7/20/8

OPTICAL FIBER
TELECOMMUNICATIONS II

OPTICAL FIBER TELECOMMUNICATIONS II

Edited by
STEWART E. MILLER
Consultant to Bell Communications Research, Inc.
Retired, Bell Telephone Laboratories, Inc.

IVAN P. KAMINOW
AT&T Bell Laboratories, Inc.
Crawford Hill Laboratory
Holmdel, New Jersey

 **ACADEMIC PRESS, INC.**

Harcourt Brace Jovanovich, Publishers

Boston   San Diego   New York
Berkeley   London   Sydney
Tokyo   Toronto

ACADEMIC PRESS, INC.
1250 Sixth Avenue, San Diego, CA 92101

United Kingdom Edition published by
ACADEMIC PRESS INC. (LONDON) LTD.
24-28 Oval Road, London NW1 7DX

Library of Congress Cataloging-in-Publication Data

Optical fiber telecommunications II/edited by Stewart E. Miller,
Ivan P. Kaminow.
  p. cm.
  Bibliography: p.
  Includes index.
  ISBN 0-12-497351-5
  1. Optical communications. 2. Fiber optics. I. Miller, Stewart
E. II. Kaminow, Ivan P., Date-   . III. Title: Optical fiber
telecommunications 2.
  TK5103.59.06763 1988   87-27049
  621.38′0414--dc 19   CIP

PRINTED IN THE UNITED STATES OF AMERICA

88 89 90 91   9 8 7 6 5 4 3 2 1

Contents

Chapter 4   Fiber Materials and Fabrication Methods

Suzanne R. Nagel

Chapter 5   Optical Fiber Cables

Charles H. Gartside, III, Parbhubhai D. Patel, and Manuel R. Santana

Chapter 6   Optical Fiber Splicing

Stephen C. Mettler and Calvin M. Miller

Chapter 7   Fiber Connectors

W. C. Young and D. R. Frey

Chapter 12   Light-Emitting Diodes for Telecommunication

Tien Pei Lee, C. A. Burrus, Jr., and R. H. Saul

Chapter 13   Semiconductor Lasers for Telecommunications

J. E. Bowers and M. A. Pollack

Chapter 14   Optical Detectors for Lightwave Communication

S. R. Forrest

Chapter 15   Integrated Optical and Electronic Devices

Kohroh Kobayashi

Chapter 20   High Performance Integrated Circuits
             for Lightwave Systems

Robert G. Swartz

Chapter 21   Introduction to Lightwave Systems

Paul S. Henry, R. A. Linke, and A. H. Gnauck

Chapter 22   Interoffice Transmission Systems

Steven S. Cheng and Eric H. Angell

Chapter 23   Terrestrial Intercity Transmission Systems

Detlef C. Gloge and Ira Jacobs

Chapter 24   Undersea Cable Transmission Systems

P. K. Runge and Neal S. Bergano

Chapter 25   Optical Fibers in Loop Distribution Systems

P. E. White and L. S. Smoot

Chapter 26   Photonic Local Networks

Ivan P. Kaminow

Contributors

Rod C. Alferness (Ch. 11), AT&T Bell Laboratories, Inc., Box 400, Holmdel-Keyport Road, Room L143, Holmdel, New Jersey 07733

William T. Anderson (Ch. 8), Bell Communications Research, Inc., 435 South Street, Room 2K168, Morristown, New Jersey 07960-1961

Eric H. Angell (Ch. 22), AT&T Bell Laboratories, Inc., 74 Fountain Avenue, Room 2C05, Ward Hill, Massachusetts 01830

Neal S. Bergano (Ch. 24), AT&T Bell Laboratories, Inc., Room 3B320, Crawfords Corner Road, Holmdel, New Jersey 07733

John E. Bowers (Ch. 13), Department of Electrical Engineering, University of California, Room 3151, Santa Barbara, California 93106

C. A. Burrus, Jr. (Ch. 12), AT&T Bell Laboratories, Inc., Box 400, Holmdel-Keyport Road, Room L247, Holmdel, New Jersey 07733

Steven S. Cheng (Ch. 22), Bell Communications Research, Inc., 435 South Street, Room 2L287, Morristown, New Jersey 07960-1961

Alan G. Chynoweth (Foreword), Bell Communications Research, Inc., 435 South Street, Room 2F383, Morristown, New Jersey 07960-1961.

Niloy K. Dutta (Ch. 17), AT&T Bell Laboratories, Inc., 600 Mountain Avenue, Room 2D301, Murray Hill, New Jersey 07974

S. R. Forrest (Ch. 14), Departments of Electrical Engineering/Electrophysics and Materials Science, Center for Photonic Technology, University of Southern California, University Park, Los Angeles, California 90089-0241

Dean R. Frey (Ch. 7), AT&T Bell Laboratories, Inc., 2000 Northeast Expressway, Room 2D48, Norcross, Georgia 30071

Charles H. Gartside, III (Ch. 5), AT&T Bell Laboratories, Inc., 2000 Northeast Expressway, Room 1D36, Norcross, Georgia 30071

Detlef C. Gloge (Ch. 23), AT&T Bell Laboratories, Inc., Room 3F632, Crawfords Corner Road, Holmdel, New Jersey 07733

A. H. Gnauck (Ch. 21), AT&T Bell Laboratories, Inc., Box 400, Holmdel-Keyport Road, Holmdel, New Jersey 07733

Paul S. Henry (Ch. 21), AT&T Bell Laboratories, Inc., Box 400, Holmdel-Keyport Road, Holmdel, New Jersey 07733

Ira Jacobs (Ch. 23), Bradley Department of Electrical Engineering, Virginia Polytechnic Institute & State University, Blacksburg, Virginia 24061

Peter Kaiser (Ch. 2), Bell Communications Research, Inc., 331 Newman Springs Road, Room 3Z231, Box 7020, Red Bank, New Jersey 07701-7020

Ivan P. Kaminow (Ch. 26), AT&T Bell Laboratories, Inc., Box 400, Holmdel-Keyport Road, Room R219, Holmdel, New Jersey 07733

R. F. Karlicek, Jr. (Ch. 16), AT&T Bell Laboratories, Inc., 600 Mountain Avenue, Room 7D308, Murray Hill, New Jersey 07974

Bryon L. Kasper (Ch. 18), AT&T Bell Laboratories, Inc., Box 400, Holmdel-Keyport Road, Room L135, Holmdel, New Jersey 07733

Donald B. Keck (Ch. 2), Applied Physics Research, Sullivan Park, Corning Glass Works, Corning, New York 14831

Kohroh Kobayashi (Ch. 15), Optical Device Research Laboratory, Opto-electronics Research Laboratories, NEC Corporation, 4-1-1, Miyazaki, Miyamae-ku, Kawasaki-city, 213, Japan

Steven K. Korotky (Ch. 9, 11), AT&T Bell Laboratories, Inc., Room 4D527, Crawfords Corner Road, Holmdel, New Jersey 07733

Tien Pei Lee (Ch. 12), Bell Communications Research, Inc., 331 Newman Springs Road, Room 3Z371, Box 7020, Red Bank, New Jersey 07701-7020

R. A. Linke (Ch. 21), AT&T Bell Laboratories, Inc., Box 400, Holmdel-Keyport Road, Room L131, Holmdel, New Jersey 07733

Ralph A. Logan (Ch. 16), AT&T Bell Laboratories, Inc., 600 Mountain Avenue, Room 1C334A, Murray Hill, New Jersey 07974

Judith A. Long (Ch. 16), AT&T Bell Laboratories, Inc., 600 Mountain Avenue, Room 7D306, Murray Hill, New Jersey 07974

D. Marcuse (Ch. 3), AT&T Bell Laboratories, Inc., Box 400, Holmdel-Keyport Road, Room L117, Holmdel, New Jersey 07733

Stephen C. Mettler (Ch. 6), AT&T Bell Laboratories, Inc., 2000 Northeast Expressway, Room 2C56, Norcross, Georgia 30071

Stewart E. Miller (Ch. 1), 67 Wigwam Road, Locust, New Jersey 07760

Calvin M. Miller (Ch. 6), AT&T Bell Laboratories, Inc., 2000 Northeast Expressway, Room 2C56, Norcross, Georgia 30071

Suzanne R. Nagel (Ch. 4), AT&T Engineering Research Center, P.O. Box 900, Princeton, New Jersey 08540

Parbhubhai D. Patel (Ch. 5), AT&T Bell Laboratories, Inc., 2000 Northeast Expressway, Room 1D40, Norcross, Georgia 30071

Dan L. Philen (Ch. 8), AT&T Bell Laboratories, Inc., 2000 Northeast Expressway, Room 1D28, Norcross, Georgia 30071

Martin A. Pollack (Ch. 13), AT&T Bell Laboratories, Inc., Box 400, Holmdel-Keyport Road, Room R241, Holmdel, New Jersey 07733

Peter K. Runge (Ch. 24), AT&T Bell Laboratories, Inc., Room 3C337, Crawfords Corner Road, Holmdel, New Jersey 07733

Manuel R. Santana (Ch. 5), AT&T Bell Laboratories, Inc., 2000 Northeast Expressway, Room 1D32, Norcross, Georgia 30071

R. H. Saul (Ch. 12), AT&T Bell Laboratories, Inc., 600 Mountain Avenue, Room 7B428, Murray Hill, New Jersey 07974

P. W. Shumate (Ch. 19), Bell Communications Research, Inc., 435 South Street, Morristown, New Jersey 07960-1961

L. S. Smoot (Ch. 25), Bell Communications Research, Inc., 435 South Street, Room 2Q286, Morristown, New Jersey 07960-1961

Robert G. Swartz (Ch. 20), AT&T Bell Laboratories, Inc., Room 4E-324, Crawfords Corner Road, Holmdel, New Jersey 07733

W. J. Tomlinson (Ch. 9, 10), Bell Communications Research, Inc., 331 Newman Springs Road, Room 3Z215, Box 7020, Red Bank, New Jersey 07701-7020

Patrick E. White (Ch. 25), Bell Communications Research, Inc., 435 South Street, Room 2Q286, Morristown, New Jersey 07960-1961

William C. Young (Ch. 7), Bell Communications Research, Inc., 331 Newman Springs Road, Room 3Z213, Box 7020, Red Bank, New Jersey 07701-7020

C. L. Zipfel (Ch. 17), AT&T Bell Laboratories, Inc., 600 Mountain Avenue, Room 2C436, Murray Hill, New Jersey 079744

Foreword

Nature appears to have looked kindly upon mankind's development of information technologies, the technologies that deal with the processing and transmission of information in the form of electronic or photonic signals. Just as society's needs for information processing and computing capacity were beginning to outstrip the capabilities of circuits based on vacuum tube switching elements, along came the discovery of the transistor. In little more than a decade this led to the first integrated circuits, which began to put almost undreamt of computing and signal processing power in the hands of the communications or computing systems designer. And, as ways have steadily been found to crowd more and more components on to a semiconductor chip, this power has escalated dramatically.

These devices ushered in the Information Age. The transistor has done for man's brain what the steam engine did for his brawn in the Industrial Age.

By the seventies, this signal processing power was beginning to place intolerable burdens on the capacity of transmission paths and telecommunications networks. Hence the natural result that major emphasis was placed on concentrating this signal processing power in large central offices or mainframe computers so as to minimize the need for using the relatively costly transmission systems. But again nature smiled and came through with the invention of the laser on the one hand and the optical fiber on the other, inventions that would tend to redress the balance between costs of signal processing and costs of signal transmission. These discoveries portended lightwave transmission capacity to keep pace with the signal handling power of silicon chips. It was the beginning of a natural partnership between solid state electronics and photonics that would substitute for the older partnership between vacuum tubes, or even electromagnetic relays, and copper wires or cables.

And yet again, nature seems to have been almost encouraging the endeavors of lightwave researchers and engineers. In the early stages many horrendous technical obstacles were expected—getting the loss of fibers to be sufficiently low, getting the lifetimes of lasers to be sufficiently long, and finding ways to splice extremely fine fibers routinely in the field under a variety of difficult environmental conditions. One after another these challenges were attacked and solved. Those of us closely involved with developing the technology kept holding our breath and marveling at what seemed to be extraordinary luck. We kept wondering when our luck would run out and we would be up against a technological impasse, but thanks to the imagination, ingenuity and determination of scientists and engineers all around the world, this has not happened. Instead, lightwave technology is "on a roll," as gamblers might say.

At the time of the earlier volume edited by Stewart Miller and myself, lightwave technology was beginning to become established in medium-haul interoffice trunks. It was based on the first generation gallium arsenide technology with wavelengths in the 0.8–0.9u range. At that time, the next generation technology based on ternary and quaternary compounds and wavelengths in the 1.3–1.5u range was just beginning to emerge in the research laboratories. Developed further, it quickly superseded the shorter wavelength technology for most applications because of the much longer spans it offered between repeaters. It thus became ideal for use in long-haul systems, both terrestrial and transoceanic submarine. Variations on the theme also saw it being developed for use in local telecommunications networks, loop feeders and local distribution right up to and on to customer premises systems. Truly, it seems that lightwave technology has turned out to be an irresistible force.

Almost without exception, lightwave is now the technology of choice for new inter-office and long-haul transmission system installations and is also preferred over copper for an increasing fraction of local feeder systems. In the competition with other technologies, it is establishing itself on the basis of hard-nosed economics in the present time frame but with the added bonus of almost unlimited potential to upgrade its signal-carrying capacity as the needs and markets build up. It is further recognized that lightwave technology is still at a relatively early stage on the learning curve and that as demand for transmission capacity continues to build, and as ways are found to increase the bandwidth of optical fibers by adding or multiplexing more and more signal channels on to them, the costs of transmission will continue to progress down the learning curve for a good many years to come.

We have become so confident of the continuing technological progress that we are beginning to take almost for granted the prospect of bandwidth as an abundant commodity. So what will be the impact of this abundant bandwidth?

One impact that we are only just beginning to wrestle with is how lightwave technology can change network topologies and architectures. Ring and gateway topologies with control located at terminals can become viable alternatives to the more familiar star topologies radiating out from central offices. In other words, lightwave makes decentralized networks and intelligence a serious alternative to centralized architectures and intelligence. A mixture of network topologies is therefore a likely outcome as networks are optimized to meet customer needs.

Another and crucial question is, what will all the abundant bandwidth be used for? While it will certainly help foster network services such as data, facsimile and image services, these require only a relatively small part of lightwave's potential bandwidth capacity. Obviously, the big customer of bandwidth can be, and I believe will be, high quality video services. It may well be that these services are necessary to justify single fiber lines to individual customers. This often presents the classic chicken-or-egg dilemma. But as one corporate telecommunications executive has put it, "you can't sell tickets for the railroad before you have built the railroad." There is clearly going to be plenty of opportunity in the years ahead for the adventurous entrepreneurs.

But the biggest obstacle currently facing the deployment of lightwave technologies and hindering customers from getting the full cost and service advantages they can offer may be not so much a technological one, perhaps not even a market one, but that which is posed by the world's fragmented telecommunications networks and industries, a condition that has only been accentuated by the break-up of the former Bell System in the U.S.A. The advance of communications technology over the last few decades has been so great and it has been made so ubiquitous that it is now everyman's technology rather than a closed preserve of the high priests of the telecommunications industries and administrations. Private networks and new services are proliferating. Faced with this proliferation, the need for universal compatibility in communications technologies, network architectures, and service protocols is more vital and urgent than ever. Lightwave technology has made the prospect of truly universal communications service much more realizable; by which anyone, anywhere, at any time will be able to communicate conveniently with anyone else, anywhere in the world, in any medium—voice, data, facsimile, image, video—and at acceptable cost. But it will take the determined and cooperative work of all those concerned with developing standards that make network engineering sense if this prospect is really to come about.

This present volume continues from where the earlier one left off. It covers the long wavelength lightwave technologies that have evolved steadily in the intervening years and the growing range of their applications.

Although this volume concentrates on this second generation of lightwave technology, there are signs of yet another generation taking shape, an even higher capability generation that will be based on the so-called coherent optical technology and on integrated optoelectronics. Perhaps this phase will be the basis of a third volume in a few years time.

Lightwave technology started with the invention of the laser a little over a quarter of a century ago. It has advanced, seemingly irresistibly, ever since. It is truly one of the great technological marvels of the 20th century, and all those who have played a part in it have a right to feel proud of what they have helped to accomplish. It is an inspiring example of what can be achieved by well supported and steadily sustained high technology research and development scientists and engineers in organizations led by managers with faith and astute vision. It has been my good fortune to have been associated with such people and most aspects of the lightwave revolution over the last quarter of a century.

ALAN G. CHYNOWETH
Bell Communications Research

Chapter 1

Overview and Summary of Progress

STEWART E. MILLER

Consultant to Bell Communications Research, Inc., Retired, Bell Telephone Laboratories, Inc.

INTRODUCTION

Telecommunications using glass fibers as the transmission media is now a major industry. Manufacturing plants devoted exclusively to fabricating optical fibers and cables are making several million kilometers of telecommunications fiber annually, which is capable of supplying fiber for a super-abundance of transmission facilities. Intercity fiber systems are in service, fiber cables interconnect the telephone switching centers in the cities, fiber feeders lead from switching centers toward the subscriber, and within buildings fiber cables interconnect switching machines. Banks are using fiber links to tie together their decentralized facilities, and there is beginning an application to interconnect computers and business equipment more generally. By the time this book appears, the installation of several glass-fiber transoceanic cables will be under way, with service commitment dates in 1988.

Almost all of this commercial realization of the potential of optical fiber telecommunications has occurred since 1979, when the book *Optical Fiber Telecommunications* (edited by S. E. Miller and A. G. Chynoweth) first appeared. So much new understanding and attractive new technology has accompanied this expansion that it is timely to produce a new volume. Most of the material in the 1979 book is applicable as a foundation for the more recent work. Also, to attempt to provide a stand-alone treatise on all of the critical elements of fiber telecommunications systems—fibers, cables, splicing and installation, lasers and LEDs, detectors, integrated optical circuitry, and a series of system design perspectives—would be impossible in a single volume of reasonable size. Nevertheless, a single source that gives most of the fundamentals will be of value for several interested groups—operating company engineers, component design and manufacturing engineers, researchers interested in the interrelation between their specialty and others in the system, and finally, students seeking background for additional contributions that they will be making.

With these objectives in mind, the present volume is designed to be a supplement to the 1979 volume of *Optical Fiber Telecommunications*. That volume and the present one are intended to provide a stand-alone source for the lightwave telecommunications field. Readers will in many cases have other earlier books that provide even more detailed information on various specialties. We have endeavored, however, to provide references to journal papers or other books to lead the reader to more detailed information where desired. These references are incorporated at the end of each chapter.

Most of the chapters can be read without detailed study of the other chapters. This makes it feasible for a reader to go directly to the material of greatest interest at the moment for a current perspective.

In the following paragraphs an attempt is made to give something of an overview of the material. This should not be regarded as an abstract, for that would require more space than we wish to take here and it would be presumptive of this author to reinterpret the authors' contributions. It might rather be seen as another view, formed in the same time frame and based on the same technology.

1.1 TELECOMMUNICATIONS FIBERS AND CABLES

Chapters 2 through 5 cover fiber fabrication processes and the formation of cables as well as a perspective on the current variety of fiber structures and the new theoretical background for understanding their performance. Perhaps the least predicted trend in the last six years is the rapidity of the movement to single-mode fibers as the medium of choice for the local (central office to subscriber) environment. The mechanical tolerances required during splicing or in joining fibers at demountable connectors are roughly a factor of 10 smaller for single-mode than for multimode fibers. Whereas it was generally known that good single-mode-fiber joints were feasible, the cost for making them was anticipated to be a real obstacle. It was also appreciated however, that the bandwidth capability for single-mode fiber is orders of magnitude larger than for multimode fiber. The long-term economic value of this characteristic has proven sufficiently attractive to move system designers to prefer single-mode fiber not only for intercity and transoceanic applications (which was fully anticipated) but also for short spans such as between central offices within cities and from the central office to the end customer or a nearby concentrator. The transmission loss of single-mode fibers is lower than for multimode, in part due to an ability to make the core (wave guiding section) with less doping of the fused silica resulting in less lightwave scattering. We will see that it has even proven practical to use an LED transmitter on single-mode fibers at hundreds of megabits/sec over spans as long as 5 km or more. Thus the inherent ability of the multimode fiber to gather more power from the LED, while still a valid asset, is not an absolute necessity. We do not conclude that multimode fibers will no longer be used. Where the bandwidth capability of multimode transmission is adequate (a bitrate-span-length product of 500 to 1000 Mb/s-km) and where cost is really critical, the multimode links will probably be economically preferred.

1.1.1 Fiber Types and Fabrication Processes

Chapter 2 provides an overview of the fiber types that currently are important. Figure 2.1 shows the index profiles of single-mode fibers—subsequent discussion will elaborate on their significance. Multimode fibers have been optimized for use in two wavelength regions—near 850 nm and near 1300 nm. International standards are being established for both single-mode and multimode fibers to try to insure interchangeability.

Chapter 4 reviews the materials science and fabrication methods associated with telecommunications fibers. A massive research and development effort has been invested in fiber making, a recognition of the long-term economic potential that the initial fiber installation represents. With labor a major contributor to installed-fiber costs, there is strong motivation for maximizing the long-range usefulness of the fiber cable; to replace it would involve major expense.

The vast majority of current fibers use lightly doped silica for both core and cladding. In Chapter 4 the material loss mechanisms are discussed: absorption and/or scattering due to impurities, damage due to radiation (ultraviolet, X rays, etc.) either manmade or natural background, hydrogen induced absorption or processing induced structural losses. After suitable modification of the processes and fiber cable design, the end result is to reach very nearly the intrinsic loss of silica in practical fiber cables. Forward-looking research seeks new materials to achieve even lower losses.

The first step in fabricating fibers is to produce a preform that has essentially the desired transverse index distribution but scaled up by a factor of several hundred. Subsequently, the preform is drawn down into the finished fiber. There are four processes for making the preforms. Corning Glass Works initiated the outside vapor deposition process (OVD), Bell Laboratories originated the modified chemical deposition process (MCVD), Phillips Research Laboratories originated the plasma activated chemical vapor deposition process (PCVD), and the Nippon Telegraph and Telephone research people first utilized the vapor axial deposition process (VAD). In each of these processes the starting materials are introduced as vapors, and to varying degrees minute glass particles are formed in the gas stream before deposition on the inside of a substrate tube or on the outside of a mandrel. The structures in which initial chemical reactions take place to create the preform are shown in Chapter 4: Fig. 4.17 for OVD, Fig. 4.14 for MCVD, Fig. 4.16 for PCVD and Fig. 4.19 for VAD. The details of the chemistry and physics for the MCVD process have been extensively published, and it is generally considered to be the best understood. In keeping with glass industry conventions, much of the Corning process has been kept proprietary. All four processes have produced top-quality fibers, and all four are considered competitive commercially.

In some applications, notably for transoceanic systems, the strength of fibers is very important. Careful studies have shown the relation between breaking strength and cracks of length less than a micron, either intrinsic or due to processing. Water vapor is known to degrade strength. By the use of hermetic coatings and prooftesting all manufactured fibers, satisfactory strength for undersea cables as well as for domestic usage has been achieved (see Chapter 4, Fig. 4.13).

Looking toward the future, one notes that theory indicates intrinsic material losses far lower than in silica at wavelengths beyond 2 microns. Chapter 2 gives the status of this research. It is not possible to foresee whether or not these potentials will be realizable in practice.

1.1.2 Connectors and Splices

Chapter 7 provides a review of the demountable connector art for both single-mode and multimode fibers. The principal message for the multimode case is that the apparent loss of the connector is so strongly dependent upon the details of the multimode fibers being connected as to require caution in assigning connector losses. Although single-mode fiber connectors must have tolerances in the submicron range, the technological situation is much simpler than for multimode connections and a budgeted connector-loss value becomes feasible for the single-mode case. Whereas field installable single-mode connectors have become commercially available, their economics must be demonstrated in the marketplace.

Chapter 6 provides a survey of permanent splicing. This technology has developed dramatically as the result of extensive development for field use and as the result of the field experience itself. Whereas losses in metallic splices were negligible, even the massive effort invested in fiber splicing leaves a finite mean loss that must be included in the transmission loss budget; splice loss does influence maximum repeater-span length in long haul systems.

In order to minimize splice losses, the manufacturing dimensional tolerances on the fibers themselves are largely set by splicing (and connection) requirements. It has proven desirable to optimize jointly the fiber, cabling and splicing technologies.

In both splices and connections, three misalignments are minimized in the designs: transverse offset, tilt (of the two fiber axes) and gap at the joint. Specialized hardware is now fully commercial and is utilized in the field to make the splices and to measure their losses (Chapter 8). Fusion splices, made by arc or flame heating of the joint, produce low-loss and high-strength joints that are no larger than the fibers themselves. Other techniques require considerably more space for the splice assembly, but in some applications may prove to have less cost per fiber joint.

Average losses under 0.1 dB have been achieved in the field for permanent splices, and single-mode connector losses averaging 0.25 dB with a standard deviation of 0.12 dB were demonstrated.

1.1.3 Chromatic Dispersion

Chapter 3 gives new detailed insight into the bandwidth capabilities of single-mode fibers. As earlier work reported, pulse spreading due to wavelength-dependent delays has two sources in single-mode fibers: (1) material dispersion intrinsic to the glass, which would be observed with a plane wave in an infinite glass medium, and (2) waveguiding dispersion, due to a partial shifting of the lightwave field distribution between core and cladding as a function of wavelength. Total pulse dispersion in the fiber we call chromatic dispersion, which increases linearly with fiber length and transmitter spectral width. In all glass fibers the glass material dispersion dominates at 820 to 900 nm and is of the order of 100 picosec./km-nm. For pure fused silica, material dispersion goes through a minimum (less than 1 picosec./km-nm) at about 1270 nm; total fiber chromatic dispersion is a minimum near 1300 nm in fibers with a simple core of constant index, the shift from 1270 to 1300 being the effect of waveguiding dispersion. Extensive use of the 1300 nm region in simple uniform-core fibers is planned for very high speed pulse transmission. For example, a rather broad transmitter spectrum (2 nm) located within 1 nm of the fiber's chromatic dispersion minimum could support a bit rate of 30 Gb/sec. for a repeater span of 100 km.

The loss in fused silica fibers, however, drops from about 0.35 dB/km at 1300 nm to about 0.2 dB/km at 1550 nm (see Fig. 2.7). This is a strong motivator for building systems at 1550 nm, but in the simple uniform-core fibers the chromatic dispersion is about 18 picosec./km-nm. The natural desire to have both lowest loss and lowest dispersion at the same transmitter wavelength led to two new single-mode fiber types: dispersion shifted fibers and what has become known as dispersion flattened fibers. (The refractive-index profiles for uniform-core, dispersion-shifted and dispersion-flattened fibers are shown in Fig. 2.1.) In both of these new fiber types, the waveguiding dispersion effect is altered so as to partially or completely cancel the glass material dispersion at the wavelengths of interest.

In the dispersion-shifted fibers, the core is made smaller and its index of refraction is usually tapered linearly from a peak value (larger than in the uniform-core fiber) to the cladding index. The result can be a shift of the fiber's total chromatic dispersion minimum to near 1550 nm.

In dispersion-flattened fibers, a series of glass layers of alternating index of refraction are used in the single-mode core-cladding structure; for example, see Fig. 3.2.6b. The net effect can be to cause the waveguiding dispersion to substantially cancel the glass material dispersion all the way from 1300 nm to 1600 nm (see curve marked (QC) in Fig. 3.2.7).

Both dispersion-shifted and dispersion-flattened fibers are now being offered commercially. Their use in systems is expected to follow. Initial experiments yielded somewhat increased loss in these fibers, in part due to increased scattering from the increased level of dopant needed to achieve the needed refractive indices. Thus, where absolutely minimum loss is more valuable than a greater permissible Gb/s-km product for the fiber, the simple fiber forms will be preferred. It is feasible to tolerate the 15–18 picosec./km-nm chromatic dispersion of the simple fibers by using a transmitter with a very narrow spectral width (to be discussed).

1.1.4 Cabling

Cabling technology has been the subject of extensive development for field use and field experience has been accumulated. Chapter 5 reviews this work. As indicated by early research experiments, it has proved feasible to produce cables in which the fiber loss is not significantly increased due to cabling. This has been achieved by using cable designs that minimize residual strain in the fibers and by carefully monitoring manufacturing parameters to achieve this design intent. A variety of specific cable designs have been produced commercially, sometimes as alternative choices for the same objective and sometimes to meet significantly different requirements. As already noted, a joint optimization of the fiber, cable and splicing technologies is needed. For long haul application, fiber count per cable is typically in the order of 10 to 20, and individual fiber splicing for minimum loss is preferred. For interoffice trunk or for distribution to the customer, fiber counts over 100 may be needed and so-called mass splicing techniques, in which numerous fibers are spliced in a single procedure, have been used. For 22- to 144-fiber cables, the overall diameter is in the 1 to 2 cm range, the weight in the 120 to 355 kg/km range, and installed bending radii as small as 12 to 43 cm are acceptable. Figure 5.14 illustrates typical histograms of loss: a mean value of 0.35 dB/km with standard deviation of 0.02 dB/km at 1310 nm, and a mean value of 0.21 dB/km with standard deviation of 0.02 dB/km at 1550 nm. Discussions of mechanical and hazard testing and undersea cable designs are also given in Chapter 5.

1.1.5 Polarization–Maintaining Fibers

Polarization-maintaining fibers are fibers that are designed to ensure that, with a linearly polarized input, the lightwave emerging from a long fiber will be essentially linearly polarized and will have a predetermined spatial orientation. In fibers without very special features this will not be the case. In nominally circularly symmetric structures there are two polarizations of the lowest order wave, and due to uncontrollable imperfections and strains these two waves are coupled. The result is usually some combination of the two waves that yields an elliptically polarized wave at the output, a result that can change with wavelength and with time due to temperature and strain variations. In systems using a photon-counter as a detector this represents no problem, but in many coherent transmission systems and in most integrated-optic devices, the arbitrary elliptically polarized wave is unacceptable. A major effort has been devoted to devising fiber structures that maintain a predetermined received polarization; several of these structures are illustrated in Fig. 2.4 and in Fig. 4.23. These fibers incorporate intentional arrangements of anisotropy in the index of refraction to give the two polarizations of the lowest-order mode (effectively two different modes) significantly different phase constants, which leads to reduced power transfer between these modes. Chapter 3 gives a discussion of the theory for several approaches. There has been some success; the undesired polarization has been suppressed by 30 dB in some experimental fibers. The losses in polarization maintaining fibers however, are still higher than in the simpler fiber structures, and no commercial telecommunications application has yet appeared. Where loss is not critical, such as in patch cords, and in fiber sensors or other devices and connecting links, this type of fiber is used. A breakthrough in providing low loss with low cost for both the fiber and its associated splicing could result in a widespread application in telecommunications; it would be attractive in coherent systems.

1.1.6 Nonlinear Effects in Fibers

Chapter 3 also summarizes some recent additions to knowledge about the effects of nonlinearity in silica fibers. The index of refraction changes slightly due to the presence of the optical field, and the effect can be undesirable or, in a special case, desirable. The more common undesirable effect is self phase modulation. When using simple binary amplitude modulation, the change in index at power levels on the order of 50 mW can increase pulse dispersion effects (single-mode fibers). In systems using phase-shift keying, the effects of fiber nonlinearity are more critical; for a multi-channel system using wavelength-division multiplexing, crosstalk between channels can become significant at individual-channel power levels on the order of 1 mW.

The Raman effect can also cause crosstalk between channels in a multi-channel wavelength-division-multiplex system. In a 50 km long single-mode fiber, significant power is transferred from one channel to another at power levels in the order of 10 mW per channel. Brillouin scattering can produce crosstalk if the fiber is used with transmission in both directions at the same time.

The beneficial effect of fiber nonlinearity is a special modal condition resulting in waves called solitons. At power levels of hundreds of milliwatts in silica single-mode fibers with negligible loss (or with gain added to simulate negligible loss), very short pulses can, in principle, propagate over very great distances without broadening. The reader is referred to Chapter 3 for a more precise discussion. Laboratory demonstration of the principle has been achieved, but no commercial application is known.

1.2 SYSTEMS

The explosive growth of applications, which resulted in the embedded complex of fiber usage referred to in the introduction, has been made possible by the research and development in the areas of lasers, detectors, receiver designs and system configurations. It is the overall guidance and motivation of the system thinking, however, that has shaped the advanced developments of the component parts. Pioneering research on the component level has in many instances provided the breakthroughs that have been the foundation for the systems, and there is a symbiotic relationship between the two disciplines that has been fruitful. Here we discuss the system viewpoint first, with the understanding that the appearance of almost every system concept as a serious practical possibility was the result of earlier device and component innovation.

The people who have led or carried out the system work have in many instances been experienced in telecommunications systems at lower carrier frequencies. This has led to a valuable perception of the key questions to ask and to the desirable forms of systems to seek. In part this background is responsible for the rapid progress—for the rapid transition from research concept to commercial reality. But even more responsible for the rapid transition has been the very nature of the new-art combination—lasers and fibers. It was recognized immediately that the small size and vast capacity of telecommunications fibers offered a revolution—fibers could be applied advantageously wherever coaxial cable had been used—in patch cords, in city streets for video, in intercity transmission and in transoceanic systems.

In Chapter 21 the principles underlying the variety of systems applications are introduced in an historical tutorial. The fundamentals are common to all applications, and we seek to provide them in a format not restricted to a specific series of trade-offs, which must be made in any commercial product offering. It is interesting to note, however, the environment in which fiberoptic telecommunications took off. With no intermediate application available for the vast bandwidths made feasible by fibers, the initial commercial value of fiber systems was their low loss. Repeater spacings could be an order of magnitude longer than with copper-based systems, with consequent savings in both initial and maintenance costs. That fact put fiber systems over the threshold (and it is still a valid driving force today). Once commercial usage had been started, the well-anticipated advantages provided by the greater bandwidth capability were utilized to yield further economic savings.

Chapters 22–26 deal with transmission systems designed to provide optimally specific types of service. Chapter 22 covers interoffice transmission systems, Chapter 23 covers terrestrial intercity systems, Chapter 24 covers undersea cable systems and Chapter 25 covers the use of optical fibers in loop transmission systems. Although the earliest interoffice systems used lasers and LEDs operating at 820 nm, essentially all transmission systems are now being designed for operation at either 1300 nm, where the fiber dispersion in simple fibers is a minimum, or near 1550 nm, where the fiber loss is a minimum. The choice of lasers or LEDs is determined by the relative importance of maximum repeater span (indicating lasers) or ruggedness, low cost and reliability (indicating LEDs). It has been found feasible to use LEDs with single-mode fibers at pulse rates up to about 600 Mb/s when the repeater span is not too long. This provides a strategy of using single-mode fibers and low-cost LEDs for initial installation (in a loop application, for example) with the prospect of later converting to laser-based repeaters for higher communication capacity when the combination of laser cost and compensating service revenues justifies the change. The initial installation of single-mode fiber for use over many years, without the need to prove in the current cost of lasers, is an attractive approach.

1.2.1 Domestic Intercity Systems

For domestic intercity systems, the current costs of lasers are fully justified, and single-mode fiber with laser transmitters are universally chosen. Digital transmission (generally with binary on-off intensity modulation) is used, with pulse rates ranging from a few hundred Mb/s to about 2 Gb/s. In maximizing channel capacity, the economics favors running a single channel at pulse rates as high as the baseband electronics can support before using multichannel wavelength division multiplexing (WDM). Some WDM has been applied, and importantly, more will be added in the future as a measure for increasing capacity on fibers already installed when increased service opportunities appear. In order to reach the status just outlined, it was necessary to get over a threshold—to prove the viability of the new art in the intercity environment. The introductory phase involved a different set of system parameters.

Initially it was the combination of low fiber loss (resulting in longer repeater spans) and the appearance of digital switching for intercity circuits that gave fiber optic intercity systems an economic advantage over coaxial systems. For very high circuit cross-section but relatively short intercity links (on the order of 200 miles), fiber systems also became more attractive than microwave radio relay or satellite systems. In the early 1980s, there began a steady growth of intercity fiber systems in the U.S., Europe, and Japan. Because the components were more advanced, the initial systems were based on multimode fibers and used repeater spans on the order of 10 km with bit rates in the 32 to 140 Mb/s range. All groups recognized from the earliest development phase that single-mode fibers were the best choice, and by about 1982 the component technologies had advanced to the point where single-mode systems could be manufactured. Since 1984 all new intercity systems have employed single-mode fibers, and repeater spans are on the order of 40 km or more.

Digital transmission is universal, and simple on-off keying of the laser with detection of the lightwave power is standard. The system hierarchies differ somewhat in the various countries, but all have a series of "levels," each characterized by a pulse rate and channel capacity. These are given in Table 23.1. In the U.S., level 1 (DS-1) is 1.544 Mb/s and 24 voice channels and level 3 (DS-3) is 44.736 Mb/s. System economics favors the use of the highest pulse rate possible on each channel, based on the state of the art for integrated circuits. At the present time the highest rate is about 2 Gb/s for commercial systems. Figure 23.12 illustrates how the cost per channel drops as the number of channels is increased on each fiber.

System requirements, general configuration of elements and supervisory planning for fiber optic intercity systems leans heavily on earlier system work. There are differences, however, made desirable by the nature of the lightwave components. For example, the multiplexers that built up the higher levels of the digital hierarchy in the past were separately packaged subsystems, which had electrical inputs and outputs and which could be used on either coaxial cable, microwave radio or satellite systems. It has proved preferable in the U.S. to combine some of the multiplexing with the fiber-span terminating equipment (terminals) of the intercity links. Lower data rate pulse streams are input directly to the transmitting terminal, and a higher level in the data-rate hierarchy emerges at the terminal output as a photon stream. An inverse arrangement is employed at the receiving end of the fiber line.

In the future we can expect to see changes as the intercity and intracity networks evolve. The distinction between intercity, intracity and local systems is perceived to be blurring with time. The forces causing the changes include the corporate evolutions following the January 1, 1984 divestiture of the Bell System and the development of new wideband services, which are made possible largely by the fiber optic art. The telecommunications systems hierarchies for all of the segments must be compatible in some sense, and much of this remains to be determined. In the future we can also anticipate a possible departure from the modulation format of on-off keying with power detection to one or more of the formats of coherent systems. Coherent systems offer increased repeater spans for intercity systems, and a convenient way of combining 10's or 100's of signals, which may prove advantageous in the local loop. Because of the need for interconnecting local loops at a distance from each other, the local loop advantages may generate a requirement for coherent intercity transmission. At the present time the state of the technology and the cost of implementing coherent systems makes intercity usage unattractive; however, the component costs can be expected to drop and the advantages may increase. The prospect for commercial usage is not clear now.

1.2.2 Undersea Cable Systems

Chapter 24 deals with fiber optic undersea cable systems. The technology used in these systems is derived from that employed in intercity systems, and experience obtained there is valuable. The major difference is in need for all components used in the undersea system to be exceptionally reliable. The costs of repairs are so large that extensive development is appropriate to avoid failures. Chapter 24 discusses these efforts.

Another difference between undersea and intercity systems is the requirements on the fiber cable. The undersea cable carries power to energize the repeaters as well as the information-carrying fibers, and it is obvious that the environment at the bottom of the ocean presents new problems. Earlier undersea coaxial cable systems provided a great deal of experience as input to the fiber optic work. Chapter 5 gives further details and gives references to more extensive discussions.

Initial systems will operate near the fiber's chromatic dispersion minimum at 1300 nm using lasers with rather broad spectra but which have been qualified for very long device lifetime. Service is scheduled in 1988 for both a trans-atlantic system (American, English and French) and a transpacific system (American-Japanese). Both systems will employ three pairs of fibers, two working pairs and one pair available as protection against failure, and will have a transmission capacity of 560 Mb/s (280 Mb/s on each fiber pair).

1.2.3 Interoffice Transmission Systems

As pointed out in Chapter 22, interoffice transmission systems differ from intercity and undersea systems as a consequence of much greater sensitivity to terminal costs. This follows because the typical length between terminal points is in the range 5–20 km. By its very nature, a local office interfaces with many more sources/receivers of the information it handles than does an intercity terminal. This potentially leads to a need for a different architecture than is ideal for the intercity network. We touch on this in subsequent paragraphs.

Fiber optic interoffice systems are digital, following the pattern of the widely used T-1 systems employed to send 24 voice channels over a single wire pair via PCM at 1.544 Mb/s. The first system was the threshold application, in which the Bell System led the world in proving in large-scale commercial usage of fiber optic telecommunication. A field experiment was held in 1976, a field trial in 1977 and commercial service with a standard commercial product offering in 1980. The system, designated FT-3, employed multimode fiber, 820 nm lasers in most links (with LEDs shown feasible for short links), and the DS-3 level of multiplexing. This provided 672 voice channels in the form of 28 DS-1 24 channel blocks.

Several years later, as single-mode fiber systems at 1300 nm became available (driven by the intercity application), single-mode systems began to be used in the interoffice application also. This form made available on a single fiber 9-12 DS-3 groups and employed bit rates of 417-680 Mb/s.

Field experience showed lower incidence of failures on the fiber optic systems than in earlier wire-pair systems, leading to operating company preference for the lightwave systems for new installations.

Current work is reexamining the question of the most desirable topology for the interoffice application. Existing systems are point-to-point, wherein all the traffic from an information source is sent to a single destination. These have grown as an economical way to increase the information capacity of a single wire-pair. With the vast bandwidth potentially available on a single fiber and as a consequence of potential new services to be offered to make use of the newly available bandwidth, it becomes appropriate to make the reexamination.

One concept currently advanced that relates to the reexamination referred to above is called SONET for synchronous optical network. According to this proposal, the multiplexing of digital voice circuits (or any other digital signal) would be done using clocks tied together throughout the country—assuring that two DS-3 bit streams originating at different locations could be combined without any further synchronization. In the present national network, this cannot be done; even though all DS-3 groups run at the nominal 44.736 Mb/s they are not at exactly the same frequency and their rates can drift with time. The SONET arrangement would facilitate the multiplexing/demultiplexing functions at all levels of the national network. Fig 22.5B and section 22.3.3 illustrate the potential impact on the interoffice architecture.

Switching of the various digital tributaries in optical bit-stream form seems feasible with current integrated optical circuitry, may prove advantageous and is facilitated by the SONET arrangement.

Other new arrangements are being considered and field tested. Architectures for interconnecting a number of geographically distributed local area networks, such as are set up in a bank or in a large office complex, are called MANs for metropolitan area networks.

Several systems concepts are being explored that use forms of coherent transmission for distributing information to a large number of local customers. These may use a "dense WDM" format in which as many as 100s of channels are combined on a single fiber, with provision for selection at the customer's premise either by the customer or by the telecommunications provider. These possibilities are far from the advanced development stage right now but represent the current thinking. In this network environment, it is acceptable to take low efficiency of optical power delivery from a source to any one destination in order to provide an attractive degree of flexibility for selective (time varying) delivery to one or several destinations. Thus, the most attractive architecture may differ significantly from that desired in intercity transmission. A very possible outcome could be some special networking configurations and the basic intercity configuration superimposed in the interoffice environment.

1.2.4 Loop Distribution Systems

The final system type we consider is the class used in the loop, which is the region between the central office and the subscriber. The largest future growth of fiber installation will be in the loop, and has already begun. Many features of the expected massive deployment, however, have not yet been decided. Our review will serve as an introduction, laying the foundation for work yet to be done.

It should be noted at the outset that the loop environment places more severe requirements on its components than any other telecommunications system. The temperature range over which loop electronics must operate reliably is -40 to $+70$ degrees centigrade and this strongly influences the configuration of loop optical systems. Components used in the loop must have far lower cost than is necessary in intercity systems; the larger volume of production associated with loop components is expected to aid considerably in achieving the cost objective.

In order to discuss specific systems and plans, we first identify the segments of the loop network with reference to Fig. 25.1. The interexchange trunks form the backbone of the interoffice network, covered in Chapter 22. The feeder cables lead from the central office toward the subscriber, and usually carry signals destined for numerous subscribers multiplexed onto a single transmission line (wire pair or fiber); the feeder cable may have several serving area interface units for connection to the distribution cables. The distribution cables lead from the serving area interfaces toward the subscriber, and may have several distribution pedestals along them for connection to drop cables. The drop cables are those that terminate on the subscriber premise. Thus, the subscriber in the present arrangement has a dedicated transmission line in both the drop and distribution cables, but at the serving area interface the subscriber's signal is usually combined with others for economical transport to/from the central office.

Two copper-pair based systems are now commonly used as feeder cables, T-1 using PCM at 1.544 Mb/s carrying 24 voice channels and a form of digital carrier that combines 96 voice pairs on 4 T-1 lines. Presently in the plant there are fiber-based digital carrier systems as feeders, using the DS-2 level (6.312 Mb/s) carrying 96 channels, frequently on a special multimode fiber (one fiber for each direction). LEDs and simple PIN detectors are used in the remote electronics because of the rugged environment. Digital carrier systems for the loop are being offered now more generally, commonly using lasers and single-mode fibers, and can be very attractive economically as a consequence of low fiber loss (permitting long repeaterless spans) and wide bandwidth.

The greater potential for fibers in the loop is to use them to carry higher information capacity signals to the subscriber. Here it is necessary to note two classes of customers—business and ordinary residence. The business customer already has used DS-1 rate channels, carried now on wire pair, and it is anticipated that larger data rates will be required. One forthcoming configuration for doing this is called ISDN–integrated services digital network. It is intended to provide simultaneously both digital service of the customer's choosing and voice service over the same access line—i.e., via one port. In the initial ISDN, 144 kb/s is a building block, arranged to provide two 64 kb/s "bearer" channels, which can be circuit or packet switched, and one 16 kb/s signaling and packet data channel. This can be provided on wire pair in the distribution and drop cables if desired. Where fiber becomes mandatory is for a proposed broadband ISDN—BISDN—which is under discussion in this country and in the international standards organization CCITT. It is not agreed just what the architecture should be for BISDN. One proposal is to provide the customer with options chosen from: fixed rate multiplexed channels at 16 kb/s, 64 kb/s, 1.544 Mb/s, 45 Mb/s and a "video channel rate" near 140 Mb/s, which could be the maximum single channel rate provided. Other proposals range up to a maximum aggregate rate near 560 Mb/s with great flexibility amongst multiplexing, packet switching and circuit switching alternatives.

It seems clear that for the short distances involved in the distribution and drop cables, it would be technically feasible to provide 600 Mb/s channels using LEDs and single-mode fiber. Later, increased capacity could be contemplated with no change in the fiber plant—merely a change of the electronics as the only loop modification. One might expect that rugged, less expensive lasers would be developed in time, which could provide multi-Gb/s rates.

It appears as though this evolution is being paced by the social innovation needed to find ways to use the very high information rates advantageously. Whereas both business and residence usage of BISDN can be anticipated, the services involved would be quite different. Whether business or residence demand will become widespread first is now a subject of debate.

1.2.5 Receiver Design

All of the systems discussed previously utilize a photon detector and baseband circuitry to recover the information on the optical carrier. Chapter 18 contains an in-depth discussion of such receivers for the special case of on-off keying of the lightwave carrier. This is the arrangement used extensively at present. Chapter 14 provides a review of the detectors used in the receiver, and Chapter 20 discusses the integrated circuits that are pacing elements, not only in the receiver and other electronic subsystems, but also in the highest-speed fiber optic systems.

1.2.5.1 Receiver Structure. Figure 18.1 shows a block diagram of the typical receiver. The wide variety of requirements that accompany receiver design for different systems makes it difficult to give details without becoming inappropriately lengthy. We point out that new receiver configurations have been devised to provide high sensitivity and wide dynamic range with respect to the level of an incoming optical signal. The transimpedance front end, (Fig. 18.6) and the improved form including active feedback (Fig. 18.15) are examples that are now widely used.

The state-of-the-art in receiver design is summarized in Table 18.4. At 10^{-9} bit-error-rate, the average received power requirement is in the range -50 to -53 dBm at 45 Mb/s, -38 dBm at 565 Mb/s (commonly achieved numbers) and -25 dBm at 8000 Mb/s (a "champion" number).

The required receiver input power level is dependent on the type of detector employed, and this choice is also controlled by the operating environment as noted below.

1.2.5.2 Detectors. Chapter 14 provides a comprehensive discussion of the semiconductor detectors found useful in fiber optic telecommunications. The types most commonly used are the PIN and the avalanching photodiode (APD). In the PIN device, incoming photons over a broad lightwave spectrum are absorbed, creating an electron-hole pair. The carriers are swept out by a bias of a few volts, creating a current flow in the baseband or microwave circuit. The APD differs in providing a region in the semiconductor wherein the carriers created by the incoming photon are accelerated and by collision create an avalanche from a single photon absorption. Typically, the semiconductors employed are Ge, InGaAs or InGaAsP, and cross sections of the structures are shown in Fig. 14.6.

The advantage of the APD is that it amplifies the received signal before the first low-frequency gain element (transistor), thereby making the receiver sensitivity less dependent on the noise properties of the transistors. Because the avalanching process is statistical, however, it does introduce some noise, and because there is some delay involved in the avalanching process the APD has bandwidth limitations that are more severe than for the PIN detector. It is found that advantages of the APD are reduced at the very highest bit rates. Note, however, that the 8000 Mb/s receiver listed in Table 18.4 used an APD.

The avalanching process is critically dependent on the APD bias voltage and is sensitive to ambient temperature changes. Thus the APD is not as well-suited to the environment of the subscriber loop, but can be very advantageous in systems providing a controlled environment.

1.2.5.3 Electronic Integrated Circuits (ICs). As previously noted, it proves advantageous economically to increase the bit rate in digital systems to provide more information capacity, rather than to use wavelength division multiplexing and a series of optical carriers. Increasing the bit rate is dependent on having electronic integrated circuits that will function at the increased speed, and the electronics tend to be the limiting factor. This subject is discussed in Chapter 20.

Silicon integrated circuits are commonly cited as being capable of bit rates up to about 2 Gb/s, but champion circuits have performed at 6 Gb/s. Intrinsic material limitations are present, however, and in parallel there are other groups of people developing GaAs ICs, because GaAs has greater intrinsic speed capabilities. Chapter 20 discusses the device structures and emphasizes that device performance cannot be weighed outside of the circuit context in which it is found.

GaAs devices have been more expensive and power consuming, but commercial devices are available and usage is growing. GaAs devices are reported to be suffering principally from an inadequate state of technological development. The competition between the silicon and gallium arsenside groups is intense.

In order to integrate electrical and 1300 or 1550 nm optical devices on the same substrate, it would be desirable to develop an integrated circuit technology compatible with InGaAs or InGaAsP. This work is in its infancy.

1.3 LASERS AND LIGHT-EMITTING DIODES (LEDs)

Second only to the fiber transmission medium, the sources used in fiber optic telecommunications systems are the elements that most importantly determine the system configuration and performance. Semiconductor injection lasers and

LEDs are universally used at the present time; their size, power output and efficiency are well matched to the need. Both lasers and LEDs are used extensively and both will continue to be important because they fill different needs as is outlined below.

The chapters that pertain to the sources and related technologies are: Chapter 12, LEDs; Chapter 13, semiconductor lasers; Chapter 16, semiconductor epitaxial growth methods; Chapter 19, lightwave transmitters; and Chapter 15, integrated optical and electronic devices.

1.3.1 LEDs

The most extensively used LED is the surface emitter, shown in cross section in Fig. 12.1. More power output and a somewhat narrower spectral width is obtained from the edge-emitting LED, shown in Figs. 12.4 and 12.5. The edge emitter has a structure more like a laser, and shares some of the laser's complexity both in fabrication and in driving-circuit requirements.

The surface emitters may be optimized for maximum power output or for maximum modulation bandwidth, but the two optimum conditions are mutually exclusive. The level of doping the semiconductor for maximum power output results in bandwidths under 100 Mb/s. However, by increasing the doping level to shorten the carrier recombination lifetime the modulation bandwidth can be increased to 500 Mb/s at the expense of reduced output power. Part of the decrease in carrier lifetime is in the form of non-radiative recombination and is an intrinsic material effect, so the power/bandwidth trade-off may be expected to be permanent unless a different material system is used.

The advantages of the surface-emitting LED are very long device lifetime, low cost and insensitivity to ambient temperature; the limitations are low power delivered into single-mode fiber (1–10 microwatts) and broad spectral width (approximately 110 nm in the 1300–1500 nm region).

The edge-emitting LED can be viewed as a modified form of laser, one in which the cavity-forming reflectors at the ends have been eliminated. Many of the laser structures described in Chapter 13 can be modified to form an edge-emitting LED. When the length of the active region behind the emitting surface is long enough, the stimulated-emission effect causes narrowing of the emission line and the device is called a superluminescent diode. Linewidths of 75 nm at 1300 nm have been observed at low output power; however, when the device is driven to high output power the linewidth may go to 100 nm or more. The advantages of the superluminescent diode are: increased power coupled into the fiber, resulting from a narrower output beam; narrower emission linewidth under some conditions; high modulation rate without compromising output power level. The disadvantages of the superluminescent diode arise from its similarity to the laser: greater structural complexity and sensitivity to ambient temperature and to current due to proximity to threshold of oscillation.

1.3.2 Lasers

Injection lasers are so important to fiber optic telecommunications that we have devoted a major chapter to this topic, Chapter 13. A tremendous research and development effort has been invested (and still continues) in order to first, solve the performance problems that the early devices exhibited, and second to refine the device to achieve more sophisticated performance properties.

We cite here a few of the problems, and the avenues to solving them. Device lifetime was marginally satisfactory in about 1979, and much further effort has gone into improving it; better material purity and cleanliness during the fabrication process were among the techniques used to achieve better life. Lasers still are not as long lived as LEDs, possibly due to their greater complexity. Early lasers exhibited more than one transverse mode, especially when driven well above threshold. In order to overcome this tendency, many varieties of transverse guidance methods were incorporated into the waveguiding active medium; several of these structures are shown in cross section in Figs. 13.13 and 13.14. Another undesired characteristic observed was output power not only at a frequency where the semiconductor gain was nearly a maximum, but also at adjacent frequencies where the longitudinal laser cavity had additional resonances and the semiconductor gain was only slightly below its maximum. These side modes gave the laser a multi-longitudinal mode output and total emission line widths on the order of 5 nm. This relatively broad spectral width, in combination with fiber dispersion, caused significant pulse broadening at the receiver of transmission links that were 10s of km long (unless the system operated near the fiber's 1300 nm dispersion minimum). In order to reduce the side modes of the laser, additional complexity was introduced into the laser structure. Whereas the simple Fabry-Perot cavity gives the same resonant transmission loss for successive longitudinal modes, distributed feedback circuits (DFB) (sketched in Fig. 13.18a) gave highly selective longitudinal resonances. The DFB laser, however, proved even more complex than initially recognized, and required considerable refinement in order to make it reproducible. External resonant cavities were used as another alternative to approximate a single longitudinal mode output; these are sketched in Figs. 13.16 and 13.17.

Occasionally, even a simple Fabry-Perot laser will oscillate in a single longitudinal mode, probably as the result of accidental imperfections introducing as little as a few tenths of a dB extra loss at the frequency of the side modes. This confuses the study of improvements; tiny loss effects mask the desired changes.

It can be appreciated from this brief summary that a large number of alternative laser structures were proposed and used experimentally. This situation has persisted, and the various commercial lasers differ significantly in their internal laser structures. The complexity has contributed to their cost, frequently in the $1000 range, for the relatively small numbers used in intercity and undersea cable systems. (For loop applications, a widely applied system would require much larger numbers, and this will make lower cost possible.)

Looking toward coherent systems for future applications, the width of the laser spectrum within a single longitudinal mode becomes significant. The laser types noted above and others have been the subject of innovation and study to try to meet the system needs. This work is currently going on, and modifications of the laser resonant circuit are still being made. For example, the addition of a long low-loss passive waveguide section within the resonant system has been shown theoretically to decrease the laser linewidth advantageously. The most attractive way to build this circuit is to integrate the passive and active section on the semiconductor chip, and new fabrication techniques are being sought to realize it in practice.

The performance numbers for injection lasers vary considerably depending on the structure, the imperfections of the particular sample and operating conditions. Increased drive yields increased power output (within an operating range) and simultaneously improves the ratio of principal-mode power to side-mode power. Increased drive also decreases the emission linewidth. Temperature changes of a fraction of a degree are frequently critical due to semiconductor gain changes, both in magnitude and in the frequency of maximum gain. From these comments, it is not surprising that the circuitry for driving a laser is complex. This subject is covered in Chapter 19.

The order of magnitude of numbers describing lasers are: length, 250 μm; at 1300 nm carrier wavelength, longitudinal modes are separated by 0.85 nm; a multi-longitudinal mode Fabry-Perot laser might have a total spectral width of 3–5 nm; a distributed feedback laser might have a linewidth of 10 to 100 MHz at a few milliwatts output power. Direct modulation of the laser is feasible at rates of at least 1.7 Gb/s and at higher rates in some commercial devices.

All of the above performance numbers have been exceeded in laboratory experiments. For example, direct modulation of a laser has been demonstrated at 16 Gb/s (Fig. 13.24), and laser linewidths well below 1 MHz have been observed in external cavity lasers. The reader is referred to Chapter 13 for a more complete view.

1.3.3 Epitaxial Growth Methods for Lightwave Devices

The initial feasibility and ultimate cost of sophisticated lasers and indeed all semiconductor lightwave devices rests on the initial production of a multilayered wafer. The starting material is a thin slice of single crystal InP (for 1300 and 1550 nm devices) on which layers of InGaAs or InGaAsP only tenths of a micron (for lasers) or several microns thick (for detectors) are grown epitaxially. Lasers and LEDs, as described in Chapters 12 and 13, require a very thin active layer and bounding layers of slightly different composition to confine the electrons and partially confine the light to this active region. The electrons that produce the gain are injected at a pn junction, which is formed near the active layer. The starting "seed" for the composite structure could be on the order of 1 cm square, whereas a typical laser or LED is on the order of 250 μm square. Thus the wafer could yield hundreds of devices by processing the entire wafer at one time, using a mask that delineates the desired structure hundreds of times. Actually, many masks and many repetitions of processing the entire wafer are needed, because the complex structure of a laser is produced by selectively etching through some of the layers and regrowing single-crystal layers of a slightly different material in patterns designed to confine the lightwave to single-mode propagation. In this complex series of steps, it is essential to avoid defects on the surfaces between the various grown layers, and the degree of success achieved dictates not only the yield of operable devices but also the level of performance found in the accepted devices. There is a great deal of empirical art as well as science involved in the processes used to make the wafers.

Four growth techniques are presently used: liquid-phase epitaxy (LPE), vapor-phase epitaxy (VPE), metalorganic chemical vapor deposition (MOCVD) and molecular-beam epitaxy (MBE). By far the most commonly used process now is LPE, whose elements are shown in Fig. 16.1. Growth of the thin layers is accomplished from melts containing the desired growth compositions in wells #1–#4, and successive layers are grown by moving the wafer in the slider portion of the carbon boat in steps from well to well. This process has proved easier to implement than the others, but is still difficult. The size of the wafers is limited in practice, and the minimum thickness of good layers is about a hundred nanometers. In VPE, diagrammed in Fig. 16.5, larger substrates can be used and thinner, uniform layers can be grown, which is being used for some devices. MOCVD is another vapor-phase process and is noted for yielding high purity materials with abrupt interfaces between layers; thickness as low as 0.8 nm has been achieved and it provides an alternative fabrication technique for devices originally pioneered with MBE. Finally, MBE is diagrammed in Fig. 16.12. As the name implies, it uses beams of molecules for the deposition and requires ultra-high vacuum. MBE is outstanding in being able to make very thin layers. It has made possible radically new types of devices called "quantum well" devices that are referred to in the discussion of detectors, Chapter 14. Initial MBE techniques had difficulty handling phosphorous, a component of AlGaAsP, although current work appears to overcome this problem. In other MBE work, attention has gone to another quaternary system, InGaAlAs, which is also of interest in the 1300–1550 nm region. Avalanche photodetectors have been made in it.

Each of the various processes has advantages and limitations. LPE is least expensive and easiest to implement, but the superior capabilities of the other processes are leading to their adoption in circumstances where their advantages pay off.

1.3.4 Reliability of Lasers and LEDs

Device lifetime of the optical elements represented one of the major uncertainties when the new art was introduced into commercial service. Significant efforts were invested to evaluate and improve it and Chapter 17 gives an overview of this work. As cited there, experience has supported the conclusion that InGaAsP lasers operated at 10°C can have a median lifetime in excess of 25 years, and InGaAsP LEDs operated at 70°C can have a median lifetime in excess of 100 years. These excellent reliability projections for lasers and LEDs are the result of optimal device design and careful controls in fabrication. Because of the unique requirements for undersea applications of lasers, much study has been undertaken to devise a sorting strategy to assure these lifetimes.

Device failure is more typically characterized by diminished light output, requiring increased drive current to maintain the needed level of power, followed eventually by burnout. The factors most typically responsible are appearance of lossy crystalline defects growing through the active region and disappearance of the current-confining capability of the material near the active region. Both effects are changes in the semiconductor. They are speeded up by elevation of the device temperature and in some circumstances by increased injection current.

In order to obtain data in shorter elapsed times, device testing is done at elevated ambient temperatures, and carefully documented methods are devised to extrapolate the observations to the temperatures expected for the actual field operation of the devices. Of course, there are uncertainties in the accelerated aging process, but experience has supported it as a valid way to produce satisfactory commercial devices.

1.4 INTEGRATED OPTICAL AND ELECTRONIC DEVICES

Integrated optical circuitry, using planar lithographic techniques to form optical circuits, goes back to 1969, when it was conceived as a much-needed replacement for optical-bench arrays of components, which were commonly used to conduct research at that time. The advantages recognized were smaller size, ruggedness, reliability and the ability to reproduce complex circuits rapidly and in an economic manner once the design and fabrication procedures had been established. The simplest forms of integrated optical circuits involve passive components or components in which the only "active" effect is the electrooptic or magnetooptic effect, and these devices advanced steadily during the 1970s. As the art advanced, the natural desire to integrate the source and/or detector with other optical and with electrical devices became technically feasible and a branch of work termed Optoelectronic Integrated Circuits (OEIC) appeared. The ultimate form of OEIC would be an optical repeater monolithically formed in a compatible set of material compositions, using lithographic techniques.

These various types of integrated optical circuits are at widely varying degrees of practicality at the present time. Chapter 9 reviews the broad concept of integrated optics. Chapters 10 and 11 review the circuits without sources or detectors integrated on the same chip, and Chapter 15 covers the OEICs.

1.4.1 Optoelectronic Integrated Circuits

The most desirable end product clearly would be the OEIC, so we consider that form first. In order to review the general fabrication procedure used and the end product that results, we can look at Fig. 9.7. The process begins with the use of a mask to create a pattern in photoresist; then a series of steps involving etching, depositing new layers, and in some instances employing diffusion (or ion implantation) ultimately create waveguiding circuits in planar form. The illustration is simplified; in semiconductors especially there are dozens of steps employed and multiple use of different masks in successive operations. We end up with a component having a width and length on the order of a centimeter or less and 0.1 cm or less in thickness.

The specific objective for the OEIC could be (1) improvement of the performance of an optical device or cluster of devices, (2) improvement of the performance of an electronic device, (3) the creation of a basically new type of device with desirable attributes and (4) lower cost. For (1), we combine the optical transmitter with its electronic driver or combine the detector with the low-noise electronic amplifier stages of the receiver; these integrations in principle enable one to reduce the parasitics, which limit pulse speed, or to reduce the noise sources at the receiver input. We shall see that these potential advantages have been only partially accomplished in practice. For (2) we consider using optical waveguides to interconnect electronic devices; a desirable thing to consider because these interconnections in electrical form are limiting features in all-electronic IC designs. We look to an open-ended future for (3); direct interactions between two or more optical waves are the type of phenomena that could prove advantageous.

Because most electronic integrated circuits are formed in silicon, we must depart from the material most commonly used in the electronic IC art. The GaAs system is now used in some high speed ICs, and that material is compatible with the GaAlAs or GaAs systems used for sources in the 800–900 nm regions of the optical spectrum. Some success has been obtained in growing GaAs crystalline layers on silicon substrates, and that material combination could become attractive by enabling the use of well established silicon ICs integrated with optical devices. For the 1300–1550 nm wavelength region, where fiber transmission is most attractive, another material system such as InGaAsP, InGaAs or InP is needed for the optical elements. Although it is possible in principle to make transistors in that system (and some have been made), that art is in its infancy; the GaAs electronic circuit art is relatively advanced as compared with InGaAsP/InP, and, as noted, even GaAs is less developed than the art for silicon.

Despite the early stage of OEIC development, numerous research successes have been achieved. For example, Fig. 15.12 shows several heterostructure bipolar transistors (HBT) integrated with a laser diode (LD) to form an integrated transmitter assembly in the 800 nm optical region. As recounted in Chapter 15, both transmitter and receiver assemblies have also been made for the 1300 nm region. Figure 15.17 shows the receiver sensitivities of some of the OEICs as points, to be compared with the dotted line below them as representative of the level achievable with separately fabricated optical and electrical elements (hybrid technology). It is clear that the potential improvement in receiver sensitivity has not yet been reached, but this art is at a very early phase. One can expect improvement as the result of further development. In addition, a practical application may take advantage of the small size and potential lower cost of OEIC without demanding better performance.

1.4.2 Electrooptic Waveguide Devices

Chapter 11 covers integrated optical devices that are passive in the sense that no optical gain or source of photons is present, but may be temporally active—changes in drive signals may produce time dependent changes in frequency, phase, intensity, polarization or direction of lightwave transmission. The most common physical property used to make those changes is the electrooptic effect, which is a change in the index of refraction of a material in proportion to an applied electric field. The electrooptic effect is very fast, occurring in less than a picosecond, and makes possible multigigabit rate devices.

This technology is very advanced, particularly as practiced in titanium-diffused lithium niobate. Research models of phase modulators, amplitude modulators, tunable frequency-selective filters, optical multiplexers, N × N optical switches and polarization controllers have all been demonstrated. The control voltages required differ depending on the bandwidth needed; a phase modulator having a bandwidth of about 5 GHz, a loss of 1.8 dB (fiber-to-fiber) and a length of 1 cm required a drive of 8 volts to produce 180 degrees phase change.

A few titanium-diffused lithium niobate waveguide devices have been offered as commercial products, and more could be made available if the need appeared. New research focuses on devices fabricated in semiconductors in order to make them integrable with sources and detectors. Some progress has been made, but much remains to be done.

1.4.3 Passive and Low-Speed Optical Components

In this section, we discuss several technologies that have in common the fact that they are entirely passive—they do not create or detect photons and they do not manipulate photonic signals in times of the order of the bit period. They do, however, fill important needs. Many were originated to fill an immediate commercial need in multimode fiber systems. As early field applications were developed in the later 1970s, system designers called for WDMs and power splitters in multimode form, and they found that the integrated optics technology could not provide them. Efforts to extend the integrated technology into multimode form were not immediately fruitful, so an alternative appeared—microoptics. As described in Chapter 10, innovative engineers devised miniature forms of lenses, mirrors, prisms, filters and gratings compatible with the very small optical beams that were available from the transmission fibers. These same elements also proved suitable for use with the beams from single-mode fibers and were used there also. The microoptic technology typically requires manual assembly of the above elements in order to produce a power divider, a multiplexing filter or other components and lacks many of the virtues of the lithographic technology; however, it provides an immediate solution.

A key element of the new microoptic technology is the GRIN-rod lens, a device that has focusing properties similar to a convex lens but has a uniform rod shape. (GRIN stands for graded index, which is suggested by the device's index of refraction profile along a radius.) A quarter-pitch transformer can be formed by proper selection of the length of the rod, and as illustrated by the central set of ray paths in Fig. 10.2, the output of a fiber is transformed into a larger beam of parallel rays. This is a building block used repeatedly in microoptic component design. For example, Fig. 10.10 shows a 4-port hybrid using four GRIN-rods and a partially reflecting planar element. As the figure shows, there are a number of individual piece parts that must be fabricated, aligned and secured in position, which limits the suitability of such components for low-cost high-volume manufacture. Figure 10.16 shows a WDM using a GRIN-rod and a tilted grating in microoptic form; the lower three fibers carry individual wavelengths to be combined and the top fiber collects the assembled set of channels. GRIN-rod lenses with extremely low aberration have been developed, and this has made it possible to make microoptic components for use with single-mode fibers.

Some combining of the planar lithographic technology with the microoptic technology has occurred. The minimum channel spacing for the filter of Fig. 10.16 is set by the properties of the grating and the physical spacing between the waves entering the GRIN-rod transformer. The size of the core and cladding of the fiber limits how closely the fibers can be assembled on the GRIN-rod, and the resulting spacing can be larger than desired, especially with single-mode fibers. Closer packing of the multiplexed channels has been achieved by the design of Fig. 10.19, where a fanout of the waves reflected from the grating allows small field spacing (20 μm) at the grating side of the integrated circuit concentrator, while providing the needed space for the fiber core and cladding (125 to 250 μm) at the other end.

Mechanically activated optical switches have been found useful where the need is for a seldom activated device and speed is not a requirement. Figure 10.22 illustrates a 4 $\times$ 4 switch using movable dielectric blocks to deflect the collimated beams from one path to another. It is apparent that this is a latching structure, which requires power only to make the change in connection. It seems unlikely that this approach will prove attractive for large N $\times$ N switches because of increasing size and limitations in the distance over which the collimated beams can be maintained free of crosstalk.

Figure 10.33 illustrates that ordered arrays of fibers can be switched in a mechanically actuated assembly; the array in the foreground of the figure can be moved as a group to connect to one of two or more other arrays in the background by means of a vertical translation. The use of precisely etched grooves for positioning the fibers in the array assures accurate alignment in the array's plane. This type of device maintains its precision after a very large number of operations, and could be useful in providing standby facilities for ribbon fiber groups, or in other architectural rearrangements of transmission facilities.

The microoptic technology permits designing and producing small numbers of needed components without the expense or delay associated with developing integrated forms of the components.

1.5 PHOTONIC LOCAL NETWORKS AND PHOTONIC SWITCHING

Chapter 26 provides a tutorial and overview of the use of lightwave technology in local area networks (LANs); also provided is an insight into the hardware aspects of photonic switching applicable to lightwave systems more generally. LANs are relatively new, but are already in use based on wire pairs or coaxial cables as the transmission medium.

The region covered by a LAN might have a diameter in the range 100 m to 10 km, might have only a few or hundreds of users connected to it and would be digital in format using rates of 10 Mb/s or higher on the network.

The functional need might arise in the home, in a factory, in a closed environment of a single business or might interconnect a more general group with a common interest. In the home, professionals are already using terminals or computers to access data or high-powered computers located in the main business building. In a factory, LANs provide process control and associated information flow. At the business location, LANs are used to interconnect computers with users in a variety of ways: word processing for secretaries, inventory control, mathematical computations for engineers and scientists, library access, data bank access generally and electronic mail. In universities, LANs provide remote classrooms for lectures and demonstrations. Entertainment or TV programming could be provided and some use of video in home shopping has occurred.

The principal advantages that fiber appears to offer for LANs are higher bit rates and longer distance coverage as a consequence of lower transmission loss, and immunity from electromagnetic interference. These advantages can translate into the ability to provide a higher bandwidth service than can realistically be provided using metallic media, or can mean increased speed or lower costs for the services already being provided.

The principal disadvantages of fiber LANs are (1) very limited range between available transmitter power and required received signal, and (2) relatively expensive and signal-consuming lightwave connectors to the common fiber medium, ultimately constrained by the limited dynamic range cited in the first instance.

1.5.1 Topology

The topology of the transmission lines interconnecting the users may be a linear bus shown in Fig. 26.3, a ring as shown in Fig. 26.1, a star as shown in Figs. 26.6 and 26.4 or as a direct interconnection of all of the users (called a mesh). The double ring, with transmission in both directions around separate facilities on the same path, has been proposed as a National Standard. It can be arranged to bypass a failed terminal. Each topology, however, has advantages and disadvantages, and all are being offered commercially and used at the present time on wire facilities. The reader is referred to Chapter 26 for more discussion.

1.5.2 Modulation Format and Access Protocols

Every specific system has, in addition to choice of topology, a choice of access protocol and modulation format. On the most extensively used LAN, known as Ethernet®, only one user has use of the bus at a given time, and a packet of data is sent once "possession" of the bus is obtained through a process of contention. The data rate is 10 Mb/s, and coaxial cable is used with very simple connecting arrangements. This system is so extensively employed that any new systems are likely to be arranged so as to at least interconnect successfully with it.

A general usage of LANs leads one to expect that two types of services may be desirable: circuit-switched service, wherein a user has total use of a channel for whatever period of time he/she chooses, and packet-switched service, which is well adapted to sending small amounts of information at intermittent times. Some LANs provide both types of service.

All of the modulation techniques developed for point-to-point transmission are candidates for use in LANs: wavelength (or frequency) division multiplex, time division multiplex and code division multiplex, which also goes by the name "spread spectrum."

1.5.3 The Outlook for the Future

There is no consensus on the best LAN arrangement, and almost all possibilities are being explored. The systems being offered commercially most commonly use wire pairs or coaxial as the medium, use a variety of topologies and sell for $200 to $1000 per access point. In a few current offerings fibers are included, usually as interconnecting point-to-point elements in a large network. Perhaps the most urgent need for LANs is by hospitals and universities, which are dependent on rapid flow of information between a large number of points located within a relatively confined area. It seems likely that the need for the functions of LANs will grow rapidly in the next five years, and that both closed systems and systems integrated with more general information transport will be useful.

Chapter 2

Fiber Types and Their Status

PETER KAISER

Bell Communications Research, Inc., Red Bank, New Jersey

DONALD B. KECK

Corning Glass Works, Corning, New York

2.1 INTRODUCTION

Optical fibers have become the preferred transmission medium for telecommunications, and different types of fibers have evolved over recent years, which best satisfied changing systems requirements and specialized applications. The trend has been to minimize the number of fiber types required for the public network in order to reduce manufacturing costs, to facilitate installation and maintenance procedures, and to assure compatibility of fibers supplied by different manufacturers. Besides long-distance transmission, which has been the primary application area, optical fibers are also used for many short-distance applications where ease of connection and compatibility with lowest-cost terminal equipment are more important than high information rate and long transmission distance. In addition, a variety of fibers have been developed for specialized applications. In this chapter, some of the more important types of optical fibers are reviewed, with emphasis being placed on telecommunication applications. Chapters 4 and 8 contain additional information on fiber designs, fabrication, and transmission characteristics, which are relevant to this chapter.

Optical fibers initially proposed for telecommunications, and the first low-loss fiber fabricated by Corning Glass Works in 1970, were single-mode (SM) fibers, since the transmission characteristics of such fibers were expected to be best suited for telecommunications. However, development efforts first concentrated on large-core multimode (MM) fibers because of difficulties encountered in coupling light into SM fiber cores only a few micrometers in diameter, and in splicing and interconnecting such fibers—not to mention the lack of suitable semiconductor light sources at that time. A near-parabolic shape of the refractive index profile of MM fibers yielded large bandwidth, and resulting graded-index fibers were used in the "first-generation" MM fiber systems, which operated in the 800 nm wavelength region. Second-generation systems took advantage of the lower loss and zero material dispersion in the 1300 nm wavelength region, resulting in graded-index MM fiber systems operating in this "second window." The growing demand for longer-distance, larger-bandwidth transmission—initially for trunk, and later for submarine cable applications— resulted in the re-introduction of SM fibers in the third-generation lightwave systems operating in the 1300 nm wavelength region. Fourth-generation systems are expected to utilize the lowest-loss 1550 nm wavelength region ("third window"), with coherent transmission eventually leading to fifth-generation lightwave systems.

In the following, we describe some of the more important fiber types, their technical capabilities, and their performance characteristics. Emphasis is placed on fibers for telecommunications, and especially single-mode fibers, whose various types are discussed first in Section 2.2. Key transmission parameters that help to categorize SM fibers are summarized in Section 2.3. They include: cut-off wavelength, mode-field diameter, dispersion and attenuation characteristics, and bend-loss behavior. Overall design trade-offs are also described. Section 2.4 presents the status of MM fibers whose use has recently tended toward short-distance applications. Section 2.5 deals with the physical and mechanical properties of optical fibers relevant to their areas of application. Other special fibers, and "mid-infrared" fibers intended for future very long distance transmission, are described in Section 2.6. Concluding this chapter, Section 2.7 discusses expected future trends.

2.2 SINGLE-MODE FIBERS

Because of their numerous advantages, single-mode fibers have become the most widely used optical transmission medium in the networks of the telecommunication companies (Kaiser, 1985; CCITT 1985). The reasons for this include:

- Assurance of the once-only installation of a new transmission medium with more than 20 years expected lifetime;
- Functionally the lowest-loss, largest-bandwidth transmission medium available;

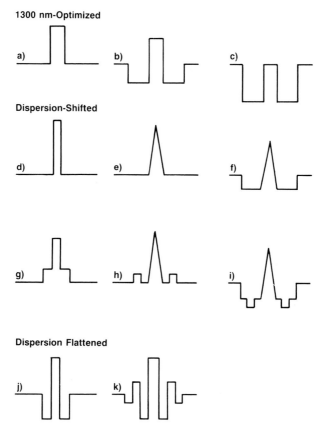

Fig. 2.1 Typical index profiles of a) 1300 nm-optimized, b) dispersion-shifted, and c) dispersion-flattened single-mode fibers.

- Substantial upgrade capability for future large-bandwidth-consuming services with either higher-speed devices, wavelength multiplexing, or coherent transmission technology;
- Superior transmission quality because of the absence of modal noise;
- Compatibility with integrated-optics technology.

Important design considerations for SM fibers are low transmission loss, suitable dispersion characteristics, low splice loss, and low macro- and micro-bending loss. With the dispersion behavior being the dominant distinguishing factor, one can identify three major SM fiber categories whose representative refractive index profiles are shown in Fig. 2.1: 1300 nm-optimized SM fibers, dispersion-shifted SM fibers, and dispersion-flattened SM fibers.

2.2.1 1300 nm-Optimized Single-Mode Fibers

The most commonly used SM fibers are near-step-index fibers, which are dispersion-optimized for operation in the 1300 nm wavelength region. The fibers are either of the "matched-cladding" (MC) or "depressed-cladding" (DC) design. In MC fibers, the region external to the core has a uniform refractive index, typically that of pure silica. In case the core itself consists of pure silica, a lower-index cladding is achieved through fluorine doping. Typical mode-field diameters (MFDs) of MC fibers are around 10 μm, and the fractional core/cladding index differences are around 0.3%. Changing user requirements have brought about additional MC designs with reduced MFDs around 9.5 μm and core/cladding index differences of 0.37% to improve the bend-loss performance in the 1550 nm region (Kanamori et al., 1986; Dixon et al., 1987).

In DC fibers, the cladding region next to the core is lower in index of refraction than that of the outer cladding region. Changing the depth and diameter of the index depression results in more degrees of freedom. Typical MFDs of DC SM fibers are 9 μm, and positive and negative fractional index differences used in commercial fibers are 0.25% and 0.12%, respectively (Jablonowski, 1986). The topographic index profile of a DC SM fiber is shown in Fig. 2.2a (Bachman et al., 1986). Both the MC and DC-SM fibers described above have become known as G.652 fibers recommended by CCITT (Table 2.1), and comprise the majority of SM fibers being used in telecommunication networks.

2.2.2 Dispersion-Shifted Single-Mode Fibers

The zero-dispersion wavelength of SM fibers can be shifted to the lowest-loss 1550 nm wavelength region by compensating the material dispersion with increased waveguide dispersion (Figs. 2.3 and 2.9). This can be achieved in a number of ways, such as through a reduction of the core diameter and an accompanying increase in the fractional index difference. The design flexibility required to achieve a desired loss, dispersion, mode-field diameter, and bend-loss characteristics, has resulted in proposals for different refractive-index profiles, a representative sample of which is shown in Fig. 2.1. Noteworthy among them are fibers with a higher-index annular ring surrounding the core structure, Fig. 2.1h, (Bhagavatula et al., 1983); and triangular-index profile fibers, Fig. 2.1f, to achieve larger modified diameters and lower loss than step-index, dispersion shifted designs, (Saifi et al., 1982; Ainslie et al., 1982). An annular index trench around the core improves the mode confinement properties (Reed et al., 1987). Similar reasoning has previously resulted in the design of a dual-step-index dispersion-shifted fiber (Ohashi, 1986). Particular attention is being paid to reducing the bend-loss sensitivity at 1550 nm. While the MFD is the most

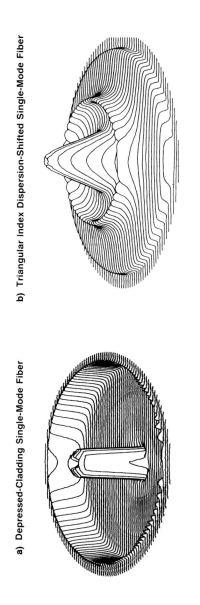

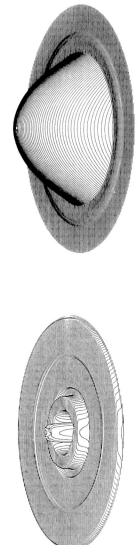

b) Triangular Index Dispersion-Shifted Single-Mode Fiber

d) Graded-Index Multimode Fiber

a) Depressed-Cladding Single-Mode Fiber

c) Quadruply-Clad, Dispersion-Flattened Single-Mode Fiber

Fig. 2.2   Topographic index profiles of different types of optical fibers (Courtesy Philips Research Laboratories and Corning Glass Works).

TABLE 2.1
CCITT Recommendation G.652

Outer Diameter	125 microns ($\pm 2.4\%$ max.)
Modefield Diameter	9 to 10 microns ($\pm 10\%$ of nominal value)
Cut-Off Wavelength	1100 to 1280 nm
1550 nm Bend Test	≤ 1 dB excess loss for 100 turns of 7.5 cm diameter
Dispersion	≤ 3.5 ps/nm $\cdot$ km between 1285 and 1330 nm
	≤ 6 ps/nm $\cdot$ km, 1270–1340 nm
	≤ 20 ps/nm $\cdot$ km at 1550 nm
	Under consideration: $1295 \leq \lambda_0 \leq 1322$ nm
	Disp. Slope $S_0 \leq 0.095$ ps/nm$^2 \cdot$ km

significant factor influencing the bend-loss behavior, it is not sufficient to characterize the bend sensitivity completely (Petermann, 1987). Other factors, such as the exact radial mode-field distribution and the cut-off wavelength also affect the observed behavior. Mode-field diameters of between 7 and 9 μm at 1550 nm wavelength are typical for the above DS fiber designs, and a suitably low cut-off wavelength generally permits systems operation in the 1300 nm window. Average losses of 0.21 dB/km achieved in a production environment (Croft *et al.*, 1985) are comparable to those of 1300 nm-optimized SM fibers. Dispersion-shifted SM fibers are presently considered for CCITT standardization.

2.2.3 Dispersion-Flattened Single-Mode Fibers

Dispersion-flattened (DF) fibers with low dispersion over the low-loss wavelength region between 1300 and 1600 nm relax the spectral requirements for light sources and allow flexible wavelength multiplexing. Dispersion-flattened SM fibers with multi-layer index profiles have been fabricated whose increased waveguide dispersion is tailored to achieve a low dispersion (e.g., < 2ps/nm $\cdot$ km) over the above wavelength region (Bhagavatula *et al.*, 1983). Typical refractive index profiles of DF fibers are shown in Figs. 2.1 and 2.2, and their dispersion characteristics in Fig. 2.3 (Cohen, 1985). Present efforts concentrate on identifying the optimum MFD and improving the bend-loss sensitivity of DF-SM fibers (Bachman *et al.*, 1987). The eventual use of DF-SM fibers in the public network will depend on their compatibility with, and cost/performance ratio relative to conventional SM fibers.

2.2.4 Polarization-Retaining Single-Mode Fibers

The polarization behavior of SM fibers plays an important role in coherent transmission systems and sensor applications. Single-mode fibers with ideal circularly-symmetric geometry propagate two orthogonal modes with identical

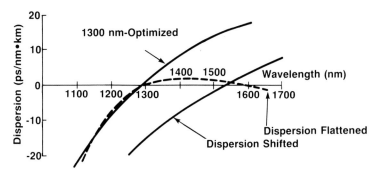

Fig. 2.3   Dispersion characteristics of different types of single-mode fibers.

propagation constants. The polarization degeneracy can be removed through geometric and/or stress-induced asymmetries, which create two preferential axes with slightly different refractive indices, and resulting different propagation constants β_x and β_y. Representative designs of polarization retaining SM fibers are shown in Fig. 2.4. An efficient way to impart a large birefringence is to insert within the cladding, and diametrically-opposed to the core, a material with thermal expansion coefficient sufficiently different from that of pure silica, such

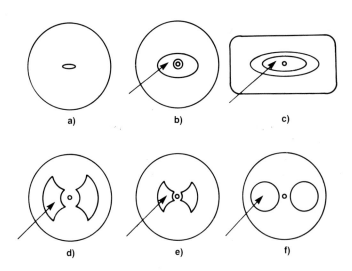

Note: Arrow Indicates Stress Inducing Member

Fig. 2.4   Representative designs of polarization-retaining single-mode fibers.

as boron-doped silica. Minimum losses achieved with such fibers are 0.22 dB/km at 1550 nm (Sasaki, 1987).

Linearly-polarized light launched with arbitrary orientation into polarization retaining fibers will propagate as an elliptically-polarized wave with beat length λ_b determined by $\lambda_b = \lambda/B = 2\pi/(\beta_x - \beta_y)$. The modal birefringence B is larger than 10^{-3}, and smaller than 10^{-9} for high- and low-birefringence fibers, respectively (Noda *et al.*, 1986). Polarization-retaining SM fibers have beat length as small as a few millimeters, while λ_b of standard SM fibers with fabrication-related birefringence can be as large as a few meters. A measure of the degree of cross coupling is the extinction ratio, ER, determined by

$$\text{ER(dB)} = 10\log(P_y/P_x) = 10\log[\tanh(hL)],$$

where P_x and P_y are the original and cross-coupled power levels measured after a fiber length L, respectively, and h is the mode coupling coefficient (Kaminow *et al.*, 1980; Noda *et al.*, 1986). The highest extinction ratios achieved were 22 dB for 26 km of fiber, and 32 dB at 1550 nm for 1 km fiber lengths.

A further improvement in the polarization characteristics of SM fibers can be obtained by providing a mechanism for preferentially attenuating one of the two linearly-polarized modes. This can be done either with diametrically-opposed absorbing regions, or by carefully constructing the azimuthal refractive index profile such that one of the LP_{01} modes is cut-off. This technique has been used to fabricate "single-polarization" fibers with extinction ratios greater than 50 dB. A recent review of the status of polarization retaining fibers has been published by Noda *et al.* (1986).

Most commercial polarization retaining SM fibers developed thus far are used for sensing applications and are operated single-mode in the 800 to 900 nm wavelength region. While some coherent transmission experiments have been performed around 1300 nm with longer lengths of polarization retaining fibers, their use in the 1300 and 1550 nm wavelength regions has been mainly confined to the fabrication of components, and as interconnection cables for single-polarization devices. Experiments have shown that the linear polarization can be recovered in conventional SM fibers through polarization transformers at the receiving end (Harmon, 1982; Okoshi, 1985). This feature may be important for future coherent communication systems, the majority of which is expected to use conventional SM fibers.

2.3 SINGLE-MODE FIBER TRANSMISSION CHARACTERISTICS

As noted earlier, the transmission characteristics of optical fibers play an important role in determining their range of applications. The following sections are therefore devoted to a more detailed discussion of some of the relevant SM and MM fiber parameters.

2.3.1 Cut-Off Wavelength

Single-mode operation occurs above the theoretical cut-off wavelength $\lambda_{c,\text{theor.}}$ determined by the core radius, a, the index of refraction, n, and the fractional core/cladding index difference, Δ, according to

$$\lambda_{c,\text{theor.}} = \frac{2\pi a n (2\Delta)^{1/2}}{V},$$

with $V = 2.405$ for step-index fibers. Since the presence of the first higher LP_{11} mode near cut-off is strongly affected by fiber length and curvature, an effective cut-off wavelength λ_c has been defined by CCITT (Recommendation G.652), which is obtained for a 2 m length of fiber containing a single, 14 cm radius loop. Recommended λ_c values range from 1100 to 1280 nm to avoid modal noise and dispersion problems in the 1300 nm wavelength region. Before stabilizing beyond a certain fiber length (Kitayama *et al.*, 1984), the measured cut-off wavelength decreases with fiber length according to

$$d\lambda_c = -m \log \frac{L}{2},$$

where m depends on the fiber type, and L is the fiber length (in meters). Representative values for m are in the 20 to 60 nm range, with MC fibers tending to be lower than DC fibers. Other than on the fiber design, the value of m depends on the cabling and deployment conditions. The characteristic relationship between theoretical, standardized, and cabled-fiber cut-off wavelength values relative to the systems operating wavelength region is depicted in Fig. 2.5. Since cabled fibers of different lengths and deployment conditions may have widely varying effective cut-off wavelengths, care must be taken that the cut-off wavelength of the shortest fiber section encountered in practical systems, such as a cable-repair section, remains below the systems operating wavelength in order to avoid modal noise (Cheung and Kaiser, 1984; Stone, 1984). Because of the length and bend dependence of the cut-off wavelength, the 2-m cut-off wavelength can be chosen higher than the systems operating wavelength provided the effective cut-off wavelength for the shortest cable length is below the operating wavelength. While the CCITT-recommended upper limit for the λ_c of the primary-coated fiber is 1280 nm, λ_c values of up to 1330 nm are being used in cabled fibers, where both the length and curvature of the fiber in the cable reduce the effective cut-off wavelength to below the systems operating wavelength. Recent experiments with one specific cable structure containing DC SM fibers showed that the effective cut-off wavelength of a 22 m cable section with a 7.5 cm diameter splice loop at each end was on the average 72 nm lower than the 2-m cut-off value of a similar set of primary-coated fibers (CCITT, 1987).

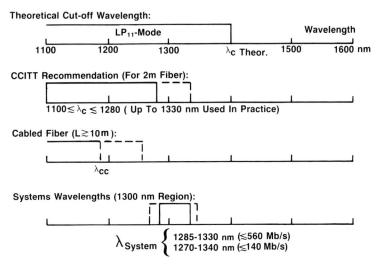

Fig. 2.5 Definitions of different types of fiber cut-off wavelengths, and relationship to systems operating wavelength range.

2.3.2 Mode-Field Diameter

The mode-field diameter (MFD) of SM fibers takes into account the wavelength-dependent field penetration into the cladding and is, therefore, a better measure of the functional properties of the fiber than the core diameter. For step-index and parabolic-index fibers operating near λ_c, the field is well approximated by a Gaussian distribution, and the MFD is simply the $1/e^2$ width of the Gaussian which best fits the power distribution. For arbitrary index distributions, however, a more general definition of the MFD has been proposed by Petermann (1983) and recommended by CCITT as discussed in more detail in Chapter 8. Using the "Petermann II" definition, good agreement is obtained between the MFDs determined for arbitrary-index fibers using different measuring techniques (Anderson *et al.*, 1987). The wavelength-dependent MFDs of some typical SM fibers are shown in Fig. 2.6.

2.3.3 Attenuation

Attenuation in state-of-the-art SM fibers is primarily caused by the intrinsic Rayleigh scattering of the doped fused silica, which decreases with the inverse 4th power of wavelength. Beyond 1600 nm, a rapidly increasing absorption is due to the intrinsic infrared tail of the Si-O and/or Ge-O vibrations. Excess

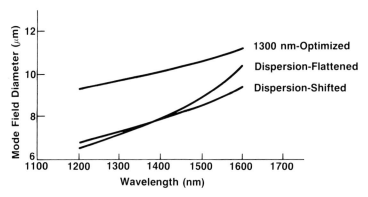

Fig. 2.6 Modefield diameter as function of wavelength for 1300 nm-optimized, dispersion-shifted, and dispersion-flattened single-mode fibers.

losses caused by waveguide imperfections and metallic impurities are negligible even in today's mass-produced fibers made by any of the commonly-used preform fabrication techniques, and minimum losses between 0.15 and 0.16 dB/km at 1550 nm wavelength have been reported (Berkey and Sarkar, 1982; Kanamori, 1986; Csencsits *et al.*, 1984). Average losses obtained in a manufacturing environment are closer to 0.35 and 0.21 dB/km at 1300 and 1550 nm, respectively, for both MC and DC fibers (Croft *et al.*, 1985; Jablonowski, 1986). Hydroxyl (OH) absorption losses at 1380 nm can be practically eliminated through suitable dehydration of the preform, but are typically in the 0.5 to 2 dB/km range. The range of spectral losses of presently manufactured SM fibers is shown in Fig. 2.7.

As noted earlier, macro- and micro-bending losses are one of the more important factors in the design of SM fibers. Generally, these losses manifest themselves as a rapid increase in attenuation beyond a critical wavelength when fibers are bent or otherwise perturbed such as with a "basket-weave test" (Kaiser *et al.*, 1979; Tomita *et al.*, 1983). This sharp loss increase (bend edge) may occur in the 1550 nm region, and may appear in cables under certain deployment conditions (Fig. 2.8, Bhagavatula *et al.*, 1985). Single-mode fibers are more susceptible to bending losses the larger the mode-field diameter, and the lower the cut-off wavelength relative to the operating wavelength. As a measure of low-loss 1550 nm operation of 1300 nm-optimized SM fibers, the excess loss of 100 turns of fiber wound loosely around a 7.5 cm diameter mandrel, should remain below 1 dB (Table 2.1). The values are derived from splice-case deployment conditions for a typical repeater span.

For short-distance applications such as in the subscriber loop, 1300 nm-optimized SM fibers can also be used in the 800 nm dual-mode wavelength

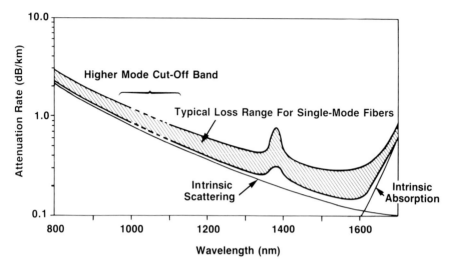

Fig. 2.7 Typical spectral loss range of commercial single-mode fibers.

region, where the effect of modal distortion can be significantly reduced by using a short section of single-mode fiber whose cut-off wavelength is below 800 nm in front of the receiver (Stern *et al.*, 1987).

2.3.4 Single-Mode Fiber Dispersion

The dispersion of SM fibers consists mainly of material and waveguide dispersion:

$$D_{tot} = D_{mat} + D_{wvg}(ps/nm \cdot km).$$

In standard SM fibers, the total dispersion is dominated by the material dispersion of fused silica, which passes through zero around 1270 nm. For germanium and several other dopants, λ_0 is shifted to increasingly longer wavelengths by higher concentrations of the dopant material (Fig. 2.9, Ainslie *et al.*, 1986). Waveguide dispersion becomes more pronounced for smaller core diameters and/or index differences, with the resulting effect of shifting λ_0 also to longer wavelengths. The dispersion around 1300 nm can be obtained from λ_0 and the dispersion slope S_0 according to (Kapron, 1987),

$$D(\lambda) = \frac{\lambda S_0}{4}\left[1 - \left(\frac{\lambda_0}{\lambda}\right)^4\right].$$

Typical values for S_0 are 0.092 for standard SM fibers, and between 0.06 to 0.08 ps/km · nm for some dispersion-shifted fibers. Alternately, a maximum disper-

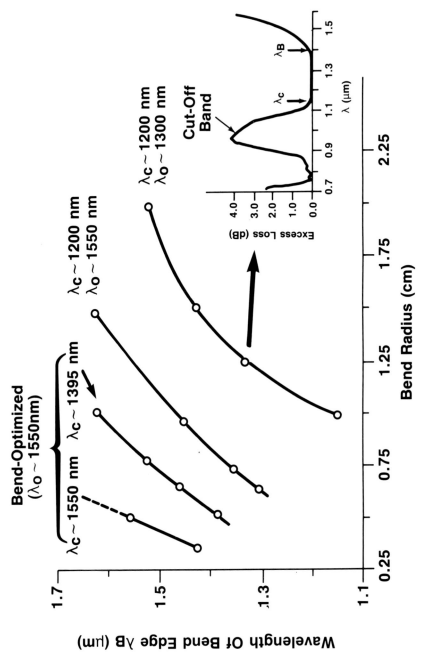

Fig. 2.8 Wavelength of bend edge versus bend radius for different types of single-mode fibers.

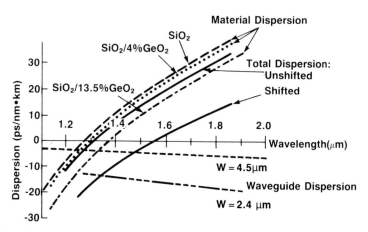

Fig. 2.9 Dispersion of single-mode fibers as function of material composition and modefield diameter (After Ainslie et al, 1986).

sion of 3.5 ps/nm · km has been specified in the 1285 to 1330 nm wavelength region (CCITT Rec. G.652, see Table 2.1). The bandwidth of SM fibers is inversely related to the dispersion as further discussed by Philen and Anderson in Chapter 8 of this book.

The low loss and long interaction lengths in SM fibers may also cause non-linear dispersion effects. For example, for specific power levels, a non-linear dispersion is generated which cancels material and waveguide dispersion. Under these conditions, a particular propagating pulse, termed soliton, does not broaden and allows long, loss-limited repeater spacings (Hasegawa, 1986).

2.3.5 Concatenation Effects

When dealing with fiber parameters, one typically refers to the characteristics of individual fibers, while the systems performance is predominantly determined by concatenated cable sections. In general, the effects of concatenation on key SM fiber parameters are straightforward, and make SM fibers easy to understand and use: Fiber, splice, and connector losses are additive, dispersion adds algebraically (with compensation occurring for positive and negative dispersion contributions), and the zero-dispersion wavelength is the length-weighted average of the individual cable sections.

2.4 MULTIMODE FIBERS

2.4.1 Fiber Types

Because of their higher loss and significantly lower bandwidth compared to SM fibers, MM fibers have increasingly been used for short-distance transmission such as in Local Area Network (LAN), data-link, and other on-premises applications where emphasis is placed on low-cost splices and connections, and on simple, low-cost source/fiber coupling schemes. Their potential use in future subscriber loop networks is still being studied. It should be noted that MM fiber interconnection cables and pigtails are used at the receiving end of SM fiber transmission systems and for measurement purposes, where single-to-multimode fiber transitions yield lower joint losses. Multimode fibers can be divided into the following major groups:

- Graded-index fibers with 50 μm core, and 125 μm cladding diameter (50/125 fibers), and with typical numerical apertures (NAs) between 0.20 and 0.24. This fiber was originally developed for telecommunication applications at 850 and 1300 nm wavelengths, and is the only MM fiber that has been standardized by CCITT (Recommendation G.651). The topographic index profile of a representative graded-index fiber is shown in Fig. 2.2.
- Graded-index fibers with 62.5 μm core and 125 μm cladding diameter, and with typical NAs of 0.26 to 0.29. This fiber was developed for longer-distance loop-feeder applications in the 850 and 1300 nm wavelength ranges, but is now used mainly in LANs.
- Graded-index fibers with 85 μm core and 125 μm outer diameter, and with typical NAs ranging from 0.26 to 0.30. This fiber was specifically developed for LANs and other short-haul systems for transmission at 850 and 1300 nm.
- Graded-index fibers with 100 μm core and 140 μm cladding diameter, and with an NA of 0.29. These fibers were developed primarily for low-cost, short-distance applications providing high coupling efficiency to 850 nm LEDs but are also operable at 1300 nm. While 100/140 fibers have also been made with plastic claddings, all-glass fibers have better handling characteristics.

Other types of MM step-index fibers are available, but have not been used in telecommunications. These include fibers with glass or silica cores of 200 μm and larger diameter, and claddings of glass or plastic; and all-plastic fibers of similar sizes that are intended for easy coupling to LEDs, and lowest-cost splicing and connectorization primarily for data link applications. All-plastic fibers offer a high degree of ruggedness and ease of handling. Their high attenuation on the order of 100 dB/km (champion data: approximately 20 dB/km) in the near-IR region, however, limits their use to distances of a few hundred meters, with primary emphasis on vehicular applications.

2.4.2 Fiber Attenuation

The range of spectral losses of commercially-available MM fibers is shown in Fig. 2.10. Multimode fiber losses have the same fundamental limitations from Rayleigh scattering and long-wavelength infra-red absorption as SM fibers. The

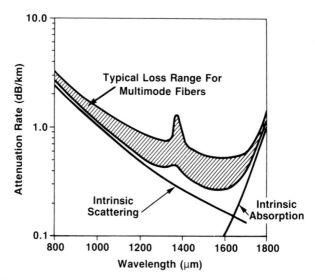

Fig. 2.10 Typical spectral loss range of commercial graded-index multimode fibers.

losses of MM fibers are generally higher than those of SM fibers, however, because of higher dopant concentrations and the resulting larger compositional fluctuation scattering, and because of possible higher-order mode losses due to core/cladding interface perturbations, or a higher sensitivity to microbending losses particularly of the higher-order modes.

Because of differential mode attenuation phenomena, the loss of MM fibers depends on the excitation conditions as well as the fiber characteristics. Unique losses are only obtained for steady-state mode-propagation conditions that establish themselves over long lengths of fibers. Steady-state conditions can be approximated either with carefully-controlled launch conditions, or by using a "mandrel-wrap" mode filter (Kaiser *et al.*, 1981). While extensive studies of these phenomena have previously been performed, the use of MM fibers for short-distance applications no longer requires the knowledge of the steady-state behavior. This puts emphasis on the transient-loss behavior, which is critically dependent on the excitation conditions.

Differential mode loss and mode coupling phenomena also affect the effective numerical aperture, effective core size, and the dispersion characteristics, and make it difficult to properly characterize the transmission characteristics of MM fibers other than in the near steady-state condition.

The microbending losses depend on the fiber parameters as follows:

$$\alpha = \frac{Ka^A}{(NA)^B d^C}(\text{dB/km}),$$

where a is the core radius, NA is the numerical aperture, and d is the fiber outer diameter. The empirical constants A, B, C, and K have values of 3.7, 7.4, 5.2, and 0.01–0.1, respectively (Olshansky, 1975). In SM fibers, micro-bending losses for most perturbations exhibit a strong wavelength dependence beyond a critical wavelength as noted earlier, and generally do not affect transmission in the 1300 nm wavelength region. In contrast, the microbending losses of MM fibers are essentially independent of wavelength and may contribute to the transmission losses particularly in tightly-packaged cable structures and/or at low temperatures.

2.4.3 Dispersion and Bandwidth

Signal degradation in MM fibers is caused by both inter-modal and material dispersion. Assuming equal power excitation of all modes, a simple expression for the rms pulse broadening in MM fibers has previously been derived (Olshansky and Keck, 1976). In practical LED-MM fiber transmission systems, however, there is a complex relationship between the fiber material dispersion, the large spectral width of the LEDs, the fiber spectral loss characteristics, and actual excitation conditions. Useful semi-empirical formulas for the bandwidth of LED-MM fiber systems in the 800 and 1300 nm wavelength regions have recently been derived by Refi (1986). As noted in this reference, while the bandwidth for short fiber lengths is strongly affected by modal distortion, for fiber sections beyond a few km's length, chromatic dispersion typically dominates. Figure 2.11 shows the representative bandwidth range of commercial 50/125 μm graded-index MM fibers as function of wavelength. The index-profile is chosen such that the bandwidth is optimized in the 1300 nm region. While, in theory, bandwidths in excess of 10 GHz km should be possible, 5 GHz km values have been achieved in practice. However, fabrication-related waveguide imperfections reduce the bandwidth of the best commercial fibers to about 1-to-2 GHz · km.

Because of profile variations from fiber to fiber, differential mode loss, mode conversion at imperfect joints, and mode coupling effects along the fiber, concatenation characteristics in MM fibers are significantly more complex than in SM fibers. Also, the cabled fiber characteristics, such as steady-state mode distribution, loss, and bandwidth may differ from those of the uncabled fiber—especially in tight buffer constructions—because of cabling-induced macro- and micro-bend effects and different mode coupling. For similar MM fibers being operated under approximately steady-state excitation conditions, the losses of individual cable sections are additive, and the bandwidth of the concatenated cable section is approximated by (Miyamoto *et al.*, 1983)

$$\mathrm{BW}(L) = \mathrm{BW}_0 (L/L_0)^\gamma$$

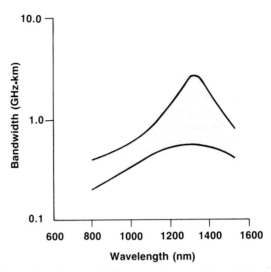

Fig. 2.11 Wavelength-dependent bandwidth range of commercial graded-index multimode fibers.

with BW_0 being the bandwidth of the section length L_0, and $0.5 < \gamma < 1.0$ representing the bandwidth-length extrapolation factor. A more recent study (Nolan et al, 1987) has shown how γ may be predicted for arbitrary mode power distributions and group delays.

2.5 PHYSICAL AND MECHANICAL PROPERTIES, RELIABILITY, AND STANDARDS

2.5.1 Geometry

Tight control of the fiber geometrical tolerances is necessary to assure good control of the waveguide characteristics, and to permit intermateability between fibers of the same generic type to achieve low joint losses. Present multimode telecommunication fibers have a standard outer diameter of 125 μm, with a tolerance of 3 μm, and a core diameter tolerance of 5 to 6%. Besides diameter tolerances, limits have been set on maximum ellipticities of core and cladding of MM fibers. Conformity with CCITT-recommended standard tolerances can be checked against a "4-Circle Tolerances Template" (CCITT Recommendation G.651). Since joint losses comprise various intrinsic and extrinsic components, including NA mismatch, an Intrinsic Quality Factor (IQF) has been proposed and accepted by CCITT as an alternate way to characterize the geometric tolerances of MM fibers. The IQF takes into account the loss compensation occurring for certain fiber parameter mismatches, and permits excursions beyond individual worst-case tolerances, while still resulting in low joint losses.

The compatibility of fibers of the same type extends also to the fiber material, and specifically to the fusion-splicing of similar fibers with different core dopant materials and levels. This results in different softening temperatures, such as for fibers with pure and Ge-doped silica cores. Careful measurements have revealed, however, that such different fibers can also be fusion-spliced with low losses provided the fusion parameters are properly adjusted. The recommended maximum core eccentricity of SM fibers has recently been reduced by CCITT from 3 to 1 μm in order to facilitate low-loss mass splicing using the outer diameter as the reference surface, and average values below 0.5 μm are now readily being achieved in commercial fibers.

2.5.2 Plastic Coatings, Materials, and Designs

Optical fibers are generally protected with plastic coatings to preserve their pristine strength and to make them handleable. Representing an integral part of the fiber, the coating may influence both the fiber characteristics and the range of application. Typical coating diameters are around 250 μm, but diameters as large as 900 μm are also being used. Preferred materials are UV-curable resins that permit higher drawing speeds—and thus lower fabrication costs—than thermally-cured materials such as silicones. Dual-layer coatings with a soft inner buffer layer and a higher-modulus outer layer generally provide more protection from lateral forces than do single-layer coatings. In dual coating designs, care has to be taken to optimize the layer thickness and Young's moduli. A high modulus of a thick outer layer may result in compressive strain on the fiber, with resulting reduction of the cut-off wavelength compared to that of the unstressed fiber, as well as excess losses at low temperatures. A too-low-modulus inner layer, on the other hand, may not permit fiber proof testing at the required stress levels without damaging this layer. In general, the coating should be easily removable —preferably by mechanical means—to allow low-cost field termination.

In specialized applications, the use of hermetic coatings leads to high-strength fibers which are highly immune against environmental influences such as stress and moisture, and which do not change the optical properties of the fibers, (Beales *et al.*, 1984; Hiskes *et al.*, 1984). Metal-copolymer dual-coated fibers have been developed as phase modulators for sensor applications (Lagakos *et al.*, 1987).

2.5.3 Fiber Strength

Depending on the area of application, fibers are generally proof-tested at specific stress levels, which assure a long service life. The strength of fibers is critically dependent on the care exerted during the preform and fiber fabrication process. Fiber strength has continuously increased over the years because of improvements made in these processes. Presently, fibers used for standard

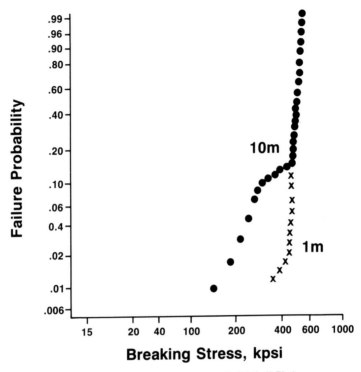

Fig. 2.12 Fiber failure probability versus breaking strength (Weibull Plot).

telecommunication cables are proof-tested at 50 Kpsi, while fibers intended for submarine cable installations are tested at 200 Kpsi and higher because of the large tensile forces encountered in this application (DiMarcello *et al.*, 1985). The ultimate strength of silica fibers for standard environmental conditions is on the order of 800 Kpsi. The fracture strength of fibers is typically represented in form of a "Weibull Plot" which depicts the probability of failure as function of instantaneous stress level (for given test lengths, say 1 or 10 m, Fig. 2.12). A sufficient total length of fiber must be tested to accurately reflect the flaw distribution of the fiber under consideration.

2.5.4 Environmental Stability and Reliability

Strength tests are carried out to eliminate weak fiber sections, and to assure the long-term reliability of optical fibers over their service life. Cable structures are designed, and installation conditions are chosen such that the long-term stress level of the installed fibers does not exceed approximately 1/3 of the proof-test level to avoid static-fatigue failures (Maurer, 1985).

Another potential failure mechanism may be due to molecular hydrogen, which can cause a loss increase in the 1300 and 1550 nm wavelength regions (Uchida and Uesugi, 1986; Beales *et al.*, 1983; Lemaire and Tomita, 1984). Some of the excess losses caused by H_2 are reversible, while others are not. Hydrogen may be generated through metallic corrosion in the cable structure due to the presence of water, or it may evolve from plastic material decomposition, including that of the fiber coating. The problem has practically been eliminated through a more careful choice of the cable materials. High levels of certain fiber dopant materials, such as phosphorous, have been found to increase the sensitivity toward hydrogen, and, therefore, have been virtually eliminated in commercial fibers.

Significant loss increase may occur in doped silica fibers (such as phosphorous-doped) due to high doses of irradiation, both of man-made and natural origin. Pure-silica-core fibers are generally less susceptible toward irradiation-induced losses. The effect of radiation is strongly dependent upon the temperature of the fiber, with low temperature being particularly detrimental. In general, both reversible and irreversible absorption centers may be created in both doped and undoped silica. Details on this subject can be found in the literature (Friebele, 1984).

2.6 SPECIAL FIBERS

2.6.1 Multi-Core and other Special Fibers

Fibers with more than one core are considered useful for spatial multiplexing instead of multi-fiber transmission (Inao *et al.*, 1979; Kashima *et al.*, 1982), or to provide controlled coupling between two or more guides for component applications. While attractive, in principle, for the low-cost mass-handling of fibers particularly in the subscriber loop (Sumida *et al.*, 1986), practical aspects such as fabrication difficulties, cross talk (Fukuma *et al.*, 1987), tolerance control (particularly in case of single-mode-fibers), and splicing, connectorization, and registration problems have precluded their practical use thus far. Two-core fibers have also been proposed as interferometric sensors, as power dividers, and wavelength-division multiplexers (Kitayama and Ishida, 1985). Fibers with active and/or non-linear properties are used for optical amplifiers through stimulated Raman scattering or 3-wave mixing (Stolen and Bjorkholm, 1982). The inclusion of neodymium, erbium, and other rare earths or transition metals into multi-component glass and/or silica fibers leads to the various properties required for fiber sensor, optical amplifier, and fiber laser applications (Payne, 1987).

Because of their compatibility with transmission fibers (particularly single-mode), components made of fibers are attractive for a variety of passive-device function applications such as couplers, star-couplers, taps, taper interferometers, wavelength-division-multiplexers, polarizers, and optical switches. The discussion of such fiber-based devices, however, goes beyond the scope of this chapter (Stewart, 1987).

2.6.2 Mid-Infrared Fibers

Because of the promise for repeaterless transmission over several 1000 kilometers distance, significant research and development effort is being spent on mid-infrared fibers whose total losses in the 2 to 4 μm wavelength region are about two orders of magnitude below those of silica fibers (Tran et al., 1984). Examples of such materials are heavy metal oxide glasses such as GeO_2 (used for fiber Raman lasers and for soliton transmission), halide glasses (e.g., Zr-based, Cd-based, and BeF_2), chalcogenide glasses (GeS) with low softening temperature, and halide crystals such as KCl. A loss of a few 0.001 dB/km is expected to be achievable with fluoride glass fibers, even though present total loss values are still in the 0.7 to 0.9 dB/km range because of metallic impurities and excess scattering due to waveguide imperfections (Kanamori et al., 1986; Tran et al., 1986). Short sections of step- and graded-index fibers made from KRS-5 (TlBr and TlI) polycrystalline material clad with KSR-6 are used for the transmission of CO_2 laser light, or for thermal radiation measurements.

Fluoride glass single-mode fibers with 10/125 μm core/cladding diameters, cut-off wavelengths λ_c of 1.20 μm, and minimum losses of 8.5 dB/km at 2.2 μm have already been fabricated (Ohishi, et al., 1986). The cut-off and zero-dispersion wavelengths of a GeO_2-core, F-doped GeO_2 cladding (10.7/98 μm) SM fiber were 1.28 and 1.72 μm, respectively (Hosaka et al., 1987). Preform and fiber fabrication process technologies still present great difficulties, and much development work remains to be done before mid-IR fibers may become available for practical use in telecommunications.

2.7 SUMMARY AND TRENDS

Research and development of optical fibers has led to a limited number of mass-produced fibers used in the public network, and a variety of fibers suited for specialized applications. Present trends are to use single-mode fibers for most telecommunication applications. Multimode fibers continue to be important for short-distance and LAN-type applications, even though SM fibers are increasingly being considered for those applications as well. Challenges ahead include the development of low-cost single-mode fibers for the subscriber loop (Kaiser and Anderson, 1986; Murata et al., 1986; Seikai et al., 1986), the further development and standardization of dispersion-shifted and dispersion-flattened single-mode fibers, and the development of low-loss and low-cost polarization-maintaining fibers for short- and medium-distance coherent transmission systems, as well as for component and sensor applications. Fibers with active and non-linear properties will expand the range of applications, specifically in the areas of fiber lasers, amplifiers, and sensors. If progress in their fabrication continues, ultra-low-loss fluoride fibers may find use for long-distance transmission in the $2-4$ μm wavelength region sometime in the future.

REFERENCES

Ainslie, B. J., Beales, K. J., Day, C. L., and Rush, J. D. (1982). The design and fabrication of monomode optical fiber. *IEEE Trans. Microwave Theory Tech.*, **MTT-30**, 1360.

Ainslie, B. J., and Day, C. R. (1986). A review of single-mode fibers with modified dispersion characteristics. *IEEE J. Lightwave Tech.*, **LT-4**, 967.

Anderson, W. T., and Lanahan, T. A. (1984). Length Dependence of the Effective Cut-off Wavelength in Single-Mode Fibers. *IEEE J. Lightwave Tech.*, **LT-2**, 238.

Anderson, W. T., Shah, V., Curtis, L., Johnson, A. J., and Kilmer, J. P. (1987). Modefield diameter measurements for single-mode fibers with non-gaussian field profiles. *IEEE J. Lightwave Tech.*, **LT-5**, 211.

Bachmann, P. K., and Lydtin, H. (1986). "Progress in the PCVD process," *OFC'86*, Atlanta.

Bachmann, P. K., Leers, D., and Wiechert, D. U. (1987). The bending performance of matched-cladding, depressed-cladding and dispersion flattened single-mode fibers. *OFC / IOOC'87*, PDP1, 5.

Beales, K. J., Cooper, D. M., and Rush, J. D. (1983). Increased attenuation of optical fibers caused by diffusion of hydrogen. *ECOC'83*, PDP.

Beales, K. J., Cooper, D. M., Duncan, W. J., and Rush, J. D. (1984). Practical Barrier to Hydrogen Diffusion into Optical Fibers. *OFC'84*, New Orleans, W15.

Berkey, G. E., and Sarkar, A. (1982). Single-mode Fibers by the OVD Process. *OFC'82*, Paper THC5, Phoenix, Arizona.

Bhagavatula, V. A., and Blaszyk, P. E. (1983). Single-mode Fiber with Segmented Core. *OFC'83*, New Orleans, Paper MF5.

Bhagavatula, V. A., Ritter, J. E., and Modavis, R. A. (1985). Bend-optimized dispersion-shifted single-mode designs. *IEEE J. Lightwave Tech.* 3, 954.

Bhagavatula, V. A., Spotz, M. S., Love, W. F., and Keck, D. B. (1983). Segmented Core Single-Mode Fibres with Low Loss and Low Dispersion. *Electron Lett.* **19**, 317.

Black, P. W. (1986). Design of fiber and cable for specialist applications. *IEEE J. Lightwave Tech.*, **4**, 1167.

CCITT (1985). COM XV-83-E, USA. Cut-off wavelength of cabled single-mode fibers.

CCITT (1987). COM XV-146-E, ATT. Relation between fiber and cable cut-off wavelength.

Cheung, N. K., and Kaiser, P. (1984). Modal noise in single-mode fiber transmission system. ECOC'84, Stuttgart.

Cohen, L. G. (1985). Comparison of single-mode fiber dispersion measurement techniques. *IEEE J. Lightwave Tech.* 3, 958.

Croft, T. D., Ritter, J. E., and Bhagavatula, V. A. (1985). Low-loss dispersion-shifted single-mode fiber manufactured by the OVD Process. *IEEE J. Lightwave Tech.* 3, 931.

Csencsits, R., Lemaire, P. J., Reed, W. A., Shenk, D. S., and Walker, K. L. (1984). Fabrication of low-loss single-mode fibers. *OFC'84*, TU13.

DiMarcello, F. V., Brownlow, D. L., Huff, R. G., and Hart, A. C. (1985). Multikilometer lengths of 3.5-GPa (500 Kpsi) prooftested fiber. *IEEE J. Lightwave Tech.* **LT-3**, 946.

Dixon, J. A., Giroux, M. S., Isser, A. R., and Van Dewoestine, R. V. (1986). Bending and Microbending Performance of Single-Mode Optical Fibers. *OFC / IOOC'87*, Reno, TUA2, 40.

Friebele, E. J., Long, K. J., Askins, C. G., and Gingerich, M. E. (1984). Radiation response of optical fibers and Selfoc lenses at 1.3 μm. *Proc. SPIE*, **506**.

Fukuma, M., Ogasawara, I., Nishimura, A., Suganuma, H., and Suzuki, S. (1987). Characterization of crosstalk in four-core graded-index multicore fiber. *OFC / IOOC'87*, Reno, WI7, 172.

Giroux, M. S., Isser, A. R., and Van Dewoestine, R. V. (1987). Bending and Microbending Performance of Single-mode Optical Fibers. *OFC / IOOC '87*, Reno, Paper TUA2.

Gloge, D., and Marcatili, E. A. J. (1973). Multimode theory of graded-core fibers. *Bell System Technical Journal* **52**, 1563–1578.

Harmon, R. A. (1982). Polarization Stability in Long Lengths of Monomode Fiber. *Electron Lett.* **18**, 1058.

Hasegawa, A. (1986). Amplification and reshaping of optical solitons in a glass fiber—IV: Use of the stimulated Raman process. *Opt. Lett.* **8**, 650.

Hattori, H., Ohishi, Y., Kanamori, T., and Sakaguchi, S. (1986). *Appl. Optics* **25**, 3549.

Hiskes, R., Schantz, C. A., Hanson, E. G., Mittelstadt, L. S., Scott, C. J., Joiner, C. S., Knutsen, G. F., and Pound, J. G. (1984). High Performance Hermetic Optical Fibers. *OFC'84*, New Orleans, W16.

Hosaka, T., et al (1985). Low-loss PANDA fiber. Nat'l Conf. IECE-J, 4.

Hosaka, T., Sudo, S., and Okamoto, K. (1987). Dispersion of pure GeO_2 glass-core and F-doped GeO_2 glass cladding single-mode optical fibre. *Electron Lett.* **23**, 24.

Inao, S., Sato, T., Sentsui, S., Kuroha, T., and Nishimura, Y. (1979). Multicore optical fiber. *OFC'79*, Washington DC, WB1, 46.

Jablonowski, D. P. (1986). Fiber Manufacture at AT & T with the MCVD Process. *JLT* **4**, 1016.

Kaiser, P., and Bisbee, D. L. (1979). Transmission losses of concatenated connectorized fiber cables. *OFC'79*, Washington DC, TU E3, 26.

Kaiser, P., French, W. G., Bisbee, D. L., and Shiever, J. W. (1979). Cabling of single-mode fibers. *ECOC '79*, Amsterdam, 7.4.

Kaiser, P. (1981). Loss measurements of graded-index fibers: accuracy versus convenience. *Opt. Fiber Measurement Symp.*, NBS.

Kaiser, P. (1985). Single-mode fiber technology for the subscriber loop. *IOOC / ECOC '85*, Venice.

Kaiser, P., and Anderson, W. T. (1986). Fiber cables for public communications: State-of-the-art technologies and the future. *IEEE J. Lightwave Tech.* **LT-4**, 1157.

Kanamori, T., and Sakaguchi, S. (1986). Preparation of elevated NA fluoride optical fibers. *Japan J. Appl. Physics* **20**, 468.

Kanamori, H., Yokota, H., Tanaka, G., Watanabe, M., Ishiguro, Y., Yoshida, I., Kakii, T., Itoh, S., Asano, Y., and Tanaka, S. (1986). Transmission characteristics and reliability of pure-silica-core single-mode fibers. *IEEE J. Lightwave Tech.* **4**, 1144.

Kapron, F. P. (1987). Chromatic dispersion format for single-mode and multimode fibers. *OFC / IOOC'87*, TUQ2, 97.

Kashima, N., Maekawa, E., and Nihei, F. (1982). New type of multicore fibers. *OFC'82*, 46.

Katsuyama, T., Matsumura, H., and Suganuma, T. (1984). Reduced pressure collapsing MCVD method for single-polarization optical fibers. *IEEE J. Lightwave Tech.* **LT-2**, 634.

Kitayama, K., Ohashi, M., and Ishida, Y. (1984). Length-dependence of effective cutoff wavelength in single-mode fibers. *IEEE J. Lightwave Tech.* **LT-2**, 629.

Kitayama, K., and Ishida, Y. (1985). Wavelength-selective coupling of two-core optical fiber: application and design. *J. Opt. Soc. Am.* **2**, 90.

Lagakos, N., Ku, G., Jarzynski, J., Cole, J. H., and Bucaro, J. A. (1985). Desensitization of the ultrasonic response of single-mode fibers. *IEEE J. Lightwave Tech.* **3**, 1036.

Lagakos, N., Ku, G., Cole, J. H., and Bucaro, J. A. (1987). High-frequency fiber optic phase modulator. *OFC / IOOC'87*, Reno, TUJ2, 77.

Lemaire, P. J., and Tomita, A. (1984). Behavior of single-mode fiber exposed to hydrogen. *ECOC'84*, Stuttgart.

Maurer, R. D. (1985). Behavior of Flaws in Fused Silica Fibers. In "Strength of Inorganic Glass." edited by C. R. Kurkjian, Plenum Publishing Corp, New York.

Miller, S. E., and Chynoweth, A. G. (1979). "Optical Fiber Telecommunications." Academic Press, Boston.

Miyamoto, M., Yamauchi, R., and Inada, K. (1983). Length Dependence of Bandwidth for Fibers with Random Axial Profile Fluctuation. *J. Lightwave Tech.* **LT-1**, 354.

Murata, H. (1986). Recent developments in vapor phase axial deposition. *JLT* **LT-4**, 1026.

Murata, H., Ogai, M., and Tachigami, S. (1986). Optical Cables and Connection Technologies for Multi-Service Subscriber Systems. *ECOC'86*, Barcelona, II/79.

Noda, J., Okamoto, K., and Sasaki, Y. (1986). Polarization-maintaining fibers and their applications. *J. Lightwave Tech.* **4**, 1071.

Nolan, D. A., Hawk, R. M., and Keck, D. B. (1987). Multimode concatenation modal group analysis., *IEEE J. Lightwave Tech.*, **LT-5**, 1727.

Payne, D. N. (1987). Special fibers and their use. *OFC / IOOC'87*, Reno, WI1, 166.

Petermann, K. (1987). Mode-field characteristics of single-mode fiber designs. *OFC / IOOC'87*, TUA1, 39.

Petermann, K. (1983). *Electron Lett.* **19**, 712.

Ohashi, M., Kuwaki, N., and Tanaka, C. (1986). Characteristics of bend-optimized convex-index dispersion-shifted fiber. *First Optoelectronics Conf.*, *OEC '86*, Tokyo, PDP, 22.

Ohishi, Y., Sakaguchi, S., and Takahashi, S., (1986). Transmission loss characteristics of fluoride glass single-mode fiber. *First Optoelectronic Conference*, *OEC'86*, PDP C11-1.

Okoshi, T. (1985). Polarization-state control schemes for heterodyne or homodyne optical fiber communications. *IEEE J. Lightwave Tech.* **3**, 1232.

Olshansky, R. (1975). Mode Coupling Effects in Graded-Index Optical Fibers. *Appl. Opt.* **14**, 935.

Olshansky, R., and Keck, D. B. (1976). Pulse Propagation in Optical Waveguides. *Appl. Opt.* **15**, 483.

Reed, W. A., Cohen, L. G., and Shang, H. T. (1987). Tailoring optical characteristics of dispersion-shifted lightguides for systems applications near 1.55-μm wavelength. *OFC / IOOC'87*, TUA5, 43.

Refi, J. J. (1986). LED bandwidth of multimode fibers as a function of laser bandwidth and LED spectral characteristics. *IEEE J. Lightwave Tech.* **4**, 265.

Saifi, M. A., Jang, S. J., Cohen, L. G., and Stone, J. (1982). *Opt. Lett.* **7**, 43.

Sasaki, Y. (1987). Long-Length Low-Loss Polarization-Maintaining Fibers. Paper WP1, OFC/IOOC-87, Reno, Nev.

Sasaki, Y., Hosaka, T., Horiguchi, M., and Noda, J. (1986). Design and fabrication of low-loss and low-crosstalk polarization-maintaining optical fibers. *IEEE J. Lightwave Tech.* **LT-4**, 1097.

Seikai, S., Kitayama, K., Uesugi, N., and Kato, Y. (1986). Design consideration on single-mode fibers suitable for subscriber cables. *IEEE J. Lightwave Tech.* **LT-4**, 1005.

Shah, V. (1987). Curvature dependence of the effective cutoff wavelength in single-mode fibers. *IEEE J. Lightwave Tech.* **4**, 35.

Stern, M., Way, W. I., Shah, M. B., Young, W. C., and Krupsky, J. W. (1987). 800-nm digital transmission in 1300-nm optimized single-mode fiber. *OFC / IOOC'87*, **MD2**, 12.

Stewart, W. J. (1987). Advances in fiber-based components. *OFC / IOOC'87*, Reno, **THB1**, 191.

Stolen, R. H., and Bjorkholm, J. E. (1982). Parametric amplification and frequency conversion in optical fibers. *IEEE J. Quantum Elec.* **QE-18**, 1062.

Stolen, R. H., Pleibel, W., and Simpson, J. R. (1984). High Birefringence optical fibers by preform deformation. *IEEE J. Lightwave Tech.* **LT-2**, 639.

Stone, F. T. (1984). Modal noise in single-mode fiber communication systems. *SPIE Conf.*, San Diego.

Sumida, S., Maekawa, E., and Murata, H. (1986). Design of bunched optical-fiber parameters for 1.3 μm wavelength subscriber line use. *IEEE J. Lightwave Tech.* **LT-4**, 1010.

Tomita, *et al.* (1983).

Tran, D. C., Sigel, G. H., and Bendow, B. (1984). Heavy metal fluoride glasses and fibers: A review. *IEEE J. Lightwave Tech.* **LT-2**, 566.

Tran, D. C., Levin, K. H., Burk, M. J., Fisher, C. F., and Brower, D. (1986). Preparation and properties of high-quality IR transmitting glasses and fibers based on metal fluorides. *Proc. SPIE* **618**, 48.

Uchida, N., and Uesugi, N. (1986). Infrared Optical Loss Increase in Silica Fibers due to Hydrogen. *IEEE J. Lightwave Tech.* **LT-4**, 1132.

Yamauchi, R., Miyamoto, M., Abiru, T., Nishide, K., Ohashi, T., Fukuda, O., and Inada, K. (1986). Design and performance of Gaussian-profile dispersion-shifted fibers manufactured by VAD process. *IEEE J. Lightwave Tech.* **LT-4**, 997.

Chapter 3

Selected Topics in the Theory of Telecommunications Fibers

DIETRICH MARCUSE

AT & T Bell Laboratories, Holmdel, New Jersey

As optical fiber telecommunications has matured, the emphasis has moved from multimode to single-mode fibers. For this reason we discuss in this chapter several properties of single-mode fibers that are related to dispersion, polarization, non-linear effects and the cutoff of the LP_{11} mode.

The chapter begins with a discussion of the effect of chromatic dispersion on pulse propagation in single-mode fibers. Chromatic dispersion is of particular interest since, in conjunction with fiber loss, it sets an upper limit on the obtainable repeater spacing for any desired bit rate. The value of chromatic dispersion depends on the operating wavelength; in pure fused silica it vanishes to first order at a wavelength of 1.27 μm. In single-mode fibers, however, dispersion is influenced by a waveguide effect making it possible to shift the desired "zero-dispersion" wavelength to coincide with the wavelength at which the fiber loss is also a minimum. This possibility motivates the design of "dispersion-shifted" fibers.

Normal, single-mode fibers actually support two modes, one in each of two orthogonal polarization states. In fibers with perfectly circular cross sections, having isotropic refractive index distributions, these two orthogonally polarized modes travel with equal phase and group velocities. Thus, light pulses that distribute their power among these two polarization states would travel in perfect synchronism. But even under such idealized conditions the state of polarization of the light at the end of the fiber cannot be predicted, since it depends on the resultant of the two principal polarization states taking their relative phases into account. The uncertain polarization can be a problem for heterodyne and homodyne detection and for coupling to integrated optics devices where the polarization of the light wave needs to be controlled. Actual fibers do not have exactly circular cross sections and usually have a small amount of anisotropy. Thus, light associated with the two polarization states travels at slightly different group velocities causing pulses to split or spread. These problems can be alleviated if one polarization state is suppressed resulting in polarization-maintaining fibers to be discussed in this chapter.

Elementary discussions of the properties of optical fibers assume that the refractive index of the fiber material at every point in space has a fixed value that may depend slightly on wavelength (chromatic dispersion) but which is assumed to be independent of the amount of light power carried in the fiber. More sophisticated studies of fiber properties need to refine this assumption and take into account that the refractive index of the fiber material does depend slightly on the value of the light energy existing in it. The nonlinearity of the refractive index gives rise to subtle effects such as Raman and Brillouin scattering, which cannot be neglected for fibers carrying power in the tens of milliwatt range.

Nonlinearities of the refractive index need not be entirely detrimental, because it is possible to utilize nonlinear effects to counteract dispersion. The interplay of these two effects can lead to stable pulse wave forms, called solitons, that, in the absence of loss, travel arbitrary distances without changing their shape or at most with periodic shape changes. Nonlinear effects, as well as the formation of solitons and their use will also be discussed in this chapter.

3.1 PULSE PROPAGATION IN SINGLE-MODE FIBERS

To be able to describe pulse propagation in single-mode fibers, we need to make a few assumptions. We begin by assuming that the fiber is excited by an externally modulated source. Thus, in the absence of modulation the light source provides a light signal ψ_0 (corresponding to a voltage in conventional transmission systems) that oscillates at a carrier frequence ω and has an amplitude $A(t)$ that exhibits random fluctuations in magnitude and phase

$$\psi_0(t) = A(t)e^{i\omega_c t}. \tag{3.1}$$

The random fluctuations of $A(t)$ give rise to a finite spectral width of the light source. (Miyagi and Nishida, 1979; Gloge, 1979). Its spectrum has an autocorrelation function of the form (Marcuse, 1980)

$$\langle \phi_0(\omega)\phi_0^*(\omega') \rangle = S(\omega - \omega_c)\delta(\omega - \omega'), \tag{3.2}$$

where $\delta(\omega - \omega')$ is the Dirac delta function and $\langle \ \rangle$ indicates an ensemble average. For simplicity, we assume that the source spectrum can be approximated by a Gaussian function (Marcuse, 1980)

$$S(\omega - \omega_c) = \frac{\hat{P}_0}{\pi^{1/2}W}\exp\left[-(\omega - \omega_c)^2/W^2\right] \tag{3.3}$$

with peak amplitude $\hat{P}_0$, W is the source half width at the $1/e$ points in units of angular frequency. The instantaneous power of the source is defined as $P_0(t) = |\psi_0(t)|^2$.

To obtain light pulses we assume that the light stream emerging from the source is modulated by Gaussian pulses of the form (Juergensen, 1978)

$$s(t) = \exp(-t^2/T^2) \tag{3.4}$$

with temporal half width T at the $1/e$ points. Thus, the light power assumes the form

$$P(t) = s(t)P_0(t). \tag{3.5}$$

The corresponding light signal is assumed to be of the form

$$\psi(t) = s^{1/2}(t)\psi_0(t). \tag{3.6}$$

Associated with the light signal $\psi(t)$ is a spectral function $\phi(\omega)$.

The light signal at any point z along the fiber at time t can now be expressed as a Fourier integral

$$\psi(z, t) = \int_{-\infty}^{\infty} \phi(\omega)\exp\left[i(\omega t - \beta z)\right] d\omega. \tag{3.7}$$

Due to dispersion, the propagation constant β of the mode in the fiber is a function of frequency. The complete form of $\beta(\omega)$ is hard to express analytically but for our purposes it is sufficient to approximate it by the following Taylor expansion ($\beta_c = \beta(\omega_c)$) (Kapron, 1977)

$$\beta = \beta_c + \dot{\beta}_c(\omega - \omega_c) + \frac{1}{2}\ddot{\beta}_c(\omega - \omega_c)^2 + \frac{1}{6}\dddot{\beta}_c(\omega - \omega_c)^3, \tag{3.8}$$

where the dots indicate differentiation with respect to the angular frequency ω.

The ensemble average of the light power is equal to the ensemble average of the square of the magnitude of $\psi(z, t)$. After considerable algebra, the average light power can also be expressed as a Fourier integral (Marcuse, 1980)

$$\langle P(z, t) \rangle = \int_{-\infty}^{\infty} G(z, t)\exp\left(i\frac{t - \dot{\beta}_c z}{T}x\right) dx \tag{3.9}$$

with the spectral function

$$G(z,t) = \frac{\hat{P}_0}{2\sqrt{\pi}} \frac{\exp(-x^2/4)\exp(-iBx^3/4)}{\left[1 + 3iBx(1 + V_s^2)\right]^{1/2}} \cdot \exp\left\{\frac{-\hat{D}^2 x^2(1 + V_s^2)}{1 + 3iB(1 + V_s^2)}\right\}.$$

(3.10)

The variable x is defined as

$$x = \omega T.$$

(3.11)

The product of the temporal pulse $(1/e)$-width with the spectral source $(1/e)$-width has been abbreviated as

$$V_s = TW.$$

(3.12)

Furthermore, we have introduced the first-order dispersion parameter

$$\hat{D} = \frac{\ddot{\beta}_c z}{2T^2}$$

(3.13)

and the corresponding second-order dispersion parameter

$$B = \frac{\dddot{\beta}z}{6T^3} = \frac{1}{3T}\frac{d\hat{D}}{d\omega}.$$

(3.14)

In general, the Fourier integral (3.9) cannot be solved analytically, but numerical solutions can readily be obtained with the fast Fourier transform algorithm (FFT).

The "zero-dispersion" point is defined as that frequency or wavelength where $\ddot{\beta} = 0$ (Payne and Grambling, 1976). At that point most dispersion effects come from the presence of the third frequency derivative $\dddot{\beta}$. Far from the "zero-dispersion" point, however, $\dddot{\beta}$ is negligibly small in most practical fibers. In that case we may set $\dddot{\beta} = 0$ and obtain the following solutions of (3.9) and (3.10),

$$\langle P(z,t) \rangle = \frac{\hat{P}_0}{\left[1 + 4\hat{D}^2(1 + V_s^e)\right]^{1/2}} \exp\left\{\frac{-\left[(t - \dot{\beta}_c z)/T\right]^2}{1 + 4\hat{D}^2(1 + V_s^2)}\right\}, \quad (3.15)$$

which shows that an initial Gaussian pulse remains Gaussian in shape but increases in width since, according to (3.13), $\hat{D}$ is linearly proportional to the length coordinate z.

Figure 3.1 shows how an initial Gaussian input pulse ($\hat{D} = 0$) broadens after traveling a distance corresponding to $\hat{D} = 3$ ($B = 0$ was assumed). It is not surprising to see the input pulse broaden due to first-order dispersion. Looking at the two pulses in Fig. 3.1, however, it appears as though the spectrum of the

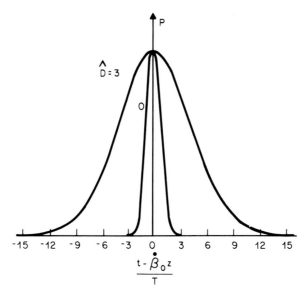

Fig. 3.1 An output pulse broadened by first-order dispersion with $\hat{D} = 3$ and $V = 0$ is compared with the corresponding input pulse (labeled $\hat{D} = 0$). (Marcuse, 1982.)

pulse might have narrowed since a broader temporal function is usually associated with a narrower spectrum. Since there is no reason to assume that the lossless, linear transmission medium (the fiber) should act as a frequency filter, narrowing of the frequency spectrum would indeed not be expected. To resolve this paradox it is helpful to consider the light signal ψ instead of the average light power $\langle P(z, t) \rangle$. Neglecting second-order dispersion, $B = 0$, it is possible to solve the Fourier integral in (3.7) for a monochromatic light source, $W = 0$: (Marcuse, 1982)

$$\psi(z, t) = \frac{A}{[1 + 2i\hat{D}]^{1/2}} \exp\left\{ \frac{-[(t - \dot{\beta}_c z)/T]^2}{2(1 + 4\hat{D}^2)} \right\} \cdot \exp\left\{ \frac{i\hat{D}[(t - \dot{\beta}_c z)/T]^2}{1 + 4\hat{D}^2} \right\}$$

$$\cdot \exp[i(\omega_c t - \beta_c z)]. \tag{3.16}$$

Figure 3.2 shows the square of the real part of $\psi \exp[-i(\omega_c t - \beta_c z)]$ as a function of $(t - \dot{\beta}_c z)/T$. Contrary to the average light power shown in Fig. 3.1, the light signal performs more rapid oscillations. The width of the sidelobes of the function in Fig. 3.2 is equal to the width of the input pulse. Thus, we see that it is only the envelope of the function in Fig. 3.2 which has become smoother and broader, and which is identical with the average power plotted in Fig. 3.1. The function itself performs rapid oscillations on a time scale commensurate with the spectrum of the input pulse.

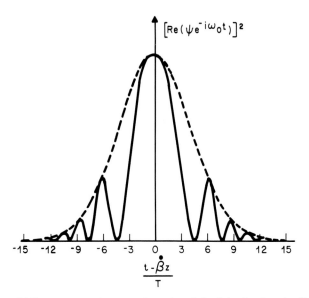

Fig. 3.2 The solid line represents the squared envelop of the light signal at the fiber output end with $\hat{D} = 3$ and $V = 0$. The dotted line is the corresponding average output power reproduced from Fig. 3.1. (Marcuse, 1982.)

We have seen that the average pulse power retains its Gaussian shape provided the second order dispersion parameter is negligible, $B = 0$. To show how second-order dispersion affects pulse propagation, Fig. 3.3 was computed from (3.9) with the help of the FFT for the input pulse of Fig. 3.1 with $\hat{D} = 0$ and $B = 1$ (Marcuse, 1980). This figure clearly shows that second-order dispersion distorts the Gaussian pulse shape. Even though the pulse remains much narrower after traveling the same distance, its shape is no longer smooth and single-peaked, (Unger, 1977).

So far we have always assumed that the light source is inherently monochromatic prior to modulation, $W = V = 0$. The modulated light has, of course, a spectrum with finite width even if $W = 0$. A finite source spectrum, $W > 0$, broadens the pulse further but does not modify the Gaussian shape of a pulse if $B = 0$. But even for $B \neq 0$ the Gaussian pulse shape tends to be restored if W and hence V is sufficiently large (Marcuse, 1980).

Equations (3.9) and (3.10) are too complicated for providing simple insights into the behavior of light pulses in fibers with first- and second-order dispersion. For this reason we introduce the concept of moments by defining

$$\langle t^n \rangle = \frac{\int_{-\infty}^{\infty} t^n P(z, t)\, dt}{\int_{-\infty}^{\infty} P(z, t)\, dt}. \tag{3.17}$$

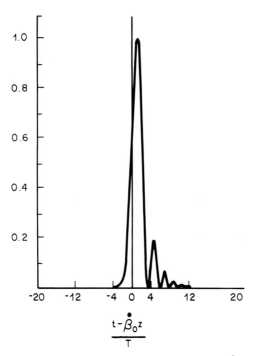

Fig. 3.3 Pulse distortion due to second-order dispersion with $B = 1$, $\hat{D} = 0$, $V = 0$. (Marcuse, 1982.)

With this definition, the variance (square of the rms deviation) of the light pulse can be written

$$\sigma^2 = \langle t^2 \rangle - \langle t \rangle^2. \tag{3.18}$$

It is possible to compute the variance from (3.9), (3.10) with the result (Marcuse, 1980)

$$\sigma^2 = \sigma_0^2 \left[1 + 4\hat{D}^2 \left(1 + V_s^2 \right) + 9B^2 \left(1 + V_s^2 \right)^2 \right] \tag{3.19}$$

or, if we substitute the values of (3.12) through (3.14),

$$\sigma^2 = \frac{T^2}{2} \left[1 + \frac{\left(\ddot{\beta}_c z \right)^2}{T^4} \left(1 + T^2 W^2 \right) + \frac{\left(\dddot{\beta}_c z \right)^2}{4T^6} \left(1 + T^2 W^2 \right)^2 \right]. \tag{3.20}$$

In (3.19), σ_0^2 is the variance of the input pulse

$$\sigma_0^2 = \frac{1}{2} T^2. \tag{3.21}$$

These expressions are sufficiently simple to yield a great deal of useful information. It should be remembered that, for Gaussian shaped pulses, that is if $\ddot{\beta}_c$ can be neglected, 95.4 percent of the pulse's energy is located inside a temporal range of width 4σ, or 99 percent is covered by a temporal range of 5.1σ. If the pulse is not Gaussian it is harder to interpret the relation of σ to the actual width of the pulse. However, it appears from Fig. 3.3 that more than 90 percent of the pulse power is contained in a time interval equal to 4σ, even in this case. Thus, we shall use

$$B_m = \frac{1}{4\sigma} \qquad (3.22)$$

as an estimate of the maximum achievable bit rate regardless of the shape of the pulse.

The expression for the variance, and hence for the maximum achievable bit rate, assumes a particularly interesting form when $V_s^2 \gg 1$, $4\hat{D}^2 V_s^2 \gg 1$ and $B = 0$. In this case we obtain from (3.20) the following expression for the full (4σ) width of the pulse in a fiber of length $z = L$

$$\Delta t = 4\sigma = 2\sqrt{2}\,\ddot{\beta}_c L W. \qquad (3.23)$$

Using the full spectral width of the source in frequency units (not angular frequency), $\Delta f = W/\pi$, and introducing the group delay in the fiber

$$\tau = \dot{\beta}_c L, \qquad (3.24)$$

we obtain from (3.23)

$$\Delta t = \sqrt{2}\,\left|\frac{d\tau}{df}\right|\Delta f, \qquad (3.25)$$

where $f = \omega/(2\pi)$ is the actual frequency. Except for the factor $\sqrt{2}$, this expression confirms the intuitive feeling that the width of an initial impulse (pulse of zero width) can be obtained by multiplying the change in group delay attributable to each spectral component of the source by the spectral width of the source. Often this expression is given in terms of wavelength with the spectral width $\Delta\lambda$ of the source expressed in wavelength units

$$\Delta t = \sqrt{2}\,\left|\frac{d\tau}{d\lambda}\right|\Delta\lambda = \sqrt{2}\,L\,|D_\lambda|\,\Delta\lambda. \qquad (3.26a)$$

The dispersion

$$D_\lambda = (1/L)\,d\tau/d\lambda \qquad (3.26b)$$

is usually measured in ps/(nm-km).

Looking at (3.20), we see that the variance of the pulse becomes infinite when the half width T of the input pulse vanishes, $T = 0$. The variance is also infinite for $T \to \infty$. Thus, σ must assume a minimum for an optimum input pulse width which we find by differentiating (3.20) with respect to T and setting the derivative equal to zero. Thus, we obtain the following implicit equation for the optimum input pulse half-width (Marcuse and Lin, 1981)

$$\frac{\left(\ddot{\beta}_c L \right)^2}{T_{opt}^4} + \frac{\left(\dddot{\beta}_c L \right)^2}{2 T_{opt}^6} \left(1 + W^2 T_{opt}^2 \right) - 1 = 0. \tag{3.27}$$

We consider the implication of this equation for several special cases. If the operating frequency is far from the zero-dispersion point, we may set $\dddot{\beta}_c = 0$ and obtain for the optimum value of the half-width of the input pulse

$$T_{opt} = \left(|\ddot{\beta}_c| L \right)^{1/2}. \tag{3.28}$$

The corresponding full width of the output pulse is

$$\Delta t_{opt} = 4\sigma_{opt} = 4 \left(|\ddot{\beta}_c| L \right)^{1/2} \left[1 + \frac{1}{2} |\ddot{\beta}_c| L W^2 \right]^{1/2}. \tag{3.29}$$

This latter equation shows that the dependence on the fiber length L of the optimum output pulse width (and hence the maximum obtainable bit rate) changes with the width of the source spectrum. For a nearly monochromatic source or for weak dispersion, $W^2 |\ddot{\beta}_c| L \ll 1$, the optimum output pulse width increases only as the square root of the fiber length L, while for a more nearly incoherent source or strong dispersion, $W^2 |\ddot{\beta}_c| L \gg 1$, the optimum output pulse width increases in direct proportion to the fiber length.

Exactly at the zero dispersion wavelength, $\ddot{\beta}_c = 0$, we obtain from (3.27) for a nearly monochromatic source, $WT \ll 1$,

$$T_{opt} = \left(\frac{|\dddot{\beta}_c| L}{\sqrt{2}} \right)^{1/3}. \tag{3.30}$$

The corresponding optimum output pulse width is

$$\Delta t_{opt} = 2\sqrt{3} \left(\frac{|\dddot{\beta}| L}{\sqrt{2}} \right)^{1/3}. \tag{3.31}$$

Thus, in this case, the optimum output pulse length grows only in proportion to the cube root of the fiber length L.

On the other hand, for a more nearly incoherent source, $WT \gg 1$, we obtain from (3.27) at the zero dispersion wavelength, $\ddot{\beta}_c = 0$,

$$T_{opt} = \left(\frac{|\ddot{\beta}_c| LW}{\sqrt{2}} \right)^{1/2} \tag{3.32}$$

with the corresponding optimum output pulse length

$$\Delta t_{opt} = 2\left(\sqrt{2}\,|\ddot{\beta}_c|\,LW\right)^{1/2}\left[1 + \frac{1}{2\sqrt{2}}|\ddot{\beta}_c|\,LW^3 \right]^{1/2}. \tag{3.33}$$

Now, the dependence of the optimum output pulse width on the fiber length changes again with changing spectral width and dispersion. For sufficiently small second-order dispersion or small source spectral width, $W^3\,|\ddot{\beta}_c|\,L \ll 1$, the optimum output pulse width is proportional to the square root of the fiber length, just as for the case of first-order dispersion and narrow source spectrum. For a source spectrum wide enough, or second-order dispersion strong enough, so that $W^3\,|\ddot{\beta}_c|\,L \gg 1$, we find once more that the optimum output pulse width grows proportional with the fiber length. But note that these last cases were already based on the assumption $WT \gg 1$.

Our discussion of the dependence of the output pulse width on the fiber length was based on assuming that the input pulse width was optimized in accordance with the condition (3.27). This means that the input pulses must be carefully adjusted to the existing fiber dispersion, fiber length, and source spectral width. If the input pulse is not optimized, and if dispersion is sufficiently strong so that the 1 in (3.20) can be neglected, then the output pulse width always grows in linear proportion to the fiber length $z = L$.

We conclude from this discussion that an increase of the optimized output pulse width less rapidly than in direct proportion to the fiber length is possible only for sufficiently narrow source spectra, that is for essentially monochromatic sources.

The amount of dispersion found in single-mode optical fibers depends on the interplay of material and waveguide dispersion (Gloge, 1971; Marcuse, 1979). For typical step-index fibers with small values of the core-cladding refractive index difference and relatively large core diameters (on the order of 10 μm), the fiber dispersion is very nearly equal to the chromatic dispersion of the fiber material, usually fused silica. Thus, we gain useful insight into pulse spreading by looking at dispersion in pure fused silica. Figures 3.4a and b show plots of the first and second (angular) frequency derivatives $\ddot{\beta}$ and $\dddot{\beta}$ of the propagation constant of a plane wave in fused silica (Kobayashi et al., 1977). By plotting these curves as functions of the difference between the actual wavelength and the zero-dispersion wavelength λ_0 we obtain dispersion curves, which are quite

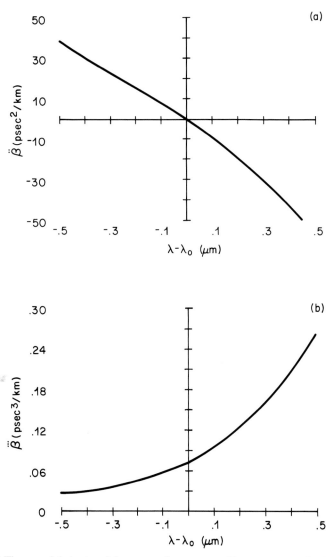

Fig. 3.4 a) The second derivative of the propagation constant with respect to angular frequency as a function of the wavelength relative to the "zero-dispersion" wavelength, b) third derivative of the propagation constant. (Note that frequency derivatives are plotted not as function of frequency but of wavelength.) (Marcuse and Lin, 1981.)

representative of typical single-mode fibers. The waveguide effect of conventional step-index fibers causes a small shift of the "zero-dispersion" point from the value $\lambda_0 = 1.273 \mu m$ of pure fused silica but the slopes and curvatures of the curves for $\ddot{\beta}$ and $\dddot{\beta}$ change only slightly. In Section 3.2 we discuss dispersion-shifted fibers, which are designed to produce a large shift of the zero-dispersion wavelength and which may also have dispersion curves of quite different shape. Thus, our present discussion does not apply to these special cases.

We use the dispersion curves of Figs. 3.4a and b to compute the optimum half width $T = T_{opt}$ of the input pulse according to (3.27) and then use this value to compute the optimum bit rate $1/(4\sigma)$ from (3.20).

Figures 3.5a through c show the optimum bit rate as a function of wavelength relative to the zero-dispersion wavelength for three different source widths. The spectral width of the source is given in wavelength units so that the corresponding width W in frequency units changes with wavelength. All three curves were computed for a fiber of 100 km length. The vertical scales are different in all three figures to adjust for the rapid decrease of the bit rate with increasing source spectral width.

Figures 3.6a through c present the optimum bit rate as a function of fiber length for three different spectral source widths and for several values of $\lambda - \lambda_0$. These curves clearly show how the length dependence (slopes of the curves) changes with the source spectral width and with $\lambda - \lambda_0$ in agreement with our discussion on the length dependence of σ.

Finally, we present the optimum bit rate as a function of source spectral width in Fig. 3.7. The Figs. 3.5 through 3.7 clearly show how large the optimum bit rate can be at the zero-dispersion wavelength and how rapidly it decreases with increasing source spectral width and with departures from the zero-dispersion point (Marcuse and Lin, 1981).

3.2 DISPERSION SHIFTED FIBERS

For the mathematical treatment of pulse dispersion presented in Section 3.1, it was natural to use the angular frequency and derivatives with respect to this variable to describe pulse dispersion. In experimental work, however, it is customary to use the wavelength and wavelength derivatives to describe dispersion. Even though the group delay, Eq. (3.24), is defined as the frequency derivative of the propagation constant, dispersion is usually defined as the quantity

$$D_\lambda = \frac{1}{L} \frac{d\tau}{d\lambda}, \tag{3.34}$$

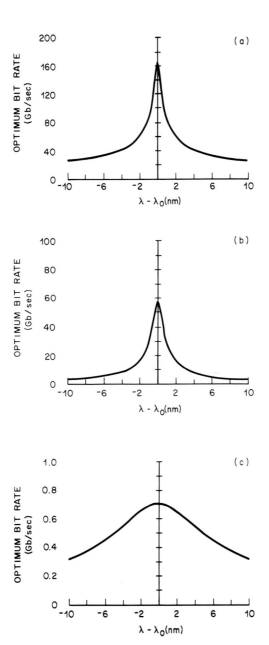

Fig. 3.5 Optimum bit rate for a fiber of 100 km length as a function of wavelength relative to the "zero-dispersion" wavelength for: a) a monochromatic source (prior to modulation), b) a source spectral width of 2 nm, c) a source spectral width of 20 nm.

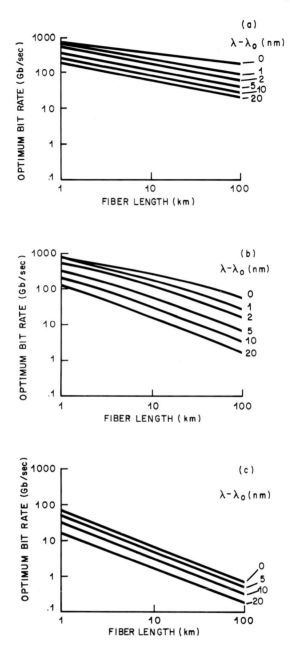

Fig. 3.6 Optimum bit rates as functions of fiber length with wavelength difference as a parameter for: a) a monochromatic source, b) a source spectral width of 2 nm, c) a source spectral width of 20 nm.

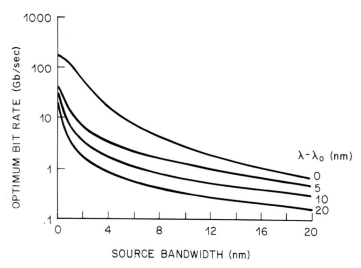

Fig. 3.7 Optimum bit rates as functions of source bandwidth.

which is the change of the pulse delay measured in picoseconds (or nanoseconds) per kilometer of fiber length and per nanometer of source spectral width. This interpretation of dispersion as pulse delay change per unit fiber length and unit source spectral width follows naturally from our discussion of pulse spreading as expressed by (3.26a). Using (c = light velocity)

$$\omega = \frac{2\pi c}{\lambda},\tag{3.35}$$

we can easily express the dispersion in terms of $\ddot{\beta}$,

$$D_\lambda = \ddot{\beta}\frac{d\omega}{d\lambda} = -\frac{2\pi c}{\lambda^2}\ddot{\beta}.\tag{3.36}$$

The specific pulse delay, τ/L, (time delay per unit fiber length) of pure fused silica as derived from Fig. 3.4 is plotted as a function of $\lambda - \lambda_0$ in Fig. 3.8a, and the corresponding dispersion function D_λ is plotted in Fig. 3.8b. This latter curve follows either as the derivative of the function plotted in Fig. 3.8a or it can be obtained from Fig. 3.4a with the help of (3.36). According to its definition, the zero-dispersion wavelength occurs at the minimum of the time delay curve, Fig. 3.8a or at the zero crossing of the dispersion curve, Fig. 3.8b.

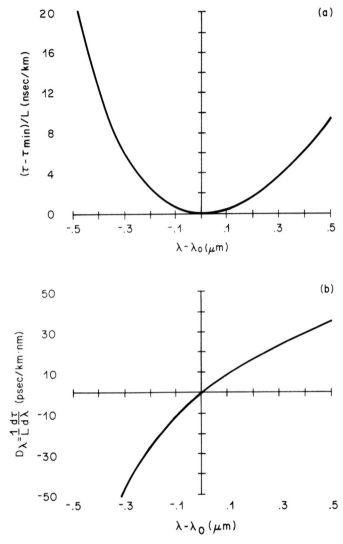

Fig. 3.8 a) Group delay per unit length as a function of relative wavelength. b) Dispersion as function of relative wavelength. These curves correspond to the dispersion curves of Fig. 3.4.

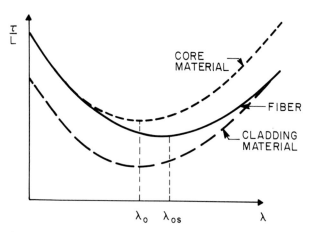

Fig. 3.9 The dotted curves show the group delay per unit length for core and cladding material of a step-index fiber as functions of wavelength. The solid curve represents the group delay per unit length of the guided mode.

For fibers with small relative index differences,

$$\Delta = \frac{n_{core}^2 - n_{cL}^2}{2n_{core}^2} \approx \frac{n_{core} - n_{cL}}{n_{core}} \tag{3.37}$$

(n_{core} = maximum index value in the fiber core, n_{cL} = cladding refractive index) the dispersion curve of the fiber mode is almost identical with the material dispersion. Fibers with small Δ require large core radii to be able to confine the guided mode reasonably tightly to the core region. If the refractive index of the core, however, becomes substantially larger than that of the cladding, that is for larger values of Δ, the time delay curves for core and cladding material separate, as shown schematically by the dotted curves in Fig. 3.9. The specific group delay curves for core and cladding materials usually do not have the same zero-dispersion wavelength, but for germania-doped silica cores the shift is sufficiently slight to be neglected in this qualitative discussion.

From the dotted curves in Fig. 3.9 we can infer the corresponding time delay curve for the guided fiber mode as follows. For short wavelengths, most of the field energy is confined inside the core so that the group delay of a pulse would be nearly the same as the group delay of a plane wave in the core material itself. At the other extreme, for long wavelengths, most of the light is spread out in the cladding. Consequently, the group delay of the fiber mode would approach the group delay of a plane wave in the cladding material. By this reasoning we may draw the qualitative specific group delay curve of the fiber mode as shown by the solid line in Fig. 3.9 (Reed *et al.*, 1986). This qualitative discussion clearly shows that the group delay minimum, that is the zero-dispersion point of the fiber, is shifted to a longer wavelength relative to the value that characterizes the fiber materials. If the shift of the zero-dispersion point is non-negligible, we speak of a dispersion shifted fiber. To ensure that the fiber can support only one single mode, the core radius must decrease with increasing refractive index n_{core} so that the V-value of the fiber (not to be confused with V_s of (3.12)), defined as

$$V = \left(2\pi n_{core} \frac{a}{\lambda} \right) \sqrt{2\Delta} \tag{3.38}$$

satisfies the condition $V < 2.4$ (for step-index fibers).

It is possible to shift the λ_0-value of single-mode fibers from the zero-dispersion point of fused silica near 1.3 μm to the 1.55 μm region, where the losses of germania-doped fibers reach a minimum (Juergensen, 1979; Ainslie et al., 1982; Jeunhomme, 1982). Such design results in an optimum single-mode fiber with the property that the points of minimum loss and minimum dispersion coincide, permitting repeater spans of the largest possible distance.

The amount of λ_0-shift that is obtainable depends on the shape of the refractive index profile of the fiber core. Figure 3.10 shows relative index profiles $\Delta(r)$ for step, trapezoidal, parabolic and triangular-shaped refractive index distributions. The zero-dispersion wavelengths for these fibers are shown as functions of the core diameter in Fig. 3.11 for several peak values of $\Delta_m = \Delta(0)$ (Reed et al., 1986). The horizontal line in each figure indicates the wavelength 1.55 μm, which the dispersion shift is intended to reach.

These figures reveal a number of interesting features. First, they show that a given value of λ_0 can be achieved with two different core radii. This behavior becomes understandable when we recall that for a fixed peak value of Δ_m a given mode spot size can be reached in two ways, if it is reachable at all. For fibers with sufficiently large core radii, the spot radius will be of the same order of magnitude as the core radius. As the core radius is allowed to shrink, the mode spot radius may shrink initially, but it will start to grow again as the core radius becomes still smaller and the fiber loses its ability to guide light. At a certain small core radius the mode spot radius reaches once more the initial value.

Of the two radii yielding the desired zero-dispersion wavelength $\lambda_0 = 1.55$ μm, the larger one is more desirable from a practical point of view, (more tightly guided mode).

Next, we see the core radii resulting in a desired value of λ_0 move to the right as we scan the figures from top to bottom. Thus, if a large core radius is desired, the fiber with triangular refractive index profile is best.

Finally, it is clear that not all the curves reach the desired value $\lambda_0 = 1.55$ μm. But with increasing values of Δ_m the maximum of each curve moves

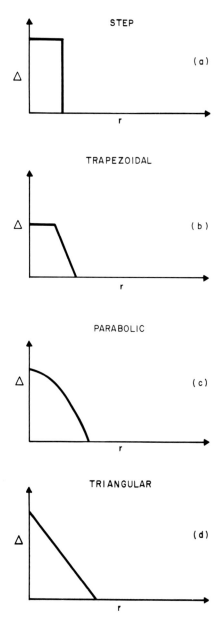

Fig. 3.10 Relative refractive index profiles of fiber cores as functions of the radial coordinate: a) step-index fiber, b) trapezoidal-profile fiber, c) parabolic-index fiber, d) triangular-index fiber.

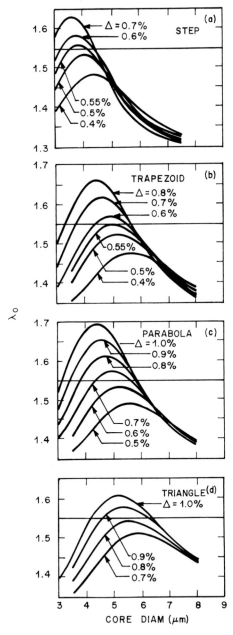

Fig. 3.11 "Zero-dispersion" wavelength as function of core diameter for the fiber types shown in Fig. 3.10. (Reed *et al.*, 1986.) Reprinted with permission from the *AT&T Technical Journal.* Copyright 1986 AT&T.

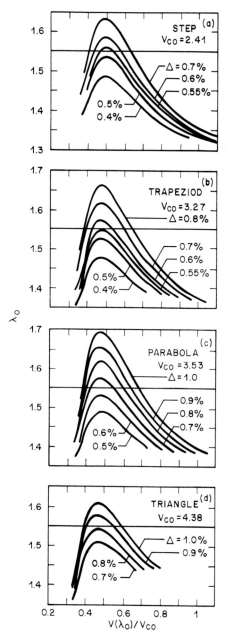

Fig. 3.12 "Zero-dispersion" wavelength as function of relative V-value for the fiber types of Fig. 3.10. (Reed *et al.*, 1986.) Reprinted with permission from the *AT&T Technical Journal.* Copyright 1986 AT&T.

upward until it coincides with $\lambda_0 = 1.55$ μm. This point is noteworthy because it yields the largest tolerance to accidental core radius variations.

Figures 3.12a through d are similar to Figs. 3.11 except that the zero-dispersion wavelength is now plotted as a function of the ratio V/V_c (Reed et al., 1986).

The parameter V is defined by (3.38). V_c is that value of V at which the next higher mode (LP_{11}) reaches its cutoff point. (For a discussion of LP_{11} cutoff see Section 3.6.) Thus, $V/V_c < 1$ is the condition for single-modeness. As V/V_c becomes smaller, the mode is more vulnerable to microbending losses. Thus, V/V_c should be as large as possible but not larger than unity. The values of V_c for each fiber type are indicated in the figure. The most remarkable feature of the Figs. 3.12 is the fact that the λ_0-curves for all fiber types have maxima at nearly the same value, $V/V_c = 0.5$.

Even more interesting than the dispersion-shifted fibers listed in Figs. 3.10 are multiply-clad fibers. Two members of this general group are sketched in Figs. 3.13. Figure 3.13a represents a "doubly-clad" refractive index profile, also known as W-fiber (Kawakami and Nishida, 1974). This name becomes more plausible if the profile is plotted symmetrically extended to negative values of r, because then it suggests the shape of the letter W. The index profile of Fig. 3.13b is a quadruply-clad fiber (Cohen et al., 1982; Francois et al., 1984). The dispersion curves of representative samples of fibers of this type are shown in Fig. 3.14 (Cohen et al., 1982). The label SC suggests single-clad, or conventional step-index fibers. The label DC stands for doubly-clad and QC for quadruply-clad fibers. Doubly and quaduply clad fibers have the remarkable property of yielding two zero-dispersion wavelengths. The amount of dispersion between these two points remains extremely small, so that these fibers are practically dispersion free over a wide range of wavelengths. The range between zero-dispersion points of the doubly-clad fiber is approximately 0.2 μm, while it is 0.3 μm for the quadruply-clad fiber. The inner cores of these fibers have diameters of approximately $2a = 8$ μm and the refractive index of the core relative to the outermost cladding is $\Delta = 0.4$ percent. In addition to low dispersion, these fibers also have low losses comparable to those of ordinary step-index fibers (Francois, 1983).

The design of dispersion-shifted fibers requires numerical analysis (Mammel and Cohen, 1982). Because of the small values of Δ it suffices to describe the guided mode in the scalar approximation. This means that instead of solving the complicated Maxwell equations, only the scalar wave equation

$$\frac{d^2\psi}{dr^2} + \frac{1}{r}\frac{d\psi}{dr} + \left(n^2(r)k^2 - \beta^2\right)\psi = 0 \tag{3.39}$$

needs to be solved. We have tacitly assumed that we are interested only in the

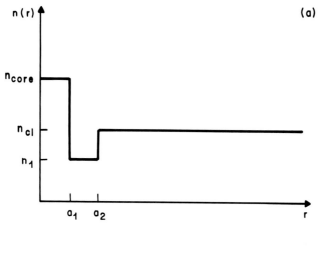

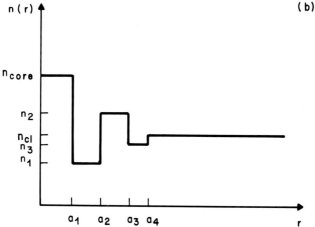

Fig. 3.13 Schematic refractive index profiles of two fiber types: a) doubly-clad or W-fiber, b) quadruply clad fiber.

mode of lowest order with azimuthal symmetry. To normalize the wave equation we introduce the dimensionless radial variable.

$$\rho = \frac{r}{b} \tag{3.40}$$

with b indicating the inner radius of the outermost (constant) cladding. Next, we introduce a dimensionless, wavelength-independent profile function $(N(\rho))$ via

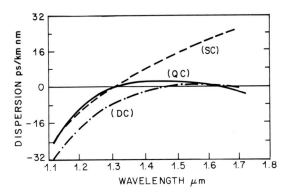

Fig. 3.14 Dispersion as a function of wavelength for an ordinary singly-clad step-index fiber (dashed curve), a doubly-clad fiber (dashed-dotted curve), quadruply clad fiber, (solid curve). (Cohen *et al.*, 1982.)

the relation

$$n^2(\rho) = n_0^2[1 - 2N(\rho)\Delta].$$ (3.41)

With these definitions, the wave equation can be written in the form of an operator eigenvalue problem

$$A\psi = \eta\psi$$ (3.42)

with differential operator

$$A = \frac{d^2\psi}{d\rho^2} + \frac{1}{\rho}\frac{d\psi}{d\rho} - V_b^2 N(\rho)$$ (3.43)

and eigenvalue

$$\eta = (\beta^2 - n_0^2 k)b^2.$$ (3.44)

The parameter V_b corresponds to (3.38) with 'a' replaced by b. The choice of the normalizing radius b is, of course, arbitrary, $b = a$ would be perfectly acceptable. For computational purposes, however, it is convenient to use a radius such that $n(r) = const$ for $r > b$.

The solution of the eigenvalue problem (3.44) can be accomplished by a large number of existing techniques and depends on the available library routines. However, acceptable solutions must satisfy the boundary condition of vanishing field at infinite radius, $\psi \to 0$ as $r \to \infty$.

To compute dispersion curves, it is necessary to solve the eigenvalue problem for a large number of different wavelengths and take the wavelength dependence on n_{core} and Δ into consideration. To achieve desired dispersion characteristics a suitable index profile must be chosen and the available free parameters varied until the desired result is obtained (Mammel and Cohen, 1982).

Computation of the dispersion (3.34) or (3.36) would seem to require two successive numerical differentiations. Because of the notoriously low accuracy of numerically computed derivatives it is advisable to obtain at least one of these derivatives by means other than numerical differentiation. For this purpose, there exists a formula (Mammel *et al.*, 1981; Snyder and Love, 1983; Eq. (33-16), p. 644), which permits the computation of the group delay by integration rather than differentiation.

$$\frac{1}{L}\tau = -\frac{\pi\lambda^2}{\beta c}\frac{\int_0^\infty \frac{d}{d\lambda}\left(\frac{n^2(r)}{\lambda^2}\right)\psi^2 r\,dr}{\int_0^\infty \psi^2 r\,dr} \tag{3.45}$$

The derivatives of the refractive index can be computed analytically by formulas derived from differentiating the Sellmeier representation of the refractive index.

This sketchy outline of the procedure required for computing fiber dispersion from analytically or experimentally given refractive index profiles shows that this is not a simple process. But, once the required programs have been written and made to work on a sufficiently fast computer, very accurate predictions of the dispersive behavior of real optical fibers can be obtained (Cohen *et al.*, 1980, 1981).

3.3 POLARIZATION MAINTAINING FIBERS

The modes of optical fibers consist of two types, $HE_{\nu\mu}$ modes and $EH_{\nu\mu}$ modes. The label ν refers to the azimuthal variation of the field while the label μ counts modes of different radial variation (Snyder and Love, 1983; Marcuse, 1982). The dominant mode of an ordinary optical fiber is designated as the HE_{11} mode. In weakly guiding fibers, characterized by the fact that $\Delta = (n^2_{core} - n^2_{cL})/(2n^2_{core}) \ll 1$, an approximate description of the mode field can be obtained by solving the scalar wave equation instead of the full set of Maxwell's equations. This dominant-mode solution is called the LP_{01} mode, a designation that serves as a reminder that the modal field is (very nearly) linearly polarized (Gloge, 1971). For the dominant mode, the exact HE_{11} mode and the approximate LP_{01} mode are equivalent. For higher order modes, the relationship is more complicated [Snyder and Young, 1978]. For $\nu > 1$, it can be shown that the $LP_{\nu\mu}$ mode is actually a linear superposition of the $HE_{\nu+1,\mu}$ and the $EH_{\nu-1,\mu}$ modes (Marcuse, 1982). These two modes have nearly the same phase velocity, that is they are nearly degenerate. Since the degeneracy is not exact, the superposition of $HE_{\nu+1,\mu}$ and $EH_{\nu-1,\mu}$ modes changes its appearance as the field travels in the fiber. For this reason, the $LP_{\nu\mu}$ modes are not really true modes of the fiber, which, by definition, must have the same cross sectional form at every point along the fiber. The LP_{01} mode, however, qualifies as a true mode since it corresponds directly to the HE_{11} mode.

The electric field of the $LP_{01}(HE_{11})$ mode has three components, E_x, E_y and E_z. Of these, either E_x or E_y predominates, while E_z, the component in the direction of the fiber axis, is very much smaller. If E_x is the most prominent field component, its magnitude is said to be of zero order, the magnitude of E_z is small of first order and E_y is small of second order (Gloge, 1971). Thus, if E_x is the dominant field component of the mode, the light is polarized in x-direction. However, in isotropic, circularly symmetric fibers, for every solution with a dominant E_x component there is a corresponding solution whose E_y component predominates. This means that a nominal "single-mode fiber" can actually support two modes that are identical except for their mutually orthogonal polarizations.

This double-modedness of ordinary single-mode fibers has important consequences. Let us assume that at $z = 0$ the mode field is excited with the x-polarization. If the fiber were a perfect cylinder this polarization would remain undisturbed throughout its length. No fiber, however, can be absolutely perfect. It is well known that any kind of fiber irregularity, such as a change in the direction of its axis, couples certain mode groups among each other (Marcuse, 1974). Other fiber imperfections consist in departures of the fiber from its nominal circular cross section. The fiber material is often assumed to be nominally isotropic. This means that the refractive index is the same regardless of the direction of polarization of the light field. In actual fibers, this is not exactly true. Such small departures from perfect circularity, or fluctuations of the anisotropy of the fiber material, couple the x-polarized mode to the y-polarized mode. This coupling is particularly effective since both modes are nominally degenerate, that is, they maintain their phase relationship forever, or in practical cases, over a relatively long distance. These conditions lead to a complete mixing of the two polarization states so that the initially linearly polarized light field quickly reaches a state of arbitrary polarization (Kaminow, 1981). For some applications the polarization state of the light does not matter. An ordinary photodetector does not distinguish polarization so that the polarization state of the light arriving at an intensity detector is of no interest. The polarization state, however, is of great concern in coherent light systems, where the incident signal is superimposed on the field of a local oscillator. The desired beats occur only among field components of the same polarization but are absent for orthogonally polarized components. Another application where polarization is of concern arises when an optical fiber is to be coupled to a modulator or other waveguide device that operates properly only if the light is linearly polarized in a given direction.

Finally, there is another drawback to loss of polarization. In slightly elliptical or slightly anisotropic fibers, the two polarization states travel at slightly different velocities. A pulse that is carried by both polarizations thus arrives at slightly different times at the receiver causing a small amount of pulse broadening.

Thus, there are good reasons why it is often desirable to use fibers that will permit light to pass through without changing its state of polarization. Much effort is directed at the design of polarization maintaining fibers (Ramaswamy et al., 1978; Kaminow and Ramaswamy, 1979). All efforts at achieving this goal are directed to breaking the degeneracy between the two mutually orthogonal polarization states. Originally it was hoped that this could be achieved by making the fiber core cross section intentionally as elliptical as possible (Love et al., 1979). We shall see, however, that such shape birefringence is not very effective. Much more successful are efforts at introducing intentional anisotropy into the fiber material (Simpson et al., 1983; Stolen et al., 1984; Varnham et al., 1983). This is achieved by utilizing the elastooptic effect, which links mechanical strain in a dielectric material to changes in the refractive index. The index anisotropy induced by strain follows the direction of the strain tensor (Born and Wolf, 1965).

Once the degeneracy between mutually orthogonal polarization states is broken, coupling between these modes is much reduced (Kaminow, 1981, 1984). To understand this effect it is important to know that coupling between two modes with propagation constants β_x and β_y is mediated by fiber imperfections whose z-dependence has a Fourier component at the spatial frequency $K = \beta_x - \beta_y$ (Marcuse, 1974). The Fourier spectra of fiber irregularities tend to have large amplitudes at low spatial frequencies and tend to drop off rapidly with increasing spatial frequencies (Krawarik and Watkins, 1978; Kaminow, 1984). Thus, large values of $\beta_x - \beta_y$ correspond to high spatial frequencies whose Fourier amplitudes are typically quite small. In this way it has been possible to construct anisotropic fibers that maintained a cross talk level between mutually orthogonal polarizations below -20 dB over several km length (Katsuyama et al., 1981; Rashleigh and Marrone, 1982).

Even more promising are fibers that are constructed in such a way that one of the two polarizations states suffers appreciable losses while the other suffers less loss (Snyder and Ruehl, 1983; Varnham et al., 1983). In this case, power converted from the desired polarization state is quickly dissipated and a true single-mode, single-polarization fiber results.

If the dominant mode of a single-mode optical fiber can exist in two polarization states with different phase velocities, the fiber has two principal transverse directions such that, in the absence of coupling, modes polarized in these preferred directions maintain their polarization. The mathematical description of the modes is simplified by orienting the coordinate system such that the transverse x and y coordinates coincide with these principal axes. The propagation constants of the x- and y-polarized modes are labeled β_x and β_y, respectively.

A fiber whose two polarization states have different propagation constants is called birefringent. The degree of birefringence is expressed as

$$B = \frac{|\beta_x - \beta_y|}{k} \qquad (3.46)$$

where, as always, $k = 2\pi/\lambda$ is the propagation constant of plane waves in free space.

Shape birefringence increases as the ellipticity of the fiber core becomes larger. In the limit of an infinitely elongated ellipse, the fiber degenerates into a slab waveguide. This case with maximum shape birefringence is also easiest to treat analytically (Kaminow and Ramaswamy, 1979). The even TE modes of a symmetric slab waveguide with core half width a, are polarized with their electric vectors parallel to the slab, which is assumed to be the y-direction. The propagation constant β_y of this mode is a solution of the eigenvalue equation (Marcuse, 1982)

$$\tan U_e = \frac{W_e}{U_e} \qquad (3.47)$$

with

$$U_e = \left[n_{core}^2 k^2 - \beta_y^2 \right]^{1/2} a \qquad (3.48)$$

and

$$W_e = \left[\beta_y^2 - n_{cL}^2 k^2 \right]^{1/2} a. \qquad (3.49)$$

The mode polarized at right angles to the TE mode is called the TM mode. Its electric vector is directed (predominantly) perpendicular to the slab. It has the propagation constant β_x, which is a solution of the eigenvalue equation (Marcuse, 1982)

$$\tan U_m = \frac{n_{core}^2}{n_{cL}^2} \frac{W_m}{U_m} \qquad (3.50)$$

with

$$U_m = \left[n_{core}^2 k^2 - \beta_x^2 \right]^{1/2} a \qquad (3.51)$$

and

$$W_m = \left[\beta_x^2 - n_{cL}^2 k^2\right]^{1/2} a. \tag{3.52}$$

An approximate expression for the birefringence can be computed from (3.47) and (3.50) in the following way. We introduce $\Delta\beta = \beta_y - \beta_x$, which is assumed to be a small quantity, and set $\beta_y = \beta$, $U_e = U$, $W_e = W$ and $\beta_x = \beta - \Delta\beta$. Next, we expand U_m, W_m and $\tan W_m$ in Taylor series neglecting terms of order smaller than $\Delta\beta$. We substitute these expansions into (3.50) and obtain with the help of (3.37) and (3.38) the following expression for the birefringence of the slab waveguide

$$kB = \Delta\beta = \frac{n_{core}^2 U^2 W^2 2\Delta}{n_{cL}^2 \beta a V^2 a(1 + W)}. \tag{3.53}$$

For most practical purposes, the case of weak guidance $(n_{core} \approx n_{cL} = n)$ is of most interest. In the weak guidance limit we obtain with $\beta \approx nk$

$$kB = \Delta\beta = \frac{(2\Delta)^{3/2}}{a}\phi_{SL} \tag{3.54}$$

with

$$\phi_{SL} = \frac{U^2 W^2}{V^3(1 + W)}. \tag{3.55}$$

A corresponding expression has been derived for fibers with slight ellipticity (Love et al., 1979)

$$e^2 = 1 - a^2/b^2, \tag{3.56}$$

whose major and minor half axes are, respectively, b and a;

$$kB = \Delta\beta = \frac{e^2}{a}(2\Delta)^{3/2}\phi_f \tag{3.57}$$

the expression for ϕ_f is more complicated than (3.55) so that we present a plot of this quantity as a function of V as the solid curve in Fig. 3.15 (Love et al., 1979). The dotted curve in this figure is a plot of (3.55). The two curves should be identical since ϕ_f is nominally independent of e. The derivation of ϕ_f, however, was limited specifically to $e^2 \ll 1$, so that exact agreement with the slab formula (that is for $e^2 = 1$) cannot be expected. It is apparent, however, that the two curves predict the same order of magnitude of shape birefringence. For a typical case, $\lambda = 1$ μm, $\Delta = 0.003$ and $a = 5$ μm, we obtain from Fig. 3.15 (with $e^2 = 1$ for the fiber) with $\phi_f = 0.15$ the following value for the shape birefringence, $B = 2.2 \times 10^{-6}$. If we define the beat wavelength L_b of the two

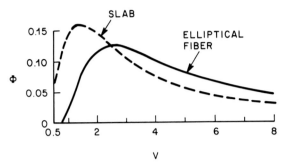

Fig. 3.15 Normalized shape birefringence for a slab waveguide (dotted curve) and an elliptical fiber (solid curve).

polarization states as

$$L_b = \frac{2\pi}{\Delta\beta}, \tag{3.58}$$

we obtain with this birefringence, $L_b = 45$ cm. The amount of birefringence is not sufficient to uncouple the two polarization states effectively. To achieve larger B requires large values of Δ, which would require impractically small core radii if single-modedness is to be preserved.

B-values that are approximately two orders of magnitude larger can be achieved by stress birefringence. To provide a basis for understanding stress birefringence, let us review a few basic facts and look at a simple example. We restrict the discussion to materials that, in the absence of stresses, are optically isotropic with a constant refractive index n. Stresses can be described by a stress tensor. For simplicity, we orient the coordinate system such that its axes coincide with the principal axes of the stress tensor so that only its diagonal elements σ_x, σ_y and σ_z are different from zero. Stress alters the refractive index of the dielectric material making it birefringent. In (originally) isotropic materials, the principal axes of the refractive index tensor coincide with those of the stress tensor so that it too can be characterized by three constants n_x, n_y and n_z. Traditionally, instead of $n_x - n$, it is $1/n_x^2 - 1/n^2$ that is related to the stress tensor. However, since we may approximate $1/n_x^2 - 1/n^2 = -2(n_x - n)/n^3$, we express the refractive index changes as follows (Born and Wolf, 1965)

$$n_x - n = -\frac{n^3}{2}\left(p_{11}\sigma_x + p_{12}\sigma_y + p_{12}\sigma_z\right) \tag{3.59}$$

$$n_y - n = -\frac{n^3}{2}\left(p_{12}\sigma_x + p_{11}\sigma_y + P_{12}\sigma_z\right) \tag{3.60}$$

$$n_z - n = -\frac{n^3}{2}\left(p_{12}\sigma_x + p_{12}\sigma_y + p_{12}\sigma_z\right) \tag{3.61}$$

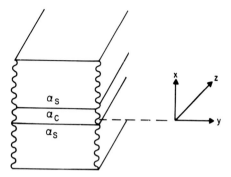

Fig. 3.16 Slab core stressed by cladding layers with different expansion coefficient.

The coefficients p_{11} and p_{12} are the stress-optical coefficients, which are also called photoelastic constants. In general, one might expect that nine different p-coefficients should appear in (3.59) through (3.61). Symmetry considerations, however, can be used to prove that of the nine possible coefficients, only two are different from each other for isotropic materials such as glass.

If the stress exists only in the $y - z$ plane, $\sigma_x = 0$, the birefringence of the material for a plane wave traveling in z-direction, follows from (3.59) and (3.60)

$$B = n_x - n_y = \frac{n^3}{2}(p_{11} - p_{12})\sigma_y. \tag{3.62}$$

We are now ready to consider a simple example of a slab core with thermal expansion coefficient α_c, that is embedded in an infinitely extended material with thermal expansion coefficients α_s, as shown in Fig. 3.16 (Kaminow and Ramaswamy, 1979). The slab is fused to the surrounding material at a temperature T_2. If the whole system is then allowed to cool to a lower temperature, $T_1 = T_2 - \Delta T$, the different expansion coefficients for the slab and for the surrounding material, cause stresses to build up in y and z direction,

$$\sigma_y = \sigma_z = (\alpha_s - \alpha_c)\Delta T \tag{3.63}$$

But there is no stress in x-direction, $\sigma_x = 0$.

For applications to polarization maintaining fibers, it is of interest to assume that the substance surrounding the slab is basically made of the same material but is modified by the addition of a small molar concentration g of a dopant with expansion coefficients α_d (for example the slab core may be made of SiO_2 and the surrounding material of SiO_2 doped with B_2O_3). Its thermal expansion coefficient can then be approximated as (Kaminow and Ramaswamy, 1979)

$$\alpha_s = \alpha_d g + (1 - g)\alpha_c \tag{3.64}$$

The birefringence of the slab is now obtained by combining (3.62), (3.63) and (3.64)

$$B = \frac{n^3}{2}(p_{11} - p_{12})g(\alpha_d - \alpha_c)\Delta T. \tag{3.65}$$

To obtain an order-of-magnitude estimate of the birefringence that can be expected in this case, we assume that the temperature difference ΔT corresponds to the difference between the softening temperature of the boron doped silica glass relative to room temperature, $\Delta T = 800°C$, $n = 1.46$, $g = 0.1$, $\alpha_c = 0.5 \times 10^{-6}C^{-1}$, $\alpha_d = 10 \times 10^{-6}C^{-1}$ and $p_{11} - p_{12} = 0.15$. With these values we obtain for the birefringence of the slab, $B = 1.77 \times 10^{-4}$. Even though this example is vastly oversimplified, it gives us an idea of the order of magnitude of stress birefringence to be expected in optical fibers and shows that it can be two orders of magnitude larger than the shape birefringence of a slab.

Stresses and hence stress-induced birefringence can be built into optical fibers by deliberately destroying the circular symmetry of the fiber cross section. This can be done in several ways. For example, a circular fiber core can be surrounded by an elliptical cladding region, or, conversely, an elliptical core could be surrounded by a circular cladding as shown in Fig. 3.17 (Ramaswamy et al., 1979; Eikhoff, 1982; Katsuyama et al., 1981). A very elongated ellipse approaches the slab assumed earlier. Since core and cladding must have different refractive indices, they are made of somewhat different materials. For example, the core could be made of pure fused silica while the cladding is made of borosilicate glass, because boron doping reduces the refractive index of fused silica. The two materials have different thermal expansion coefficients so that stresses are introduced and frozen into the glass when the molten fiber solidifies. As the stresses lack circular symmetry, since the structure itself is asymmetric, the fiber material becomes birefringent. The ellipticity of either the core or the cladding cross sections is achieved by careful control of temperature and pressure during the collapse of the MCVD preform tube. Separate control of the core and cladding geometries is made possible by the fact that pure silica and boron doped silica have different softening temperatures (Katsuyama et al., 1981). Boron doping has the added advantage of yielding a particularly large difference of the expansion coefficients of core and cladding materials. A disadvantage of this method, however, is the relatively high loss of boron-doped silica. For this reason, other methods of producing stress birefringence in fibers have been invented.

A particularly successful structure is shown in Fig. 3.18 (Shibata et al., 1983b; Varnham et al., 1983). It consists of a Ge-doped core in a fused silica cladding. Added to these standard fiber components are two stress members made of borosilicate glass. Again, the asymmetry of the structure causes stress birefrin-

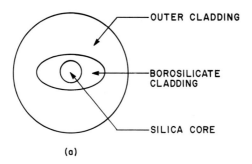

(a)

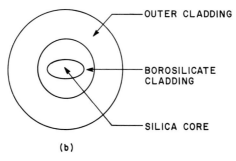

(b)

Fig. 3.17 Stress induced by: a) elliptically deformed cladding, b) elliptically deformed core.

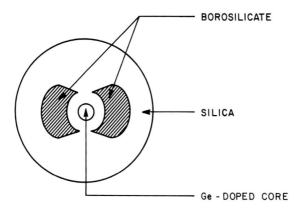

Fig. 3.18 "Panda" or "bow-tie" fiber with borosilicate stress members embedded in cladding.

88   Dietrich Marcuse

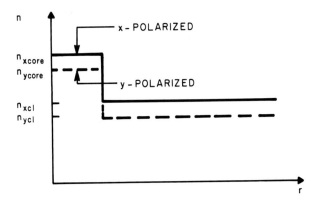

Fig. 3.19 Refractive index profile of birefringent step-index fiber. The solid curve shows the refractive index as "seen" by the x-polarized mode, the dotted curve is the refractive index as "seen" by the y-polarized mode.

gence due to the different thermal expansion coefficients of borosilicate glass and fused silica. This design has the advantage that the somewhat lossier borosilicate glass is farther removed from the fiber core as in the designs shown in Fig. 3.17. A compromise must be reached by placing the stress members at a sufficient distance from the core to prevent degradation of the low fiber loss and yet placing it close enough to the core to yield sufficiently high stress birefringence. The structure shown in Fig. 3.18 has been made by producing the germania-doped core and boron doped stress members separately by the VAD method. The resulting rods are pulled to the desired diameter and are placed inside a silica tube such that the core rod occupies the center and the stress members assume their respective positions. The remaining space is filled up with glass rods made of pure silica. The whole arrangement is then fused together to become the fiber perform from which the polarization maintaining fiber is pulled. Fibers of this type have been given whimsical names such as "PANDA" [Shibata *et al.*, 1983] and "Bow-tie" [Birch *et al.*, 1982] fibers.

For fibers of the type shown in Fig. 3.17, birefringence values as high as $B = 7 \times 10^{-4}$ have been reported. For the structure shown in Fig. 3.18 reported birefringence values range from $B = 1.2 \times 10^{-4}$ to $B = 5.3 \times 10^{-4}$.

In principle, truly single mode operation is possible with stress birefringent optical fibers. To see this, we show schematically in Fig. 3.19 the refractive index distributions as "seen" by an x- and a y-polarized mode [Snyder and Ruehl, 1983]. Because the values of $n_{xcore} - n_{xcL}$ are typically quite small (on the order of 0.005), it is safe to assume that the stress induced birefringence of core and cladding are the same. Each mode has an effective refractive index defined as $n_{eff} = \beta/k$, where β indicates the modal propagation constant of either the x- or y-polarized mode. As long as $n_{eff} > n_{xcL}$, modes of both polarization are truly guided. But, once the y-polarized mode has an effective refractive index such that $n_{yeff} < n_{xcL}$, this mode becomes a leaky wave. This leakage phenomena is subtle. If the y-polarized mode had no electric field components in x and z direction it would be "unaware" of the higher refractive index value n_{xcL} and no leakage could occur. We have already mentioned, however, that even the electric field of the y-polarized mode has a small z-component and an even smaller x-component. These small field components couple the y-polarized to the x-polarized mode. Since both guided modes have different phase velocities, such coupling would not be very effective. In addition to the guided modes discussed so far, there are so-called radiation modes whose effective refractive index lies below the cladding index value. Since the radiation modes form a continuum, there is always one x-polarized radiation mode whose effective refractive index matches exactly that of the y-polarized guided mode. Coupling to radiation modes causes power to be carried away from the fiber core and hence results in radiation losses. Thus, coupling of the y-polarized guided mode with $n_{yeff} < n_{xcL}$ to x-polarized radiation modes having the same effective refractive index, is the mechanism by which the y-polarized mode loses power. These power losses can be substantial so that, effectively, only the x-polarized mode is truly guided. Any power coupled into the y-polarized mode dies out before reaching the receiver.

Figure 3.20 [Snyder and Ruehl, 1983] shows the normalized power loss coefficient α (multiplied by the core radius and divided by $\Delta^{5/2}$) as a function of the normalized birefringence nB/Δ for cases, $n_x = n_z$ and $n_y = n_z$, where n_z is the value of the refractive index in the direction of the fiber axis. To obtain the loss in dB/km, when α is measured in μm, the value read off the vertical axis of Fig. 3.20 must be multiplied by the factor 4.3×10^9. For a fiber with $a = 5\ \mu$m, $\Delta = 0.003$, the leakage loss can reach values on the order of 400 dB/km. This loss value corresponds to a normalized loss of unity on the scale of Fig. 3.20. On average, the leakage losses will be somewhat lower but still considerable relative to the fiber losses, which are typically below 1 dB/km.

The design of a polarization maintaining fiber, following the principle just outlined, has the disadvantage that the x-polarized mode must be operated at very small V-values so that it is only extremely weakly guided. This restriction stems from the fact that, typically $nB \ll \Delta$ so that small values of V are required to ensure $n_{yeff} < n_{xcL}$. Such weak guidance requires that the fiber remain extremely straight to avoid bending losses. It has been demonstrated, however, [Simpson et al., 1983; Varnham et al., 1983] that fibers operating with V-values below $V = 2$ still yield very considerable loss discrimination between x- and y-polarized modes. This occurs because uncontrolled but inevitable microbending losses are very sensitive to the actual V-values of each mode. Due to its design we have $V_x > V_y$, making the y-polarized mode more susceptible to excess loss. Thus, even though the fiber is not operating in the regime $n_{yeff} < n_{xcL}$, it

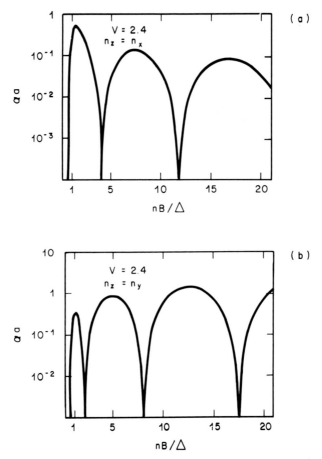

Fig. 3.20 Normalized leakage loss as a function of the normalized birefringence for a y-polarized mode in a fiber with index profile shown in Fig. 3.19. a) $n_z = n_x$, b) $n_z = n_y$. (Snyder and Ruehl, 1983.)

satisfies the requirement that the loss of the y-polarized mode substantially exceeds the loss of the x-polarized mode. For practical applications, it behaves as though only one polarization can propagate.

3.4 SOLITONS

The distance that can be bridged by an optical fiber, before regeneration of the signal becomes necessary, is determined by the fiber loss and by its dispersion. Fiber losses have been reduced to about 0.2 dB/km at 1.55 μm wavelength. Thus, if the receiver can detect signals that are 20 dB below the input signal, a distance of 100 km can be spanned without need for signal amplification. Once the repeater span is thus determined, the maximum signaling rate depends on fiber dispersion. We have seen in Section 3.1 that, with a monochromatic light source, data rates on the order of 100 Gb/sec are feasible if the fiber is operated at the zero-dispersion wavelength.

In principle, one could imagine inserting optical amplifiers into the fiber cable, thus extending the range over which a signal can be detected. Dispersion, however, would still seem to set an upper limit on the maximum spacing between regenerative repeaters if high data rates are required. To people used to thinking of optical fibers as linear transmission media, this conclusion seems final and inevitable. But fibers are not exactly linear waveguides. The small amount of nonlinearity, in conjunction with some linear dispersion, permits certain special pulse shapes to establish themselves, which, in a lossless but dispersive fiber, can travel arbitrary distances without changing shape. These special waveforms, called solitons, must carry a given amount of light energy that depends on the pulse duration, the fiber nonlinearity and dispersion [Hasegawa and Kodama, 1981]. Solitons do not solve the problem of fiber losses, but they do conquer the limits set on achievable data rates by fiber dispersion. For solitons to become useful, some kind of optical amplifier would be needed to counter fiber losses and thus maintain the critical pulse energy. The amplifier problem may well be solvable with the help of the Raman effect [Stolen and Ippen, 1973] so that optical communication by means of solitons may permit the transmission of light signals across the Atlantic or Pacific Oceans without regenerative repeaters. Even though this possibility is still speculative, it is sufficiently exciting to warrant taking a look at solitons in optical fibers [Hasegawa, 1983, 1984; Mollenauer et al., 1985, 1986].

We derive the differential equation for solitons in two stages. First, we neglect dispersion and concentrate on the contribution of the nonlinearities of the fiber material [Stolen and Lin, 1978]. Next, we neglect nonlinearities and derive a differential equation for the pulse envelope in a dispersive medium. Finally, we combine the two effects into one equation. The derivation is simplified by considering the fiber mode as a plane wave. The connection with the actual fiber mode will be made by introducing an effective area covered by the mode field. The plane wave assumption is not essential, but it simplifies the derivation of the soliton equation and highlights its principal features without being burdened by additional steps needed to accommodate waveguide effects.

A (plane) wave propagating in z-direction can be described as a solution of the wave equation

$$\frac{\partial^2 E}{\partial z^2} = \frac{1}{c^2} \frac{\partial^2 (n^2 E)}{\partial t^2} \tag{3.66}$$

with c indicating the speed of light in vacuum. The refractive index n is assumed to be nonlinear, that is, it depends slightly on the square of the magnitude of the electric field

$$n = n_0 + n_2 |E|^2. \tag{3.67}$$

The electric field E varies rapidly as a function of z with propagation constant β_0 and as a function of time t with angular frequency ω_0. The problem is simplified if we introduce the slowly varying envelope function $\phi(z, t)$ of the pulse

$$E = \phi(z, t) e^{i(\omega_0 t - \beta_0 z)}. \tag{3.68}$$

When taking the second derivatives of E with respect to z and t we neglect second derivatives of ϕ since, due to the slower variation of this function, they are much smaller than the terms $\omega_0^2 \phi$ and $\beta_0^2 \phi$. Using $\beta_0 = n_0 \omega_0/c$ we obtain, by substitution of (3.67) and (3.68) into (3.66),

$$\frac{\partial \phi}{\partial z} + \dot{\beta}_0 \frac{\partial \phi}{\partial t} = -i \frac{n_2}{n_0} \beta_0 |\phi|^2 \phi. \tag{3.69}$$

The step from (3.66) to (3.69) involves two additional approximations. Because n_2 is very small, the square of this quantity is being neglected. For the same reason, we also keep only the leading term, $-\omega^2 |\phi|^2 \phi$, of the second time derivative $\partial^2 [|\phi|^2 \phi]/\partial t^2$ since the products of all additional terms with n_2 are negligibly small. Finally, we have written $n/c = \dot{\beta}_0$, where the dot indicates the derivative with respect to the angular frequency ω_0. This last step is exact since n_0 was assumed to be independent of ω_0.

Equation (3.69) describes the envelope of a pulse propagating in a nonlinear, nondispersive medium, but this equation does not yield soliton solutions. Solitons can exist if dispersion is added to the nonlinearity. To see how dispersion affects the pulse envelope we derive the corresponding differential equation for a linear but dispersive medium. A plane wave with angular frequency ω can be described as

$$E_p = A e^{i(\omega t - \beta z)}, \tag{3.70}$$

while a pulse is obtained as a superposition of infinitely many plane waves

$$E = \int_{-\infty}^{\infty} A(\omega) e^{i(\omega t - \beta z)} \, d\omega. \tag{3.71}$$

Since the medium is dispersive, the propagation constant β is a nonlinear function of ω. Except in the immediate vicinity of the zero-dispersion point, an excellent approximation is provided by

$$\beta = \beta_0 + \dot{\beta}_0 (\omega - \omega_0) + \frac{1}{2} \ddot{\beta}_0 (\omega - \omega_0)^2. \tag{3.72}$$

The dots indicate derivatives with respect to ω taken at the central (carrier) frequency ω_0. Using (3.72), we can now rewrite (3.71) such that it assumes the form (3.68) with

$$\phi(z, t) = \int_{-\infty}^{\infty} A(u)\exp\left\{ i\left[(t - \dot{\beta}_0 z)u - \frac{1}{2}\ddot{\beta}_0 zu^2 \right] \right\} du. \qquad (3.73)$$

The integration variable u is $\omega - \omega_0$. It is easy to see by substitution of (3.73) that the pulse envelope satisfies the differential equation

$$\frac{\partial \phi}{\partial z} + \dot{\beta}_0 \frac{\partial \phi}{\partial t} = \frac{i}{2}\ddot{\beta}_0 \frac{\partial^2 \phi}{\partial t^2}. \qquad (3.74)$$

This differential equation for the pulse envelope is exact, provided the pulse travels in a medium that exhibits only first order dispersion as expressed by (3.72).

The left-hand sides of (3.69) and (3.74) are identical. Each describes a traveling pulse of an arbitrary shape.

$$\phi = f(t - \dot{\beta}_0 z) \qquad (3.75)$$

The right-hand side of (3.69) modifies the traveling wave to account for the nonlinearity of the medium, while the right-hand side of (3.74) modifies it to account for dispersion. If both effects are weak, as they are in typical optical fibers, the two effects can be taken to be additive so that we obtain the soliton equation for waves in a nonlinear and dispersive medium

$$\frac{\partial \phi}{\partial z} + \dot{\beta}_0 \frac{\partial \phi}{\partial t} = \frac{i}{2}\ddot{\beta}_0 \frac{\partial^2 \phi}{\partial t^2} - i\frac{n_2}{n_0}\beta_0 |\phi|^2 \phi. \qquad (3.76)$$

From its derivation it is clear that (3.76) is only approximately valid provided that nonlinearity and dispersion are weak. The equation, however, yields the desired soliton solutions.

It is customary to convert the soliton equation (3.76) to a normalized form by introducing dimensionless variables

$$t' = \frac{1}{\tau}(t - \dot{\beta}_0 z) \quad \text{and} \quad z' = \frac{|\ddot{\beta}_0|}{\tau^2}z \qquad (3.77)$$

and a new envelope function

$$u(z', t') = \tau \left[\frac{n_2\beta_0}{n_0 |\ddot{\beta}_0|} \right]^{1/2} \phi. \qquad (3.78)$$

With this transformation, (3.76) assumes the form

$$\frac{\partial u}{\partial z'} = \frac{i}{2} \frac{\ddot{\beta}_0}{|\ddot{\beta}_0|} \frac{\partial^2 u}{\partial t'^2} - i |u|^2 u. \tag{3.79}$$

We shall see that τ in (3.77) is closely related to the pulse width.

The simplest solution of the soliton equation (3.76) is known as a soliton of first order

$$\phi(z, t) = \phi_0 \frac{e^{iaz}}{\cosh\left(\dfrac{t - \dot{\beta}z}{\tau}\right)}. \tag{3.80}$$

It is easy to see that (3.80) is a solution of (3.76) provided that

$$a = \frac{\ddot{\beta}_0}{2\tau^2} \tag{3.81}$$

and that the amplitude of the soliton is given by

$$|\phi_0|^2 = -\frac{n_0}{n_2} \frac{\ddot{\beta}_0}{\beta_0 \tau^2}. \tag{3.82}$$

The soliton (3.80) shows that the width τ of the square magnitude of the pulse is independent of z. This means that the soliton travels in the dispersive fiber without changing its shape.

The amplitude equation (3.82) is interesting for two reasons. First, it shows that the amplitudes of solutions of the soliton equation are not arbitrary. This is a general property of nonlinear differential equations. Thus, if a soliton of first order exists, its amplitude is uniquely specified by the nonlinearity of the medium represented by n_2, but the amount of dispersion as indicated by the value of $\ddot{\beta}_0$ and by the pulse width as specified by τ.

Of equal importance is the second observation. Solitons can only exist if $\ddot{\beta}_0/n_2 < 0$. Thus, nonlinearity and dispersion must cooperate in the right sense. Since n_2 of fused silica is positive [Stolen and Lin, 1978], we see from Fig. 3.4 that the fiber must be operated at wavelengths that are longer than the zero dispersion wavelength if we want to excite solitons.

The full width Δt at half maximum of the soliton pulse power (proportional to $|\phi|^2$) is related to τ by

$$\Delta t = 1.76\tau. \tag{3.83}$$

The energy carried by the soliton is

$$W_e = \sqrt{\frac{\epsilon_0}{\mu_0}} \, A_{eff} \frac{n_0^2 \, |\ddot{\beta}|}{n_2 \beta_0 \tau} \tag{3.84}$$

and the power at the peak of the pulse is

$$P_{max} = \frac{W_e}{2\tau}. \tag{3.85}$$

A_{eff} is an effective area of the mode supported by the fiber [Mollenauer *et al.*, 1986]. It is not known precisely from our derivation but reasonable estimates are easily made.

Equation (3.84) shows that for given dispersion and nonlinearity the product of pulse energy and pulse width is constant, a shorter soliton pulse requires more energy. This may seem paradoxical but can be explained as follows. For a given amount of dispersion, a short pulse spreads more rapidly than a longer pulse. To counteract this tendency requires a larger contribution from the nonlinearity, that is more field intensity.

The soliton equations (3.76) or (3.79) have an infinite number of solutions yielding solitons of higher order. Each has a definite amplitude but the higher-order solutions are not as simple as the first-order soliton (3.80). All solitons of higher order change their shape periodically along the fiber. Thus, they too are pulses that do not spread continuously due to dispersion, but their widths pulsate and their shapes are very complicated. Nevertheless, higher order solitons may also be useful for data transmission over long distances, [Doran and Blow, 1983].

Having developed the theory of the soliton, it is now easy to give a qualitative explanation of why these wave forms escape pulse spreading due to fiber dispersion. It is well known that pulses do not necessarily always becomes longer as they travel in dispersive media. For example, if the pulse is chirped—if the carrier frequency changes so that it is lower at one end of the pulse and higher at the other end—the pulse may either spread more rapidly than an unchirped pulse, or it may even contract temporarily until a minimum pulse width is reached [Marcuse, 1981]. The direction in which the chirp affects the pulse depends on the direction of the chirp and the sign of $\ddot{\beta}$. Such compression of chirped pulses has been used to great advantage to generate extremely short light pulses starting with longer pulses, [Nakatsuku *et al.*, 1981; Shank *et al.*, 1982].

It is easy to show qualitatively that the fiber nonlinearity imposes a chirp on the pulse. The phenomenon is known as self-phase modulation [Meinel, 1983;

Tomlinson et al., 1984]. By substitution we see that

$$\phi = f \exp\left\{ -i \frac{n_2}{n_0} \beta_0 f^2 z \right\} \tag{3.86}$$

is a solution of (3.69) where f is an arbitrary real function $f = f(t - \dot{\beta}_0 z)$. Combining (3.68) and (3.86) we see that we are dealing with a pulse whose instantaneous carrier frequency—the time derivative of the phase angle of the complex function $E(z, t)$—is

$$\omega = \omega_0 - 2 \frac{n_2}{n_0} \beta_0 f \frac{\partial f}{\partial t} z. \tag{3.87}$$

Thus, we see that the pulse has become chirped by the nonlinearity of the medium. This observation makes it plausible that the dispersion of the fiber can lead to compression of the self-chirped pulse. If self chirping and dispersion act in the proper direction and have the proper magnitudes, the pulse compression exactly cancels the tendency of dispersion to broaden the pulse. As a result, the pulse maintains its width exactly. We can now also understand why solitons of higher order pulsate. Here, self chirping and pulse compression are not perfectly balanced at every instant. Instead, the pulse shrinks and expands periodically depending on whether the forces causing pulse compression or pulse spreading momentarily have the upper hand.

To convey an idea as to the order of magnitude of peak pulse power required for solitons, we consider an example. For fused silica the nonlinear coefficient of the refractive index is* $n_2 = 6.1 \times 10^{-19}$ cm$^2/V^2$. We assume that the effective mode area has a radius $r = 5$ μm, so that $A_{eff} = 7.85 \times 10^{-7}$ cm^2. At $\lambda = 1.5$ μm we see from Fig. 3.4 that $\ddot{\beta} = -20$ psec^2/km $= -2 \times 10^{-28}$ sec^2/cm. With $n_0 = 1.46$ and $\lambda = 1.5$ μm, we have $\beta = 6.12 \times 10^4$ cm^{-1}. Finally, let the pulse width be $\Delta t = 10$ psec. With these values we compute from (3.84) and (3.85) a peak pulse power of 375 mW. The required peak power decreases in inverse proportion to the square of the pulse width, thus a 10 times longer pulse requires only 3.75 mW peak power. The required peak power also decreases linearly with decreasing values of $|\ddot{\beta}|$. Thus, placing the operating wavelength closer to the zero-dispersion point reduces the required peak power. This example shows that the power requirements for solitons are easily within range of existing laser sources.

To excite a first-order soliton, the input pulse to the fiber must have the shape and amplitude of the wave form defined by (3.80) through (3.82). If the energy is too low, no soliton will form and the light pulse will spread according to the rules explained in Section 3.1. If the energy is too high, or the pulse too broad or

---

* $n_2 = 5.5 \times 10^{-14}$ cm^2/(statvolt)^2 = 6.1×10^{-19} cm$^2/V^2$. Note that n_2 as given by Stolen and Lin, 1978 is twice as large since they define $n = n_0 + \frac{1}{2} n_2 |E|^2$.

too narrow, a soliton (possibly of higher order) with the proper energy and pulse shape will form, but the remaining energy will be spun off as a spurious pulse. Since the spurious pulse travels in the fiber along with the soliton, but with different dispersion, its presence can lead to undesirable signal distortion.

Once a soliton has formed, it will be attenuated by the fiber loss mechanisms. To characterize the degree of attenuation it is useful to introduce the quantity

$$z_0 = \frac{\pi}{2} \frac{\tau^2}{|\ddot{\beta}|} \approx 0.5 \frac{(\Delta t)^2}{|\ddot{\beta}|}, \tag{3.88}$$

which is closely related to $1/a$ of (3.81). For fiber loss coefficients α that obey $z_0 < 1/\alpha$ the soliton will adjust its width so that, according to (3.83) and (3.84), the product $W_e \Delta t$ remains constant. If the loss coefficient is large so that $z_0 > 1/\alpha$ the soliton does not preserve its unique character but becomes an ordinary light pulse subject to dispersion spreading. This discussion shows that soliton transmission over large distances requires repeated amplification of the light pulse to preserve the solitons.

An ideal solution to the optical amplifier problem is provided by the Raman effect [Stolen and Ippen, 1973]. This is a special kind of parametric process in which one of the three waves needed in ordinary parametric amplifiers is replaced by vibrations of the molecules of the dielectric medium. Thus, a Raman amplifier requires a pump at an angular frequency ω_P that is slightly higher than the signal frequency ω_S. The difference between these two frequencies corresponds to a vibrational frequency of the molecules of the glass. Phase matching of the two optical waves is not required. In fused silica, Raman amplification takes place over a relatively broad band of frequencies centered at a frequency 450 cm^{-1} below the pump frequency. Thus, all that is required to turn the communications fiber into a distributed amplifier is a continuously operating optical pump wave whose power must be in the range between 30 and 100 mW. To amplify a signal at 1.55 μm requires a pump at 1.45 μm wavelength.

Hasegawa, 1984 and Mollenauer et al., 1986 have studied communications systems using the Raman effect to amplify the soliton pulses. They propose a communications system in which a pump wave is injected into the fiber through a frequency selective directional coupler every 20 to 60 km. The directional couplers are tuned to the pump frequency but do not couple out light at the soliton frequency. The pump waves are sent in both directions. The Raman effect works even if pump and signal waves travel in opposite directions. To avoid interaction between solitons and to avoid timing jitter introduced by fundamental spontaneous emission processes, Mollenauer et al., 1986 suggest that the spacing T between pulses be $T = 10 \times \Delta t$. Table 3.1 lists design and performance parameters of several proposed soliton systems. In this table, L is

TABLE 3.1
Design Examples of Soliton Based, Single Channel, High Bit Rate Systems

| $A_{eff} = 25 \ \mu m^2$; $\ddot{\beta} = -2.4$ psec^2/km | | | | | |
Design No.	L (km)	z_0 (km)	Δt (psec)	P_{max} (mw)	R (GHz)	Z (km)
1	30	6	5.5	50	18	1600
2	40	11	7.5	27	13	2200
3	50	17	9.4	18	10.6	2700
4	(~ 50)	30	12.3	10	8.1	3600
5	(~ 50)	50	16	6.1	6.2	4700
6	(~ 50)	100	22.6	3.0	4.4	6600

the spacing between injection points of the bidirectional Raman pump waves, z_0 is defined by (3.88), $R = 1/(10 \times \Delta t)$ is the maximum available signal bandwidth and Z is the total length of the fiber.

Mollenauer *et al.*, 1986 found by computer simulation that a resonance instability develops if $z_0 = L/8$. In this case the soliton energy and width tend to fluctuate wildly. The table shows that this condition is avoided in all listed designs.

A soliton communication system has several advantages over one using ordinary pulses, even at the zero-dispersion wavelength. The Raman pumps required for providing the gain to cancel out fiber losses need not be sophisticated, tightly controlled single wavelength lasers. They and their power control circuitry are expected to be considerably cheaper than the regenerative repeaters needed to restore the shape of the ordinary pulses. The capacity of the system can be increased by wavelength multiplexing. The Raman bandwidth is sufficiently wide to accommodate several channels with a single pump laser. In a system using wavelength multiplexing, however, it may be necessary to use several pumps at slightly different wavelengths to smooth out the gain spectrum over the utilized wavelength range. But wavelength multiplexing in a soliton system is still very much simpler than in a conventional system where at each repeater the various channels would first have to be filtered out, their pulses regenerated and finally, for each channel a tightly controlled signal source would be needed to send out anew all the different wavelengths. The error rate in a soliton system would be very low since the received pulses are very much stronger than those of a conventional system, which allows the pulses to be attenuated as much as possible to increase the repeater spans.

3.5 DETRIMENTAL NONLINEAR EFFECTS

We have seen that solitons can exist if the nonlinearity of the fiber material cooperates with fiber dispersion such that the detrimental influences of both effects cancel each other to permit a pulse to propagate without deformation in the absence of loss. A first-order soliton can exist only if the pulse carries enough energy to satisfy the condition (3.84). Similar conditions, requiring even higher pulse energies, apply to solitons of higher order. For the formation of solitons the fiber nonlinearity is beneficial. But if the pulse energy (in conjunction with pulse width and fiber dispersion) is insufficient for a soliton to form, nonlinear effects may still be noticeable and may have a detrimental influence on optical communications.

In addition to the nonlinearity appearing on the right-hand side of (3.69), we now also include fiber loss by adding a term with the power loss coefficient α. Thus, (3.69) becomes

$$\frac{\partial \phi}{\partial z} + \dot{\beta}_0 \frac{\partial \phi}{\partial t} = -\frac{\alpha}{2}\phi - i\frac{n_2}{n_0}\beta_0 |\phi|^2 \phi \qquad (3.89)$$

with the solution

$$\phi = fe^{-\frac{\alpha}{2}L_z} \exp\left[-i\frac{n_2}{n_0}\beta_0 f^2 L_z \right] \qquad (3.90)$$

taken at an effective length L_z defined by

$$z = L_z = \frac{1 - e^{-\alpha z}}{\alpha}. \qquad (3.91)$$

As before, f is an arbitrary function of $t - \dot{\beta}_0 z$, $f = f(t - \dot{\beta}_0 z)$. The spectrum of the light wave is the Fourier transform of the field function (3.68),

$$F(\omega - \omega_0) = e^{-\frac{\alpha}{2}z} e^{-i\beta_0 z} \frac{1}{2\pi} \int_{-\infty}^{\infty} fe^{-i\Delta\psi} e^{-i(\omega - \omega_0)t} \, dt \qquad (3.92)$$

with

$$\Delta\psi = \frac{n_2}{n_0}\beta_0 f^2 L_z. \qquad (3.93)$$

For Gaussian shaped pulses,

$$f(\tau) = B \cdot \exp\left[-\left(\frac{\tau}{T}\right)^2 \right], \qquad (3.94)$$

the integral in (3.92) can be solved by expanding the exponential function $\exp(-i\Delta\psi)$ in a Taylor series and integrating this series term by term. Thus, we obtain the spectrum of the light wave in the nonlinear fiber in the following form

$$F(\omega - \omega_0) = e^{-\frac{\alpha}{2}z} e^{-i\beta_0 z} \frac{BT}{2\sqrt{\pi}} \sum_{n=0}^{\infty} \frac{(-i\Delta\psi_m)^n}{n!\sqrt{2n+1}} \exp\left[-\frac{T^2(\omega - \omega_0)^2}{4(2n+1)} \right] \qquad (3.95)$$

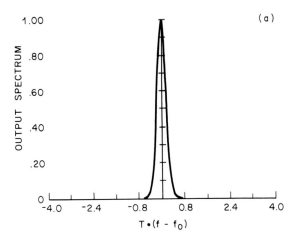

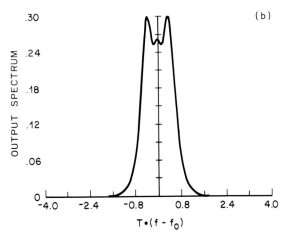

Fig. 3.21 Spectra of Gaussian pulse deformed by self-phase modulation. a) undeformed spectrum of input pulse, b) deformed spectrum with $\Delta\psi_m = \pi$ [see Eq. (3.96)], c) $\Delta\psi_m = 2.5\pi$, d) $\Delta\psi_m = 4.5\pi$.

with

$$\Delta\psi_m = \frac{n_2}{n_0}\beta_0 B^2 L_z. \tag{3.96}$$

The square magnitude of this spectral function is shown in Fig. 3.21 for several values of $\Delta\psi_m$ [Stolen and Lin, 1978].

Obviously, the nonlinearity of the fiber cannot be neglected once the original spectrum of the pulse is significantly deformed. To estimate the power level at

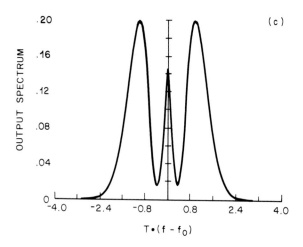

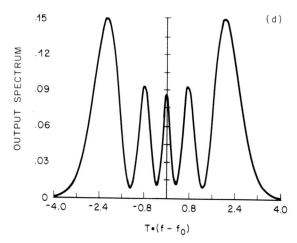

Fig. 3.21 (*Continued.*)

which the nonlinearity becomes noticeable, we assume that $\Delta \psi_m$ should not exceed the value $\Delta \psi_m = 2$. This criterion leads to the restriction that the maximum pulse power should not exceed the value [Stolen and Lin, 1978]

$$P_{max} = \frac{n_0^2 A_{eff}}{n_2 \beta_0 L_z \sqrt{\dfrac{\mu_0}{\epsilon_0}}}$$

(3.97)

with $\sqrt{\mu_0/\epsilon_0} = 377$ ohm. Using again the numerical values of the example used in the preceding section [between Eqs. (3.87) and (3.88)], we obtain from (3.97) $P_{max} = 55$ mW. In this example we have used a loss of $\alpha = 0.2$ dB/km so that the effective length for $z \rightarrow \infty$ becomes $L_z = 1/\alpha = 21.7$ km. Thus, if the peak power of a light pulses in our example approaches 55 mW, pulse broadening will no longer be governed just by dispersion, but will be modified due to self-phase modulation, [Potasek et al., 1986].

Self-phase modulation due to the fiber nonlinearity expressed by (3.67) is even more detrimental if, instead of pulse modulation, phase modulation of the optical signal is being used. Since the phase angle of the optical signal carries the information, phase changes due to nonlinearity of the refractive index can be very detrimental. Here we must distinguish between the case of self modulation by a single light wave or cross modulation due to the existence of other light waves as, for example, in wavelength division multiplexing (WDM), [Chraplyvy et al., 1984].

At first glance it may seem surprising that cross modulation should be twice as effective as self modulation [Stolen and Bjorkholm, 1982]. To see that this is the case let us consider a light wave made up of two components E_1 and E_2 at frequencies ω_1 and ω_2 such that the total light field is given as $E = E_1 + E_2$. The fiber nonlinearity considered here is caused by a polarization of the medium that is proportional to the third power of the electric field vector

$$\hat{P} = \chi E^3 = \chi\left(E_1^3 + 3E_1^2E_2 + 3E_1E_2^2 + E_2^3\right). \tag{3.98}$$

A nonlinearity of this kind gives rise to polarization of the medium at the frequencies ω_1, ω_2, $\omega_1 \pm 2\omega_2$, $\omega_2 \pm 2\omega_1$ $3\omega_1$ and $3\omega_2$. If the two signals are expressed as $E_1 = \sqrt{P_1}\cos(\omega_1 t)$ and $E_2 = \sqrt{P_2}\cos(\omega_2 t)$, the polarization at frequency ω_1 becomes

$$\hat{P}_1 = \tfrac{3}{4}\chi(P_1 + 2P_2)\sqrt{P_1}\cos(\omega_1 t). \tag{3.99}$$

There are other components oscillating at frequencies ω_2, $3\omega_1$ and $3\omega_2$, etc., but these do not contribute to the polarization wave at ω_1. The numerical coefficients appearing in (3.99) originate from the expansions $\cos^3(x) = [3\cos(x) + \cos(3x)]/4$ and $\cos^2(x) = [1 + \cos(2x)]/2$ with $x = \omega t$. It is apparent that the term P_1 inside the parenthesis in (3.99) is the contribution of the signal at frequency ω_1 itself, while the term $2P_2$ is the contribution to the signal at frequency ω_1 stemming from the presence of the signal at frequency ω_2. Thus, the first term leads to self-phase modulation, while the second term leads to cross-phase modulation, whose effectiveness is obviously twice as large.

The phase change due to fiber nonlinearity, as given by (3.93), can be expressed in terms of power if we use

$$P_j = \frac{n_0}{2}\sqrt{\frac{\epsilon_0}{\mu_0}}\,A_{eff}\,|E_j|^2. \tag{3.100}$$

The subscript j stands for 1 or 2, that is, it indicates the wave that is responsible for the refractive index change. A_{eff} is the effective area of the mode. With $|E|^2 = f^2$ we obtain from (3.93) for the phase change suffered by the wave at frequency ω_1

$$\Delta \psi_1 = 2(1 + \delta_{2j}) \frac{n_2}{n_0^2} \frac{\beta_{01} P_j L_z}{A_{eff}} \sqrt{\frac{\mu_0}{\epsilon_0}} . \tag{3.101}$$

The Kronecker delta symbol δ_{2j} accounts for the fact that cross-phase modulation is twice as effective as self-phase modulation.

If the power P_2 in channel 2 is constant, cross-phase modulation simply causes a constant phase shift which usually does no harm. Even if the power in channel 2 fluctuates on a short term basis, these fluctuations average out if the light waves in channels 1 and 2 are propagating in opposite directions. In most cases, however, the signals are propagating in the same direction. In this case, fluctuations of the power in one channel travel (almost) in synchronism with the signal in the other channel and can give rise to undesirable phase fluctuations. Power fluctuations in channel 2 may be caused either by intensity modulation of the light in that channel or they may be due to unavoidable random fluctuations. To handle this latter case more adequately, we consider the rms fluctuations σ_{ψ_1} of the phase of the wave in channel 1 that is caused by power fluctuations in other channels. If the fluctuations of several channels are uncorrelated, we obtain from (3.101) [Chraplyvy et al., 1984]

$$\sigma_{\psi_1} = K \left\{ \sigma_{P_1}^2 + 4 \sum_{j=2}^{N} \sigma_{P_j}^2 \right\}^{1/2} \tag{3.102}$$

with

$$K = 2 \frac{n_2}{n_0^2} \frac{\beta_{01} L_z}{A_{eff}} \sqrt{\frac{\mu_0}{\epsilon_0}} , \tag{3.103}$$

where σ_P^2 indicates the variance of the power fluctuations in the jth channel.

Using once more $n_2 = 6.1 \times 10^{-19}$ $(cm/V)^2$, $n_0 = 1.46$, $\beta_0 = 6.12 \times 10^4$ cm^{-1}, $A_{eff} = 7.85 \times 10^{-7}$ cm^2, $L_z = 2.17 \times 10^6$ cm, $\sqrt{\mu_0/\epsilon_0} = 377$ ohm, the coefficient K of (3.103) assumes the value $K = 0.037$ mW^{-1}. The rms value σ_{ψ_1} of the phase is measured in radians.

An rms power fluctuation of $1/2$ mW in one cross channel would cause an rms phase fluctuation $\sigma_{\psi_1} = 0.037$ radians $= 2.1°$. Such a slight phase angle fluctuation can probably be tolerated. If many channels are wavelength multiplexed on the same fiber, however, their combined effect would become significant. Power fluctuations of this magnitude would easily arise if one or more of the adjacent channels were intensity modulated. Laser light, though, suffers from intensity fluctuations even if the laser is nominally operating under cw conditions. Self-phase modulation of a single channel due to random intensity fluctuations of the light source do not cause significant phase modulation, but if many channels are wavelength multiplexed onto the same fiber, their random intensity fluctuations could become a problem if phase shift keying is used.

In principle, the Raman effect also contributes to self- and cross-phase modulation in wavelength multiplexed fiber systems [Chraplyvy et al., 1984]. The mechanism for Raman-induced phase modulation is as follows. Any light wave traveling in the fiber serves as a pump for Raman-induced gain at longer wavelength and Raman-induced loss at shorter wavelength. This gain or loss can be regarded as a change in the imaginary part of the refractive index of the fiber material. The Raman-induced loss or gain is wavelength dependent. Knowing this dependence permits us to predict the corresponding change of the real part of the refractive index via the Kramers-Kronig relations. However, this Raman-induced index change is one order of magnitude smaller than the nonlinear index change expressed in (3.67).

As a final example of undesirable nonlinear effects in fibers, we consider Raman-induced cross talk between two channels operating at different wavelengths. Channel 1 at wavelength $\lambda_1 < \lambda_2$ acts as a pump for channel 2. Thus, channel 2 gains power at the expense of channel 1. The resulting problem is illustrated in Fig. 3.22 [Chraplyvy and Henry, 1983]. Figure 3.22a shows the bit pattern (pulses) that is launched into the fiber. Because of the Raman-induced power exchange, the bit pattern at the receiver assumes the appearance of Fig.

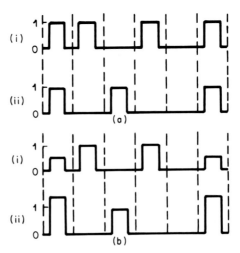

Fig. 3.22   a) Bit pattern in channels 1 and 2 with no stimulated Raman action. b) bit pattern in channels 1 and 2 with stimulated Raman action. (Chraplyvy and Henry, 1983.)

3.22b. The pulses in channel 2 gain power if they overlap with a corresponding pulse in channel 1. As shown in the figure, equal group velocity is assumed for both channels. Even if the group velocities were slightly different, the same problem would arise if successive "ones" in channel 1 overlap a "one" in channel 2. Such overlap could exist for an effective length $1/\alpha$ (α = fiber power loss) even if the group velocities were slightly different in both channels. The power gain in channel 2 would not lead to errors at the receiver, but the decrease in power in channel 1 would certainly increase the bit error rate in that channel.

The interplay of power in the two channels is described by the following set of differential equations

$$\frac{dP_1}{dz} = -\frac{\lambda_2}{\lambda_1}\frac{\gamma}{2A_{eff}}P_1P_2 - \alpha_1 P_1 \tag{3.104}$$

$$\frac{dP_2}{dz} = \frac{\gamma}{2A_{eff}}P_1P_2 - \alpha_2 P_2. \tag{3.105}$$

P_1 and P_2 indicate the power in channel 1 and 2. The Raman gain coefficient γ ($\gamma = 7 \times 10^{-12}$ cm/W) is divided by 2 to account for randomization of the polarization of the guided waves. A_{eff} is the effective area of the fiber mode and α_1 and α_2 are the power loss coefficients. The factor λ_2/λ_1 appearing in (3.104) is larger than unity if channel 2 experiences gain, that is if $\lambda_2 > \lambda_1$. This factor accounts for the fact that the signal in channel 1, the Raman pump, loses power not only to the signal in channel 2 but also supplies power to the vibration of the glass molecules that serve the function of the idler of the parametric Raman process.

Figure 3.23 [Chraplyvy and Henry, 1983], which was computed by solving the coupled differential equations (3.104) and (3.105), shows the output power in both channels as a function of input power, assuming $P_1(0) = P_2(0)$, for a fiber of length $L = 50$ km, with power loss $\alpha_1 = \alpha_2 = 0.25$ dB/km and an effective mode area $A_{eff} = 50.3$ μm^2 (corresponding to a mode radius of $r = 4$ μm). The wavelength in channel 1 is $\lambda_1 = 1.5$ μm. The wavelength in channel 2 is 100 nm longer. This wavelength shift corresponds to a frequency shift of 450 cm^{-1}. The dotted curve in Fig. 3.23 indicates the output power in the absence of Raman intreraction but in the presence of loss. The figure shows that channel 2 gains power at the expense of channel 1. In fact, for $P_1(0) > 0.05$ W increasing input power in channel 1 no longer increases the output power of channel 1. However, already for $P_1(0) > 0.01$ W = 10 mW, channel 1 loses power due to Raman interaction with the neighboring channel 2 so that its bit error rate deteriorates.

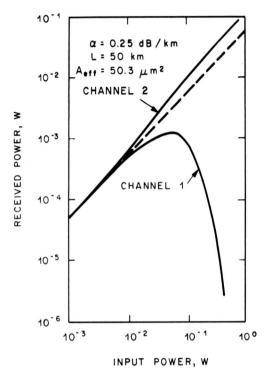

Fig. 3.23 Received power as function of input power for channel 1 (pump) and channel 2 in the presence of stimulated Raman interaction in a fiber of 50 km length with 0.25 dB/km loss. The input powers in both channels are assumed to be equal. The broken line represents the power in both channels in the absence of Raman interaction. The Raman gain coefficient is -7×10^{-12} cm/W. (Chraplyvy and Henry, 1983.)

These examples show that the nonlinear behavior of the fiber material can cause detrimental effects for power levels exceeding approximately 10 mW.

Finally, a few words should be said about stimulated Brillouin scattering, which is adequately treated in many publications [Cotter, 1983]. Brillouin scattering is a special case of Raman scattering, where the material vibration takes the form of an acoustical density wave. Brillouin backscattering can become a problem at power levels as low as 1 mW for strictly monochromatic light waves. Fortunately, Brillouin scattering decreases with increasing spectral width of the light source. Another way of suppressing Brillouin backscattering consists in reversing the phase of successive transmitted symbols, such as used in pulse code modulation or phase shift keying. In the latter case, if "ones" and "zeros" are transmitted by signals that are 180° out of phase, Brillouin backscattering would be eliminated. Thus, even though Brillouin backscattering has the lowest threshold of all nonlinear effects, it can be counteracted by suitable means such as using a light source of relatively broad spectral width or by using specially designed modulations schemes.

3.6 DISCUSSION OF THE LP_{11} MODE CUTOFF

Except for the guided mode of lowest order (the LP_{01} mode), all other modes fail to be guided by the fiber core if the V-value [see (3.38)] drops below a certain cutoff value that is characteristic for each mode. Cutoff in optical fibers is different from the cutoff phenomenon familiar in metallic microwave waveguides. If a metallic waveguide is tapered smoothly and continuously in the z-direction, the guided mode carries power up to a cutoff point z_c where the wave is completely reflected. Beyond the cutoff point the mode field decays exponentially as a function of z. By contrast, cutoff in optical fibers is a less dramatic event. If cutoff is accomplished by reducing the fiber core radius continuously, the mode does not cease to transport power in the direction of the fiber axis. Instead, the field detaches itself from the fiber core at the point where the core radius reaches the cutoff value for the mode in question. Instead of being reflected, the wave simply merges with the radiation field and carries radiative power at a slight angle to the fiber axis. In a cutoff metallic waveguide the wave becomes evanescent in the z-direction. By contrast, the fiber mode, which is evanescent in the radial direction prior to cutoff, becomes a traveling wave in the radial (as well as the longitudinal) direction beyond the cutoff point.

A mode beyond cutoff is often called a leaky wave [Snyder and Love, 1983, Marcuse, 1981]. This term is justified by the observation that light power traveling in the fiber core does not escape from it abruptly but, instead, leaks out steadily. This process can best be visualized with the help of a slab waveguide model using geometrical optics. Before cutoff is reached, light is trapped inside the slab core by total internal reflection. As the frequency of the light is lowered, the effective ray angle relative to the core-cladding interface increases until total internal reflection can no longer be sustained. At that point some light power leaks out of the core on each bounce along the ray's zigzag path. The leakage rate increases with increasing ray angle, that is, with decreasing light frequency. In round optical fibers the picture is not always that simple since geometric optics does not completely account for the cutoff phenomenon. Here, a tunneling phenomenon analogous to the tunneling effect in quantum mechanics takes place for helical waves [Snyder and Love, 1983]. But the fact remains that beyond cutoff light power leaks out of the core, justifying the descriptive term "leaky wave" [Tomita and Cohen, 1985].

The cutoff frequency or wavelength of the LP_{11} mode is an important parameter for characterizing an optical fiber. For wavelengths shorter than the cutoff wavelength of the LP_{11} mode, the fiber supports at least two guided

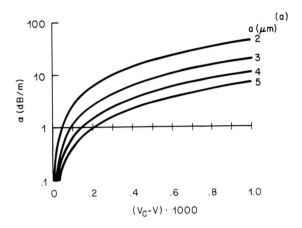

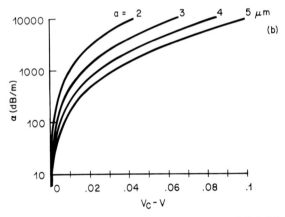

Fig. 3.24 Leakage loss of the LP mode in a fiber with infinitely extended cladding as a function of V relative to the cutoff value V. a) and b) represent the same functions on different horizontal and vertical scales.

modes (four if the polarization degeneracy of each mode is taken into account). It would appear that the cutoff frequency or wavelength could easily be observed by measuring fiber loss as a function of wavelength with the help of an incoherent light source, which is capable of exciting all guided modes. For a fiber core embedded in an infinite cladding the loss would shoot up dramatically as the wavelength increases through cutoff. The leakage loss of the LP_{11} mode in a fiber with infinitely extended core is shown in Fig. 3.24 as a function of the difference between the cutoff V-value, V_c, and the actual value of V [Snyder and

Love, 1983]. The wavelength is assumed to be $\lambda = 1.3$ μm for $V = V_c$. The various curves in the figures are drawn for different values of the fiber core radius a. The two parts of the figure, a and b, show the same loss curves on different scales. It is apparent that, in most practical cases, the leakage loss exceeds 10 dB/m for $V_c - V > 0.001$. Such a rapid loss increase should permit a determination of the cutoff wavelength with high precision.

In actual fibers, the leakage losses of the cutoff LP_{11} mode are much lower since the fiber cladding modifies the guided mode field. In most practical fibers the refractive indices of the fiber cladding and its outside plastic jacket are not exactly identical. Thus, substantial reflection may occur at this boundary for rays at grazing incidence. As a result, the intensity of the electromagnetic field at the cladding-jacket boundary assumes very small values. Fig. 3.25a and b show the field distributions ψ of the LP_{11} mode in a fiber with cladding-to-core radius ratio $b/a = 10$ below and above cutoff. The dotted line represents the mode before reaching cutoff. This curve was computed from the usual theory that assumes an infinitely extended cladding. Thus, the field does not assume zero values at the cladding-jacket boundary. The solid curves, computed from the field function listed in the appendix, represent the mode beyond cutoff whose field is assumed to be zero at the cladding-jacket boundary. The figure shows that inside the core the field does not look much different before and after reaching cutoff. It is apparent, however, that the cladding-jacket boundary has a considerable influence on the shape of the field and influences its behavior.

In the appendix we list formulas, valid near cutoff, for the loss coefficient of the LP_{11} mode in a fiber with finite cladding and an infinitely extended jacket. If the jacket index is higher than the cladding index, the field with β near $n_{cL}k$ becomes a radiation field inside the jacket, carrying power away from the fiber core and cladding. Because of the high loss of the jacket material, we may assume that all this energy is lost. If the jacket has a lower refractive index than the cladding, the light field inside the jacket is evanescent. With a lossless jacket, no additional losses would occur. However, the typical high jacket losses will impart substantial losses to any field reaching into the jacket, even if it is evanescent.

Loss curves for both cases are shown in Figs. 3.26 and 3.27. In Fig. 3.26 the jacket is assumed to have a higher refractive index n_j than the cladding. In this case no total internal reflection occurs at the cladding-jacket boundary, permitting radiation to escape into the jacket. The mode loss in dB/m is shown as a function of several variables. Figure 3.26a shows the loss as a function of the excess of V over the theoretical cutoff value (for infinitely extended cladding), $V_c = 2.405$, for several values of the jacket-cladding index difference, $n_j - n_{cL}$. Comparison with Fig. 3.24 shows that the mode losses are very much lower if the finite cladding thickness is taken into account. In all cases we assumed that the wavelength at cutoff is $\lambda = 1.3$ μm. For $n_j - n_{cL} = 0.05$ the mode loss remains

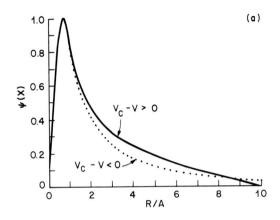

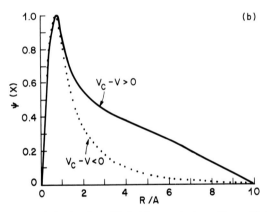

Fig. 3.25 Field function of the barely guided LP_{11} mode in a fiber with infinitely extended cladding (broken curve) and with finite cladding of radius $b = 10a$ just beyond cutoff (solid curve). a) $|V_c - V| = 0.03$, b) $|V_c - V| = 0.1$.

below 10 dB/m for $V_c - V < 0.03$. For an infinite cladding 10 dB/m loss already occurs at $V_c - V = 0.0012$. On the other hand, it is important to note that the mode experiences substantial losses even for $V > V_c$ contrary to the fiber model with infinite cladding. This shows that the presence of the cladding modifies the cutoff V-values that would result from observing the loss increase of the LP_{11} mode near its theoretical cutoff point.

Figure 3.27 displays LP_{11} mode losses for a fiber whose jacket has a lower refractive index than the cladding. In this case total internal reflection occurs at the cladding-jacket boundary so that much lower losses are to be expected. The loss is now caused by the dissipation of the evanescent field in the lossy jacket

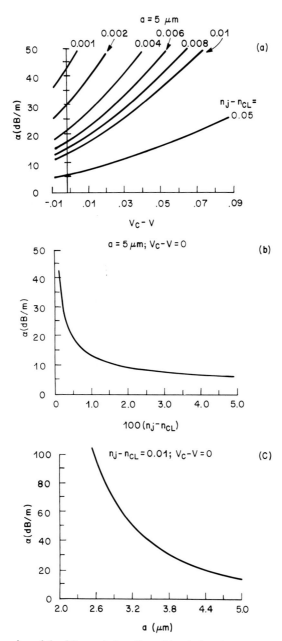

Fig. 3.26   Leakage loss of the LP_{11} mode in a fiber with a jacket whose refractive index is slightly larger than the cladding, a) as function of V, b) as function of the jacket index, c) as function of the core radius.

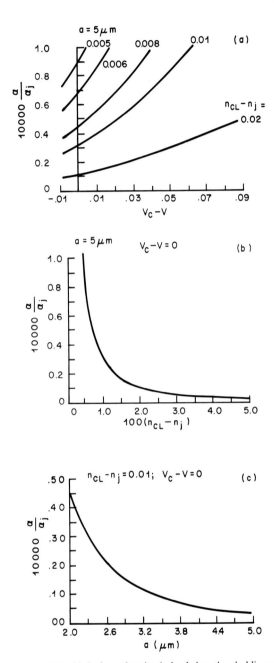

Fig. 3.27 Similar to Fig. 3.26 with jacket refractive index below the cladding value.

material. The curves hold only in the limit of moderate jacket losses. This means that the loss coefficient α_j of the jacket material must be much smaller than the propagation constant in the jacket $\alpha_j \ll n_j k = 2\pi n_j/\lambda$. This requirement, however, is easily satisfied in most practical plastic jacket materials unless they appear black at the operating wavelength. If we assume $\alpha_j = 10$ dB/mm = 10,000 dB/m we find from Fig. 3.27a, for $V_c - V = 0.03$ and $n_j - n_{cL} = 0.02$, a mode loss of 0.2 dB/m. This loss is substantially lower than the corresponding loss for the example of a fiber whose jacket index exceeds the cladding index. Thus, it is even more difficult to measure the cutoff of the LP_{11} mode in a fiber with plastic jacket of lower index than the cladding.

In practice, it is desirable to infer the value V_c for long fibers from measurements of short fibers [Anderson and Lenahan, 1984, Katsuyama et al., 1976, Murakami et al., 1979]. Experimenters often resort to bending the fiber to reduce the influence of the finite cladding and thus obtain more reproducible values for the experimentally determined cutoff value V_c of the LP_{11} mode. Bending of the fiber has two consequences. First, any light that is trapped in low-loss cladding modes experiences increased bending losses and thus disappears from the cladding more rapidly. In addition, bending distorts the effective refractive index distribution of the fiber, shifting the point at which the evanescent field changes to a radiation field radially inwards.

The influence of bending on fiber losses is schematically shown in Figs. 3.28a and b. Figure 3.28a shows the refractive index profile of a straight fiber whose jacket has a slightly larger refractive index than the cladding. The guided mode is very nearly cutoff so that the value of the effective refractive index β/k is just slightly above the refractive index of the cladding. At $r = b$ the evanescent cladding field turns into the radiative field tail existing in the jacket. Loosely speaking, it is at this point that radiation begins.

A bent fiber can be described approximately as a straight fiber with a deformed effective refractive index profile as shown in Fig. 3.28b. The effective index of the deformed fiber is [Marcuse, 1976]

$$n_R(r) = n(r)\left[1 + \frac{r}{R}\cos\phi\right], \qquad (3.106)$$

where R is the radius of curvature of the fiber axis and ϕ is the azimuthal coordinate. Figure 3.28b shows a cut through the refractive index profile at $\phi = 0$. Because of the bend, the evanescent cladding field changes into a radiation field at the point where $n(r) = \beta/k$ that now occurs somewhere in the cladding, as indicated in Fig. 3.28b. The magnitude of the bending loss depends on the radial distance from the core boundary $r = a$ to the point at which β/k becomes equal to the cladding refractive index. Keeping the measured loss constant and increasing the radius of curvature (so that the distorted index

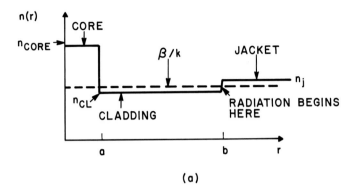

(a)

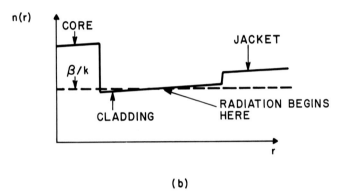

(b)

Fig. 3.28 a) Schematic of the refractive index distribution of a straight step-index fiber with effective refractive index close to the cutoff, β/k only slightly larger than the cladding index n_{cL} ; b) distorted refractive index profile that simulates the influence of a bent fiber axis.

profile becomes more nearly straight) keeps the point at which radiation begins roughly in the same radial position. But at the same time, the value of β/k approaches more and more the constant value n_{cL} of the straight fiber. Thus, if we plot the V-value for a given, fixed curvature loss as a function of curvature and extrapolate this plot to the point $1/R = 0$, we obtain a good approximation to the true cutoff V-value of the LP_{11} mode, [Lazay, 1980].

3.6.1 Appendix

In this appendix we sketch the method that was used for computing the losses of the LP_{11} mode just beyond cutoff in a fiber with lossy jacket.

We work with the well known weak guidance approximation [Gloge, 1971]. For a fiber with a refractive index profile as shown in Fig. 3.28a, the scalar field solution can be expressed as

$$\psi = AJ_1(\kappa r)\cos\phi\, e^{-i\beta z} \qquad 0 \le r \le a \tag{3.107}$$

$$\psi = \left[BH_1^{(2)}(\sigma r) + CH_1^{(1)}(\sigma r) \right]\cos\phi\, e^{-i\beta z} \qquad a \le r \le b \tag{3.108}$$

$$\psi = DH_1^{(2)}(\rho r)\cos\phi\, e^{-i\beta z} \qquad r \ge b \tag{3.109}$$

with

$$\kappa = \sqrt{n_{core}^2 k^2 - \beta^2} \tag{3.110}$$

$$\sigma = \sqrt{n_{cL}^2 k^2 - \beta^2} \tag{3.111}$$

$$\rho = \sqrt{n_j^2 k^2 - \beta^2}\,. \tag{3.112}$$

The Bessel function of the first kind, $J_1(\kappa r)$, represents a cylindrical standing wave in the fiber core. The Hankel functions $H_1^{(2)}(\sigma r)$ and $H_1^{(1)}(\sigma r)$ of the second and first kind represent radial outgoing and incoming traveling waves in the cladding and $H_1^{(2)}(\rho r)$ represents an outgoing cylindrical wave in the fiber jacket. If $n_j < n_{cL}$ then $\rho = -i\gamma$ (γ is real and positive in a lossless jacket) and the radiation field in the jacket becomes an evanescent field.

The problem is now solved in two steps. First, we assume that the refractive index step at $r = b$ is large enough to permit us to set $\psi = 0$ at $r = b$. This approximation leads to

$$C = -\frac{H_1^{(2)}(\sigma b)}{H_1^{(2)}(\sigma b)}B. \tag{3.113}$$

The requirement of continuity of ψ and $d\psi/dr$ at $r = a$ results in the eigenvalue equation (N_0 and N_1 are Neumann functions of order 0 and 1)

$$\sigma J_1(\kappa a)\left[N_1(\sigma b)J_0(\sigma a) - J_1(\sigma b)N_0(\sigma a) \right]$$
$$= \kappa J_0(\kappa a)\left[N_1(\sigma b)J_1(\sigma a) - J_1(\sigma b)N_1(\sigma a) \right]. \tag{3.114}$$

The second step consists in computing the amplitude of the reflected and transmitted waves at $r = b$ in terms of the incident wave amplitude B from the requirement that ψ and $d\psi/dr$ be continuous at $r = b$. The amplitude C of the reflected wave is of no further interest but the amplitude of the transmitted wave is

$$D = \frac{\sigma\left[H_1^{(1)'}(\sigma b)H_1^{(2)}(\sigma b) - H_1^{(1)}(\sigma b)H_1^{(2)'}(\sigma b) \right]}{\sigma H_1^{(1)'}(\sigma b)H_1^{(2)}(\rho b) - \rho H_1^{(1)}(\sigma b)H_1^{(2)'}(\rho b)}B \tag{3.115}$$

The prime indicates differentiation of the function with respect to the whole argument.

The power loss coefficient α of the mode is now obtained as the ratio

$$\alpha = \frac{\Delta P_r}{P_z}, \tag{3.116}$$

where ΔP_r represents the radial power outflow per unit fiber length

$$\Delta P_r = ib \int_0^{2\pi} \left(\psi^* \frac{\partial \psi}{\partial r} - \psi \frac{\partial \psi^*}{\partial r} \right) d\phi = \frac{16(\rho + \rho^*)b |B|^2}{\pi |1 + i\rho b|^2 T_1} \tag{3.117}$$

and P_z is the power carried by the mode in z-direction

$$P_z = i \int_0^{2\pi} d\phi \int_0^b \left(\psi^* \frac{\partial \psi}{\partial z} - \psi \frac{\partial \psi^*}{\partial z} \right) r \, dr = \frac{2\pi\beta |B|^2}{\sigma^2 T_1} \left\{ \frac{8}{\pi^2} \left[1 - \pi J_1(\sigma b) N_1(\sigma b) \right] \right.$$

$$- 2a^2 T_2 \frac{J_0(\kappa a)}{J_1(\kappa a)} \left[\left(\frac{2\sigma^2}{\kappa a} + (\kappa^2 - \sigma^2) \frac{J_0(\kappa a)}{J_1(\kappa a)} \right) T_2 + \frac{2\kappa}{a} J_1(\sigma b) N_1(\sigma a) \right]$$

$$\left. + 4a\sigma T_2 J_1(\sigma a) N_1(\sigma b) \right\} \tag{3.118}$$

with the abbreviations

$$T_1 = J_1^2(\sigma b) + N_1^2(\sigma b) \tag{3.119}$$

and

$$T_2 = N_1(\sigma b) J_1(\sigma a) - J_1(\sigma b) N_1(\sigma a). \tag{3.120}$$

The value for ρ appearing in (3.117) follows from the solution of the eigenvalue equation (3.114) and from (3.112). If $n_j < n_{cL}$ we relate $\rho = -i\gamma$ to the power loss coefficient α_j of the jacket material

$$\rho + \rho^* = \frac{\gamma - \gamma^*}{i} = \frac{2n_j k}{\mathrm{Re}(\gamma)} \alpha_j. \tag{3.121}$$

REFERENCES

Ainslie, J. B., Beales, K. J., Day, C. R., and Rush, J. D. (1982). The design and fabrication of monomode optical fiber. *IEEE J. Quant. Electr.* QE-18, 514–522.

Anderson, W. T., and Lenahan, T. A. (1984). Length-dependence of the effective cutoff wavelength in single-mode fibers. *IEEE J. Lightwave Tech.* L.T-2, 238–242.

Birch, R. D., Payne, D. N., and Varnham, M. P. (1982). Fabrications of polarization maintaining fibres using gas-phase etching. *Electr. Lett.* 18, 1030–1030.

Born, M., and Wolf, E. (1965). "Principles of Optics." Pergamon Press, New York.

Chraplyvy, A. R., and Henry, P. S. (1983). Performance degradiation due to stimulated raman scattering in wavelength-division-multiplexed optical-fibre systems. *Electr. Lett.* **19**, 641–643.

Chraplyvy, A. R., Marcuse, D., and Henry, P. (1984). Carrier-induced phase noise in angle-modulated optical-fiber systems. *IEEE J. Lightwave Tech.* LT-2, 6–10.

Cohen, L. G., Mammel, W. L., and Presby, H. M. (1980). Correlation between numerical predictions and measurements of single-mode fiber dispersion characteristics. *Applied Optics* **19**, 2007–2010.

Cohen, L. G., Mammel, W. L., and Lumish, S. (1981). Dispersion and bandwidth spectra in single-mode fibers. *IEEE J. Quant. Electr.* QE-18, 49–53.

Cohen, L. G., Mammel, W. L., and Jang, S. J. (1982). Low-loss quadruple-clad single-mode lightguides with dispersion below 2 ps/km nm over the 1.28 μm—1.65 μm wavelength range. *Electr. Lett.* **18**, 1023–1034.

Cotter, D. (1983). Stimulated Brillouin scattering in monomode optical fibers. *J. Opt. Commun.* **4**, 10–19.

Doran, N. J., and Blow, K. J. (1983). Solitons in optical communications. *IEEE J. Quant. Electr.* QE-19, 1883–1888.

Eikhoff, W. (1982). Stress-induced single-polarization single-mode fiber. *Optics Letters* **7**, 629–631.

Francois, P. L. (1983). Zero dispersion in attenuation optimized doubly clad fibers. *IEEE J. Lightwave Tech.* LT-1, 26–37.

Francois, P. L., Bayon, J. F. and Alard, F. (1984). Design of monomode quadruple-clad fibers. *Electron. Letters.* **20**, 688–689.

Gloge, D. (1971). Weakly Guiding Fibers. *Applied Optics* **10**, 2252–2258.

Gloge, D. (1979). The effect of chromatic dispersion on pulses of arbitrary coherence. *Electr. Letters* **15**, 686.

Hasegawa, A., and Kodama, Y. (1981). Signal transmission by optical solitons in monomode fiber. *Proc. IEEE* **69**, 1145–1150.

Hasegawa, A. (1983). Amplification and reshaping of optical solitons in glass fiber. *Opt. Lett.* **8**, (650–652).

Hasegawa, A. (1984). Numerical study of soliton transmission amplified periodically by stimulated raman process. *Applied Optics* **23**, 3302–3309.

Hosaka, K., Okamotot, K., Sasaki, Y., and Edahiro, T. (1981). Single-mode fiber with asymmetrical refractive index pits on both sides of the core. *Electr. Lett.* **17**, 191–193.

Jeunhomme, L. (1982). Single-mode fiber design for long haul transmission. *IEEE J. Quant. Electr.* QE-18, 727–732.

Juergensen, K. (1978). Gaussian pulse transmission through monomode fibers, accounting for source linewidth. *Applied Optics* **17**, 2412–2415.

Juergensen, K. (1979). Dispersion minimum of monomode fibers. *Applied Optics* **18**, 1259–1261.

Kaminow, I. P., and Ramaswamy, V. (1979). Single-polarization optical fibers: slab model. *Appl. Phys. Lett.* **34**, 268–270.

Kaminow, L. P. (1981). Polarization in optical fibers. *IEEE J. Quant. Electr.* QE-17, 15–22.

Kaminow, I. P. (1984). Polarization maintaining fibers. *Applied Scientific Research* **41**, 257–270.

Kapron, F. P. (1977). Maximum information capacity of fiber-optic waveguides. *Electr. Letters* **13**, 96–97.

Katsuyama, Y., Tokuda, M., Uchida, N., and Nakahara, M. (1976). New method for measuring *V*-value of a single-mode optical fibre. *Electr. Lett.* **12**, 669–670.

Katsuyama, R., Matsumura, H., and Suganuma, T. (1981). Low-loss single-polarization fibres. *Electr. Letters* **17**, 473–474.

Kawakami, S., and Nishida, S. (1974). Characteristics of a doubly clad optical fiber with a low-index cladding. *IEEE J. Quant. Electr.* QE-10, 879–887.

Kobayashi, S., Shibata, S., Shibata, N., and Izawa, T. (1977). Refractive index dispersion of doped fused silica. *Digest of Int. Conf. on Integr. Optics and Optical Fiber Commun.* Tokyo paper B8-3.

Krawarik, P. H., and Watkins, L. S. (1978). Fiber geometry specifications and its relation to measured fiber statistics. *Applied Optics* **17,** 3984–3989.

Lazay, P. (1980). Effect of curvature on the cutoff wavelength of single mode fibers. *Technical Digest*-Symposium on Optical Fiber Measurements, Boulder, Colorado, 93–95.

Love, J. D., Sammut, R. A., and Snyder, A. W. (1979). Birefringence in elliptically deformed optical fibres. *Electr. Letters* **15,** 615–616.

Mammel, W. L., Cohen, L. G., and Lumish, S. (1981). Improving propagation characteristics in single-mode optical fibers with computer-aided analysis using wave equation techniques. *SPIE* **294,** 26–33.

Mammel, W. L., and Cohen, L. G. (1982). Numerical prediction of fiber transmission characteristics from arbitrary refractive-index profiles. *Applied Optics* **21,** 699–703.

Marcuse, D. (1974). "Theory of Dielectric Optical Waveguides." Academic Press, New York.

Marcuse, D. (1976). Field deformation and loss caused by curvature of optical fibers. *J. Opt. Soc. Am.* **66,** 311–320.

Marcuse, D. (1979). Interdependence of waveguide and material dispersion. *Applied Optics* **18,** 2930–2932.

Marcuse, D. (1980). Pulse distortion in single-mode fibers. *Applied Optics* **19,** 1653–1660.

Marcuse, D. (1981a). Pulse distortion in single-mode fibers. 3: chirped pulses. *Applied Optics* **20,** 3573–3579.

Marcuse, D. (1981b). "Principles of Optical Fiber Measurements." Academic Press, New York.

Marcuse, D., and Lin, C. (1981). Low dispersion single-mode fiber transmission—the question of practical versus theoretical maximum bandwidth. *IEEE J. Quant. Electr.* QE-17, 860–877.

Marcuse, D. (1982). "Light Transmission Optics." 2nd edition. Van Nostrand Reinhold, New York.

Meinel, R. (1983). Generation of chirped pulses in optical fibers suitable for an effective pulse compression. *Opt. Commun.* **47,** 343–346.

Miyagi, M., and Nishida, S. (1979). Pulse spreading in single-mode optical fiber due to third-order dispersion: effect of optical source bandwidth. *Applied Optics* **18,** 2237–2240.

Mollenauer, L. F., Stolen, R. H., and Islam, H. (1985). Experimental demonstration of soliton propagation in long fibers: loss compensated by raman gain. *Optics Letters* **10,** 229–231.

Mollenauer, L. F., Gordon, J. P., and Islam, M. N. (1986). Soliton propagation in long fibers with periodically compensated loss. *IEEE J. Quant. Electr.* QE-22, 157–173.

Murakami, A., Kawana, A., and Isuchiya, H. (1979). Cut-off wavelength measurement for single-mode optical fibers. *Applied Optics* **18,** 1101–1105.

Nakatsuka, H., Grishkowsky, D., and Balant, A. C. (1981). Nonlinear picosecond-pulse propagation through optical fibers with positive group velocity dispersion. *Phys. Lett.* **47,** 910–913.

Payne, D. N., and Gambling, W. A. (1976). Zero material dispersion in optical fibers. *Electr. Letters* **12,** 549–550.

Potasek, M. J., Agrawal, G. P., and Pinault, S. C. (1986). Analytical and numerical study of pulse broadening in nonlinear fibers. *J. Opt. Soc. Am. B.* **3,** 205–211.

Ramaswamy, V., Kaminow, I. P., and Kaiser, P. (1978). Single polarization optical fibers: exposed cladding technique. *Appl. Phys. Lett.* **33,** 814–816.

Ramaswamy, V., Stolen, R. H., Divino, M. D., and Pleibel, W. (1979). Birefringence in elliptically clad borosilicate single-mode fibers. *Applied Optics* **18,** 4080–4084.

Rashleigh, S. C., and Marrone, M. J. (1982). Polarization holding birefringent fibers. *IEEE J. Quant. Electr.* QE-18, 1515–1524.

Rashleigh, S. C., Burns, W. K., Moeller, R. P., and Ulrich, R. (1982). Polarization holding in birefringent single-mode fibers. *Optics Letters* **7,** 40–42.

Reed, W. A., Cohen, L. G., and Shang, H. T. (1986). Tailoring optical characteristics of dispersion-shifted lightguides for lightwave systems applications near 1.55 μm wavelength. *OFC/IOOC, 1987*, Reno, Nev.

Shank, C. V., Fork, R. L., Yen, R., Stolen, R. H., and Tomlinson, W. J. (1982). Compression of femtosecond optical pulses. *Appl. Phys. Lett.* **40**, 761–763.

Shibata, N., Sasaki, Y., Okamoto, K., and Hosaka, T. (1983). Fabrication of polarization maintaining and absorption-reducing fibers. *IEEE J. Lightwave Tech.* LT-1, 38–43.

Simpson, J. R., Stolen, R. H., Sears, F. M., Pleibel, W., MacChesney, J. B., and Howard, R. E. (1983). A single-polarization fiber. *IEEE J. Lightwave Tech.* LT-1, 370–374.

Snyder, A. W., and Young, W. R. (1978). Modes of optical waveguides. *J. Opt. Soc. Am.* **68**, 297–309.

Snyder, A. W., and Ruehl, F. (1983). Single-mode single-polarization fibers made of birefringent material. *J. Opt. Soc. Am.* **73**, 1165–1174.

Snyder, A. W., and Love, J. D. (1983). "Optical Waveguide Theory." Chapman and Hall, London.

Stolen, R. H., and Ippen, E. P. (1973). Raman gain in glass optical waveguides. *Appl. Phys. Lett.* **22**, 276–278.

Stolen, R. H., and Lin, C. (1978). Self-phase modulation in silica optical fibers. *Phys. Rev. A* **17**, 1448–1453.

Stolen, R. H., and Bjorkholm, J. E. (1982). Parametric amplification and frequency conversion in optical fibers. *IEEE J. Quant. Electr.* QE-18, 1062–1072.

Stolen, R. H., Pleibel, W., and Simpson, J. R. (1984). High-birefringence optical fibers by preform deformation. *IEEE J. Lightwave Tech.* LT-2, 639–641.

Tomita, A., and Cohen, L. G. (1985). Leaky-mode loss of the second propagating mode in single-mode fibers with index well profiles. *Applied Optics* **24**, 1704–1707.

Tomlinson, W. J., Stolen, R. H., and Shank, C. V. (1984). Compression of optical pulses chirped by self-phase modulation in fibers. *J. Opt. Soc. Am. B* **1**, 139–149.

Unger, H. G. (1977). Optical pulse distortion in glass fibers at the minimum dispersion wavelength. *Arch. Electr. Uebertragung (AEU)* **31**, 518–519.

Varnham, M. P., Payne, D. N., Birch, R. D., and Tarbox, E. J. (1983). Single-polarization operation of highly birefringent bow-tie optical fibers. *Electr. Lett.* **19**, 146–247.

Chapter 4

Fiber Materials and Fabrication Methods

SUZANNE R. NAGEL

AT&T Bell Laboratories, Murray Hill, New Jersey

4.1 INTRODUCTION

Lightguide structures can be utilized to transmit light signals from a variety of sources over the entire optical spectrum. In general, the thrust has been on amorphous materials being formed into low loss fibers, although crystalline and hollow waveguide structures are possible. For lightwave telecommunications systems, the driving force has been to develop long lengths of practical, reliable, high bandwidth, low loss transmission media designs that operate in the near infrared (0.6–1.6 μm), where high speed solid state light sources and detectors have also been perfected. Glass fiber lightguides based on silica glass with small amounts of dopants to principally vary the optical properties have become the material system of choice for optical fibers for telecommunication applications. They have demonstrated low optical loss and have dispersion characteristics that allow high bandwidths to be achieved. A variety of optimized fiber designs have been realized, as discussed in Chapter 2. These glasses have excellent chemical durability, high intrinsic tensile strength, and demonstrated long term reliability. Equally important, tremendous advances have occurred in the evolution of large-scale manufacturing technology for now standard designs, which have realized the requisite optical, dimensional, and mechanical properties, while steadily decreasing the fiber cost, essential for the economic viability of lightwave telecommunications systems. Thus, high silica fibers have not only demonstrated theoretical performance, but have also met the technological and engineering requirements for lightwave communications applications.

Four major vapor phase manufacturing techniques have emerged for fabricating high silica telecommunications glasses. The focus in the development of these techniques has been to evolve fundamental understanding of the fabrication process, then scale-up the size and rate of fabrication of the precursor glass from which fiber is drawn, while realizing optimized designs with high performance and yield. All techniques have in common the fiber drawing and coating step, where the goal has been to maximize the draw speed while realizing controlled fiber and coating dimensions, excellent mechanical properties and reproducible low loss and dispersion. The emergence of single mode fibers as the preferred telecommunication fiber design, coupled with the emphasis on cost reduction, has resulted in further extensions on these basic techniques as well as examination of new approaches such as sol-gel processing to prepare the glasses.

A number of other aspects of high silica fibers continue to be studied. Attention is currently focused on defects and their relationship to optical loss, since further scale-up in processing as well as long term optical reliability are impacted by such effects. In particular, the interaction of hydrogen and radiation with defects in lightguides is an important reliability concern, and enhanced fiber designs are required to extend their useful temperature range and allow application in harsh environments. Mechanical property understanding and optimization continues. Very high tensile strength fibers in long lengths are required for some applications. Better understanding of static fatigue is necessary for realistic mechanical reliability assessment. Hermetic coatings, which can improve the long term fatigue performance of fibers as well as act as a barrier to H_2, are beginning to emerge, and improved organic coatings for lower and higher temperature use are under study. On another front, fibers with controlled birefringence to allow their use in applications where the phase or optical polarization of the light signal is important are under investigation, and a variety of design and fabrication approaches are being explored.

In addition to high performance, high silica lightguides, a number of other technological thrusts for communications fibers are underway. For lower bandwidth, short distance communications systems where other lightguide features are of concern, both plastic and plastic-clad-silica fibers are being explored, and materials and fabrication technology has emerged. Lower loss transmission media are also of concern, and both silicate and halide material systems, which hold the prospect for even lower losses than silica, are being explored. In particular, much recent research has focused on heavy metal fluoride materials and fiber fabrication, since their loss minimum in the mid-infrared is projected to be one to two orders of magnitude lower than silica.

The primary focus of this chapter will be on high silica telecommunications fiber. Chapters 2 through 12 of *Optical Fiber Telecommunications* [Miller and Chynoweth, 1979] provided extensive background and information on materials, design, fabrication and property concerns, while Chapter 2 of this book has reviewed fiber types and their performance status. Section 4.2 reviews optical properties of materials with a focus on high silica glasses, and includes discussion of radiation and hydrogen induced effects. Section 4.3 examines the current understanding of mechanical properties of silica based fibers, including fatigue effects. Section 4.4 focuses on high silica preform fabrication techniques for conventional fiber designs, while Section 4.5 reviews the associated fiber drawing and coating of high silica fibers. In Section 4.6, approaches for fabrication of birefringent silica fibers will be discussed. Section 4.7 briefly examines the status of plastic fibers, while Section 4.8 reports on the current status of midinfrared fibers. The chapter will close with a discussion of future directions for lightguide transmission.

4.2 OPTICAL PROPERTIES OF MATERIALS AND LIGHTGUIDES—GENERAL CONSIDERATIONS

Light guidance, pulse dispersion and optical attenuation are the critical optical properties of concern for the transmission media. Light guidance relies on the tailoring of the refractive index profiles of core and cladding compositions at a given wavelength. Minimization of pulse distortion depends not only on the fiber refractive index profile but also on the dispersion of the core and cladding materials, i.e., the variation of the refractive index with wavelength. Attenuation of the signal intensity is determined by interaction of photons with the material through absorption and scattering processes, and by the details of the fiber design. The thrust of fiber optimization for telecommunications has been to choose materials and designs that achieve light guidance with minimal loss and pulse distortion over very long distances. In general, for telecommunications applications, fibers are operated at power levels where the optical properties are linear, i.e., neither the loss or the dispersion characteristics vary with signal intensity. Thus, non-linear properties, reviewed in Chapter 5 of the first book, will not be discussed. Long-term stability of optical loss is of particular concern, and aging effects due to interactions with radiation or hydrogen in the fiber environment are two key areas of concern.

4.2.1 Refractive Index and Material Dispersion

The retractive index, n, of materials is a key property for realizing a waveguide structure, and is a measure of the ratio of the speed of light in vacuum, c, relative to its speed in the material. The refractive index varies with wavelength, λ, and thus gives rise to dispersion. Frequently, the value of the refractive index of glass, n_d, is quoted at a standard wavelength of the sodium d line at 589.3 nm, and the Abbe number, V_d, is used to classify the dispersion of optical

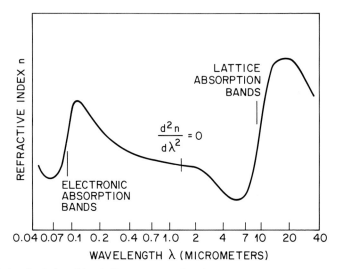

Fig. 4.1 Refractive index of fused silica versus wavelength.

glasses:

$$V_d = n_d - 1/(n_f - n_c), \tag{4.1}$$

where n_f and n_c are the index values at 479.99 nm and 656.27 nm, respectively.

An excellent discussion of the refractive index of glasses and dispersion is contained in Rawson (1980) and Fanderlik (1983), while extensive data on the effects of glass composition on index and dispersion can be found in glass textbooks such as Morey (1954), Bansal and Doremus (1986), as well as in glass manufacturer's handbooks and catalogs. For SiO_2, $n_d = 1.456$, and normalized index differences, $\Delta = (n_{core} - n_{clad})/n_{core}$ of 1–2% for multimode and 0.3 up to 1% are typical for single mode structures.

The wavelength dependence of the refractive index arises from the interaction of light with intrinsic material absorption bands in the ultraviolet and infrared. Figure 4.1 schematically illustrates the refractive index versus wavelength for fused silica as an example of such an effect. An important component of delay distortion in lightguides results from the wavelength dependence of the refractive index and is often characterized by the material dispersion parameter, M:

$$M = -\frac{\lambda}{c} \frac{d^2 n}{d\lambda^2}. \tag{4.2}$$

Figure 4.2 illustrates the relationship among refractive index, pulse delay and material dispersion. The material dispersion coupled with the use of sources of

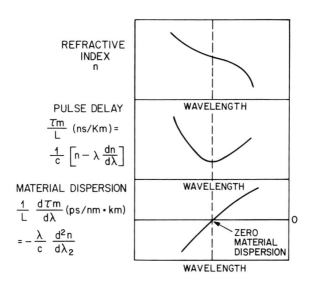

Fig. 4.2 Relationship of refractive index versus wavelength behavior of glasses to the resultant material dispersion. At any wavelength, a pulse will be delayed by a time τ_m per unit length L according to the equation shown. The material dispersion is a measure of this delay with respect to wavelength, and is minimized at the zero material dispersion wavelength.

non-zero spectral width make important contributions to the pulse spreading in both single-mode and multimode fibers. Payne and Gambling (1975) were the first to recognize that because the material dispersion went to zero at ~ 1.3 μm for high silica glasses, greatly reduced pulse spreading due to material dispersion effects could be realized. The wavelength λ_o corresponding to zero material dispersion is a function of composition, and such effects have now been extensively evaluated by a number of investigators for lightguide glasses. For pure silica, $M = O$ at $\lambda_o = 1.274$ (Malitson, 1965). GeO_2 additions increase λ_o, and for pure GeO_2, $\lambda_o = 1.738$ (Fleming, 1984). P_2O_5 additions have little impact on $M = O$ relative to SiO_2 (Payne and Hartog, 1977); B_2O_3 (Fleming 1976; Sladen et al., 1977) and F (Fleming, 1978) slightly decrease this wavelength.

Adams (1981) has presented an excellent review and summary of how the refractive index variations of single-mode and multimode fibers impact their resultant total dispersion and bandwidth. For multimode fibers, the variation in material dispersion as a function of composition gives rise to profile dispersion, as described by Olshanky and Keck (1976). This must be considered for determining the optimal graded index profile parameter α for a given core-clad multimode system. For high silica single-mode fibers, the composite material dispersion of the core/clad structure along with associated profile dispersion for the specific refractive index profile is used to balance waveguide dispersion in

order to design dispersion optimized fibers, with zero dispersion at one or more
wavelengths.

The refractive index data for glasses is most commonly fit to a three term
Sellmeier equation of the form:

$$[n(\lambda)]^2 - 1 = \sum_{i=1}^{3} A_i \lambda^3 / (\lambda^2 - l_i^2), \tag{4.3}$$

where A_i are constants related to material oscillator strengths, and l_i are the
corresponding oscillator wavelengths (Sutton and Stavroudlis, 1961). Material
dispersion data have been reported by a number of investigators for high silica
bulk glasses (Malitson, 1965; Fleming, 1976, 1978, 1984; Kobayashi et al., 1977)
and for heavy metal fluoride glasses (Bendow et al., 1981; Mitachi and Miyashita,
1983; Brown and Hutta, 1985). Other workers have determined values directly
from fibers by measuring the total transit time of a short pulse as a function of
wavelength (Luther-Davies et al., 1975; Miyashita et al., 1977; Payne and
Hartog, 1977; Cohen and Lin, 1977; Lin et al., 1978; and Horiguchi et al.,
1979). As pointed out by Fleming (1976), it is important to take into account
that the refractive index and its variation in glasses is markedly impacted by
processing and thermal history. The rapid quenching of optical fibers often
freezes in a higher temperature structure whose index is different from bulk
annealed glasses.

Wemple (1979) proposed a model for material dispersion that provided a
good fit to experimental refractive index data for optical fiber materials, and
elucidated the role played by bond strength, lattice structure, chemical valence,
average energy gap and atomic mass of materials. He was able to use broad
trends in electronic and phonon oscillator strengths to derive simple expressions
for predicting the material dispersion and the wavelength for zero material
dispersion. Nassau used these concepts to calculate projected values of λ_0 for a
wide variety of oxides, halides, and mixtures (1981a, b). As new materials are
examined for lightguide structures, the index and material dispersion characteristics will play a key role in optimizing the future designs.

4.2.2 Intrinsic Optical Loss

The potential for low-loss transmission arises from fundamental properties
of lightguide materials as well as design considerations, as discussed in Chapter
2. Electronic transitions from valence to conduction bands cause absorption at
short wavelengths, typically in the far ultraviolet, which decay exponentially
with increasing wavelength; the tail of this absorption is commonly referred to as
the Urbach edge (Urbach, 1953). Strong absorptions from polar cation-anion
modes of lattice vibration occur in the infrared, and also decay exponentially

with decreasing wavelength. Rayleigh scattering is the third intrinsic loss phenomena, associated with small-scale refractive index variations due to density and compositional fluctuations; it is further exacerbated by "freezing in" a high temperature structure in the process of quenching or forming a fiber lightguide. The magnitude of this loss mechanism is largest at short wavelengths, where the wavelength of light is similar to the fluctuation length, and decays as λ^{-4}. The sum of these three intrinsic effects results in a "transmission window" where the lowest losses are projected for a given material system.

The intrinsic ultraviolet (UV) absorption behavior of amorphous materials is not totally understood theoretically, and experimental measurements of the fundamental Urbach edge in such materials are difficult due to the effects of very low levels of impurities. The most definitive work in this area was reported by Schultz (1977) for high purity silica glasses. In pure silica, the UV contribution to absorption is negligible at wavelengths in the near IR but the addition of modifiers typically shifts the edge to longer wavelengths. In GeO_2-doped lightguide glasses, the ultraviolet absorption edge is associated with Ge^{4+} and described by:

$$\epsilon'_\lambda = \epsilon_0 \exp(E/E_0), \tag{4.4}$$

where ϵ'_λ is the extinction coefficient in units of dB/km/ppmw Ge, $\epsilon_0 = 1.474 \times 10^{-11}$, $E_0 = 0.268$ eV, and E is the photon energy level at wavelength λ. Absorption due to Ge^{2+} was shown to have negligible effect for transmission wavelengths. Miya et al. (1979a) expressed the UV loss, L_{UV} in dB/km, of GeO_2-SiO_2 glasses in terms of the mole fraction of GeO_2, X_{GeO_2}:

$$L_{UV} = \left[\frac{1.542 X_{GeO_2}}{44.6 X_{GeO_2} + 60} \right] \times 10^{-2} \exp(4.63/\lambda). \tag{4.5}$$

Thus, as the GeO_2 content increases, the UV absorption at a given wavelength increases. At the far infrared portion of the optical frequency range, materials have intense single-phonon absorption bands due to lattice vibrations. The exponential tail of these bands results from multiphonon excitations, which have overtone and combination bands of far-infrared fundamental vibration frequencies, arising from anharmonic coupling of the various vibrational modes (Lines, 1984). For high silica glasses, Osanai et al. (1976) showed that the absorption wavelengths of the fundamental vibrations of B-O, P-O, Si-O and Ge-O bonds occur at 7.3, 8.0, 9.0 and 11.0 μm, respectively. Izawa et al. (1977a) examined the absorption spectra in the 3–25 μm wavelength region and delineated the effects of overtone and combination bands associated with the SiO_4 tetrahedral structural building block, as well as the effect of dopants on such spectra. B_2O_3, used in early lightguide compositions, was shown to have an absorption that extended into the low-loss window at wavelengths greater than 1.2 μm. The role of contributions of the P-O phonon absorption could not be clearly delineated in silica based glasses due to the complications of P-OH absorptions (Irven, 1981).

For the very lowest loss lightguide compositions, the use of boron-doping is totally avoided, while phosphorus is typically used only at very low levels, and avoided in single-mode core compositions. Miya (1979a) expressed the contribution of the infrared absorption loss, L_{IR}, in dB/km, for GeO_2-SiO_2 glasses as:

$$L_{IR} = 7.81 \times 10^{11} \exp(-44.48/\lambda), \qquad (4.6)$$

where he assumed that for all silica low-loss lightguide compositions, the IR loss edge component is due to Si-O vibrations. Pure GeO_2 (Osanai et al., 1976) and non-oxide glasses have infrared absorption edges that occur further into the UV, as reviewed by Goodman (1980), Gannon (1980), and Drexhage et al. (1981). These glasses are currently the focus of much experimental work for lower loss as well as longer wavelength infrared transmitting glasses; the exact position of the multiphonon edge must be considered in order to project minimum loss.

The third intrinsic loss mechanism is due to Rayleigh scattering of light due to fluctuations in the dielectric constant. Such fluctuations can arise from both density and concentration fluctuations. Density fluctuations are minimized when a glass is in thermal equilibrium; however, the high temperature quenching associated with glass formation "freezes-in" nonequilibrium fluctuations corresponding to the glass structure at some higher temperature corresponding approximately to its glass transition temperatures, T_g. The temperature at which the structure is frozen in is often referred to as the fictive temperature, T_f, and is a function of the specific thermal processing and cooling rate. In addition, as glass structures become more complex, such as by doping SiO_2 to change its refractive index, scattering associated with compositional fluctuations must also be considered. The intrinsic scattering coefficients of glassy materials may be calculated from first principles (Lines, 1984, 1986) by considering local density fluctuations, polarizability, electronic or ionic carrier concentration, molecular orientation for materials composed of anisotropic molecules, and concentration fluctuation effects. However, more typically the values of the Rayleigh scattering for glasses are determined experimentally, and take into account the thermal history of the sample, which affects the measured scatter. In silica-based glasses, there is variation in the reported scattering coefficients, but the effect of processing on the resultant scattering is not well documented. Glasses of pure silica or slightly P_2O_5-doped silica have the lowest reported values of scattering achieved in lightguides, while additions of GeO_2 increase the scattering coefficient of SiO_2 (Olshansky, 1981; Miya, 1979a, b).

In general, the Rayleigh scattering contribution to loss, L_R, in dB/km, is expressed as:

$$L_R = A\lambda^{-4}, \qquad (4.7)$$

where A is the scattering coefficient with units of dB/km-μm^{-4}. In high silica-based lightguide glasses in the 1.3–1.6 μm wavelength region, this is the dominant intrinsic contribution to lightguide loss with A values ranging from 0.7–1.0 for single-mode, and 1–1.6 for multimode fibers. Other bulk glasses whose intrinsic scattering coefficients are reported to be less than fused silica include K$_2$O-SiO$_2$ (Schroeder et al., 1973), ZnCl$_2$, BeF$_2$ (Van Uitert and Wemple, 1978), Na$_2$O-B$_2$O$_3$-SiO$_2$ (Tynes et al., 1979) and Na$_2$O-Al$_2$O$_3$-SiO$_2$ (Pinnow et al., 1975). Although such glasses offer the potential for lower losses than silica at equivalent wavelengths, if other intrinsic and extrinsic effects are not dominant, low-loss fibers have not been reported. In general, the concentration flucuations in glasses are thought to increase to the fourth power of the anion valence and between linearly and the square of the cation valence, and lower density fluctuations result for glasses with lower transition temperatures. Thus, combinations of these two effects can result in lower scattering than silica (Lines and Nassau, 1986; Lines, 1986), such as those projected for fluoroberyllates. Heavy metal halides have achieved scattering coefficients equivalent to silica (Tran et al., 1983) but have lower projected total losses since their infrared absorption edges occur at longer wavelengths. Thus, lightguides based on other families of glasses as alternatives to high silica lightguides continue to interest researchers.

A detailed treatment of general considerations for low-loss glasses was recently been put forth by Lines (1984, 1986), which considers the underlying intrinsic mechanisms as well as available data in the literature. A general relationship for minimum attenuation, α_{min}, was expressed as:

$$\alpha_{min} \propto (1 - \Lambda)^2 T_F D^f Z_A^g Z_c^h / (\mu^2 C^2), \qquad (4.8)$$

where Λ is a photoelastic factor, which is a measure of the fractional change of bond polarizability; T_F is the temperature at which the structure is frozen in; D is a dimensionless density factor; Z = anion valence; Z = cation valence; μ = the relevant reduced mass, c is a parameter related to cation-anion bond strengths, lengths and valence of the constituent atoms, and varies from 0.3 to 1 depending on composition; and f, g and h, are approximately 3, 4 and 1, respectively. This basic relationship led the author to conclude that the desirable physical and chemical attributes leading to a low-loss amorphous material were low valency, particularly anion valency; low fictive temperature; open structures; large reduced mass and a low value of $(1 - \Lambda)$. In addition, his formalism indicated that cations with shallow d-electron cores should be avoided to achieve low loss. Figure 4.3 shows a schematic of projected losses versus wavelength for a number of materials based on such considerations. While single crystal materials have the lowest projected loses, in general they have not been pursued for long length, low-loss telecommunications due to the difficulties of processing and overcoming defect-induced losses.

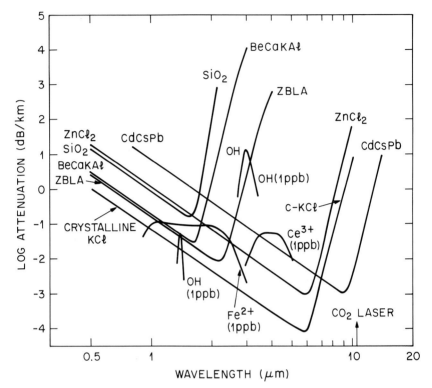

Fig. 4.3 Theoretical loss versus wavelength for a variety of possible low-loss materials. The BeCaKAl represents a fluoride glass with those cations; ZBLA represents a zirconium-barium-lanthanum-aluminum fluoride glass; CdCsPb represents a chloride glass with those cations. Absorption due to select impurities is also depicted.

4.2.3 Extrinsic Loss Mechanisms

While the fundamental structure of materials gives rise to intrinsic attenuation as a function of wavelength, a wide variety of phenomena can give rise to additional loss. Because these effects can in principle be overcome by controlled processing of the materials, they are considered extrinsic loss mechanisms.

The most important source of extrinsic loss typically arises from cation impurities, as reviewed in Chapter 7, Miller and Chynoweth (1979). In general, transition metal impurities have very strong, broad absorption bands in glass and must be reduced to the part per billion level (pbb) in order to avoid loss contributions. Newns *et al.* (1973) documented the optical loss coefficients for transition metals for sodium calcium silicate glasses, while Schultz (1974) examined their effects in pure silica. The oxidation state of the impurity was shown to be important, and controlled oxidation-reduction (redox) conditions during multicomponent glass melting was used to minimize such effects (Beales and Day, 1980). In high silica glasses, vapor phase processing technology has resulted in glasses where such impurity absorption has been essentially eliminated. In the case of longer wavelength materials, both transition metal and rare earth impurities such as Pr, Nd, Sm, Eu, Tb and Dy are important. For fluorozirconate glasses, which are being exploited for mid-infrared transmission, Ohishi et al. (1983) have reported the absorptivities of such ions as a function of wavelength. In some cases, sub-ppb purity levels in the resultant fibers are required to eliminate such effects.

Another key impurity that causes high absorption is OH. Water incorporated in glass as OH has its fundamental stretching vibration at 2.7–3.0 μm, depending on the specific glass composition. For silica-based glasses, the overtone absorption bands of the fundamental vibration at 2.73 μm have been examined in detail (Kaiser, 1973; Keck et al., 1973a). For pure silica, the first overtone occurs at 1.38 μm, with 1 ppmw Si-OH resulting in added losses of 48 dB/km, while the combination band at 1.24 μm causes 2.5 dB/km/ppmw. These peaks are sufficiently broad that they can affect loss in the low-loss regions at 1.3 and 1.55 μm, and thus severely limit the achievable loss characteristics if not reduced to the ppb level. The addition of GeO_2 to silica broadens the OH peak to the long wavelength side (Shibata et al., 1980a; Stone and Walrafen, 1982), with Ge-OH having its first overtone at 1.41 μm. Additions of P_2O_5 to silica glasses can give rise to a P-OH band whose fundamental absorption occurs at ~ 3.05 μm (Mita et al., 1977). A broad overtone in the 1.5–1.6 μm region occurs whose magnitude scales with P_2O_5 content (Edahiro et al., 1979a; Blankenship et al., 1979; Irven, 1981). In heavy metal halide glasses, the fundamental OH absorption, which occurs at ~ 2.9 μm is quite important, as well as combination bands at 2.25 and 2.45 μm (Mitachi and Miyashita, 1982). This places extremely stringent OH control requirements on such fibers, since the minimum loss is at ~ 2.5 μm.

As materials are examined for mid-infrared (2–12 μm) applications, many other types of impurities absorptions must be considered, particularly those associated with small accounts of oxygen, hydrogen, sulfer and carbon. In general, as these materials are being examined in greater detail for lightguide applications, critical impurity requirements are being defined. For example, for the chalcogenides whose minimum transmission is in the 5 μm region, the 6.32 and 2.76 water bands, carbon band at 4.94 μm, SH and SeH fundamental, overtone and combination bands, and an oxide band at 7.9 μm have been shown to cause deleterious absorption (Kanamori et al., 1984a, b).

Another source of loss is that associated with non-impurity defect absorptions. Such defect centers can have optical absorptions that effect transmission loss in the optical fiber, or act as precursor sites, which give rise to absorption when they interact with radiation or hydrogen. Such defects have been best documented for silica and high silica glasses, but in most cases the relationship between the electronic structure of the defect and any resultant optical absorption is not well understood. Multivalent cations can give rise to oxidation states, which can affect the optical absorption. Schultz (1977) reported the fundamental Urbach edge for Ti^{3+} and Ge^{2+} in the binary silicate glasses and concluded small amounts of Ti^{3+} limited the near infrared transmission of Ti-doped silica but that the Ge^{2+} edge did not extend into the wavelength region of concern. In heavy metal fluoride glasses, the Zr^{3+} or Hf^{3+} absorption edge can extrapolate to mid-IR wavelengths, and effect ultra-low-loss transmission (Tanimura et al., 1985; Griscom, 1985). The valence state of cations is impacted by the specific processing conditions, and often the high temperatures associated with glass forming and fiber drawing make it difficult to avoid such effects at the ppm level.

A number of workers have reported on the effect of drawing on the resultant optical attenuation. Kaiser (1974) reported drawing induced coloration of pure silica fibers resulting in a 0.63 μm absorption band, and others looked at the fundamental silica defect centers, their removal by chemical treatment (Levin and Pinnow, 1983), and the impact of fiber drawing on their concentration (Friebele et al., 1976, 1985a; Hibino et al., 1983, 1985, 1986; Kawazoe, 1986). Other workers reported drawing losses associated with germanium in multimode fibers (Yoshida et al., 1977; Hooper, 1983) and single-mode fiber (Ainslie et al., 1982a, b). UV light, such as from curing lamps during drawing if coatings do not block the radiation can also cause Ge-related defect absorption (Blyler et al., 1980), while other studies have shown the role of the processing conditions on the draw induced loss (Hibino et al., 1986). The origin of silica and germanium defect associated losses in optical fibers are difficult to study since many of the defects are not paramagnetic, and thus one must infer a defect from the characteristics of the optical spectra. Ainslie et al. (1982a, b) showed that the losses in single-mode fibers were absorptive in nature, while in the multimode case, scattering losses can also occur if bubbles are formed to the volatilization of GeO and formation of O_2 at high temperatures.

A number of studies have looked at defects in optical fiber materials and tried to correlate the defect concentration and luminescence behavior to optical absorption, as recently reviewed by Griscom (1985, 1986) and Friebele and Griscom (1986). The bulk of such studies have been directed at understanding the radiation performance of fibers (as reviewed in the next section) and focused on electron spin resonance (ESR) techniques to identify paramagnetic defects. To the degree other defects are non-paramagnetic, they have remained elusive. Further studies are examining new techniques such as optically detected magnetic resonance as well as Raman spectroscopy and hydrogen treatment as a probe to understand such defects. The radiation studies indicated species such as Cl_2 and O_2, which are used during processing, can effect both the optical loss and fiber radiation and H_2 sensitivity.

Other extrinsic sources of loss include processing related bubbles and inhomogenities that can lead to scattering, or the presence of crystals or phase separation, a particularly important consideration for new glass candidates. Yet another source of loss relates to the fiber design that can lead to microbending and macrobending sensitivity, as well as radiative losses due to diameter fluctuations or microdeformation.

The optimization of silica-based fibers has resulted in as-drawn fibers exhibiting very low losses, essentially equivalent to that projected from intrinsic loss considerations, with trace amounts of OH related absorption at ~ 1.39 μm frequently remaining in commercial fibers. Studies are focusing on controlling the sensitivity of loss to processing conditions, as well as subsequent interaction with H_2 or radiation in the environment. The challenge for new fiber materials is to demonstrate similar control of these extrinsic loss mechanisms.

4.2.4 Radiation Induced Losses

Silica-based lightguides can experience increases in attenuation due to the interaction of radiation from a variety of potential sources with the glass matrix. Such effects are important for the long-term reliability of lightguides, as well as for fiber use in applications where radiation might be present. Excellent reviews of the basic radiation induced effects in glasses and fibers (Friebele, 1979a; Friebele and Griscom, 1979; Friebele et al., 1984, 1985b; Dianov et al., 1983), along with the basic associated induced defects, which can interact with radiation (Friebele and Griscom, 1986; Asamov, 1983; Griscom, 1985; Robertson, 1985; Kawazoe et al., 1985c, 1986; Kawazoe, 1985), have been made. Specific studies have analyzed germanium related-defects (Friebele et al., 1974; Kawazoe, 1985; Kawazoe et al., 1985b, 1986a; Iino et al., 1985; Nakahara et al., 1986; Chamulitrat et al., 1986; Simpson et al., 1986), and phosphorus defects (Griscom et al., 1983; Kawazoe et al., 1986b), along with the various defects associated with SiO_2 (Friebele et al., 1976, 1979; Friebele and Gingerich, 1980; Hibino et al., 1986; Nagasawa et al., 1984a, b, 1986b; Vitko, 1978; see also review articles cited above). It is beyond the scope of this section to discuss such effects in detail, and the interested reader is encouraged to examine the literature to derive more specific insight into these complex phenomena. This section will briefly summarize some of the important findings for high silica fiber performance in radiation environments.

There are a wide range of possible sources of radiation with different energies and dose rates. Natural sources of radiation include radiative isotopes, which occur in soils and materials in the terrestrial and undersea environment, as well as extraterrestrial sources such as cosmic rays, solar energy and radiation, and Van Allen Belt radiation. In addition, fibers may be used in a number of nuclear environments such as found in reactors, medical instrumentation, accelerators and electron beams, and would experience exposure in the event of a nuclear blast. Unusual conditions, such as high activity geological formations or nuclear dump sites, must be considered as well.

Two basic classes of radiation are important in assessing degradation of optical properties of glass: particles such as electrons, protons, neutrons, alpha and beta rays; and electromagnetic radiation such as ultraviolet light, X rays and gamma rays. Since radiation must reach the optically active region of the fiber to induce loss, alpha and beta radiation can be ignored. When ionizing radiation interacts with the glass matrix, two effects can occur depending on the specific nature of the radiation: ionizations of electrons or direct displacement of atoms due to elastic scattering, with ionization predominating over displacement. The electrons and holes that are produced by the radiation results in electrons moving through the glass matrix, where they can be trapped at pre-existing or newly created defects to form radiation induced defect centers, which are optically absorbing; others recombine with the positive charged holes. When the energy is very high, a secondary electron cascade can be produced by Knock-on collision with bound electrons (Friebele and Griscom, 1979).

As mentioned in the previous section, a wide variety of defects can exist in the glass network, and give rise to optical attenuation due to radiation exposure. These range from Schottky, Frenkel and strained bond defects occurring due to rapid quenching and the random nature of the glass network, to substitutional, interstitial, multivalent cation, and non-stoichiometric defects. For high silica fibers, a large number of defects associated with the basic Si, Ge, P, B and O ions have been documented. The concentration of these defects is intimately tied to the specific glass and lightguide composition as well as to the fiber design and the detailed processing conditions and history of the glass. At this time, it is impossible to predetermine or predict what the radiation response of a given lightguide fiber will be, except in the broadest terms.

In addition to these basic fiber characteristics, the induced radiation effects are influenced by a large number of variables. These include the nature and energy of the radiation, the dose and the dose rate; the wavelength of operation; the temperature, length and uniformity of fiber exposed; the light injection conditions and intensity; the previous radiation and even hydrogen treatment history of the fiber (see Nagasawa, 1985; Nakahara et al., 1968); and radiation exposure. These many variables further underscore the need to have detailed radiation performance for the fiber as well as to be able to define the details of the fiber operational environment.

When fibers are exposed to radiation, the induced losses are composed of a permanent and transient component. Thus, the losses due to steady-state exposures versus loss at a given time after exposure due to recovery mechanisms must be considered, depending on the nature of the application and radiation exposure. In some cases, photobleaching due to the transmission of power through the fiber (Friebele and Gingerich, 1981) or exposure to ambient light can enhance the recovery of electron-hole recombination characteristics of the fiber, which also impacts performance and must be taken into account during characterization. While it is beyond the scope of this section to detail the complex radiation effects in fibers, some basic characteristics can be described. For silica-based fibers, the majority of the defects that lead to radiation induced loss have absorption, which occurs in the UV region of the spectrum. An exception to this case is the P_1 defect associated with phosphorus, which has an absorption peak in the region of 1.5 μm (Griscom *et al.*, 1983). From a compositional perspective, the very best radiation resistance has been reported for pure silica glasses; germanium doped silica has much greater UV-induced absorptions, but because the tails of the absorptions decay with increasing wavelength, in the longer wavelength region of 1.3–1.6 μm these fibers are only slightly worse than silica. Phosphorus-doped fibers typically show the highest induced losses, especially at 1.55 μm. When recovery characteristics are important, however, some phosphorus is beneficial, as discussed by Friebele *et al.* (1984). Thus, while general compositional trends can be considered, the optimization of a fiber design for a

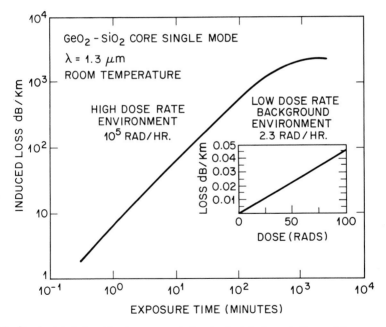

Fig. 4.4 Steady-state induced loss in a representative standard single-mode fiber at 1.3 μm at room temperature for two different radiation environments using Co^{60} radiation, after Nagel (1986). Saturation effects occur at high doses, while at low dose rates, the induced loss usually shows a linear relationship with dose.

given application must consider the complex radiation response of the fiber in the given environment.

Figure 4.4 illustrates the induced loss for a single wavelength of operation for a germanium silicate core single-mode fiber, typical of that used in current telecommunications applications; it serves to illustrate loss phenomena for two extremes of radiation environments. Typical doses expected in the terrestrial or

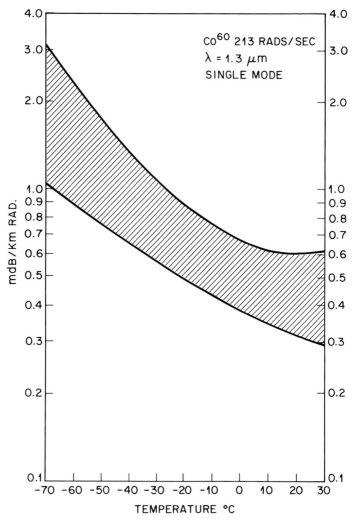

Fig. 4.5 Representative induced loss sensitivity versus temperature for single-mode fibers at 1.3 μm. Data were determined for dose rates of 213 rads/sec of Co^{60} radiation up to a total dose of 65,000 rads. The observed induced loss is a linear function of total exposure, after Nagel (1986). For higher doses, saturation effects are expected to occur.

undersea environment are 0.1–1 rad/year, so are much lower than the low dose rate depicted in the figure. For the low dose rate and total dose case, the induced loss is linear versus time. This linear regime is often used to calculate a sensitivity factor in units of milli dB/km-rad, in order to predict the total induced loss for a given total radiation exposure over the fiber service lifetime. For the high dose rate and total dose case, such as might be encountered in a nuclear reactor, the induced loss shows saturation effects, attributed to a steady state being established between the creation and annihilation of absorbing defects. It should be pointed out that this figure serves to illustrate basic phenomena, but it is critical to characterize the sensitivity of fibers with radiation and dose rates that are as close as possible to real service conditions in order to have an accurate measure of the sensitivity.

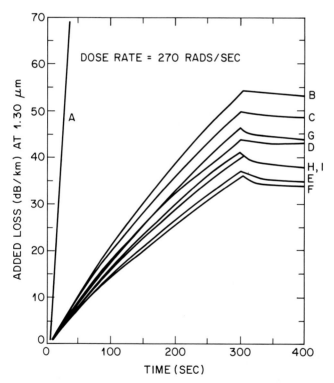

Fig. 4.6 Induced loss versus time at room temperature for GeO_2-SiO_2 fibers with varying effective phosphorus concentrations, after Mies and Soto (1985). Dose rates of 270 rads/sec for 300 sec, with Co^{60} radiation, were used with some subsequent monitoring of recovery. The curves of (a) represent 1.3 μm losses, where losses due to Ge-related defects dominate, while (b) illustrates the sensitivity of loss at 1.53 μm to additional loss caused by phosphorus-related defects.

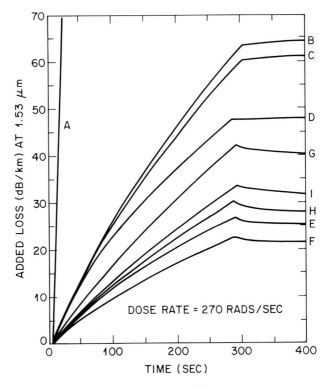

Fig. 4.6 (*Continued.*)

Figure 4.5 illustrates the temperature dependence of the radiation induced losses, again at a single wavelength, representative of a number of germanium doped silica fiber designs. Typically, the induced losses are higher as temperature decreases due to slower recovery processes at low temperatures. Friebele *et al.* (1985b), have pointed out the importance of small variations in dopant concentration on the detailed radiation response, particularly at low temperatures, once again pointing out the importance of detailed characterization of specific fiber designs and the environmental conditions in which it will be used.

There is a need for specific studies that relate radiation performance to the variations in processing conditions and composition. An example of such studies is illustrated in Fig. 4.6, after Mies and Soto (1985), where a series of germania doped silica single-mode fibers, which had different effective phosphorus concentrations (calculated on the basis of mode field power and phosphorus concentration as a function of radial position), were examined. These studies enabled them to correlate the radiation response of fiber to a quantitative compositional variation, thus providing insight into the critical role of composition. They then correlated this data to lower dose rate data in order to derive long-term radiation induced loss penalty values for 1.3 and 1.55 μm designs in an undersea environment. More recently, Watanabe *et al.* (1985) reported excellent improved performance for SiO_2 core fibers based on compositional optimization.

While these examples only illustrate a few aspects of radiation induced effects, they reflect the tremendous advances that have occurred in understanding and characterizing radiation induced losses. The radiation behavior of fibers is a complex function of many variables that must be taken into account. While all fibers darken or incur induced losses to some degree when exposed to radiation, under normal operating conditions in telecommunications applications, such induced losses are well within allocated system loss budgets and are not a cause for concern. As fiber implementation is being planned for harsher radiation environments, careful fiber optimization will be necessary to assure adequate performance. From an ongoing research perspective, there is a great deal of work required to fully define and understand such effects, with the possibility that defects can be eliminated through controlled processing, or "deactivated" by selective dopants.

4.2.5 Hydrogen Induced Loss

Hydrogen permeation into silica-based fibers can give rise to increases in optical attenuation, and thus it has implications on the long-term reliability of the transmission medium. Many studies have documented the nature of such loss increases, and long-term predictions as well as strategies to minimize such effects have emerged.

The relatively open network of high silica glasses allows hydrogen to permeate them quite rapidly, even at low temperatures. Solubility, permeability, and diffusion data have been extensively documented (Lee *et al.*, 1962, 1966; Lee, 1963; Shackleford *et al.*, 1972; Shelby, 1977). At room temperature, Raman scattering has been used to study the dissolution process; it was shown that the hydrogen dissolves as a molecule, occupies interstices in the glass, and binds weakly to the silica (Hartwig, 1976). Stone *et al.* (1982), using H_2 in silica as a Raman gain media, reported a new set of infrared absorption peaks between 1.08 and 1.24 μm, which they attributed to the first overtone spectra of H_2. The fundamental vibration of solid H_2 occurs at 2.42 μm (Kumar *et al.*, 1982), and had been reported for bulk silica at this same wavelength (Mattern, 1976). A number of studies have examined the details of the induced infrared absorption spectra and attributed it to the lack of inversion symmetry at the solution site, as first proposed by Vitko *et al.* (1976). Thus, the infrared activity is due to H_2 overtones and combination bands of molecular H_2 vibrations with the SiO_4

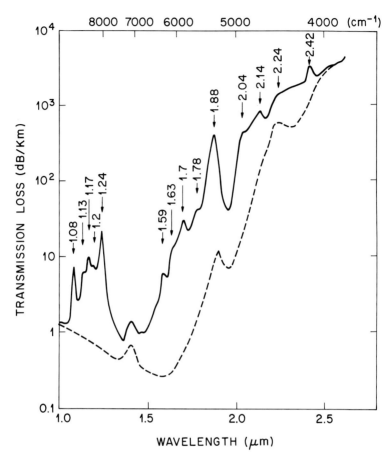

Fig. 4.7   Added loss (−) due to hydrogen absorption relative to hydrogen free spectra (--) for silica fibers, after Mochizuki *et al.* (1984). The peak at 2.42 μm is due to the fundamental H_2 vibration with its overtone at 1.24 μm, other associated combinations are also shown. Data represents equilibration in 1 atm H_2 at room temperature.

vibrations; the relative intensities and wavelengths of these absorption bands have now been extensively documented (Stone *et al.*, 1982; Mochizuki *et al.*, 1983, 1984; Beales *et al.*, 1983a, b). Overtones occur at 1.245 μm and 0.851 μm, with a number of additional combination bands, as shown in Figure 4.7 (after Mochizucki *et al.*, 1984).

At 20°C, the magnitude of the loss increase in the two long wavelength low-loss operating windows is 0.3 dB/km/atm H_2 at 1.3 μm and 0.6 dB/km/atm H_2 at 1.55 μm (Beales *et al.*, 1983a, b). The solubility increases linearly with increasing pressure up to relatively high pressures (Hartwig, 1976) and decreases weakly with increasing temperature (Shelby, 1977; Shackleford *et al.*, 1972). At room temperature, when hydrogen is present, it has a diffusion coefficient of 1.7×10^{-11} cm^2/sec, and takes ~ 300–500 hours to equilibrate with the fiber through diffusion. Moreover, these losses due to interstitial hydrogen are reversible, i.e., when the hydrogen atmosphere is removed, out-diffusion occurs (Beales *et al.*, 1983a, b). This type of loss increase occurs in all silica fibers since the solubility is roughly the same, independent of composition. The absolute magnitude of losses will depend on hydrogen partial pressure and temperature.

In addition to absorption effects due to molecular hydrogen, additional permanent loss increases due to chemical reaction of the H$_2$ with the glass network to form OH, Ge-related defect absorption, and other absorbing species have been documented (Uesugi *et al.*, 1983a, b; Mochizuki *et al.*, 1983, 1984; Beales *et al.*, 1983; Rush *et al.*, 1984; Uchida *et al.*, 1983, 1986; Lemaire and Tomita, 1984; Pitt and Marshall, 1984; Tomita and Lemaire, 1984; Ohmori *et al.*, 1983). Since the first report of such losses by Uchida *et al.* (1983) in part of a field installed cable in Japan containing graded index multimode germanium phosphosilicate core fiber, extensive studies have been made in detail to determine the nature of the chemical reactions in the glass, the origin and causes of hydrogen in the operating environment, and the long-term reliability character-istics of fibers exposed to a hydrogen environment.

The source of hydrogen in cables can be corrosion or electrolytically assisted corrosion of the metallic cable elements; evolution of H$_2$ from the fiber coating material, particularly some silicones; the complex interaction of the cable structure materials to produce H$_2$; natural background hydrogen, which is ~ 10^{-6} atm; and even radiolytic hydrogen when fibers are exposed to such radiation as gamma rays (Mochizuki *et al.*, 1983; Uesugi, 1983b; Mies *et al.*, 1984; Murakami *et al.*, 1983, 1984, 1985; Hinchliffe *et al.*, 1984; Barnes *et al.*, 1985; Pitt and Marshall, 1984; Friebele *et al.*, 1985). Because of such effects, it is critical to evaluate the transient and permanent loss increases that might impact long-term fiber performance, as well as understand the basic loss mechanisms.

The loss increases in silica fibers exposed to hydrogen are a function of wavelength, partial pressure of hydrogen, time, temperature, and fiber composi-tion and design. Two general approaches have been taken to evaluate the permanent loss increases due to chemical reactions with hydrogen in the fiber, as reviewed by Rush *et al.* (1984). The first approach involves testing the fibers in a hydrogen atmosphere at higher temperatures in order to monitor the chemical reaction growth kinetics for the various attenuation mechanisms, then apply Arrhenius analysis of the temperature dependence in order to predict the performance at lower temperatures characteristic of the service environment. For a complete picture, the time and pressure dependence is also required. The second approach involves studying the pressure dependence of loss as a function of time at a temperature characteristic of the operating environment. This type of approach in general takes considerably longer time, but makes no assumptions that the higher temperature loss mechanism is equivalent to that at lower operating temperatures. In situ monitoring of the loss increase using discrete sources or a monochromator allows the kinetics of the reactions to be readily studied (Tomita and Lemaire, 1985). Higher temperature testing is further complicated by the potential for coating degradation and microbending, so carefully designed experimental apparatus and fiber handling are required. Analysis of the data requires that H_2 reversible losses be separated from permanent loss effects; this is often accomplished by out-diffusing the test atmosphere hydrogen, or subtracting out such effects from the measured spectrum.

In general, the experimental studies suggest that the permanent loss increases due to hydrogen in silica fibers are due to reaction of the hydrogen at defect sites in the glass, although there is little direct evidence correlating species to changes in optical absorption (Iino, 1985; Nakahara, 1986). The behavior of multimode fibers, typically GeO_2-SiO_2 co-doped with varying concentrations of P_2O_5 is markedly different than conventional GeO_2-SiO_2 core single-mode fibers with various cladding compositions. The many studies have underscored the complexity of the induced loss as a function of composition and fiber design, making it critical to consider specific test data for a given fiber type.

The general features of hydrogen induced losses in fibers can be summarized as follows:

1. Hydrogen permeates the lightguide structure and gives rise to strong absorptions due to molecular hydrogen, which are quantified as a function of H_2 pressure and temperature (due to changes in solubility). These losses are reversible.

2. When hydrogen permeates the lightguide, chemical reactions at defect sites can take place, giving rise to permanent changes in loss if the hydrogen is present long enough for reaction to take place.

3. The permanent induced losses are a function of time, temperature, hydrogen pressure, composition and fiber design. The loss increases initially roughly linearly with time, then gives evidence of approaching some saturated value, presumed to reflect depletion of available defect sites for reaction.

In germanium-doped multimode fibers, especially those co-doped with phosphorus, the predominant loss mechanism is due to OH growth. The Ge-OH band at 1.41 μm typically preferentially grows relative to the 1.38 μm Si-OH

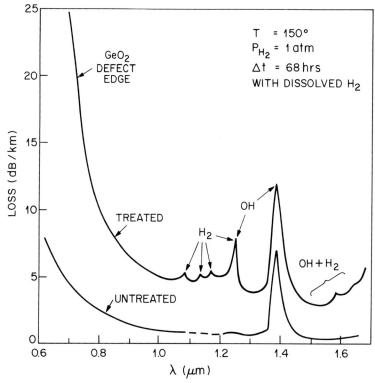

Fig. 4.8 Example of loss increases in germanium doped silica single-mode fibers heat treated in a hydrogen atmosphere, showing the germanium related defect absorption edge, some growth of OH, and H₂ absorption peaks, after Tomita and Lemaire (1985).

band, and a long wavelength loss tail grows markedly, attributed to fundamental and combination bands of the network with OH. While the magnitude of such effects increases with increasing P_2O_5 content for high P_2O_5 concentrations, 0.1 wt% P_2O_5 fibers show less OH growth than P_2O_5 free fibers (Rush *et al.*, 1984), suggesting a complex dependence of reactive defects on composition.

In contrast, GeO_2 doped single-mode fibers characteristically show a short wavelength absorption loss edge, which extrapolates to long wavelengths and is the dominant loss increase mechanism. While OH growth occurs with the shape of the peak growing in proportion to the original peak, the magnitude of the growth is much less than the multimode case, and does not significantly contribute to added loss in the two operating long wavelength windows. Figure 4.8 illustrates the three loss mechanisms for single-mode fiber. The specific nature of the short wavelength loss is not well understood. Other studies have shown that fibers with small amounts of P_2O_5 in the cladding have less OH growth than P_2O_5 free fibers (Lemaire and Tomita, 1984).

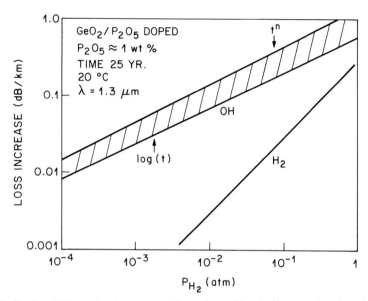

Fig. 4.9 Predicted 25-year loss increases at 1.3 μm for multimode fiber as a function of partial pressure of H_2, after Rush *et al.* (1984). Two different time dependencies for permanent losses are shown.

At this time, much further work is required to understand in detail the nature of the reactive defects in silica fibers and their relationship to processing conditions, composition and design. Nevertheless, based on tests for various fiber designs as a function of time, temperature and hydrogen pressure, a number of investigators have analyzed such data in order to make long-term predictions on the increased loss that might occur in a hydrogen environment (Rush *et al.*, 1984; Tomita and Lemaire, 1985; Fox and Stannard-Powell, 1983; Pitt and Marshall, 1984; Plessner and Stannard-Powell, 1984; Noguchi, 1985). Figure 4.9 shows twenty-five-year reliability projections of Rush *et al.* (1984) for loss increases due to H_2 and OH growth at 1.3 μm in a multimode fiber at 20°C as a function of hydrogen partial pressure, using a square root H_2 pressure dependence for the permanent loss contribution, determined experimentally. At low pressures, permanent losses dominate. Figure 4.10 shows the analysis of Tomita and Lemaire (1985) for a single-mode fiber design at 1.3 μm for a twenty-year time period. A more complex pressure dependence was reported, and the maximum tolerable hydrogen pressure versus temperature is plotted for four maximum tolerable loss increases for the time period. At low temperatures, the tolerable values are only weakly temperature dependent and are determined by molecular H_2 contributions to loss, while at higher temperature, the permanent loss mechanism dominates, and the effects are strongly temperature dependent.

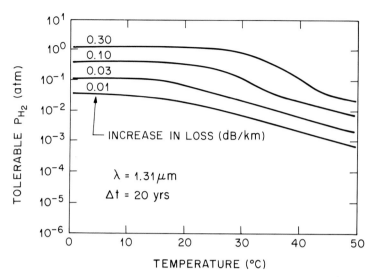

Fig. 4.10 Predictions for tolerable hydrogen pressure versus temperature for standard single-mode fiber at 1.3 μm to maintain the added loss, $\Delta\alpha_{tot}$, below some tolerable level for 20 years, after Tomita and Lemaire (1985).

More recently, an absorption peak at 1.52 μm has been reported for single-mode fibers (Sakaguchi *et al.*, 1985; Ogai *et al.*, 1986; Lemaire *et al.*, 1986; Nagasawa *et al.*, 1986; Blankenship *et al.*, 1987), attributed to the Si-H overtone of the ~ 4.5 μm band. The peak has been observed in both GeO$_2$-doped and silica core fibers and it is transient. Nagasawa *et al.* (1986) indicated it only forms in oxygen rich silica glasses, and attributed formation to reaction at peroxy radicals, while others (Ogai *et al.*, 1986; Lemaire, 1986; Blankenship *et al.*, 1987) have shown its growth is directly correlated with the disappearance of the 0.63 μm draw-induced peak when hydrogen is permeated into the lightguide structure; subsequently, the 1.52 μm peak disappears as SiOH grows. The peak is thought to occur in the SiO$_2$ core or cladding, and is only observed in specific fiber designs. Further work, however, is required to clarify both the specific mechanism and the importance of such a transient mechanism on long-term performance in the 1.55 μm window, since only very low levels of H$_2$ are required for its formation (Lemaire, 1986).

Yet another hydrogen-related phenomena has been recently reported by Iino *et al.* (1986, 1987) and Ogai *et al.* (1987) associated with the diffusion of alkali impurities from substrate tubing into the active region of fibers. Enhanced hydrogen-induced loss occurs, especially at 1.55 μm, pointing to the subtle role impurity ions can have in forming reactive defects.

Considerable progress has been made in a very short time period to understand and quantify hydrogen-induced loss effects in fibers. A number of strategies, as reviewed by Rush *et al.* (1984), have emerged to account for and minimize such effects. The recent work on hermetic coatings offers the potential to render fibers impermeable to hydrogen (Beales *et al.*, 1984), especially in harsh environments, but further study is necessary to prove-in the efficacy of such an approach. Work is continuing to further refine long-term reliability models to assure the required fiber performance, as well as to understand the underlying reactive defects.

4.3 MECHANICAL STRENGTH AND STATIC FATIGUE

The mechanical properties of optical fibers are critical for high yield lightguide manufacture as well as for subsequent cabling, installation and service lifetime considerations. The mechanical properties of inorganic glasses in general have recently been critically reviewed (Kurkjian, 1985) and those of optical fibers were further discussed in Chapter 12, Miller and Chynoweth (1979) and elsewhere (Midwinter, 1979; DiMarcello *et al.*, 1985). This section will briefly review the current understanding of the mechanical properties of silica fibers, as well as some of the outstanding issues and ongoing directions for long-length, high-strength optical fiber.

Silica glass fibers, at normal operating temperature, behave as ideal elastic bodies up to their breaking strength. The theoretical strength of silica glasses is estimated by considering the cohesive bond strength of the constituent Si-O atoms, with values as high as 3 million psi (20 GPa) postulated. Proctor *et al.* (1967) first reported measurements as high as 2 million psi (13.8 GPa) for short lengths of silica fibers at liquid nitrogen temperatures, where the effect of moisture was eliminated from the test environment, thus demonstrating the intrinsic high strength of silica. The realizable strength of fibers is typically much lower, and is impacted by many environmental and extrinsic factors, which are now reasonably well understood. At ambient temperatures and moisture levels, the intrinsic strength is reduced to values on the order of 800 ksi (5.56 GPa) due to fatigue during dynamic tensile testing, which will be discussed in more detail later.

Typically, however, long lengths of fibers exhibit a distribution of much lower strengths, and this variability of strength is understood in terms of the classic Griffith model of brittle fracture in which surface flaws, mechanically or chemically induced, act as local stress concentrations leading to failure at a lower fracture stress than the intrinsic level (Griffith, 1920). Fracture mechanics theory is applied to fibers to account for stresses at the tip of the crack or flaw, as well as for different crack geometries and loading situations. When stresses in the vicinity of the crack reach a critical value, fracture occurs. The stress causing failure is inversely proportional to a critical flaw size, a_{cr}, as shown in Fig. 4.11, after Krause, *et al.* (1986). For a flaw, which has an atomically sharp tip with a

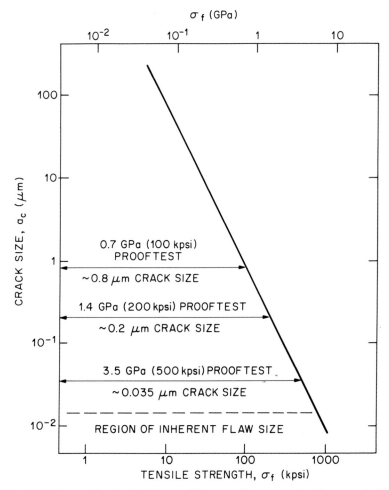

Fig. 4.11   Crack size associated with failure at different tensile stresses, after Krause *et al.* (1986).

crack plane perpendicular to the fiber axis, the critical size leading to failure is related to the fracture stress by:

$$a_{cr} = \left(YK_{IC}\right)^2/\sigma_f, \qquad (4.9)$$

where Y = geometrical constant = 0.806 and K_{IC} = critical stress intensity factor = 0.789 MN/m$^{3/2}$ (0.72 ksi-in$^{1/2}$) for silica. Many extensions on the original Griffith model have been made to describe the fracture of brittle materials, particularly glass (see, for example, Wiederhorn, 1969, 1974; Evans and

Wiederhorn, 1974; Evans and Johnson, 1975; Ritter *et al.*, 1978; Kurkjian, 1986).

The statistical variability of strength observed in silica glass lightguides is interpreted as due to a random distribution of surface flaws introduced during the various stages of fiber fabrication. In a given length, the largest or most severe flaw will lead to failure. As the length increases, there is an increasing probability of encountering increasingly larger flaws. Weibull's (1939) statistical analysis is commonly used to study this strength distribution, as well as strength versus length characteristics of fiber (Maurer, 1975). The cumulative failure probability, F, for fiber of test length L, subjected to an applied tensile stress σ, is given by:

$$F(\sigma) = 1 - \exp[-L\sigma^m] \qquad (4.10)$$

$$lnln[1/(1 - F)] = ln L + m \, ln \, \sigma. \qquad (4.11)$$

The value of m, the Weibull probability slope obtained from a plot of $lnln[1/(1 - F)]$ versus $ln \, \sigma$, is indicative of the distribution of flaw sizes for a given test length. The coefficient of variation of the measured strength, ν is related to m by $\nu = 1.25/m$ and is an indication of the uniformity of flaw sizes. The Weibull model assumes a single failure mode, while frequently, depending on the test length, more than one mode of failure is observed in fibers. Weibull analysis is still used with the data split into segments, each of which are fit to the Weibull equation. Strength for different gauge lengths L_1 and L_2 can be compared using the relationship:

$$\sigma_{f1}/\sigma_{f2} = (L_2/L_1)^{1/m}, \qquad (4.12)$$

where σ_{f1}, and σ_{f2} are strengths for equal F, and this relationship is sometimes used to predict the strength of long lengths of fiber from short length data, but if the distribution is not unimodal, this must be taken into account.

Figure 4.12, from Krause *et al.*, illustrates an example of such Weibull analysis and strength versus length characteristics for high quality silica fiber, derived from short length tensile tests and prooftest data. It illustrates the probability of failure due to a critical flaw at a given tensile stress as the length is increased. Very low ν values are achieved for short 10 m test lengths, and Kurkjian and Paek (1983) have shown that the strength is single valued, indicative of "perfect" flaw free fibers as drawn. All of the variation could be accounted for by diameter variations. At the length increases to 1 km, the distribution becomes bimodal with the lower strength, high variation mode attributed to extrinsic flaws. As the test length is further increased, the distribution is dominated by the presence of extrinsic flaws. This basic statistical feature of glass fiber strength has led to proof strength screening of all fiber as an essential manufacturing step to eliminate the weakest links, which might lead to

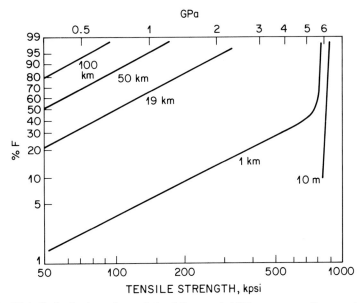

Fig. 4.12 Wiebull distributions of cumulative failure probabilities versus tensile stress for high strength fibers of different gauge lengths, after Krause *et al.* (1986).

failure in subsequent use. The source of these extrinsic flaws has been extensively investigated, and fiber drawing techniques, which minimize or eliminate their formation, have been achieved, as reviewed by DiMarcello *et al.* (1979, 1985).

In addition to the reduction and variation in strength due to the presence of flaws in long lengths of fiber, time dependent failure can occur, attributed to subcritical crack growth, the slow growth of flaws in an active environment. The mechanisms for subcritical crack growth have been critically reviewed (Pavelchuck and Doremus, 1976; Adams and McMillan, 1977; Wiederhorn, 1975, 1978). The most commonly accepted is that of stress corrosion, where stress enhances the interaction of water with a silica surface via the reaction (Charles, 1958a, b; Charles and Hillig, 1962; Hillig and Charles, 1965):

$$Si-O-Si + H_2O \rightleftarrows 2SiOH. \qquad (4.13)$$

This reaction is enhanced in the vicinity of the crack tip and slow crack growth occurs in an active environment under an applied stress, leading to failure when the flaw reaches a critical size.

Two types of time-dependent failure can occur in glass fibers, depending on the nature of the applied stress. Dynamic fatigue occurs in an active environment when the applied stress is constantly increasing, such as in a standard tensile test or when a fiber is rapidly loaded. As the stress is applied, a flaw grows due to interaction with moisture and failure occurs when the flaw size reaches its critical size. Thus, the breaking stress will be a function of the load rate as well as the relative humidity and temperature. Fibers tested in ambient conditions exhibit lower strengths than those equilibrated and tested in dry environments or at low temperatures, where the immobility of the moisture prevents the enhancement of crack growth. Static fatigue occurs when the fiber is in an active environment with an applied stress at some level lower than that leading to fast fracture; moisture interacts with the surface of the glass causing cracks to grow with time, leading to failure when the crack reaches a critical size for the applied stress.

Many formulations have been proposed to describe the fatigue phenomena in glasses in general, and in optical fibers specifically. The exact details and micromechanical nature of the subcritical crack growth mechanism continues to be debated, particularly for high strength fibers, and is beyond the scope of this chapter to discuss (for review see Kurkjian, 1985). Very frequently, a fracture mechanics model is used to describe the crack velocity, V, by the empirical relationship:

$$V = AK_I^n, \tag{4.14}$$

where A is a materials constant, K_I is the stress intensity factor and n is the stress corrosion susceptibility factor.

The original model of Charles (1958a, b) or the fracture mechanics analog of Evans (1974) are used to relate the subcritical crack growth to time to failure for a given applied stress. For static fatigue, the time to failure when a static stress, σ_{fs}, is applied is given by

$$\log t_{fs} = -n \log \sigma_{fs} + \log K_s \tag{4.15}$$

and for dynamic fatigue, the time to failure, t_{fd}, at a failure stress, σ_{fd}, is

$$\log t_{fd} = -n \log \sigma_{fs} + \log K_d, \tag{4.16}$$

where the static constant K_s and dynamic constant K_d are related by the expression (Davidge, et al., 1973)

$$\log K_d = \log K_s + \log(n + 1). \tag{4.17}$$

In these relationships, the n derived from static or dynamic testing is predicted to be interchangeable, and thus dynamic tests are frequently used to predict the long-term static fatigue performance of fibers. However, this is not always experimentally observed.

Many studies of dynamic and static fatigue of fused silica have been made (Proctor et al., 1967; Weiderhorn and Bolz, 1970; Ritter and Sherborne, 1971; Wang and Zupko, 1978; Kalish and Tariyal, 1978; Gulati et al., 1979; Krause,

1979, 1980; Craig *et al.*, 1982; Chandan and Kalish, 1982; Donaghy and Nicol, 1983; Sakaguchi, 1982, 1984; Dabbs and Lawn, 1985; Maurer, 1986; Duncan *et al.*, 1981, 1986), and the fatigue phenomena of high strength silica fibers is far more complex than the basic fracture mechanics models would predict. A wide variation in the stress corrosion susceptibility factor reported in such works exists, ranging from $n = 36-41$ for direct measurements of crack velocity on bulk samples, to most measurements on pristine silica fibers showing empirical values ranging from $18-24$, with $n \simeq 20$ the most common value. A number of different effects have been observed. The measured n value can vary depending on the type of flaw [Gulati *et al.*, 1979; Craig *et al.*, 1982; Donaghy and Nicol, 1983; Maurer, 1986]. Treatment in different environments, such as soaking in water before the test, results in different n values (Krause, 1980). Other work has documented the dependence of the measured n on the chemical environment, relative humidity and temperature (Kalish and Tariyal, 1978; Sakaguchi *et al.*, 1982, 1984; Duncan *et al.*, 1981, 1986; Krause, 1980). At longer test times, typical of static versus dynamic testing, a transition in the fatigue behavior of high silica fibers has been reported showing much more rapid degradation in strength, with $n \simeq 7$; the effect can depend on the particular fiber coating system used (Wang and Zupko, 1978; Krause, 1979, 1980; Chandan and Kalish, 1982). Other work has suggested that a static fatigue limit may occur at very long times, or a slowing of the normal fatigue process (Krause, 1980; Maurer, 1986). Work by Dabbs and Lawn (1985), where controlled micro-indentation flaws were introduced into fibers, try to account for differences in fatigue based on different crack geometries and residual stresses at the crack tip. Thus, the specifics of the fatigue process in silica fibers continues to point to the need to understand microscopically the nature of the fatigue processs for flaws and flaw-free fiber, and the associated impact of coating systems and the enviroment on long-term performance. Because of these various effects, caution must be taken in using short-term fatigue data on short gauge length samples to extrapolate the long-term performance of long-length fiber. Typically, the most conservative estimate of projected lifetime is made for a given application because of these ambiguities.

As a result of the statistical nature of strength, along with such considerations of fatigue, prooftesting of all fiber has been adopted in order to guarantee that a fiber will have a minimum strength acceptable for a minimum lifetime in a given application. In prooftesting, the fiber is subjected to a tensile stress greater than that expected in any subsequent step during cable manufacture, installation and service. Figure 4.11 shows a representation of the flaw sizes eliminated by prooftesting at two typical levels, 100 ksi and 200 ksi. A variety of silica fiber prooftester designs and approaches have been reported (Justice and Gulati, 1978; Tariyal *et al.*, 1977; Craig *et al.*, 1983; Duncan, 1981; Mitsunaga *et al.*, 1982). Care must be taken to consider and minimize any crack growth during

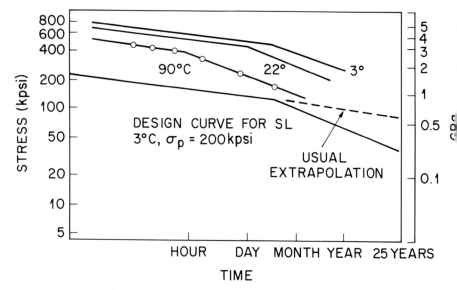

Fig. 4.13 Static fatigue design diagram, after Krause *et al.* (1986).

the unloading portion of the prooftest, in order to assure that damage is not occurring due to the test itself (Evans and Fuller, 1975; Tariyal and Kalish, 1978).

Static fatigue design diagram analysis is typically used for determining the fiber characteristics for a given application. The higher the prooftest level, the greater the safety factor built into projected fiber performance, but practical issues of manufacturing yield must also be considered when setting the prooftest level. Figure 4.13, after Krause *et al.* (1986), shows an example of a conservative design diagram used for fiber which has been prooftested at 200 ksi. For this case, static fatigue data for fiber in water over the temperature range 3–90°C has been considered, including the transition in static fatigue characteristic of the particular coating system used. Fiber in water represents the worst case environment for fiber in regard to the dependence of static fatigue on humidity. The design curve for the twenty-five-year service lifetime is constructed based on the initial prooftest level of 200 ksi. For this example, characteristic of an undersea environment at 3°C, all of the calculated stresses that are expected to be imposed upon the fiber during cable installation, possible recovery for repair, and ocean bottom lifetime are well below the design curve, thus assuring the mechanical reliability for the design time period. This analysis includes the transition in fatigue and thus represents a very conservative estimate relative to the normal linear extrapolation. More generally, for a constant applied stress, as long as the applied stress for a given time is below the design curve for the prooftested fiber, failure is not predicted to occur. Prooftesting; conservative lifetime projections; and coating, handling and cabling techniques which minimize applied stress (as described in Chapter 5), can all assure adequate reliability for a given application.

Other studies have been directed at trying to improve the fatigue resistance of silica-based fibers. The organic coating typically used to buffer and protect fibers are permeable to water, and thus the glass fiber surface equilibrates to the ambient environment, which contain moisture that contributes to the fatigue process. An approach to improving the fatigue resistance involves using "hermetic" coatings, which prevent moisture from reaching the silica fiber surface. A number of investigators have explored ductile metal coatings, which can elastically and plastically deform under applied stress; early work reported the achievement of markedly improved fatigue resistance in short lengths, (Pinnow et al., 1979). However, long lengths of pinhole-free metal coatings proved difficult to fabricate; their high coefficient of expansion and the relatively thick coatings required due to the coating application technology caused unacceptable microbending induced losses; upon occasion, a very low unexplained failure mode was noted.

Other workers have focused on amorphous ceramic coatings (Hiskes et al., 1984; Duncan et al., 1981; Chandhari and Schultz, 1986; Beales et al., 1984; Maurer, 1986), such as silicon oxynitride, silicon nitride, silicon carbide, titanium carbide and carbon. Such coatings are brittle ceramic materials and thus their mechanical properties and crack growth characteristics must be considered. Control of porosity is essential to achieve impermeability, and application technology that results in pinhole and defect-free coatings is critical. Typically, coatings on the order of 200–1000 Å have been reported, and with their relatively low expansion coefficients, no added loss is observed. The most success has been achieved with chemical vapor deposition approaches, which do not appear to damage the underlying silica during deposition. Well-adhering coatings are deemed desirable, as well as good environmental stability. Using such approaches, n values in the range of 50–200 have been reported, but such coatings have characteristically reduced the ~ 800 ksi initial ambient strength to values in the range of 300–500 ksi. A number of explanations have been put forward to explain both these phenomena, but further study is required to fully account for the meaning of such results. The achieved n values may be characteristic of the coating or composite coating—fiber system and the associated crack growth, while the lower strength may be related to a number of effects such as mechanical properties of and stresses in the coating as well as the chemical bonding. Most of the reported data has been determined from dynamic tests, and further study of long-term static fatigue is required.

The higher n values achieved to date hold promise for improved fatigue performance of fibers despite the reduction in initial strength, particularly for applications in harsh environments. Test methods must be evolved to assure 100% hermeticity screening, since a single pinhole or defect could obviate the benefit of such coatings. These coatings have also been reported to minimize hydrogen diffusion into the fiber (Beales *et al.*, 1984), and thus may offer additional reliability in this regard. Continued research in this area is expected to further improve the fatigue performance of silica, but additional studies are required to understand the detailed characteristics and performance of such fiber.

While tremendous progress has been made in understanding the mechanical properties of silica fibers and in achieving long length, high strength reliable transmission media, ongoing studies continue to address many of the outstanding questions in regard to strength, crack growth, fatigue and microscopic models, which account for the detailed behavior of such fibers.

4.4 GENERAL APPROACHES TO SILICA GLASS FABRICATION

The proposal by Kao and Hockham (1966) to use silica-based fibers as the transmission medium in lightwave communication systems and the realization of 20 dB/km losses in a lightguide structure using vapor phase processing by Corning in 1970 (Kapron *et al.*) are two important milestones in lightguide research. Much of the early work focused on fiber performance at short wavelengths (0.6 μm–1.0 μm), where solid-state sources and detectors were available. While a variety of fiber properties were recognized as important, the early emphasis concentrated on materials and processes that had sufficient control of impurities to approach achieving theoretical low-loss performance. As propagation theory was developed (for review see Chynoweth and Miller, 1979; Midwinter, 1979; Adams, 1981), the ability to fabricate graded-index multi-mode fiber with bandwidth optimized structures for use with available light sources was also deemed important. These two considerations, loss and graded index, played a critical role in the evolution of current manufacturing methods, as well as the preference for high silica glass over more conventional multicomponent glasses. Further studies of materials suggested that the lowest losses occurred at longer wavelengths (1.3–1.6 μm), and sources and detectors simultaneously evolved for these wavelengths. This further favored high silica glasses over multicomponent glasses. As the various aspects of technology (such as lasers, coupling techniques, splicing and connectorization) advanced, techniques for higher bandwidth, lower loss single-mode fibers became practical, and fabrication technology focused on realizing optimized single-mode designs. Fiber strength, dimensional control, reliability, and reproducible high yield fabrication assumed increasing importance. The major manufacturing approaches to be described in the following sections emerged due to their demonstrated ability to achieve all requisite requirements.

One thrust of early work was to improve conventional glass compositions and processing approaches, where raw materials were melted and quenched to form a bulk glass, then drawn into fiber. Soda-lime-silica and sodium borosilicate glasses were explored, with the primary objective of eliminating transition metal impurity absorption. Excellent reviews of the early work on fiber optics detail these approaches (Maurer, 1973b; French et al., 1975; Beales et al., 1976; Black, 1976; Beales and Day, 1980). Special purification techniques were devised to provide high purity starting materials, and improved techniques of melting, homogenizing and refining bulk glass were reported. Rod-and-tube and double crucible drawing techniques were used to make fibers. Diffusion of glass constituents between the core and cladding were developed to achieve graded index structures (Koizumi et al., 1974). Indeed, much progress was reported and ultimately reasonable losses were achieved (Beales et al., 1980; 1985).

A second approach involved nonconventional vapor phase preparation of high silica glass preforms with only small amounts of dopants to vary the refractive index. These preforms could then be drawn into fibers. Volatile silicon compounds could be converted to SiO_2 by high temperature vapor deposition techniques, and suitable silica dopant sources were also examined. Flame hydrolysis, CVD, and high temperature oxidation were explored (for review, see Beales and Day, 1980). While the first low-loss fiber utilized TiO_2 to increase the index of the core (Kapron et al., 1970), GeO_2-doped silica emerged as the preferred composition. GeO_2 was an excellent glass former in its own right with volatile compounds, which could be used for vapor phase deposition, and good intrinsic loss characteristics. TiO_2-doped silica had deleterious absorption associated with the defect Ti^{3+} (Carson and Maurer, 1973), while Ge^{2+} defect, although present in Ge-doped silica, did not have an associated deleterious absorption edge (Schultz, 1977). Small amounts of B_2O_3 were added to aid in multimode processing, which was later replaced by P_2O_5 for long wavelength optimized designs, where the B-O absorption edge limited loss. P_2O_5 also allow lower scattering to be achieved (Payne and Gambling, 1974). While silica cladding was most commonly used, cladding compositions based on doping with B_2O_3 (Van Uitert et al., 1973) or fluorine (Mulich, 1975; Abe, 1976) were also used, expecially to fabricate SiO_2 core structures, since these binary glasses had a lower refractive index than SiO_2. Tremendous progress in realizing very low loss and high bandwidth designs occurred.

As technology matured, an additional focus of vapor phase processing studies concentrated on process scale-up in order to improve the economics of manufacture. Studies of process limits explored higher deposition rates and larger preform sizes. Variations on the basic vapor phase processes are now directed at high rate high efficiency methods of adding cladding to the basic optically active core-clad structure in order to further improve economics. All vapor phase processes, however, have finite deposition rates and more recently, sol-gel methods of silica preparation are being explored as the next generation process technique in order to overcome the basic deposition rate limit of vapor phase processing.

4.4.1 Modified Chemical Vapor Deposition Process

The modified chemical vapor deposition (MCVD) process was invented at AT & T Bell Laboratories (MacChesney and O'Connor, 1980) and was first disclosed in 1974 (MacChesney et al., 1974). It involves vapor phase deposition of high optical quality material on the inner surface of a tube, collapsing this composite to form a preform rod with the core/clad structure, then drawing the preform into fiber. A variation on the process, plasma MCVD (Jaegar et al., 1980; Fleming et al., 1982), uses an atmospheric plasma to enhance deposition rate and efficiency. Recent reviews (Nagel et al., 1982; 1983; 1985; Jablonowski, 1985) have detailed the process mechanisms and performance. This section will briefly review some of the key elements of this process.

MCVD is the simplest and perhaps most flexible of the lightguide processes. It starts with a tube, typically commercial silica, which provides part of the cladding in the lightguide structure, and also acts as a containment vessel for the deposition process. The tube is characterized and selected for dimensions, siding, cross-sectional area and uniformity, and cleaned prior to use. The deposition station consists of a glass working lathe, a chemical delivery system, and associated computer control console. The entrance end of the tube is mounted in one of two synchronously rotating chucks of the lathe and coupled to the chemical delivery system via a rotating joint. The other end of the tube is flared and fused to a larger tube mounted in the second chuck; this large tube serves to collect unincorporated material resulting from the deposition process and is coupled to a chemical scrubbing system. After setup of the tube, it is rotated and fire polished by means of a traversing heat source. Next, the deposition phase of the process begins.

The basic deposition process predominantly involves the high temperature homogeneous gas phase oxidation of volatile vapor delivered compounds that are deposited as submicron particles via thermophoresis and fused to a clear glass film. The deposition process uses controlled amounts of chemical reagents entrained in a gas stream by passing carrier gases such as O_2 or He through liquid dopant sources such as $SiCl_4$, $GeCl_4$ or $POCl_3$ or direct proportionation of gaseous dopants such as SiF_4, BCl_3, and CCl_2F_2. As in all forms of vapor phase processing, this method of delivery acts as a purification step relative to transition metal impurities which might be contained in raw materials, and are characterized by much lower vapor pressures. The chemical gas mixture is injected into the rotating tube where a hot zone is traversed along the length of the tube by an external moving heat source, typically an oxyhydrogen burner.

The temperature of the hot zone is controlled via optical pyrometry monitoring and feedback to a flame temperature controller. Layer by layer of material is deposited and sufficient heat from the moving heat source results in the sintering of the deposit as the hot zone passes over it.

In MCVD, first high purity cladding is deposited, then core. This cladding serves a number of functions: it acts as a barrier to in-diffusion of impurities, particularly OH, into the active region of the lightguide; it insures low cladding losses for any power which propagates in the cladding; lastly, it minimizes any scattering losses that might occur due to tubing defects or interfacial irregularities at the tubing inner surface. For single-mode fibers, the deposited cladding can also serve the additional function of allowing more complex, dispersion optimized designs to be made, by tailoring the cladding index profile. Core deposition ensues next, involving 30–70 layers for multimode structures, versus one to several layers for single-mode structures. The deposited cladding can be a variety of index matched or depressed compositions in the $F-SiO_2-GeO_2-P_2O_5$ system, where small amounts of P_2O_5 are sometimes used predominantly to decrease the deposition temperature. Core compositions are typically GeO_2-SiO_2, where small amounts of P_2O_5 are used for graded index multimode fibers. The dimensions and refractive index profile of the eventual fiber structure is built up by depositing sucessive layers of controlled composition to the desired thickness, then collapsing the composite tube plus deposit to a solid preform rod. The total number of deposited layers is chosen on the basis of starting tube dimensions, deposition rate, profile complexity and fiber design.

Excellent understanding of all aspects of the process has been documented, as reviewed by Nagel et al. (1985). As depicted in Fig. 4.14, the deposition phase of the process involves several interactive steps: the reaction of the precursor chemicals to nucleate particles in the gas phase, which grow by Brownian coagulation; the deposition of these particles, which are thermophoretically driven to and deposited on the tube wall; then the subsequent consolidation of particles to a bubble-free film by viscous sintering. The details of each of these interactive steps plays a critical role in determining the characteristics of the resulting waveguide material.

A number of investigations (Walker et al., 1979, 1980a; Simpkins et al., 1979) conclusively demonstrated thermophoresis as the deposition mechanism in MCVD. Thermophoresis refers to the phenomena in which particles suspended in a gas with a temperature gradient experience a net force in the direction of decreasing temperature. More energetic small gas molecules impacting the particle on its high temperature side than those on the low temperature side result in the particle acquiring a velocity in the direction of cooler temperature. In MCVD, the tube is heated by a traversing oxhydrogen torch, which gives rise to a variable temperature field within the tube. Under typical operating conditions, the gas flow in the tube in laminar and as the cool gas enters the

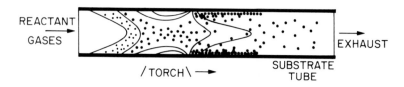

1. REACTION AND PARTICLE NUCLEATION
2. PARTICLE GROWTH - BROWNIAN COAGULATION
3. DEPOSITION - THERMOPHORESIS
   EFFICIENCY α $(1 - Te/T_{rxn})$
4. CONSOLIDATION - VISCOUS SINTERING

Fig. 4.14 Schematic representation of the deposition phase of the MCVD process at any point in time. Oxidation reactions to nucleate particles occurs within the tube when the gas mixture reaches a critical temperature. Particles continue to grow as they are transported along streamlines in the tube, and some fraction deposit by thermophoresis at the tube wall while others are swept out in the exhaust. The particles deposited at some previous time are simultaneously consolidated via viscous sintering in the region of the torch.

heated zone; it starts to increase in temperature. When the moving gas front reaches a critical temperature, T_{rxn}, which is sufficiently high enough for the oxidation reaction to take place (typically ~ 1200°C), particles are nucleated and rapidly grow to submicron size. The detailed composition and size of the particles is strongly effected by the gas composition, chemical equilibrium, particle growth dynamics and transport phenomena. Once the particle is formed, it moves in a trajectory determined by the thermophoretic temperature gradient. Initially, particles tend to move inward since the tube wall is hotter than the gas. Downstream from the torch, the tube wall is cooler and the drift inward is reversed, resulting in particles moving to the wall where they are deposited. Walker *et al.* (1979) has shown that the efficiency, E, of particle incorporation can be simply described by:

$$E \sim 0.8\left[\left(1 - \left(T_e/T_{rxn}\right)\right)\right], \qquad (4.18)$$

where T_e is the downstream temperature to which the gas and tube wall thermally equilibrate. T_e strongly depends on the torch traverse length and velocity, and tube wall thickness, and weakly on the flow rate and tube radius. Only certain trajectories result in deposition. The length over which deposition takes place depends on Q/α_g, where Q is the total volumetric flow and α_g is the thermal diffusivity of the gas mixture. This deposition length results in a finite entry taper at the beginning of deposition, which can be minimized by ramping the initial torch velocity.

The composition of the particles and resulting deposited layer is critical to control of the process, and many studies have examined the details of the high temperature chemistry and mass transport during MCVD, especially in regard to germanium incorporation. In general, the $SiCl_4$ and $POCl_3$ are completely oxidized at high temperatures in MCVD (Wood *et al.*, 1978; French *et al.*, 1978; Powers, 1978), while the $GeCl_4$ oxidation and incorporation are strongly affected by unfavorable thermodynamic equilibrium of the reaction (Wood *et al.*, 1981, 1982, 1987; Kleinert *et al.*, 1980):

$$GeCl_4(g) + O_2(g) \rightleftarrows GeO_2(s) + 2Cl_2(g). \qquad (4.19)$$

The large amounts of chlorine generated by the oxidation of $SiCl_4$ further shift this equilibrium to the left. A simple model for the incorporation of GeO_2 has been presented (Wood *et al.*, 1987), which takes into account the control of kinetics at low temperatures, and the control of thermodynamic equilibria at high temperatures. The model quantitatively can predict GeO_2 incorporation by including the dynamics of particle formation, the role of phosphorus, and also accounts for the characteristic layer structure in MCVD.

Other studies have been directed at fluorine incorporation, first reported in MCVD by Abe (1976). The fluorine incorporation into the glass deposit has been shown to depend on the concentration of SiF_4 in the high temperature gas stream as well as on the temperature at which the deposited particulate layer is consolidated (Walker *et al.*, 1983). The resultant refractive index depression was found to be proportional to the 0.25 power of the partial pressure of SiF_4, predicted for the equilibrium

$$SiF_4(g) + 2SiO_2(s) \rightleftarrows 4SiO_{1.5}(s), \qquad (4.20)$$

where the SiF_4 concentration depends on the equilibria

$$3SiF_4(g) + SiO_2(s) + 2Cl_2(g) \rightleftarrows 4SiF_3Cl(g) + O_2(g) \qquad (4.21)$$

$$Cl_2(g) \rightleftarrows 2Cl(g). \qquad (4.22)$$

Thus, there is a finite amount of fluorine that can be incorporated into the glass, corresponding to a maximum relative index depression of $\sim 0.6\%$. In MCVD, SiF_4 is the preferred dopant source for maximum deposition rate and efficiency, since all other dopants form SiF_4 at high temperatures at the expense of SiO_2.

Consolidation of the deposited material is also critical to the MCVD process and the fusion mechanism has been numerically modelled and experimentally shown to be viscous sintering of thin particulate layers (Walker *et al.*, 1980b). Kosinksi *et al.* (1981) showed the difference in sintering characteristics across a layer as a function of P_2O_5 dopant concentration. In general, control of pore size and aggregates during deposition are important to avoid bubble growth during consolidation and subsequent thermal processing such as collapse. A hot zone sufficient to properly sinter the deposited material is also required. Consolidation is typically not a rate limiting step during MCVD deposition with proper hot zone optimization.

OH contamination during MCVD has been the subject of many studies directed at minimizing such effects. There are two primary sources of OH contamination in fibers made by MCVD: (1) OH incorporated from hydrogenic impurity species that enter the gas stream during various phases of processing, and (2) OH that thermally diffuses from the substrate tube into the active region during deposition, collapse and fiber drawing (Osanai, 1978). A variety of approaches are used to minimize these effects.

The amount of OH incorporated during processing is controlled by the equilibria (Wood et al., 1979; Wood and Shirk, 1981; Walker et al., 1981)

$$H_2O + Cl_2 \rightleftarrows 2HCl + \tfrac{1}{2}O_2 \qquad (4.23)$$

$$H_2O + [Si\text{-}O\text{-}Si]_{solid} \rightleftarrows 2[SiOH]_{solid}. \qquad (4.24)$$

H_2O is incorporated into the glass as OH, while HCl is not. The resultant equilibrium concentration of SiOH in the glass is described by

$$C_{SiOH} \alpha \left[P_{H_2O}^i \right] \left[P_{O_2} \right]^{1/4} \left[P_{Cl_2} \right]^{-1/2}, \qquad (4.25)$$

where $P_{H_2O}^i$ is the initial partial pressure of H_2O in the gas stream from all sources, and P_{O_2} and P_{Cl_2} are the partial pressures of oxygen and chlorine, respectively. This equilibrium not only pertains to MCVD but is also applicable to dehydration and sintering of OVD and VAD, to be discussed in the following sections. Figure 4.15 shows the dependence of SiOH incorporation on the chlorine and oxygen partial pressures when 10 ppm H_2O by volume is present in the process atmosphere. During deposition, 3–20% Cl_2 is typically present in the atmosphere due to the oxidation of the chloride reactants, resulting in typically a 4000:1 incorporation ratio due to favorable equilibrium. While O_2 pressures could be reduced and Cl_2 further raised to impact this equilibria, this could adversely affect germanium incorporation. More typically, very low H_2O contamination levels are maintained through chemical purification and leak tight delivery systems, as reviewed by Nagel et al. (1985). During collapse, it is critical to introduce a chlorine atmosphere, especially during single-mode fabrication where any hydrogen impurities during collapse can easily diffuse into the entire core region. Pearson (1980) demonstrated low OH single-mode fibers using controlled collapse atmospheres.

In-diffusion of OH from the substrate tube can also add to loss and is commonly prevented by depositing low OH cladding as a buffer. Kosinski et al. (1982) quantified the role of barrier layer thickness on resultant OH levels in multimode fibers for various OH level tubes. Substantial amounts of deposited cladding is necessary for low OH single-mode fibers, due to the power in the cladding; the exact amount will depend on the substrate tube type and dimensions, as well as the detailed processing conditions. Nagel et al. (1982b, 1985),

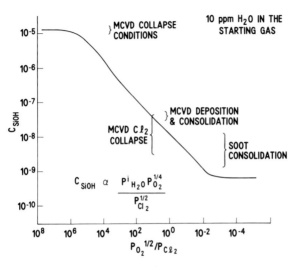

Fig. 4.15 SiOH incorporation sensitivity to H_2O, Cl_2 and O_2 partial pressures, after Walker *et al.* (1981). Representative conditions for MCVD and soot processes are illustrated when 10 ppm H_2O vapor is present in the gas stream.

incorporated a variety of OH control approaches and demonstrated OH peaks less than 0.05 dB/km for both single-mode and multimode fibers with MCVD.

High deposition rate scale-up approaches have been demonstrated for multimode and single-mode fibers by considering all aspects of the deposition process (Simpson *et al.*, 1980, 1981; Walker *et al.*, 1982a). In these approaches, hot zone geometry was extended to assure complete reaction and sintering; entry taper was minimized; cooling or gas thermal conductivity was optimized, and in general, conditions that resulted in low-loss deposition at rates as high as 2.3 g/min were reported. In another extension of MCVD, an atmospheric oxygen plasma was used to create a high temperature fireball inside the tube to enhance reaction and thermophoretic deposition (Fleming *et al.*, 1981a, b, c; Fleming, 1983; Nagel *et al.*, 1985). Deposition rates greater than 5 g/min for low-loss single-mode fibers were reported, with much higher incorporation efficiencies than in MCVD.

The second major step in the MCVD process for preform preparation involves collapse of the composite tube to form a solid preform rod. Typically, a number of very high temperature (2000–2300°C) passes with the external heat source are made to reduce the diameter. Details of the collapse process have been numerically modelled and experimentally verified (Walker *et al.*, 1982b; Geyling *et al.*, 1983), and the role of surface tension driven viscous flow and pressure differential on the collapse rate and stability were elucidated for the two very different cases of single-mode and multimode collapse. The derived model provides key guidelines for understanding, controlling and optimizing collapse to achieve stability and maximum rate of collapse, with the maintenance of slight positive internal pressures key to achieving high circularity, particularly for more fluid deposits.

In detail, all aspects of the MCVD process physics and chemistry, as well as control approaches have been documented. Excellent performance has been achieved for a number of fiber designs ranging from 1.3 μm dispersion, bend, and loss optimized structures (Pearson et al., 1982a, b) to high 0.3 NA, 62.5 μm core multimode fibers (Klein et al., 1983), as reviewed in detail by Nagel et al. (1985) and Jablonowski (1985). MCVD has been the predominant technique used to make polarization maintaining fibers, as described in Section 4.6. While the process relies on a tube, this relatively cheap source of silica can be used to advantage, not only to provide cladding material for the optically inactive region of the fiber, as in conventional MCVD, but also by overcladding using rod-in-tube. Such an approach was shown by Pearson (1981) to yield 40 km low-loss single fiber, simply by increasing the volume of deposit in the original preform. In general, MCVD can be scaled up to greater than 100–200 km preform sizes by using overcladding approaches coupled with initial starting tube sizes which result in high deposition rates and efficiencies. Addition of overcladding in a secondary step is an approach under examination in all processes. Critical issues are dimensional control, eccentricity, yield, strength and resultant optical properties. MCVD continues to be a process practiced by many manufacturers throughout the world. It is the most extensively and quantitatively documented of the processes in regard to process control, and the simplest to implement conceptually and from an equipment perspective.

4.4.2 Plasma Activated Chemical Vapor Deposition (PCVD) Process

The PCVD process was invented by Phillips Research Laboratories and first disclosed in 1975 (Koenings et al., 1975). Deposition is similar to MCVD in that it occurs within a silica tube using chloride precursor reagents to deposit high purity glass, followed by collapse to form a preform rod with the requisite core/clad structure. The chemical reaction, however, is initiated by a non-isothermal microwave plasma operating at low pressures. A number of papers have reviewed the current understanding of the method, process mechanisms as well as performance (Geittner et al., 1976, 1986; de Wert et al., 1985; Bachmann et al., 1982a, b, 1983a, b, 1986; Lydtin, 1986).

A schematic of PCVD deposition is shown in Fig. 4.16. A non-rotating silica tube is positioned within a furnace held at temperatures in the range of 1000–1250°C. A moving microwave resonator operating at a frequency of 2.45 GHz with powers up to 6 kW is used to generate a plasma within the tube,

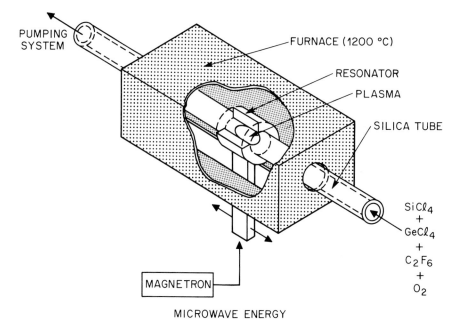

PUMPING SYSTEM

FURNACE (1200 °C)

RESONATOR

PLASMA

SILICA TUBE

$SiCl_4$
$+$
$GeCl_4$
$+$
C_2F_6
$+$
O_2

MAGNETRON

MICROWAVE ENERGY

Fig. 4.16 Schematic representation of the PCVD process.

which is maintained at pressures of 10–25 mbars using a rotary pump. The resonator traverses the length of the tube within the furnace at speeds up to 8 m/min. These speeds are achievable since the microwave energy is coupled directly into the plasma rather than the tube, thus thermal inertia effects are avoided. Vapors of reagents such as $SiCl_4$, $GeCl_4$ and C_2F_6 in an oxygen carrier gas are proportioned by electronic flow controllers and injected into the tube. Clear glass material is deposited via a heterogeneous reaction directly on the tube wall, with the reaction initiated only within a tube diameter of where the plasma is present, due to the excitation and formation of highly reactive free radicals. Reaction is kinetically inhibited away from the plasma; thus, no soot particles are formed by homogeneous gas phase reaction. Typically, thousands of thin layers are deposited, which allows very accurate compositional control. After deposition is complete, the tube is collapsed in a similar manner to MCVD, and an index dip is avoided by using C_2F_6 etching technique during collapse (Peelen *et al.*, 1984).

Excellent understanding of the deposition process has evolved, allowing high efficiency of dopant incorporation, high rates, reproducible index profiles and excellent bandwidth and loss to be achieved by this process. Early work (Kuppers *et al.*, 1979) underscored the importance of substrate temperature control to achieve low-loss, bubble free material. In particular, substrate temperatures of ~ 1200°C were high enough to avoid deleterious chlorine incorporation, which caused bubbling, yet low enough to avoid thermal oxidation of the reactants to form particles (Koel, 1983).

Other studies have focused on deposition efficiency and dopant incorporation. While the incorporation efficiency is ~ 100% for SiO_2, it is typically on the order of 80–90% for GeO_2, much higher than for other processes (Bachmann et al., 1982b). The GeO_2 incorporation was found to be weakly dependent on such process variables as total pressure, microwave resonator velocity, microwave power, substrate temperature and flow rate. Substrate temperature control within 10°C is important, and smooth uniform profiles are achieved due to the thousands of deposited layers.

Fluorine doping in PCVD has also been extensively investigated (Bachmann, 1982a). Earlier, Kuppers et al. (1978) had reported very efficient doping of F using SiF_4 in PCVD, but by using C_2F_6 as a fluorine source, index differences up to 2%, with 57% incorporation efficiency could be achieved. Such efficiencies and fluorine dopant levels are unattainable with other processes, due to the differences in process chemistry. Although losses were high in early fibers, it was shown that very small amounts of GeO_2 codoping markedly improved the optical properties of both single-mode and multimode fibers (Bachmann et al., 1986).

Fluorine doping has the additional advantage of making the process less sensitive to OH incorporation. Without fluorine present, hydrogen impurities are incorporated at ratios of 1:80, imposing very stringent requirements on the purity levels of all starting materials as well as the maintenance of a leak tight delivery system (Lennartz et al., 1983). Very small amounts of F markedly decrease this sensitivity, and OH peaks heights as low as 0.1 dB/km have been achieved (Bachmann, 1983a, b; Peelan et al., 1984; Lydtin, 1986). An additional benefit of the use of C_2F_6 during collapse is to produce smooth profiles without a central refractive index dip, due to an etching mechanism (Peelen, 1984).

The progress in PCVD processing has demonstrated its ability to manufacture high quality multimode and single-mode structures at practical deposition rates. Like MCVD, the tube supplies a large fraction of the cladding in the waveguide structure, and larger preforms are achieved by rod-in-tube overcladding of the preform (deWert et al., 1985; Lydtin, 1986). Preform sizes > 200 km are projected. Champion data include a 8GHz-km multimode fiber (Peelen et al., 1984) and highly controlled single mode profiles for dispersion flattened designs with dispersion < 2ps/nm-km between 1.3 and 1.6 µm (Bachmann et al., 1986). Higher fluorine doping levels than in any other process as well as smoother complex refractive profiles are unique features of PCVD. Continued efforts on large-scale manufacturing will determine the ultimate cost effectiveness of the process.

4.4.3 Outside Vapor Deposition (OVD) Process

OVD is a process invented and developed by Corning Glass Works, and its details have remained largely proprietary to that company. Many patents and reviews have provided insight into the basic process (Keck et al., 1973b; Maurer, 1973a; Schultz, 1974; Schultz and Maurer, 1975; Schultz, 1979a, b; Blankenship and Deneka, 1982; Morrow and Schultz, 1983; Sarkar and Schultz, 1983; Deneka, 1985; Morrow et al., 1985; VanDewoestine and Morrow, 1986).

The OVD process is based on flame hydrolysis, which is used to generate submicron size glassy particles, which are deposited onto a rotating mandrel to build up a soot boule or preform that is subsequently dehydrated, consolidated and drawn into fiber. The process steps are shown schematically in Fig. 4.17, after Schultz, 1979b. First, submicron (~ 0.25 μm) soot is formed by flame hydrolysis reactions, typically using a methane-oxygen flame, which is directed at a rotating low expansion target rod or mandrel such as Al_2O_3 or graphite. The flame issues from a torch or burner that traverses the length of the mandrel, and soot is deposited by thermophoresis and built up, layer by layer, into a cylindrical porous soot boule. In a variation, the torch can remain stationary while the mandrel moves. Typically, the average pore size is ~ 0.3 μm with overall densities of 15–25% of the bulk glass density. Torch design has remained proprietary but in general consists of chemical reagents issuing from a center oriface with additional rings of ports for shield and fuel gases. Chemical delivery of the reagents such as $SiCl_4$ and $GeCl_4$ can be accomplished in a number of ways (Morrow et al., 1985) using vapor entrainment with oxygen or inert carrier gas, direct chemical vaporization, or metering pumps with subsequent vaporization (Blankenship, 1979; Blankenship and Deneka, 1982).

In OVD, the composition of the deposit is controlled by changing the gas composition with time in order to build up the desired refractive index profile as a function of radial position. Typically, both the core and the cladding are deposited, and up to 1000 deposited layers have been reported. No discussion of the impact of the torch design, flow conditions and other process variables on the resultant chemical composition of doped-SiO_2 soot has been reported, but these variables are thought to effect the GeO_2 incorporation efficiency and the specific composition of the particles. Particles of different sizes and GeO_2 content have been reported (Carrier, 1980; Morrow et al., 1985) but the process parameters, which give rise to such effects have not been delineated.

After deposition, in the typically described OVD configuration, the mandrel is removed prior to sintering. Great care must be used to avoid defects at the inner surface, which could impact the resultant fiber performance. The boule is then placed in a sintering furnace where it is dehydrated and consolidated to form a bubble-free glass boule. The sintering of the porous soot boule is achieved by suspending it and passing it vertically through a controlled temperature zone,

A. SOOT DEPOSITION

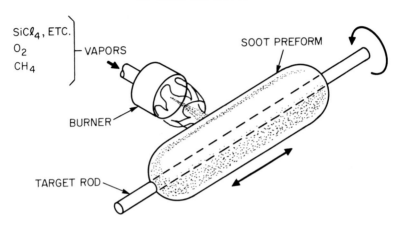

B. SINTERING   C. FIBER DRAWING

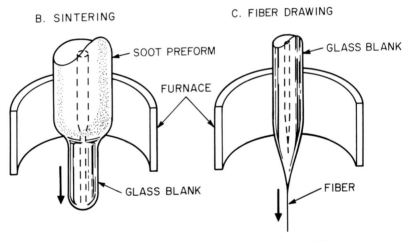

Fig. 4.17 Schematic representation of the OVD process, after Schultz (1979b).

refractory muffle furnace, operated at ~ 1500°C. The viscous sintering of OVD soot has been kinetically modelled and experimentally verified (Scherer, 1977, 1979a, b). During this step, controlled atmospheres of Cl_2-He mixtures are used to reduce the resultant hydroxl level (DeLuca, 1976; Powers, 1979; Aronson et al., 1979). Sintering in such an atmosphere takes advantage of the same H_2O-HCl chemical equilibrium described for MCVD, as depicted in Fig. 15. Such treatment reduces the hydroxyl level from as high as 200 ppm to the ppb level in currently manufactured fiber (Sarkar and Schultz, 1983). However, in using such a Cl_2-containing atmosphere, caution must be taken since GeO_2 can be removed as $GeCl_4$. Use of some O_2 minimizes such effects and trapped gases such as O_2 and Cl_2 must also be minimized to prevent bubble formation (Powers, 1979). Well-controlled sintering conditions yield low loss, low OH, controlled profile fiber.

The removal of the target rod allows delivery of drying gases through the center hole, thus allowing most efficient removal of OH from the optically active region of the resultant fiber. However, the existence of a hole in the center of the large boule after consolidation creates problems due to thermal expansion mismatch effects. The difference in thermal expansion of the core relative to the cladding puts this hole into tension, which can result in stress-induced cracking if special steps are not taken to relieve the stress. Early on, this greatly limited the compositions and designs of fibers that could be made, especially high NA fibers. By using controlled chemical compositions, lower stress levels were reported to improve this constraint. Using a stress balancing concept developed by Gulati and Scherer (1978), the expansion mismatch across the boule could be minimized, allowing 0.3 NA fibers to be made.

Another approach utilized in OVD involves closing the hole in the boule during the consolidation step (Scherer, 1979; Blankenship, 1982) by making use of forces generated during this step. In some cases, P_2O_5 was deliberately increased at the inner surface, forming a more fluid layer during consolidation, thus enhancing the surface tension forces to cause hole closure. However, the use of P_2O_5 in the core is not suitable for very low-loss single-mode fibers, and is not known to be currently in use in OVD production. Hole closure during sintering has the additional advantage that the central dip is minimized due to the lower temperatures used during sintering.

More recently, fluorine doping in the OVD process has been reported (Berkey, 1984) in order to fabricate depressed cladding fiber designs. Fluorine is introduced into the consolidation atmosphere, but the preferred chemical reagent for such doping was not specified. Complex cladding structures were also reported (Berkey, 1984; Bhagatavula et al., 1984). In this case, very sharp profiles were realized, and a complex series of steps are believed to be practiced involving multiple deposition and consolidation to control the fluorine doping and prefent fluorine diffusion and penetration into the core, but process details were not revealed. Polarization maintaining fibers have also been achieved (Morrow and Schultz, 1983) using a stress rod design. Dispersion-shifted designs using either triangular cores or depressed cladding structures have been optimized and show excellent performance (Bhagavatula et al., 1984). Thus, a variety of fiber designs have been reported.

Another important aspect of the OVD process is the achievable deposition rates. In OVD, the deposition rate and efficiency increases with the square root

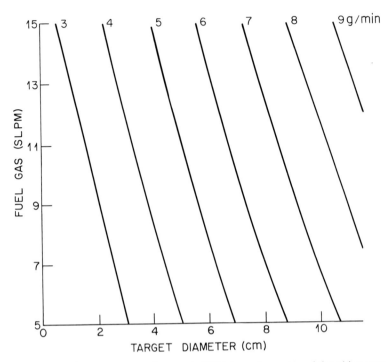

Fig. 4.18 Fuel gas flow versus target diameter in OVD for various achieved deposition rates, after Blankenship and Deneka (1982).

of the target size, for a fixed fuel gas flow, as shown in Fig. 4.18 after Blankenship and Deneka (1982). Rate also scales with fuel gas flow, where it is assumed the reactant flow increases proportionally to accomplish this. Average rates of 4 gm/min to produce 50 km boules have been reported with preform sizes of 1.8 kg and maximum deposition rates of ~ 10 g/min. Because the efficiency and rate scale with target size, the lowest efficiencies result for the most highly doped portion of the core. For high NA multimode fibers based on GeO_2, this implies very low incorporation efficiencies, which could impact process economics. No deposition rate limit has yet been reported for the OVD process, and research and development continue to report improvements in this area. A number of proprietary approaches to process scale up are underway with preform boules capable of yielding 250 km, and variations on the basic process, such as using multiple torch deposition, and secondary overcladding of a sintered preform among the approaches considered.

Fiber production has been described in terms of a balanced line (Sarkar and Schultz, 1983) with 100,000 km/year/line typical for 1983, which was projected to increase by a factor of 2.5 by 1985. However, no definition of the number of stations for the various process steps, which constitute a balanced line, were detailed. Excellent performance characteristics have been achieved for a wide variety of fiber designs (Morrow et al., 1985; DeWoestine, 1986; Deneka, 1985; Croft *et al.*, 1985).

Currently, the OVD process is in large-scale manufacture by Corning Glass Works and has also been implemented through ventures in a number of countries throughout the world. It continues to improve in its process economics and performance, and is a well-proven large-scale manufacturing technique.

4.4.4 Vapor Axial Deposition (VAD) Process

The development of VAD was the last vapor phase process to be announced (Izawa *et al.*, 1977b, c). It is similar to OVD in that it is based upon the flame hydrolysis process, using a flame to react halide reactants with oxygen and water to form oxide particles. Typically, an oxhydrogen flame is used in contrast to methane in OVD. It differs from OVD in that deposition occurs on the end of a rotating cylindrical bait rod to build up a soot boule with time. In this configuration, the soot form has no central hole or mandrel that must be removed, and many problems associated with thermal expansion mismatch and cracking during consolidation are avoided. Conceptually, it can be viewed as a continuous process, although as practiced, deposition, consolidation and fiber drawing are achieved in separate steps. Reviews (Niezeki *et al.*, 1985; Murata, 1985, 1986) have detailed the current understanding and recent progress using this technique. Some of the basic findings are summarized here.

The VAD apparatus consists of a rotating bait rod that is withdrawn from the deposition zone by means of a lead screw as shown in Fig. 4.19. Halide reagents such as $GeCl_4$ and $SiCl_4$ are entrained in a vapor and fed into a glass torch consisting of a series of concentric rings. The central ring typically contains both reagents, while in one configuration the second ring has $SiCl_4$. Fuel gases and shield gases are fed through additional rings. A large number of torch configurations have been used, and an optimal torch design is critical in determining the resultant composition and distribution of dopants depositing on the rotating boule surface. The torch is directed at the surface of the rotating bait rod or boule in order to collect the particles formed in the flame and to build up a controlled geometrical and compositional soot boule. In the case of multimode fiber fabrication, only the core composition is typically deposited along with a very small amount of primary cladding. For single-mode fiber fabrication, the core is deposited end on, while up to five cladding torches can be used to deposit cladding material from the side. After deposition, the resultant cylinder is consolidated, usually in two steps. First, the soot boule is treated in a halide containing atmosphere, using chlorine or thionyl chloride at low temperatures, typically below 1200°C (Sudo *et al.*, 1978; Edahiro *et al.*, 1979b; Moriyama

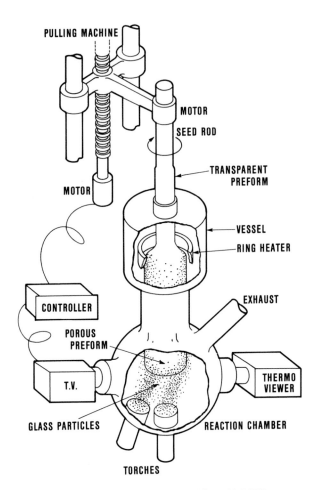

PULLING MACHINE

MOTOR

SEED ROD

TRANSPARENT
PREFORM

MOTOR

VESSEL

RING HEATER

EXHAUST

CONTROLLER

POROUS
PREFORM

T.V.

THERMO
VIEWER

GLASS PARTICLES

REACTION CHAMBER

TORCHES

Fig. 4.19   Schematic of VAD apparatus, after Izawa and Inagaki (1980).

et al., 1980; Hanawa *et al.*, 1980). The halide vapors react with hydroxyl contained in the relatively wet soot boule to form HCl, which is removed by a flowing gas stream. Next, further heating to temperatures in the range of 1500°C result in consolidation of the boule to a solid, bubble-free glass rod. Removal of Ge in the form of $GeCl_4$ is minimized by low or no chlorine during this step. The resultant rod is typically stretched or elongated, such as by using an oxyhydrogen torch, to reduce the diameter of the rod. In the most common process configuration, a commercial fused silica tube is then used to overclad the rod so that the desired clad/core ratio is achieved. During the stretching and over cladding operations, care must be taken to avoid OH contamination of the active region of the resultant fiber. Additional cladding can also be achieved by subsequent by subsequent soot deposition (Tamaru *et al.*, 1980). The resultant preform, which can be large enough to yield more than 100 km of fiber (Kawachi *et al.*, 1981a), is drawn into fiber.

It is important ot understand that in VAD, profile and compositional control is achieved spatially in a very different manner than the other processes, where the ratio of constituents in the gas stream are varied with time. The process as shown in Fig. 4.19 involves deposition in the lower zone of the apparatus, and simultaneous consolidation in the upper electric furance zone, when the boule has reached a sufficient length that it enters this zone. Although VAD is frequently depicted as occurring in this manner, there are many operational parameters that must be optimized to allow uniform preform fabrication. Deposition rate at the growing boule face must be matched to the required feed rate through the dehydration and consolidation zone to realize low OH, bubble-free preforms. As shrinkage of the boule begins, the pulling rate must be adjusted to maintain the required fixed distance between the torch and growing boule surface.

The deposition takes place within an enclosed glass container, which provides a protective atmosphere and allows for precise control of the reactant and exhaust flows. The burner projects through the wall of the container, and in the typical configuration, the angle of the torch relative to the boule surface can be adjusted. A controlled exhaust pressure is maintained by means of a pressure monitor coupled to a buffer tank, and such control is essential for minimizing extraneous soot build up in the chamber or upstream from the growing boule surface. A key factor in the fabrication process includes the maintenance of constant process conditions by regulation of raw material flows, exhaust gas flow, flame temperature and the resulting surface temperature of the growing boule. The rotation rate and position of the boule surface relative to the torch are also important. Fluctuations in this position can cause variations in the index profile along the axial direction, since GeO_2 incorporation can change. A computer-aided refractive index profile monitor (Chida *et al.*, 1983) has been developed to achieve the necessary control for high bandwidth multimode structures. A pyrometer is used to measure the temperature distribution at the surface of the boule. By knowing the relationship between this temperature distribution and the GeO_2 incorporation, and by taking into account profile changes during dehydration and consolidation, the process can be operated in a feedback mode to control growth conditions.

The design and the precise construction of the oxyhydrogen soot torch is essential to the development of the desired compositional distribution and thus profile in the preform. A numbr of descriptions of torch design have been proposed for VAD (Sudo *et al.*, 1981a, b; Ishida *et al.*, 1982; Imoto and Sumi, 1981). Reoptimization of the torch design is necessary as one wants to change average boule size and deposition rate. The exact preferred torch design typically has remained proprietary to the various manufacturers of fiber by VAD. The complexity of profile control, since it is impacted by the many other process variables, also factor into this approach, and burners that result in excellent properties have been devised.

Much study has been undertaken to understand the mechanisms by which the profile is actually achieved in VAD; the process is complex with the understanding by no means complete. Burner configuration, its relative position and angle to the growing boule face, the flow rates of the reagents and fuel gases, the shape of the growing face, and the flame and surface temperature are all important in determining the end profile, and small fluctuations in any of these variables can greatly influence profile control. Sudo *et al.* (1981a) investigated the effect of mixing in the flame on profile formation by varying the ratio of chemical flows through two adjacent orifices in the VAD torch, and demonstrated the importance of mixing on achieving profile control. Additional experiments demonstrated that surface temperature is also important in determining the resultant profile (Sudo, 1981b). Work reported by Sanada *et al.* (1979) examined the effect of varying the hydrogen to oxygen ratio in the burner on the resultant profile shape, as well as total GeO_2 incorporation. Introduction of Cl_2 into the flame also severely alters the resultant profile and index, (Kuwahara *et al.*, 1981).

Edahiro *et al.* (1980) first showed that the GeO_2 concentration in the deposited particles in VAD was strongly dependent on the substrate temperature during deposition, and further work by Sudo *et al.* (1981a, b) attempted to clarify the detailed mechanisms. Figure 4.20, after Edahiro *et al* (1980), shows the complex relationship of the GeO_2 concentration in the resultant boule as a function of both surface temperature as well as H_2 flow, which is related to flame temperature. For fixed-flame temperatures of $1200°C$ and $1300°C$, the GeO_2 concentration increased with increased substrate temperature, but at flame temperatures of $1400°C$, the GeO_2 concentration decreased when the substrate temperature exceeded $700°C$. Studies by Kawachi *et al.* (1980, 1981b) and Edahiro *et al.* (1980) have offered an explanation for this complex behavior. For the typical conditions used in VAD, GeO_2 particles are not thought to form in the flame, in contrast to SiO_2 particles which are. Vapor phase GeO_2 is carried to the substrate where cooler temperatures exist. When the substrate temperature is sufficiently low, crystalline GeO_2 solidifies on the surface. At higher temperatures above approximately $500°C$, however, amorphous GeO_2 condenses. This process will be markedly dependent on the specific flow conditions and temperature distribution. The oxidation to form GeO_2 may not be complete, particularly at high temperatures. On consolidation, homogenization of the GeO_2-SiO_2 takes place. Although the various studies have attempted to quantify such effects, the complexity of interrelated mechanisms makes it difficult to understand or

GeO$_2$ (Mol %)

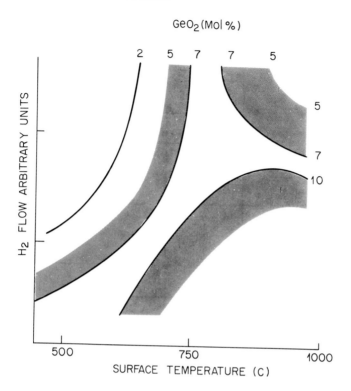

Fig. 4.20   Contour map of GeO$_2$ concentration in GeO$_2$-SiO$_2$ soot boules versus flame and substrate temperature, where the numbers represent the mole percent GeO$_2$, after Sudo *et al.* (1981).

predict. Experimental conditions have been identified that allow control over the index profile and composition in the resultant fiber, and sophisticated control schemes are necessary to operate the process in a feedback control mode. Nonetheless, excellent profile control and bandwidth have been achieved.

Dehydration of VAD soot boules to achieve very low OH levels has been another aspect of processing that has been studied and reported extensively. In VAD, as in OVD, formation of the soot boule takes place in an atmosphere of H$_2$O resulting from the flame combustion products, and hundreds of ppm hydroxl ion are incorporated into the soot boule. When the boule is heated to consolidation temperatures of ~ 1500°C, if a dry inert gas is used, a considerable portion of the hydroxl is removed. Early VAD fibers, which did not receive special dehydration treatment, contained 10–100 ppm OH in the resultant fiber. Ultimately, total elimination of OH was reported (Hanawa *et al.*, 1980; Moriyama *et al.*, 1980). The dehydration step in VAD takes advantage of the same chemical equilibrium described for MCVD. Because this step is separate from deposition, very low O_2 pressures as well as a Cl_2 atmosphere can be used to control the final OH concentration.

Another aspect of concern during dehydration and consolidation is the potential for increased attenuation and profile distortion due to deleterious effects associated with Cl_2. Ishida *et al.* (1982) reported that if partial pressures of chlorine above 10^2 pascals were used, the resultant loss was significantly increased. The increase in wavelength independent loss is presumably due to the formation of pores, which act as Mie scattering sites. Thus, great care must be taken to identify dehydration conditions that achieve low OH, low loss, and do not radically alter the resultant refractive index profile in a manner that can be compensated for, since GeO_2 dopant removal can occur during the high temperature dehydration/sintering step (Kuwahara et al., 1981).

A number of studies have been directed at high rate multimode VAD, with deposition rates as high as 4.5 g/min reported for conventional torches (Niizeki, 1982). Multiflame VAD torches (Suda *et al.*, 1984, 1985; Kawazoe *et al.*, 1985; Miyamoto *et al.*, 1985; Danzuka *et al.*, 1985) as well as chemical modification of the flame such as by using $SiHCl_3$ (Danzuka *et al.*, 1984; Suda *et al.*, 1985) have been used to further increase deposition rates to 6.5 g/min. While optimization of the torch flow characteristics and its associated impact on deposition can improve deposition rate, the detailed understanding of the physics and chemistry of optimal torch design in flame process is complex.

Single-mode fiber fabrication, although in many ways similar to multimode VAD, has some important differences. The basic deposition and consolidation chambers are the same. Similar controls must be used to achieve stable flames, profile control, and boule sizes. A modified burner, capable in some cases of producing small diameter core boules, is directed at the rotating bait to deposit the core material of the desired single-mode composition and profile. After the core boule has stabilized and grown to some finite length, a cladding torch (or torches) is directed from the side at the boule to deposit material, as shown in Fig. 4.21, after Miya *et al.* (1984). Care must be taken to minimize interference among the torches. In addition, soot density control of the deposited material is important to avoid subsequent cracking, especially during sintering, in the resultant soot form. The cladding torches are stationary relative to the moving boule. Thus, a finite and constant diameter is achieved at any vertical position, and at some distance from the growing core surface, the final diameter of the soot boule is achieved. This distance is analogous to the entry taper in MCVD, and thus represents non-usable material. The more torches that are added to build up cladding, the longer this length becomes. Typically, GeO_2-SiO_2 core and SiO_2 cladding are deposited. For the single-mode case, sufficient cladding must be deposited in order to avoid OH contamination due to water introduced on stretching or overcladding, and to avoid loss increases due to tubing or surface irregularities.

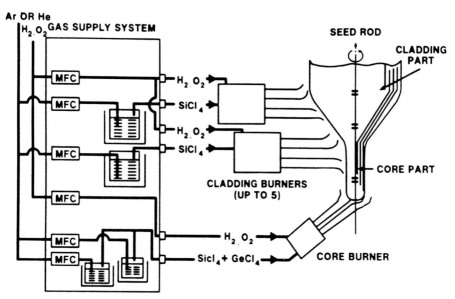

Fig. 4.21 Single-mode VAD preform fabrication, after Miya *et al.* (1984).

Deposition rates for single-mode fibers of 6 g/min have been reported (Shioda *et al.*, 1983). In this case, five cladding deposition burners were used. Deposition rate was found to increase with target size, in similar fashion as in OVD. The efficiency of deposition for each burner was maximized by varying the fuel gas flow, raw material flow and burner geometry. In this manner, totally synthetic VAD performs can be made without the use of an overcladding tube.

VAD has also recently reported success in making single-mode fiber structures doped with fluorine (Miyamoto *et al.*, 1983; Kanamori *et al.*, 1983; Kyoto *et al.*, 1984; Kokura *et al.*, 1984). Doping during both deposition and consolidation has been examined with a variety of freon sources, as well as SF_6. Deposition rate and efficiency decrease when used in the flame as contrasted to using a fluorine during consolidation. Using such F-doping, both GeO_2 core, depressed cladding designs and pure SiO_2 core, F-doped cladding structures have been made.

As reviewed by Murata (1985, 1986) and Niizeki *et al.* (1985), tremendous understanding of the VAD process has evolved and its suitability for large-scale manufacture of a variety of multimode and single-mode designs has been demonstrated. For single-mode fibers, it can be practiced as a hybrid process in a sense, with the core and inner cladding fabricated by VAD, and the outer cladding deposited in a separate step from the side, or supplied via a tube. Very large preforms, OH free fiber, and excellent losses have been shown. It remains the process of choice for Japanese manufacturers, and is being practiced in other parts of the world through ventures.

4.4.5 Process Comparisons

The four processes described have all demonstrated basically equivalent performances for both multimode and single-mode standard designs. While champion results have been achieved for a given property or feature, there is no set of process or performance advantages that has emerged to clearly favor one process over the next. Table 4.1 compares and contrasts some of the essential process features. All use electronic grade chemicals for vapor delivery of dopants to achieve low transition metal impurities, but the details of OH control are markedly different. All require chemical delivery systems with controlled chemical ratios. All have a common drawing step requiring similar process control. In all cases, some form of scrubbing and collection of the by-products of the chemical reactions is necessary. Oxyhydrogen fuel is necessary for deposition and collapse in MCVD, for collapse in PCVD and for deposition in VAD; OVD typically uses methane. Electrical energy is required to drive the plasma source in PCVD and the plasma enhanced MCVD process, to run the high temperature sintering furnaces in OVD and VAD, and for the fiber drawing heat source. MCVD requires a lathe and associated traverse and flame control apparatus; PCVD requires sophisticated equipment for rapidly traversing the microwave resonator, and pumps to maintain a vacuum; VAD requires a deposition apparatus with controlled pressure and feed mechanism coupled with carefully designed torches and sintering chambers; OVD requires traverse control, carefully designed torches and sintering chambers, and some kind of enclosed clean environment during deposition. MCVD involves the simplest apparatus that is readily available commercially, while the other processes in general utilize specialized systems developed in a proprietary manner by the practitioners, and thus are more difficult to implement without fabrication experience.

All processes have reported high NA multimode designs as well as matched and depressed index cladding single-mode structures, and each has perfected fabrication technology to achieve good control. The highest overall incorporation efficiencies have been reported for the plasma processes, with PCVD able to incorporate high amounts of fluorine. Since the inside processes typically rely on a silica tube, all fluorine doped claddings such as reported for VAD are more difficult, although they can be achieved using an F-doped tube. While F-doping has been achieved in VAD and OVD, standard designs have avoided its use. In soot processes, however, F-doping of the cladding during consolidation without diffusion into the core is difficult, but can be accomplished by careful control or with multiple deposition and consolidation steps. Complex F-doped profiles are more difficult to make.

Since MCVD and PCVD are inside processes, scale-up is limited by the finite tube size, while in principle, OVD and VAD deposition can continue indefinitely if deposition and subsequent sintering can be controlled. Overcladding

TABLE 4.1
Process Comparisons for Major High Silica Fiber Preform Fabrication

Feature	MCVD	PCVD	OVD	VAD
Basic Approach	Deposition of high purity material inside tube, which becomes outermost cladding	Deposition of high purity material inside tube, which becomes outermost cladding	Deposition of core; then cladding material on removable mandrel	End on deposition of core material on bait rod; deposition of clad from side
Major Process Steps	Tube set up; simultaneous deposition and sintering of porous material; tube collapse; draw	Tube set up; direct deposition of glass; tube collapse, draw	Mandrel set up; deposition of soot; mandrel removal; dehydration and sintering, sometimes with hole closure; draw	Bait rod set up; deposition of soot; dehydration; high temperature consolidation; sometimes stretch; overclad with tube or more soot, which must be consolidated; draw
Reaction Mechanism	High temperature homogeneous gas phase oxidation of chlorides to form particles	Microwave plasma initiated oxidation of chlorides	Flame hydrolysis of chlorides to form oxide particles, with methane fuel	Flame hydrolysis of chlorides to form oxide particles, using O_2/H_2 flame
Deposition Mechanism	Thermophoretic deposition of particles on tube wall	Heterogeneous surface nucleation at tube wall	Thermophoretic deposition of particles on mandrel	Thermophoretic deposition of particles on end of bait rod or boule

TABLE 4.1
(*Continued.*)

Feature	MCVD	PCVD	OVD	VAD
Consolidation	Viscous sintering of particles simultaneous with deposition of layer	No consolidation necessary; substrate temperature control to assure deposition of vitreous layer	Separate viscous sintering step of soot form after deposition	Separate viscous sintering step of soot form after deposition
Refractive Index and Profile Control	Controlled chemical composition of each layer by controlled flow ratios; controlled collapse to minimize central index dip.	Controlled chemical composition of each layer by controlled flow ratios; controlled collapse to minimize central index dip.	Controlled chemical composition of each layer by controlled flow ratios; controlled consolidation to minimize Ge removal; center index dip minimized during consolidation or draw	Radical variation of index complex function of diffusion in flame and condensaton of GeO_2; controlled by control of variables such as fuel mixture, torch design, substrate temperature, torch position, chemical flows; features no index dip
Transition Metal Impurity Control	Vapor phase delivery of electronic grade chloride dopants; tube prevents ingress of contaminants from atmosphere	Vapor phase delivery of electronic grade chloride dopants; tube prevents ingress of contaminants from atmosphere	Vapor phase delivery of electronic grade chloride dopants; clean deposition environment required	Vapor phase delivery of electronic grade chloride dopants; deposition inside glass containment vessel
OH Control	Low hydrogen levels in starting materials; chlorine atmosphere during deposition and collapse; leak tight delivery	Very low hydrogen levels in starting materials; F found to reduce OH incorporation; leak tight delivery	Chlorine atmosphere during sintering to remove OH from soot	Low temperature dehydration in Cl_2 atmosphere to remove OH prior to consolidation
GeO_2 doping	Determined by high temperature particle growth and equilibrium considerations	Determined by low pressure chemical reaction and deposition position relative to	Determined by flame chemistry, surface temperature of boule and condensation; can	Determined by flame chemistry, surface temperature of boule and condensation; can

TABLE 4.1
(Continued.)

Feature	MCVD	PCVD	OVD	VAD
F-doping	Determined by high temperature equilibria and consolidation temperature. SiF$_4$ preferred dopant.	Determined by low pressure chemical reaction. C$_2$F$_6$ preferred dopant. Highest F incorporation level of all processes.	Low efficiency incorporation during hydrolysis; F-atmosphere during consolidation typically used.	Low efficiency incorporation during hydrolysis; F-atmosphere during consolidation typically used.
Scale up	Larger tubes and higher deposition rates for larger preforms; overcladding to further increase size	Larger tubes and higher deposition rates with increased microwave power to make larger preforms; overcladding to further increase size	Scaled-up torches; deposition rate increases with increased target size results in larger preform; multiple torches possible, can add cladding in secondary step	Scaled-up torches, for core and for cladding; multiple torches for cladding; larger preform sizes; can add cladding in secondary step

techniques with the former, however, are capable of achieving very large preforms (~ 200 km) similar to those under development with VAD and OVD. Tube versus alternative high rate overcladding methods are under investigation by a number of workers, and to the degree process information is available, all seem to be competitive techniques.

Further process work continues to address performance, advanced designs and scale-up of fabrication. Enhanced designs such as lower loss silica core fibers or improved radiation performance fiber have been demonstrated. Advanced fabrication techniques are addressing the formation and control of defects to further improve performance. It is expected that no dominant process will emerge, but that the individual practitioners will continue to advance the specific implementation of their process to maintain competitiveness and/or to enhance or optimize specific performance characteristics.

4.4.6 Sol-Gel Process for Silica Preparation

The general principal of glass formation via a sol-gel route involves introducing glass precursor chemicals into a solution that undergoes chemical reactions to form a gel; this gel can then be dried and after solvent is removed, a solid glass body results. The sol-gel process for making high-silica glass and fibers for lightwave applications has received increasing attention for a number of reasons: it permits lower temperature processing of silica; it has the potential for very high purity; it allows the casting and shaping of bodies; and it can overcome the deposition rate limitations of vapor phase processing of silica. Two basic approaches have been used to prepare the glasses: (1) polymerization of alkoxides to form a monolithic gel (Fleming, 1976; Fleming et al., 1976; Susa et al., 1982; Harmer et al., 1982; Puyane et al., 1982; Sudo et al., 1983; Matsuyama et al., 1984), or (2) gelation of aqueous colloidal sols (Rabinovich et al., 1982, 1983, 1984; Johnson et al., 1983; Wood et al., 1983; Scherer and Luong, 1984). The gel can be directly formed into a body and dried and sintered to a final shape, or glass particles can be formed by drying and sintering, then used to make glass. While this approach is still in the earliest stages, a number of researchers have demonstrated its potential, and it holds great promise for future technology. One of the major obstacles to this technique has been cracking on drying, as well as the tendency to form bubbles at very high temperatures.

Very early work on gel-derived silica glasses for lightguides was reported by Fleming et al. (1976), where doped powders were formed via sol-gel synthesis, then fed through a plasma torch to fabricate glass boules. By control of particle size, volatile dopants such as GeO_2 were able to be efficiently incorporated into the resultant glass despite the high temperature processing. In an investigation directed at developing high rate deposition of multimode fiber (Sudo et al.,

1983), a GeO_2-SiO_2 gel was formed by hydrolyzing $Si(C_2H_5O)_4$ and $Ge(C_2H_5O)_4$ alkoxides, then dried and sintered to form 100–300 μm particles. The particles were fed through an oxyhydrogen flame in a configuration similar to VAD where they were deposited and melted to form a boule at a deposition rate of 5 g/min. Using a rod-in-tube technique, fiber was drawn and losses of 7 dB/km at 0.8 μm was achieved.

A second approach involved using Ge-Si alkoxide sols to form layers on the inside of a SiO_2 tube (Harmer et al., 1982; Puyane et al., 1982). Each layer was gelled, dried and cured by a fast thermal treatment. After deposition and sintering, the tube was collapsed and drawn into fiber with losses of 22 dB/km at 0.85 μm.

Silica core rods for optical fibers were formed by hydrolyzing $Si(OCH_3)_4$ in alcohol to form a gel that was cast into a mold, slowly dried (> 1 week) and sintered (Susa et al., 1982). This was then inserted into a silica tube with a lower index borosilicate deposit on its inside and a fiber was drawn by the rod-in-tube method. Losses as low as 6 dB/km were achieved at 0.85 μm. Using a similar approach, Shibata and Nakahara (1985) made SiO_2 rods, where they carefully controlled the gel particle size to overcome cracking problems and sintered in a Cl_2 atmosphere to remove OH. They then used VAD to deposit a F-doped cladding and the resultant fiber had losses as low as 1.8 dB/km at 1.6 μm with 0.1 ppm OH (~ 5 dB/km peak height at 1.39 μm).

Other workers have focused on using colloidal approaches to form high silica bodies (Rabinovich et al., 1982, 1983; Johnson et al., 1980). Fumed silica was dispersed in a controlled pH water solution where gel is formed by hydrogen bonding. In contrast to alkoxide gel, colloidal gels have less shrinkage on drying and thus can more readily be dried without cracking. To further minimize cracking problems, a double dispersion technique was reported in which the resultant gels exhibited very low shrinkage on drying, attributed to formation of an interlocking aggregate structure with large pores, which provide low imped-ance diffusion paths for the removal of water during drying. After casting and drying, the gel bodies can be sintered to transparent bodies at temperatures in the range of 1300–1500°C, and also dehydrated and/or fluorinated using a controlled atmosphere. Large silica bodies have been reported using this tech-nique, demonstrating the ability to overcome the cracking problems and exces-sive drying times required for alkoxide gels. Fluorine doped substrate tubes were also reported (MacChesney et al., 1985) where dehydration and fluorination of the cast gel was used to achieve a depressed index difference relative to silica, Δ^-, of 0.4%. The tube was then used to make a silica core, depressed cladding structure with a depressed deposited clad/core ratio of 4.5. Losses of 0.28 dB/km at 1.55 μm were reported for this structure; if it had been made in a conventional silica tube, four times the deposited cladding would have been required to avoid leaky-mode losses.

While the sol-gel process has not yet achieved fibers with the performance of vapor phase processes, it has already demonstrated critical milestones of improved loss as well as its efficacy for making large silica bodies. The consolidation and sintering step can take advantage of the same technology already developed for soot-derived bodies, and thus offers the potential for low OH, low-loss glasses without the rate limitations of vapor phase processing.

4.5 FIBER DRAWING AND COATING

The second major step in the fabrication of lightguide is fiber drawing and it is common to all the major lightguide fabrication processes. (For more detailed review, see Blyler and DiMarcello, 1980; Chapter 9, Miller and Chynoweth, 1979; DiMarcello et al., 1985.) Fiber drawing and the associated fiber coating step is critical in determining the resultant optical, dimensional, and mechanical properties of the final lightguide. Significant progress has been made in highly controlled fiber drawing techniques with the advent of lightguides for optical communication systems, where very stringent requirements must be met.

Figure 4.22 shows a schematic of some of the essential features of a state-of-the-art fiber drawing facility, for drawing silica-based fibers from glass preforms. A tall main frame is used for supporting the associated equipment and control instrumentation. The precursor glass preform is fed into a high temperature furnace by means of a preform feed mechanism. Alignment and centering of the preform relative to the furnace is critical, and can be accomplished by manual or automated alignment techniques. For silica fiber drawing, glass viscosity considerations require draw temperatures in the range of 1950–2300°C. A variety of heat sources have been examined to achieve these temperatures (for review see Miller and Chynoweth, 1979; DiMarcello et al., 1985), with graphite resistance or inductively heated zirconia-type furnaces most commonly used. The tip of the preform softens as it is fed into the high temperature furnace and both gravity and an applied tensile force causes the glass to "neck down" to a small diameter fiber. The shape of the neck-down region is determined by a variety of factors, including the thermal gradient in the furnace and the draw forces (Paek and Runk, 1978). A controlled tensile force is sustained by using a fiber pulling capstan or some other source of tension. The preform feed rate and capstan rotation rate determine the draw-down ratio from preform to fiber. Typically, preforms ranging from 10–70 mm are drawn down to fiber in the 100–225 μm range, with 125 μm the most commonly used fiber outer diameter. Fiber diameter control is most commonly achieved by varying the draw speed while feeding the preforms at a fixed rate through a constant temperature heat source. Typically, the fiber diameter is measured and controlled by means of a contactless technique based on shadow graphs or forward- or back-scattered signals. The diameter data is collected at a point just below the furnace where

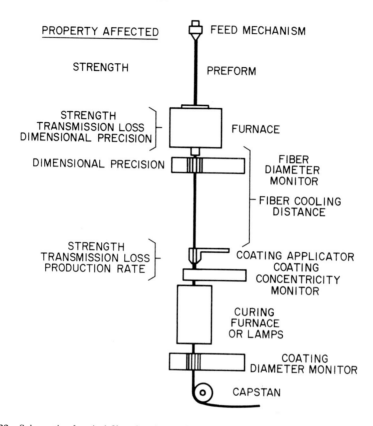

Fig. 4.22 Schematic of optical fiber drawing and coating apparatus.

the final drawn down diameter has been achieved. Smithgall *et al.* (1977) has achieved 0.1 μm precision and accuracy in diameter measurement over the range of 50 to 150 μm by means of a high speed forward scattering diameter measurement technique that samples at a rate of 1 KHz. Control of fiber diameter is achieved by means of a feedback control loop. Excellent fiber diameter control and response, with mean diameter variations < 0.1 μm and standard deviations of 0.25 μm, have been achieved by adjusting the draw velocity with a four element control loop (Smithgall, 1979). Such control is essential to minimize the effect of fiber diameter fluctuations on such properties as microbending sensitivity and splicing loss.

In-line coating application is a key step in silica lightguide drawing. The coating plays two essential roles: it provides mechanical protection of the fiber surface and buffers the fiber from externally induced microbending effects. Typically, such coatings are applied as liquids by means of coating applicator, which the fiber passes through. Rapid solidification is achieved by on-line thermal or ultraviolet curing; the coating must solidify before contact with any associated fiber drawing mechanism that might damage the surface of the glass fiber. Coating diameter control and centering of the fiber within the coating are essential, and frequently, coating geometry is monitored and controlled. Extruded jackets such as nylon are applied off-line. A catenary loop is often used to synchronize the speed of the fiber take-up mechanism with the draw speed: this allows the fiber to be wound at low tension independent of the applied draw tension. Frequently, the fiber draw environment is controlled by use of filtered clean air or clean rooms in order to achieve high strength. Prooftesting to test the tensile strength of fiber can be done in-line or off-line.

The focus on fiber drawing research and development has been to realize fiber drawing and coating procedures that eliminate or minimize the introduction of flaws during all stages of fiber processing, which might degrade the initial strength. DiMarcello et al. (1979) summarized and demonstrated the conditions required for the fabrication of long lengths of high-strength fiber. These included the use of high quality tubes or substrate materials without any inclusions or defects; preparation and handling of the glass precursor in a clean environment to avoid surface damage; a clean heat source, which operates in an environment free of contaminants; relatively high drawing temperatures; uniformly applied coating whose application does not damage the fiber surface; and drawing and coating in a clean, particle-free environment.

Typically, strengths on the order of 50–100 ksi (0.35–0.7 GPa) are deemed suitable for many lightwave applications. A number of high strength applications for lightguides have evolved, however, such as undersea cable systems, missile guidance systems, and variety of military harsh in-service environment applications. For such use, strengths of 200–300 ksi (1.4–2.1 GPa) are necessary coupled with very low losses and dimensional precision. Recent results have shown that such properties can be simultaneously realized with high yield (Brownlow et al., 1982; Sakaguchi et al., 1983) and that high strength, low loss fusion splicing (Krause et al., 1981) can be used to assemble long lengths of high performance fiber (Runge et al., 1982).

Other properties of concern that can be effected during fiber drawing are loss and dimensions. Drawing induced losses have been reported involving absorption, scattering and microbending. Although there have been many studies to examine the effect of draw speed, temperature, and tension on both loss and strength in a variety of fiber designs and compositions, the detailed mechanisms of such loss increases are only qualitatively understood. However, procedures that reproducibly give low loss with high strength have evolved.

Coating materials and their uniform application play a critical role in fiber loss and strength. The primary function of the coating is to protect the lightguide from any external effects, which could reduce the intrinsic strength by abrading or contaminating the fiber surface. The coating should not contain any particles that might damage the surface (Blyler and DiMarcello, 1980). Coatings also reduce the susceptibility of a given fiber design to microbending losses (Gardner, 1975; Gloge, 1975). Typically, the choice of a fiber coating system considers such material properties as the glass transition temperature, elastic modulus, stability, toughness, adhesion, moisture resistance, index of refraction, and strippability (DiMarcello et al., 1985). Dual structures combining a low modulus primary buffer surrounded by a high modulus rigid structure are frequently used to optimize the coating structure (Gloge, 1975; Naruse et al., 1977; Santana et al., 1982). More recently, fiber coating materials that do not evolve hydrogen over their lifetime have become important for reliability considerations.

As lightguide process technology has been introduced into large-scale manufacture, fiber draw speed is an important step in determining the process economics. Paek and Schroeder (1981) reported the important parameters for high speed drawing. These included minimizing the shear forces in the coating applicator and control of the fiber temperature entering the application to avoid coating minicus collapse. As recently reviewed by Paek (1986), much progress has been reported in high speed drawing and coating, with fiber fabrication speeds in the range of 10–12 m/sec achieved (Wagatsuma et al., 1983; Paek and Schroeder, 1984) coupled with well centered coatings, high strength, and low loss (Paek and Schroeder, 1984).

Thus, tremendous progress has been made in fiber drawing and coating technology, ranging from understanding the basic processes that impact fiber strength, diameter control and optical attenuation, to implementation of advanced control technology to realize high throughput with excellent yield of controlled property fiber. Further research and development continues to address very high strength, long length fiber; formulation and application of new organic coating materials, which allow either higher or low temperature use of fiber without optical or mechanical degradation of properties; hermetic coatings for improved fatigue resistance in harsh environments; and special coatings such as required for fiber sensors and other new applications.

4.6 POLARIZATION MAINTAINING FIBER FABRICATION

Polarization maintaining fibers refer to a class of single-mode fibers that can preserve a state of linear polarization over a given length, typically long lengths. An ideal circular single-mode fiber can support two independent degenerate modes of orthogonal polarization, HE_x and HE_y, which propagate independently with the electric field of propagating light being a simple linear superposition of the two polarization eigenmodes. However, "real" single-mode fibers have internal imperfections such as slightly elliptical cores, and can experience external perturbations, which cause the two modes to propagate with different phase velocities, B_x and B_y, related to modal birefringence, B,

$$B = \frac{\lambda(B_x - B_y)}{2\pi}. \tag{4.26}$$

This is further complicated by perturbations that can cause coupling between the two modes. These effects lead to polarization mode dispersion and instabilities in the output polarization. Excellent reviews of the origin and control of such effects have been made (Okoshi, 1981; Kaminow, 1981; Payne et al., 1982; Rashleigh, 1983; Noda et al., 1986).

Two basic approaches have been utilized to address such effects: low birefringence and high birefringence fibers. For low birefringence fibers, controlled fabrication techniques are utilized to make highly circular cores with minimum stress in order to minimize the difference in propagation constants (Schneider et al., 1978; Norman et al., 1979). Another approach to further decreasing birefringence involves spinning the preform during drawing to average the residual birefringence (Barlow et al., 1981). While conventional single-mode fibers exhibit very low modal birefringence, the output polarization is typically unstable due to environmental and external perturbations. These changes, however, typically occur with very slow time constants, and can be compensated for by a polarization controller at the output end of the fiber for applications such as coherent detection, which requires such control.

A second approach, which has resulted in a variety of specialized fabrication techniques, is to deliberately make fibers with sufficiently high birefringence so that there is minimal coupling between the two linear modes, even when subjected to external perturbations. In the extreme, fibers that support only one of the two fundamental modes can be designed and made, termed single polarization fibers or polarizers (Okoshi et al., 1982; Simpson et al., 1983; Snyder and Ruhl, 1983; Varnham et al., 1983).

Birefringence in fibers can be produced by two basic mechanisms: geometrical shape birefringence and stress induced birefringence. More insight into these mechanisms can be obtained from the cited review articles. This in turn has led to three basic fabrication approaches, which are sometimes combined: fabricating fibers with elliptical cores or non-symmetric cores; non-symmetric, high stress claddings; and side pit or side tunnel structures that utilize axially non-symmetric refractive index distributions having two pits on either side of the core. A number of fabrication approaches have been taken to realize such approaches, as summarized in Fig. 4.23 after Noda et al. (1986).

The earliest approach, depicted in (a) involved grinding flats on an MCVD substrate tube prior to deposition, depositing cladding and core, then allowing surface tension, during the collapse phase of the process, to round the outer

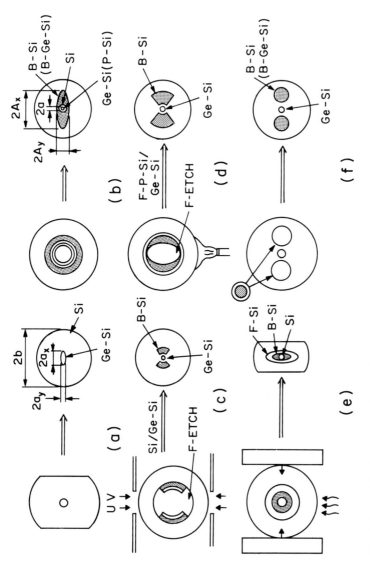

Fig. 4.23   Fabrication processes for high birefringence fibers, after Noda *et al.* (1986).

surface of the preform (Stolen *et al.*, 1978). Depending on the viscosity characteristics of the deposited core and cladding material relative to the outer silica tube,
elliptical or circular cores, surrounded by elliptical claddings are achieved. A
high viscosity core material such as SiO_2 results in a circular core. A variation
on this technique, called the "exposed cladding" technique (Ramaswamy *et al.*,
1978; Kaminow *et al.*, 1979), involved grinding flats or slots on the circular
MCVD preform after collapse. This also allowed fibers approaching the idealized "slab" geometry to be fabricated (Kaminow and Ramaswamy, 1979).
Dyott and Schrank (1982) fabricated elliptical core, D-shaped fibers for exposing the guiding region for coupling, as well as for alignment purposes; in this
case they ground a flat on one side of the MCVD preform, then drew fiber.

A second fabrication approach involves reduced pressures or vacuum collapse
during MCVD, illustrated in Fig. 4.23b. First proposed by Ramaswamy *et al.*
(1979) and extended by Katsuyama *et al.* (1983, 1984) to define the relationship
of processing conditions, viscosity and composition on the resultant structure, a
variety of optimized structures have made with both circular and elliptical cores
and high stress inner claddings. Katsuyama *et al.* (1981) was the first to report
the addition of an intermediate layer to avoid losses at long wavelengths due to
B-O absorption.

Yet another approach to forming high stress asymmetric structures involved
etching techniques, again using the MCVD process. In a lithographic technique
(Stolen *et al.*, 1982), a borosilicate cladding was deposited, followed by a resist
that was selectively exposed and developed to allow selective etching of the
borosilicate. Next, a silica cladding and GeO_2–SiO_2 core were deposited and
collapsed to form the asymmetric structure depicted in Fig. 4.23c. Another more
practical technique (Birch *et al.*, 1982) utilized high temperature gas phase
etching. In this case, cladding material was deposited, then the lathe rotation
was stopped and two burners, on opposite sides of the tube, were traversed while
passing a chemical F_2 containing etchant gas through tube to selectively remove
material. Additional material was then deposited and the composite collapsed to
form the "bow-tie" structure shown in Fig. 4.23d. A rod-in-tube technique
reported by Hosaka *et al.* (1981) and Shibata *et al.* (1983), where boron-doped
rods were used, produced similar structures. Al_2O_3 high expansion stress rods
using the rod-in-tube technique have also led to bow-tie structures (Morrone
et al., 1984).

A preform deformation technique was reported (Stolen *et al.*, 1984) and is
depicted in Fig. 4.23e. Cladding and core compositions were deposited by
conventional MCVD and collapsed to a preform rod, which was then deformed
by local heating and squeezing from the sides to form a rectangular (elliptical)
preform. By using a high viscosity core material, and fluid high stress cladding
composition, an asymmetric stress cladding was formed while maintaining a
circular core. The fiber is drawn at a temperature low enough to prevent

rounding during drawing, resulting in a fiber whose shape facilitates location of the principal axes, and increases the resistance to polarization breakdown, since the fibers tend to bend only along one axis. Recently, VAD PANDA fibers of similar shape have been reported, fabricated by grinding off the preform cladding, then drawing at low temperatures (Okamoto *et al.*, 1985). The last method, depicted in Fig. 4.23f, has been described as the "pit-in-jacket" method, as proposed and reviewed by Sasaki *et al.* (1984). Because of the excellent polarization and loss results, it has been called the PANDA fiber (polarization maintaining and absorption reducing). It is a hybrid MCVD-VAD process in which a VAD single-mode preform is first fabricated, then side holes are drilled into which stress applying rods, made by MCVD, are inserted and the composite is drawn into fiber. This approach allows very symmetric and uniform structures to be realized, critical for high birefringence in long lengths. If the holes are not filled and positive pressure control during drawing is used, the side tunnel fiber proposed by Okoshi *et al.* (1982) can be made. For some applications, preservation of circular birefringence is important (Payne *et al.*, 1982). A novel helical core circularly birefringent fiber structure was reported by Varnham *et al.* (1985) using a rod and tube technique continued with spinning during draw.

Polarization maintaining fibers can be used in a variety of applications ranging from short length components to longer lengths used in sensors or other applications, where the phase is important. Dramatic progress has been made in realizing long length, low loss, and excellent polarization characteristics. It is expected that processing approaches will play a critical role in the availability and economics of such fibers.

4.7 POLYMERIC OPTICAL FIBERS

Polymeric or plastic optical fibers (POF), while typically exhibiting much high transmission losses than silica-based fibers, have specialized uses in communications systems. Directed at short wavelengths, short distance/low bandwidth applications, the high numerical aperatures of these step index structures, as well as their large cores allow easy coupling, splicing and connectorization.

An excellent review of these fibers has been given by Glen (1986). Plastic fiber development work began in the late sixties, and was directed at fibers for display and automotive purposes. Hager *et al.* (1967) reported on fibers with a poly-methyl methacrylate (PMMA) core, clad with a coextruded fluorinated acrylic polymer, with NA = 0.54 and losses on the order of 500 dB/km. These fibers were commercialized as Crofon[TM] by Dupont, and lower losses of 300 dB/km were demonstrated (Schleinitz, 1977) by using deuterated PMMA. In the early 1970s, a continuous casting process was developed for PMMA, and the Eska[TM] family of fibers with losses as low as 110 dB/km were marketed by Mitsubishi.

Extensive work has been done by Kaino *et al.* (1981, 1982, 1984) to address loss mechanisms in PMMA and polystyrene core fibers, and to develop fabrication approaches, which would allow achievement of even lower loss for communications applications.

The basic loss mechanisms in polymeric optical fibers have been recently reviewed (Kaino, 1985; Glen, 1986). As with glasses, both intrinsic and extrinsic effects must be considered. In general, amorphous polymers have been considered, and three basic systems investigated for the core material: PMMA, polystyrene (PS) and polycarbonate polymers. Rayleigh scattering occurs due to density fluctuations and fluctuations due to anisotropic structure of the polymer. Absorption is due to the long wavelength tail of the intrinsic absorption, and is generally quite small in 0.5–0.9 μm range where such fibers might operate. Vibrational absorptions in the infrared result in higher order harmonics, with basic building block overtone bands of the C—H bond having a dominant effect. A number of extrinsic mechanisms are critical to the realizable loss. These include absorptions due to transition metal and organic contaminants, as well as OH overtone bands, which overlap those of the C—H bands; scattering due to dust contamination and microvoids; fluctuations in core diameter; orientational birefringence and core-cladding boundary imperfections.

The very best results have been achieved by Kaino and coworkers, as summarized in a paper that discussed preparation of such fibers (Kaino *et al.*, 1984). They utilized a completely enclosed system for purification, polymerization and fiber extrusion to fabricate highly pure polymers. Distillation was used to purify the monomers until no visible scattering was detected at He–Ne wavelengths. They then polymerized the monomer thermally, and coextruded fiber with a fluorinated acrylic polymer cladding. With such an approach, they were able to achieve minimum losses of 55 dB/km at 0.56 μm for PMMA and 114 dB/km at 0.67 μm for polystyrene. To lower the contribution of C—H absorption to the losses, they used the approach of Schleinitz (1977) to perdeuterate PMMA, creating shifted C-D absorptions, and achieved 20 dB/km at 0.68 μm; they predict achievable losses as low as 10 dB/km if structural imperfection losses can be overcome.

The progress in perfecting high quality plastic optical fibers has been dramatic, and additional work is focusing on their environmental performance, particularly at higher temperatures. Although low losses have been achieved, absorbed water can increase these losses, particularly at longer wavelengths. Chemical durability and mechanical properties must also be improved for many proposed applications. Much of the recent work has been addressing such concerns as summarized in recent reviews (Glen, 1986; Blyler, 1986). The very lowest loss fibers are unlikely to ever be as low cost as high silica fibers because of the specialized processing required. Progress continues, however, and such fibers are already implemented in a number of short distance data link applications, where their larger cores and ease of connection, coupled with availability of low cost LED devices allow economical, low performance systems to be implemented.

4.8 ULTRA-LOW-LOSS / MID-INFRARED FIBERS

The incentive for seeking very low-loss fiber materials begins with the birth of the modern day lightwave communications revolution, stimulated by the invention of the laser and the search for a suitable transmission medium. While a number of technical milestones focused much of the effort on silica-based glasses, any number of amorphous, crystalline and even liquid low-loss materials have been considered. High silica glass fibers, as described in the previous sections, became the preferred transmission media for telecommunication systems, because of the dramatic demonstration of very low loss and other optimal properties, coupled with the rapid evolution of practical fabrication technology. In 1979, Miya *et al.* achieved fibers with losses thought to be at the theoretical limit for silica. Nevertheless, many researchers continued to ask the question if even lower losses than silica could be achieved. A number of workers reported silica-based glasses with lower scattering than pure silica (Schroeder *et al.*, 1973; Pinnow *et al.*, 1975; Tynes *et al.*, 1975). Glass forming BeF_2 and $ZnCl_2$ were predicted to have losses of 10^{-2} dB/km at 1.05 μm and 10^{-3} dB/km around 3.5–4.0 μm, respectively (Van Uitert and Wemple, 1978). Goodman (1978) considered halides, chalcogenides and heavy metal oxides based on their longer wavelength absorption edge and Pinnow *et al.* (1978) suggested polycrystalline materials based on bromides, iodide and chlorides could transmit at longer wavelengths and possibly be low-loss materials. Others examined single crystal fibers such as AgBr (Bridges *et al.*, 1980), KRS-5 (Mimura, 1980), CsBr and CsI (Okamura *et al.*, 1980). In 1975, Poulain *et al.* reported the discovery of heavy metal halide glasses, based principally on large amounts of ZrF_4 (> 50 mol%) and since then, a large number of heavy metal fluoride compositions have been studied. These glasses were projected to have losses of 0.01–0.001 dB/km in the 2–5 μm spectral region (Gannon, 1980; Shibata *et al.*, 1981). Chalcogenides glasses were also projected to have losses of 0.01 dB/km (Shibata *et al.*, 1980b; Dianov, 1982). Oxide glasses based on GeO_2 were investigated and reported to have theoretical losses as low as 0.1 dB/km (Takahashi *et al.*, 1982). Thus, there has been tremendous activity in looking for lower loss infrared transmitting materials. Excellent general reviews (Goodman, 1978; Miyashita and Manabe, 1982; Iwasaki, 1986; Harrington, 1986), as well as reviews on heavy metal fluorides (Drexhage *et al.*, 1981; Mitachi *et al.*, 1982, 1984; Tran *et al.*, 1984; Drexhage, 1984; Comyns, 1986) and chalcogenides (Savage *et al.*, 1980; Kanamori, 1984a, b) have discussed the potential and progress in lightguides based on such compositions.

Despite the many possibilities for ultra-low-loss transmission, the most progress has been achieved in heavy metal fluoride fiber fabrication. It is beyond the scope of this section to discuss all the mid-IR fiber work, and only a brief examination of the current status of the heavy metal fluorozirconates will be given.

The most common family of glasses investigated for ultra-low-loss fiber applications are the fluorozirconates, with zirconium fluoride as the major constituent and fluorides of barium, lanthanum, aluminum, gadolinium, sodium, lithium and sometimes lead added as modifiers and stabilizers. Alkali additions are particularly important for stability as well as to improve the glass working characteristics. In addition, HfF_4 can be substituted for ZrF_4 to vary the refractive index. Extensive work has been done in two systems: zirconium-barium-lanthanum-aluminum fluoride (ZBLA) and zirconium-barium-gadolinium-aluminum fluoride (ZBGA). Sodium fluoride is frequently added to further stabilize ZBLA, to form ZBLAN glasses, and core-clad refractive index differences are achieved by varying the sodium fluoride level or partially substituting hafnium tetrafluoride for zirconium. In the case of ZBGA, the aluminum fluoride content is typically varied. Given the complexity of these glasses, many compositional variations are possible to achieve index differences, and extensive compositional work has resulted in a broad range of compositional variation aimed at forming compatible core/clad compositions and optimized glass forming characteristics, particularly aimed at minimizing the tendency to crystallize on cooling. Often, the difference between the crystallization temperature, T_{xt1} and the glass transition temperature, T_g, is used as a measure of the working range, with values of 80–150°C being typical, and maximum values most desirable. Extensive work on glass formation in a variety of systems has been reported, as reviewed by Drexhage (1984), and Tran et al. (1984).

The basic challenge for fluorozirconate-based fibers, as for all ultra-low-loss materials candidates, is to realize a fabrication technology which overcomes extrinsic absorption and scattering phenomena, and realizes designs with controlled profiles, dimensions, and mechanical properties in long lengths. The basic glass-forming properties of the fluorozirconates present difficult challenges. The glasses must be rapidly quenched to avoid crystallization while reheating, such as often required for fiber drawing, can cause the nucleation and growth of crystals. The glasses are typically very fluid at their liquidus temperature, and exhibits a very rapid decrease in viscosity above the glass transition temperature, resulting in the sensitivity to crystallization and a very narrow range of temperature within which fiber drawing must be controlled.

Additional fabrication problems relate to the reactivity of the raw materials and resultant glasses and melts with water and oxygen, requiring specialized processing conditions. Such contaminants not only result in deleterious absorption in the mid-infrared but may be the source of nucleation sites, which increase the susceptibility to crystallization and thus scattering-induced losses. In addition, high purity preparative techniques in regard to transition metals and specific rare earths are required. While vapor phase processing of silica glasses was able to take advantage of the large differential vapor pressure of reagents used to make the glasses relative to impurities contained in the raw materials, similar vapor phase processing advantages do not exist at present for the fluorozirconate-based glasses. Few volatile compounds are available and transport characteristics of dopants and certain critical impurities are similar. While organometallic liquid compounds are possible, additional issues related to carbon, oxygen and hydrogen contamination must be overcome for vapor phase processing. The relative complexity of the glass compositions add another severe constraint to vapor phase approaches, and to date no progress has been reported using such an approach. Rather, the focus has been on purification of the individual raw materials and subsequent high purity bulk melting of glasses from which fiber is then drawn. A variety of approaches to purification of the individual raw materials have been taken, such as sublimation, vapor phase halogenation followed by fluorination, other vapor phase purification reactions, solvent extraction and ion exchange techniques.

Two basic approaches have been used for melting the fluorozirconate glasses. The first involves converting oxide raw materials to fluorides by heating in the presence of ammonium bifluoride (NH_4F-HF), then fusing. At temperatures between 150 to 400°C, a sequence of chemical reactions takes place to fluorinate the oxides, and large excesses of NH_4F-HF are required to completely convert the melt. This process requires working in a fume hood due to the corrosive reaction products, but represents a relatively inexpensive method to prepare the glasses. In the second technique, anhydrous fluorides are mixed and directly fused at temperatures in the range of 800–1000°C in an inert or controlled atmosphere, typically within a dry box. Prolonged melting can result in ZrF_4 volatization. Typically, reactive atmosphere processing (RAP) is used to minimize oxygen and OH contamination, while preventing the reduction of the zirconium fluoride. CCl_4, CF_4, NF_3, SF_6, Cl_2, HF-controlled atmospheres have been reported where these gases act as oxidizing agents, and can have the additional benefit of maintaining transition metal and rare earth impurities in their highest oxidation state, which typically have lower absorption at longer wavelengths. The procedures used for melting must be carefully controlled to avoid particles and other contaminants, and any gases must be dry and particle free. Melting is typically done in Pt, Au or vitreous carbon crucibles, and care must be taken that the reactive atmosphere does not attack the crucible material.

After melting, both crucible drawing and preform techniques have been used to prepare lightguide structures. Crucible drawing is attractive, since it has the potential for being a continuous process for fiber drawing. To achieve core-clad fiber structures, usually two concentric crucibles containing glasses of different compositions are used. A number of variations on conventional double crucible drawing have been proposed (Tran *et al.*, 1984; Iwasaki, 1986) to overcome some of the crystallization and deformation issues that arise in these fluid glasses. Crystallization initiated in the region of the orifice of the crucible from which the fiber is drawn continues to be an obstacle, and controlled single-mode fiber structures, necessary for ultimate use in long distance applications, are difficult to achieve. Another approach has been to make preforms. Figure 4.24a depicts one process approach, a built-in casting technique developed by Mitachi *et al.* (1981). The cladding glass is first poured into a heated mold, which is maintained near the glass transition temperature. The mold is then inverted causing unsolidified glass to flow out, thus creating a hollow region in the center. The cladding glass is then poured and the entire structure is annealed to produce a clad-core preform. Uniformity, crystallization and contamination at the core-clad interface are drawbacks to this technique. Tran *et al.* (1982) implemented a rotational casting technique, depicted in Fig. 4.24b, which resulted in greater uniformity and better interfacial quality. In this process, the glass is poured into a gold-coated cylindrical mold preheated to T_g and then rotated at speeds of greater than 3000 revolutions/minute, resulting in highly uniform glass tubes. Core glass is then poured into the resulting tube to achieve a preform with a well-defined core-clad interface. In both cases, the achievable preform sizes are limited to less than 1 km, it is very difficult to achieve small cores necessary for single-mode structures, and interfacial crystallization often results. Nevertheless, the best results have been achieved using such techniques, with champion loss values as low as 0.7–0.9 dB/km reported for short lengths, typically ~ 30 m (Tran, 1986; Yoshida, 1986).

Fiber drawing from a preform requires stringent control of the hot zone and temperature, since reheating the glass and dwell time in the draw furnace can lead to nucleation and growth of crystals, particularly from pre-existing defects. Typically, very narrow hot zones are used to minimize such effects. In addition, since the glasses are hydroscopic and reactive with water and oxygen at higher temperatures, controlled atmospheres must be used. Fiber drawing temperatures in the range of 300–400°C are used. Coating presents additional problems due to these low draw temperatures, and often Teflon^{TM} is co-extruded to provide protection. Surface crystallization often degrades the achievable fiber strength, and reaction with atmospheric water, which permeates the coating further degrades the fiber surface. Hermetic coatings are felt to be essential to protect long lengths of fiber over their lifetimes. The intrinsic strength of the fluorozirconates is thought to be on the order of 400–500 Kpsi (2.8–3.5 GPa) although further studies in the area of mechanical properties of these relatively new glasses are necessary (Mecholsky *et al.*, 1983).

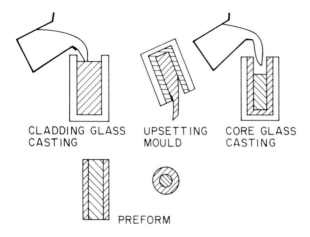

CLADDING GLASS
CASTING

UPSETTING
MOULD

CORE GLASS
CASTING

PREFORM

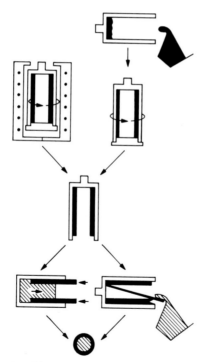

Fig. 4.24   Schematic representation of heavy metal fluoride glass preform fabrication using (a) the built-in casting technique, after Mitachi *et al.* (1981) and (b) the rotational casting technique, after Tran *et al.* (1982).

Other studies have examined fiber designs required to optimize the dispersion and bending performance of such fibers at the minimum loss wavelength (Walker *et al.*, 1986). For fluorozirconate fiber designs, dispersion optimization with < 1 psec/nm-km can be achieved with modest relative index differences, while also optimizing the designs for macrobending and microdeformation loss. In particular, the correlation length of the microdeformation plays a critical role in the required design parameters, with long microbending correlation lengths limiting design options consistent with dispersion optimization. Further fabrication studies of long lengths of such fibers will be required to determine if such optimized designs can be achieved in practice, and could present an obstacle in the attainment of ultra-low loss in long lengths.

Fiber fabrication studies of such glasses is in their earliest phase. Many technologically difficult challenges must be met in order to realize the potential of these lightguide materials. If such technical obstacles can be overcome, the relative advantages of lower loss will have to be determined relative to other telecommunication system requirements such as long term reliability and cost. Nevertheless, these new materials offer exciting prospects for even longer repeater spacings to further extend the future options in lightwave systems.

4.9 SUMMARY / FUTURE DIRECTIONS

Large-scale manufacturing techniques are now in place to make high quality, long length silica fibers economically. Studies continue to further improve the economics and control of these techniques, as well as look to new silica processing approaches that will offer further benefits, such as sol-gel. New fiber designs with tailored dispersion characteristics or enhanced performance require further extensions of the processing technology. New material systems that have the potential for even lower losses are under examination to further extend the system performance limits. Figure 4.25, after Broer and Cohen (1986), shows one representation of the potential repeater spacing-bit rate limits of system performance that could be achieved with optimized fibers, directly modulated lasers and detectors operating at the sensitivity limit. For long haul telecommunication trunking systems, these features have made fiber the transmission media of choice. For silica, Gbit/s transmission rates over greater than 100 km distances have already been achieved; the tremendous interest and potential for new ultra-low-loss materials is evident from this diagram. As the information capacity of lightwave systems is exploited, however, it appears that silica-based glass fibers have advantage in terms of the achievable unrepeated distance at the higher bit rates due to dispersion limited performance. For telecommunications in the loop and distribution portion of the network, fiber performance and cost now allow lightwave transmission to be an economical alternative to copper

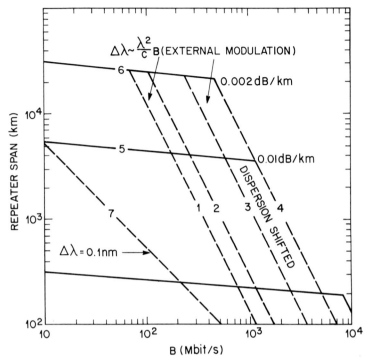

Fig. 4.25 Achievable repeater spacing versus bit rate for silica, heavy metal halide and chloride low-loss fiber systems, after Broer and Cohen (1986). The bottom curve represents silica fibers at 1.55 μm with a 0.2 dB/km loss, 60 dB system budget, a modulation broadened linewidth of 1 ps/km · nm, and coherent detection. Curves 1 and 3 refer to chlorides operating at 6 μm for a modulation broadened laser linewidth with a dispersion value of 10 and 1 ps/nm · km, respectively, while curves 2 and 4 refer to heavy metal fluorides with dispersion values of 20 and 1 ps/nm · km, respectively. Curve 7 illustrates a non-dispersion optimized fluoride operating with a laser with 0.1 nm linewidth.

transmission. For these applications, fiber upgradability to higher bit rates is an attractive feature.

Other studies have been directed at developing a variety of special fibers for both telecommunication and newer applications. These include improving the radiation and mechanical performance of the fibers so they can be used in harsh environments or applications that have extremely stringent reliability considerations, and controlling the polarization characteristics so that a whole new family of optical fiber based sensors can be realized. Another exciting area of research that has been beyond the scope of this chapter to discuss is fiber devices such as couplers, splitters, active devices and sensors, as well as a variety of other fiber components.

Fundamental understanding of defects, whether optical or mechanical, and their impact on performance and reliability will continue to be a thrust of basic studies on silica, as well as a necessity for the successful development of new materials.

The past twenty years has shown dramatic progress in the development of fiber technology; it is expected that the future will continue to extend the performance limits as well as use fibers in a variety of new applications, some yet to be determined.

REFERENCES

Abe, K. (1976). Fluorine doped silica for optical waveguides. *Proc., 2nd Eur. Conf. Opt. Comm.*, Paris, France, 59–61.

Adams, M. J. (1981). "An Introduction to Optical Waveguides," John Wiley and Sons, New York.

Adams, R., and McMillan, P. W. (1977). Review, static fatigue in glass. *J. Mat. Sci.* **12**, 643–657.

Ainslie, B. J., Beales, K. J., Cooper, D. M., Day, C. R., and Rush, J. D. (1982a). Drawing-dependent losses in dispersion-shifted monomode fibers. *Tech. Dig., Opt. Fib. Comm., THEE6*, Phoenix, Ariz., 66–67.

Ainslie, B. J., Beales, K. J., Cooper, D. M., Day, C. R. and Rush, J. D. (1982b). Drawing-dependent transmission loss in germania-doped silica optical fibers. *J. Non-Cryst. Sol.* **47** (2), 243–246.

Aronson, B. S., Powers, D. R. and Sommer, R. G. (1979). Chlorine drying of a doped deposited silica preform simultaneous to consolidation. *Tech. Dig., Top Meeting, Opt. Fiber Comm.* 54–55.

Asamov, A. V. (1983). A new concept of the mechanism of the formation of radiation paramagnetic color centers in vitreous silica. *Fiziki Khimiya Stekla* **9** (5), 569–583. Translated by Plenum Press, 1984.

Bachmann, P., Hübner, H., Lennartz, U., Steinbeck, E., and Ungelenk, J. (1982a). Fluorine doped single mode and step index fibers prepared by the low pressure PCVD process. *Proc., 8th Eur. Conf. Opt. Comm.*, Cannes, 66–69.

Bachmann, P., Geittner, P., and Wilson, H. (1982b). The deposition efficiency for the GeO_2-doped in optical fiber preparation by means of low pressure PCVD. *Proc., 8th Eur. Conf. Opt. Comm.*, Cannes, 614–617.

Bachmann, P., Leers, D., Lennartz, M., and Wehr, H. (1983a). Preparation of single mode fibres by the low pressure PCVD process. *Proc., 9th Eur. Conf. Opt. Comm.*, Geneva, 5–9.

Bachmann, P., Geittner, P., Hermann, W., Lydtin, H., Rau, H., Ungelenk, J., and Wehr, H. (1983b). Recent progress in the preparation of GI and SM-fibres by means of the PCVD process. *Proc., 4th Integer. Optics and Opt. Fib. Comm. Conf.*, Tokyo, Post Deadline, 29A5–3.

Bachmann, P. K., Geittner, P., Leers, D., and Wilson, H. (1986). Loss reduction in fluorine doped SM and high NA PCVD fibers. *J. Lightwave Tech.* **LT4** (7), 813–817.

Bansal, N. P., and Doremus, R. H. (1986). "Handbook of Glass Properties." Academic Press, Inc.,

Barlow, A. J., Ramskov-Hansen, J. J., and Payne, D. N. (1981). Birefringence and polarization mode dispersion in spun single mode fiber. *Appl. Opt.* **20**, 2962–2968.

Barnes, S. R., Riley, S. P., and Wolfe, S. V. (1985). Hydrogen evolution behavior of silicone coated optical fibers. *Electron Lett.* **21** (6), 712–713.

Beales, K. J., Day, C. R., Duncan, W. J., Midwinter, J. E., and Newns, C. R. (1976). Preparation of sodium borosilicate glass fibre for optical communication. *Proc. Inst. Elec. Eng.* **123**, 591–596.

Beales, K. J., and Day, C. R. (1980). A review of glass fibers for communications. *Phys. Chem. Glasses* **21**, 5–20.

Beales, K. J., Day, C. R., Dunn, A. G., and Partington, S. (1980). Multicomponent glass fibers for optical communications. *Proc. IEEE* **68** (10), 1191–1194.

Beales, K. J., Cooper, D. M., Rush, J. D., Fox, M., Plessner, K. W., and Stannard-Powell, S. J. (1983a). Increased attenuation of optical fibers caused by diffusion of hydrogen. *Proc., 9th Eur. Conf. Opt. Commun.*, Geneva, Post Deadline.

Beales, K. J., Cooper, D. M., and Rush, J. D. (1983b). Increased attenuation in optical fibres caused by diffusion of molecular hydrogen at room temperature. *Electron. Lett.* **19** (22), 917–919.

Beales, K. J., Cooper, D. M., Duncan, W. J., and Rush, J. D. (1984). Practical barrier to hydrogen diffusion in optical fibers. *Tech. Dig., Opt. Fib. Commun. Conf.*, New Orleans, Post Deadline, W15-1-4.

Beales, K. J., Carter, S. F., France, P. W., and Partington, S. (1985). Multicomponent glass optical fibres with a lowered water content and reduced loss. *J. Non-Cryst. Sol.* **70**, 253–262.

Bendow, B., Brown, R. N., Drexhage, M. G., Loretz, T. J., and Kirk, R. L. (1981). Material dispersion of fluorozirconate-type glasses. *Appl. Opt.* **20**, 3688–3690.

Berkey, G. E. (1984). Fluorine doped fibers by the outside vapor deposition process. *Tech. Dig., Opt. Fiber Comm. Conf.*, New Orleans, MG-3.

Bhagavatula, V. A., Spotz, N. S., and Love, W. F. (1984). Dispersion shifted segmented core fibers. *Opt. Lett.* **9**, 186–188.

Birch, R. D., Payne, D. N., and Varnham, M. P. (1982). Fabrication of polarization maintaining fibers by gas phase etching. *Electron Lett.* **18** (24), 1036–1038.

Black, P. W. (1976). Fabrication of optical fiber waveguide. *Electr. Comm.* **51** (1), 4–11.

Blankenship, M. G. (1979). System for delivering materials to deposition site on optical waveguide blank, U.S. Pat. 4,173,305.

Blankenship, M. G., Keck, D. B., Leven, P. S., Love, W. F., Sarkar, A., Schultz, P. C., Sheth, K. D., and Siegfried, R. N. (1979). High phosphorus containing P_2O_5–GeO_2–SiO_2 optical waveguides. *Tech. Dig., Opt. Fib. Comm. Conf.*, Washington, D.C., PD 3–1.

Blankenship, M. G., and Deneka, C. W. (1982). The outside vapor deposition method for fabricating optical waveguide fibers. *IEEE J. Quant. Electronics* **QE-18**, 1418–1423.

Blankenship, M. G. (1982). Current status of outside vapor deposition process and performance. *Tech. Dig., Opt. Fiber Comm. Conf.*, Phoenix, 8–9.

Blankenship, M. G., Morrow, A. J., and Powers, D. R. (1987). Short-term transient attenuation in single mode fibers due to hydrogen. *Tech. Dig., Opt. Fiber Comm. Conf.*, Reno, A3.

Blyler, L. L., Jr., and DiMarcello, F. V. (1980). Fiber drawing, coating and jacketing. *Proc. IEEE* **68**, 1194–1197.

Blyler, L. L., DiMarcello, F. V., Simpson, J. R., Sigety, E. A., Hart, A. R., and Foertmeyer, V. A. (1980). UV-radiation induced losses in optical fibers and their control. *Proc., XII Inter. Cong. on Glass*, Albuquerque, 165–170.

Blyler, L. L., Jr., Cogan, K. A., and Ferrara, J. A. (1987). Plastics and polymers for optical transmission, *MRS Symposium Proceedings, Vol. 88, Optical Fiber Materials and Processes*, 3–10.

Bridges, T. J., Haziak, J. S., and Strand, A. R. (1980). Single crystal AgBr infrared optical fibers. *Opt. Lett.* **5**, 85–86.

Broer, M. M., and Cohen, L. G. (1986). Heavy metal glass fiber lightwave systems. *J. Lightwave Tech.*, **LT4**(10), 1509–1513.

Brown, R. N., and Hutta, J. J. (1985). Material dispersion in high quality heavy metal fluoride glasses. *Appl. Opt.*, **24** (24), 4500–4503.

Brownlow, D. L., DiMarcello, F. V., Hart, A. C., Jr., and Huff, R. G. (1982). High strength multikilometer lightguides for undersea applications. *Proc., 8th Eur. Conf. Opt. Comm.*, Cannes, Post Deadline Paper.

Carrier, G. B. (1980). Characterization of glasses and ceramics with the analytical electron microscope. *Proc., XII Intern. Cong. on Glass*, Albuquerque, 15–20.

Carson, D. S., and Maurer, R. D. (1973). Optical attenuation in titania-silica glasses. *J. Non-Cryst. Solids* **11**, 368–380.

Chamulitrat, W., Kevan, L., Schwartz, R. N., Blair, R., and Tangonan, G. L. (1986). Radiation damage of germanium doped silica glasses: spectral simplification by photo- and thermal bleaching, spectral identification and microwave saturation characteristics. *J. Appl. Phys.* **59** (8), 2933–2939.

Chandan, H. C., and Kalish, D. (1982). Temperature dependence of static fatigue of optical fibers coated with a UV-curable polyurethane acrylate. *J. Amer. Ceram. Soc.* **65** (3), 171–173.

Charles, R. J. (1958a). Static fatigue of glasses, I-II. *J. Appl. Phys.* **29** (11), 1549–1560.

Charles, R. J. (1985b). Dynamic fatigue of glass. *J. Appl. Phys.* **29** (12), 1657–62.

Charles, R. J., and Hillig, W. B. (1962). In "Symposium on the Mechanical Strength of Glass and Ways of Improving It." *Union Scientifique Continentale du Verre*, Charleroi, Belgium, 517–527.

Chaudhari, R. and Schultz, P. (1986). Hermetic coating on optical fibers. *SPIE*, Vol. 717, *Reliability Considerations in Fiber Optic Applications*, 27–29.

Chida, K., Nakahara, M., Sudo, S., and Inagaki, N. (1983). On-line monitoring technique of the refractive index profile in the VAD process. *J. Lightwave Tech.* **LT-1** (1), 56–60.

Cohen, L. G., and Lin, C. (1977). Phase delay measurements in the zero dispersion wavelength region for optical fibers. *Appl. Opt.* **16**, 3136–3139.

Comyns, A. (1986). Fluoride glasses for fiber optics. *Chemistry in Britain* 47–52.

Craig, S. P., Duncan, W. J., France, P. W., Snodgrass, J. E. (1982). The strength and fatigue of large flaws in silica optical fiber. *Proc. 8ᵗʰ Eur. Conf. Opt. Fibre. Comm.*, Cannes, France, 205–208.

Craig, S. P., Dunn, P. L., Rush, J. D., Smith, D. G., and Beales, K. J. (1983). Simultaneous multilevel proof-testing of high strength silica fibre. *Elect. Lett.* **19** (14), 516–517.

Croft, T. D., Ritter, J. E., and Bhagavatula, V. A. (1985). Low loss dispersion shifted single mode fiber manufactured by the OVD process. *J. Lightwave Tech.*, **LT3** (5), 931–935.

Dabbs, T. P., and Lawn, B. R. (1985). Strength and fatigue properties of optical glass fibers containing microindentation flaws. *J. Am. Ceram. Soc.* **68** (11), 563–569.

Danzuka, T., Yokota, H., Tsuchiya, I., and Takimoto, H. (1984). Investigation of $SiHCl_3$ on high speed deposition of VAD. *Nat. Conv., Opt. Wave, IECE J.* **416**, 2–160.

Danzuka, T., Yokota, H., and Tsuchiya, I. (1985). High speed deposition of graded index fiber by the multiflame vapor axial deposition method. *Tech. Dig., Opt. Fib. Comm. Conf.*, New Orleans, **MF5**, 16–17.

Davidge, R. W., McLaren, J. R., and Tappin, G. (1973). Strength-probability-time (SPT) relationship in ceramics. *J. Mat. Sci.* **8**, 1699–1705.

DeLuca, R. D. (1976). Method of making optical waveguides, U.S. Pat 3,933,454.

Deneka, C. W. (1985). "Manufacturing in Corning Glass Works in Opt. Fiber Comm." Vol. 1, ed. by T. Li, Academic Press, New York, 271–296.

de Wert, H. P. J., van der Hulst, V., Kuyt, G., and Peelen, J. (1985). Production of single mode fibres by the PCVD process. *Phillips Telecom. Rev.* **43** (1), 50–56.

Dianov, E. M., Kornienko, L. S., Nikitin, E. P., Rybaltovskii, A. O., Sulimov, V. B., and Chernov, P. V. (1983). Radiation-optical properties of quartz glass fiber-optic waveguides (review). *Sov. J. Quantum Electron.* **13** (3), 274–289.

Dianov, E. M. (1982). Materials for infrared low loss fibers in advances in IR fibers. *Tech. Dig., SPIE*, Paper 320-04.

DiMarcello, F. V., Hart, A. C., Williams, J. C., and Kurkjian, C. R. (1979). "High Strength Furnace-Drawn Optical Fibers in Fiber Optics: Advances in Research and Development." ed. by B. Bendow and S. S. Mitra, Plenum, New York, 125–135.

DiMarcello, F. V., Brownlow, D. L., and Shenk, D. S. (1981). Strength characterization of multikilometer silica fiber. *Tech. Dig., San Francisco, Int. Conf. IOOC*, 26–27.

DiMarcello, F. V., Kurkjian, C. R., and Williams, J. C. (1985). "Fiber Drawing and Strength Properties in Opt. Fiber Comm." Vol. 1, Ed., T. Li, Academic Press, New York, 179–248.

Donaghy, F. A., and Nicol, D. R. (1983). Evaluation of the fatigue constant n in optical fibers with surface particle damage. *J. Am. Ceram. Soc.* **66** (8), 601–604.

Drexhage, M. G., Quinlan, K. P., Moynihan, C. T., and Boules, M. S. (1981). "Fluoride Glasses for Visible to Mid-IR Guided Wave Optics, in Physics of Fiber Optics." ed. B. Bendow and S. S. Mitra, *Amer. Cer. Soc.*, Columbus, Ohio, 57–73.

Drexhage, M. G. (1984). "Heavy Metal Fluoride Glasses, in Treatise on Materials Sciences and Technology." Vol. 26: Glass IV, ed. M. Tomozawa and H. Doremus, Academic Press, New York, 151– .

Duncan, W. J., France, P. W., and Beales, K. J. (1981). Effect of service environment on proof testing of optical fibres. *Proc., 7th Eur. Conf. Opt. Commun.*, Copenhagen, 4.5-1–4.5-4.

Duncan, W. J., France, P. W., and Craig, S. P. (1986). "The Effect of Environment on the Strength of Optical Fiber in Strength of Inorganic Glasses." Ed., C. R. Kurkjian, Plenum Press, NY, 309–328.

Dyott, R. B., and Schrank, P. F. (1982). Self-locating elliptically cored fiber with an accessible guiding region. *Electron Lett.* **18** (22), 980–81.

Edahiro, T., Horiguchi, M., Chida, K., and Ohmori, Y. (1979a). Spectral loss characteristics of GeO_2–P_2O_5–doped silica graded index fibres in long-wavelength band. *Electron. Lett.* **15** (10), 274–275.

Edahiro, T., Kawachi, M., Sudo, S., and Takata, H. (1979b). OH reduction in VAD optical fibers. *Electron. Lett.* **15** (16), 482–483.

Edahiro, T., Kawachi, M., Sudo, S., and Tamaru, S. (1980). Deposition properties of high silica particles in the flame hydrolysis reaction for optical fiber fabrication. *Jap. J. Appl. Phys.* **19**, 2047–2054.

Evans, A. G. (1974). Slow crack growth in brittle materials under dynamic loading conditions. *Int. J. Fracture* **10**, 251–259.

Evans, A. G., and Wiederhorn, S. M. (1974). Prooftesting of ceramic materials—an analytical basis for failure prediction. *Int. J. Fract.* **10**, 379–392.

Evans, A. G., and Johnson, H. (1975). The fracture stress and its dependence on slow crack growth. *J. Mat. Sci.* **10** (2), 214–222.

Fanderlik, I. (1983). "Optical Properties of Glass." Elsevier Scientific Publishing Co., New York.

Fleming, J. W. (1976). Material and mode dispersion in GeO_2-B_2O_3-SiO_2 glasses. *J. Am. Ceram. Soc.* **59**, 503–507.

Fleming, J. W., Jaegar, R. E., and Miller, T. J. (1976). Optical glass and its production, US Pat. No. 3,954,431.

Fleming, J. W. (1978). Material dispersion in lightguide glasses. *Electron. Lett.* **14** (11), 326–328.

Fleming, J. W., and O'Connor, P. B. (1981a). "High Rate Lightguide Fabrication Technique, Phys. Fib. Opt. Adv. in Ceram." Vol. II, ed. B. Bendow and S. S. Mitra, Plenum Press, New York, 21–26.

Fleming, J. W., and Raju, V. R. (1981b). Low optical attenuation fibers prepared by plasma enhanced MCVD. *Tech. Dig., IOOC*, WD2, San Francisco.

Fleming, J. W., and Raju, V. R. (1981c). Low loss single mode fibers prepared by plasma enhanced MCVD. *Elect. Lett.* **17** (23), 867–868.

Fleming, J. W., Jr., MacChesney, J. B., and O'Connor, P. B. (1982). Optical fiber fabrication by a plasma generator. U.S. Pat. No. 4,331,462.

Fleming, J. W. (1983). Status and Prognosis for Plasma MCVD. *Tech. Dig., Opt. Fib. Comm. Conf.*, New Orleans, LA. WG1, 88–90.

Fleming, J. W. (1984). Dispersion in GeO_2-SiO_2 glasses. *Appl. Opt.* **23** (24), 4486–4493.

Fox, M., and Stannard-Powell, S. J. (1983). Attenuation changes in optical fibers due to hydrogen. *Electron Lett.* **19**, 916–917.

France, P. W., Paradine, M. J., Reeve, M. H., and Newns, G. R. (1980). Liquid nitrogen strengths of coated optical glass fibers. *J. Mat. Sci.* **15**, 825–830.

French, W. G., MacChesney, J. B., and Pearson, A. D. (1975). Glass fibers for optical communication. *Ann. Rev. Mat. Sci.* 373–394.

French, W. G., Pace, L. J., and Foertmeyer, V. A. (1978). Chemical kinetics of the reactions of $SiCl_4$, $SiBr_4$, $GeCl_4$, $POCl_3$, and BCl_3 with oxygen. *J. Phys. Chem.* **82**, 2191–2194.

Friebele, E. J., Griscom, D. L., and Sigel, G. H. (1974). Defect centers in a germanium doped silica core fiber. *J. Appl. Phys.* **45**, 3424–3428.

Friebele, E. J., Sigel, G. H., and Griscom, D. L. (1976). Drawing induced defect centres in fused silica core fibre. *Appl. Phys. Lett.* **28**, 516–518.

Friebele, E. J. (1979). Optical fiber waveguides in radiation environments. *Optical Eng'g.* **18** (6), 552–561.

Friebele, E. J., and Griscom, D. L. (1979). "Radiation Effects in Glass in Treatise on Materials Science and Technology." Vol. 17: Glass II, M. Tomozawa and R. H. Doremus, Ed., Academic Press, New York, 257–351.

Friebele, E. J., Griscom, D. L., Stapelbroek, M., and Weeks, R. A. (1979). Fundamental defect centers in glass: The peroxy radical in irradiated high purity fused silica. *Phys. Rev. Lett.* **42**, 1346–1349.

Friebele, E. J., and Gingerich, M. E. (1980). Radiation induced optical absorption bands in low loss optical fiber waveguides. *J. Non-Cryst. Solids* **38–39**, 245–250.

Friebele, E. J., and Gingerich, M. E. (1981). Photobleaching effect in optical fiber waveguides. *Appl. Opt.* **20**, 3448–3452.

Friebele, E. J., Askins, C. G., Gingerich, M. E., and Long, K. J. (1984). Optical fiber waveguides in radiation environments II. *Nucl. Inst. Meth. in Phys. Res.* **B1**, 355–369.

Friebele, E. J., Griscom, D. L., and Marrone, M. J. (1985a). The optical absorption and luminescence bands near 2 eV in irradiated and drawn synthetic silica. *J. Non-Crys. Solids* **71**, 133–144.

Friebele, E. J., Long, K. J., Askins, C. G., Gingerich, M. E., Marrone, M. J., and Griscom, D. L. (1985b). Overview of radiation effects in fiber optics. *SPIE, Radiation Effects in Optical Materials* **541**, 70–88.

Friebele, E. J., and Griscom, D. L. (1986). Color centers in glass optical fiber waveguides in defects in silica. *MRS* **61**, 319–332.

Gannon, J. R. (1980). Optical fiber materials for operating wavelengths longer than 2 μm. *J. Non Cryst. Solids* **42**, 239–245.

Gardner, W. B. (1975). Microbending loss in optical fibers. *Bell Syst. Tech. J.* **54**, 457–465.

Geittner, P., Küppers, D., and Lydtin, H. (1976). Low loss optical fibers prepared by plasma-activated chemical vapor deposition (CVD). *Appl. Phys. Lett.* **28** (11), 645–646.

Geittner, P., Hagemann, H. J., Warnier, J., and Wilson, H. (1986). PCVD at high deposition rates. *J. Lightwave Tech.* **LT4** (7), 818–822.

Geyling, F. T., Walker, K. L., and Csencsits, R. (1983). The viscous collapse of thick-walled tubes. *J. Appl. Mech.* **50** (2), 303–10.

Glen, R. M. (1986). Polymeric optical fibre. *Chemitronics* **1**, 98–105.

Gloge, D. (1975). Optical fiber packaging and its influence on fiber straightness and loss. *Bell. Syst. Tech. J.* **54**, 245–262.

Goodman, G. H. L. (1978). Devices and materials for 4 μm-band fiber optical communications. *Solid State Electron. Devices* **2**, 129–37.

Griffith, A. A. (1920). The phenomena of rupture and flow in solids. *Phil. Trans. Roy. Soc. London*, 163.

Griscom, D. L., Friebele, E. J., Long, K. J., and Fleming, J. W. (1983). Fundamental defect centers in glass: electron spin resonance and optical absorption studies of irradiated phosphorus-doped silica glass and optical fibers. *J. Appl. Phys.* **54** (7), 3743–3762.

Griscom, D. L. (1985). Nature of defects and defect generation in optical glasses. *SPIE, Radiation Effects in Optical Materials* **541**, 38–59.

Griscom, D. L. (1986). Point defects in amorphous SiO_2: What have we learned from 30 years of experimentation?. *Mat. Res. Soc. Symp.* **61**, 213–221.

Gulati, S. T., and Scherer, G. W. (1978). Design of optical waveguide blanks from residual stress point of view. *Proc., 4th Eur. Conf. Opt. Comm.,* Genoa, 37–46.

Gulati, S. T., Helfinstine, J. D., Justice, B., McCartney, J. S., and Runyan, M. A. (1979). Measurement of stress corrosion constant *n* for optical fibers. *Bull. Am. Ceram. Soc.* **58** (11), 1115–1117.

Hanawa, F., Sudo, S., Kawachi, M., and Nakahara, M. (1980). Fabrication of completely OH free, VAD fibre. *Electron. Lett.* **16** (18), 299–700.

Harmer, A. L., Puyane, R., and Gonzalez-Oliver, (1982). The sol-gel method for optical fiber fabrication. *Proc. Int. Conf. Opt. Fiber. Commun.* (IFOC) 40–44.

Harrington, J. A. (1986). A look at the future of infrared fibers. *Optics News* **5**, 23–29.

Hartwig, C. M. (1976). Raman scattering from hydrogen and deuterium dissolved in silica as a function of pressure. *J. Appl. Phys.* **47**, 956–959 (1976).

Hibino, H., and Hanafusa, H. (1983). ESR study on E′-centers induced by optical fiber drawing process. *Jap. J. Appl. Phys.* **22**, 766–768.

Hibino, Y., Hanafusa, H., and Sakaguchi, S. (1985). Formation of drawing-induced E′ centers in silica optical fibers. *Jap. J. Appl. Phys.* **24** (9), 1117–1121.

Hibino, Y., Hanafusa, H., and Yamamoto, F. (1986). Oxygen influence on drawing induced defect centres and radiation-induced loss in pure silica optical fibers. *Electron lett.* **22** (8), 434–435.

Hillig, W. B., and Charles, R. J. (1965). In "High Strength Materials," Ed. V. F. Zackay, Wiley & Sons, New York, 682–705.

Hinchliffe, J. D. S., Ashpole, R. S., Powell, R. J. W., and Lewis, D. L. (1984). Hydrogen in optical cables. *IEE Colloquim,* London, Dig. No. 1984/74, paper 8.

Hiskes, R., Schantz, C. A., Hanson, E. G., Mittelstadt, L. S., Scott, C. J., Joiner, C. S., Knutsen, G. F., and Pound, J. G. (1984). High performance hermetic optical fibers. *Tech. Dig., Opt. Fib. Commun. Conf.,* New Orleans, PD-WI6.

Hooper, E. A. (1983). Effects of draw conditions on loss in multimode fibers. *Tech. Dig., Opt. Fib. Comm.,* New Orleans, LA, WC5, 82–85.

Horiguchi, M., Ohmori, Y., and Miya, T. (1979). Evaluation of material dispersion using a nanosecond optical pulse radiator. *Applied Optics* **18**, 2223–2228.

Hosaka, T., Okamoto, K., Miya, T., Sasaki, Y., and Edahiro, T. (1981). Low loss single polarization fibers with asymmetrical strain birefringence. *Electron. Lett.* **17** (15), 530–531.

Iino, A., Tamura, T., Orimo, K., Kamiya, T., Ogai, M. (1985). Defect studies in MCVD fibers. *Proc., IOOC-ECOC,* Venezia, 527–530.

Iino, A., Tamura, J., Orimo, K., and Okubo, K. (1986). Contamination free single mode fibers. *Tech. Dig., Opt. Fib. Comm. Conf.,* Atlanta, TUI3.

Iino, A., Ogai, M., Matsubara, K. (1987). Mechanism of infrared loss increase caused by impurities contained in silica tubes. *Tech. Dig., Opt. Fiber Comm. Conf.,* Reno, WA5.

Imoto, K., and Sumi, M. (1981). Modified VAD method for optical fiber fabrication. *Electron. Lett.* **17**, 525–526.

Irven, J. (1981). Long wavelength performance of $SiO_2/GeO_2/P_2O_5$ core fibers with different P_2O_5 levels. *Electron. Lett.* **17** (1) 2–3.

Ishida, K., Imoto, K., and Suganuma, T. (1982). Transmission characteristics of VAD fibers using a multiple burner with a single source gas nozzle. *Fiber and Integrated Optics* **4** (2), 191–202.

Itoh, H., Ohmori, Y., and Nakahara, M. (1984). Chemical change from diffused hydrogen gas to hydroxyl ion in silica glass optical fibers. *Electron. Lett.* **20** (3), 140–142.

Iwasaki, H. (1986). Development of IR optical fibers in Japan. *SPIE Vol. 618, Infrared Optical Materials and Fibers*, **IV**, 2–9.

Izawa, T., Shibata, N., and Takeda, A. (1977a). Optical attenuation in pure and doped fused silica in the IR wavelength region. *Appl. Phys. Lett.* **31** (1), 33–35.

Izawa, T., Kobayashi, S., Sudo, S., and Hanawa, F. (1977b). Continuous fabrication of high silica fiber preform. *Tech. Dig., Int. Conf. Integ., Opt. Opt. Fiber Commun.*, 375–378, Tokyo.

Izawa, T., Miyashita, T., and Hanawa, F. (1977c). Continuous optical fiber preform fabrication method, U.S. Pat. 4,062,665.

Izawa, T., and Inagaki, N. (1980). Materials and processes for fiber preform fabrication-vapor-phase axial deposition. *Proc. IEEE* **68** (10), 1184–1187.

Jablonowski, D. P. (1985). MCVD fiber manufacture. In "Advances in Optical Fiber Communication." Ed. T. Li, Academic Press, New York, 249–269.

Jaeger, R. E., MacChesney, J. B., and Miller, T. J. (1978). The preparation of optical waveguide preforms by plasma deposition. *Bell Syst. Tech. J.* **57** (1), 205–210.

Johnson, D. W., Jr., Rabinovich, E. M., MacChesney, J. B., and Vogel, E. M. (1983). Preparation of high-silica glasses from colloidal gels. II: sintering. *J. Am. Ceram. Soc.* **66** (10), 688–693.

Justice, B., and Gulati, S. (1978). Tensile tester for long optical fibers. *Bull. Am. Ceram. Soc.* **57** (2), 217–219.

Kaino, T., Jujiki, M., and Nara, S. (1981). Low-loss polystyrene core-optical fibers. *J. Appl. Phys.* **52** (12), 7061–7063.

Kaino, T., Kaname, J., and Nara, S. (1982). Low loss poly(methyl methacrylate-d5) core optical fibers. *Appl. Phys. Lett.* **41** (9), 802–804.

Kaino, T., Fujiki, M., and Jinguiji, K. (1984). Preparation of plastic optical fibers. *Rev. Elect. Comm. Lab.* **32** (3), 478–488.

Kaino, T. (1985). Absorption losses of low loss plastic optical fibers. *Jap. J. Apl. Phys.* **24** (12), 1661–1665.

Kaiser, P. (1973). Spectral losses of unclad fiber made from high grade silica. *Appl. Phys. Lett.* **23**, 478–479.

Kaiser, P. 91974). Drawing induced coloration in vitreous silica fibers. *J. Opt. Soc. Am.* **64** (4), 475–481.

Kalish, D., and Tariyal, B. K. (1978). Static and dynamic fatigue of a polymer-coated fused silica optical fiber. *J. Am. Ceram. Soc.* **61** (11–12), 518–523.

Kaminow, I. P., and Ramaswamy, V. (1979). Single polarization optical fibers: slab model. *Appl. Phys. Lett.* **34**, 268–270.

Kaminow, I. P., Simpson, J. R., Presby, H. M., and MacChesney, J. B. (1979). Strain birefringence in single polarization germanosilicate optical fibers. *Electron. Lett.* **15**, 677–679.

Kaminow, I. (1981). Polarization in optical fibers. *IEEE J. Quantum Electronics* **QE17** (1), 15–22.

Kanamori, H., Yoshioka, N., Kyoto, M., Watanabe, M., and Tanaka, G. (1983). Fluorine doping in the VAD method and its application to optical fiber. *Proc., 9th Eur. Conf. Opt. Comm.*, Geneva 13–16.

Kanamori, T., Terunuma, Y., and Miyashita, T. (1984a). Preparation of chalcogenide optical fiber, Rev. ECL **32** (3), 469–477.

Kanamori, T., Terunuma, Y., Takahashi, S., and Miyashita, T. (1984b). Chalcogenide glass fibers for mid-infrared transmission. *J. Lightwave Tech.* **LT-2** (5), 605–613.

Kao, K. C., and Hockmam, G. A. (1966). Dielectric fiber surface waveguides for optical frequencies. *Proc. IEEE* **133** (7), 1151–1158.

Kapron, F. P., Keck, D. B., and Maurer, R. D. (1970). Radiation losses in glass optical waveguides. *Appl. Phys. Lett.* **17**, 423–425.

Katsuyama, T., Matsumura, H., and Suganuma, T. (1981). Low loss single polarization fibers. *Electron Lett.* **17** (13), 473–474.

Kastsuyama, T., Matsumura, H., and Suganuma, T. (1983). Low loss single polarization fibers. *Appl. Opt.* **22** (11), 1741–1747.

Katsuyama, T., Matsumura, H., and Suganuma, T. (1984). Reduced pressure collapsing method for single-polarization optical fibers. *J. Lightwave Tech.* **LT2** (5), 634–639.

Kawachi, M., Sudo, S., Shibata, N., and Edahiro, T. (1980). Deposition properties of SiO_2–GeO_2 particles in the flame hydrolysis reaction for optical fiber fabrication. *Jap. J. Appl. Phys.* **19** (2), 69–71.

Kawachi, M., Tomoru, S., Yasu, M., Horiguchi, M., Sakaguchi, S., and Kimura, T. (1981a). 100 km single mode VAD fibers. *Electron Lett.* **17**, 57–58.

Kawachi, M., Sudo, S., and Edahiro, T. (1981b). Threshold gas flow rate of halide materials for the formation of oxide particles in the VAD process for optical fiber fabrication. *Jap. J. Appl. Phys.* **20** (4), 709–712.

Kawazoe, H., Orimo, K., and Iino, A. (1985a). High-rate deposition VAD process using a double flame burner. *Proc. Opt. Fiber. Comm. Conf.*, San Diego, **MF2**, 14–15.

Kawazoe, H., Yamane, M., and Watanabe, Y. (1985b). Intrinsic and photon induced defects associated with Ge in optical fiber. *J. de Physique* **C8** 651–655.

Kawazoe, H. (1985c). Effects of modes of glass formation on structure of intrinsic or photon induced defects centered on III, IV or V cations in oxide glasses. *J. Non-Cryst. Solids* **71**, 231–243.

Kawazoe, H., Watanabe, Y., Shibuya, K., and Muta, K. (1986a). Drawing or radiation induced paramagnetic defects associated with Ge in SiO_2:GeO_2 optical fiber. *Mat. Res. Soc. Symp. Proc.* Vol. **61**, 349–357.

Kawazoe, H., Kohketsu, M., Watanabe, Y., Shibuya, K., and Muta, K., 349–357 (1986b). New phosphorus oxygen hole center in γ-irradiated. *Mat. Res. Soc. Symp. Proc.* Vol. **61**, 339–347.

Keck, D. B., Maurer, R. D., and Schultz, P. C. (1973a). On the ultimate lower limit of attenuation in glass optical waveguides. *Appl. Phys. Lett.* **22**, 307–309.

Keck, D. B., Schultz, P. C., and Zimar, F. (1973b). Method of forming optical waveguide fibers. U.S. Patent 3,737,292.

Klein, A. A., Nguyen, Q. D., and Shang, H. T. (1983). Mass production of a large core high NA fiber design. *Tech. Dig., Opt. Fib. Comm.*, New Orleans, LA, **WG4**, 90–93.

Kleinert, P., Schmidt, D., Kirchhof, J., and Funke, A. (1980). About oxidation of $SiCl_4$ and $GeCl_4$ in homogeneous gas phase. *Kristall and Technik.* **15** (9), 85–90.

Kobayashi, S., Shibata, S., Hibata, N., and Igawa, T. (1977). Refractive index dispersion of doped fused silica. *Proc. Int. Conf. Opt. Fiber Commun.*, Tokyo, Japan, 390–312.

Koel, G. J. (1983). Technical and economic aspects of the different fiber fabrication processes. *Ann. Telecommun.* **38**, 1–2, 36–46.

Koenings, J., Küppers, D., Lydtin, H., and Wilson, H. (1975). Deposition of SiO_2 with low impurity content by oxidation of $SiCl_4$ in nonisothermal plasma. *Proc. 5th Intern. Conf. CVD*, 270–281.

Koizumi, K., Ikeda, Y., Kitano, I., Furukawa, M., and Sumimoto, T. (1974). New light focusing fibers made by a continuous process. *Appl. Opt.* **13**, 255–260.

Kokura, K., Yoshida, K., Iino, A., and Orimo, K. (1984). Fluorine doping in the consolidation process of VAD soot preform for single mode fibers. *Tech. Dig., Opt. Fib. Comm. Conf.*, New Orleans, **MG6**, 22–23.

Kosinski, S. G., Soto, L., Nagel, S. R., and Watrous, T. (1981). Characterization of germanium phosphosilicate films prepared by modified chemical vapor deposition. *Cer. Bull.* **60** (8), 860.

Kosinski, S. G., Nagel, S. R., Lemaire, P. J., and Stone, J. (1982). Effect of cladding on OH content of MCVD lightguides. *Cer. Bull.* **61** (8), 822.

Krause, J. T. (1979). Transitions in static fatigue of fused silica fiber lightguides. *Proc., 5th Eur. Conf. Opt. Comm.*, Amsterdam, **19**, 1–4.

Krause, J. T. (1980). Zero stress strength reduction and transitions in static fatigue of fused silica fiber lightguides. *J. Non-Cryst. Sol.* **38–39**, 497–502.

Krause, J. T., Kurkjian, C. R., and Paek, U. C. (1981). Tensile strengths 74 GPa for lightguide fusion splices. *Electron Lett.* **17**, 812–813.

Krause, J. T., Meade, D. A., and Shapiro, S. (1986). Assuring mechanical reliability of high strength fiber and cable for SL. *Proc., Suboptic Intern Conf., Opt. Fiber. Sub. Telecommun. System,* Paris, 117–122.

Kumar, C. K. N., Nelson, E. T., and Karl, R. J. (1981). Vibrational overtone absorption in solid hydrogen. *Phys. Rev. Lett.* **47**, 1631–1633.

Küppers, D., Koenings, H., and Wilson, H. (1978). Deposition of fluorine doped silica layers from a $SiCl_4/SiF_4/O_2$ gas mixture by the plasma-CVD method. *J. Electrochem. Soc.* **125** (8), 1298–1302.

Küppers, D., Koenings, J., and Wilson, H. (1979). Influence of the substrate temperature on the deposition properties for the plasma activated chemical vapor deposition process. *Proc., 5th Eur. Conf. Opt. Comm.,* Amsterdam, **3.5**-1–4.

Kurkjian, C. R., and Paek, U. C. (1983). Single-valued strength of "perfect" silica fibers. *Appl. Phys. Lett.* **R42** (3), 251–253.

Kurkjian, C. R., Ed. (1985). "The Strength of Inorganic Glass," Plenum Press, New York.

Kuwahara, T., Watanabe, M., Suzuki, S., and Sudo, S. (1981). Refractive index profile formation mechanism on VAD fiber. *Proc., 7th Eur. Conf. Opt. Commun.,* Copenhagen, **2.2**-1–4.

Kyoto, M., Kanamori, H., Yoshioka, N., and Tanaka, G. (1984). Fluorine doping in the VAD sintering process. *Tech. Dig., Opt. Fiber Comm. Conf.,* New Orleans, **MG5**, 22–23.

Lee, R. W., Frank, R. C., and Swets, D. E. (1962). Diffusion of hydrogen and deuterium in fused quartz. *J. Chem. Phys.* **36**, 1062.

Lee, R. W. (1963). Diffusion of hydrogen in natural and synthetic fused quartz. *J. Chem. Phys.* **38**, 448–455.

Lee, R. W., and Fry, D. L. (1966). A comparative study of the diffusion of hydrogen in glasses, *Phys. Chem. Glasses* **7**, 19–28.

Lemaire, P. J., and Tomita, A. (1984). Behavior of single mode MCVD fibers exposed to hydrogen. *Proc. 10th Eur. Conf. Opt. Comm.,* Stuttgart, 306–307.

Lennartz, M., Rau, H., Trafford, B., and Ungelenk, J. (1983). Hydroxyl incorporation in SiO_2/GeO_2. Prepared by the *PCVD Process Proc., 9th Eur. Conf. Opt. Comm.,* Geneva, 21–24.

Levin, P. S., and Pinnow, D. A. (1983). Chemical treatment of the 630 nm defect in silica fibers. *Tech. Dig., Opt. Fib. Comm. Conf.,* New Orleans, 46–47.

Lin, C., Cohen, L. G., French, W. G., and Foertmeyer, V. A. (1978). Pulse delay measurements in the zero material dispersion region for germanium and phosphorus-doped silica fibers. *Electron. Lett.* **14**, 170–172.

Lines, M. E. (1984). The search for very low loss fiber optic materials. *Science* **226**, 663–668.

Lines, M. E. (1986). Ultralow-loss glasses. *Ann. Rev. Mat. Sci.* **16**, 113–135.

Luther-Davies, B., Payne, D. N., and Gambling, W. A. (1975). Evaluation of material dispersion in low-loss phosphosilicate core optical fibers. *Opt. Commun.* **13**, 84–88.

Lydtin, H. (1986). PCVD: A technique suitable for large-scale fabrication of optical fibers. *J. Lightwave Tech.* **LT-4** (8), 1034–1038.

MacChesney, J. B., O'Connor, P. B., DiMarcello, F. V., Simpson, J. R. and Lazay, P. D. (1974). Preparation of low loss optical fibers using simultaneous vapor phase deposition and fusion. *Proc. 4th Intern. Cong. Glass,* 6.40–6.44.

MacChesney, J. B., and O'Connor, P. B. (1980). Optical fiber fabrication and resultant product, U.S. Patent 4,217,027.

MacChesney, J. B., Johnson, D. W., Jr., Lemaire, P. J., Cohen, L. G., and Rabinovich, E. M. (1985). Depressed index substrate tubes to eliminate leaky mode losses in single mode fibers. *J. Lightwave Tech.* **LT-3** (5), 942–945.

Malitson, I. H. (1965). Interspecimen comparison of the refractive index of fused silica. *J. Opt. Soc. Am.* **55** (10), 1205–1209.

Marrone, M. J., Rashleigh, S. C., and Blaszyk, P. E. (1984). Polarization properties of birefringent fibers and stress rods in the cladding. *J. Lightwave Tech.*, **LT-2** (2) 155–160.

Matsuyama, I., Susa, K., Satoh, S., and Suganuma, T. (1984). Synthesis of high purity silica by the Sol-Gel method. *Bull. Am. Ceram. Soc.* **63** (11), 1408–1411.

Mattern, P. L. (1976). Induced infrared absorption in H$_2$ containing vitreous silica. *Bull. Am. Phys. Soc.* **21**, 226.

Maurer, R. D. (1973a). Method of forming an economic optical waveguide fiber, U.S. Pat. 3,737293.

Maurer, R. D. (1973b). Glass fibers of optical communication. *Proc., IEEE* **61** (4), 452–62.

Maurer, R. D. (1975). Strength of fiber optical waveguides. *Appl. Phys. Lett.* **27** (4), 220–221.

Maurer, R. D. (1986). Behavior of flaws in fused silica fibers, In "Strength of Inorganic Glasses." Ed. C. R. Kurkjian, Plenum Press, New York, 291–308.

Mecholsky, J. T., Lau, J., MacKenzie, J. D., Tran, D. and Bendow, B. (1983). Fracture analysis of fluoride glass fibers. *Proc., 2nd Intern. Symp. Halide Glass,* Troy, New York, Paper 32.

Midwinter, J. E. (1979). "Optical Fibers for Transmission." John Wiley & Sons, Ltd., New York.

Mies, E. W., Philen, D. L., Reents, W. D., and Meade, P. A. (1984). Hydrogen susceptability studies pertaining to optical fiber cables. *Proc., Opt. Fiber. Comm. Conf.*, New Orleans, Post Deadline, WI-5.

Mies, E. W., and Soto, L. (1985). Characterization of the radiation sensitivity of single mode optical fibers. *Proc. IOOC-ECOC,* Venezia, 255–258.

Miller, S. E., and Chynoweth, A. G., Eds. (1979). "Optical Fiber Telecommunication." Academic Press, New York.

Mimura, Y., Okamura, Y., Komazawa, Y., and Ota, C. (1980). Growth of fiber crystals for infrared transmission. *Jap. J. Appl. Phys.* **19**, L269–L272.

Mita, Y., Matsushita, S., Yanase, T., Nomura, H. (1977). Optical absorption characteristics of hydroxyl radicals in phosphorus-containing silica glass. *Electron Lett.* **13** (2), 55–56.

Mitachi, S., Miyashita, T., and Kanamori, T. (1981). Fluoride glass cladded optical fibers for mid-infrared ray transmission. *Electron. Lett.* **17**, 591–592.

Mitachi, S., and Miyashita, T. (1982). Preparation of low loss fluoride glass fiber. *Electron Lett.* **18** (4), 170–171.

Mitachi, S., and Miyashita, T. (1983). Refractive index dispersion for BaF$_2$-GdF$_3$-ZrF$_4$-AlF$_3$ glasses. *Appl. Opt.* **22**, 2419–2425.

Mitachi, S., Ohishi, Y., and Takahashi, S. (1984). Preparations of fluoride optical fiber. *Rev. ECL* **32** (3), 461–468.

Mitsunaga, Y., Katsuyama, Y., Kobayashi, H., and Ishida, Y. (1982). Failure prediction for long length optical fiber based on prooftesting. *J. Appl. Phys.* **53** (7), 4847–4853.

Miya, T., Teranuma, Y., Hosaka, T., and Miyashita, T. (1979a). Ultra low-loss single mode fibers at 1.55 μm. *Rev., ECL* **27** (7–8), 497–505.

Miya, T., Teranuma, Y., Hosaka, T., and Miyashita, T. (1979b). Ultra low-loss single mode fibers at 1.55 μm. *Electron. Lett.* **15**, 106–108.

Miya, T., Nakahara, M., and Inagaki, N. (1984). VAD single mode fiber fabrication techniques. *Rev. ECL* **32** (3), 411–417.

Miyamoto, M., Akiyama, M., Shioda, T., Sanada, K., and Fukuda, O. (1983). Fabrication and transmission characteristics of VAD fluorine doped single mode fibers. *9th Eur. Conf. Opt. Comm.,* Geneva, 9–12.

Miyamoto, M., Ohashi, T., Yamauchi, R., Fukada, O., and Inada, K. (1985). High rate deposition of VAD graded index fibers. *Tech. Dig. Opt. Fib. Comm. Conf.*, San Diego, MF4 16–17.

Miyashita, T., Horiguchi, M., and Kawana, A. (1977). Wavelength dispersion in a single mode fiber. *Electron. Lett.* **13**, 227–228.

Miyashita, T., and Manabe, T. (1982). Infrared optical fibers. *IEEE Trans. Mic. Theory & Tech.* **MTT30** (10), 1420–1438.

Mochizuki, K., Namihira, Y., and Yamamoto, H. (1983). Transmission loss increase in optical fibers due to hydrogen permeation. *Elect. Lett.* **19** (18), 743–745.

Mochizuki, K., Namihara, Y., Kuwazura, M., and Iwamoto, Y. (1984). Behavior of hydrogen molecules absorbed on silica in optical fibers. *IEEE, J. Quant. Electron* **QE20** (7), 694–697.

Morey, G. W. (1954). "The Properties of Glass, Second Edition." Reinhold Publishing Corp., New York.

Moriyama, T., Fukuda, O., Sanada, K., Inada, K., Edahiro, T., and Chida, K. (1980). Ultimately low OH content VAD fibers. *Electron. Lett.* **16** (18), 693–699.

Morrow, A. J., and Schultz, P. C. (1983). Recent advances in the outside vapor deposition (OVD) process. *LIA*, Vol. 40, ICALEO, 3–6.

Morrow, A. J., Sarkar, A., and Shults, P. C. (1985). "Outside Vapor Deposition in Opt. Fiber Comm." Vol. 1, Ed. T. Li, Academic Press, New York, 65–94.

Mühlich, A., Rau, K., Simmat, F., and Treber, N. (1975). A New doped fused silica as bulk material for low loss optical fibers. *1st Eur. Conf. on Opt. Comm.*, London, Post Deadline.

Murakami, Y., Uesugi, N., Noguchi, K., and Mitsunaga, Y. (1983). Optical fiber loss increase in the infrared wavelength region induced by electric current. *Appl. Phys. Lett.* **43**, 896–897.

Murakami, Y., Noguchi, K., Ishihara, D., and Negishi, Y. (1984). Fiber loss increase due to hydrogen generated at high temperatures. *Electron Lett.* **20**, 226–228.

Murakami, Y., Noguchi, K., Useugi, N., Ishihara, K., and Negishi, Y. (1985). Optical fiber loss increase in the infrared wavelength region due to hydrogen molecules induced by electrolysis. *Trans. IECE*, Japan **68** (2), 65–70.

Murata, H. (1985). "Optical Fiber communications." Vol. 1., Ed. T. Li, Academic Press, New York, 297–352.

Murata, H. (1986). Recent developments in vapor phase axial deposition. *J. Lightwave Tech.* **LT4** (8), 1026–1033.

Nagasawa, K., Tanabe, M., Yahagi, K., Iino, A., and Kuroha, T. (1984a). Gamma-ray induced absorption band at 770 nm in pure silica core optical fibers. *Jap. J. Appl. Phys.* **23** (5), 606–611.

Nagasawa, K., Tanabe, M., and Yahagi, K. (1984b). Gamma-ray-induced absorption bands in pure silica core fibers. *Jap. J. Appl. Phys.* **23** (12), 1608–1613.

Nagasawa, K., Hashi, Y., Ohki, Y., and Yahagi, K. (1985). Improvement of radiation resistance of pure silica core fibers by hydrogen treatment. *Jap. J. Appl. Phys.* **24** (9), 1224–1228.

Nagasawa, K., Todoriki, T., Fujii, T., Ohki, Y., and Hama, Y. (1986a). The 1.52 μm absorption band in optical fibers induced by hydrogen treatment. *Jap. J. Appl. Phys.* **25** (10), 853–855.

Nagasawa, K., Hashi, Y., Ohki, Y., and Yahagi, K. (1986b). Radiation effects on pure silica core optical fibers by γ-rays. Relation between 2 eV band and non-bridging oxygen hole centers. *Jap. J. Appl. Phys.* **25** (3), 464–468.

Nagel, S. R., MacChesney, J. B., and Walker, K. L. (1982a). An overview of the modified vapor deposition (MCVD) process. *IEEE J. Quantum Electron.* **QE18**, 459–476.

Nagel, S. R., Kosinski, S. G., and Barns, R. L. (1982b). Low OH MCVD optical fiber fabrication. *Cer. Bul.* **61** (8), 822.

Nagel, S. R. (1983). Advances in the MCVD process rate and fiber performance. *Tech. Dig., IOOC*, Tokyo, Japan, 2–3.

Nagel, S. R., MacChesney, J. B., and Walker, K. L. (1985). "Modified Chemical Vapor Deposition in Optical Fib. Comm." Vol. 1, Ed. T. Li, Academic Press, New York, 1–64.

Nagel, S. R. (1980). Reliability issues in optical fibers. *SPIE Proceedings*, Vol. **717**, 8–20.

Nakahara, M., Ohmori, Y., Itoh, H., Shimizu, M., and Inagaki, N. (1986). Formation and disappearance of defect centers in GeO_2-doped silica fibers with heat treatments. *J. Lightwave Tech.* **LT4** (2), 127–132.

Naruse, T., Sugawara, Y., and Masuno, K. (1977). Nylon jacketed optical fiber with silicone buffer layer. *Electron. Lett.* **13**, 153–154.

Nassau, K. (1981a). The material dispersion zero in infrared optical waveguide materials. *Bell Syst. Tech. J.* **60**, 327–337.

Nassau, K. (1981b). Materials dispersion zero in glass mixtures. *Electron Lect.* **17**, 768–769.

Nassau, K., and lines, M. E. (1986). Calculation of scattering and dispersion-related parameters for ultra low-loss optical fibers. *Opt. Eng.* **25** (4), 602–607.

Newns, G. R., Pantelis, P., Wilson, J. L., Uffen, R. W. J., and Worthington, R. (1973). Absorption losses in glasses and optical waveguides. *Optoelectronics* **5** (1973), 289–296.

Niizeki, N., Inagaki, N., and Edahiro, T. (1985). In "Optical Fiber Communications." Vol. 1, Ed. T. Li, Academic Press, New York, 97–178.

Noda, J., Okamoto, K., and Sasaki, Y. (1986). Polarization maintaining fibers and their applications. *J. Lightwave Tech.* **LT4** (8), 1071–1089.

Noguchi, K., Shibata, N., Uesugi, N., and Negishi, Y. (1985). Loss increase for optical fibers exposed to hydrogen atmosphere. *J. Lightwave Tech.* **LT3** (2), 236–243.

Norman, S. R., Payne, D. N., Adams, M. J., and Smith, A. M. (1979). Fabrication of single mode fibers exhibiting extremely low polarization birefringence. *Electron Lett.* **15** (11), 309–331.

Ogai, M., Iino, A., Matsubara, K., and Katsuhiko, O. (1986). Absorption peak at 1.52 μm in silica fiber. *Proc. 12th Eur. Conf. Opt. Comm.*, Barcelona, 7–10.

Ogai, M., Iino, A., and Matsubara, K. (1987). Behavior of alkali impurities and their adverse effect on germania doped silica fibers. *Tech. Dig., Opt. Fiber Comm. Conf.*, Reno, PD3.

Ohishi, Y., Mitachi, S., Kanamori, T., and Manake, T. (1983). Optical absorption of 3d transition metal and rare earth elements in zirconium fluoride glasses. *Phy. Chem. Glasses*, **24** (5) 135–140.

Ohmori, Y., Itoh, H., Nakahara, M., and Inagaki, N. (1983). Loss increase in silicone coated fibers with heat treatment. *Electron. Lett.* **19**, 1006–1008.

Okamoto, K., Hosaka, T., and Noda, J. (1985). High birefringence polarizing fiber with flat cladding. *J. Lightwave Tech.* **LT3** (4), 759–762.

Okamura, Y., Mimura, Y., Komazawa, Y., and Ota, C. (1980). CsBr and CsI crystal fibers for infrared transmission. *Paper Tech. Group OQE, IECE*, Japan, 25–30.

Okoshi, T. (1981). Single polarization single mode optical fibers. *IEEE J. Quant. Electron.* **QE17** (6), 879–884.

Okoshi, T., Oyamada, K., Nishomura, M., and Yokota, H. (1982). Side tunnel fiber: an approach to polarization maintaining optical waveguide scheme. *Electron. Lett.* **18** (19), 824–826.

Olshansky, R., and Keck, D. B. (1976). Pulse broadening in graded-index optical fibers. *Appl. Opt.* **15**, 483–491.

Olshansky, R. (1981). Optical properties and waveguide materials 1.2 μm to 1.8 μm, In "Physics of Fiber Optics, Advances in Ceramics." Vol. 2, Ed. B. Bendow and S. S. Mitra, Plenum Press, New York, 40–46.

Osanai, H., Shioda, T., Moriyama, T., Araki, S., Horiguchi, M., Izawa, T., and Takata, H. (1976). Effects of dopants on transmission loss of low-OH content optical fibers. *Electron. Lett.* **12**, 549–590.

Osanai, H. (1978). Fabrication of ultra-low-loss, low OH content high silica optical fibers. *Ext. Abstr., Fall Meeting, Electrochem. Soc.*, Pittsburgh, 367–369.

Paek, U. C., and Runk, R. B. (1978). Physical behavior of the neck-down region during furnace drawing of silica fibers. *J. Appl. Phys.* **49**, 4417–4422.

Paek, U. C., and Schroeder, C. M. (1981). High speed coating of optical fibers with UV curable materials at a rate of greater than 5 m/sec. *Appl. Opt.* **20**, 4028–4034.

Paek, U. C. (1986). High-speed high-strength fiber drawing. *J. Lightwave Tech.* **LT-4** (8), 1048–1060.

Pavelchuk, E. K., and Dorelmus, R. H. (1976). Static fatigue in glass—a reappraisal. *J. Non-Cryst. Solids* **20**, 305–321.

Payne, D. N., and Gambling, W. A. (1974). New silica-based low loss optical fiber. *Electron. Lett.* **10** (15), 289–290.

Payne, D. N., and Gambling, W. A. (1975). Zero material dispersion in optical fibers. *Electron. Lett.* **11**, 176–178.

Payne, D. N., and Hartog A. (1977). Determination of the wavelength of zero material dispersion in optical fibers by pulse-delay measurements. *Electron. Lett.* **13**, 627–269.

Payne, D. N., Barlow, A. J., and Ramskov Hansen, J. T. (1982). Development of low and high birefringence optical fibers. *IEEE J. Quant. Elect.* **QE-18** (4), 477–488.

Pearson, A. D. (1980). Hydroxyl contamination of optical fibers and its control in the MCVD process. *Tech. Dig., Sixth Eur. Conf. on Opt. Fib. Comm.*, York, England, 22–25.

Pearson, A. D. (1981). Fabrication of single mode fiber at high rate in very long lengths for submarine cable. *Tech. Digest, IOOC*, San Francisco, CA, WA3, 86–87.

Pearson, A. D., Lazay, P. D., and Reed, W. A. (1982a). Fabrication and properties of single mode optical fiber exhibiting low dispersion, low loss, and tight mode confinement simultaneously. *Bell Syst. Tech. J.* **61** (2), 262–66.

Pearson, A. D., Lazay, P. D., Reed, W. A., and Saunders, M. J. (1982b). Bandwidth optimization of depressed index single mode fiber by means of a parametric study. *Conf. Proc. 8th Europ. Conf. on Opt. Commun.*, Cannes, France, 93–97.

Peelen, J., Pluijms, R., and Koel, G. (1984). Dipless multimode and monomode fibers manufactured by the PCVD process. *Proc., Opt. Fib. Commun. Conf.*, New Orleans, **TUM-3**, 68–69.

Pinnow, D. A., van Uitert, L. G., Rich, T. C., Ostermayer, F. W., and Grodkiewicz, W. H. (1975). Investigation of soda-alumino silicate glass systems for application to fiber waveguide. *Mat. Res. Bull.* **10**, 133–146.

Pinnow, D. A., Gentile, A. L., Standlee, A. G., and Timper, A. (1978). Polycrystalline fiber optical waveguide. *Apl. Phys. Lett.* **33**, 28–29.

Pinnow, D. A., Robertson, G. D., and Wysocki, J. A. (1979). Reductions in the static fatigue of silica fibers by hermetic jacketing. *J. Appl. Phys.* **34** (1), 17–19.

Pitt, N. J., and Marshall, A. (1984). Long term stability of single mode optical fibers exposed to hydrogen. *Electron Lett.* **20** (12), 512–514.

Plessner, K. W., and Stannard-Powell, S. J. (1984). Attenuation/time relation for OH formation in optical fibers exposed to H_2. *Electron. Lett.* **20** (6), 250–252.

Poulain, M., Poulain, U., Lucas, J., and Brun, P. (1975). Verres fluores are tetrafluorure de zirconium proprietes optiques d'un verre dope au Nd^{3+}. *Mat. Res. Bull.* **10**, 243–246.

Powers, D. L. (1978). Kinetics of $SiCl_4$ oxidation. *J. Am. Ceram. Soc.* **61** (7–8), 295–297.

Powers, D. R. (1979). Method of making dry optical waveguides, U.S. Patent 4,165,223.

Proctor, B. A., Whitney, I., and Johnson, J. W. (1967). The strength of fused silica. *Proc. Royal. Soc.* **297A**, 534–557.

Puyane, R., Harmer, A. L., and Gonzalez-Oliver, C. J. R. (1982). Optical fibre fabrication by the Sol-Gel method. *Proc., 8th Eur. Conf., Opt. Comun.*, Cannes, C-24.

Robertson, J. (1985). Defect mechanisms in a-SiO_2. *Philos. Mag.* **52** (3), 371–377.

Rabinovich, E. M., Johnson, D. W., Jr., MacChesney, J. B., and Vogel, E. M. (1982). Preparation of transparent high silica glass articles from colloidal gels. *J. Non-Cryst. Solids* **47**, 435–439.

Rabinovich, E. M., Johnson, D. W., Jr., MacChesney, J. B., and Vogel, E. M. (1983). Preparation of high silica glasses from colloidal gels: I, preparation for sintering and properties of sintered glasses. *J. Amer. Ceram. Soc.* **66** (10), 683–688.

Rabinovich, E. M., MacChesney, J. B., Johnson, D. W., Jr., Simpson, J. R., Meagher, B. W., DiMarcello, F. V., Wood, D. L., and Sigety, E. A. (1984). Sol-Gel preparation of transparent silica glass. *J. Non-Cryst. Solids* **63**, 155–161.

Ramaswamy, V., Kaminow, I. P., Kaiser, P., and French, W. G. (1978). Single polarization optical fibers: exposed cladding technique. *Appl. Phys. Lett.* **33**, 814–816.

Ramaswamy, V., Stolen, R. H., Divine, M. D., and Pliebel, W. (1979). Birefringence in elliptically clad borosilicate single mode fibers. *Appl. Opt.* **18** (24), 4080–4084.

Rashleigh, S. C. (1983). Origins and control of polarization effects in single mode fibers. *J. Lightwave Tech.* LT1 (2), 312–321.

Rawson, H. (1980). "Properties and Applications of Glass." Elsevier Scientific Publishing Company, New York.

Ritter, J. E., and Sherburne, L. L. (1971). Dynamic and static fatigue of silicate glasses. *J. Am. Ceram. Soc.* **54** (12), 601–605.

Ritter, J. E., Sullivan, J. M., and Jakus, K. (1978). Application of fracture mechanics theory to fatigue failure of optical glass fibers. *J. Appl. Phys.* **49** (9), 4479–4782.

Runge, P. K., Brackett, C. A., Gleason, R. F., Kalish, D., Lazay, P. D., Meeker, T. R., Ross, D. G., Swan, C. B., Wahl, A. R., Wagner, R. E., Williams, J. C., and Jablonowski, D. P. (1982). 101-Km undersea system experiment at 274 Mb/s. *Tech. Digest, Opt. Fiber Comm.*, Phoenix, PD7, 1–2.

Rush, J. D., Beales, K. J., Cooper, D. M., Duncan, W. J., and Rabone, N. H. (1984). Hydrogen related degradation in optical fibers-system implications and practical solutions. *Br. Telecom. Tech. J.* **2** (4), 84–93.

Sakaguchi, S., Nakahara, M., and Tajima, Y. (1984). Drawing of high strength long length optical fibers. *J. Non Cryst. Solids* **64**, 173–183.

Sakaguchi, S., Itoh, F., Hanawa, F., and Kimura, T. (1985). Drawing induced 1.53 μm wavelength optical loss in single mode fibers drawn at high speeds. *Appl. Phys. Lett.* **47** (4), 344–346.

Sakaguchi, S., Sawaki, S., Abe, Y., and Kawasaki, T. (1982). Delayed failure in glass. *J. Mat. Sci.* **17**, 2878–2886.

Sakaguchi, S., Hibino, Y., and Tajima, Y. (1984). Fatigue in silica glass optical fibers. *Rev. Elec. Commun. Lab.* **32** (3), 444–451.

Sanada, K., Shioda, T., Moriyama, T., Inada, K., Kawachi, M., and Takata, H. (1979). Refractive index profile of the graded index fibers made by the VAD method. *Proc., 5th Eur. Conf. Opt. Comm.*, Paper 1.23 Amsterdam.

Sanada, K., Moriyama, T., Shroda, T., Fukuda, O., Inada, K., and Chida, K. (1981). Behavior of GeO_2 in dehydration process of VAD method. *7th Eur. Conf. Opt. Comm.*, Copenhagen, 2.2-1–4.

Santana, M. R., Lovelace, C. R., Hart, A. C., DiMarcello, F. V., and Blyler, L. L. (1982). Transmission and performance of three optical fiber coatings in a ribbon structure. *Tech. Dig., Opt. Fiber Comm. Conf.*, WCCA4, 40–41.

Sarkar, A., and Schultz, P. C. (1983). Recent advances in the outside vapor deposition process. *Tech. Digest, 4th Intern. Conf. Integr. Optics and Opt. Fib. Commun.*, Tokyo, 27A2-2.

Sasaki, Y., Hosaka, T., and Noda, J. (1984). Fabrication of polarization-maintaining optical fibers with stress-induced birefringence. *Rev. ECL* **32** (3), 452–460.

Savage, J. A., Webber, P. J., and Pitt, A. M. (1980). The potential of Ge-As-Se-Te glasses for 3–5 μm and 8–12 μm infrared optical materials. *Infrared Physics.* **20**, 313–320.

Scherer, G. W. (1977). Sintering of low density glasses, I, II, III. *J. Am. Ceram. Soc.* **60**, 5–6, 236–246.

Scherer, G. W. (1979a). Sintering in homogeneous glasses: application to optical waveguides. *J. Non-Cryst. Solids* **34**, 239–256.

Scherer, G. W. (1978b). Thermal stresses in a cylinder: application to optical waveguide blanks. *J. Non Cryst. Solids* **34**, 223–238.

Scherer, G. W., and Luong, J. C. (1984). Glasses from colloids. *J. Non-Cryst. Solids* **63**, 163–172.

Schneider, H., Harms, H., Papp, A., and Aulich, H. (1978). Low birefringence in single mode fiber; preparation and polarization characteristics. *Appl. Opt.* **17**, 3035–3037.

Schroeder, J., Mohr, R., Macedo, P. B., and Montrose, C. J. (1973). Rayleigh and Brillouin scattering in $K_2O–SiO_2$ glasses. *J. Am. Ceram. Soc.* **56**, 510–513.

Schultz, P. C. (1974). Optical absorption of the transition elements in vitreous silica. *J. Am. Ceram. Soc.* **57** (7), 309–313.

Schultz, P. C., and Maurer, R. D. (1975). Preparation and properties of low loss glass waveguides. *J. Solid State Chem.* **12**, 176.

Schultz, P. C. (1977). Ultraviolet absorption of titanium and germanium in fused silica. *Proc. Int. Congr. Glass, 11.* Vol. **3**, 155–162.

Schultz, P. C. (1979a). Progress in optical waveguide process and materials. *Appl. Optics* **18** (21), 3684–3693.

Schultz, P. C. (1979b). "Vapor Phase Materials and Processes for Glass Optical Waveguides in Fiber Optics: Advances in research and development." Plenum Press, New York, 3–30.

Schultz, P. C. (1980). Fabrication of optical waveguides by the outside vapor deposition process. *Proc. IEEE* **68** (10), 1187–1190.

Shackelford, J. F., Studt, P. L., and Fulrath, F. M. (1972). Solubility of gases in glass II. He, Ne and H_2 in fused silica. *J. Appl. Phys.* **43** (4), 1619–1626.

Shelby, J. E. (1977). Molecular diffusion and solubility of hydrogen isotopes in vitreous silica. *J. Appl. Phys.* **48** (8), 3887–3894.

Shibata, N., Kawachi, M., and Edahiro, T. (1980a). Optical loss characteristics of GeO_2 content silica fibers. *Trans. IECE*, Japan **63** (12), 837–841.

Shibata, S., Terunuma, Y., and Manabe, T. (1980b). Ge-P-S chalcogenide glass fibers. *Jap. J. Appl. Phys.* **19**, L603–L605.

Shibata, S., Horiguchi, M., Jinguji, K., Mitachi, S., Kanamori, K., and Manabe, T. (1981). Prediction of loss minima in the infrared optical fibers. *Electron. Lett.* **17**, 775–777.

Shibata, N., Sasaki, V., O Kamoto, K., and Hosaka, T. (1983). Fabrication of polarization maintaining and absorption-reducing fibers. *J. Lightwave Tech.* **LT1** (1), 38–43.

Shibata, S., and Nakahara, M. (1985a). Fluorine and chlorine effects on radiation induced loss for GeO_2-doped silica optical fibers. *J. Lightwave Tech.* **LT3** (4), 860–863.

Shibata, S., and Nakahara, M. (1985b). Low OH content fiber fabrication by particle size control Sol-Gel method. *Proc., 11th Eur. Conf. Opt. Comm.,* Venice, 3–6.

Shioda, T., Miyamoto, M., Kasaka, K., Sanada, K., and Fikada, O. (1983). High speed production of VAD single mode fibers. *Proc., 4th Intern. Conf. on Integ. Opt. Opt. Fiber Comm.,* Tokyo, 10–11.

Simpkins, P. G., Kosinski, S. G., and MacChesney, J. B. (1979). Thermophoresis: the mass transfer mechanism in modified chemical vapor deposition. *J. Appl. Phys.* **50** (9), 5676–5681.

Simpson, J. R., MacChesney, J. B., and Walker, K. L. (1980). High rate MCVD. *J. Non-Cryst. Sol.* **38**, 831–836.

Simpson, J. R., MacChesney, J. B., Walker, K. L., and Wood, D. L. (1981). "MCVD Preform Fabrication at High Deposition Rates, in Phys. Fib. Opt., Advances in Ceramics." Vol. II, Ed. B. Bendow and S. S. Mitra, 8–13, Plenum Press, New York.

Simspon, J. R., Stolen, R. H., Sears, F. M., Pliebel, W., MacChesney, J. B., and Howard, R. E. (1983). A single polarization fiber. *J. Lightwave Tech.* **1** (2), 370–374.

Simpson, J. R., Ritger, J., and DiMarcello, F. V. (1986). UV radiation induced color centers in optical fibers. *Mat. Res. Soc. Proc.* **61**, 333–338.

Sladen, F. M. E., Payne, D. N., and Adams, M. J. (1977). Measurement of profile dispersion in optical fibers: a direct technique. *Electron Lett.* **13**, 212–213.

Smithgall, D. H., Watkins, L. S., and Frazee, R. E. (1977). High speed non-contact fiber diameter measurement using forward scattered light. *Appl. Opt.* **16**, 2295–2402.

Smithgall, D. H. (1979). On the control of the fiber drawing process. *Tech. Dig., Top. Meeting, Opt. Fiber Transmission,* WF4, 70–71.

Snyder, A. W., and Rühl, F. (1983). Single mode, single polarization fibers made of birefringent material. *J. Opt. Soc. Am.* **73** (9), 1165–1174.

Stolen, R. H., Ramaswany, V., Kaiser, P., and Pliebel, W. (1978). Linear polarization in birefringent single mode fibers. *Appl. Phys. Lett.* **33** (8), 699–701.

Stolen, R. H., Howard, R. E., and Pliebel, W. (1982). Substrate tube lithography for optical fibers. *Electron. Lett.* **18** (18), 764–765.

Stolen, R. H., Pliebel, W., and Simpson, J. R. (1984). High birefringence fibers by preform deformation. *J. Lightwave Tech.* **LT2** (5), 639–641.

Stone, J., and Walrafen, G. E. (1982). Overtone vibrations of OH groups in fused silica fibers. *J. Chem. Phys.* **76** (4), 1712–1722.

Stone, J., Chaplyvy, A. R., and Burrus, C. A. (1982). Gas-in-glass—a new Raman-gain medium molecular hydrogen in solid silica optical fibers. *Opt. Lett.* **7**, 297–299.

Suda, H., Chida, K., and Sudo, S. (1984). High rate fabrication technique for VAD optical fiber preforms. *Rev., ECL,* **32** (3), 418–424.

Suda, H., Shibata, S., and Nakahara, M. (1985). Double flame VAD process for high rate optical preform fabrication. *Electron. Lett.* **21** (1), 29–30.

Sudo, S., Kawachi, M., Edahiro, T., Izawa, T., Shioda, T., Gotoh, H. (1978). Low OH content optical fibre fabricated by vapor phase axial deposition method. *Electron Lett.* **14** (17) 534–535.

Sudo, S., Edahiro, T., and Kawachi, M. (1980). Sintering process of porous preforms made by a VAD method for optical fiber fabrication. *Trans. IECE Jap.* **63** (10), 731–737.

Sudo, S., Kawachi, M., Edahiro, T., and Chida, K. (1981). Transmission characteristics of long length VAD fibers. *Trans IECE Jap.* **64** (3), 175–180.

Sudo, S., Kawachi, M., Suda, H., Nakahara, M., and Edahiro, T. (1981). Refractive index profile control techniques in the vapor-phase axial deposition method. *Trans. IECE Jap.* **64** (8), 536–543.

Sudo, S., Nakahara, M., and Inagaki, N. (1983). A novel high rate fabrication process for optical fiber preforms. *Tech. Digest, 4th intern. Conf., Integ. Optics and Opt. Fib. Comm.,* 14–15.

Susa, K., Matsuyama, I., Satoh, S., and Suganuma, T. (1982). New optical fibre fabrication method., *Electron. Lett.* **18** (12), 499–500.

Sutton, L. E., and Stavroudis, O. N. (1961). Fitting refractive index data by least squares. *J. Opt. Soc. Am.* **51**, 901–905.

Takahashi, H., Sugimoto, I., Sato, T., and Yoshida, S. (1982). GeO_2-Sb_2O_3 glass optical fibers for 2–3 μm fabricated by VAD method in advances in IR fibers. *Tech. Dig., SPIE,* 320–321.

Tamaru, S., Yasu, M., Kawachi, M., and Edahiro, T. (1980). VAD single mode fibre with 0.2 dB/km loss. *Electron. Lett.* **17** (2), 92–93.

Tanimura, K., Sibley, W. A., Suscavage, M., and Drexhage, M. (1985). Radiation effects in fluoride glasses. *J. Appl. Phys.* **58**, 4544–4552.

Tariyal, B. K., Kalish, D., and Santanya, M. R. (1977). Proof testing of long length optical fibers for a communications cable. *Bull. Am. Cer. Soc.* **56** (2), 204–205.

Tariyal, B. K., and Kalish, D. (1978). "Mechanical Behavior of Fibers in Fracture Mechanics of Ceramics," Eds. Brandt, R. C., Hasselman, D. P. H., and Lange, F. F., Vol. 3, Plenum Press, New York, 161–165.

Tomita, A., and Lemaire, P. J. (1984). Observation of a short wavelength loss caused by hydrogen in optical fibers. *Proc., 10th Eur. Conf. Opt. Commun.,* Stuttgart, Post Deadline.

Tomita, A., and Lemaire, P. J. (1985). Hydrogen induced loss increases in germanium doped single mode fibers: long term predictions. *Electron Lett.* **21** (2), 71–72.

Tran, D. C., Fischer, C. F. and Sigel, G. H. (1982a). Fluoride glass preforms prepared by a rotational casting process. *Electron Lett.* **18**, 657–658.

Tran, D. C., Sigel, G. H., Levin, K. H., and Ginther, R. J. (1982b). Rayleigh scattering in ZrF_4-based glasses. *Electron. Lett.* **18**, 1046–1048.

Tran, D. C., Sigel, G. H., and Bendow, B. (1984). Heavy metal fluoride glasses and fibers. *J. Lightwave Tech.* **LT-2** (5), 566–586.

Tran, D. C. (1986). Low optical loss fluoride glass waveguides. *Tech. Dig., Opt. Fib. Comm.*, Atlanta, 20–21.

Tynes, A. R., Pearson, A. D., and Northover, W. R. (1979). Rayleigh scattering losses in soda borosilicate glasses. *J. Am. Ceram. Soc.* **62** (7–8), 324–26.

Urbach, F. (1953). The longwave length edge of photographic sensitivity and of the electronic absorption of solids. *Phys. Rev.* **92**, 1324.

Uchida, N., Uesugi, N., Murakami, Y., Nakahara, M., Tanifuji, T., and Inagaki, N. (1983). Infrared loss increase in silica optical fiber due to chemical reaction of hydrogen. *Proc., 9th Eur. Con. Opt. Comm.*, Geneva, Post Deadline.

Uchida, N., and Uesugi, N. (1986). Infrared optical loss increase in silica fibers due to hydrogen. *J. Lightwave Tech.* **LT4** (8), 1132–1138.

Uesugi, N., Kuwabara, T., Koyamada, Y., Ishida, Y., and Uchida, N. (1983a). Optical loss increase of phosphor-doped silica fiber at high temperature in the long wavelength region. *Appl. Phys. Lett.* **43**, 327–328.

Uesugi, N., Murakami, Y., Tanaka, C., Ishida, Y., Mitsumaga, Y., Negishi, Y., and Uchida, N. (1983b). Infrared loss increase in silica fiber in cable filled with water. *Electron Lett.* **19** (19), 762–764.

Van Dewoestine, R. V., and Morrow, A. J. (1986). Developments in optical waveguide fabrication by the outside vapor deposition process. *J. Lightwave Tech.* 1020–1025.

Van Uitert, L. G., Pinnow, D. A., Williams, J. C., Rich, T. C., Jaeger, R. E., and Grodkiewicz, W. H. (1973). Borosilicate glasses for fiber optic waveguides. *Mat. Res. Bull.* **8** (4), 469–476.

Van Uitert, L. G., and Wemple, S. H. (1978). $ZnCl_2$ glass: a potential ultralow-loss optical fiber material. *Appl. Phys. Lett.* **33** (1), 57–58.

Varnham, M. P., Payne, D. M., Birch, R. D., and Tarbox, E. J. (1983). Single polarization operation of highly birefringent bow tie optical fibers. *Electron Lett.* **19**, 246–247.

Varnham, M. P., Birch, R. D., and Payne, D. W. (1985). Helical-core circularly birefringent fibres. *Proc. IOOC-ECOC*, Venezia, 135–138.

Vitko, J. (1978). ESR studies of hydrogen hyperfine spectra in irradiated silica. *J. Appl. Phys.* **49** (11), 5530–5535.

Vitko, J., Hartwig, J. M., and Mattern, P. L. (1978). Interaction of dissolved hydrogen with a vitreous silica host. *Proc., Intern. TOP Conf. on the Physics of SiO_2 and Its Interfaces*, New York, 215–221.

Wagatsuma, M., Chida, K., and Kimura, T. (1983). High speed coating of optical fibers using pressurized dies. *Proc., IOOC*, 27A4-3, 22–23, Tokyo, Japan.

Walker, K. L., Homsy, G. M., and Geyling, F. T. (1979). Thermophoretic deposition of small particles in laminar tube flow. *J. Colloid. & Interface Science* **69** (1), 138–147.

Walker, K. L., Geyling, F. T., and Nagel, S. R., (1980a). Thermophoretic deposition of small particles in the modified chemical vapor deposition (MCVD) process. *J. Amer. Cer. Soc.* **63** (9–10), 552–558.

Walker, K. L., Harvey, J. W., Geyling, F. T., and Nagel, S. R. (1980b). Consolidation of particulate layers in the fabrication of optical fiber preforms. *J. Amer. Cer. Soc.* **63** (1–2), 96–102.

Walker, K. L., MacChesney, J. B., and Simpson, J. R. (1981). Reduction of hydroxyl contamination in optical fiber preforms. *Tech. Dig., Third Int. Conf. on Integ. Optics and Opt. Fib. Commun.*, San Francisco, CA, 86–88.

Walker, K. L., and Csencsits, R. (1982a). High rate fabrication of single mode fibers. *Tech. Dig., Opt. Fiber Comm., PDI*, 1–2, Phoenix, Ariz.

Walker, K. L., Geyling, F. T., and Csencsits, R. (1982b). The collapse of MCVD optical preforms. *Conf. Proc., 8th Eur. Conf. on Optic Commun.* Cannes, France, 61–65.

Walker, K. L., Csencsits, R., and Wood, D. L. (1983). Chemistry of fluorine incorporation in the fabrication of optical fibers. *Tech. Dig., Opt. Fiber Comm.*, TUA7, New Orleans, LA, 36–37.

Walker, K. L., Broer, M. M., and Carnevale, A. (1986). Dispersion shifted fiber designs with low bending loss for infrared materials. *SPIE* **618**, 17–24.

Wang, T. T., and Zupko, H. M. (1978). Long term mechanical behavior of optical fibers coated with a UV curable epoxy acrylate. *J. Mat. Sci.* **13**, 2241–2248.

Watanabe, M., Matsui, K., Yokota, H., Kanamori, H., Tanaka, G., and Isoya, J. (1985). The reduction of the radiation induced loss increase of pure silica core single mode fiber. *Proc., IOOC-ECOC*, Venezia, 259–262.

Wemple, S. H. (1979). Material dispersion in optical fibers. *Appl. Opt.* **18** (1), 31–35.

Weibull, W. (1939). A statistical theory of the strength of materials. *Proc. Royal Swed. Instit. for Eng. Res.*, No. 151.

Wiederhorn, S. M. (1969). Fracture energy of glass. *J. Amer. Cer. Soc.* **52**, 99.

Wiederhorn, S. M. and Bolz, L. H. (1970). Stress corrosion and static fatigue of glass. *J. Am. Ceram. Soc.* **53**, 543–548.

Wiederhorn, S. M. (1974). Subcritical crack growth in ceramics. In "Fracture Mechanics of Ceramics," Vol. 2, Eds., Brandt, R. C., Hasselman, D. P. H., and Lange, F. F., Plenum Press, New York, 613–646.

Wiederhorn, S. M. (1975). Crack growth as an interpretation of static fatigue. *J. Non-Cryst. Solids* **19**, 169–181.

Wiederhorn, S. M. (1978). Mechanisms of subcritical crack growth in glass. In "Fracture Mechanics of Ceramics," Eds., Brandt, R. C., Hasselman, D. P. H., and Lange, F. F., Plenum Press, New York, 549–580.

Wood, D. L., MacChesney, J. B., and Luongo, J. P. (1978). Investigation of the reactions of $SiCl_4$ and O_2 at elevated temperatures by infrared spectroscopy. *J. Mat. Sci.* **13**, 1761–1768.

Wood, D. L., Kometani, T. Y., Luongo, J. P., and Saifi, M. A. (1979). Incorporation of OH in glass in the MCVD process. *J. Am. Ceram. Soc.* **62** (11–12), 638–639.

Wood, D. L., and Shirk, J. S. (1981). Partition of hydrogen in the modified chemical vapor deposition process. *J. Am. Ceram. Soc.* **64** (6), 325–327.

Wood, D. L., Walker, K. L., Simpson, J. R., MacChesney, J. B., Nash, D. L., and Angueira, P. (1981). Chemistry of the MCVD process for making optical fibers. *Conf. Proc., Seventh ECOC*, Copenhagen, Denmark, 1.2-1.2–4.

Wood, D. L., Walker, K. L., Simpson, J. R., and MacChesney, J. B. (1982). Reaction equilibrium and resultant glass compositions in the MCVD process. *Tech. Dig., Opt. Fib. Comm.*, Phoenix, Ariz, TUCC4.

Wood, D. L., Rabinovich, E. M., Johnson, D. W., Jr., MacChesney, J. B., and Vogel, E. M. (1983). Preparation of high-silica glasses from colloidal gels: III, infrared spectrophotometric studies. *J. Am. Ceram. Soc.* **66** (10), 693–699.

Wood, D. L., Walker, K. L., MacChesney, J. B., Simpson, J. R., and Csencsits, R. (1987). Germanium chemistry in the MCVD process for optical fiber fabrication. *J. Lightwave Tech.* **LT5** (2), 277–285.

Yoshida, K., Sentsui, S., Shii, H., and Kuroha, T. (1977). Optical fiber drawing and its influence on fiber loss. *Int. Conf. Int. Opt. and Opt. Fiber Comm.* 27A3-5, 16–17.

Yoshida, S. (1986). Progress in fiber preparation in Japan. Nato Advanced Research Workshop, Halide Glasses for Infrared Fiber Optics, Vilamoura, Portugal.

Chapter 5

Optical Fiber Cables

CHARLES H. GARTSIDE III

AT&T Bell Laboratories, Transmission Media Laboratory, Norcross, Georgia

PARBHUBHAI D. PATEL

AT&T Bell Laboratories, Transmission Media Laboratory, Norcross, Georgia

MANUEL R. SANTANA

AT&T Bell Laboratories, Transmission Media Laboratory, Norcross, Georgia

5.1 INTRODUCTION

Although optical fiber cables have been commercially manufactured since 1980, the technology has undergone significant changes due to the evolution of fiber designs, system electronics, and markets. Beginning in 1980, multimode fiber was first used in interoffice trunks and a few long-haul networks. Systems initially operated at about 825 nm at 45 Mb/s. The fiber loss was between 3 and 3.5 dB/km and the maximum repeater spacing was about 6 km. High fiber count cables, with between 60 and 144 fibers, were expected to meet the trunk market. In less than a year, improved fiber designs and long wavelength sources made operation at 1300 nm possible. With losses in the 1 dB/km range, repeater spacing tripled. Transmission rates also increased to 90 Mb/s, cutting the fiber count in cables in half. These early systems were usually installed in underground ducts, minimizing the field hazards and simplifying the cable design requirements.

The introduction of single-mode fiber, beginning in 1983, was the next very important shift in the technology. Although multimode fiber is still used in buildings and in local area networks, just about all outside plant cable is now single-mode. The initial market was long-haul systems. With typical losses of about 0.5 dB/km at 1310 nm, the repeaterless spans were about 30 km. Transmission rates were between 400 and 600 Mb/s. In current production cable, the loss is less than 0.4 dB/km, and the span length is about 40 km. System upgrading is also an important design variable. At 1310 nm, the systems can be upgraded by increasing the transmission rate with current technology lasers. Rates up to about 2 Gb/s will be introduced in 1987. Systems can also be upgraded with multiplexing at 1550 nm. Narrow line width lasers will be necessary, since the dispersion is high at 1550 nm in most fiber designs. Typical fiber losses at the long wavelength are less than 0.25 dB/km. Low loss at 1550 nm is an important cable design requirement. With the high transmission rates, long-haul cables generally contain a modest number of fibers—typically between 20 and 30. Long-haul systems impose additional field hazards on the cable—in particular, rodent and lightning damage in open country. The divestiture of AT & T resulted in an unexpected demand for high fiber count long-haul cables. Business arrangements between common carriers called for the sharing of the fiber within one cable among multiple owners. In addition, a few networks were installed in which fibers were sold to private customers. As a result, long-haul cables with as many as 96 fibers have been manufactured and installed. Transmitting at 1310 nm at 2 Gb/s, these cables have the capacity of about one-and-a-half million voice circuits.

In addition to overtaking the long-haul market, single-mode fiber was rapidly applied in central office trunks. Fiber counts are generally in the 24- to 36-fiber range. Within metropolitan areas, interoffice trunks can usually be installed without repeaters. Here, long duct pulls are common and small diameter cables make installation much easier. In the loop market, single-mode fibers are already used extensively in loop feeders, linking central offices with remote terminals. A few distribution networks have also been installed, directly linking large business customers. Both large cables with more than 100 fibers and small distribution cables are needed. The progression of fiber directly to the customer in the loop is still in its infancy. In fact, the question of multimode or single-mode fiber is not entirely settled, due to the high cost of single-mode electronics. In progressing from long-haul to trunk to loop applications, the cable requirements change. Long-haul networks require the lowest loss cable and splicing and long cable lengths, maximizing span lengths. Individual fiber splicing is used exclusively. In the loop, splicing time and the human factors in working with the media are extremely important. Rearrangement or "churning" of the outside plant is common and must be accommodated in cable and apparatus designs. The lowest loss is not essential.

The evolution of fiber and cable clearly shows that cable designs must be flexible. Important design variables are: fiber count, optical loss, environmental stability, mechanical protection, cable handling, fiber splicing, and size and weight, where their relative importance is dictated by the particular application. Optical cable designs, which are currently available, have been derived from the early multimode designs. In all cases, the fibers are loosely packaged. In the late 1970s, there were two leading cable designs: ribbon and loose tube. In the ribbon cable, the fibers are assembled into 12-fiber ribbons and a maximum of 12 ribbons are packaged in a common reinforced sheath. The main features of this design are its high packaging density and the option of factory connectorization, thus allowing for easy installation and splicing. In the loose-tube cable, the fibers are enclosed in tubes that are stranded around a central strength member and then followed by a polymeric sheath. With this design, large fiber strain relief is possible. Two additional designs are now widely used: slotted core and loose fiber bundles. In the slotted-core design, the fibers are inserted in grooves or slots in a preformed core containing a central strength member. In the loose-fiber-bundle design, the fibers are packaged into units that are bound by color-coded binders, and the entire core is enclosed in a single, loose-fitting tube, followed by a reinforced sheath. All four designs are available with single-mode as well as multimode fibers.

In this chapter, the construction and the optical and mechanical performance of these four leading designs for outside plant applications are summarized. The primary emphasis is on single-mode cable used in North America and Europe. Tight buffer cables will also be reviewed briefly. These designs are used in undersea crossings and tactical battlefield communications, providing extremely robust cables for difficult environments.

5.2 DESIGN CONSIDERATIONS

Fiber and cable designs are strongly interrelated in providing good optical and mechanical performance. Materials must be environmentally stable over many years of service. The fiber must be prooftested, ensuring against failure due to installation or in-service stresses. The overall mechanical design requirements are dictated by the outside plant environment—over which the designer has practically no control.

5.2.1 Fiber Design

5.2.1.1 Single-mode. Two single-mode fiber designs are currently in mass production—matched clad and depressed clad fibers. The matched clad fiber has a simple step index profile, with a core diameter of about 8.7 μm and an index difference (Δ) of about 0.3%. The depressed clad fiber has a core diameter of about 8.3 μm, a total index difference of about 0.37%, and an inner cladding depression of about 0.12%. The actual fiber parameters vary slightly, depending on the manufacturer. The fibers have outer cladding diameters of 125 μm and are optimized for zero dispersion at about 1300 nm. Although both fiber designs perform very well in cable, the depressed clad fiber is more resistant to macrobending and microbending, due to tighter mode confinement with its smaller core and larger index difference. At 1310 nm, either fiber is insensitive to bending loss effects. Above about 1500 nm, however, either microbending or macrobending may cause large loss increases. Both fiber and cable design are very important in this long wavelength region. Mean losses for production fiber are typically 0.35 and 0.21 dB/km at 1310 and 1550 nm, respectively.

5.2.1.2 Multi-mode. Multi-mode fibers are commonly manufactured with core diameters of 50 or 62.5 μm. Finely graded parabolic index profiles provide low dispersion and high bandwidths. For the 50 μm core fiber, the maximum index difference varies between 1 and 1.3%, depending on the manufacturer. The 62.5 μm core fiber generally has an index difference of 2%. The 50 μm fiber was initially used in trunk routes, while the 62.5 fiber is intended in loop or LAN systems using LED sources, due to its high NA and coupling efficiency. Good quality fiber has losses in the range of 2.5 to 3.0 dB/km at 850 nm and 0.5 to 1 dB/km at 1300 nm.

5.2.1.3 Coatings. Due to their small diameter and brittle material properties, glass fibers must be protected by polymeric coatings immediately after drawing. Single or dual layers of UV-curable acrylate materials are commonly used. The dual coatings consist of a soft inner layer and a harder outer shell. The elastic modulus for the inner layer is usually less than 7 MPa, while it's about 350 MPa for the outer layer. The outer diameter is about 250 μm, allowing for high fiber draw speeds and low cost. The thin dual-coating structure buffers the fiber from microbending and also minimizes the effect of thermal contraction on the fiber at low temperature.

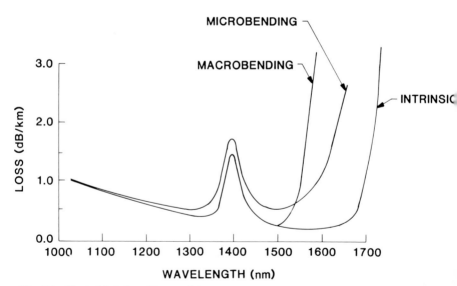

Fig. 5.1 Single-Mode Loss Characteristics

5.2.1.4 Bending Loss.   The transmission characteristics of optical fibers are greatly influenced by fiber design and by the packaging environment. The interaction between fibers and the cable structure can result in random perturbations of the fiber axis, allowing optical energy to radiate through the cladding. This loss mechanism is known as microbending (Gardner, 1975; Gloge, 1975), or macrobending (Marcuse, 1982), depending on the bend spectrum of the fiber axis perturbations. The microbending effect is characterized by a small amplitude, on the order of a nanometer, and a period of critical spatial frequency on the order of a millimeter. Figure 5.1 illustrates the loss mechanisms that can affect a single-mode fiber. Microbending loss is dependent on wavelength. For a given mechanical bend spectrum, the added loss can be minimal at 1310 nm, but substantial at 1550 nm. On the other hand, the wavelength dependence of macrobending is even more dramatic. There is no effect at short wavelengths; however, the macrobending loss is pronounced at about 1500 nm. Both effects are strongly fiber-design dependent.

Microbending loss has been the subject of many theoretical treatments. Although the analytical models cannot generally be applied directly in cable design, they are valuable in interpreting cabling results and in comparing fiber and cable designs. Microbending losses are caused by the coupling between the guided and radiation modes. Microbending is analogous to the common resonance phenomena in which the forcing frequency approaches the system's natural frequency. In a single-mode fiber, the spatial frequency of the bends must be near

$$\overline{\Omega} = \beta_g - n_2 k, \tag{5.1}$$

where β_g and $n_2 k$ are propagation constants of the guided and radiation modes, respectively. The cladding refractive index is n_2. Assuming that the core-cladding boundary of a fiber with random bends is described by

$$r(\phi, z) = a + f(z)\cos\phi \tag{5.2}$$

coupled mode theory predicts the power loss coefficient, 2α, (Marcuse, 1976):

$$2\alpha = \frac{\gamma_0^2 \gamma_1^2 J_0^2(\kappa_0 a) J_1^2(\kappa_1 a)\left[k^3\langle F^2(\overline{\Omega})\rangle\right]}{2a^2 n_2^2 k^5 J_1^2(\kappa_0 a)|J_0(\kappa_1 a) J_2(\kappa_1 a)|}, \tag{5.3}$$

where:

a = core radius

n_1 = refractive index of core

n_2 = refractive index of cladding

ν = azimethal mode number

β_ν = propagation constant

$$k = 2\frac{\pi}{\lambda} = \text{free-space propagation constant}$$

$$\lambda = \text{wavelength}$$

$$\overline{V} = \frac{2\pi a}{\lambda} n_1 (2\Delta)^{1/2}$$

$$\kappa = (n_1^2 k^2 - \beta_\nu^2)^{1/2}$$

$$\gamma^a = [V^2 - (\kappa a)^2]^{1/2}$$

$$J_\nu(X) = \text{Bessel Function.}$$

The term $\langle F^2(\overline{\Omega}) \rangle$ represents the convolution of the mechanical bend spectrum and the mode coupling. The bend spectrum is described by a correlation function $R(u)$. The Fourier spectrum of distortion function $f(z)$ is

$$F(\theta) = \lim_{L \to \infty} \frac{1}{\sqrt{L}} \int_0^L f(z) e^{-i\theta z} \, dz \qquad (5.4)$$

and the ensemble average of the square of $F(\theta)$ can be expressed as the Fourier transform of the correlation function $R(u)$ of $f(z)$:

$$\langle F^2(\theta) \rangle = \int_{-\infty}^{\infty} R(u) e^{-i\theta u} \, du. \qquad (5.5)$$

Although equation (5.3) is complex, the attenuation coefficient can be computed for a given bend spectrum. Unfortunately, it is usually unknown. In order to practically apply equation (5.3) assumptions on the bend spectrum must be made. Marcuse (1984) assumed that the autocorrelation function of $f(z)$ is Gaussian:

$$R(u) = \sigma^2 e^{-(u/L_c)^2}, \qquad (5.6)$$

where σ is the rms deviation of the distortion and L_c is the correlation length. With the added assumptions that the bound mode is Gaussian and the power is coupled to a discrete set of cladding modes, the computations for a step index single-mode fiber are straightforward. Typical results are plotted in Figs. 5.2 and 5.3. In Fig. 5.2, the loss coefficient is plotted versus wavelength for various correlation lengths. The core radius is 5 μm, Δ is 0.3% and σ is 1 nm. The loss and its wavelength dependence are strong functions of L_c. For large values of

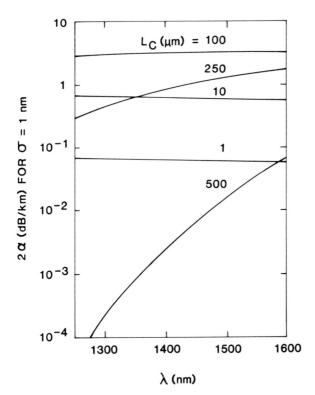

Fig. 5.2 Microbending Loss vs Wavelength for Varying Correlation Length (Lc), a Step-Index Fiber with a Core Radius, $a = 5$ μm, $\Delta = 0.3\%$ and $\sigma = 1$ nm

L_c, such as 500 μm, the loss coefficient is small but strongly dependent on wavelength. As L_c decreases, however, the loss increases greatly and the wavelength dependence is reduced. For $L_c = 100$ μm, the loss coefficient is large, about 5 dB/km. As L_c continues to decrease, 2α goes through a maximum and then decreases. For other values of σ, 2α is proportional to σ^2. Based on the trends predicted in Fig. 5.2 and experimental data, the correlation length in cables is generally greater than about 500 μm. In Fig. 5.3, the effect of core radius is shown for $L_c = 500$ μm and $\sigma = 1$ nm. For each core radius, Δ is adjusted to maintain single-mode operation ($V = 2.405$). Near 1300 nm, the effect of core radius is small, since the field is tightly bound near the cutoff wavelength. At about 1550 nm, however, small reductions in core radius result in large improvements in microbending performance.

Rather than evaluate equation 5.3, semi-empirical approximations can often be used. If a simple power law spectrum is assumed and the results of the computation are empirically fitted to the V number, the following approxima-

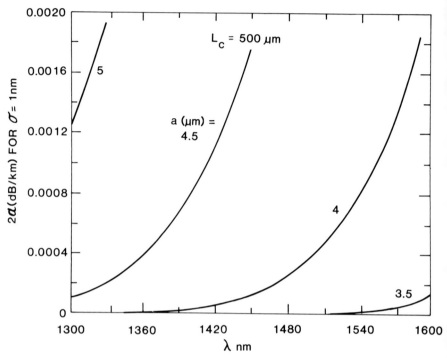

Fig. 5.3 Microbending Loss vs Wavelength for a Step-Index Fiber with a Varying Core Radius (a), $L_c = 500 \ \mu m$, $\sigma = 1$ nm and $V = 2.405$

tions are obtained (Marcuse, 1976):

$$2\alpha = \frac{n_2 k^4 \langle F^2(\overline{\Omega}) \rangle}{(m-1)(m-2)} \left(\frac{\overline{\Omega}}{k} \right)^3 \left\{ V + 7.58 \times 10^{-5} \frac{m^9 V^M}{e^{1.7m}} \right\} \quad 1 < V < 2.405 \quad (5.7)$$

$$2\alpha = \frac{\langle F^2(\overline{\Omega}) \rangle}{2n_2^2 a^6 k^2} \left[9(V - 1.87) - \frac{0.616}{(V - 2.405)^{1/2}} \right] \quad 2.405 < V < 3.832, \quad (5.8)$$

where m is the power law exponent. For a given cable structure, the bend spectrum parameters m and $\langle F^2(\overline{\Omega}) \rangle$ are constant and effect of wavelength on attenuation can be quickly evaluated.

Using a power law bend spectrum, Petermann (1977) has also derived an approximate expression for the attenuation coefficient:

$$2\alpha = K(kn_1\omega)^2 (kn_1\omega^2)^{2p}. \quad (5.9)$$

Here, K is a constant, ω is the spot size and p is the exponent in the power law. The Petermann's treatment is based on deterministic arguments and is less rigorous than coupled-mode theory. Experimental data (Hornung, 1982) has shown the exponent to be in the 4 to 5 range. This simple expression is also valuable in comparing fiber designs, in that the spot size is a convenient parameter. In multimode fiber, the loss mechanisms are similar. The flow of energy between the modes, however, must be included. Mode-coupling is treated as a diffusion phenomena within the core. Gloge (1975) and Olshansky (1975) have developed simplified models, taking into account the mechanical deformation of the fiber and coating. The loss coefficient is given by

$$2\alpha = K\left(\frac{a^4}{\Delta^3 b^6} \right). \tag{5.10}$$

Here, b is the outer cladding radius. In comparing 50 μm fibers with 1 and 1.3% Δs, the higher Δ fiber is about half as susceptible to microbending. A 62.5 μm fiber with $\Delta = 2\%$ has only about one-third the sensitivity of a 50 μm fiber with $\Delta = 1\%$. In most practical cases, the microbending loss in multimode fibers is not strongly wavelength dependent.

5.2.1.5 Hydrogen. In selecting fiber coating and cable materials, careful consideration must be given to hydrogen gas generation. It is now widely recognized that molecules such as hydrogen can diffuse into optical fibers. An optical fiber that has been saturated in an atmosphere of hydrogen will show a broad loss increase of about 6 dB/km at 1240 nm, and 0.2 dB/km at 1310 nm due to absorption by molecular hydrogen (Beales *et al.*, 1983; Mies 1984). This loss increase is reversed when the hydrogen atmosphere is removed and the absorbed hydrogen outdiffuses. Hydrogen gas that has diffused into the fiber may also chemically react with constituents of the fiber. This can result in a permanent loss increase at 1380 nm (Si-OH), 1410 nm (Ge-OH), and 1600 nm (P-OH). The sensitivity of fibers to these loss increases has also been linked to high levels of phosphorus in the fibers (Uesugi *et al.*, 1983). Fibers that contain abnormally high levels of phosphorus (8%) tend to react with hydrogen at elevated temperatures to produce permanent loss increases. The 1410 and 1600 nm absorption bands overlap in the long wavelength region to cause unacceptable loss increases at 1550 nm (Nakahara *et al.*, 1983; Tanaka *et al.*, 1984). Both temporary and permanent loss increases can be prevented through careful cable design, resulting in negligible hydrogen gas generation. For example, a combination of materials to be avoided in sheath construction is aluminum and steel, since they can generate hydrogen by electrolytic corrosion. Negligible amounts of hydrogen is generated with a copper-steel construction. Other materials to be avoided are certain silicone fiber coating compounds. They can degrade with aging, forming hydrogen as a byproduct. The problem of hydrogen gas release can be eliminated through material selection (Philen and Gartside, 1984). Individual materials and complete cables are generally tested in accelerated aging tests. Satisfactory long-term field experience has also been obtained with most production cable designs.

5.2.2 Fiber Strength

A glass fiber under tension behaves as a perfect elastic body with an elastic modulus of 73.8 GPa. Although the pristine strength of the glass fiber in short length is over 4.8 GPa (above 7% strain), the practical stress limit for a longer length is much lower (about 0.5% strain) due to microscopic flaws. Therefore, optical fibers are proof-tested during manufacture, guaranteeing a safe operating stress. Typically, fibers are proof-tested to 0.35 GPa for terrestrial applications and 1.38 GPa for undersea applications.

The presence of randomly distributed flaws makes fiber strength a statistical quantity. Furthermore, the flaw size grows under the influence of time, stress and humidity. For a constant stress, this phenomenon is commonly known as static fatigue (Miller and Chynoweth, 1979). A minimum time to failure (t_f), for a proof-tested fiber (σ_p), using the power law for crack growth (Evans, 1974) and applied tensile stress σ_a, is given by

$$t_f = B\left(\frac{\sigma_a}{\sigma_p}\right)^{-n} \sigma_p^{-2}, \tag{5.11}$$

where

$$B = 2\left[AY^2(n-2)K_{IC}^{(n-2)}\right]^{-1}. \tag{5.12}$$

K_{IC} is the critical stress-intensity factor in tension [for fused silica $K_{IC} = 0.789 \text{MPa}\sqrt{m}$], Y is a geometric constant ($Y = 1.24$), n is the stress-intensity factor exponent (for ambient condition, $23°C$ and 45% RH, $n = 20$; for wet condition, $23°C$ and 97% RH, $n = 14$), and A is known from static fatigue data (in GPa-s units) by

$$\text{Log } A = 3.289n - 10.05. \tag{5.13}$$

Figure 5.4 shows two curves of minimum time to failure vs. stress ratio (σ_a/σ_p) for 0.35 and 1.38 GPa proofstress fibers in a wet environment. The results show that a smaller stress should be applied for a longer life span. Safety factors for short- and long-term loading are used in cable design. Short-term loading is generally limited to about 60% of proof stress. The long-term limit is about 20% of proof stress. In practical applications, the cables and fibers are also subjected to bending. Small bends usually cause high tensile stress over part of the cross section. In practice, the static fatigue curves developed for pure tension are often

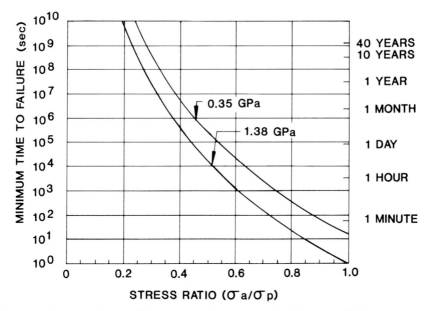

Fig. 5.4 Minimum Time to Failure vs Stress Ratio (σ_a/σ_p) for 0.35 and 1.38 GPa Proof-tested Fiber in a Wet Environment

applied to bending by assuming that the maximum bending stress is equivalent to a tensile stress (Patel *et al.*, 1981). Since the stress distribution is non-uniform, this approach is conservative in failure time estimates.

5.2.3 Cable Tension, Weight and Handling

In addition to providing good optical performance, the physical design of a cable must protect the fiber during installation and service. The three most important design parameters are tensile load rating, diameter, and bend radius. They are all derived from the outside plant requirements. Tensile load ratings are specified for short- and long-term loading. For the short term, a 2700 N limit is typical and adequate for cable installation in underground ducts or by direct plowing. A typical long-term limit is 600 N, which may be applied for many years. These load ratings are based on the fiber proof-stress level, cable tensile stiffness (expressed in N/% of strain), strain relief, and appropriate safety factors for static fatigue. In most cable designs, the fibers are intentionally longer than the cable sheath, providing strain relief when the cable is tensioned. The amount of excess length is a critical design parameter, which varies with cable design (on the order of tenths of a percent). Although large strain relief is desirable, the excess fiber length must be accommodated within the cable structure without inducing bending losses.

Physical size and weight are also very important. Cable diameters are generally between 10 and 30 mm and weights are generally between 75 and 600 Kg/km. The weight is approximately proportional to diameter squared (area). Cables are commonly available in long reel lengths (nominally 2 to 5 km), allowing for long underground pulls, which minimize splicing. During installation, the cable must be pulled under tension through manholes and around sheaves, and the bend radius must be set to prevent sheath damage or kinking. Under load, the radius is usually limited to 20 times the cable diameter. The radius of coils to store slack cable in manholes must also be manageable. For slack cable, it is usually 10 times the cable diameter. In general, smaller and lighter cables are easier to handle, require less space, and longer lengths may be installed with a considerable savings to the customer.

5.3 CABLE CONSTRUCTION

During the last decade, a large variety of lightguide cable designs have been developed. Those used in the largest volume of applications are the loose tube, loose fiber-bundle, ribbon and slotted core designs. Although each design has its own advantages, they all have two common characteristics. The fibers are loosely packaged in a unit construction and the core is filled. The loose construction allows for free fiber movement in the core. Fiber stresses can be relieved, and microbending losses are minimized. Early cable designs were air core constructions and pressurization systems were used to keep the fibers dry. Filled core designs are now used almost exclusively, greatly simplifying cable installation and maintenance. One distinguishing feature is strength member placement. In the loose-tube and slotted-core designs, the strength members are placed in the core. They are included in the sheath in the loose-fiber-bundle and ribbon designs.

5.3.1 Loose Tube

In the loose tube design (Oestreich et al., 1981; Bark and Oestreich, 1978; Jackson, et al., 1977), a filled loose tube is extruded over a fiber bundle, with typically up to 12 fibers per tube. Multiple tubes are stranded around a center strength member forming the cable core, and the cable is completed with a polyethylene sheath. A typical loose-tube cable is illustrated in Fig. 5.5. The tube is filled with a soft thixotropic material allowing for free fiber movement in the tube. This most common tube materials are nylon, polypropelene, or dual extrusions of polymers. The tube od varies from about 1 to 3 mm. The fibers are generally slightly longer than the tube by between 0 and 0.1%. The excess length in a single tube is illustrated in Fig. 5.6. The fiber takes a periodic shape such as

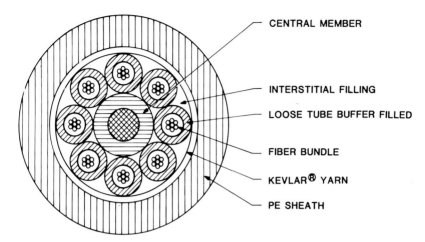

CENTRAL MEMBER

INTERSTITIAL FILLING

LOOSE TUBE BUFFER FILLED

FIBER BUNDLE

KEVLAR® YARN

PE SHEATH

Fig. 5.5   Loose Tube cable

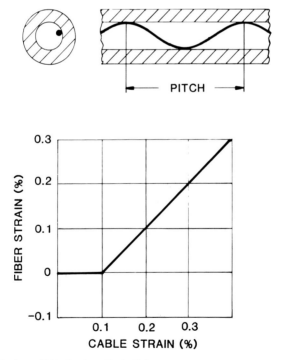

Fig. 5.6   Fiber Strain vs Cable Strain—Single Tube

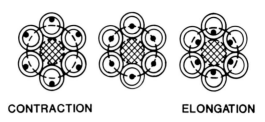

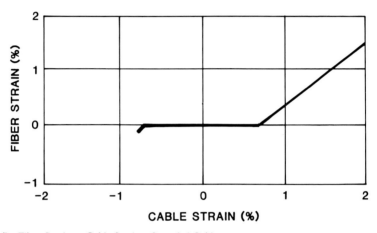

Fig. 5.7   Fiber Strain vs Cable Strain—Stranded Cable

a sinusoid or helix. If the excess length exceeds about 0.1%, high microbending losses can occur. In stranding the tubes around the central strength member, either full helical or S-Z stranding may be used. The stranding pitch is generally about 10 cm. In addition to the central strength member, DuPont Kevlar® or glass yarns are sometimes stranded over the cable core. The short stranding pitch and the free radial fiber movement within the tubes provide axial strain relief of applied tensile or thermal contraction loads. This is illustrated in Fig. 5.7. Depending on the tube dimensions and the stranding pitch, strain relief values of up to about 1% are possible. Various size cables are fabricated with multiple tubes. Typical cable dimensions range from about 12 to 21 mm, depending on fiber count. The physical design parameters for loose tube and other cables are summarized in Table 5.1.

5.3.2 Loose Fiber-Bundle

The basic cable construction (Patel and Gartside, 1985; Gartside and Cottingham, 1986a) is shown in Fig. 5.8. Up to 12 fibers are assembled in a unit or bundle, which is identified with a color-coded binder. Several units are then

TABLE 5.1
Physical Design Summary

Cable Diameter (mm)				
Fiber Count	22	44	72	96	144
Lose Tube	12	14	16	19	21
Loose Fiber-Bundle	10	10	13	13	—
Ribbon	13	13	13	13	13
Slotted Core	15	17	17	—	—
Weight (kg / km)					
Fiber Count	22	44	72	96	144
Loose Tube	120	175	225	300	355
Loose Fiber-Bundle	120	120	135	135	—
Ribbon	135	135	135	135	135
Slotted Core	190	300	300	—	—
Installation Bend Radius (cm)					
Fiber Count	22	44	72	96	144
Loose Tube	26	30	33	39	43
Loose Fiber-Bundle	20	20	26	26	—
Ribbon	26	26	26	26	26
Slotted Core	30	34	34	—	—

assembled into a cable core. A large diameter filled tube is extruded over the core. The tube diameter is typically between 6 and 8 mm. As with the loose-tube cables, an essential design feature is the free movement of the fibers in the filling material. The core undulates in the tube to accommodate the excess fiber length, as illustrated in Fig. 5.6 for an individual fiber. This provides low optical loss and excellent mechanical performance. The cable is completed with a reinforced

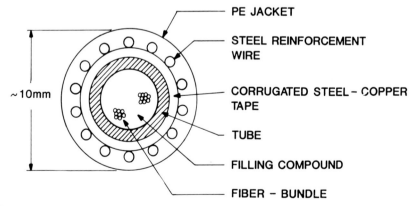

PE JACKET

STEEL REINFORCEMENT WIRE

CORRUGATED STEEL – COPPER TAPE

TUBE

FILLING COMPOUND

FIBER – BUNDLE

~10mm

Fig. 5.8 Loose-Fiber-Bundle Cable with Rodent and Lightning Protection

sheath. The sheath illustrated is reinforced with steel wires and includes a corrugated copper-steel laminate for rodent and lightning protection. In the standard sheath, the reinforcing elements are applied helically in two layers with opposite stranding directions providing a torque-balanced cable, i.e., easy to handle with no twist under tensile load. The helical pitches of the strength member layers are set to provide good bending performance. Steel wire or non-metallic strength elements are used. The core construction allows for very compact and low-weight cables, as illustrated in Table 5.1. For example, the cable diameter is about 10 mm for fiber counts between 4 and 48 and about 13 mm for between 50 and 96 fibers.

The core design also provides a unique feature during splicing. Cable preparation is very simple. The end of the cable can be stripped back in about five minutes. After exposing and cutting the sheath reinforcement, the core tube is ring cut. The entire sheath then slides off of the core, exposing the fiber units. The color-coded binders provide unit identification. The core design provides easy access to the individual fibers—a particularly important feature in loop-distribution networks.

5.3.3 Ribbon Cable

As illustrated in Figs. 5.9 and 5.10, the lightguide ribbon cable is based on a ribbon that is manufactured by packaging 12 fibers between two adhesive-backed polyester tapes (Saunders and Parham, 1977; Gagen and Santana, 1979; Eichenbaum and Santana, 1982). The ribbon structure was originally developed for multimode fiber and allowed factory-installed connectors and rapid splicing. The design was extended to include single-mode fiber in 1984 (Gartside and Baden, 1985). Up to 12 ribbons may be stacked into a rectangular array for fiber counts as high as 144 per cable. After twisting the ribbon stack to provide good bending performance, a loose plastic tube is extruded over the core. The tube is filled with a soft water-blocking compound. The filling compound allows for free movement of the ribbons within the core. This loose construction is essential in providing low optical loss. The ribbons are slightly longer than the tube, assuring that the fibers will be stress-free. Finally, a reinforced high-density

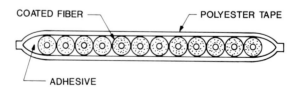

Fig. 5.9 Adhesive Sandwich Ribbon (ASR)

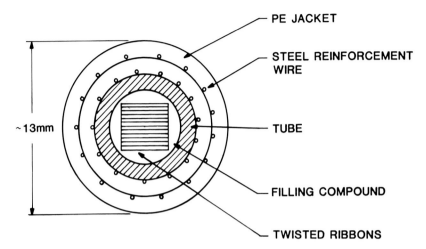

PE JACKET

STEEL REINFORCEMENT WIRE

~13mm

TUBE

FILLING COMPOUND

TWISTED RIBBONS

Fig. 5.10   Ribbon Cable with a Steel-reinforced Crossply Sheath

polyethylene sheath is applied over the core tube. The complete cable has an outer diameter of 13 mm.

For installations requiring maximum repeater spacing, minimum splice loss is obtained using individual fiber splicing. Within the splice case, each ribbon is handled with the same flexibility as a loose-tube or loose-fiber-bundle unit. Simply by peeling apart the polyester tapes, individual fibers may be spliced using fusion, bonded or mechanical techniques. The major advantages of ribbon cable are its compact size and rapid mass splicing. In most metropolitan and loop installations, where splice loss requirements are less critical, splicing time can be greatly reduced using factory-connectorized ribbons. Silicon array connectors are applied to each ribbon in the factory, and high-productivity splicing is accomplished by simply joining two arrays together in the field. Twelve fibers are spliced simultaneously. Mean splice loss values of about 0.4 dB with single-mode fibers have been obtained under a large variety of field conditions. With array splicing, 144 single-mode fibers can be spliced in less than three hours by craft personnel.

In addition to the adhesive-backed ribbon structure, other ribbon designs are available. The fibers are arranged in a linear array and a plastic material is extruded over it (Katsuyama *et al.*, 1985) or the fibers are bonded together with a UV-curable material, similar to the fiber coating (Nirasawa *et al.*, 1985). A variety of cable structures have been proposed. In one case, about 100 fibers are packaged in a tube and multiple tubes are stranded together, forming the cable core. Cables with fiber counts exceeding a thousand have been proposed.

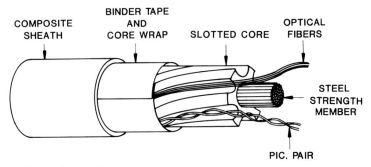

Fig. 5.11   Slotted Core Cable

5.3.4 Slotted Core

The basic slotted-core construction is illustrated in Fig. 5.11. A slotted profile is extruded over a steel strand using a hard plastic to form the basic core (Hope *et al.*, 1981). The steel strand provides tensile load capability and limits compressive strains at low temperatures. During the extrusion process, the core is oscillated downstream of the extrusion head and, in this manner, an oscillated lay is formed and frozen into the core. The use of an oscillated lay structure eliminates the need for either a rotating die, or rotating take-ups, and, therefore, allows high manufacturing speeds. The primary advantages of the design are easy manufacture and versatility. The core is also filled with a soft, water-blocking material. Hydroscopic powders have also been used.

Core designs with various size and numbers of slots have been developed. Up to about 10 fibers may be laid in each slot. Once the fibers are laid in the slot, a binder covers the slot and the sheath is applied. Instead of fibers, plastic-insulated copper pairs can also be laid into the slot. For example, a 6-slot structure with a bonded aluminum polyethylene sheath typically contains 24 fibers and 2 copper pairs for an overall diameter of 15 mm. A larger 8-slot structure accommodates 48 fibers and 3 copper pairs within a diameter of 17 mm. Without copper pairs, cables with as many as 72 fibers have been manufactured with the 8-slot core. In addition to the single slotted core, cables are also manufactured in a unit construction. Each unit has a V groove construction (deVecchis *et al.*, 1983) and multiple units are assembled to form the cable core.

5.4 OPTICAL PERFORMANCE

5.4.1 Attenuation

The four filled loose-structure cable designs have excellent optical performance with both multimode and single-mode fiber. With either fiber type, cabling results in no added fiber loss. In Fig. 5.12, loss distributions for production

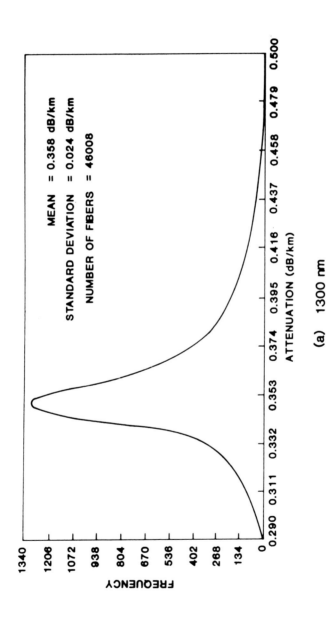

Fig. 5.12   Optical Performance—Loose Tube Cable

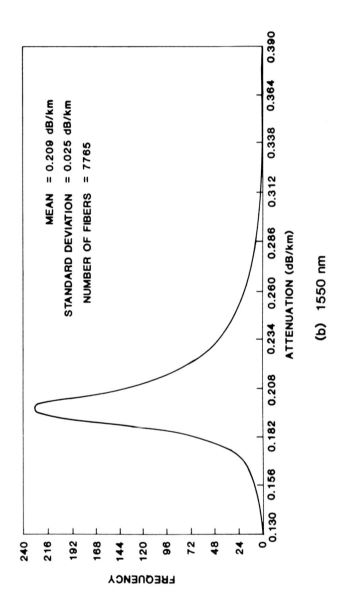

MEAN = 0.209 dB/km
STANDARD DEVIATION = 0.025 dB/km
NUMBER OF FIBERS = 7765

(b) 1550 nm

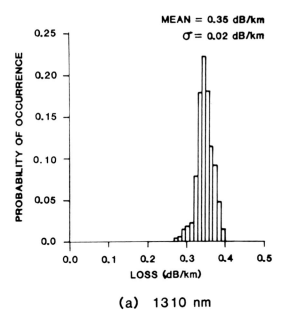

(a) 1310 nm

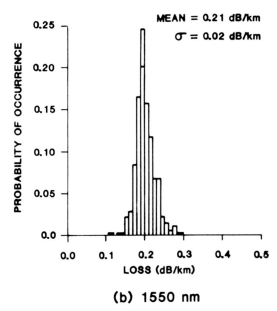

(b) 1550 nm

Fig. 5.13   Optical Performance—Loose-Fiber-Bundle Cable

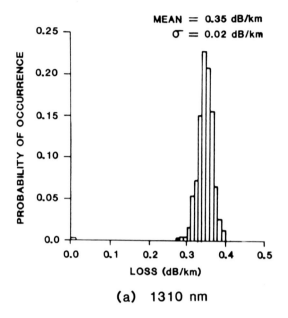

(a) 1310 nm

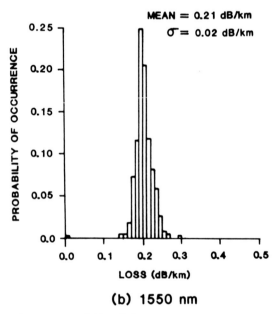

(b) 1550 nm

Fig. 5.14 Optical Performance—Ribbon Cable

single-mode loose-tube cable are illustrated (Bark and Lawrence, 1985). The results are shown for both 1300 and 1550 nm. The fiber was of a matched clad design with dual UV coatings. At 1300 nm, the mean loss is 0.36 dB/km with a standard deviation of 0.02 dB/km. The mean loss and standard deviation at 1550 nm are 0.21 and 0.03 dB/km. In selecting the fiber, only fibers with losses less than 0.4 dB/km at 1300 nm were used.

Optical loss histograms are shown for production loose fiber-bundle and ribbon cable in Figs. 5.13 and 5.14, respectively. For both designs, the depressed-clad fiber was used and identical performance was obtained. Results are presented at both 1310 and 1550 nm. The mean cable loss is 0.35 dB/km at 1310 nm, with a standard deviation of 0.02 dB/km. The maximum individual fiber loss is 0.4 dB/km. The mean loss and standard deviation at 1550 nm are 0.21 dB/km and 0.02 dB/km, indicating excellent upgrade capability at the longer wavelength. At both wavelengths, the loss distributions are tight and the cabling process induces no added loss above the intrinsic loss. Excellent results have also been obtained with the slotted core design (McKay, 1986).

The low production cable losses are primarily due to the loose construction employed. With the combination of high quality cable and low-loss splicing, repeaterless span lengths between 55 and 60 km are possible. In many applications, particularly short path length loop systems, higher loss fiber can be used effectively. More typical mean loss values are 0.4 dB/km and 0.25 dB/km at 1310 and 1550 nm, respectively.

5.4.2 Environmental Stability

The effect of daily and seasonal temperature variations on optical performance is a very important design consideration. For cables installed in underground ducts or directly buried in the ground, the temperature range is small, typically 0 to 30°C. The temperature range is much greater in aerial plant, -25 to 65°C for most regions of the country and -40 to 75°C in extreme climates. Optical performance is evaluated in environmental cycling tests performed on complete cables. The cycle generally consists of a high temperature exposure and subsequent exposure to low temperatures down to -40°C. The high temperature exposure is selected to accelerate aging effects, which occur in service and the entire cycle provides much more severe conditions than will be experienced in the field. Hence, thermal cycling tests are valuable in accessing cable performance under extreme conditions and in evaluating long-term performance and product quality.

Test results for a loose-tube cable using single-mode matched clad fiber are shown in Fig 5.15. The cable included 48 fibers in six buffer tubes. These plots also show the effects on performance of high temperature exposures. Results are shown for no high temperature exposure, after five days at 66°C, and after five

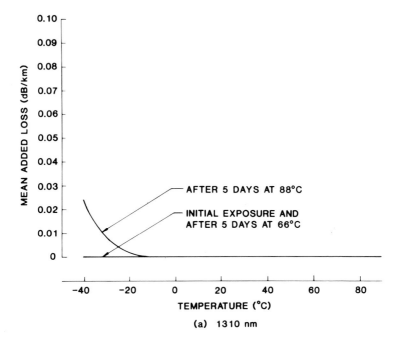

(a) 1310 nm

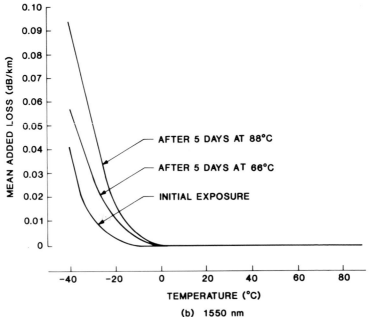

(b) 1550 nm

Fig. 5.15 Environmental Performance—Loose Tube Cable

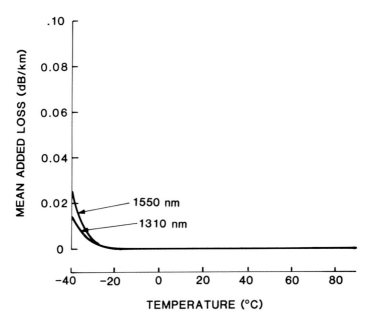

Fig. 5.16   Environmental Performance—Loose-Fiber-Bundle Cable

days at 88°C. At 1310 nm, the loss changes are very small, ~ 0.02 dB/km for the worst case. At 1550 nm, however, the initial change at −40°C is 0.04 dB/km, while after the 88°C exposure, the mean change is about 0.09 dB/km. The data shows the importance of high temperature aging on cable performance. The microbending sensitivity is much greater at 1550 nm as compared to 1310 nm. The increased sensitivity at 1550 nm is expected from theoretical considerations, since the optical power is less tightly bound to the core at the longer wavelength.

Environmental test results for a 72-fiber single-mode loose-fiber-bundle cable are shown in Fig. 5.16. At both 1310 and 1550 nm, the mean loss change is very small, less than 0.03 dB/km, over the entire temperature range. The results are shown after high temperature aging at 88°C. For any individual fiber, there is no change in loss above about −20°C. Below −20°C, the acrylate fiber coating materials stiffen, which can result in microbending loss. In Fig. 5.17, test results are shown for 48-fiber ribbon cable. The results are similar to those for the loose-fiber-bundle cable. Both cables were manufactured with the dual UV-coated depressed clad fiber. In Fig. 5.18, environmental test results are shown for the slotted core design (McKay, 1986). At 1300 and 1550 nm, the mean added losses at −40°C are .03 and .05 dB/km, respectively. This cable was manufactured with dual-coated depressed clad fibers and also exhibits excellent environmental performance.

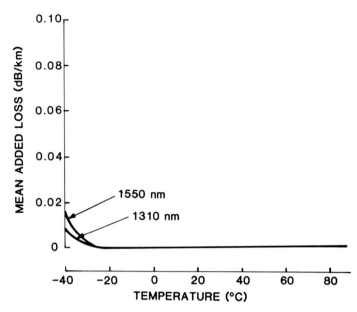

Fig. 5.17 Environmental Performance—Ribbon Cable

With multimode fiber, the added loss at $-40°C$ is generally a few tenths of a dB/km, after high temperature aging. It is nearly constant with wavelength. As with single-mode, there is no performance penalty above about $-20°C$. The increased sensitivity is intrinsic with multimode fiber and the four designs perform equally well. Early in the development of optical cable, the added loss due to low temperature exposure was large, more than one dB/km for either fiber type. During recent years, however, fiber and cable designs, materials and manufacturing processes have matured to the point that low temperature performance is not a serious limitation. For most installations, no loss budget penalty is necessary in engineering cable spans.

5.5 CABLE INSTALLATION

5.5.1 Underground Plant

The three basic installation environments for terrestrial applications are underground ducts, direct buried, and aerial. Underground conduits are the prevalent plant for central office trunks, especially in metropolitan areas. The standard four-inch ducts (Hale *et al.*, 1980) used in copper plant can accommodate up to four one-inch innerducts or duct liners. This approach provides efficient use of duct space, allowing for future growth. Figure 5.19 shows the relationship

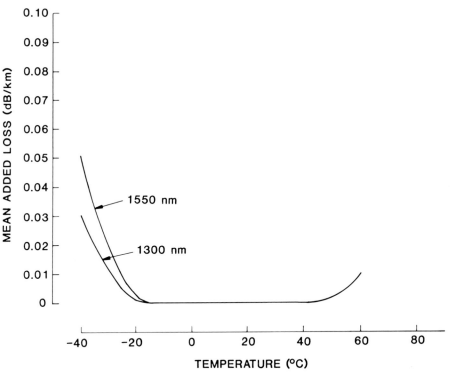

Fig. 5.18   Environmental Performance—Slotted Core Cable

between pulling length and cable diameter. A 2700 N load rating was used for all cables. The plot is derived from empirical data obtained from many underground pulls and is plotted for cables pulled into 25 mm innerduct in a duct run containing 160° of turn per 300 m of duct length and for a coefficient of friction of 0.1. The figure shows that the effect of increasing cable diameter is to reduce the pull lengths because of the added weight as the diameter increases. In addition, there is a further penalty resulting from the interaction between the cable's reel set and the inside wall of the innerduct. This interaction is more critical as the cable diameter increases and approaches the inner duct diameter. The data clearly demonstrates that small, lightweight cables offer significant advantages in underground plant. Lightning or rodent hazards are minimal in ducts. Adjacent copper cables or other structures generally attract lightning strokes in metropolitan areas. Sometimes ingress of contaminated water is a concern. A standard sheath with plastic jacket (Cornelison and Fleck, 1984; Gagen and Santana, 1979) provides adequate protection.

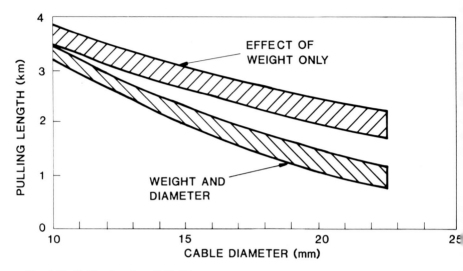

Fig. 5.19   Pulling length vs Cable Diameter

5.5.2 Buried and Aerial Plant

In long-haul terrestrial networks, direct-buried construction is used for most routes. Aerial plant is most common in suburban areas and is the least expensive way of installing telephone cables. Unlike underground plant, directly buried and the aerial plant are prone to human and natural hazards, such as dig-ups, vandalism, lightning, and rodents (Connolly *et al.*, 1970; Cogelia *et al.*, 1976; Sunde and Trueblood, 1949; Uman, 1969; Fisher *et al.*, 1971). There is no cost-effective defense against dig-ups or vandalism through cable design; however, protection against lightning and rodents is possible.

Based on AT & T coaxial cable experience, lightning damage is the second most important hazard category after the contractor damage. Figure 5.20 shows a map of the USA with estimated lightning exposure factors for buried cables. The highest probability of lightning damage is in the southeast and midwest. Rodent damage, on the other hand, is primarily caused by gnawing animals such as gophers and squirrels. Gophers, particularly the plains pocket gopher (*Geomys busarius*), will destroy unprotected cables buried less than six feet below ground. The rodent-prone areas encompass most of the states west of the Mississippi River and sections of Georgia, Florida and Alabama as shown in Fig. 5.21.

5.5.2.1 Lightning Protection.   The prediction of lightning damage has been the subject of a number of analytical and experimental studies (Fisher *et al.*, 1971; Sunde and Trueblood, 1949; Uman, 1969). The electrical energy from a

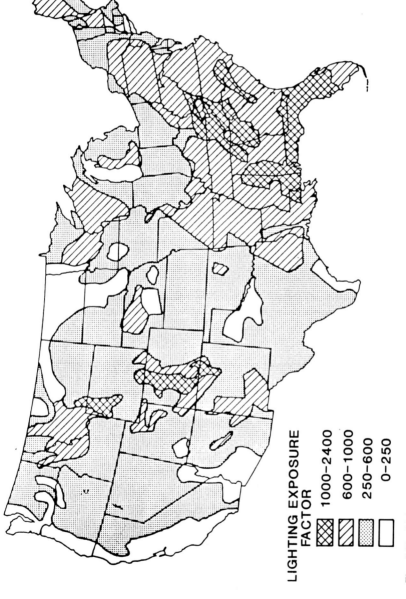

Fig. 5.20 Lightning Prone Areas in USA

LIGHTING EXPOSURE
FACTOR

1000–2400
600–1000
250–600
0–250

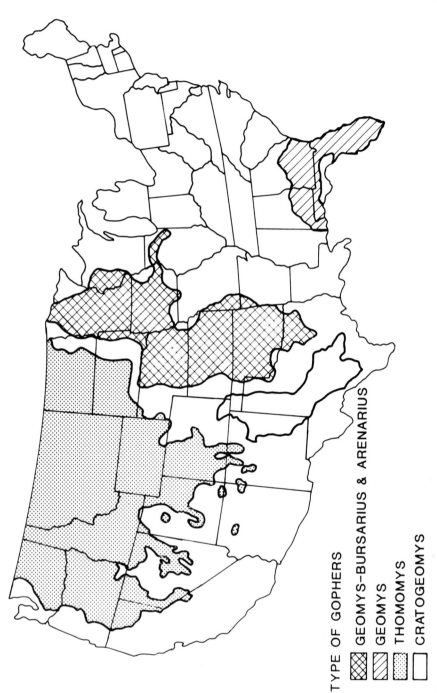

TYPE OF GOPHERS

GEOMYS-BURSARIUS & ARENARIUS

GEOMYS

THOMOMYS

CRATOGEOMYS

Fig. 5.21   Rodent Prone Areas in USA

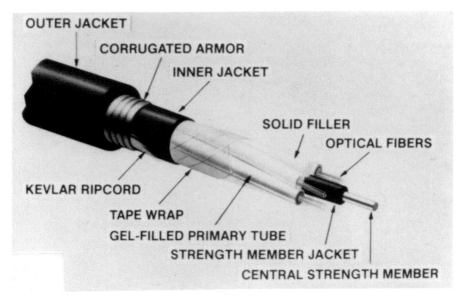

Fig. 5.22   Fiberoptic Cable-Corrugated Armor

lightning bolt flows through a path of least resistance, causing damage by passing through the cable. Since conductors attract lightning, a non-metallic cable construction would provide excellent lightning protection; however, there is no protection from the rodent hazard. An alternative economical approach is to add enough conductive layers in the sheath to dissipate the electrical and thermal energy. The most common conductive materials used are aluminum, steel, stainless steel, or stainless steel and copper composites. Figure 5.22 illustrates a typical steel-armored design. The armoring layers are corrugated for lateral strength and bending flexibility. They are also coated with an adhesive that bonds to the outer jacket. The stainless steel and copper composite provides both corrosion protection and high conductivity.

A lightning simulation test, known as the sand-box test, has been developed by AT & T Bell Laboratories (Fisher *et al.*, 1971). Large peak currents with discharge times simulating actual lightning strikes are applied to a cable sample in wet sand. The moisture is important. In addition to electrical heating, cables can be damaged by the large pressures developed during rapid vaporization of the water. This is known as the "steamhammer effect" (Reynolds *et al.*, 1986). Figure 5.23 is a photograph of four different sheath constructions after the lightning test. The samples were exposed to a peak current level of 80 KA, which is representative of a severe lightning strike. Peak lightning currents of

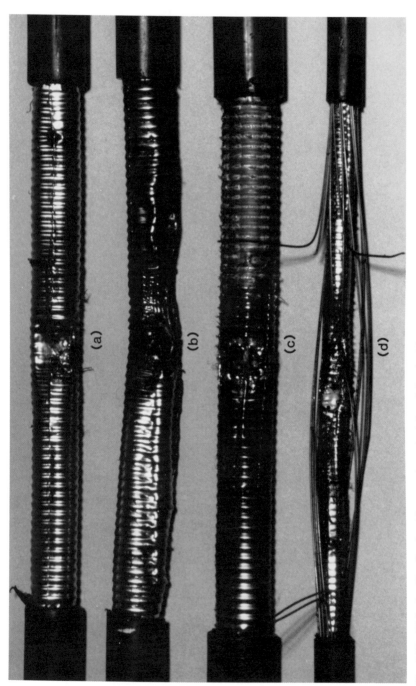

Fig. 5.23 Lightning Test Results: (a) a crossply steel-reinforced sheath with an oversheath consisting of 304 stainless steel over copper armor; (b) the same sheath and oversheath without the copper; (c) a steel-reinforced sheath with an internal layer of carbon steel armor; and (d) a smaller diameter

more than 80 KA are uncommon (Cianos and Pierce, 1972). The samples included:

(a) a crossply steel reinforced sheath with an oversheath consisting of 304 stainless steel over copper armor; the armor layers are 0.13 mm thick;

(b) the same sheath and oversheath without the copper;

(c) a steel-reinforced sheath with an integral 0.15 mm layer of carbon steel armor;

(d) a smaller diameter steel-reinforced sheath with a stainless steel-copper laminate armor.

In Fig. 5.23, the polyethelene jackets have been removed, exposing the armor. The oversheath with only the stainless armor (design (b)) experienced the greatest damage, although the fibers were untouched. The other designs experienced less damage. Without the copper layer, carbon steel armor exhibits better performance than the stainless steel due to its higher conductivity. Although the steel armored designs provide adequate protection, the addition of copper results in an extra margin of protection.

5.5.2.2 Rodent Protection. Rodents have been known to damage cables that are installed aerially, buried directly or in a plastic conduit. No commercial non-metallic designs have yet been developed that can resist the persistent gnawing of gophers and squirrels. The geographic areas requiring lightning protection also coincide with the gopher-prone sections of the country. Therefore, a common sheath design for both the lightning and rodent protection is desirable. An attacking rodent will penetrate all cable materials except metal. Early work by AT & T Bell Laboratories (Connolly and Cogelia, 1970), demonstrated that carbon steel does not provide a permanent barrier to rodent intrusion. After exposure to moisture, carbon steel corrodes and loses its protection against repeated attacks from rodents, whereas stainless steel affords a more permanent protection (Southwell *et al.*, 1976; Gerhold and McCann, 1976; Fink *et al.*, 1982). Gopher testing is conducted by the Denver Wildlife Research Center of the U.S. Department of the Interior's Fish and Wildlife Service. Figure 5.24 shows the catastrophic result of gopher attack on unprotected cables. Figure 5.25 is a photograph of protected cables after rodent exposure. These tests demonstrate that metal armors with stainless steel and copper layers provide excellent protection against both rodent and lightning hazards.

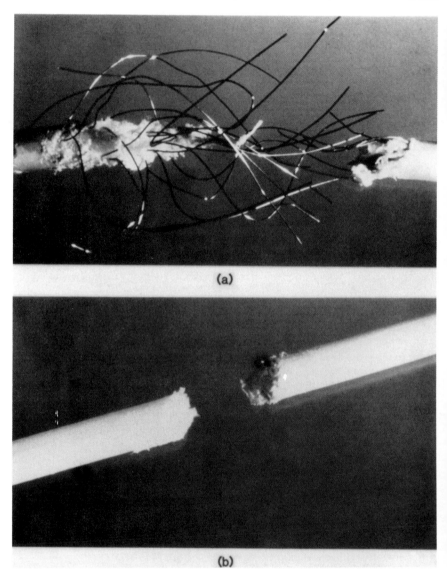

Fig. 5.24   Gopher Damage in unprotected Sheath: (a) Steel-reinforced Sheath, and (b) Non-metallic Sheath

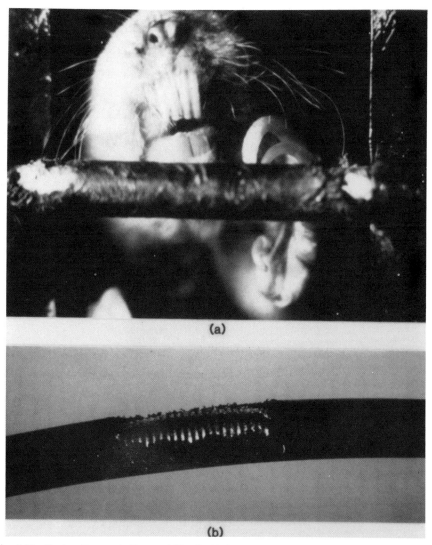

Fig. 5.25 Gopher Damage in Protected Sheath: (a) Gopher Attacking Cable in a Laboratory, and (b) Damaged Cable After Seven Days

5.5.3 Special Protection

For special applications such as river crossings, submarine, long aerial span along high tension wires, and extreme temperature or chemical environment, custom-tailored protection can be provided. In many instances, this involves wire armor for great tensile, crush, and abrasion resistance or hermetic seals for high temperature or extreme chemical environments. Usually, this type of protection is provided over a standard terrestrial cable.

5.6 MECHANICAL AND HAZARD TESTING

Optical cables undergo extensive testing to evaluate their mechanical performance (Venkatesan and Korbelak, 1982; Cornelison and Fleck, 1984). The Electronics Industries Association (EIA) has provided a number of standard tests that the manufacturers and customers use to qualify and select the product, respectively. In addition, specialized tests are often devised by manufacturers to evaluate design limits. As an illustration of specialized testing, plow simulation tests will be discussed.

5.6.1 Industry Standards

Some of the most important EIA tests adopted as standard in the optical cable industry are combined tension and bending, compression loading, twist, low and high temperature bending, cyclic flex, and cyclic impact. In all of these tests, the conditions imposed are much more severe than those expected in the field. A summary of the test conditions is given in Table 5.2.

In the tensile loading and bend test, a 150 m section of a test cable is looped between a set of sheaves and is subjected to 2700 N tension. The sheave

TABLE 5.2
ELA Test Requirements

Test	ELA Designation	Test Condition
Tensile Loading and Bending	FOTP-33	Tensile Load = 600 lbs Bend Radius = 20 × Cable OD
Compressive Loading	FOTP-41	Linear Load = 1000 lbs Over length = 4 in.
Twist	FOTP-85	Twist Angle = ±180° Sample Length = 13 ft Cycles = 10
Low and High Temperature Bending	FOTP-37	Bend Radius = 10 × Cable OD[†] No. of Wraps = 4 Temperature = −20°F and 140°F
Cyclic Flex	FOTP-104	Bend Radius = 10 × Cable OD[†] Arc = 180° Cycles = 25
Cyclic Impact	FOTP-25	Impact = 52 in.-lb No. of Impacts = 25

[†]15 × Cable OD for Rodent Lightning Protected Sheath

diameter is equal to 20 times that of the cable diameter. The compression test applies a 4,450 N force for a ten minute period over a 100 mm section of cable. In the twist test, a 4 m length of cable sample is rotated $\pm 180°$ about its axis. The cable must endure a minimum of 10 cycles lasting a total time of 10 minutes. The capability of the cable to endure bending at extreme temperatures is evaluated in the low and high temperature bend test; a cable sample is conditioned at $-30°$ and $60°C$ for four hours and wrapped four times on a mandrel whose diameter is 10 times that of the cable diameter. In the cyclic flex test, the cable sample is bent around a mandrel with a diameter equal to 10 times that of the cable diameter. The test is performed at a rate of 30 cycles per minute for a total of 25 cycles. In all of the above EIA tests, there should be no sheath failure. Maximum added loss, measured on a kilometer length, should be less than 0.2 dB.

5.6.2 Plow Simulation

Two simulation tests were developed to evaluate the ability of various cable structures to withstand abuse during plowing. During cable plowing, a rapid acceleration of the plow can result in high cable tensions. With rapid vertical plow motion, the cable can be severely deformed by concentrated lateral loading at the trailing edge of the plow chute. The complex loading, resulting from simultaneous plow acceleration and vertical motion, can cause cable damage and fiber breakage. The passage of a cable through a plow chute under tension is simulated in the first test. A cable sample is tensioned around a 18-cm diameter pulley with a discontinuous ramp. The ramp simulates the trailing edge of the plow. The springs on one end of the test setup allow the cable to slide past the sharp ($\sim$ 3 mm radius) ramp discontinuity while under tension. Cable tension and fiber continuity are recorded during the test as the load is increased. This being a destructive test, loading progresses until all the fibers are broken. A typical plot of the percentage of broken fibers versus sheath load is shown in Fig. 5.26. Results are presented for the ribbon, loose-tube and loose-fiber-bundle cable designs. Even though the loading conditions are very severe, all fibers remain continuous until the sheath load reaches about 4,450 N for the loose fiber-bundle cable. Above 4,450 N, the sheath begins to collapse as the cable passes over the sharp edge under the combination of tensile and lateral loads. For the loose-tube design, the unit tubes are crushed between the central strength member and the sheath, resulting in fiber breakage at about 2,200 N. For the ribbon cable, no fibers break, even for loads up to 8,900 N. This test demonstrates the advantage of strength member placement in the sheath, rather than in the core. It should be noted, however, that the objective of this test is to determine failure limits and all three cables perform very well in the field.

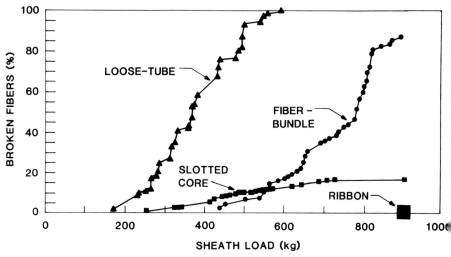

Fig. 5.26 Plow Simulation—Sharp Edge Test Results

In the dynamic squeeze test, a cable sample is repeatedly pulled through rollers over a 10-foot test length. Rapid downward movement of a plow chute into rock or compacted soil, combined with tractor acceleration, can result in cable squeezing. This problem has also been encountered in certain aerial installations (Kameo *et al.*, 1981). The purpose of this test is to quantitatively evaluate the susceptibility of various cable structures to dynamic squeezing loads. Figure 5.27 shows a schematic of the test apparatus with a cable sample squeezed between movable and guide rollers. The roller separation is expressed

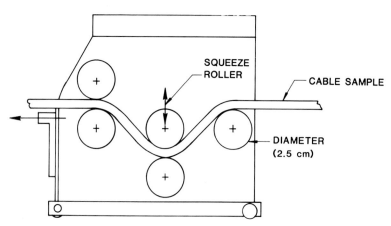

Fig. 5.27 Plow Simulation—Dynamic Squeeze Test

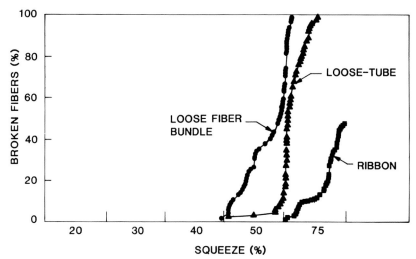

Fig. 5.28   Plow Simulation—Dynamic Squeeze Test Results

in % squeeze (% of the cable diameter). Cable pulling tension and fiber continuity are recorded during each segment of a test, lasting about 600 seconds at a given reduction in cable diameter. The cable diameter is progressively squeezed and cumulative fiber breakage is recorded. The test is concluded when all the fibers or the sheath have been broken. The percentage of broken fibers as a function of percent squeeze (cumulative squeeze time) is shown in Fig. 5.28. All three designs survive a typical abuse of about 30% squeeze. In the loose-fiber-bundle cable, the diameter must be reduced by more than 50% before the fibers are damaged. The loose-tube cable exhibits comparable performance and the ribbon cable performance is exceptional.

5.7 TIGHT BUFFER CABLE DESIGNS

As its name implies, the optical fibers in a tight-buffer core are encapsulated or buffered with relatively stiff elastomers; thus, the fibers are prevented from moving about when outside forces are exerted on them. This requires a careful balancing between the material and design selection and the structural and environmental requirements, since the added margin of safety of strain relief is not available. Particular attention must be given to fiber strength and thermally induced losses. This is not to say that tight-buffer designs are inferior to their loose-buffer counterparts; in fact, tight-buffer undersea cables are the strongest cables available and tight-buffer military cables with stable optical performance at −55°C are commercially available.

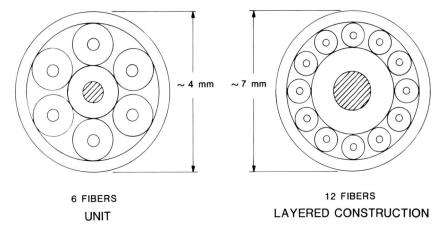

6 FIBERS   12 FIBERS
UNIT   LAYERED CONSTRUCTION

Fig. 5.29   Stranded Tight Buffer

5.7.1 Stranded Cable

Stranded cable has been used successfully with both multimode or single-mode fiber, primarily in Japan (Uchida *et al.*, 1981). The fiber is buffered with a thick layer of silicone and a nylon outer jacket. The overall diameter of the structure is about 0.9 mm. The six buffered fibers are stranded around steel or FRP strength members, forming units, and the units are assembled into a cable core, as illustrated in Fig. 5.29. The units are wrapped with a plastic yarn, providing additional buffering. The unit diameter is about 4 mm. For small cables, less than 12 fibers, a layered construction is used. The diameter of the strength member buffer is sized to accommodate the number of fibers in the core. The cable is generally completed with a laminated aluminum polyethelene (**LAP**) sheath. Both filled and air core construction have been used. By carefully controlling manufacturing parameters, i.e., tensions, excellent optical performance has been obtained. The structures are also very stable environmentally, the strength elements limit thermal expansion and contraction. Although excellent performance has been reported in Japan, the designs have not gained wide acceptance in North America or Europe. The buffers are expensive to apply, and the resulting cable structure is large, compared with the loose designs.

5.7.2 Undersea Cable

Because of their high information-carrying capacity and their low-loss, single-mode fibers are well suited for long-haul transmission systems such as undersea links. Because of their long span lengths and their unaccessibility after placement, undersea systems require long cable lengths with minimum splices. In

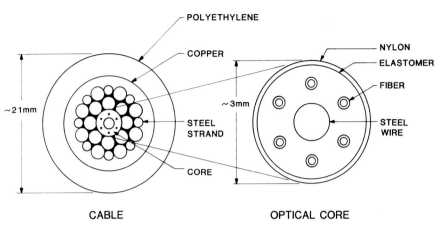

POLYETHYLENE

COPPER

NYLON

ELASTOMER

FIBER

~21mm

~3mm

STEEL
STRAND

STEEL
WIRE

CORE

CABLE   OPTICAL CORE

Fig. 5.30   Undersea Cable

addition, these cables require high-strength fibers (typically 1.38 GPa) having very low loss. Exceptional reliability and strength are of major importance. As an example, to recover a cable from the ocean floor, it must be strong enough to support its own weight in a catenary between the surface and the bottom and also withstand dynamic stresses from ship motion in response to wave action. A breaking strength equal to the weight of 19 km of cable has been found to be satisfactory. For typical undersea cables, the strength is 96 to 135 kN. Gleason *et al.* (1978) presented a detailed treatment of undersea cable design.

Figure 5.30 shows a cross-sectional view of a typical undersea cable design. The core is composed of a steel wire over which the fibers are helically applied within a polyester-elastomer encapsulant followed by a high-modulus polymeric sheath, typically nylon. This core design is rugged and easier to handle during later cable manufacturing operations. The core is placed in the center of a steel strand followed by a metallic conductor, typically copper, and a high-voltage insulation layer, usually of low-density polyethylene. The strength of this cable is about 110 kN without any further armoring. Depending on special hazards near the shore site, the basic cable design can be further protected by single-wire or double-wire steel armor.

Although undersea cable designs are extremely rugged and of a tight-buffer design, the optical transmission results are excellent. An 18.25-km cable containing six depressed-index cladding fibers had a mean loss of 0.37 dB/km at 1310 nm and 0.27 dB/km at 1550 nm (Adl *et al.*, 1984). The loss of this cable was also tested in an ocean-simulating facility and its loss was found to be insensitive to pressure variations up to 69 Pa, to temperature variations over the range 3°C to 30°C and to tension variations up to 80 kN. With these excellent results, applications such as a 147-km repeaterless span operating at 3 Mb/s at 1550 nm between a shore terminal building in Okinawa and a moored offshore platform are practical (Chu *et al.*, 1985).

5.7.3 Military

The market for military fiber-optic cable is in its infancy with applications for tactical battlefield communication leading the way. Military applications range from conventional tactical communication to more exploratory uses such as missiles guided by a trailing fiber. The requirements for tactical communication cables are now in the process of review by representatives of both industry and the armed services (Kalomiris, 1984).

The basic building block of a tactical cable is the tightly-buffered radiation-hard multimode fiber with exploratory work with single-mode fibers now in progress in the industry. The fiber has a 50 μm graded index core with transmission capability at 850 and 1300 nm. The bandwidth requirements are modest, 400 MHz-km. The proof-test level is 0.69 GPa. Radiation hardening is the only unusual fiber requirement. The cable is of a non-metallic construction with a 1780 N load rating. It has a maximum diameter of 6 mm to allow for rapid deployment in the field. The environmental requirements are from $-55°$ to 85°C. In addition to these overall requirements, the cable is specified in detail in DoD-STD-1678.

The material selections and structural design details are left up to the cable industry designers. Interestingly, the end-point requirements are such that the now available cable designs all follow the design pattern of Fig. 5.31. The fiber is buffered to the required 1-mm diameter using a polymeric material that varies with the different manufacturers but is commonly nylon or a polyester-elastomer buffer. Two fibers are stranded together forming the center of the cable over which aramid yarn or fillers are stranded to attain circularity. The cable is completed with the application of a reinforced polymeric sheath, typically of flame-retardant polyurethane and epoxy glass rods or aramid yarns for rein-

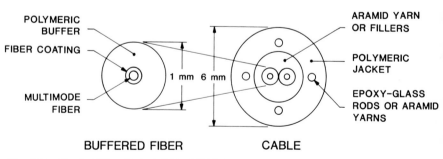

Fig. 5.31 Tactical Fiber-Optical Cable (TFOCA)

forcement. Jenkins *et al.* (1985) and Bark *et al.* (1984) provide more specific details on the different manufacturers' designs. Typical average losses at room temperature are 3.5 dB/km at 850 nm and 1.0 dB/km at 1300 nm, with added losses of less than 0.5 dB/km over the $-55\,^{\circ}C$ to $+85\,^{\circ}C$ temperature range.

5.8 FUTURE DIRECTIONS

As the long-haul and central office trunk cable markets saturate, the future of the fiber-optic cable industry depends on penetrating the loop distribution network. Optical-fiber systems should offer an economic solution to the growth of voice frequency and data services, while providing enough bandwidth for future enhanced services. Currently available cable designs have evolved to satisfy the requirements of point-to-point networks. Besides the obvious goal of lower costs, a further design evolution is needed to satisfy the reconfiguring needs of the loop plant while offering reliable service. Some areas where future work will probably be directed are: larger fiber count cables (200 to 1000 fibers), fiber tapering and rearrangement.

Coherent detection is a polarization-sensitive process that offers significant improvements in repeater spacing and information capacity. Improvements in devices and the elimination or compensation of the polarization drift that occurs in present cables will have to be achieved, perhaps through cable design changes, before this technology can enter the network.

In the more distant future, the promise of optical fibers of new composition with losses of 0.001 dB/km in the 2 to 10 μm range should provide new challenges in cable design. One could envision repeaterless spans across the ocean with few splices since any splice loss would greatly affect span loss. Thus, extremely long cable lengths would be needed using novel designs with zero packaging and environmental added loss. In addition, these new fiber materials pose new mechanical protection challenges since they are hygroscopic and the fiber's integrity is greatly affected by the presence of water.

REFERENCES

Adl, A., Chien, T. M., Chu, T. C. (1984). Design and testing of the SL cable. *IEEE Journal on Selected Areas in Communications* Vol. **SAC-2**, No. 6, 864–872.

Bark, P. R., and Oestreich, N. (1978). Stress-strain behavior of optical fiber cables. *Proc. Int. Wire & Cable Symp., 27th*.

Bark, P. R., Lawrence, D. O., Oestreich, U., and Mayr, E. (1984). Composite buffering for high performance fiber cables. *Proc. Int. Wire & Cable Symp., 33rd*, 98–101.

Bark, P. R., and Lawrence, D. O. (1985). How Singlemode cable performs its tasks. *Telephone Engineer and Management*.

Beales, K. J., D. M. Cooper and J. D. Rush (1983). Increased attenuation in optical fibers caused by diffusion of molecular hydrogen at room temperature. *Electronic Letters* **19**, 917.

Chu, T. C., Marra, L. J., and Stix, R. K. (1985). Mechanical architecture of a 147 kilometer repeaterless fiber optic undersea cable system. *Proc. Int. Wire & Cable Symp., 34th*, 346–354.

Cianos, N., and Pierce, E. T. (1972). A ground-lightning environment for engineering usage. Stanford Research Institute, prepared under contract L.S-2817-A3 for McDonnell Douglas Astronautics Co.

Cogelia, N. J., Lavoie, G. K., and Glahn, J. F. (1976). Rodent biting pressure and chewing action and their effects on wire and cable sheath. *Proc. Int. Wire & Cable Symp., 25th,* 117–124.

Connolly, R. A., and Cogelia, N. J. (1970). The gopher and buried cable. *Bell Telephone Laboratories Record.*

Cornelison, K., and Fleck, M. (1984). Applications and comparative performance of lightwave cable sheaths. *Proc. Int. Wire & Cable Symp., 33rd,* 130–140.

deVecchis, M., Demey, J. P., Hulin, J. P., Personne, Jr., and Staath, J. C. (1983). Cylindrical V-grooved non-metallic optical fiber cable. *Proc. Int. Wire & Cable Symp., 32nd,* 215–219.

Eichenbaum, B. R., and Santana, M. R. (1982). Design and performance of a filled high fiber count multi-mode optical cable. *Proc. Int. Wire & Cable Symp.*

Evans, A. G. (1974). Slow crack growth in brittle materials under dynamic loading conditions. *Int. J. Fract.* **10,** 251.

Fink, J. L., Escalante, E., and Gerhold, W. F. (1982). Corrosion Evaluation of Underground Telephone Cable Shielding Materials. Published by U.S. Department of Commerce.

Fisher, E. L., Kelch, E. C., and Bishop, W. F. (1971). The effect of lightning arcing currents on telephone cables. *Proc. Int. Wire & Cable Symp., 20th,* 285–292.

Gagen, P. F., and Santana, M. R. (1979). Design and performance of a crossply lightguide cable sheath. *Proc. Int. Wire & Cable Symp., 28th,* 391–395.

Gardner, W. B. (1975). Microbending loss in optical fibers. *Bell System Technical Journal* **54,** No. 1, 457–465.

Gartside, C. H., III, and Baden, J. L. (1985). Single-mode ribbon cable and array splicing. *Telephony* 80.

Gartside, C. H., III, and Cottingham, C. F. (1986a). Production and field experience with AT & T lightpack™ cable. Optical Fiber Communication Conference.

Gerhold, W. F., and McCann, J. P. (1976). Corrosion evaluation of underground telephone cable shielding materials. Paper No. 31, National Association of Corrosion Engineers, Houston, Texas.

Gleason, R. F., Mondello, R. C., Fellows, B. W., and Hadfield, D. A. (1978). Design and manufacture of an experimental lightguide cable for undersea transmission systems. *Proc. Int. Wire & Cable Symp., 27th,* 864–872.

Gloge, D. (1975). Optical-fiber packaging and its influence on fiber straightness and loss. *Bell System Technical Journal* **54,** No. 2, 245–262.

Hale, A. L., Pope, D. L., and Rutledge, D. R. (1980). Lightguide cable installation in underground plant. *International Fiber Optics and Communications, 1980–1981 Handbook and Buyers' Guide* 71.

Hope, T. S., Williams, R. J., and Abe, K. (1981). Developments in slotted core optical fiber cables. IProc. Int. Wire & Cable Symp. 226.

Hornung, S., and Doran, N. J. (1982). Monomode fibre microbending loss measurements and their interpretation. *Optical and Quantum Electronics* **14,** 359.

Jackson, L. A., Reeve, M. H., and Dunn, A. G. (1977). Optical fibre packaging in loose fitting tubes of oriental polymer, *Optical and Quantum Electronics* **9,** 493.

Jenkins, A. C., Lovelace, C. R., Reynolds, M. R., and Kalomiris, V. E. (1985). A tactical fiber optic cable. *Proc. Int. Wire & Cable Symp., 34th,* 82–87.

Kalomiris, V. E. (1984). Tactical fiber optic system requirements. *Proc. Int. Wire & Cable Symp., 33rd,* 388–394.

Kameo, Y., Horima, H., Tanaka, S., Ishida, Y., and Koyamada, Y. (1981). Jelly-filled optical fiber cables. *Proc. Int. Wire & Cable Symp., 30th,* 236–243.

Katsuyama, Y., Hatano, S., Kokubun, T., and Hogari, K. (1985). Design and performance of several-hundred-core high-density optical fiber ribbon cable. *IOOC-ECOC '85 Technical Digest*, 375–378.

Marcuse, D. (1976). Microbending loses in single-mode, step-index and multimode, parabolic-index fibers. *BSTJ* **55**, No. 7, 937.

Marcuse, D. (1982). "*Light Transmission Optics.*" 2nd Edition. Van Nostrand, Reinhold, New York.

Marcuse, D. (1984). Microdeformation losses of single-mode fibers. *Applied Optics* **23**, No. 7, 1082.

McKay, G. (1986). Private communication with C. H. Gartside.

Mies, E. W. (1983). Hydrogen susceptibility studies pertaining to optical fiber cables. Optical Fiber Communication Conference, Paper No. W13-3.

Miller, S. E., and Chyanoweth, A. G. (1979). "Optical Fiber Telecommunications." Academic Press, New York.

Nakuhara, M. Y. Ohmori and H. Itoh (1983). ESR study on loss increase of 500°C heat-treated Ge-doped optical fibers. *Electronic Letters* **19**, 1004.

Nirasawa, N., Yamazaki, Y., Tanaka, S., Suzuki, S., and Oagasaware, I. (1985). Design of fiber tape with improved lateral pressure resistance. *IOOC-ECOC '85 Technical Digest*, 379–382.

Oestreich, U., Zeidler, G. H., Bark, P. R., and Liertz, H. M. (1981). High fiber count cables of the mini-bundle design. *Proc. Int. Wire & Cable Symp., 30th*, 255–258.

Olshansky, R. (1975). Distortion loses in cabled optical fibers. *Applied Optics* **14**, No. 5, 20.

Patel, P. D., Chandan, H. C., and Kalish, D. (1981). Failure probability of optical fibers in bending. *Proc. Int. Wire & Cable Symp., 30th*, 37–44.

Patel, P. D., and Gartside, C. H., III (1985). Compact lightguide cable design. *Proc. Int. Wire & Cable Symp.*

Petermann, K. (1977). Fundamental mode microbending loss in graded-index and W fiber. *Optical and Quantum Electronics* **9**, 167.

Philen, D. L., and Gartside, C. H., III (1984). Prevention of hydrogen gas induced loss in optical fibers by proper lightguide cable design. *Proc. Int. Wire & Cable Symp.*

Reynolds, M. R., Arroyo, C. J., and Kinard, M. D. (1986). Primary rodent and lightning protective sheath for lightguide cable. *Proc. Int. Wire & Cable Symp., 35th.*

Saunders, M. J., and Parham, W. L. (1977). Adhesive Sandwich Optical Fiber Ribbon. *Bell System Technical Journal* **56**, No. 6, 1013–1014.

Southwell, C. R., Bultman, J. D., and Alexander, A. L. (1976). Corrosion of metals in tropical environments—final report of 16 year exposure. *Materials Performance* **15**, No. 7, 9–25.

Sunde, E. D., and Trueblood, H. M. (1949). Lightning current observations in buried cable. Bell System Technical Journal **28**, 278–302.

Tanaka, S., M. Kyoto, M. Watanak, and H. Yokota (1984). Hydroxyl group formation caused by hydrogen diffusion into optical glass fibre. *Electronic Letters* **20**, 284.

Uchida, N., Ishida, Y., and Ishahara, K. (1981). Single-mode and graded-index multimode optical cables for use in long wavelength transmission systems. *International Conference Communications*, *IEEE*.

Useugi, N., T. Kuwabara, Y. Koyamada, and N. Uchida (1983). Optical loss increase of phosphor-doped silica fiber at high temperature in the long wavelengths region. *Applied Physics Letters* **43**, 327.

Uman, M. A. (1969). "Lightning." McGraw-Hill, New York, New York.

Venkatesan, P. S., and Korbelak, K. (1982). Characterization of ruggedized fiber optic dual wavelength cables. *31st International Wire & Cable Symposium Proceedings, 358–370.*

Chapter 6

Optical Fiber Splicing[1]

STEPHEN C. METTLER

AT & T Bell Laboratories, Norcross, Georgia

CALVIN M. MILLER

AT & T Bell Laboratories, Norcross, Georgia

6.1 INTRODUCTION

Efficient, cost effective, field connection of optical fibers has been one of the fundamental problems implicit in the deployment of lightguide systems. The design of early fibers, cables and systems was strongly influenced by the need to ease the time, effort and expense of splicing. Although all splicing problems have a variety of solutions that have been demonstrated in the laboratory, few solutions meet today's optical fiber system requirements of performance, cost and productivity in the field. Efforts continue in this area and significant advances are needed before optical communications realizes its full potential. The recent trend toward the use of single-mode fibers in all areas of the telecommunications network has heightened interest in splicing and connecting problems. In this chapter we will examine the current theory and practice of both multimode and single-mode fiber splicing.

The desirable properties of optical fibers (small size, low loss, wide bandwidth) must be retained by having connection technology with comparable characteristics. The loss associated with copper splices is so low that it is difficult to measure directly. Optical fiber splices have at least a factor of 10 higher loss, and

[1] Most of the material for this chapter was taken with permission from Miller, Mettler and White, 1986.

connectors have close to 100 times more loss than their copper counterparts. Although bandwidth is negligibly affected in both glass and copper connections, the size of glass connections is much greater than the medium itself (except for fusion splices) making transitions bulky and vulnerable. While copper wires, cables and systems are designed and optimized largely without regard for how the media will be connected, manufacturing tolerances on fiber are largely set by connection considerations and most fiber cable designs are influenced by splicing considerations with some fiber cables specifically designed to ease splicing difficulties in the field (Saunders and Parham, 1977; deVecchis *et al.*, 1978; Jocteur and Carratt, 1980). Because of the low loss of the medium, a much higher percentage of the overall fiber system loss budget is allocated to connections than for copper systems. This higher connection loss in fiber systems has prompted systems designers to minimize the number of connections, thereby increasing the length of cables, installation difficulties, cable load requirements, etc. Independent optimization of fiber, cable or connection design leads to less than optimum system performance.

6.1.1 Loss Contributors in Multimode Fiber Splices

The predominant method for connecting optical fibers (single-mode or multimode) has been with simple butt-joint connections. Splicing operations fall into three main categories: 1) fiber end preparation, 2) fiber alignment, and 3) alignment retention. Fiber ends must be generated that are smooth and perpendicular to the fiber axis. Mechanical alignment of the fiber cores in three axes must be accomplished to micron tolerances for multimode fibers (and submicron tolerances for single-mode fibers). Index-of-refraction matching material is usually added between the fiber ends to optically couple the light from one fiber to the other in non-fusion splices. Contamination of the splice region must be avoided. Some splicing techniques retain alignment mechanically and others heat and fuse the fibers together, which requires accurate control of fusion conditions to minimize core deformation loss. Sensitivities to the various operations involved in splicing differ; however, any of them, if not properly performed, can result in extremely high connection loss.

Loss in multimode, butt-joint fiber connections (splices and connectors) is caused by differences in the fibers being connected (intrinsic parameters) and the quality of the connection (extrinsic parameters). Extrinsic losses can be minimized by proper connection design and depend on fiber end quality, alignment (transverse, angular, and longitudinal), cleanliness, degree of index matching in the connection and core deformation (fusion splices). Alignment is the most important extrinsic loss contributor. Figure 6.1 shows the alignment parameters that occur in fiber connections: transverse offset normalized to core radius, r_o/a; end separation normalized to core radius, z_o/a; and angular misalignment (tilt) normalized to acceptance angle, θ_t/θ_a. Transverse alignment is the most signifi-

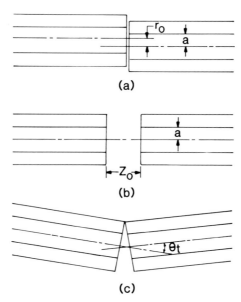

Fig. 6.1 Alignment errors in optical fiber connection. (a) Transverse offset, r_0/a. (b) End separation, z_0/a. (c) Angular misalignment, θ_t/θ_a.

cant extrinsic parameter in butt-joint fiber connections. In order to achieve low splice loss the gap between the ends of the fibers in the splice must be filled with index matching material or the ends must be in physical contact. The peak index-of-refraction at the center of the fiber core is approximately 1.47 and the loss due to Fresnel reflection at each glass/air interface averages 0.16 dB giving 0.32 dB total loss for the connection.

Retention of fiber alignment is vitally necessary to assure low splice loss over the service life of the splice. Indeed, many splicing techniques (bonding, fusion, etc.) concentrate on this aspect of splicing to such an extent that additional constraints are placed on end preparation and fiber alignment. The achievement of stable splice loss over the service life of the splice is not assured by the demonstration of low splice loss at room temperature.

Loss caused by differences in the fibers being connected is called intrinsic loss. The primary intrinsic factors are core radius, a, maximum index difference between core and cladding, Δ, profile shape parameter, g, core eccentricity, e, and core ellipticity, ϵ. In a realistic manufacturing environment, tolerance limits on intrinsic parameters must be set because variations can cause significant performance degradation. Variations in a and Δ have only slight effects on attenuation and bandwidth but can cause large splice losses. Bandwidth, however, is extremely sensitive to g, and therefore its tolerances are usually set directly from bandwidth considerations.

Splice loss caused by differences in Δ, *a* and *g* cannot be reduced by careful connection design. Losses due to variations in these intrinsic parameters typically average 0.05 to 0.2 dB per connection. These losses are due to differences in the total number of modes that the two fibers will support and therefore tend to be asymmetrical. Fortunately, intrinsic and extrinsic losses generally combine to give a total splice loss in multimode fibers that is less than the algebraic sum of the losses due to the individual factors. Intrinsic losses due to outer diameter variations and core concentricity can be minimized or eliminated by proper connection design. In general, these parameters can be controlled well in large scale manufacture, and do not constitute serious problems in multimode fiber connections.

6.1.2 Loss Contributors in Single-Mode Fiber Splices

Basic operations for single-mode fiber splicing are the same as for multimode fiber splicing; however, in general, each operation must be performed more accurately and the final loss in the splice is more dependent on this accuracy. The beam width or mode field diameter (MFD) is the only intrinsic parameter mismatch that cannot be eliminated by proper splice design. Unlike multimode fiber splices, losses due to intrinsic parameter mismatch in single-mode fiber splices usually average only ~ 0.01 dB for the normal range of MFD variation unless splicing fibers from different manufacturers. Intrinsic and extrinsic losses in single-mode fibers are approximately symmetrical, that is, losses in a splice in the direction of increasing mode field diameter are theoretically (and practically) the same as in the direction of decreasing mode field diameter for small (realistic) MFD mismatch. Also, intrinsic and extrinsic losses combine to give a total splice loss equal to the algebraic sum of the two losses to first order.

In contrast to multimode fiber splices, fiber core eccentricity is extremely important in single-mode fiber splices because of sensitivity to transverse offset. A great effort has been expended on reducing the effects of this parameter.

6.2 SPLICE LOSS THEORY

A splice is the dielectric interface between two optical fibers. Any index-of-refraction mismatch at any point in this interface will produce reflection and refraction of the light incident at that point. Light reflected or refracted at an angle greater than the critical angle at that point results in local splice loss. In multimode fibers, splices can also affect the loss of the receiving fiber (length dependence) and subsequent splices (cascade effect). These effects and the local loss at the splice all depend on the modal power distribution (MPD) entering the splice as well as the intrinsic and extrinsic factors listed in the previous section. Length dependence and the cascade splice loss effects are extremely difficult to calculate (see Section 6.2.3) because the loss and coupling coefficients of the individual modes in a multimode fiber are not known. The total effect of a splice in a multimode system is therefore difficult to accurately characterize. Single-mode splice loss theory, on the other hand, is straightforward compared to the multimode case.

6.2.1 Multimode Splice Theories

Most multimode fibers are either parabolic or step-index, that is, they have radially symmetric core index-of-refraction profiles of the form given by Gloge and Marcatili (1973) with a profile parameter, g, of 2 or ∞. These fibers have numerical aperatures (NAs) of:

$$NA(r) \sim \sqrt{2\Delta}\left[1 - \left(\frac{r}{a}\right)^{g}\right]^{1/2}. \tag{6.1}$$

NA decreases monotonically from its maximum at the center of the core to zero at the core cladding boundary. The NA gives the acceptance angle for each point on the core, that is, the maximum angle θ with respect to the longitudinal axis of the fiber at which light rays can propagate and still be guided by total internal reflection (guided or bound modes). Any light entering the fiber at that point on the core at angles greater than $\theta_a = \arcsin[NA(r)]$ is very rapidly lost through radiation or leaky modes. The models presented in this chapter assume that the fiber carrying light into the splice (the transmitting fiber) contains all rays at any point on its core within the NA at that point. From an electromagnetic theory point of view, this means that all power is propagating in bound modes. If this condition is satisfied, each point on the core of the transmitting fiber can be considered a point source with light distributed over the acceptance angle at that point.

For a perfect splice between identical, perfectly aligned fibers with perfect ends and no gap, the angle θ does not change across the interface between the two fiber end faces. Use of index-of-refraction matching material in the splice approximates this condition and most models and theories assume perfect matching. The presence of an air gap usually causes ~ 0.32 dB loss due to Fresnel reflection except in some special cases which will be discussed later. For non-identical or misaligned fibers the acceptance angle (NA) at the corresponding, contiguous point on the face of the receiving fiber core determines what fraction of the light from the point source is launched into propagating modes in the receiving fiber, and how much is lost through radiation and leaky modes. If the NA of the receiving fiber is equal to or greater than that of the transmitting fiber at a given point on the core, all light incident from the transmitting fiber at that point is assumed received and the point transmission is 100%. If the NA of the receiving fiber at that point is less than that of the transmitting fiber, the rays at angles greater than θ_a cannot propagate in the receiving fiber and are lost through radiation or leaky modes. The light lost is proportional to the ratio of the NAs squared (Miller and Mettler, 1978).

To calculate splice loss the amount of power propagating in the transmitting fiber and the angular distribution at each radial point on the core must be known. The fractional splice loss is simply the summation or integral of the product of the point transmission ratios and the point power distribution over the receiving fiber core divided by the total power in the transmitting fiber. Because graded index fiber modal power distributions are not known, assumed power distributions are used. For geometric optics models, the integrals involved in performing this calculation for parabolic fibers can only be solved in closed form for certain special cases. The completely general case of mismatch in all intrinsic and extrinsic parameters must be performed by numerical methods. One minor limitation of geometric optics models is that none of them successfully model longitudinal offset effects beyond a few core radii (8 to 10) of offset. Electromagnetic (EM) theory, on the other hand, provides closed form solutions for the bound-mode to bound-mode coupling coefficients for both intrinsic and extrinsic mismatches for any desired modal power distribution. Calculating the many thousands of coupling coefficients in this case requires a very powerful computer and often the simpler geometric optics models provide useful splice loss calculations with much less time and effort.

Geometric optics models using both uniform and Gaussian modal power distributions have been proposed. The Gaussian model (Miller and Mettler, 1978; Mettler, 1979) gives very good agreement with measured splice losses both for individual parameter mismatches and for combinations of both intrinsic and extrinsic mismatches, which are extremely difficult to calculate any other way. The accuracy of this approach and its results have been verified for parabolic fibers by careful comparison to many splice loss measurements.

The initial EM theory of fiber splicing was published by Kashima in 1981. A more complete analysis (White and Mettler, 1983) gives single term expressions for mode coupling coefficients for the case of either transverse offset or core parameter mismatch for parabolic-index multimode fibers. These coupling coefficients are used to calculate loss and mode mixing in splices. The theory agrees with geometric optics results for a uniform modal power distribution (geometric optics and EM theory must agree in this limit). Also, by comparing the results of this theory with a measurement of splice loss as a function of transverse offset, an indication of the modal power distribution in a fiber is given. Figure 6.2 shows loss versus transverse offset for several power distributions, the Gaussian splice loss model, and experimental data (Peckham *et al.*, 1984). The theoretical power distribution for "steady state" due to microbending (Marcuse, 1973) gives excellent agreement with this data. The agreement

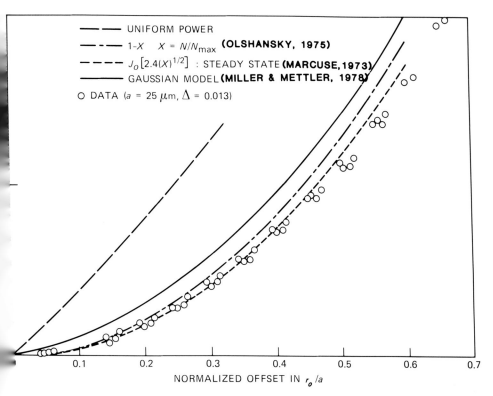

Fig. 6.2   Splice loss versus transverse offset for uniform power, $1 - N/N_{max}$, steady state, and Gaussian model compared to measurement data (from White and Mettler, 1983).

between the data and this analysis using the steady-state power distribution confirms both the distribution and the splice loss theory.

For fibers with no transverse offset, the individual mode fields are insensitive to small changes in the normalized frequency, V, and self-coupling dominates (i.e., mode $l_T \cdot m_T$ of the transmitting fiber couples primarily with mode $l_R m_R$, $R = T$, of the receiving fiber). Therefore, splice loss due to parameter mismatch is caused only by the slight field mismatch unless the ranges of the bound mode spectra of the two fibers differ, i.e., if the normalized frequencies of the fibers are such that the modes in the highest-order mode group of the transmitting fiber are not bound modes in the receiving fiber. Because self-coupling dominates, most of the power in the highest-order modes couples to leaky/radiation modes in the receiving fiber and is therefore lost. Parameter mismatch splice loss is therefore caused primarily by the difference between the bound mode volumes of the two fibers, which is a function of V. Both Δ and radius mismatches can therefore be expressed simply as V mismatch. The EM splice loss theory predicts that parameter mismatch should cause a wavelength dependent splice loss fluctuation. This prediction has been experimentally verified.

6.2.2 Single-Mode Splice Theory

Both geometric optics and electromagnetic theory were used in multimode fiber splice analyses; however, for single-mode fibers, core sizes are so small that geometric optics is invalid. For normal single-mode fibers at 1.3 μm it has been shown (Marcuse, 1977) that the fundamental modal field near cutoff is approximately Gaussian. For triangular-profile dispersion-shifted fibers, designed to be used at 1.55μ, this Gaussian approximation may or may not be valid, depending on the particular fiber design. This is an area of current research; however, the Gaussian approximation has been used with very good results for some triangular-profile fibers.

The Gaussian beam model calculates the amplitude of the fundamental mode in the receiving single-mode fiber assuming that the modal field is Gaussian. Splice loss calculations are therefore reduced to evaluating the coupling between mismatched Gaussian beams. A general equation has been derived (Nemoto and Makimoto, 1979) using Gaussian beams to calculate the splice loss between non-identical single-mode fibers with longitudinal, transverse and angular misalignment and end angle. This equation is in particularly useful form because it depends only on the mode field radius, w, of each fiber rather than the difficult to measure core refractive index profile parameters. The splice loss for the general case is (Nemoto and Makimoto, 1979):

$$\Gamma(\text{dB}) = -10 \log \frac{16 n_{co}^2 n_o^2}{\left(n_{co} + n_o\right)^4} \frac{\sigma}{q} \exp\left(-\frac{pu}{q}\right), \qquad (6.2)$$

where $p = (k_g w_T)^2/2$, $q = G^2 + \dfrac{(\sigma + 1)^2}{4}$

$u = (\sigma + 1)F^2 + 2\sigma FG(\sin \theta_t)(\cos \gamma) + \sigma\left(G^2 + \dfrac{\sigma + 1}{4}\right)\sin^2 \theta_t$

$F = r_o/k_g w_T^2$, $\quad G = z_o/k_g w_T^2$

$\sigma = (w_R/w_T)^2$, $\quad k_g = 2\pi n_o/\lambda$

n_{co} and n_o are the index of refraction values of the fiber core and the gap between the fiber ends, respectively;

w_T and w_R are the mode field radii (radius at which the near field power falls to $1/e^2$ of its maximum value) of the transmitting and receiving fibers, respectively;

λ is the wavelength of light used, and

r_o, z_o, and θ_t are the transverse, longitudinal, and angular (tilt) misalignments, respectively, as shown in Fig. 6.1;
γ is the angle between the direction of tilt and the plane containing the transverse offset and the fiber axis. The cos γ term was not included in the original expression.

This expression, with all splice loss components present, is simple enough to calculate on a hand calculator. It has been experimentally verified by extensive laboratory measurements (Kummer and Fleming, 1982).

A second analytical approach calculates the coupling coefficient between the transmitting fiber bound-mode and the radiation modes of the receiving fiber, which provides complementary information to the previous approach by calculating how much bound-mode power is lost in the splice (White and Kuhl, 1983). Coupling coefficients for this loss mechanism have been developed for microbending loss in fibers and can be applied directly to calculate splice loss. The microbending splice loss theory uses a perturbation analysis (Marcuse, 1976) that was originally intended to calculate the loss induced by small-amplitude, short-period, randomly-distributed fiber axis deviations, called microbends. Core distortion effects in fusion splices can be calculated using this microbending theory.

In the case of dispersion shifted fibers with triangular profile designs, the equivalent step-index profile of the fiber can be used to calculate fusion loss sensitivity of fiber parameters. Fusion splice loss caused by core distortion for dispersion shifted fibers should be about a factor of four greater than typical step-index single-mode designs. Such an increase in fusion splice loss for triangular profile fibers has been reported (McCartney et al., 1983).

6.2.3 Applications of Splice Loss Models and Theories

Splice loss models can be used to determine the effects of the different intrinsic parameters on splice loss. This information can then be used to establish manufacturing tolerances on these parameters. Normalized core diameter and Δ (or NA) mismatches cause approximately the same added loss and are usually given equal emphasis in manufacturing. Core eccentricity, e, can cause significant loss by contributing to transverse offset, but the effect of ellipticity is small compared to a, e and Δ and its tolerance is relatively lenient. Both the uniform power model (Thiel and Davis, 1976) and the Gaussian model (Kummer and Mettler, 1981) have been used with computer simulation of randomly spliced fibers to numerically derive expressions for statistical splice loss distributions due to normal distributions of a, Δ and g.

The traditional approach to setting manufacturing tolerances on fiber core parameters is to consider the effect on splice loss of each parameter independently. This approach, however, does not take into account the compensating or

TABLE 6.1
Compatability of Multimode Fibers
Splice Loss (dB) for Different Core Sizes and NAs

Receiving Fiber	Transmitting Fiber			
	50 μm	62.5 μm	85 μm	100 μm
50 μm	0	1.6	3.0	4.7
62.5 μm	0	0	0.85	2.1
85 μm	0	0.02	0	0.85
100 μm	0	0	0	0

Fiber Characteristics	
Core/Cladding Diameter	NA
50/125 μm	0.23
62.5/125 μm	0.29
85/125 μm	0.275
100/140 μm	0.29

compounding effect of simultaneous parameter deviations in multimode fibers. An alternate approach (Peckham *et al.*, 1984) to fiber intrinsic parameter specification is to use a fiber "intrinsic quality factor" (IQF) specification, which is a measure of the effect of combined fiber parameter deviations on splice loss. Thus, requirements can be relaxed on each individual fiber parameter while maintaining the same mean splice loss performance resulting in higher product yield. Alternatively, improved system splice loss can be realized while maintaining the same yield. The IQF can be used on a statistical basis to optimize yield and splice loss. The results of these statistical studies have been used to recommend reasonable manufacturing tolerances to the CCITT and to determine acceptable loss values to be used by splicing crews to accept or reject (and remake) splices in the field.

The Gaussian multimode fiber splice loss model has been used to determine the effects of mixing fiber types in systems. Due to the current lack of standardization, fibers with core id/fiber od of 100/140, 85/125, 62.5/125 and 50/125 may be spliced together in the field. Table 6.1 shows the splice losses predicted by the Gaussian model for splices between these fibers. These values were experimentally verified.

6.3 SPLICE LOSS MEASUREMENTS AND SYSTEM CONSIDERATIONS

From a system point of view, splice loss is not merely the amount of light lost immediately at the splice, but also the total effect of the splice on the end-to-end system loss. This section contains a discussion of these effects, splice loss measurement techniques, and additional system considerations.

6.3.1 Factors Affecting Multimode Splice Loss Measurements

The modal power distribution (MPD) at the input to a splice can significantly affect the "local" loss at the splice; therefore, to measure the loss of a splice, the proper MPD is required. Because each multimode fiber has a different equilibrium MPD (EMPD) and, because low-loss telecommunications fibers have extremely low mode coupling, "steady-state" is almost never achievable in the laboratory or field. Lasers and LEDs excite extremely different MPDs in multimode fibers (Cherin et al., 1986); therefore, splices and connectors close to sources may behave differently than if measured using an EMPD. "Underfilled" and "overfilled" conditions can also be generated by launching with a lens NA that does not match the fiber NA and a fiber loss value close to steady-state can be generated by making the launch NA approximately equal to the fiber NA. It has been shown, however, that adjusting the launch NA cannot duplicate a long input fiber for making splice loss vs transverse offset measurements (Kummer, 1980). Despite the difficulty of generating, verifying, and realistically characterizing the EMPD, it remains the goal for splice loss measurements, especially when comparisons are being made between various splicing techniques or loss measurement data.

If the power received by the second fiber is not properly distributed in the EMPD for that fiber then mode coupling and differential mode attenuation, if they are present in the fiber, will tend to redistribute the power (Kawakami and Tanji, 1983). This process couples some of the power being redistributed to radiation or leaky modes resulting in a distributed loss over the "coupling length" of the fiber as a result of the splice. Of course, if no mode mixing is present in the fiber, a steady-state power distribution will never be established in the fiber and this loss will not occur. Measurements of this effect depend strongly on the fibers used, especially on the amount of mode mixing and differential mode attenuation present. Fortunately, as fiber manufacturing and splicing technology has improved, mode mixing and differential mode attenuation have been reduced and this effect has decreased significantly as a splice loss contributor.

The loss of a given splice may be increased by the presence of other splices (cascade effect) closer to the source (Kummer, 1979). That is, the modal power distribution is shifted to higher order modes by the presence of the first splice, and the second splice generally has higher attenuation because the input modal distribution has increased power in the higher order modes. This effect is small ($\sim 10\%$ additional splice loss in the worst case of many splices close together) and is negligible for low loss splices and systems with relatively long lengths between splices (> 500 m).

6.3.2 Single-Mode Splice Loss Measurements

Compared to multimode measurements, single-mode splice loss measurements can be made more easily, more accurately and with greater repeatability because: (1) the proper MPD is easily generated by simply attenuating the LP_{11} mode, (2) there are no significant effects due to transmitting or receiving fiber length, (3) there is no significant cascade splice loss, and (4) intrinsic parameter mismatch losses are usually smaller and add linearly to alignment losses to first order. A small loop approximately one inch in diameter (depending on cut-off λ) placed before and after the splice, assures single-mode operation by attenuating the LP_{11} mode. Variations in mode field radii in typical single-mode fibers account for only a small component of average splice losses in the field (< 0.01 dB). The only other major concern is the frequency spectrum of the source. LED sources can provide enough power for single-mode splice measurements; however, the increased spectral width can give anomalous results in some cases.

6.3.3 Laboratory Splice Loss Measurements

The break and splice measurement technique provides the most repeatable splice loss measurement configuration; however, its use is limited to the laboratory. In this technique, light from a source is coupled through the fiber to a detector. A relatively straight section of fiber is laid out and the power, P_0, through the fiber is recorded. The fiber is broken and spliced back together without disturbing the source-to-fiber or fiber-to-detector coupling and with as little disturbance to the fiber as possible. Splice loss in dB is equal to $-10 \cdot \log(P_1/P_0)$, where P_1 is the power level after splicing. This highly repeatable splice loss measurement technique is excellent for comparing two or more fiber splicing techniques and measuring splice loss caused by extrinsic factors.

The effects of longitudinal offset on dry splice loss are complicated by the fact that if the fiber end faces are mirror smooth and parallel to each other, the dry gap between the ends constitutes a low-finesse Fabry-Perot etalon (Wagner and Sandahl, 1982). The interference effects in the etalon produce the splice loss behavior shown in Fig. 6.3, which can complicate dry splice loss measurements and active alignment of dry splices.

6.3.4 Field Splice Loss Measurements

Most installers of optical fiber systems measure splice losses as splices are being made and, finally, total span loss in order to guarantee a given span loss budget. Accurate individual splice loss measurements are not required; however, individ-

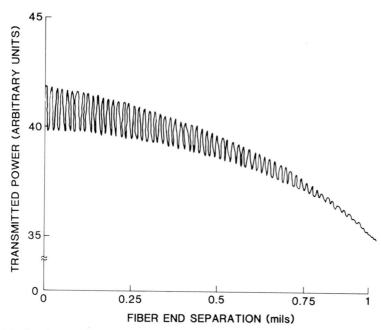

Fig. 6.3   Interference effects in dry single-mode fiber splices.

ual field splice loss measurements must be at least accurate enough to qualify a given splice as being acceptable in the system. It is highly desirable that individual splice loss measurement errors be random so that total span loss can be accurately predicted statistically and so that highly accurate splice loss averages are available to the system designer and splice hardware designer.

The three major categories of field splice loss measurement techniques are 1) far-end transmission measurements, 2) locally measured scattering and bending loss measurements and 3) optical time domain reflectometry (OTDR). All three methods are also being used for active splice alignment for single-mode fibers.

Power guided through the transmitting fiber, the splice, and the receiving fiber can be detected at the far-end and used to qualify splices in the field. If the span is spliced sequentially, this measurement confirms that each added cable section contributes the budgeted loss to the span.

Two techniques for measuring splice loss at the splice location in the field are 1) collecting and measuring a portion of the light lost in the splice (radiation and leaky modes), and 2) collecting and measuring a portion of the light transmitted through the splice and contained in the core of the receiving fiber (bound modes). Detecting power lost in a splice for splice loss measurements by sensing the light scattered from the splice is a relatively early technique

(Kohanzadeh, 1976; Tynes and Derosier, 1977). A technique for collecting and detecting a portion of the radiation and LP_{11} scattered light in a single-mode splice is used in a field splice loss test-set (Miller, 1985).

6.3.4.1 Local Light Injection.   Light injection at the splice input simplifies some problems associated with field splicing. It eliminates (1) having to move the light source from fiber-to-fiber at the central office, (2) communication between the splicers and the central office and (3) sequential splicing from the central office. Laboratory and field experiments have been reported that have given promising results (DeBlok and Matthijsse, 1984). A disadvantage of random splicing is that no indication of system continuity is present until the total end-to-end span loss is measured. Sequential splicing from one end of a span assures the quality of the span as splices are made.

The advantages of light injection are significant; however, the technical difficulties are also significant. Some methods for light injection consist of bending the fiber around a small radius ($\sim$ 4 mm) mandrel and focusing light on the bend. Aside from the possibility of fiber breakage, injecting enough light to use local scattering splice loss measurement is difficult because a signal level of approximately 34 dB below 1 milliwatt (-34 dBm) is required. The LSIM (low stress injection module), a microbend device providing this power level with no damage to the fibers, has been demonstrated, field tested and used as a light source to measure rotary splice loss in several installations (Aberson and White, 1986). This makes splice installation and measurement a true one-man one-point operation, independent of sources and detectors at either end of the system and allowing installation procedures to be optimized for local conditions with no requirements for sequential installation.

6.3.4.2 Optical Time Domain Reflectometry (OTDR).   OTDR has been used to estimate splice loss for multimode and single-mode fibers in the field. The OTDR requires access to only one end of the fiber under test and is usually required for fault location; therefore, its use is convenient and economical. The basic arrangement for OTDR measurements (Personick, 1977) consists of a pulsed laser and detector connected to the same end of a fiber through an optical coupler. A narrow optical pulse is launched into the fiber. The back-scattered signal versus time (Fig. 6.4) represents backscattering versus length along the fiber. The fraction, s, of power lost due to Rayleigh scattering that is backscattered and captured by the fiber is a function of Δ and g for multimode fibers and of the mode field radius for single-mode fibers.

The OTDR measurement is (Kummer *et al.*, 1982)

$$L_{TR}^{o} = \left(\frac{L_{TR} + L_{RT}}{2} \right) + 5 \log\left(\frac{s_R}{s_T} \right). \tag{6.3}$$

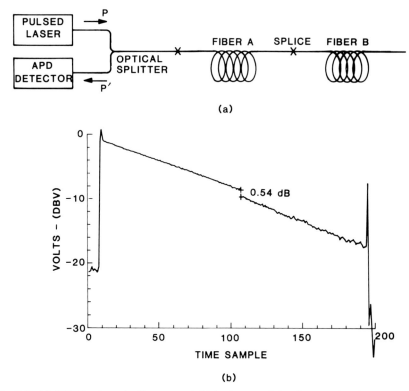

(a)

(b)

Fig. 6.4 (a) OTDR schematic diagram, and (b) backscattered signal versus time for a high loss fusion splice (from Kummer *et al.*, 1982).

This value is equal to the average of the transmission splice loss values in the two propagation directions, L_{TR} and L_{RT}, plus a term that depends on the ratio of the scattering coefficients of the two fibers. This scattering term is a measurement artifact that can lead to erroneous splice loss results amounting to plus or minus five tenths of a dB, typically.

Because this measurement artifact term is random, average OTDR results will be accurate while individual splice loss measurements will be in error by as much as five tenths of a dB. The dependence of the scattering on mode field radius causes this error of ± 0.5 dB in any individual measurement because of MFR mismatch between the fibers, despite the excellent parameter control on MFR in current production fibers. This error severely limits the use of OTDR for qualifying acceptable splices in the field, especially low loss single-mode fiber splices.

6.3.5 System Considerations

The very high bandwidth inherent in fiber optic systems results in most system designs having maximum span length limited by span loss. The importance of splice loss in total span loss, and thus on maximum span length, can be determined by examining typical system loss budgets. Because most loss components in fiber optic systems are characterized by a distribution of losses rather than a single value, the most commonly used system design method is a statistical approach (Buckler, 1983; Buckler and Meskell, 1983) that assures, at the 98% confidence level, that no path will exceed the design loss limit.

A sample single-mode system operating in the 1300 nm wavelength region is shown in Fig. 6.5 representing a fiber optic telecommunication trunk system, the most common type being installed today. Single-mode fibers have typical losses of 0.3–1.0 dB/km (0.4 dB/km used in calculations) in the 1300 nm wavelength region. This low fiber loss increases the percentage of span loss taken up by splices. The example system is implemented with passive splices (mean loss of 0.5 dB/splice) and optimized splices (0.1 dB/splice). The figure shows that active alignment reduces the contribution of splice loss to 6% to 32% of the total span loss. This type of system is often used in the long haul trunk network where cable lengths are 4.0 km or longer.

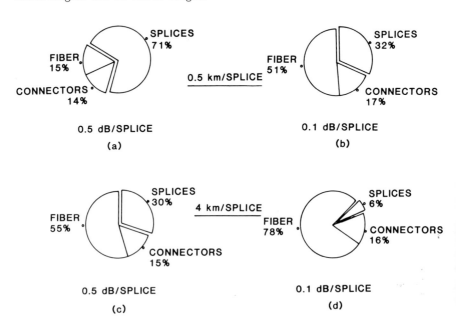

Fig. 6.5 Components of splice loss for single-mode fiber systems with: (a) 0.5 km/splice and 0.5 dB/splice, (b) 0.5 km/splice and 0.1 dB/splice, (c) 4 km/splice and 0.5 dB/splice, and (d) 4 km/splice and 0.1 dB/splice.

The most commonly referenced parameter to characterize splices is the mean splice loss. However, the splice loss terms in the system loss budget-splice loss (mean and standard deviation) and added splice loss due to environment (mean and standard deviation)-show that any system design that overlooks environmental terms is obviously incomplete.

6.3.5.1 Reflections from Splices (Judy and Neysmith, 1985). Optical energy may be reflected back toward its source by non-fusion fiber optic splices or connectors. If these reflections are large enough they can cause problems by: (1) interacting with very high bit rate single-longitudinal-mode lasers (Vodhamel and Ko, 1984; Cassidy, 1984), (2) limiting the performance of same-wavelength duplex transmission systems (Drake, 1981), or (3) interfering with OTDR test sets.

The amount of optical energy reflected from a non-fusion splice depends on three factors: (1) the quality of the fiber end surface, (2) the closeness of the match of the index matching material to the fiber's index of refraction, and (3) the spacing between the two fiber ends. In particular, the degree of polish on the end of a fiber has a substantial effect on reflection even when the end is immersed in an oil whose index of refraction matches the fiber extremely well. The effect of these reflections, if less than -30 dB, have been found to be unimportant for high bit rate, direct detection systems. Coherent and full duplex systems require lower levels of reflections and splices and connectors with less than -45 dB reflection are being designed.

6.3.5.2 Modal Noise. Splices and connectors can also generate modal noise that affects the bit error rate in a fiber optic system. Constructive and destructive interference between individual modes in a multimode fiber (speckle pattern) occurs due to phase variations caused by their different propagation times in the fiber. The speckle pattern shifts in time due to instabilities in the laser itself (mode hopping, etc.) or any movement of the fiber as might be caused by air currents or mechanical vibrations. If this pattern is the input power distribution for an imperfect splice (transverse offset, core diameter mismatch, etc.), then the splice loss varies in time as the pattern moves into and out of the receiving fiber's cone of acceptance. This phenomenon has been called modal noise (Epworth, 1978), and can cause significant noise under certain conditions. Modal noise is statistical in nature and many theoretical and experimental studies have been carried out and published (Daino et al., 1979; Rawson et al., 1980; Peterman, 1980).

Splices and connectors in single-mode digital systems can also generate modal noise (Duff et al., 1985). Although the modal noise phenomenon has been extensively studied and can cause difficulties under some circumstances, in general, it can easily be eliminated in all practical optical fiber systems (Sears et al., 1985).

6.4 SPLICING HARDWARE

This section presents a splicing hardware classification system, general splicing requirements, fundamental principles of connecting fibers and some examples of equipment and hardware available to accomplish splicing operations. While concentrating on splices (permanent connections), principles and much of the hardware described can also be applied to optical fiber connectors.

6.4.1 Classifications, Requirements and Fundamental Operations

The classification system shown in Fig. 6.6 has been devised to simplify the discussion of basic principles and hardware for optical fiber splicing and to assist those involved in connection design and evaluation. Alignment is classified to be

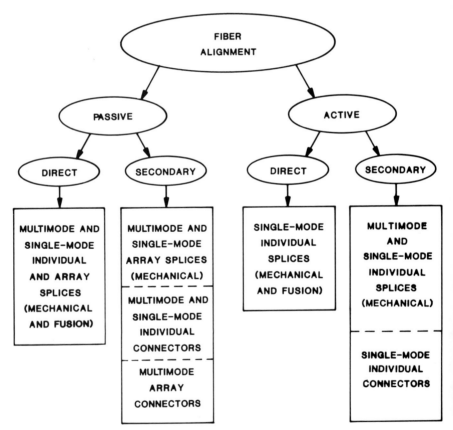

Fig. 6.6 Fiber alignment classifications. Passive alignment is adequate for most multimode applications and active alignment is required for most single-mode connections. Alignment retention technique for splices is shown in parentheses.

active when the connection is optimized by adjusting for maximum transmission (minimum loss), and passive when it is not. In addition, passive or active alignment is classified as either direct, if the bare fiber is aligned, or secondary (Bowen, 1981), if the bare fiber is mounted in a supporting structure and then aligned in a secondary operation. Specific examples of splicing hardware for each of the types of splices in Fig. 6.6 will be given.

Splicing requires the "lowest environmentally stable loss in the shortest time with the least trouble and expense." Needless to say, these factors are not mutually independent and, in addition, the relative importance of cost and performance depends drastically on the application as discussed in Section 6.1.2. The next sections discuss the fundamental operations: 1) end preparation, 2) core alignment, and 3) alignment retention. The design service life of splices is typically 20 to 40 years, which places severe requirements on mechanical and environmental stability. The thermal properties of materials used in splices play a dominant role in the loss variations with temperature. Analytical methods combined with environmental testing can insure good long-term environmental stability. Lack of mechanical or environmental stability can easily result in high connection loss and system failure.

6.4.2 End Preparation

Fiber end preparation begins with the removal of the protective coating. Techniques for removing the coating depend on the coating material and the strength requirements of the stripped fiber. Normal mechanical and chemical stripping techniques are, in general, well known or specified by the fiber manufacturer; however, special techniques needed to remove the coating for high strength fusion splicing will be briefly discussed. Most end making techniques use either scoring and breaking or grinding and polishing.

The advantages of scoring and breaking for end preparation for bare fibers are simplicity and speed and the high quality of the resulting ends. The primary disadvantages are that small fibers have to be handled, mechanical damage is difficult to avoid, and cleaning scored and broken fiber ends is difficult because rubbing the fiber can produce static charge, which attracts dust, etc. The mechanical theory of scoring and breaking is straightforward (Johnson and Holloway, 1966; Gloge et al., 1973; Miller and Chynoweth, 1979) and has been experimentally verified (Sakamoto et al., 1978).

The earliest and simplest method of scoring and breaking is a hand-held technique in which the fiber is stripped of its protective coating and placed on a flexible, straight support (typically steel shim stock approximately 0.01 inches thick). The fiber is then scored with a diamond or carbide wedge, held against the flexible support and bent until the fiber fractures. This simple score and break technique or some variation has been used almost universally for obtaining relatively high quality ends quickly in the laboratory and factory for fiber measurements.

The elimination of bending (Chesler and Dabby, 1976) has resulted in smaller end angles (Gordon et al., 1977). End angle appears to depend on torsional components imparted to the fiber during clamping, bending and pulling (Saunders, 1979). Hand-held score and break tools are available that use both bending (Furakawa, 1983; Fulenwider and Dakss, 1977) and straight pulls (Thomas and Betts, 1981). Desktop tools are also available (Sieverts, 1983).

End angles can be controlled easily with grinding and polishing, depending on the accuracy of the holding fixture. Flatness is more difficult to achieve because the various materials that support the glass—adhesives, metals, ceramics, plastic, etc.—have different grinding rates. This is partially due to the difference in hardness of the materials but also due to the material removal mechanism (Nelson and Ahmed, 1982). Lubricants (water, generally) serve to smooth the abrasive process and help to dissipate heat. Hardware for grinding and polishing ranges from simple hand-held tools and fixtures to automatic machines.

Sawing has been used for end preparation for an array splicing technique used in France (Bouvard and Hulin, 1981). Although sawing can give very small end angles and flat surfaces, it typically takes longer than other methods and leaves deeper scratches.

6.4.3 Active and Passive Alignment

Manufacturing tolerances and the relatively large core size of multimode fibers allow passive techniques to be used for core alignment. The small core size of single-mode fibers, however, requires the use of active alignment if splice losses of less than 0.1 dB are desired.

6.4.3.1 Passive Direct Alignment. Direct fiber alignment using passive techniques usually falls into one of two categories—V-groove or tube. The V-groove geometry (Someda, 1973) provides alignment along two lines of contact with the fiber. V-grooves provide a high degree of horizontal confinement; however, vertical misalignment, x, due to fiber radius, a, difference is amplified an amount dependent on the groove angle.

$$x = \frac{a_{max} - a_{min}}{\sin \theta}, \tag{6.4}$$

where θ = V-groove half angle. This transverse offset is usually small compared to multimode fiber core radius with the excellent fiber diameter control presently available; however, it can contribute significant splice loss when used with single-mode fibers. The trapezoidal groove geometry (Schroeder, 1978; Kurokawa et al., 1980) and the "three-rod" alignment technique (O'Hara, 1975) are also basically V-groove splices.

Tight fitting glass (Murata *et al.*, 1975) and metal (Derosier and Stone, 1973) tubes were suggested for direct passive fiber alignment many years ago. Metal tubes are difficult to fabricate with smooth, accurately controlled inner surfaces but smooth glass tubes can be made very accurately. The fiber must fit tightly in the tube for good alignment; however, fiber outer diameter and glass tube inner diameter tolerances must be taken into consideration.

6.4.3.2 Passive Secondary Alignment. Additional elements can be applied to the fiber to increase the size and ruggedness of the structure for ease of handling and end preparation. These secondary elements are then aligned using an overlapping matching surface. Structures in common use include grooved silicon chips (Miller, 1975), steel (Murata, 1976) and glass (Miller, 1984) tubes, and molded plastic (Kurokawa *et al.*, 1980), ceramic (Nawata *et al.*, 1979) and glass (Miller *et al.*, 1985) ferrules. With proper design, the fiber coating can be terminated within the secondary elements so that only the fiber end face remains exposed. In addition, secondary elements can be designed to terminate a single fiber or a group of fibers quickly and most secondarily aligned splices are reenterable. Individual fiber splices using passive secondary alignment usually use glass, ceramic or plastic ferrules to hold the fibers. Glass provides the highest accuracy and best environmental performance (Miller, 1984); however, depending on the application, plastics and ceramics can also give excellent results. Array splicing can be extremely cost effective for cables designed appropriately (Miller, 1973; Miller and Schroeder, 1976; Jocteur and Carrott, 1980; deVecchis *et al.*, 1978). Alignment accuracy with the silicon chip array splice has been good enough to be adopted for single-mode fiber splicing (Baden *et al.*, 1986). Average single-mode array splice loss of 0.3 dB has been achieved in field installations throughout the country.

6.4.3.3 Active Alignment. Active alignment is used primarily with single-mode fibers (Fig. 6.6). Active alignment implies a splice geometry that allows adjustment of the splice for minimum loss allowing fiber tolerances to be relaxed. Although all active alignment methods are more complicated and more time consuming than passive methods, the reduced loss for single-mode splices is often required.

Active alignment can be accomplished by either maximizing the light transmission through the splice by detecting and adjusting the splice for maximum transmitted power (at the far end or at the splice) or minimizing the light lost at the splice by detecting and adjusting the splice for minimum scattering. The ability to use the detected signal for accurate field splice loss measurements is also an important consideration. In an alternate approach (Tatekura *et al.*, 1982), not active in the sense of splice transmission, UV light is used to excite the Ge compounds in the core of the single-mode fiber causing them to fluoresce so that visual alignment of the cores can take place. Another highly accurate

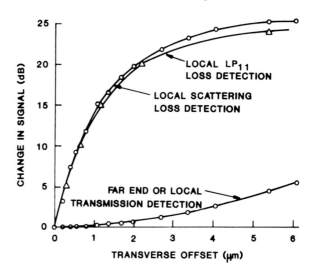

Fig. 6.7 Comparison of sensitivity to transverse offset for far-end or local transmission detection, local scattering loss detection, and local LP_{11} mode loss detection.

alignment technique (Fujikura, 1985) uses a focusing method to illuminate and magnify the core of the single-mode fiber and a microprocessor and video camera to align the cores and calculate splice loss.

Some techniques for monitoring the level of transmitted power at the splice (local detection) involve bending the fiber around a small radius (1–5 mm) and monitoring a portion of the power lost in the bend. The change in transmission signal (local or far end) as a function of transverse offset for single-mode fibers is shown in Fig. 6.7. Although this sensitivity is high relative to multimode fibers, the slope of the curve is zero for zero transverse offset, making highly accurate alignment (to less than 1 μm) extremely difficult. Additional disadvantages of the fiber bending local transmission technique include the possibility of fiber breakage, restrictions on fiber coatings, limited loss measurement capability and greatly reduced alignment sensitivity compared to local splice loss detection.

Active alignment is more attractive with local splice loss detection using the two techniques of local splice loss measurement discussed on page 275. Since losses below 0.1 dB are often required for single-mode fiber splices, alignment precision in the tenths of a micron range is needed. When the splice is adjusted for maximum transmission, the scattered or lost power at the splice is a minimum. Figure 6.7 compares far-end or local transmission detection, local scattering and local LP_{11} detection sensitivity to transverse offset. The null signal using loss detection changes rapidly around zero offset so that the sensitivity to offset is increased by more than two orders of magnitude over transmission detection. This increased sensitivity to offset is a significant advantage of local loss detection. The signal level with local loss detection is much lower (approximately 30 to 40 dB) than local transmission detection, which is not a problem if the input fiber is connected to a laser at the far end, but requires significantly more locally-injected power compared to local transmission detection. The use of local light injection makes it possible for one person to perform all the splicing activities. Light can be launched and detected and the splices actively tuned and measured with no outside assistance. This greatly accelerates (and reduces the cost of) the splicing process.

6.4.4 Alignment Retention Methods

Alignment retention is the most important of the fundamental operations. Adhesives, mechanical clamping and fusion have been used to maintain alignment. The alignment retention operation should not increase the connection loss and must maintain stable alignment over both time and the range of environmental conditions expected.

Adhesives play a major role in optical fiber connection hardware and few connection techniques totally avoid their use. However, the vast differences in material properties, especially thermal coefficient of expansion, between adhesives and glass make reliance on thick cross-sections of adhesives for primary alignment retention hazardous under variable environmental conditions. Thin adhesive cross-sections are often used for attaching fibers to supporting structures for mechanical splices and for protecting (recoating) fusion splices. In most cases the long term physical retention mechanism is either mechanical clamping or fusion.

6.4.4.1 Mechanical Alignment Retention. Mechanical methods often used for alignment retention include clamps, screws, springs, etc. Long-term variations in the loss of mechanical optical fiber splices can be caused by either degradation of the index matching material or extrinsic alignment changes due to thermal variations. Analytical methods (finite element analysis) and knowledge of the material properties of the structure in combination with the results given in the previous sections for sensitivities of loss to alignment parameters can be used to "design in" good environmental performance. Environmental testing is always necessary to verify the design and to quantify the degree of environmental variation of splice loss for inclusion in the system loss budget.

Splices are usually index matched to provide low loss and stable environmental performance. The best approach is the use of a soft, pliable index matching material that can withstand large strains. Silicone gels have been found to be ideal index matching materials (Mettler and Gotthardt, 1981); elongations of several hundred percent are obtainable, the index of refraction can be made very close to n_{co}, wetting to glass is excellent (chemical adhesion can be obtained by priming), high temperature chemical stability of silicones is excellent, and the

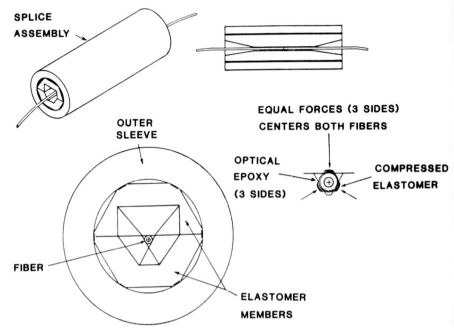

Fig. 6.8 Elastic splice. Resilient cylindrical members allow splicing of fibers with different diameters (from Knecht *et al.*, 1982).

optical properties of silicones in the wavelength region $0.6 < \lambda < 1.6$ μm are good. The gel can be applied as a liquid with the advantages of wicking and then be heat cured.

Changes in the alignment of fiber ends in a mechanical connection due to environmental variations can be a serious problem requiring that the splice geometry itself be optimized for stability. For the purpose of mechanical analysis, splice designs fall into two categories of fiber alignment retention; point clamping and continuous gripping. Designs using point clamping can be analyzed using general closed-form mathematical methods (Hardwick, 1985), while designs that employ continuous gripping require detailed finite element analysis techniques for evaluation (Hardwick and Reynolds, 1985).

6.4.4.2 Examples of Direct Alignment. Figure 6.8 shows a typical passive direct field splicing technique for single fibers (Knecht *et al.*, 1982), which can accommodate fibers with different diameters and is small and easy to use in the field. Environmental variation is 0.25 dB over a temperature range of $-30°$C to $85°$C and this variation may not be suitable for some applications.

A passive direct splice for arrays of fibers is shown in Fig. 6.9 (Hardwick and Davies, 1985). Two 12-fiber ribbons are simultaneously ground and polished

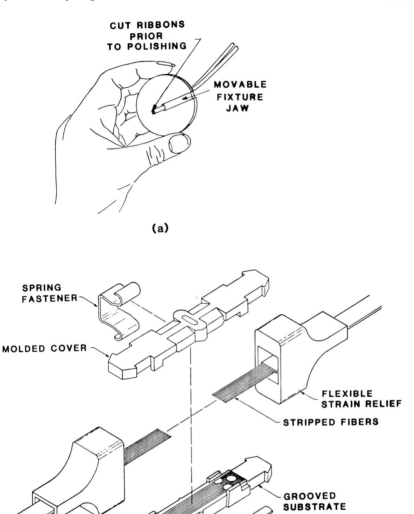

Fig. 6.9 Passive direct ribbon splicing. (a) Polishing method. (b) Components. (c) Assembled splice. (d) Field repair loss histogram (from Hardwick and Davies, 1985).

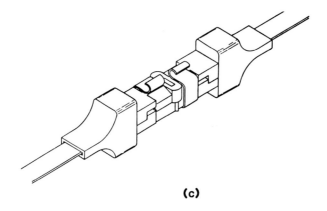

(c)

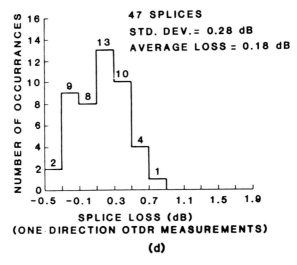

SPLICE LOSS (dB)
(ONE DIRECTION OTDR MEASUREMENTS)

(d)

Fig. 6.9 (*Continued.*)

and then laid in guides. Vacuum holds the fibers in the grooves while the cover plate is applied. Index gel is added through a slot in the cover plate and spring clips hold the assembly together and provide a mechanical force to keep fibers seated in the grooved substrate. The reported thermal variation of only 0.1 dB over the $-40°C$ to $77°C$ temperature range indicates good environmental performance. The assembly time of 20 minutes for 12 fibers makes this tech-

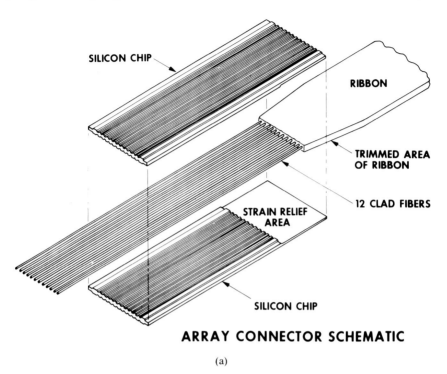

SILICON CHIP

RIBBON

TRIMMED AREA
OF RIBBON

STRAIN RELIEF
AREA

12 CLAD FIBERS

SILICON CHIP

ARRAY CONNECTOR SCHEMATIC

(a)

Fig. 6.10 Silicon chip array splicing. (a) Schematic. (b) Array assembly tool. (c) Finished array splice (from Friedrichsen and Gagen, 1981).

nique useful for rapid splicing or restoration in the case of cable dig-ups, lightning hits, etc.

6.4.4.3 Examples of Secondary Alignment. The silicon chip array splice (Miller, 1975), shown in Fig. 6.10, is an example of a passive secondary splice. After stripping the ribbon and coating material from the fiber ends, the fibers are laid into a comb, which matches the spacing of the fibers in the ribbon and the spacing of the grooves in the silicon chips. After the top chip is applied and epoxied, the front face is ground and polished. Most of the fiber splices using linear arrays have been connected in manholes (fabrication usually occurs in the factory) under normal field conditions without microscopes and without handling individual fibers. The average splice loss has been 0.12 dB for multimode fibers and 0.32 dB for single-mode fibers (Gartside and Baden, 1985) with silicone gel matching material. Splicing time is approximately one minute/fiber splice (plus travel and manhole preparation time).

The splice shown in Fig. 6.11 can provide either active or passive secondary alignment (Aberson and Yasinski, 1984). It uses precision glass capillary tubes,

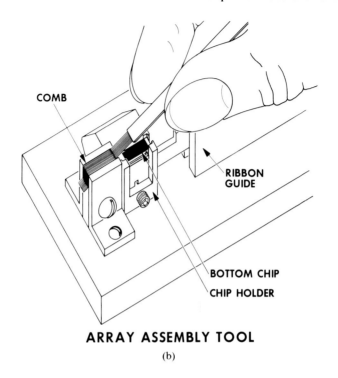

ARRAY ASSEMBLY TOOL

(b)

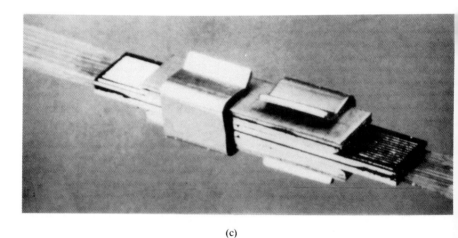

(c)

Fig. 6.10 (*Continued.*)

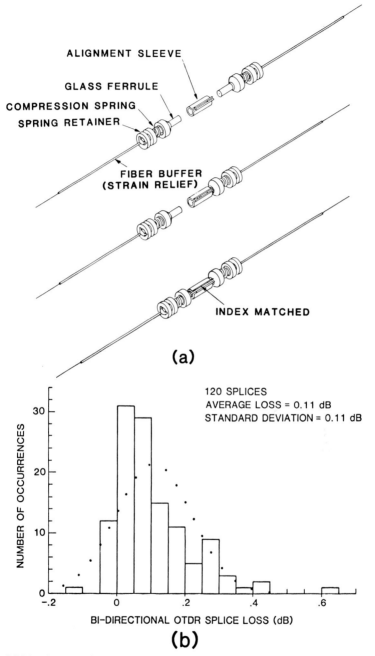

(a)

120 SPLICES
AVERAGE LOSS = 0.11 dB
STANDARD DEVIATION = 0.11 dB

(b)

Fig. 6.11 Multimode mechanical splice. (a) Glass capillary tubes (ferrules) are used to terminate each fiber and are aligned using an alignment sleeve. (b) Splice loss distribution (from Aberson and Yasinski, 1984).

grinding and polishing for end preparation, and an alignment sleeve of metal or plastic for aligning the glass tubes. Passive assembly of the splice gives a mean splice loss of less than 0.2 dB for 50 μm core diameter fiber. Local detection can be used allowing the splice to be tuned by rotating the terminations for minimum loss giving average splice loss of 0.05 dB for identical fibers. The loss histogram shown is for a field trial using production fibers. Environmental characteristics show less than 0.05 dB variation in loss occurs over the temperature range of $-40°$F to $+170°$F.

An active secondary alignment splice for single-mode fibers called the rotary splice (Miller *et al.*, 1985) is shown in Fig. 6.12(a). Fibers to be spliced are first terminated in precision glass capillary tubes using UV curable adhesive. The

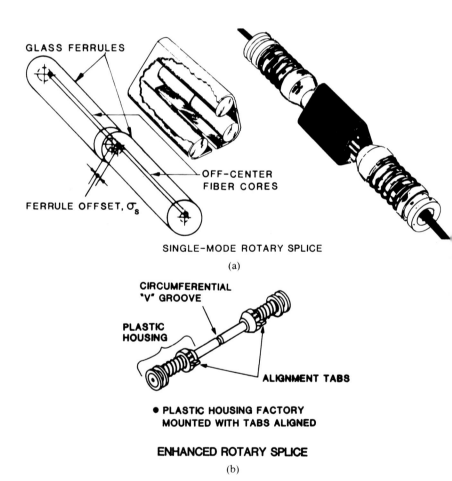

GLASS FERRULES

OFF-CENTER
FIBER CORES

FERRULE OFFSET, σ_s

SINGLE-MODE ROTARY SPLICE

(a)

CIRCUMFERENTIAL
"V" GROOVE

PLASTIC
HOUSING

ALIGNMENT TABS

● PLASTIC HOUSING FACTORY
MOUNTED WITH TABS ALIGNED

ENHANCED ROTARY SPLICE

(b)

Fig. 6.12 (a) Rotary splice for single-mode fibers and enhanced rotary mechanical splice.

fibers are prepared by grinding and polishing and gel is used for index matching. Alignment accuracies on the order of hundredths of a micron are obtainable with splice parts accurate to 1 or 2 μm, while allowing for total mechanical back-up, by using a modified three-glass-rod alignment sleeve. This sleeve has a built-in offset so that as each ferrule is rotated within the sleeve, the two circular paths of the centers of each core cross each other. A simple algorithm allows near-perfect alignment and a strong metal spring provides positive alignment retention. Field splice loss results closely match laboratory results (0.03 dB mean and 0.015 dB standard deviation) indicating that the rotary splice technique is extremely insensitive to the skill level of the splicer and to the harsh field environment.

A second-generation rotary splice—called the enhanced rotary mechanical splice (ERMS)—achieves low average single-mode splice loss of 0.20 dB without tuning. Splicing procedures and the tolerances of the glass tubes have been optimized to achieve these results; however, the splice can optionally be tuned to give splice losses below 0.05 dB with active alignment. The three largest offset-producing effects in glass-tube splices are: 1) eccentricities in the bore of the glass tubes, 2) oversize diameter of the bore, and 3) eccentricities of the fiber core. The ERMS totally eliminates the offset resulting from the first factor by using the scored continuous glass ferrule with factory mounted alignment tabs shown in Figure 6.12(b). The ferrule is broken in half in the field to produce a set of matched glass ferrules that are realigned using the tabs after the fibers are inserted and polished. Stability is the same as the standard rotary splice and installation time is reduced by a factor of two (five minutes total assembly time).

6.4.4.4 Fusion Splicing.   Fiber fusion results in the smallest size and potentially the most environmentally stable method of alignment retention. The protective coating is removed from the fibers, the fiber ends are prepared by scoring and breaking, and heat is applied until the fibers melt and fuse together (Bisbee, 1976). Electric arc and gas flame are the predominantly used heat sources. CO_2 lasers have also been used as the heat source for fusion in the laboratory but have not been found to be practical in the field. Fusion splices can be strong, especially with flame fusion, and losses can be low. The necessity of removing the fiber coating along with the handling and aligning of bare fibers can cause strength degradation, which can only be detected and eliminated by proof-testing. If the fiber splice is proof-tested, recoated and reinforced properly (Ogai et al., 1981), environmental properties are excellent.

Fusion differs from mechanical alignment retention techniques in that some splice loss increase is unavoidable because fibers ends are heated to the softening point where some degree of core deformation must take place (Payne and McCartney, 1984). In addition, arc fusion times longer than one or two seconds usually result in a phenomenon known as self-centering (Tsuchiya and

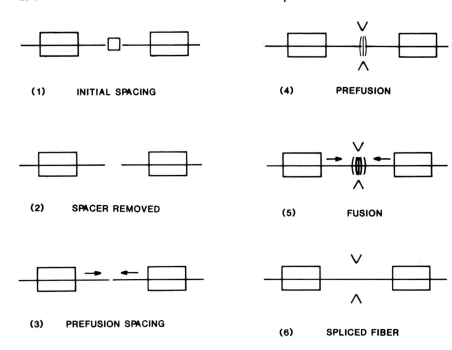

Fig. 6.13 Steps involved in most automatic fusion machines (from Sakamoto *et al.*, 1978).

Hatakeyama, 1977), which causes the fibers to more or less align relative to the fiber outer diameter because of the surface tension of the molten glass between the two fibers. Proper control of the critical fusion parameters must be maintained to repeatably fabricate low-loss, fusion splices.

The greater strength of flame fusion splices assures use of this technique for ocean cable systems. Splice strengths using a hydrogen chorine flame can match the original fiber strength under ideal conditions (Krause and Kurkjian, 1984). Strengths of oxy-hydrogen flame fusion splices can be 600 kpsi and higher. The coating stripping operation is a critical consideration when high strength splices are required. Any mechanical contact with the fiber or any sliding of the coating against the fiber will result in damage to the surface. Usually, hot acids are needed (H_2SO_4 at 200°C is used for an epoxy coating) and strengths of flame fusion splices are strongly dependent on acid temperature. Various rinses are then needed to neutralize the acid left on the fiber and adsorbed by the remaining coating. Flame fusion equipment consists mostly of laboratory or factory set-ups with almost no automation.

6.4.4.5 Fusion Hardware.   Present-day fusion equipment generally operates in the manner shown in Fig. 6.13. First, an aligned reference position for the fiber ends is established. Early machines (Sakamoto, *et al.*, 1978) used a spacer of known thickness as shown in the figure, however, most machines now require that the fiber ends be positioned and aligned close together. The machine sequence then separates the ends and a low temperature arc is initiated to prefuse the fiber ends. As the ends are heated the fibers are brought into contact, the arc temperature is increased and a prescribed amount of extra fiber (pressing stroke, usually ~ 10 microns) is fed into the melted joint. The quality of the splice depends on the accuracy and repeatability of these steps and the end angles of the fibers. Reducing the fusion time to a minimum and optimizing the remaining fusion parameters has resulted in practically eliminating self-centering by surface tension.

A large number of arc-fusion splicing machines are commercially available. A completely automatic fusion splicing machine (Fujikura, Ltd., 1984) uses direct core detection of a magnified (250 ×) cross-sectional view as shown in Fig. 6.14(a). A high-resolution TV camera and video signal processor aligns the fibers and controls all the prefusion and fusion operations. After fusing the

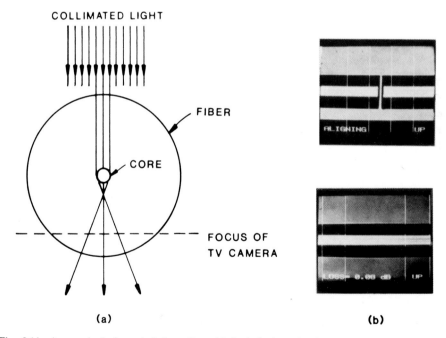

(a)   **(b)**

Fig. 6.14   Automatic single-mode fusion splicer. (a) Optical schematic. (b) Direct core observation display. (c) Splice loss histogram for identical fibers in the laboratory (from Fujikura, 1984).

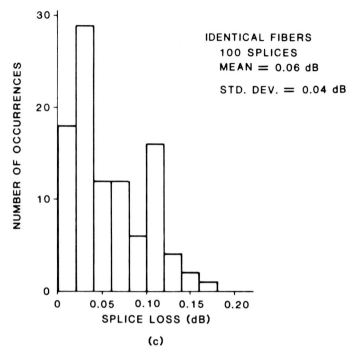

Fig. 6.14 (*Continued.*)

splice, the microprocessor measures core offset and tilt, calculates an estimated splice loss and displays the value on the TV monitor. This splicing machine does not require local injection or local detection and corrects internally for eccentric fibers. Fibers are offset so that movement due to surface tension effects returns the cores to an aligned condition (Kawata *et al.*, 1984). Results for identical fibers are typically 0.06 dB mean loss. This machine operates quickly and gives extremely repeatable, low loss results in the laboratory. Almost all sensitivity to operator skill has been eliminated.

Japanese investigators have reported success in the laboratory (Tachikura, 1984) and in a field test (Kawase *et al.*, 1982) in which 5-fiber ribbons were array fusion welded in manholes and on a pole. The fibers were placed in a region between the electrodes, where the temperature profile was nearly constant. Work is underway to extend fusion to 10-fiber ribbons and, in locations where fusion is allowed, this technique could be attractive.

Arc and flame fusion splices must be recoated and reinforced after proof-testing. Hydrofluoric acid etching can partially restore fiber strength (Ogai *et al.*, 1982). Split molds are used for recoating with UV curable materials (Hart and Krause, 1982), silicones or nylon to obtain a coated fiber the same size as the original fiber. In cases where size is not a constraint, a steel pin surrounded by a heat shrink tube is used for reinforcement. The five-fiber mass fusion welded ribbon is reinforced by sandwiching it between flat glass-ceramic substrates with a hot-melt film surrounding the fibers (Kawase *et al.*, 1982). Fusion splices must be proof-tested, recoated and/or reinforced properly or splice failure during thermal cycling can occur (Stueflotten, 1982).

6.5 CONCLUSION

The theories and principles presented here are a brief summary of a more complete discussion found in *Optical Fiber Splices and Connectors* (Miller, Mettler and White, 1986). These concepts provide a sound basis for establishing guidelines for splice and connector design and for evaluating present and future designs. The next challenge is to reduce the cost and complexity of connection hardware so that the full potential of optical fiber communications can be extended to the widest possible market. This will require ingenuity and advances in both materials and fabrication techniques especially for single-mode connections.

REFERENCES

Aberson, J. A., and White, I. A. (1986). Low-stress high-efficiency local-power injection, *Proceedings of IWCS*, Reno, Nevada, 334.

Aberson, J. A., and Yasinski, K. M. (1984). Multimode mechanical splices, *Proceedings of Tenth ECOC*, Stuttgart, Germany, 182.

Baden, J. L., Aberson, J. A., and Swiderski, M. J. (1986). Mass splicing of single-mode fibers, *Proceedings of OFC '86*, Atlanta, Georgia, 52.

Bisbee, D. L. (1971). Optical fiber joining technique, *Bell Syst. Tech. J.*, **50**, 3153–3158.

Bouvard, A., and Hulin, J. P. (1981). Results on field operation of a mass splicing process, *Proceedings of Seventh ECOC*, Copenhagen, Denmark, Paper 7.3.

Bowen, T. (1981). Optical fiber connectors and splices, Professional Program Session Record, Electro '81, New York, New York, Paper 8a/3.

Buckler, M. J. (1983). Design and engineering of a lightguide route, *Proceedings of the NEC*, Oak Brook, Illinois, 404–408.

Buckler, M. J., and Meskell, D. J., Jr. (1983). A new method for lightguide route engineering, *Proceedings of Globecom '83*, San Diego, California, 1155–1159.

Cassidy, D. L. (1984). Influence on the steady-state oscillation spectrum of a diode laser for feedback of light interacting coherently and incoherently with the field established in the laser cavity, *Appl. Opt.* **23**, 2070.

Cherin, A. H., Head, E. D., White, I. A., and Mettler, S. C. (1986). Characterization of source power distributions in multimode fiber by a splice offset technique, *J. of Lightwave Tech.* **LT-4**, 259.

Chesler, R. B., and Dabby, F. W. (1976). Simple testing methods give users a feel for cable parameters, *Electronics* **49**, 90.

Daino, B., DeMarchis, G., and Piazzolla, S. (1979). Analysis and measurement of modal noise in an optical fibre, *Elect. Lett.* **15**, 755.

DeBlok, C. M., and Matthijsse, P. (1984). Core alignment procedure for single-mode-fibre jointing, *Elect. Lett.* **20**, 109–110.

Derosier, R. M., and Stone, J. (1973). Low-loss splices in optical fibers, *Bell Syst. Tech. J.* **52**, 1229.

deVecchis, M., LeNoane, G., and Hulin, J. P. (1978). Experimental results of cylindrical V-grooved structure optical cables laid in ducts and spliced, *Proceedings of Fourth ECOC*, Genova, Italy, 218–221.

Drake, M. D. (1981). Low reflectance terminations and connections for duplex fiber-optic telecommunication links, *Appl. Opt.* **20**, 1640.

Duff, D. G., Stone, F. T., and Wu, J. (1985). Measurements of modal noise in single-mode lightwave systems, *Proceedings of OFC '85*, San Diego, California, Paper Tu01.

Epworth, R. E. (1978). The phenomenon of modal noise in analogue and digital optical fibre systems, *Proceedings of the Fourth ECOC*, Genova, Italy, 492.

Fujikura, Ltd. (1984). Abstract of the FSM-20 Single-Mode Fusion Splicer, Tokyo, Japan.

Fulenwider, J. E., and Dakss, M. L. (1977). Hand-held tool for optical-fibre end preparation, *Elect. Lett.* **13**, 578.

Furukawa Electric Company, Ltd. (1983). Stripper S-211 and breaker S-3111. *Product Bulletin*, Tokyo, Japan.

Gartside, C. H., III, and Baden, J. L. (1985). Single-mode ribbon cable and array splicing, *Proceedings of OFC '85*, San Diego, California, Paper WU2.

Gloge, D., and Marcatili, E. A. J. (1973). Multimode theory of graded core fibers, *Bell Syst. Tech. J.* **52**, 1563.

Gloge, D., Smith, P. W., Bisbee, D. L., and Chinnock, E. L. (1973). Optical fiber end preparation for low-loss splices, *Bell Syst. Tech. J.* **52**, 1579–1588.

Gordon, K. S., Rawson, E. G., and Nafarrate, A. B. (1977). Fiber-break testing by interferometry: a comparison of two breaking methods, *Appl. Opt.* **16**, 818–819.

Hardwick, N. E. (1985). Longitudinal stresses in mechanical splices. Unpublished paper.

Hardwick, N. E., and Davies, S. T. (1985). Rapid ribbon splice for multimode fiber splicing. *Proceedings of OFC '85*, San Diego, California, Paper TUQ27.

Hardwick, N. E., and Reynolds, M. R. (1985). Finite element analysis of thermal stresses in the single-mode bonded splice, *Opt. Lett.* **10**, 241–243.

Hart, A. C., Jr., and Krause, J. T. (1982). Technique for coating high strength lightguide fusion splices, *Proceedings of OFC '82*, Phoenix, Arizona, Paper ThAA3.

Jocteur, R., and Carratt, M. (1980). A new fiber ribbon cable, *Proceedings of Sixth ECOC*, York, England, 342–343.

Johnson, J. W., and Holloway, D. G. (1966). On the shape and size of the fracture surfaces, *Philos. Magn.* **14**, 731–743.

Judy, A. F., and Neysmith, N. E. S. (1985). Reflections from polished single-mode fiber ends, *Proceedings of SPIE Conference*, San Diego, California, 146–151.

Kashima, N. (1981). Transmission characteristics of splices in graded-index multimode fibers, *Appl. Optics* **20**, 3859.

Kawakami, S., and Tanji, H. (1983). Evolution of power distribution in graded-index fibres, *Elect. Lett.* **19**, 100.

Kawase, M., Tachikura, M., Nihei, F., and Murata, H. (1982). Mass fusion splices for high density optical fiber units, *Proceedings of Eighth ECOC*, Cannes, France, Paper AX-5.

Kawata, O., Hoshino, K., Miyajima, Y., Ohnishi, M., and Ishihara, K. (1984). A splicing end inspection technique for single-mode fibers using direct core monitoring, *J. of Lightwave Tech.* **LT-2**, 185–190.

Knecht, D. M., Carlsen, W. J., and Melman, P. (1982). Fiber optic field splice, *Proceedings of SPIE*, Los Angeles, California, 44–50.

Kohanzadeh, Y. (1976). Hot splices of optical wave guide fibers, *Appl. Opt.* **15**, 793.

Krause, J. T., and Kurkjian, C. R. (1984). Intrinsic glass strength achieved in fiber splices, Post Deadline Paper at OFC '84, New Orleans, Louisiana, Paper WI7.

Kummer, R. B. (1979). Observation of a cascade effect for optical fiber splices, *Proceedings of OFC '79*, San Francisco, California, Paper ThE2.

Kummer, R. B. (1980). Lightguide splice loss-effects of launch beam numerical aperture, *Bell Syst. Tech. J.* **59**, 441–447.

Kummer, R. B., and Fleming, S. R. (1982). Monomode optical fiber splice loss—combined effects of misalignment and spot size mismatch, *Proceedings of OFC '82*, Phoenix, Arizona, Paper THAA1, 44.

Kummer, R. B., and Mettler, S. C. (1981). Effects of lightguide manufacturing tolerances and misalignment on splice loss, *Proceedings of IOOC '81*, San Francisco, California, 92–94.

Kummer, R. B., Judy, A. F., and Cherin, A. H. (1982). Field and laboratory transmission and OTDR splice loss measurements of multimode optical fibers, *Proceedings of the Symposium on Optical Fiber Measurements*, Boulder, Colorado, 109–121.

Kurokawa, T., Yoshizawa, T., and Katayama, Y. (1980). Precisely moulded plastic splices for optical fibres, *Elect. Lett.* **16**, 911–912.

Marcuse, D. (1973). Losses and impulse response of a parabolic index fiber with random bends, *Bell Syst. Tech. J.* **52**, 1423.

Marcuse, D. (1976). Microbending losses of single-mode, step-index and multimode, parabolic-index fibers, *Bell Syst. Tech. J.* **55**, 937–955.

Marcuse, D. (1977). Loss analysis of single-mode fiber splices, *Bell Syst. Tech. J.* **56**, 703.

McCartney, D. J., Payne, D. B., and Wright, J. V. (1984). Analysis of splices in shifted zero dispersion monomode fibre, *Elect. Lett.* **20**, 78–80.

Mettler, S. C. (1979). A general characterization of splice loss for multimode optical fibers, *Bell Syst. Tech. J.* **58**, 2163.

Mettler, S. C., and Gotthardt, M. R. (1981). Index-of-refraction matching materials for optical fiber splicing, *Proceedings of IOOC '81*, San Francisco, California, Paper WC4.

Miller, C. M. (1973). U.S. Patent 3,864,018, Method and means for splicing arrays of optical fibers, filed October 1973, issued February 1975.

Miller, C. M. (1975). A fiber-optic cable connector, *Bell Syst. Tech. J.* **54**, 1547.

Miller, C. M. (1984). Single-mode fiber splicing, *Proceedings of OFC '84*, New Orleans, Louisiana, Paper TUM5.

Miller, C. M. (1985). Rotary splice aligns optical fibers quickly, accurately, *Design News*, June 17, 1985, 128–129.

Miller, C. M., and Mettler, S. C. (1978). A loss model for parabolic-profile fiber splices, *Bell Syst. Tech. J.* **57**, 3167.

Miller, C. M., and Schroeder, C. M. (1976). Fiber optic array splicing, Conference on Laser and Electrooptical Systems, San Diego, California, Paper ThE8.

Miller, C. M., DeVeau, G. F., and Smith, M. Y. (1985). A simple, high performance mechanical splice for single-mode fibers, *Proceedings of OFC '85*, San Diego, California, Paper MI2.

Miller, C. M., Mettler, S. C., and White, I. A. (1986). "Optical Fiber Splices and Connectors," Marcel Dekker, Inc., New York.

Miller, S. E., and Chynoweth, A. G. (1979). "Optical Fiber Telecommunications," Academic Press, New York, 458–460.

Murata, H. (1976). Broadband optical fiber cable and connecting, *Proceedings of Second ECOC*, Paris, France, Paper VI.1.

Murata, H., Inao, S., and Matsuda, Y. (1975). Connection of optical fiber cable, *Proceedings of First Topical Meeting on Optical Fiber Transmission*, Williamsburg, Virginia, Paper WA5.

Nawata, K., Iwahara, Y., and Suzuki, N. (1979). Ceramic capillary splices for optical fibres, *Elect. Lett.* **15**, 470–472.

Nelson, J., and Ahmed, W. (1982). Optimum surface finish of fiber-optic termination cuts down on optical losses, *Communications News* **19**, No. 10.

Nemoto, S., and Makimoto, T. (1979). Analysis of splice loss in single-mode fibers using a Gaussian field approximation, *Opt. and Quan. Elect.* **11**, 447.

Ogai, M., Nishimura, M., Kamikura, Y., Tachigama, S., Miyauchi, M., and Matsumoto, M. (1981). Injection molding of fusion spliced fiber, *Proceedings of IOOC '81*, San Francisco, California, Paper WC3.

O'Hara, S. (1975). Status of fiber transmission system research in Japan, *Proceedings of First Topical Meeting on Optical Fiber Transmission*, Williamsburg, Virginia, Paper ThA2.

Payne, D. B., and McCartney, D. J. (1984). A comparative study of monomode splicing techniques, *Proceedings of ICC '84*, Amsterdam, Netherlands, 1075–1079.

Peckham, D. W., Mettler, S. C., and Kummer, R. B. (1984). A systematic approach to specifying multimode fiber manufacturing tolerances, *Proceedings of the Symposium on Optical Fiber Measurements*, Boulder, Colorado, 73–76.

Personick, S. D. (1977). Photon probe—an optical-fiber-time-domain reflectometer, *Bell Syst. Tech. J.* **56**, 355–366.

Peterman, K. (1980). Nonlinear distortions due to fibre connectors, *Proceedings of the Sixth ECOC*, University of York, United Kingdom, 80–83.

Rawson, E. G., Goodman, J. W., and Norton, R. E. (1980). Experimental and analytical study of modal noise in optical fibers, *Proceedings of the Sixth ECOC*, University of York, United Kingdom, 72–75.

Sakamoto, K., Miyajiri, T., Kakuzen, H., Hirai, M., and Uchida, N. (1978). The automatic splicing machine employing electric arc fusion, *Proceedings of Fourth ECOC*, Genova, Italy, 296.

Saunders, M. J. (1979). Torsion effects on fractured fiber ends, *Appl. Opt.* **18**, 1480–1481.

Saunders, M. J., and Parham, W. L. (1977). Adhesive sandwich optical fiber ribbons, *Bell Syst. Tech. J.* **56**, 1013–1014.

Sears, F. M., White, I. A., Kummer, R. B., and Stone, F. T. (1985). Probability of modal noise in single-mode lightguide systems, *Proceedings of IOOC/ECOC '85*, Venice, Italy, 823–826.

Schroeder, C. M. (1978). Accurate silicon spacer chips for an optical fiber cable connector, *Bell Syst. Tech. J.* **57**, 91–97.

Sieverts, Kabelverk (1983). Product Bulletin, Cutting Device for Optical Fibres, Sundbyberg, Sweden.

Someda, C. G. (1973). Simple low-loss joints between single-mode optical fibers, *Bell Syst. Tech. J.* **52**, 583.

Stueflotten, S. L. (1982). Protection of optical fiber arc fusion splices, *J. Opt. Comm.* **3**, 19–25.

Tachikura, M. (1984). Fusion mass-splicing for optical fibers using electric discharges between two pairs of electrodes, *Appl. Opt.* **23**, 492–498.

Tatekura, K., Yamamoto, H., and Nunakawa, M. (1982). Novel core alignment method for low-loss splicing of single-mode fibres utilizing UV-excited fluorescence of Ge-doped silica core, *Elect. Lett.* **18**, 712–713.

Thiel, F. L., and Davis, D. H. (1976). Contributions of optical-waveguide manufacturing variations to joint loss, *Elect. Lett.* **12**, 340.

Thomas and Betts (1981). Universal Fiber Optic Cleaving Tool, Product Bulletin, Raritan, New Jersey.

Tsuchiya, M., and Hatakeyama, I. (1977). Fusion splices for single-mode optical fibers, *Proceedings of OFC II*, Williamsburg, Virginia, Postdeadline Paper, PD1.

Tynes, A. R., and Derosier, R. M. (1977). Low-loss splices for single-mode fibres, *Elect. Lett.* **13**, 673.

Vodhanel, R. S., and Ko, J.-S. (1984). Reflection induced frequency shifts in single-mode laser diodes coupled to optical fibres, *Elect. Lett.* **20**, 973.

Wagner, R. E., and Sandahl, C. R. (1982). Interference effects in optical fiber connections, *Appl. Opt.* **21**, 1381–1385.

White, I. A., and Kuhl, J. F. (1983). Microbending theory of single-mode fiber splices—butt joint and fusion, *Proceedings of Fourth IOOC*, Tokyo, Japan, 402.

White, I. A., and Mettler, S. C. (1983). Modal analysis of loss and mode mixing in multimode parabolic index splices, *Bell Syst. Tech. J.* **62**, 1189.

Chapter 7

Fiber Connectors

W. C. YOUNG

Bell Communications Research, Inc., Red Bank, New Jersey

D. R. FREY

AT & T Bell Laboratories, Inc., Norcross, Georgia

7.1 INTRODUCTION

In recent years the state of the art of optical fiber technology has progressed such that the achievable attenuation levels for the fibers are very near the limitations due to Rayleigh scattering. As a result, optical fibers, and particularly single-mode fibers, can be routinely fabricated to attenuation levels in the range of about 0.5 dB/km at 1300 nanometers, and 0.25 dB/km at 1550 nanometers and employing these fibers in lightwave systems requires precise connectors. Considering the small size of the fiber cores, less than 10 microns in diameter for single-mode fibers and less than 100 microns for multimode fibers, it is not surprising that these connectors must provide precision alignments or they can easily introduce high optical losses. Furthermore, since single-mode fibers have practically unlimited bandwidth, they have recently become the favorite choice for most of the lightwave systems presently being designed for telecommunication networks, and perhaps in the future may be used in local area networks as well. To provide low-loss connectors, particularly for these single-mode fibers, alignment accuracies in the sub-micron range are required, and these sub-micron alignments must be both reliable as well as cost-effective. Achieving these goals is presently the challenge facing the connector technologist.

In this chapter we will review the fundamental technology used in the design of demountable connectors. In particular, since single-mode fibers in the future will more than likely dominate lightwave systems and will also require the greatest precision, most of our attention will be directed toward the particular problems encountered in the interconnection of these fibers.

We begin by defining the term optical fiber connector. Generally, the term connector is used when referring to the jointing of two fibers in a manner that permits and expects unjointing by its design intent. Connectors are usually used for terminating components, for system configuration, testing, and maintenance. Generally, the connectors are either factory-installed or field-installed depending on the particular application. In this sense, the term field usually refers to an environment other than the connector factory. In the case of single-mode fibers, however, the required sub-micron alignment tolerance has, until recently, dictated that the connector installation be done in the connector factory. This limitation has hampered some single-mode fiber installation, but field installations, of single-mode connectors, are becoming common practice.

7.2 CONNECTOR APPLICATIONS

7.2.1 General

Optical fiber connectors are indispensable passive devices for:

- telephone networks,
- local area networks (LAN),
- data transmission,
- connections between computers,
- industrial applications and
- military applications.

Connectors are used for connecting the fiber cable to terminal and line equipment, to access fiber paths for testing, and for network rearrangements. The terminal and line equipment includes active devices (sources and detectors) as well as passive devices (attenuators, branching devices, wavelength division multiplexers, and couplers). These devices are often furnished with fiber pigtails terminated with a connector plug on the end of the fiber, but they are also furnished with connector couplings as an integral part of the housing or mounting.

By far, most of the optical transmission systems in place and those being installed are for telephony in long-haul routes, interoffice trunks, and distribution feeder routes. The early systems were multimode, but essentially all new telephony systems are likely to be single-mode. These are fiber dominated systems, of many kilometers, with only 4 to 6 connectors per span. Applications are emerging, however, for commercial distribution and in the near future, distribution to residences, which will be shorter in length, require less fiber per span, and thus be more connector intensive. In these latter applications, connectors will be needed to connect terminal apparatus to the network, and to interconnect the distribution terminal apparatus.

The next largest application for fiber optic systems, which is a large user of connectors, is LANs for university and industrial campuses and for large buildings for data and video transmission. In these applications, many terminals are connected to each other in a network and to central computing facilities. The use of fiber optics in industry for data collection and manufacturing control is in its infancy but will expand rapidly over the next few years with many applications for connectors. The same can be said for military applications where fiber optic systems are immune to electro-magnetic interference. These types of systems tend to be shorter in length with a high concentration of high quality and rugged connectors.

Optical fiber systems are poised for rapid expansion in many areas. Connectors will become ubiquitous, which will increase the already present pressure for connector designs that are easier and faster to install, at lower prices, while maintaining high performance. This will be a real challenge for connector industry.

7.2.2 Single Fiber Connectors

Essentially, all of the connectors in use today are single fiber designs using one of the principles discussed in Section 7.4. Bidirectional systems on a single fiber are beginning to be used for low bit rate data, but high speed systems (40 mbits and

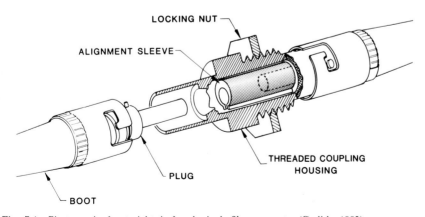

Fig. 7.1 Photograph of a straight tip ferrule single fiber connector (Carlisle, 1985).

Fig. 7.2 Photograph of a conical ferrule single fiber connector (Young and Curtis, 1983).

higher) typically use unidirectional transmission. An example of a straight ferrule single fiber connector is shown in Figure 7.1 and an example of a conical ferrule is shown in Figure 7.2.

A number of duplex connector designs have been developed, but as yet none are widely used. This may change in the near future, particularly in LANs with ring structures. An ANSI (American National Standards Institute) committee is developing standards for the architecture, protocols, and hardware (including a duplex connector) for these LANs. These duplex connectors will mate directly with connectorized optical data links (transmitters and receivers). An example of a duplex connector to simplex connector is shown in Figure 7.3.

Ruggedized duplex and multiple fiber connectors are also being developed, and are in use, for military applications. These connectors often have the added requirement of being hermaphroditic, so that all connectorized cable ends will mate together.

Japan is considering a ribbon cable structure for distribution applications using a five- or ten-fiber ribbon. A linear array connector, which uses two precision alignment pins, has been demonstrated for connecting the fiber ribbons to themselves and to a transition piece, which connects the ribbon to individual fibers. The insertion loss of this linear array connector is somewhat higher than the loss of a single fiber connector but acceptable for the particular application.

7.2.4 Backplane Connectors

Generally, optical connectors have not been used in the conventional backplane applications where the connector is inserted blind, which is common for electrical connectors. AT & T has been unique in using the biconic connector in a blind insertion backplane application for plug-in regenerators for both multimode and single-mode fibers. AT & T has also introduced a multi-fiber circuit pack to backplane optical connector using a linear array of fibers sandwiched between etched silicon chips. The arrays contain 12 to 18 fibers and fit into an area $0.5'' \times 0.6''$ on the backplane. Since the density of optical circuits is increasing, we can expect to see a proliferation of connectors for these applications in the next few years. An example of an array type backplane connector is shown in Figure 7.4. This connector replaces some of the electrical contacts in a commonly used electrical backplane connector.

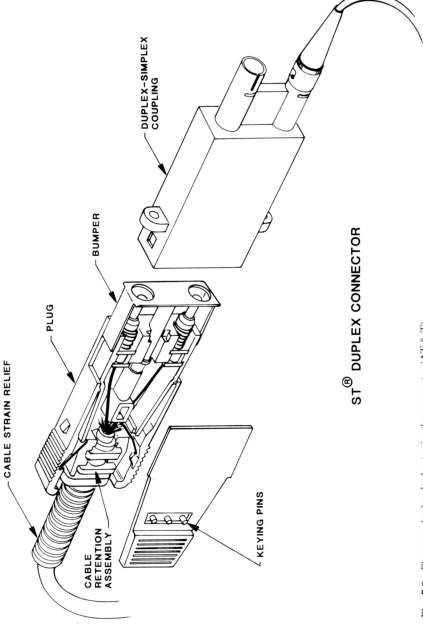

Fig. 7.3 Photograph of a duplex-to-simplex connector (AT & T).

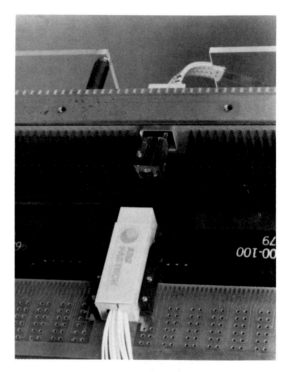

Fig. 7.4   Photograph of an array backplane connector (AT & T).

7.3 CONNECTOR LOSS PHENOMENA

It is common practice to separate the factors that cause optical losses into two categories, as detailed in Table 7.1. The first category, which includes factors intrinsic to the fiber such as lateral offset between cores, longitudinal offset (end-gap), angular misalignments (tilt), end-face quality, and reflections, results from inaccuracies related directly to the connector design and manufacturing control. The second category, intrinsic factors, are related directly to the particular properties of the two optical fibers that are actually being interconnected. These factors include mismatches in core radii, index profiles, and core ellipticity.

In addition to these factors, the coupling efficiency of a connector may also depend on the center wavelength and coherence length of the optical source, and in the case of multimode fibers, on the relative spacing of the optical source and the connector as well as the mode-mixing and differential mode attenuation characteristics of the fibers. For example, if the source is an incoherent LED and the length of fiber before the connector is sufficiently long so that the modal

TABLE 7.1
Factors Causing Optical Losses

Extrinsic Factors

Offset (lateral core misalignment)
End-gap (longitudinal offset)
Tilt (angular misalignment)
End-face quality
Reflections

Intrinsic Factors

Mismatch in fiber core diameters
Mismatch in index-profiles
Core ellipticity

power distribution has reached a steady-state condition, then the effects due to offsets are less than when the length of fiber is sufficiently short so that the steady-state mode distribution is not achieved and an over-moded state still exists. Therefore, in the case of multimode fibers and incoherent sources, the relative locations of the connectors in a particular fiber span must be considered when predicting the optical losses of connectors. Because of these mode distribution effects, a common method of evaluating connectors is to use an over-moded launch followed by a mode-filter that selectively filters the higher-order modes, thereby creating an approximate steady-state mode distribution. A further discussion on these launch conditions and their consequences will follow. For now, we will consider two distinct launch conditions for graded-index multimode fibers, namely, the uniform launch and steady-state launch conditions, and their effect on coupling efficiency.

7.3.1 Extrinsic Loss Factors

Various experiments (Kummer and Fleming, 1982; Chu and McCormick, 1978) and analytical models (Miller and Mettler, 1978; Nemota and Makimoto, 1979) have been used to quantify the effect that extrinsic factors have on the coupling losses of both single-mode and multimode fiber joints. Based on these studies, we have plotted the coupling loss as a function of lateral, longitudinal, and angular offset for both graded-index multimode fiber and for single-mode fiber in Fig. 7.5 and Fig. 7.6, respectively. In the multimode fiber's case we have shown the effect of the misalignments for both uniform and steady-state modal power distributions. From these curves it can be seen that as the quality of the alignment improves the effect of the higher-order modes being present diminishes. It should be re-emphasized that it is necessary to know the mode

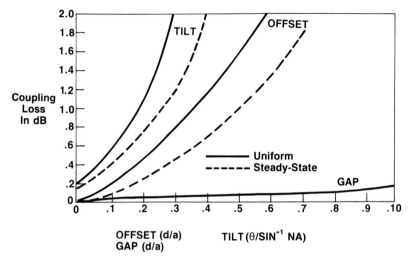

Fig. 7.5 Coupling loss as a function of normalized lateral, longitudinal, and angular offset for graded-index multimode fibers.

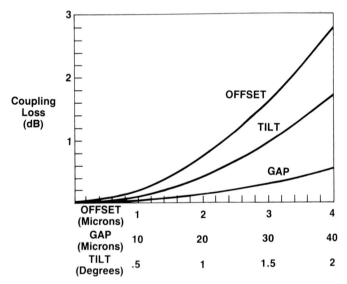

Fig. 7.6 Coupling loss as a function of lateral, longitudinal, and angular offset for single-mode fiber having a mode field diameter of 10 microns.

Fig. 7.7 Micro-interferogram of connector/fiber end-face.

distribution that will be present in a particular application of multimode connectors in order to correctly predict the performance of a particular connector design. Also, from these two figures it can be seen that the coupling loss is significantly more sensitive to lateral misalignments, for both single-mode and multimode fibers, than either longitudinal or angular misalignments. For example, for low-loss joints (0.5 dB), lateral offsets must be controlled to sub-micron accuracies for single-mode fibers and to micron accuracies for multimode fibers.

In regards to end-face quality, connectorization usually employs polishing procedures for end-face finishing. These polishing procedures usually require special tools for controlling end-face angles, end-face profiles, and sometimes for length control as well. In some cases the ends are polished slightly concave so that the fibers in the assembled connector cannot make contact with each other. In other cases, the tools are designed to create a flat, and sometimes even a convex surface, to achieve fiber end-face contact. Finally, the end-faces can also be polished at a slight angle to maximize the return loss (minimize the reflected power) of the connector. The surface finish that is generally acceptable for low-loss connectors is about 0.025 microns for non-index matched connectors and can be as much as 0.18 microns for connectors employing index-matching fluids. The end-face shown in Fig. 7.7 is more than acceptable for a low-loss non-index matched connector. The surface finish for the end-face in the photograph is about 0.025 microns and the flatness as shown in the micro-interferogram is less than 0.25 microns.

The final extrinsic factor to be discussed is the effect of reflections on coupling efficiency. For connectors having an air gap between the adjacent fibers, Fresnel reflections exist, and in this case, for the two air/glass interfaces the loss amounts to about 0.31 dB for silica fibers. Furthermore, when these gaps are small, ($\lesssim 10$ microns for LED sources and significantly larger for LD sources), interference

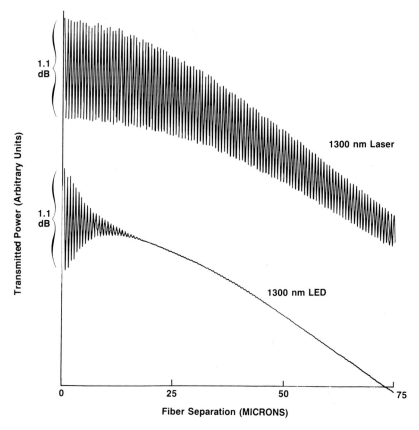

Fig. 7.8 Transmission versus fiber end separation for single-mode fiber at a wavelength of 1300 nanometers.

effects can exist that may have an additional affect on the coupling efficiency (Wagner and Sandahl, 1982), and while maximum coupling efficiency can be achieved when the fiber end-faces are in contact, small gaps can cause multiple reflections and the resulting loss can be as high as 0.6 dB. Recent work (Shah *et al.*, 1987) has shown that the polishing of fiber end-faces may substantially increase the refractive index of the surface from 1.47 to about 1.6. This region of higher index, about 0.15 microns deep, results in interference effects that can cause reflection losses of greater than 1.1 dB, which is greater than the maximum specified loss for some connectors. Figure 7.8 shows the effect of small changes in the air gap between two single-mode fibers on the coupling efficiency of a connector having polished end-faces. It should be noted that these reflections can be minimized by using an index-matching medium (fluid, gel, etc.)

between the fiber end-faces, and while this approach is quite common in the splicing of fibers, due to considerations like cleanliness and contamination it is usually not used in demountable connectors. In cases where reflections must be minimized, however, the connectors can be polished at an angle such that the reflected power is not guided by the fiber.

7.3.2 Intrinsic Loss Factors

Like extrinsic factors, intrinsic factors can also have a large effect on the coupling efficiency of fiber connectors. While extrinsic factors have a less significant effect on the loss of multimode fiber connectors than on single-mode fiber connectors, intrinsic factors have the opposite effect. That is, when evaluating the coupling efficiency of multimode fiber connectors, it is important that the characteristics of the fibers on either side of the connector be considered, and the direction of propagation through the joint must also be taken into account. In the case of the single-mode fiber, differences in the characteristics of the fibers have a practically insignificant effect on the coupling efficiency, particularly when compared with the effect caused by sub-micron-type lateral misalignments. Also, for practical single-mode fiber connectors, the law of reciprocity applies and therefore, the coupling efficiency is independent of the direction of propagation through the connector.

Again, various experiments (Kaiser *et al.*, 1980) and analytical models have been used to quantify these effects. First, in the case of multimode fiber joints, we have summarized in Fig. 7.9 the dependence of the coupling loss on

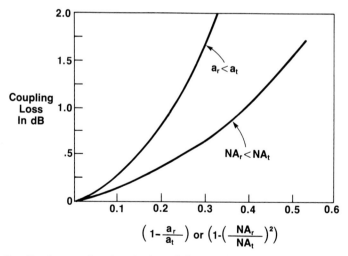

Fig. 7.9 Coupling loss as a function of mismatch in numerical aperture and core diameters for graded-index multimode fibers.

mismatches in numerical apertures and core radii. It should be noted that this figure shows the dependence of loss as the optical power propagates from larger values of numerical aperture to smaller values, and from larger core radii to smaller core radii. When the direction of propagation is from smaller to larger NAs and core radii, there is no associated loss.

In the case of single-mode fiber connectors, it has been shown (Marcuse, 1977) that the fields of most single-mode fibers being used today, those designed for use at 1300 nanometers, are nearly Gaussian and therefore the coupling losses for these connectors can be calculated by evaluating the coupling between two misaligned Gaussian beams. Based on this model the following general equation has been derived (Nemota and Makimoto, 1979) for calculating the coupling loss between single-mode fibers (Fig. 7.10) that have unequal mode-field diameters (intrinsic factor) and lateral, longitudinal, and angular offsets as well as reflections (extrinsic factors):

$$\text{Coupling Loss} = -10 \log\left[\left(\frac{16n_1^2 n_2^2}{(n_1 + n_2)^4}\right) \frac{4\sigma}{q} \exp\left(\frac{-\rho u}{q}\right)\right] (dB)$$

where $\rho = (kw_1)^2, \qquad q = G^2 + (\sigma + 1)^2$

$u = (\sigma + 1)F^2 + 2\sigma FG \sin\theta + \sigma(G^2 + \sigma + 1)\sin^2\theta$

$F = x/kw_1^2, \qquad G = z/kw_1^2$

$\sigma = (w_2/w_1)^2, \qquad k = 2\pi n_2/\lambda$

n_1 = refractive index of fibers

n_2 = refractive index of medium between fibers

λ = wavelength of source

x = lateral offset

z = longitudinal offset

θ = angular misalignment

w_1 = spot size, transmitting fiber ($1/e$ power, radius)

w_2 = spot size, receiving fiber ($1/e$ power, radius).

It should be noted that the equation does not consider interference effects that may exist in connectors having fiber end-face separations and that do not employ exact index-matching fluids or gels.

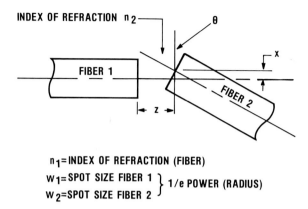

Fig. 7.10 Typical offsets for butt-jointed fibers.

Using this general equation, which has been found to have very good correlation with various experimental investigations, the coupling loss due to intrinsic factors can be determined. Assuming that no losses are present due to extrinsic factors, the equation reduces to the following expression:

$$\text{Coupling loss} = -10 \, \text{Log} \left[4 (W_2/W_1 + W_1/W_2)^{-2} \right] (dB).$$

Using this expression, the loss due to a 10% mismatch in mode-field diameters (intrinsic factor), a typical value for today's fibers, and with no other factors present, is calculated to be 0.05 dB. It can also be seen in this expression, that contrary to multimode fibers, the loss is independent of the direction of propagation through a single-mode fiber connector.

7.4 CONNECTORS DESIGNS

Low-loss practical connectors for lightwave systems, both multimode and single-mode, require precise alignments to control the extrinsic factors discussed in Section 7.3.2. These connectors must also provide precise alignments under various operating conditions such as shock, vibration, repetitive engagement and separation, as well as changes in environmental conditions such as temperature and humidity. To be practical, these connectors must also permit easy and simple installation outside the environment of the connector factory. To achieve these goals, considerable work has been directed at two types of connector designs, the butt-joint design and the expanded-beam design.

7.4.1 Butt-joint Connectors

In the discussion of butt-joint connectors, we will concentrate on the requirements of single-mode fiber connectors, since the goal, precise alignment, is the

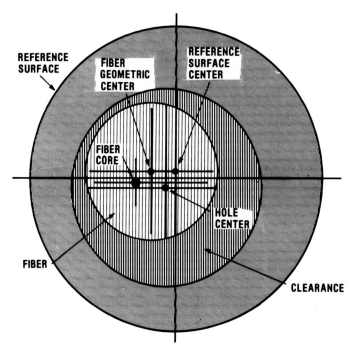

Fig. 7.11 Accumulation of inaccuracies in a typical field-installable butt-jointed type connector plug.

same as that for multimode fiber except that the required alignment accuracy is extended to sub-micron accuracies instead of micron-type accuracies.

As previously noted, the most critical and also the most difficult alignment to achieve is the control of lateral offset between adjacent fiber cores. The resulting lateral offset in a connector assembly can be influenced by the fiber's core/cladding offset and outside diameter variations, and in the connector, by the hole diameter variation and the hole-to-reference surface offset. In evaluating and designing various connector parts, the combined effect of these inaccuracies must be completely understood and controlled. To explain this, we can use the field-installable butt-jointed type connector plug shown schematically in Fig. 7.11. In this example, the reference surface may be either the outside diameter of a cylindrical ferrule, the tapered alignment surface of a biconic connector plug, both popular designs, or any other type of alignment element. From this schematic, it is obvious that a simple linear summation of these offsets does not represent a true measure of the expected offset between the fiber-core axis and the reference-surface axis, which is of course the predominate factor to be minimized. A more realistic expected offset can be statistically calculated using

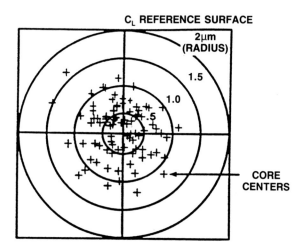

GIVEN: NUMBER OF CONNECTORS = 100
          HOLE ECC, MEAN = 0.5µm, SIGMA = 0.2µm
          DIAMETRAL CLEARANCE, MEAN = 0.7µm, SIGMA = 0.2µm
          FIBER ECC, MEAN = 0.3µm, SIGMA = 0.1µm

RESLUTS: CORE TO REFERENCE SURFACE C$_L$
          MEAN = 0.6µm
          SIGMA = 0.3µm

Fig. 7.12 Scatter plot of the calculated locations of fiber core centers in a field-installable connector plug.

appropriate distributions for offsets of the hole-to-reference-surface axis, and the fiber-core-to-cladding axis, as well as the variations in the outside diameter of the fiber and the diameter of the hole in the ferrule or plug. The appropriate distributions for these factors are clearly dependent upon the fabrication processes as well as any sorting processes that may be employed in selecting the fibers and connectors. With this in mind, a Monte Carlo simulation, based on distributions representing good-quality fibers and the connectors, can be used to calculate the expected core-to-reference-surface offset for a particular set of connector plugs.

A scatter plot of the calculated locations of these core centers, assuming fiber-core-to-cladding offsets having a mean of 0.3 and a standard deviation of 0.1 microns, hole offsets having a mean of 0.5 and a standard deviation of 0.2 microns, and diametrical clearances between the fiber and hole having a mean of 0.7 and a standard deviation of 0.2 microns is shown in Fig. 7.12. The mean and standard deviation for this particular connector model was calculated to be 0.6 and 0.3 microns, respectively. This type of distribution has been routinely achieved in laboratories by using special connector trimming techniques, high

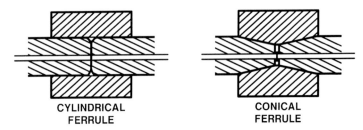

CYLINDRICAL
FERRULE

CONICAL
FERRULE

Fig. 7.13 Examples of two popular alignment mechanisms used in fiber-optic connectors.

quality fibers, and by measuring and matching the fiber and hole diameters prior to installing the connectors (Young, 1984).

After determining the distribution of the fiber-core-to-reference-surface offsets, we must next consider the ferrule-to-ferrule (plug-to-plug) alignment mechanism. The two types of connectors shown in Fig. 7.13 use two plugs and an alignment sleeve and are presently the most popular types of butt-joint connector designs being used in both single-mode fiber and multimode fiber systems. The resulting lateral offsets between fibers in these types of connectors can be represented schematically as shown in Fig. 7.14. With the calculated distribution for the core-to-reference-surface offset from above, and typical offset effects attributed to the connector alignment sleeves, (a mean of 0.5 and a standard deviation of 0.2 microns) the distribution of the expected offset for adjacent fiber cores in a connection can also be calculated. Using these distributions and a Monte Carlo simulation, the mean and standard deviation for the core-to-core offset was calculated to be 0.9 and 0.4 microns, respectively.

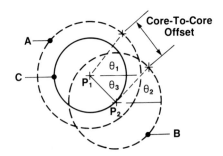

Radius A = Core/Taper Offset, Plug 1
Radius B = Core/Taper Offset, Plug 2
Radius C = Sleeve Offset

Fig. 7.14 Schematic representation of the combined offsets in a typical butt-jointed type connector assembly.

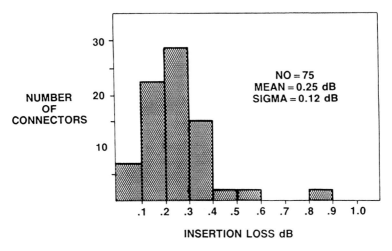

Fig. 7.15 Insertion loss data for low loss butt-jointed single-mode connectors.

Applying the general loss equation in Section 7.3.2, the distribution of the expected loss for these connectors, employing single-mode fibers operating at 1300 nanometers, was calculated to have a mean of 0.25 dB and a standard deviation of 0.1 dB. This calculation assumed an angular offset of 0.8 degrees and fiber end-face contact, which is typical for many connector designs. Experiments (Young, 1984) using fibers and connectors representing the assumptions used in this example have resulted in similar distributions for insertion loss measurements (see Fig. 7.15). It should also be mentioned that many effects have not been considered in this example such as fiber end-face quality, precise circularity and straightness of the reference surfaces (hole, fibers, periphery of the ferrules), centering effects of epoxies (if any are used), static friction, etc.

In the previous example, it was assumed that the fiber core could be located on the axis of the reference surface to less than one micron as required by single-mode fibers. When this is not possible or practical, a keying feature is sometimes used. In this technique, the axial alignment of the fiber core results in a distribution having a minimum and maximum offset, which is larger than that allowed for low loss connectors. In order to tolerate this larger offset, a key is positioned on the plug, which is referenced to the angular position of the core offset. The cross-hatched patterns in Fig. 7.16 show the locations of the fiber cores in these two types of plugs. This keyed-offset-type of plug usually requires a more complex assembly procedure involving special reference adjustment connectors and tools, as well as a keyed alignment sleeve housing. Connectors based on both of these techniques are presently being used successfully in many lightwave systems.

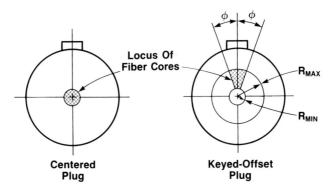

Fig. 7.16 Comparison of axially centered and keyed-offset plugs.

The above examples were used to show the complexity that one must consider when designing and characterizing a connector. The same philosophy applies whether the connector will be used with multimode or single-mode fibers; the only difference is that the clearances and accuracies can be traded-off for cost-effectiveness and ease of assembly.

7.4.2 Expanded-Beam Connectors

A typical expanded-beam connector (Nica and Tholen, 1981; Carroll, 1985), schematically shown in Fig. 7.17, consists of an optical element that collimates the beam radiating from the transmitting fiber, or focuses the expanded-beam onto the core of the receiving fiber. Two possible advantages over the butt-joint

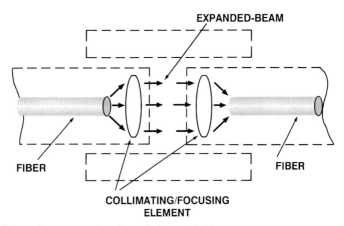

Fig. 7.17 Schematic representation of a typical expanded-beam type connector.

connector can be readily seen from this schematic. First, engagement and separation of the connector takes place within the expanded beam, meaning that the connector may be much less dependent on lateral alignments and may also be more tolerant to dirt or any other contamination on the end-faces, and therefore, exhibit more repeatable coupling efficiencies. A second advantage is that optical processing elements, to perform functions such as beam-splitting, switching, etc., can be inserted into the expanded beam.

A further analysis of the expanded-beam design reveals that the demanding submicron and micron-type alignments for single-mode and multimode fibers, must still be satisfied. Although the expanded beams do not require alignments to these accuracies, the fibers must be positioned with respect to the axes of the collimating and focusing element to these same demanding tolerances. Also, since angular misalignments of the expanded-beam cause the focusing point to be displaced laterally from the fiber core, the angular alignment of the expanded beam is critical. For these reasons, the major trade-off in butt-joint and expanded-beam connector designs is in the control of either lateral alignments or angular alignments. Furthermore, with these expanded-beam designs, even when perfect alignment is achieved, losses due to aberrations of optical elements may also be incurred.

Even with the advantages that expanded-beam connectors apparently have despite their increased complexity, the majority of connectors used today are of the butt-joint type. However, interest in expanded-beam connectors is increasing, partially due to the advancement in the molding of glass aspheric lenses, and partially due to the increased anticipation in the optical processing of the signal, such as wavelength division multiplexing and switching. The use of glass aspheric lenses instead of spherical lenses, can reduce the loss by as much as 0.4 dB (Carroll, 1985), while also allowing highly reliable anti-reflection coatings. Also, like butt-joint connectors, expanded-beam connectors must provide field-installability and cost effectiveness, even though they may contain more parts and may be more complex than butt-joint type connectors.

7.5 OPTICAL PROPERTIES AND MEASUREMENTS

The first goal of an optical fiber connector is to connect two fibers together in a manner that best approaches an optically transparent joint. Any inaccuracies in the joint may result in the loss of optical power (insertion loss), reflections propagating backward along the fiber to the optical source (return loss), and/or optical power transfer between modes (modal noise). Additionally, multi-fiber connectors may introduce crosstalk between the fibers in the joint. The level to which these properties are controlled is a primary factor in selecting a particular connector design philosophy as well as selecting a particular test schedule that is required to assure quality performance. This level is dependent on the optical system requirements and cost budget.

7.5.1 Insertion Losses

Unfortunately, there exists no unique insertion loss value that is valid for all multimode fiber connector applications. As mentioned in Section 7.3, the optical source, the number of joints and their location along the fiber, and the mode-mixing properties and differential mode attenuation of the particular fibers all play an important role in the performance, or more correctly, the apparent performance of the connector. Besides the previously mentioned effect that the actual incident mode distribution, at the connector, has on the insertion loss (due to extrinsic loss factors), for special cases additional effects can also be realized. For example, consider the case when a long length of fiber exists after a connector and there is also a parameter mismatch between the transmitting and receiving fibers. In the case when the parameter mismatches cause the transmitting fiber to underfill the numerical aperture of the receiving fiber (i.e., receiving fiber having a larger NA and/or core size), the actual loss of the receiving fiber may be much less than the steady-state loss of the fiber that is usually measured in fiber characterization. In fact, when the extrinsic loss of the connector is very small, the decrease in fiber loss may be greater than the connector loss and result in an apparent negative loss for the connector. Furthermore, this redistribution of modal power may cause a substantial change in the incident mode distribution at the next connector and result in a different connector loss than the expected steady-state loss. For these reasons and others, care must be taken when assigning a value of insertion loss for multimode fiber connectors; therefore, when one evaluates connectors, or plans a fiber transmission system, it is necessary to have a thorough understanding of the total system configuration.

Compared to multimode systems, it is straightforward to evaluate and predict the insertion loss of single-mode fiber connectors. Therefore, it is usually possible to include the measured value of insertion loss, of single-mode fiber connectors, directly to the system loss budget. Since only one mode is propagating, the source and length dependence associated with multimode fibers is no longer a concern. The only condition that must be satisfied is that only the fundamental mode propagates both before and after the connector. With single-mode fibers, however, there exists a wavelength at which the first higher-order mode is cutoff or sufficiently attenuated so that above this wavelength the fiber only guides the fundamental mode (Katsuyama et al., 1976), and below this wavelength more than one mode can propagate. Theoretically, this cutoff wavelength is defined as the wavelength at which the mode index equals the cladding index. Unfortunately, the theoretical cutoff wavelength is very difficult to measure since various loss mechanisms such as microbending, curvature, and stress that exist in

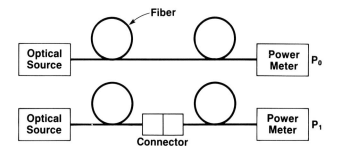

IL = –10 Log P1/P0 (dB)

Fig. 7.18 Ideal connector insertion loss measurement method.

deployed systems, strongly increase the loss of the first high-order mode. Consequently, the theoretical cutoff wavelength has little significance in practical fiber systems and a more useful concept of defining an effective cutoff wavelength has been adopted. The effective cutoff wavelength is defined as the wavelength above which the power in the first higher-order mode is below a specified level, when a specified length of fiber is deployed with a controlled radius of curvature (CCITT G.652 recommends a 2-meter length and radius of curvature equal to 14 centimeters). Therefore, it should be noted, that for single-mode fiber joints having sufficiently large offsets, it is possible to launch power into the first higher-order mode, and if the length of fiber between consecutive joints is not sufficiently long enough or curved enough to attenuate the higher-order mode, then degradation in system performance due to modal noise may result. Of course, for this to happen the second joint would also have to have a substantial offset as well. Usually, with well designed fibers (cutoff wavelength) and low-loss joints, this type of modal-noise problem is avoided.

Finally, the measurement of a connector's insertion loss can be accomplished in many different ways. Ideally, but usually not practical, the insertion loss of a connector is measured by cutting a length of fiber into two lengths and restoring continuity by installing a connector, as shown in Fig. 7.18. The insertion loss is then given by:

$$IL = -10 \operatorname{Log} P1/P0\, (dB).$$

Many other measurement methods can be found in documents provided by various national and international standards organizations such as the Electronic Industries Association (EIA), Washington DC, and the International Electronics Commission (IEC), Geneva, Switzerland.

7.5.2 Return Loss

Connectors in fiber spans can sometimes cause reflections that result in the return of optical power along the input fiber (return loss). In laser systems, this reflected power may cause system degradation (Cheung, 1984). The magnitude of this degradation is dependent on such factors as the location of the reflection with regards to the laser, the type of system (analog or digital), laser structure and of course, the amount of power that reaches the laser. As previously mentioned, most non-fusion splicing techniques employ some form of index-matching material that nearly eliminates reflections from the fiber joint. No gap exists between fused fibers and therefore there is no reflection. With connectors, however, reflections can become a concern particularly in the case of the first connector in the span, which is usually located close to the optical source. Various approaches exist to minimize the power penalty resulting from the reflections. First, an index-matching material can be used. In this case the return loss is usually greater than 30 dB. A second approach is to have fiber end-face contact without index-matching material (a typical design goal for some connectors). With this approach return losses of 25 to 40 dB have been measured. It should be noted that when small air gaps exist between fiber endfaces, multiple-beam interference can cause the return loss to be as low as 8.8 dB. Also in the case of index-matching, non-perfect matches can also result variations in return losses as great as 20 dB (for example 28 to 48 dB). Finally, if required, return losses of 45 dB or higher can be achieved by using end-faces that have a face angle of about 5 to 15 degrees. It should be noted that the first two approaches also reduce the insertion loss of the connector, while the last approach usually increases the insertion loss while also increasing the complexity of the connector.

7.5.3 Modal Noise

As previously mentioned, with regards to multimode fiber systems some amount of redistribution of power among the propagating modes usually exists at all discontinuities in a fiber span such as connectors. Interference effects between the propagating modes at these discontinuities can cause modal noise, and as a result, cause system degradation. When a significant amount of power is present in the first higher-order mode in a short section of fiber between two connectors, similar interference effects can also occur in single-mode fiber systems (Cheung, 1984). Consequently, modal noise in single-mode fiber systems may occur due to short connectorized patch-cords and laser pigtails. To avoid this problem, it is important to properly specify the cutoff wavelength of the fibers used in these applications, such that the operating wavelength is well above the effective cutoff wavelength for fibers deployed in these short lengths. In long lengths of fiber between connectors, any power launched into the first higher-order mode is usually sufficiently attenuated and modal noise is therefore not a problem.

7.6 MECHANICAL PROPERTIES

The applications of connectors in an optical system expose them and the attached fiber to handling and potential mechanical damage. The sheathed fiber cable is usually firmly attached to the connector, and between the connector and the cable there is a transition piece that limits the bending strain of the fiber. In use, the fiber cable may be pulled axially or at an angle to the connector and when uncoupled the end of the cable and the connector plug may fall to the floor or swing and strike a hard object. A connector must be sufficiently rugged to withstand these exposures. Further, the connectors are usually located in inhabited space, but may be located in remoted buildings, huts, or cabinets on the ground, on poles or underground in manholes. These exposures require the connector to be insensitive to climatic variations and vibrations. Therefore, connectors are usually subjected to a number of mechanical tests to assure that they will perform properly in an optical system application. The number and type of tests required will depend on the particular applications. Table 7.2 below lists typical tests.The tests listed in Table 7.2 have been standardized by the Electronic Industries Association (EIA), Washington D.C. under the category of Fiber Optic Test Procedures (FOTP). Briefly, these tests are:

Mechanical Endurance. This test measures the effects of uncoupling and recoupling the connector on insertion loss. The connectors are uncoupled and recoupled several hundred times and the insertion loss is measured after a given number of recouplings during the test and after completion of the test. A failure occurs when the allowable increase in loss is exceeded.

Cable Retention. In this test the connector is fixed and the cable is pulled until the fiber in the cable is broken. A failure occurs when the fiber breaks below the required tension limit.

Flexure. In this test the connector is tilted $90°$ in a plane, both directions, from vertical with a weight attached to the cable. The tilting is repeated a prescribed number of times. A failure occurs if the fiber or the cable sheath breaks, or when the allowable increase in insertion loss is exceeded.

TABLE 7.2

Mechanical Endurance
Cable Retention
Flexure
Impact
Vibration
Temperature Cycling
High Humidity

Impact. In this test the connector is dropped from a given height onto a hard surface a prescribed number of times. The insertion loss after the impacts is compared to the loss before the test. A failure occurs when the loss after the impact exceeds an allowable specified value.

Vibration. In this test the connector, with a fiber cable attached, is vibrated transverse and parallel to the axis of the connector. The frequency range, amplitude, and the number of cycles are determined by the connector application. A failure occurs when the connector's insertion loss exceeds the specified limit.

Temperature Cycling. In this test the connector with a fiber cable attached is exposed to temperature variations following a prescribed cycle. The temperatures are chosen to simulate the worst conditions of an application and to determine the effect of temperature cycling on the optical and mechanical characteristics of the connector. Indication of failure are broken parts, inability to uncouple or couple, and excessive increases in optical loss. Temperature cycling tests can be used to determine what loss factors should be included in system margins for environmental effects.

High Humidity. This test exposes the connector to high humidity in order to determine the effects of moisture and heat on the materials used in the connector.

In addition to the list, connectors may be tested to determine their resistance to other factors such as: dust, industrial environment, corrosion (salt spray), and flammability. Procedures for these have also been standardized by the EIA as well as other standards organizations.

7.7 FUTURE DIRECTIONS

Although much progress has been made in the past 10 years in optical fiber connector technology, new applications and requirements continue to appear. One area is in single-mode connectors where the required sub-micron precision usually results in a substantial increase in cost over multimode connectors. Various approaches are taken to reduce their cost, such as from reducing their insertion loss requirements to designing multi-fiber connectors, where the cost of the precision interface is shared among the fibers. Other requirements are also appearing that demand higher performance from the connector design such as high return loss (low reflection) and polarization alignment. Today, the lack of connector interface standards can sometimes make the use of connectors, particularly single-mode fiber connectors, a troublesome process. Various organizations throughout the world are very active in generating these required standards, and those required in the future, and the results of their work should soon reach the marketplace and user.

REFERENCES

Carlisle, A. W. (1985). Small-size high-performance lightguide connector for LANs, *Proc. of OFC '85*, San Diego, CA, February 1985.

Carroll, J. P., Messbauer, F. B., and Whitfield, C. H., *et al.* (1985). Design considerations of the expanded beam lamdek single-mode connector, *Proc. of FOC/LAN '85*, San Francisco, CA, September 1985.

Cheung, N. K. (1984). Reflection and modal noise associated with connectors in single-mode fibers, *SPIE Proc.* Vol. **479**, 56–59.

Chu, T. C., and McCormick, G. R. (1978). Measurement of loss due to offset, end separation and angular misalignment in graded-index fibers, *Bell System Tech. J.* **57**, No. 3, 595–602.

Kaiser, P., Young, W. C., Cheung, N. K., and Curtis, L. (1980). Loss characterization of biconic single-fiber connectors, Tech. Digest-Symposium on Optical Fiber Measurements, 1980, *NBS Pub.* **597**, 73–76.

Katsuyama, Y., Tokuda, M., Uchida, N., and Nakahara, M., *et al.* (1976). New method for method for measuring V-value of a single-mode optical fiber, *Electron. Lett.* **12**, 669.

Kummer, R. B., and Fleming, S. R. (1982). Monomode optical fiber splice loss, *Proc. Optical Fiber Communication* 44.

Marcuse, D. (1977). Loss analysis of single-mode fiber splices, *Bell System Tech. J.* **56**, 703–17.

Miller, C. M., and Mettler, S. C. (1978). A loss model for parabolic profile fiber splices, *Bell System Tech. J.* **57**, No. 9, 3167–80.

Nemota, S., and Makimoto, T. Analysis of splice loss in single-mode fibers using a Gaussian field approximation (1979). *Optical Quantum Electronics* **11**, 447.

Nicia, A., and Tholen, A. (1981). High-efficient ball-lens connector and related functional devices for single-mode fibers, *Proc. 7th European Conference on Optical Communications*.

Shah, V., Young, W. C., and Curtis, L. (1987). Large fluctuations in transmitted power at fiber joints with polished endfaces, *Proc. OFC/IOOC'87*, Reno, Nevada.

Wagner, R. E., and Sandahl, C. R. (1982). Interference effects in optical fiber connections, *J. of Applied Optics* Vol. **21**, No. 8, 1381–85.

Young, W. C. (1984). Single-mode fiber connectors—design aspects, *Proc. 17th Annual Conn. and Interconn. Tech. Symposium*, Electronic Connector Study Group 295–301.

Young, W. C., and Curtis, L. (1983). Low-loss field-installable biconic connectors for single-mode fibers, *Proc. OFC'83*, New Orleans, Louisiana.

Chapter 8

Optical Fiber Transmission Evaluation

DAN L. PHILEN

AT&T Bell Laboratories, Norcross, Georgia

WILLIAM T. ANDERSON

Bell Communications Research, Morristown, New Jersey

8.1 INTRODUCTION

Optical fiber measurements owe their great importance to the various areas where the measurement data is used. They offer valuable feedback to the fiber designer in improving and testing new fiber designs. They provide data that the systems designer requires in calculating the parameters for a practical transmission system. Finally, they are used as checks during fiber manufacture for insuring the uniformity and quality of product.

Historically, optical fiber measurements have been a story of innovation and intense development. This has largely been pushed ahead by the need for more accurate measurements as fiber technology has developed. Several years ago attenuation measurement accuracies of 0.1 dB were questionable and some questioned the need for such accuracy. Today, glass fiber purity is so high that if the ocean were as pure as optical fiber glass, then the bottom could be seen at its deepest point. Not surprisingly, attenuation accuracies of a few hundredths of a dB are commonplace today. Likewise, measurements that were the accepted technique of a few years ago are no longer used. For example, the first refractive index profiles were measured by grinding and polishing wafers cut from preforms. Today single-mode fiber profiles are routinely measured in fiber form from a short sample of fiber, using the refracted-near-field technique.

Looking at the section on fiber characterization in the previous edition of this book shows techniques that, while classic measurement techniques, are simply not applicable today. Also, measurements that were in their infancy then (for example, optical time domain reflectometry), have developed into routine measurement procedures. OTDR is a commercial field portable measurement technique widely used for fiber installations today, and dispersion measurement instruments are now being commercially produced and are beginning to be used in the field.

Fiber measurements can be grouped into two broad classes: geometric measurements that refer to the physical properties of the fiber such as cladding diameter, core diameter, and core concentricity error; and transmission measurements that refer to the optical properties of the fiber, such as attenuation and bandwidth. An exhaustive discussion of every fiber measurement technique is not the purpose of this chapter. Since several are well established and have been exhaustively described elsewhere, we will mention them here for completeness, but concentrate on those measurements that are especially suited to single-mode fibers, and those that are common to both single- and multimode fibers.

8.2 GEOMETRIC MEASUREMENTS

Determination of the physical properties such as cladding diameter, core diameter, core concentricity error, and refractive index and numerical aperture, are necessary to fiber measurements to assure splicing compatibility and uniformity between fibers. These are generally referred to as geometrical parameters; and to get an idea of the accuracy involved, consider that typical copy paper is about 100 μm thick. Typical fiber diameters are 125 μm, or about the same thickness as a piece of copy paper. The core diameter of typical multimode fibers is 50 μm, or half the paper thickness, and single-mode fibers have core diameters of about 10 μm, or one tenth the paper thickness. The accuracy required for measurements is about one tenth of a μm; or one thousandth of the paper thickness.

8.2.1 Refractive Index Profiling

8.2.1.1 Refracted Near-Field Method. A method that has gained widespread use, is the refracted near-field method. This method, originally proposed by Stewart (1977), uses light that is refracted out of the fiber core rather than transmitted through it. The profile shape is measured by moving a small spot (relative to the core size) across the fiber core and detecting the refracted light that escapes from the core-cladding boundary. When light is focused onto the fiber core with a cone of light that is larger than the acceptance angle of the fiber, some of the light remains in the fiber core, some is lost to radiation modes, and some is lost to leaky modes. To avoid having to correct for leaky modes, an

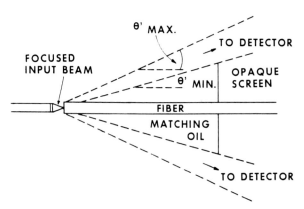

Fig. 8.1 Optical arrangement for refractive index profiling by the Refracted Near Field method.

opaque mask is used to prevent the leaky modes, which escape at angles below a certain minimum (θ_{min}), from reaching the detector. The minimum escaping angle, θ_{min}, corresponds to a certain input angle, θ'_{min}, and all light of $\theta' > \theta'_{min}$ must reach the detector. Input apertures are used to limit the numerical aperture of the input beam to obtain the proper $\theta' = \theta'_{max}$, as shown in Fig. 8.1.

In practice, the fiber is immersed in index matching oil to prevent reflection at the outer cladding boundary, and measurements of the detected light power as a function of input beam position, $P(r)$, are made. We know θ_{min} by the size of the mask in front of the detector, and we know θ'_{max} from the incident light numerical aperture. The refractive index of the core is obtained from the following formula:

$$n(r) = n_2 + n_2\cos\theta_{min}(\cos\theta_{min} - \cos\theta'_{max})\left[\frac{P(a) - P(r)}{P(a)}\right],$$

where $P(r)$ is the detected light power, $P(a)$ is determined from the $P(r)$ curve as the input beam is focused into the cladding, and n_2 is the refractive index of the cladding. The above equation can be rewritten as:

$$n(r) = k_1 - k_2 P(r),$$

where k_1 and k_2 are constants that can be determined by calibration, since $n(r)$ is directly proportional to $P(r)$ (White, 1979). The refracted near field method has gained widespread acceptance for measuring index profiles because of its high resolution (about 0.3 μ), and the ease of sample preparation (Saunders, 1981).

8.2.1.2 Transverse Interferometry. A method that has been used because of its ease of sample preparation is transverse interferometry. This method is similar to the slab interferometry mentioned extensively in the previous edition of this book, and is no longer used because of its difficult sample preparation. The reader is referred to that work for a detailed explanation of the technique. Basically, the interferogram is obtained by measuring the refractive index profile transverse to the fiber rather than end on, otherwise the technique is the same. The great advantage is that no grinding or polishing of the sample is required, and core diameter, core concentricity error, core noncircularity, and the refractive index profile and delta (Δ) are easily obtained with one measurement technique (Boggs, 1979).

8.2.2 O.D. Measurement and Control

The outside diameter of the fiber is usually monitored during fiber drawing to maintain uniform control on the diameter, and several novel techniques have been developed for this measurement. Probably the most reliable and widely used is the forward-scattering technique developed by Watkins *et al.* (1974). This method uses interference from a laser beam striking the fiber at a right angle to the fiber axis. Interference is observed from a ray striking the cladding-air boundary and a ray passing through the fiber. The forward scattered light has an angular distribution between 6° and 68° and is an interference pattern with rapid fluctuations that contains the diameter information to be processed by a microcomputer. Resolution near 0.25 μ has been shown using this technique, and locking the fringe pattern to a feedback loop has made this method particularly suited to on-line diameter control of fiber drawing.

8.2.3 Optical Microscopy

Other measurements may be used to verify the diameter control after fiber drawing, and the most common method is to use a microscope to determine the outside diameter in several orientations. This is time consuming, however, and involves some operator interpretation on the placement of a cursor. The procedure can be automated to make it faster and less prone to operator error (Marcuse, *et al.*, 1979).

A recent development in microscopy is a "shearing microscope," which shears (displaces) the image of the fiber end face for more accurate measurements. The technique provides a means, in a single system, for inspecting the end faces of multimode and single-mode fibers and accurately measuring dimensional parameters. These parameters include cladding diameter, cladding non-circularity, core diameter, and core concentricity error. Two identical images of the fiber to be measured are sheared relative to one another until the edges of the images are exactly touching. The degree of shear is measured by a very accurate strain gauge system and translated into dimensional units that are displayed on a digital indicator. By rotating sheared images, the varying degrees of concentricity, or variations in concentricity are highlighted and can be accurately and precisely measured.

In measuring core diameter, the precise measurement required is achieved by varying the intensity of the core image as it is displayed on a video monitor. The shear measurements are taken between pre-determined intensity percentage values and in this way, those fibers that show no clearly defined boundary between the core and cladding can be measured with great accuracy and consistency. Accuracy of 0.1 μm has been routinely demonstrated using this technique.

8.2.4 Numerical Aperture

There are three parameters that can be used to describe the difference in refractive index between the core and cladding in multimode fibers. They are: (1) the normalized refactive index difference, delta (Δ); (2) the maximum theoretical numerical aperture calculated from Δ; (3) and the measured numerical aperture from the far-field measurement. Any multimode fiber could be described by any one of these three parameters and ideally, the other two would be known. In practice, there is some uncertainty in making the transition from Δ to theoretical NA to measured NA. In addition, the term "numerical aperture" is often used imprecisely to describe an optical waveguide. The precise terms of "acceptance angle" and "radiation angle" would be preferred, but the term numerical aperture has gained widespread usage.

The theoretical NA is calculated from Δ by the following relation:

$$\mathrm{NA}_{th} = \sqrt{n_1^2 - n_2^2},$$

where n_1 = core refractive index, and n_2 = cladding refractive index.

The normalized refractive index difference, Δ, is defined as

$$\Delta = \frac{n_1 - n_2}{n_2}.$$

Rearranging and substituting, we get

$$\mathrm{NA}_{th} = n_2\sqrt{(1 + \Delta)^2 - 1}$$

$$\approx 1.45\sqrt{(1 + \Delta)^2 - 1},$$

where Δ is measured at 0.63 μm with the Refracted-Near-Field technique, or at 0.54 μm by transverse interferometry.

The measured NA is determined at 0.85 μm according to EIA test method EIA-455-47 by measuring the far-field radiation pattern of the fiber, and is defined by the following relation:

$$\mathrm{NA}_{meas} = \sin \theta_5,$$

where θ_5 is the half angle of the radiation pattern scanned in the far-field, and taken at the 5% intensity point.

The value of NA-measured will be less than NA-theoretical because of the change in wavelength (n varies with λ), and because of the way the values are defined. The theoretical value is defined at the baseline of the refractive index profile; and there may be some judgement involved in determining the peak value, while the measured value is arbitrarily defined at the 5% point of the radiation pattern above the baseline.

At the current state of NA measurements, there is about a 5% difference in NA calculated from Δ as measured by refracted near field, and the lower measured NA value from the far-field data. There is a small correction to the wavelength from 0.63 μm to 0.85 μm, but the largest factor in making the two measurements agree is in the difference in the definitions. Extensive work by EIA standards groups has resolved this issue and a mapping function has been formulated to correlate the two methods, EIA-455-177.

8.3 MODE FIELD DIAMETER—SINGLE-MODE FIBERS

The difference between the mode field diameters (MFDs) of two single-mode fibers is a measure of the degree of optical compatibility of the two fibers in regard to splice loss. The splice loss is a slowly varying function of MFD with the fibers in common use today, and fibers with nearly equal MFDs can be spliced with low losses, while fibers whose MFDs differ cannot be spliced without some added loss. The lowest achievable splice loss between two fibers with MFDs $2w_1$ and $2w_2$ can be predicted as

$$\alpha_{\text{splice}} = 20 \log \left[\frac{w_1^2 + w_2^2}{2w_1 w_2} \right].$$

This loss model was first proposed for step-index fibers whose fields have a radial dependence that is nearly Gaussian (Marcuse, 1977). For step-index fibers, the Gaussian approximation leads to a definition of the MFD that maximizes the efficiency of launching a Gaussian beam onto the fiber (Marcuse, 1978). That is, w is chosen to maximize

$$I = \frac{\left[\int_0^\infty E_m(r) E_g(r) r \, dr \right]^2}{\int_0^\infty E_m^2(r) r \, dr \int_0^\infty E_g^2(r) r \, dr},$$

where $E_m(r)$ are the fields in the fiber, and $E_g(r)$ is the Gaussian function, $\exp(-r^2/w^2)$.

8.3.1 Measurement Methods

The measurement of the MFD can be made several different ways. Five techniques are currently in widespread use:

1. Scanning the fiber near fields by focusing a greatly magnified image of the fiber core either onto an infrared vidicon or onto the plane of a scanning small-area detector (Murakami et al., 1979).

2. Measuring the transmission through a splice as a function of offset, from which the two dimensional autocorrelation of the near fields can be obtained (Streckert, 1980).

3. Measuring the fiber far fields any of three ways:
   a. Scanning a small area detector in the far-field (Hotate et al., 1979);
   b. Measuring transmission through apertures in the far-field as a function of aperture radius (Nicolaisen et al., 1983);
   c. Measuring transmission past a knife edge in the far-field as a function of lateral position (Otten et al., 1986).

All these techniques have certain advantages, and there appears to be little chance that any one of them will be widely accepted as a standard, or reference technique, by which the others could be judged. Therefore, consistency among the techniques is essential if the mode field diameter is to be used as a measure of the splice compatibility of two fibers.

The degree of consistency achievable with these five techniques is a direct result of the choice of a definition for the mode field diameter and depends on the type of fiber measured. The Gaussian definition leads to small but systematic differences among the measurement techniques for quasi-step index fibers, especially at longer wavelengths (Anderson, 1984; Franzen, 1985). Differences of greater than 10% have been observed for non-step-index dispersion-shifted fibers (Auge et al., 1985), whose fields are substantially non-Gaussian. Since these inconsistencies are attributable to deviations from perfect Gaussian fields, the computational method used to fit the measured data to a Gaussian function is also important. For example, the near-field and far-field scan measurements are consistent if the overlap integral is minimized, but any other method of fitting the data, such as interpolating to find the $1/e$ point, eliminates this consistency. Therefore, when using the Gaussian definition, care must be taken to specify completely the method of fitting used, and the range of data that is used in the fit. While the Gaussian definition leads to adequate control of splice loss for quasi-step-index fibers, it is not always applicable.

An alternative definition based on the second moment of the far fields, frequently called the "Petermann 2" definition (Petermann, 1983; Pask, 1984),

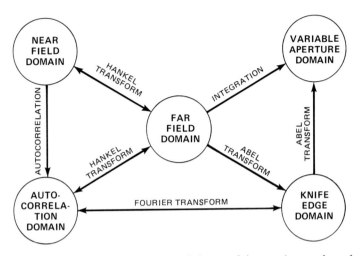

Fig. 8.2 Five-mode field diameter measurement techniques and the mapping transforms from one method to another.

can, in principle, solve this problem. The MFD is defined as

$$2w = \frac{2}{\pi W},$$

where

$$W^2 = 2\frac{\int_0^\infty q^3 F^2(q)\,dq}{\int_0^\infty q F^2(q)\,dq},$$

where $q = 1/\lambda \sin \theta$, F^2 is the far-field power, and the integration is assumed to be truncated at some large enough angle to satisfy the above integral. The five measurements are related by several transforms to a good approximation, and are shown in Fig. 8.2. Unlike the Gaussian definition, the second moment definition can be mapped from one measurement domain to another using the transform relations shown in Fig. 8.2. In this way, consistency between the measurement methods can be maintained. Initial measurement comparisons have shown that consistency is achieved (Anderson *et al.*, 1987).

8.4 ATTENUATION MEASUREMENTS

8.4.1 The Cutback Method

The measurement of spectral attenuation by the cutback technique is the oldest and conceptually, the simplest fiber measurement. The measurement technique as applied to multimode fibers is covered in some detail elsewhere (Miller *et al.*,

1979), but it has more recently been applied to single-mode fiber characterization. A typical measurement system is shown in Fig. 8.3. A tungsten-halogen lamp and monochromator provide a low-power but wavelength-tunable source, and the output beam is chopped so that synchronous detection may be used. A lock-in amplifier and a current-sensitive preamplifier detect the output of an unbiased PIN photodiode. The photodiode converts optical power to electrical current, which is transformed to voltage by the preamplifier, making the output of the lock-in amplifier directly proportional to the received optical power.

For single-mode fibers, a 2 to 4 cm loop is usually inserted near the launch end to suppress the higher-order modes, so that the measurement will be accurate in the upper portion of the multimode regime. For multimode fibers, the launch conditions must be controlled so that an equilibrium mode distribution is simulated, either by controlling the launch spot diameter and numerical aperture to 70% of the fiber core diameter and numerical aperture, or by overfilling the fiber core and numerical aperture with a mode scrambler and a mandrel wrap mode filter to strip out the higher-order modes. Additionally, a cladding mode stripper may be required if the fiber coating has a refractive index lower than that of the silica cladding. The typical epoxy acrylate coatings most commonly used, however, do not require additional cladding mode stripping.

In operation, one first scans the desired wavelengths and measures the transmitted power, $P_1(\lambda)$. The fiber is then broken, typically one to two meters from the launch, and the desired wavelengths are again scanned and the power $P_2(\lambda)$ measured. The attenuation is found as the ratio of these two powers,

$$\alpha = 10 \log \frac{P_1(\lambda)}{P_2(\lambda)}.$$

The repeatability of the cutback technique is dependent primarily on the ability to couple the fiber output to the detector repeatably. Three different coupling techniques are in use: splicing to a large core multimode fiber pigtail, imaging the fiber end onto a small area detector, or coupling directly to a large area detector. Since the dominant source of noise is detector dark current, using a large area detector may significantly reduce the dynamic range of the system. Standard deviation of .01 to .04 dB and a dynamic range of 20 to 35 dB are typical for this technique when applied to single-mode fibers.

8.4.2 Substitution Method

For installed spans or connectorized cables, a destructive technique such as the cutback method is inappropriate. In these cases, the attenuation by substitution or insertion loss technique is commonly employed. The launch and detector couplings are made through connectors, and unlike the cutback technique, both

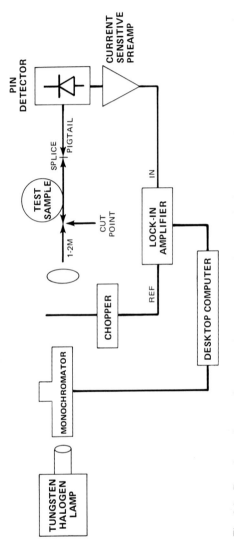

Fig. 8.3 Experimental set-up for attenuation measurements by the cutback technique. This same apparatus can be used to measure cut-off wavelength.

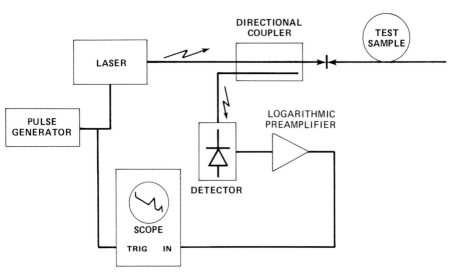

Fig. 8.4   Block diagram of an optical time domain reflectometer to measure fiber attenuation, splice loss, and splice location.

launch and detector couplings must be reproducible. Also, for an installed span, the source and detector are separated so that either two detectors or a source with excellent long-term stability is required. A systematic bias can be introduced by using two detectors that are not accurately cross-calibrated or by using fiber jumpers that differ optically from the fibers in the installed span. The former factor can be eliminated by using a second source with the second detector so that bidirectional measurements can be made and averaged, but this increases the cost and difficulty of the measurement. The second factor, however, may complicate field measurements since the type of fiber used for test set pigtails may not be known by the user. These factors clearly make the characterization of installed spans using the attenuation by substitution technique less accurate than the characterization of fiber and cable using the cutback technique.

8.4.3 Backscatter Methods

A third commonly used attenuation measurement technique is backscatter, which employs an optical time domain reflectometer (OTDR) (Miller *et al.*, 1979). A block diagram of a typical OTDR is shown in Fig. 8.4. A narrow (typically, 10 to 100 ns) pulse is launched onto the fiber. Scattering along the fiber causes a small fraction of the power to be lost, and a small fraction of the total scattered power is captured by the fiber and propagates in the reverse direction. This power is coupled onto the detector, and the received signal is displayed on an oscilloscope. Since the propagation velocity of the pulse is known *a priori* with considerable accuracy, the oscilloscope time-base can be accurately recalibrated into distance along the fiber by

$$z = 2v_g t = \frac{2ct}{n_g},$$

where n_g is the group index. This backscatter signal $P_B(t)$ is related to the power propagating in the fiber $P(z)$ as

$$P_B(t) = P(z)S_c R,$$

where S_c is the scattering coefficient and R is the fraction of the total scattered light that is captured by the fiber and returned to the detector. If the fiber is uniform along its length, then S_c and R are constant, and the attenuation can be found from the slope of the displayed trace on the oscilloscope (because of the logarithmic amplification). At discontinuities, such as splices, the loss at the discontinuity can be found from the change in vertical level of the trace, but this procedure is only accurate if the fibers before and after the splice are identical.

Unlike other measurement techniques, OTDRs can locate faults, measure splice losses, and resolve the attenuation over different parts of an installed span. OTDRs are also single-ended, so the measurement of installed spans requires only one craftsman rather than the two required by the substitution method. OTDRs, however, have three significant disadvantages. First, the resolution of 1300 nm single-mode OTDRs is currently about 100 m, making it impossible to resolve the first one or two splices in an installed span. Better lasers and detectors should improve the resolution as the technology continues to advance. Second, when measuring installed spans, the scattered and capture coefficients vary somewhat from fiber to fiber, so that the losses measured using this technique are not always accurate. In principle, averaging two bidirectional measurements can eliminate this effect, but only at the cost of doubling the time and effort required to make the measurement. Finally, when measuring spans that contain highly reflective splices or connectors, the large reflected pulse may be re-reflected at the OTDR. If there is significant power in the reflected pulse, "ghosts" may be observed because multiple pulses are propagating simultaneously. This phenomenon requires some judgment and experience by the operator to determine those parts of the trace that are real and those that are spurious.

8.5 MULTIMODE BANDWIDTH

The measurement of bandwidth in multimode fibers is a measure of the impulse response of the fiber. It is defined as the baseband modulation frequency that gives a 3 dB reduction in the receiver's baseband output, as compared to the zero frequency (DC) response. A pulse traveling along a fiber is broadened by intermodal, material, and waveguide dispersion, making bandwidth measurements on multimode fibers complicated by the same problems as attenuation measurements. The intermodal dispersion depends on the power distribution among the various modes, the distribution of power that is launched into the modes, the differential mode attenuation, and the mode coupling. Many attempts have been made to extract a length scaling factor that will give an appropriate bandwidth vs. length scaling factor; however, no universal constant has been forthcoming because of these problems. Therefore, tight control of the launching conditions is necessary for good accuracy and precision on multimode bandwidth to give an overfilled launch condition so that all the modes are excited.

8.5.1 Time Domain Measurements

The first widely used fiber bandwidth measurement is referred to as a "time domain" measurement because it uses pulsed lasers for the measurement. While this is a time domain measurement, Fourier transform theory is employed to convert the time (impulse) information to frequency (transfer) response information. The transfer function is obtained by the ratio of the Fourier transforms, and by its inverse, the impulse response back in the time domain. The relationship between the impulse response and the transfer function is shown in Fig. 8.5.

A short pulse, assumed to be Gaussian in shape, is launched into a fiber from a pulsed laser source such as a diode laser. The output pulse is not a pure Gaussian so the following relationships are approximate for a real fiber since they assume Gaussian pulse shapes.

To obtain the impulse response, the input pulse must be deconvolved from the output pulse. A short strap is used to obtain the input pulse and the output pulse is measured through the long length of fiber. In practice, however, the long length is measured first, and then the fiber is broken for the short strap measurement to avoid disturbing the launching conditions. The detector and other electronics response are canceled out since these effects are superimposed on both the input and output pulses; and if the input pulse is attenuated so that it is about the same amplitude as the output pulse, then any non-linear effects in the detector are minimized. An experimental apparatus is shown in Fig. 8.6.

The input and output pulse Fourier transforms are related by the following equation:

$$P_2(f) = H(f)P_1(f),$$

where $P_1(f)$ and $P_2(f)$ are the Fourier transforms of the input $P_1(t)$ and output $P_2(t)$ in the time domain. $H(f)$ is the transfer function, also known as the baseband frequency response. The transfer function can be written in polar

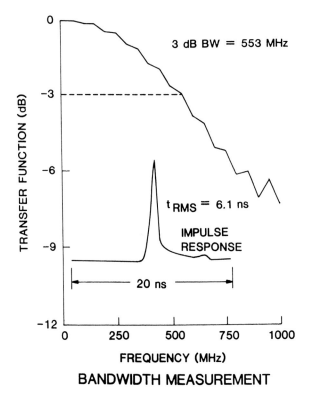

Fig. 8.5 Relationship between impulse response (time) and transfer function (frequency) to measure the 3 dB bandwidth.

coordinates as (Franzen *et al.*, 1982)

$$H(f) = M(f)e^{i\theta(f)},$$

where $M(f)$ and $\theta(f)$ are functions of (f).

The bandwidth is determined from the ratios of the fast Fourier transforms of the input and output waveforms by measuring the magnitude of the transfer function $M(f)$, and the phase response $\theta(f)$. The impulse response is just the inverse of the transfer function.

Bandwidth or frequency response, and rms pulse broadening can now be related in simple terms. If the input pulse is represented by

$$P_1(t) = e^{-4\ln w(t/\tau_1)^2}$$

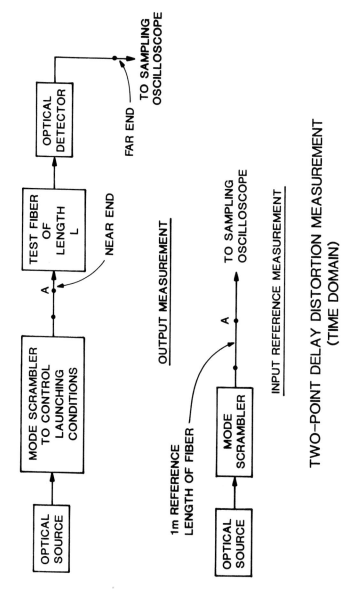

Fig. 8.6   Schematic arrangement of a bandwidth measurement in the time domain.

and the output pulse is

$$P_2(t) = e^{-4 \ln 2(t/\tau_2)^2},$$

then these are Gaussians having full width at half maximum (FWHM) of τ_1 and τ_2. The 3 dB fall-off point is

$$f_{3\,\text{dB}} = \frac{(0.44)}{\sqrt{(\tau_2^2 - \tau_1^2)}}$$

$$= \frac{0.44}{\tau_T},$$

where τ_T is the half width broadening of the fiber. This is the factor we would have determined in the time domain had we deconvolved the input pulse from the output pulse. The rms pulse duration for a Gaussian is related to the FWHM by

$$\sigma = 0.43\tau.$$

Using these definitions along with the definition for the 3 dB frequency, the 3 dB fall-off in frequency is related to the rms pulse broadening, σ_T, by

$$f_{3\,\text{dB}} = \frac{0.19}{\sigma_T}.$$

8.5.2 Frequency Domain Measurements

Time domain measurements are generally used in the laboratory or factory with relatively short fiber lengths where the total bandwidth may be in the GHz range. Frequency domain measurements are generally used in the field where cabled spans may be long and the total bandwidth may be a few hundred MHz. Also, the simpler more robust equipment used for frequency domain measurements makes this a widely used field measurement technique. Instruments that measure high frequencies directly in the frequency domain are unfortunately not commonplace in the optical laboratory. There are, however, instruments designed for working with radio frequency and microwave electronics that can be adapted to optical fiber measurements.

The simplest approach to measuring the bandwidth in the frequency domain is to modulate a laser with a rf sweep generator and detect the output with a wideband, linear, rf detector. The transfer function is obtained by dividing the output frequency response with the fiber in place, by the input response with the fiber removed. This method is sensitive to the signal-to-noise ratio and harmonic distortion introduced by the optical source.

Spectrum analyzers used in radio frequency analysis have choices of gain, averaging, sweep rate, and detection bandwidth. These narrow band devices thus make good detectors for frequency domain measurements. Maximum use of a spectrum analyzer, however, is made by locking the laser transmitter source to the local oscillator of the spectrum analyzer. The output frequency then varies as the sweep of the spectrum analyzer. Such devices are known as tracking generators because the output tracks the input sweep. These are probably the most widely used instruments for frequency domain measurements because of their simplicity of construction and their commercial availability; and they are limited only by how accurately the lock between the tracking generator and spectrum analyzer can be maintained. As long as we are interested only in the magnitude of the transfer function these methods are adequate. Other more sophisticated methods have been used to extract phase information (Marcuse, 1981; Cancellieri et al., 1984; and Day, 1983).

8.6 SINGLE-MODE DISPERSION

In single-mode fibers pulse broadening is caused by waveguide dispersion and material dispersion. Material dispersion is primarily caused by the dependence of refractive index with wavelength of the fiber core, and depends on the core dopants used, and is generally "fixed" by other considerations. Waveguide dispersion depends on the shape of the core and cladding profile and can be modified to give the desired total dispersion vs wavelength curve. The total dispersion is the parameter of interest, and is sometimes referred to as intramodal dispersion, since there is only one mode, or "chromatic dispersion." For the purposes of this discussion we will use the simple term "dispersion" to mean the total dispersion, since that is what is actually measured.

The dispersion is given by

$$D(\lambda) = \left(\frac{1}{L} \right) \left(\frac{\partial \tau}{\partial \lambda} \right)$$

and is usually expressed in picoseconds per nanometer-kilometer. The total pulse broadening over a length of fiber is given by

$$\Delta\tau = D(\lambda)L\,\Delta\lambda,$$

where D is the dispersion, L is the fiber length, and $\Delta\lambda$ is the wavelength spread of the source.

The pulse broadening in single-mode fibers is usually too small to measure directly. However, the variation in pulse delay over a wide wavelength range is easier to determine. At a point where the waveguide and material dispersion cancel each other, the pulse delay curve will go through a minimum. When a curve of the Sellmeier type is fitted to the pulse delay data, the derivative of the calculated curve is the dispersion, and the point on the curve where the dispersion changes sign (goes through zero) is called the "zero dispersion wavelength."

The most general equation for fitting to the delay data is a five term Sellmeier equation of the type

$$\tau = A + B\lambda^4 + C\lambda^2 + D\lambda^{-2} + E\lambda^{-4}.$$

For most applications, however, a three term Sellmeier equation of the form

$$\tau = A + B\lambda^2 + C\lambda^{-2}$$

works just as well, and has been recommended by the EIA for fitting dispersion data of fibers that have zero dispersion wavelengths near 1.31 μm. For fibers having their zero dispersion wavelength near 1.55 μm, the five term equation should be used for the best fit to the data, and while the following derivation is for the three term equation, the procedure is the same in solving the five term one.

The dispersion at any wavelength is the derivative of the fitted curve

$$\frac{\partial \tau}{\partial \lambda} = 2B\lambda - 2C\lambda^{-3}.$$

Solving for the wavelength, where the dispersion is zero, we get

$$\lambda_0 = \left(\frac{C}{B} \right)^{1/4}.$$

For some systems calculations, it is desirable to know the dispersion slope (the derivative of the dispersion curve), and inserting the value of λ_0 into the derivative of the dispersion equation gives the value of the slope at the zero dispersion wavelength; S_0.

$$S_0 = 8B$$

S_0 is expressed in $ps/(nm^2 - km)$.

8.6.1 Dispersion vs Bandwidth

Unlike multimode fiber bandwidth, the bandwidth of a single-mode fiber depends on the input source spectral width in addition to the fiber dispersion. The calculated bandwidth of a single-mode fiber would seem to go to infinity at the zero dispersion wavelength, except that higher order terms, the dispersion slope, limit the bandwidth to finite values. The bandwidth can be given by

$$BW \leq \frac{1}{4} \left[2(DL\,\Delta\lambda)^2 + \left(SL(\Delta\lambda)^2 \right)^2 \right]^{-1/2},$$

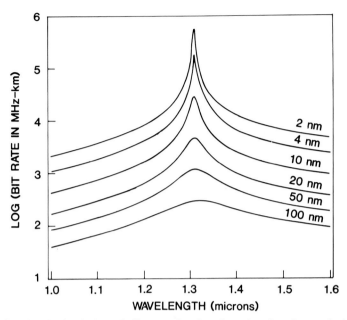

Fig. 8.7   Bandwidth of a single-mode fiber as a function of wavelength and spectral width of the source.

where $\Delta\lambda$ is the spectral width of the source, and S is the dispersion slope at the wavelength of interest.

$$S = 2B - 6C\lambda^{-4}$$

From the bandwidth equation, it is seen that near the zero dispersion wavelength the bandwidth is determined by the dispersion slope ($D \approx 0$), and at wavelengths away from λ_0 the bandwidth is determined by the dispersion ($D \gg S$). This function along with its dependence on the source width is plotted in Fig. 8.7 for a fiber having λ_0 near 1.31 μm.

8.6.2 Experimental Methods

8.6.2.1 The Fiber-Raman Laser.   The fiber-Raman laser was the first technique to give accurate dispersion data, and it is the method by which all the other techniques have been compared. This method uses narrow pulses of different wavelengths to measure the transit time through a fiber (the pulse delay curve). The technological difficulty, however, is in generating the narrow pulses over a wide spectral range in the 1.3 to 1.7 μm region (Lin *et al.*, 1978; Cohen *et al.*, 1977).

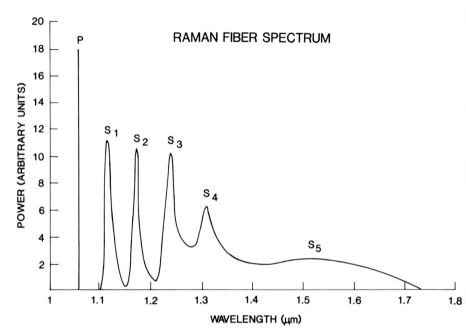

Fig. 8.8 Spectral output of a Raman fiber. P is the pump laser wavelength at 1.064 μm, and S1, S2, etc. are the Raman shifted components at longer wavelengths.

When a laser is focused into a single-mode fiber, light is scattered by the Raman effect into a weak component at slightly longer wavelengths than the "pump" laser. The scattered light and the incident light travel together down the fiber, and the scattered light component increases in intensity because it is being pumped by the incident light. The Raman scattered light eventually increases in intensity so that another Raman scattered component grows, and this process may be repeated several times to give five or six Raman components. Because the Raman shifted light is broader than the initial pump light, each successive shifted component is broader than the one before, and eventually they overlap to form a continuum. The Raman fiber acts like a one-pass amplifier sometimes found in high power lasers; and therefore, it is also called a Raman-fiber laser. Figure 8.8 shows the spectrum from a fiber with Raman shifted output.

Lasers of the Nd:YAG type are well suited as pump lasers for this application since they provide high power Q-switched pulses with a reasonable repetition rate. The Q-switched pulses are too long for accurate delay measurements, and modelocked pulses are used because they are only a few hundred picoseconds in duration. The modelocked pulses, however, are too low in power for adequate Raman gain, and they are modulated at about a hundred megahertz. The solution is to run the Q-switch from the modelocker timer (Lin *et al.*, 1978) by having the modelocked pulse start a count-down circuit that divides the modelock frequency by a preset factor to run the Q-switch. The modelocker and Q-switch are now linked together through the timing circuit, resulting in the laser operating like a simple Q-switched laser, but with several modelocked pulses in each Q-switched pulse. All of the modelocked pulses have the same time relationship for all the Q-switched pulses, and normal sampling electronics can select a modelocked pulse from the pulse train.

The Raman fiber's output vs wavelength is shown in Fig. 8.8, but the output vs time is a series of modelocked pulses modulated at the Q-switched rate. This sounds more complicated than it is, but what results is every laser designer's dream—a narrow pulse, multiple wavelength (tunable), high intensity source. Wavelengths are selected using a monochromator, and the pulses are separated using sampling electronics, as shown in the experimental arrangement for such an apparatus in Fig. 8.9.

The delay curve is obtained by scanning the monochromator and measuring the pulse transit time (Fig. 8.10). A curve fit to the measured delay is calculated using a least squared fitting technique, and the coefficients determined. The zero dispersion wavelength, dispersion, and dispersion slope are easily calculated as previously discussed.

8.6.2.2 Other Transit Time Techniques. The fiber-Raman technique is the most versatile dispersion measurement technique, but it is also complex and uses high power lasers. Other techniques have been developed using diode lasers and have gained widespread use because of their inherent simplicity and potential for field measurements.

One approach is to use five or six diode lasers at widely spaced wavelengths to cover the wavelength range of interest and measure the transit time (Lin *et al.*, 1983; Modavis *et al.*, 1984). The diode lasers give narrow pulses of a few hundred picoseconds, but a precise determination of the wavelength of each laser is needed. The disadvantage to this technique is that the wavelengths are fixed, each different wavelength requires a different laser, and the wavelength drift with time must also be determined because the emission wavelength drifts with laser age. The advantage is that the equipment is fairly compact and robust so a portable field measurement could be made.

Instead of measuring the pulse transit time, one could measure the apparent "optical length" of the fiber. The group velocity is a function of wavelength; thus, the optical length measured by timing a propagating pulse will also be the same function of wavelength

$$L(\lambda) = \left[\frac{c}{n_g(\lambda)} \right] \Delta t.$$

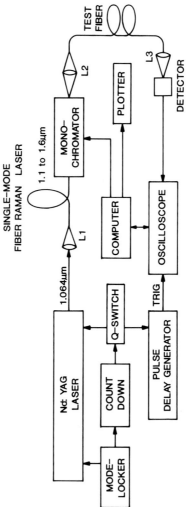

Fig. 8.9 Experimental arrangement of a Fiber-Raman laser to measure dispersion by the pulse delay method.

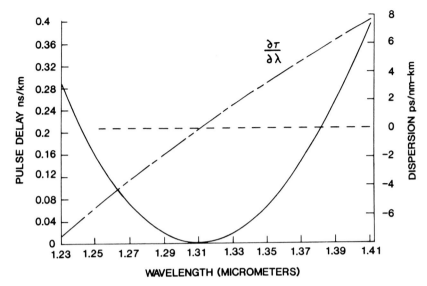

Fig. 8.10   The pulse delay curve (solid line) and its derivative, the dispersion (dashed line). The zero dispersion wavelength is the point where the dispersion curve crosses zero.

Phase detection techniques allow the optical length measurement to be made without having to produce a narrow pulse. A CW diode laser is sinusoidly modulated and the phase of the transmitted wave compared to a reference phase to determine the length. By measuring at various wavelengths and converting to the corresponding transit time, the delay curve is obtained as before.

Instead of a laser, a light emitting diode (LED) has also been used with phase techniques. A CW modulated LED having a broad spectral distribution is passed through a monochromator to select a narrow wavelength band and the optical length of the fiber determined at different wavelengths by the phase method.

Another LED-phase technique has been introduced using a differential phase technique. A special monochromator is modulated at a few hundred hertz over a narrow spectral range, and the optical phase difference is measured at each end of the modulation, i.e., at 1305 nm and 1315 nm. The detected signal is directly proportional to dispersion, and since it is modulated at a few hundred hertz, the 1/frequency noise common in other DC-phase detection schemes is greatly reduced. For example, the dispersion at 1310 nm is determined directly by interpolating between 1305 and 1315 nm. This has the advantage of measuring the dispersion directly; thus, no fitting function to the group delay, or assumption about the dispersion curve, is needed (Barlow and MacKenzie, 1987).

There are currently several commercial instruments using these techniques and comparisons with the fiber-Raman technique give results in good agreement between the various methods.

8.6.2.3 Interferometric Techniques. Interferometric techniques measure the phase delay of a CW light beam as it traverses a short length of fiber. A diagram of such an experiment is shown in Fig. 8.11 with the associated delay curve, and good agreement between this method and the traditional pulse delay method has been shown (Saunders *et al.*, 1984; Sears *et al.*, 1984). Since this technique uses short lengths of fiber, the variation of dispersion with length may be examined. Whereas other techniques measure the average dispersion over the entire length of fiber, the interferometric method measures a short length and infers the dispersion for the long length. Obviously, there are advantages and disadvantages to each method, depending on the application involved.

The interferometric technique is able to use short lengths of fiber (about 0.5 meter), since the time delay corresponding to the measured phase shift is so much shorter than in the pulse delay method, and both Mach-Zehnder (Saunders *et al.*, 1984; Sears *et al.*, 1984) and Michelson (Bomberger *et al.*, 1981) interferometers have been used with good results. The increased time resolution is because one fringe shift in the interferometer corresponds to a small length change in one arm of the interferometer. A typical measurement is made by scanning the interferometer with the fiber in one arm, and with the fiber removed, and the path difference between the two arms produces a phase difference $\Delta\theta$. The maximum fringe visibility condition is

$$\left(\frac{d}{d\lambda}\right)(\Delta\theta) = 0.$$

If the delay path is adjusted an amount δl when the fiber is removed, then the above becomes

$$n(\lambda) = \left(\frac{\delta l}{l}\right) + 1,$$

where $n(\lambda)$ is the index of refraction of the LP_{01} group delay per unit length times c, and l is the fiber length. The interferometer is scanned for each wavelength and the index determined from the measured δl.

8.7 CUTOFF WAVELENGTH—SINGLE-MODE FIBERS

At wavelengths longer than the cutoff wavelength, the fiber supports only one mode (actually, a degenerate mode group). The higher order modes, however, become highly attenuated at wavelengths substantially below the theoretical cutoff wavelength. By convention, the "effective" cutoff wavelength is defined as

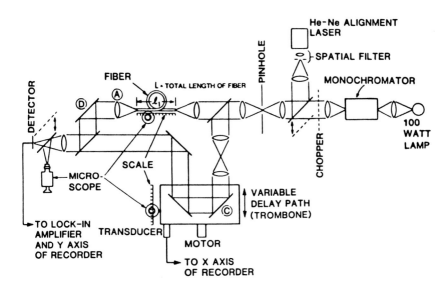

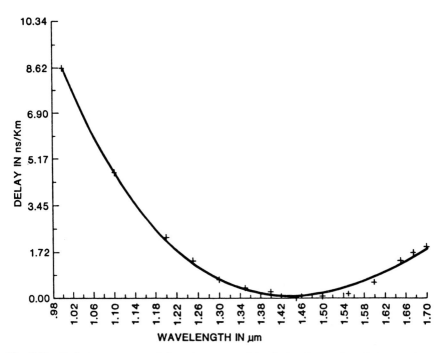

Fig. 8.11 Optical arrangement of dispersion measured by the interferometric technique, and the delay curve.

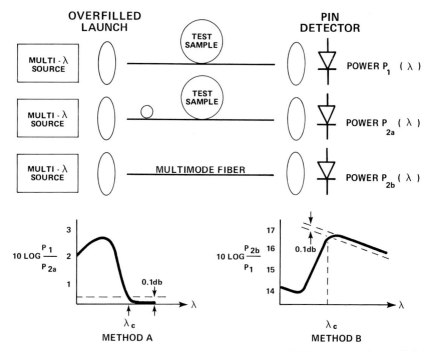

Fig. 8.12 Two techniques for cut-off wavelength measurements. Method A is sometimes called the bend technique and method B, the power step technique.

the wavelength at which the higher order mode attenuation exceeds some chosen level; and therefore, higher order modes cease to propagate significant distances. This is an appropriate definition for telecommunication systems, where the distance between mode coupling perturbations, such as splices, are typically a few hundred meters, and the adverse effects of multimode propagation, modal noise and distortion, can be avoided if the higher order modes are sufficiently attenuated. For short distance applications, such as fiber sensors, there may be different requirements.

The attenuation level that defines the cutoff wavelength depends on the measurement technique. CCITT and EIA specify a measurement sample two meters long with a 28 cm diameter loop (CCITT Recommendation G.652 and EIA-455-80). Figure 8.12 shows the cutoff wavelength measurement using two possible methods. The apparatus required for cutoff measurements is identical to that shown in Fig. 8.3 for the spectral attenuation measurement, and these two measurements are frequently made simultaneously. The power transmitted by the test sample, $P_1(\lambda)$, is larger when the higher order modes propagate. To find the transition between multimode and single-mode regimes, the power is compared to one of two references: either a multimode fiber, which establishes the spectral response of the launch and detection components; or the test sample with a small diameter (typically 2 to 4 cm) loop inserted to suppress the higher-order modes. In either case, cutoff is identified as the wavelength above which the normalized power transmitted by the test sample approaches a linear region to within 0.1 dB. An interlaboratory comparison has verified that these two methods give nearly equal values for the cutoff wavelength (Franzen, 1985). Either of these methods equate cutoff with about 20 dB of attenuation for the higher-order mode, although this attenuation may be attributable to either the sample length or the bend, depending on the fiber design.

Obviously, these measurement conditions have been chosen for convenience, not in consideration of the real usage of the fibers. Since both the length and the bend condition of the fibers significantly alter the measured cutoff wavelength (Anderson *et al.*, 1984; Shah, 1984), relating measurements made using this arbitrarily chosen length and bend condition do not relate simply to the design of telecommunications systems. To avoid modal noise and distortion, the attenuation of the higher-order mode must be large over the shortest distance between mode coupling perturbations, such as splices. While this distance is typically kilometers, it can be substantially shorter for jumpers and pigtails, for stub cables, under certain difficult installation conditions or for repair sections. This measurement method with different sample deployment conditions could also be used to determine the cutoff wavelength of cabled fiber. The user would specify a sample length and deployment condition consistent with his application, and the results of this measurement would be easier to apply to the design of systems. Users and suppliers should be cautious, however, about arbitrarily increasing the maximum allowable cutoff wavelength, since this may preclude operation in wavelength regions that may have some value for future use. Because the expected life of installed fiber and cable is long and the technology of optical sources and systems is still evolving rapidly, the future uses of cables installed today cannot be completely predicted.

REFERENCES

Stewart, W. J. (1977). A new technique for measuring the refractive index profiles of graded optical fibers, *Tech. Digest of the 1977 Int. Conf. on Integrated Optics and Optical Fiber Communication*, July 18–20, Tokyo, Japan.

White, K. I. (1979). Practical application of the refracted near-field technique for the measurement of optical fiber refractive index profiles, *Opt. Quant. Electron.* **11**, 185.

Saunders, M. J. (1981). Optical fiber profiles using the refracted near-field technique: A comparison with other methods, *Applied Optics* **20**, 1645.

Boggs, L. (1979). Rapid automatic index profiling of whole fiber samples. part I, *BSTJ* **58**, 867.

Watkins, L. S. *et al.* (1974). Scattering from side-illuminated clad glass fibers for determination of fiber parameters, *J. Opt. Soc. Am.* **64**, 767.

Marcuse, D. (1979). Automatic geometric measurements of single mode and multimode optical fibers, *Applied Optics* **18**, 402.

Marcuse, D. (1977). *Bell Syst. Tech. J.* **56**, 703.

Marcuse, D. (1978). *J. Optical Soc. Am.* **68**, 103.

Murakami, Y. (1979). *Applied Optics* **18**, 1101.

Streckert, J. (1980). *Opt. Lett.* **5**, 505.

Hotate, K. (1979). *Applied Optics* **18**, 3265.

Nicolaisen, E. (1983). *Electron. Lett.* **19**, 27.

Otten, W. G. (1986). *Tech. Digest, OFC 86*, Paper TUL30.

Anderson, W. T. (1984). *IEEE J. Lightwave Tech.* **LT-2**, 191.

Franzen, D. L. (1985). *IEEE J. Lightwave Tech.* **LT-3**, No. 5, 1073.

Auge, J. (1985). *Tech. Digest, SPIE Symp.*

Petermann, K. (1983). *Electron. Lett.* **19**, 712.

Pask, C. (1984). *Electron. Lett.* **20**, 144.

Anderson, W. T. (1987). Mode-field diameter measurements for single mode fibers with non-Guassian field profiles. *L. Lightwave Tech.* **5**, 211.

Miller, S. E. (1979). "Optical Fiber Telecommunications," Academic Press, New York, Section 11.2.3.1.

Miller, S. E. (1979). "Optical Fiber Telecommunications," Academic Press, New York, Section 11.4.4.

Franzen, D. L. (1982). Measurement of optical fiber bandwidth in the time domain, NBS Special Publication **1**, No. 637, 47.

Marcuse, D. (1981). "Principles of Optical Fiber Measurement," Academic Press, New York.

Cancellieri, G. (1984). "Measurements of Optical Fibers and Devices: Theory and Experiments," Artech House, Inc., Dedham, Mass.

Day, G. W. (1983). Measurement of optical fiber bandwidth in the frequency domain, NBS Special Publication **2**, No. 637, 57.

Lin, C. *et al.* (1978). Pulse delay measurements in the zero material dispersion region for germanium and phosphorus doped silica fibers, *Electron. Lett.* **14**, 170.

Cohen, L. G. (1977). Pulse delay measurements in the zero material dispersion wavelength region for optical fibers, *Applied Optics* **16**, 3136.

Lin, C. (1983). Chromatic dispersion measurements in single mode fibers using picosecond InGaAsP injection lasers, *Bell Syst. Tech. J.* **62**, 457.

Modavis, R. A. (1984). Multiple wavelength system for characterizing dispersion in single mode optical fibers, *Tech. Digest Sym. Optical Fiber Measurements*, NBS Special Publication 683.

Barlow, A. J., and MacKinzie, I. (1987). Direct measurements of chromatic dispersion by the differential phase technique, *Tech. Digest, OFC 87*, paper TUQ1.

Saunders, W. J. (1984). Precision interferometric measurement of dispersion in short single mode fibers, *Tech. Digest Symposium on Optical Fiber Measurements*, NBS Special Publication No. 683.

Sears, F. M. (1984). Interferometric measurements of dispersion spectra variations in a single mode fiber, *J. Lightwave Tech.* **LT-2**, 181.

Bomberger, W. D. (1981). Interferometric measurement of dispersion of a single mode optical fiber, *Electron. Lett.* **17**, 495.

CCITT Recommendation G.652 and EIA-455-80.

Franzen, D. L. (1985). *IEEE J. Lightwave Tech.* **LT-3**, No. 1, 128.

Anderson, W. T. (1984). *IEEE J. Lightwave Tech.* **LT-2**, 238.

Shah, V. S. (1984). *Tech. Digest, Symposium on Optical Fiber Measurements* 7.

Chapter 9

Integrated Optics: Basic Concepts and Techniques

W. J. TOMLINSON

Bell Communications Research, Red Bank, New Jersey

S. K. KOROTKY

AT&T Bell Laboratories, Holmdel, New Jersey

9.1 INTRODUCTION

The basic motivation for integrated optics is to do for optical circuits what integrated electronics has done for electrical circuits—replace a set of large individually-fabricated elements, that are individually-interconnected, with a single chip, in which all the miniaturized circuit elements and their interconnections are fabricated at the same time (Miller, 1969). A major concept of integrated optics is that in all the elements and interconnections the optical signals are confined in compact optical waveguides, and thus the field is often referred to as "guided wave optics" and the devices as photonic circuits.

In the years since this general concept was originally formulated, a number of integrated optical fabrication techniques have been developed, and a variety of discrete devices has been demonstrated. These individual discrete guided-wave devices have been shown in many cases to have advantages over conventional bulk optical devices. In more recent years the integrated optics fabrication technology has advanced to the point that not only is it now possible to make high-performance individual devices, but reasonably complex assemblies of devices have now been demonstrated (e.g., an 8×8 matrix switch integrating 64 2×2 interconnected switch elements on a single chip (Granestrand *et al.*,

1986)). In addition, in recent years optical communications technology has advanced, evolving toward high-speed single-mode systems, and there now appears to be an increasing need for such components.

This chapter is an introduction to the basic concepts of integrated optics, and to the basic technology that has been developed to realize those concepts. It is intended to serve as a preface to the discussion of passive integrated optical components in the following chapter, and to the more detailed discussion of active integrated optical components in Chapter 11. More detailed discussions of the basic concepts and techniques can be found in (Tamir, 1979), (Iizuka, 1983), (Hunsperger, 1984), and (Hutcheson, 1987).

9.2 BASIC CONCEPTS

The basic concepts of integrated optics are conveniently illustrated in Fig. 9.1, which is a schematic drawing of an integrated optical four-channel wavelength-division multiplexer.

9.2.1 Planar Patterns of Channel Waveguides

Integrated optics is based on the fact that one can produce regions of increased refractive index in the surface of a planar substrate, and that these regions will guide light in the same way as the core of an optical fiber. These guides are used both to form the various circuit elements (both active and passive), and to make the connections between elements, as shown in the figure.

9.2.2 Controlled Interactions between Guides

Because the locations of the waveguides are rigidly fixed in the substrate, and their characteristics and relative positions can be precisely controlled during

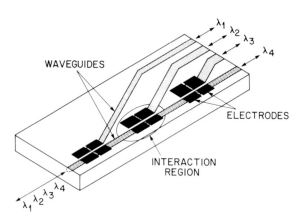

Fig. 9.1 Schematic drawing of an integrated optical wavelength-division multiplexer. This device illustrates some of the basic concepts of integrated optics.

manufacture, the waveguides can be brought near each other, such that they will interact in well-defined ways. For example, in the device illustrated in Fig. 9.1, the waveguides in the interaction regions have been constructed such that light of a particular wavelength entering the interaction region in the first of two guides will transfer to the second guide, but light at other wavelengths will remain in the first guide. For waveguides made on suitable substrates (e.g., one exhibiting a large electrooptic effect) one can use an electric field to control the interaction. In the device illustrated in Fig. 9.1, a voltage applied to the electrodes can control the wavelength selected by the coupler. With appropriately designed structures, voltage-controlled devices can serve as phase modulators, amplitude modulators, polarization controllers, tunable wavelength filters, frequency shifters, and very-high-speed broadband spatial switches. Because the light and the controlling electric field are confined to small lateral dimensions, the voltage swing required to change the device state can be low.

9.2.3 Integration of Multiple Elements in a Single Circuit

A key feature of integrated optical and integrated electronic circuits is the integration of multiple circuit elements, and their interconnections, on a single chip. In the device illustrated in Fig. 9.1, three individual filter elements are combined to make a one-piece four-channel wavelength-division multiplexer. This offers the possibility of both lower cost, because the elements and their interconnections are all made at the same time, eliminating separate fabrication and interconnection processes; and higher performance, because the interconnections are short, efficient (low loss), reproducible, and robust.

9.2.4 Possibility of High-Volume Low-Cost Manufacture

The processes used to fabricate integrated optical circuits are essentially the same as are used in the fabrication of integrated electronic circuits (e.g., photolithography, diffusion, evaporation, etching, etc.), and, as with integrated electronics, many circuits can be made simultaneously on a large wafer, and then cut apart. Thus there is the possibility that in large volumes they can be manufactured at very low unit cost. (Integrated optical and electronic circuits share a common problem, in that as the cost of the circuit decreases, the cost of packaging and interconnection becomes dominant.)

9.3 BASIC DEVICES

9.3.1 Waveguides

9.3.1.1 Channel waveguides. The geometry of the waveguides employed in integrated optical circuits depends on the particular materials system, fabrication technique, and application, but most guides used for telecommunications

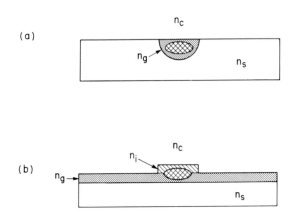

Fig. 9.2 Schematic cross sections of channel waveguides: a) embedded strip guide; b) ridge-loaded guide. The crosshatched regions indicate the areas where the guided light is concentrated.

applications can be described as channel guides of either the embedded strip or ridge-loaded type.

The embedded strip guide, as illustrated by the schematic cross section in Fig. 9.2a, consists of a substrate of index n_s, an embedded guiding region of raised index n_g (analogous to the core of a fiber), and a cover region of index n_c (frequently air). In dielectric materials, the embedded region is typically formed by doping the substrate material with an impurity (usually by diffusion or by ion exchange). In semiconducting materials the embedded region is typically formed in an epitaxial crystal growth process with a composition that differs from the substrate.

To enable light to be coupled efficiently between integrated optical circuits and single-mode fibers, channel waveguides are commonly designed to support a single spatial mode, with a mode size matched to that of the fibers. For compatibility with typical single-mode fibers designed for the 1.3–1.5 μm wavelength region, this requires a waveguide index difference of $(n_g - n_s)/n_s \approx$ 0.4%, and a guide with transverse dimensions of $\approx$ 8 μm. Most of the doping processes used to fabricate channel guides in dielectric materials can give guides that are reasonably well matched to fibers, with coupling losses of $\leq$ 1 dB commonly reported. In semiconducting materials it is frequently difficult to control the fabrication processes so as to obtain such a small index difference reproducibly. Typically the index change is large ($\sim$ 3%) and the waveguide and mode sizes are small ($\sim$ 1 μm). Consequently, butt coupling losses are typically somewhat higher than for doped guides. In addition, the large index difference between glass fibers and semiconducting materials results in a reflection loss ($\sim$ 0.7 dB), unless some type of antireflection coating is used. In semiconducting materials, embedded guides are most commonly used for lasers, as described in Chapter 13.

The ridge-loaded guide, illustrated in Fig. 9.2b, consists of a substrate of index n_s, a guide layer with a higher index n_g, and a layer of index n_i ($n_i \leq n_g$) in the form of a pattern of ridges. A feature of this design is that it provides a separation of functions. The patterned layer defines the waveguide widths and circuit pattern. The guiding layer can then be optimized for other functions (e.g., optical gain or electrooptic effect) and can be made using materials and/or processes that are not suitable for making narrow channels. The effective index difference of the guided mode (somewhat analogous to $n_g - n_s$ for embedded guides), which determines the transverse width of the guided mode, is primarily determined by the geometry of the cover layer, and small index differences can be obtained reproducibly. Ridge-loaded guides are commonly used in integrated optical devices made on semiconducting substrates.

9.3.1.2 Planar waveguides. Virtually all integrated optical devices used in optical fiber telecommunications make use of channel waveguides of the types just described, but there is another whole field of integrated optics based on planar or slab waveguides. In such guides light is confined (guided) in the direction normal to the plane, but free to spread in the plane. Most of the applications of planar waveguides in integrated optical devices are for signal processing. For example, the most extensively developed planar integrated optical device is a wideband rf spectrum analyzer, which includes two planar waveguides lenses and a planar acoustooptic beam deflector. Since planar waveguides have found relatively little application in telecommunications devices, we will not consider them any further in this chapter. Descriptions of such waveguides and their applications can be found in (Iizuka, 1983).

9.3.2 Waveguide Bends

To provide interconnections between circuit elements and between the circuit and the system it is to be used with, curved waveguides are required. For example, in the circuit illustrated in Fig. 9.1, the separation between the guides in the interaction regions is of the order of 10 μm, but to be able to attach fibers to the four guides on the right-hand end, it is desirable to have the guides separated by at least 125 μm. A typical bend structure used for this purpose is illustrated in Fig. 9.3 and is referred to as a waveguide "S-bend." To obtain a compact circuit layout, bends with very small radii of curvature are desirable, but the minimum usable bend radius is severely limited by radiation loss. The bending loss depends on the radius of curvature, the length of the bend, and the modal characteristics of the waveguide.

To a good approximation, the attenuation coefficient for planar waveguide bends is of the form

$$C_1 e^{-C_2 R}, \tag{9.1}$$

WAVEGUIDE S-BEND

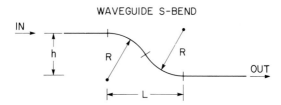

Fig. 9.3 Waveguide s-bend.

with C_1 and C_2 independent of the bend radius R (Marcatili and Miller, 1969). The coefficient C_2, and hence the bend loss, is a super-linear function of the mode confinement, and consequently it is desirable to operate the waveguide close to the cutoff of the second order mode, because this provides the maximum confinement possible in a single-mode guide.

For Ti:LiNbO$_3$ single-mode waveguides, where the index change is ~ 0.5%, the mode size is about 7 μm at 1.3 μm wavelength, and a bend radius of 20 mm can be used with negligible bend loss. For semiconductor waveguides based on compositional modifications of the crystal, the index change is ~ 3%. The waveguide and mode size are correspondingly smaller and bend radii of 1 mm or smaller are possible.

9.3.3 Y Branches

One of the most basic of optical elements for constructing more sophisticated circuits is an optical power divider, or splitter. Such an element can be used to distribute an optical signal to several other functional elements. In Fig. 9.4 we illustrate a waveguide Y-branch, which is commonly used as a power splitter. The Y-branch structure used as a splitter consists of one input and two output waveguides. For a perfectly symmetric branch, the input power will divide equally between the two output branches. For typical Ti:LiNbO$_3$ single-mode branches it is possible to obtain splitting ratios in the range 0.47–0.53 by careful control of the lithography. Radiation loss caused by the redirection of the optical

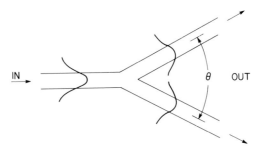

Fig. 9.4 Y-branch waveguide splitter.

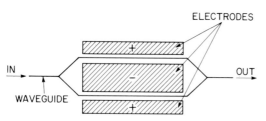

Fig. 9.5 Y-branch interferometric modulator.

field can be kept low by using a branching angle appropriate to the waveguide confinement. A general rule to ensure low radiation loss in the branch is that the branch angle should be much smaller than the ratio of the wavelength to the mode size. For Ti:LiNbO$_3$ waveguides at 1.3 μm, the branching angle should be less than 1°. For optimum performance the branch should be constructed with gradual waveguide bends as discussed before.

9.3.4 Interferometers

To illustrate an active integrated optic device, we consider a basic waveguide intensity modulator—the Y-branch interferometer, frequently referred to as a waveguide Mach-Zehnder interferometer. The device, shown in Fig. 9.5, consists of single-mode waveguides. Light entering the input waveguide is first divided at a Y-branch junction. From that point, equal amounts of light enter the two waveguide arms of the interferometer, where the two signals propagate independently. Provision is made in the arms to permit adjustment of the local index of refraction—usually via the electrooptic effect, so as to control the difference of the optical path lengths of the two arms.

If the optical path lengths of the two arms are identical, the optical fields impinging on the outgoing Y-junction arrive in phase with an optical field distribution that is symmetric about the axis of the output waveguide, and the light is coupled into the output waveguide.

If now the optical path lengths are changed so as to introduce a 180° phase shift between the optical fields entering the second Y-branch, then the field impinging on the outgoing waveguide is anti-symmetric about the waveguide axis. This anti-symmetric, i.e., odd, profile is orthogonal to that of the mode the waveguide supports. Thus, the output field radiates into the substrate. More generally, the transmission of the interferometer is proportional to the cos^2 of the differential phase shifts in the two arms.

9.3.5 Directional Couplers

One of the most versatile of integrated optic elements is the evanescent-wave directional coupler. As illustrated in Fig. 9.6, it consists of two single-mode

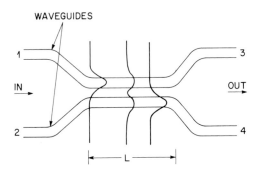

Fig. 9.6 Waveguide directional coupler.

waveguides that are brought close together so that light propagating in either guide experiences the presence of the other guide. If light is launched into one waveguide, then as the light propagates down the interaction region, light is gradually transferred from the input waveguide into the second waveguide. This process occurs along the entire interaction length of the coupler. If the waveguides are identical, then the light crossing over at various positions along the guide adds constructively. Ultimately 100% of the input light crosses over to the second waveguide. The process is reciprocal and, if the interaction region is long enough, the light will crossover from the second guide back to the first.

The spatial period of the crossover oscillation, which is related to the coupling per unit length, depends on the degree to which the tail of the mode of a waveguide penetrates into the other waveguide. By choosing a coupler length of less than one coupling length, the directional coupler can be used to tap off a small portion of the optical signal for processing at different locations on the chip. As will be discussed in Chapter 11, the directional coupler can also be designed for active operation so that the crossover efficiency can be adjusted electrically. This provides the basis of both fast optical amplitude modulators and spatial switches.

9.4 WAVEGUIDE MATERIAL SYSTEMS AND FABRICATION TECHNIQUES

9.4.1 Introduction

Integrated optical waveguide circuits have been fabricated in several different materials, each with its own particular features. In this section we describe some of the fabrication steps common to all materials systems (photolithography and fiber attachment), and then briefly review the characteristics of the most important materials systems and fabrication techniques.

9.4.1.1 Waveguide Lithography. Just as for electronic integrated circuits, the fabrication of optical waveguides is preceded by the design and construction of a device mask. This mask is the two-dimensional "street map" intended to delineate the path light will take in the guiding layer of the structure. It typically consists of a quartz plate onto which a metallic and photosensitive polymer layer have been deposited. The photoresist is exposed by a focused electron-beam under computer control in order to write the waveguide map. After developing the exposed photo-resist, the metal is etched through the selectively opened areas. This mask then serves as an original to be copied to the optical substrate in a similar process. The details depend upon whether the waveguide is physically formed by doping or etching as discussed below.

9.4.1.2 Fiber Attachment. For most telecommunications applications of integrated optical circuits, the input and/or output signals are on fibers, and thus it is important to be able to achieve these interconnections with low cost and low loss. To date, the almost universally used technique has been to manually align the fibers (using multi-axis micropositioners) to maximize the coupling efficiency, and then to cement the fiber in position (typically using uv-cured cements). While this approach has been shown to be capable of achieving low-loss connections, significant improvements and innovations are still necessary to achieve a low-cost manufacturable process necessary for the practical application of integrated optic circuits.

9.4.2 Lithium Niobate

Lithium niobate ($LiNbO_3$) is the most popular dielectric material for use in integrated optics. This is because large optical quality crystals can be routinely grown and the material has a reasonably large electro-optic figure of merit, i.e., the change of index with a change of applied electric field strength. In addition, the index change necessary to produce waveguiding is achieved by locally doping the crystal, and the resulting guides have low losses (less than a few tenths of a dB/cm). In the most developed approach, the crystal is doped by titanium using thermal diffusion. In another method, hydrogen ions (protons) are exchanged for a fraction of the lithium ions of the crystal. We outline the general aspects of these two methods next.

9.4.2.1 Ti Indiffusion. The general procedure for fabricating waveguides by the diffusion of titanium (Ti) into lithium niobate ($LiNbO_3$) is presented schematically in Fig. 9.7. First, photoresist is spun onto the prepared crystal surface. Next, the device mask is brought into contact with the coated surface and the sandwich is exposed by ultraviolet light through the mask. The photoresist is then developed, leaving openings in the photoresist layer. The size of the openings determines the width of the optical waveguide. Afterward, a thin

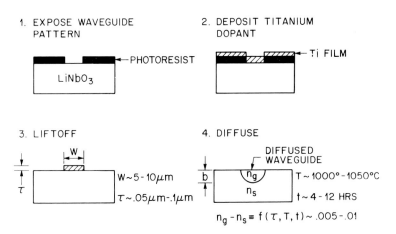

Fig. 9.7 Ti:LiNbO$_3$ waveguide fabrication steps.

(~ 0.05–0.10 μm) layer of titanium is deposited on the surface. The thickness of the titanium film is used to control the index change of the waveguide. A solvent is used to dissolve the remaining photoresist, lifting off the unneeded titanium, and leaving a bar of titanium on the surface of the LiNbO$_3$ crystal. This metal is then diffused into the crystal at approximately 1050°C for six hours, which are values typical for operation at 1.3 μm wavelength.

By varying the diffusion time and temperature, the diffusion depth can be controlled. The atmosphere of the diffusion furnace is carefully controlled to avoid the out-diffusion of Li during the heat treatment. Using the titanium thickness, the initial strip width, the diffusion time and temperature, the size and index change of the waveguide are controlled. Through these fabrication parameters, the designer can control the size and degree of confinement of the waveguide mode to optimize its performance for a given application. Typical values for 1.3 μm are a strip width of 7 μm and a titanium thickness of 900 Å.

After the diffusion, the surface of the crystal is often coated with a thin insulating buffer layer of SiO$_2$ and metal electrodes are placed over the active sections using a lithographic process similar to the waveguide fabrication. The purpose of the buffer layer is to eliminate metallic loading losses of the optical wave (Otto and Sohler, 1971, Masuda and Koyama, 1977). Next, the endfaces are carefully polished and anti-reflection coatings are deposited if required. Finally, single-mode fibers are attached to permit connection into the fiber system.

9.4.2.2 Proton Exchange.   Guides can also be formed in lithium niobate by an ion exchange process (Jackel *et al.*, 1982), in which Li$^+$ ions diffuse out of the crystal and are replaced by H$^+$ ions (protons). The basic steps in the fabrication

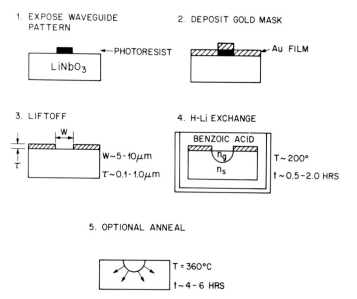

Fig. 9.8 Steps in the process of forming waveguides in LiNbO$_3$ by the proton exchange process.

process are illustrated in Fig. 9.8. A metal mask (typically made of Au) is applied to the surface of a crystal in the same way as the Ti pattern is applied for making Ti-indiffused waveguides, except that in this case the metal covers the entire surface of the crystal *except* where one wants to form guides. The crystal is then placed in an acid bath (typically benzoic acid, at ~ 250°C, for a few hours), and the ion exchange takes place through the openings in the metal mask. After the exchange is completed, the metal mask is removed. In some cases, the sample is annealed to cause further diffusion of the protons and Li$^+$ ions, and thus modify the refractive-index profile of the guides. Buffer layers, electrodes, and fibers, are applied in the same manner as for Ti-indiffused guides.

Proton exchange can produce a rather large refractive index increase (5% vs. 0.5% for Ti indiffusion), and this is attractive for use in devices where one wants guides with small-radius bends. Proton exchange only increases the extraordinary index, however, with the ordinary index actually decreased slightly. This means that proton-exchanged guides will only guide light of a single polarization, which can be an advantage or a disadvantage, depending on the application. Since the proton exchange process is carried out at a substantially lower temperature than Ti indiffusion, the two processes can be combined, with Ti indiffusion used to form conventional guides and elements, and proton exchange used to form single-polarization elements, or structures requiring very large index changes.

In the early experiments on proton exchange in lithium niobate, there was some evidence of instabilities of the resulting guides. These were evidently caused by a phase separation of a thin surface layer, which resulted in changes in the propagation constant of the guided mode and in scattering losses. Since then a number of techniques have been developed to avoid this problem. The proton exchange process has not been as thoroughly developed as Ti indiffusion, but it is currently a subject of active research in a number of laboratories around the world.

9.4.3 Ion exchange in glass

Waveguides can be formed in glass substrates by ion exchange processes very similar to the proton exchange process (Findakly, 1985). The process is basically the same as that illustrated in Fig. 9.8, except that the ions being exchanged are different. Most glasses that have been used for this process contain Na^+, which is exchanged for K^+, Ag^+, or Tl^+ from the exchange bath containing salts (typically nitrates) of one or more of these ions. An alternative approach reported recently is to use a base glass containing K^+ ions, which are exchanged for Cs^+ ions (Ross et al., 1986).

Since glasses exhibit a negligible electrooptic effect, the resulting guides are not generally useful for active devices, but they have a number of features that make them attractive for passive devices. Glass substrates are less expensive, more rugged (partially because they can be made thicker without incurring excessive material cost), and readily available in larger sizes, than single-crystal lithium niobate. Since the glass is not birefringent, the waveguides can be made relatively insensitive to the polarization of the light propagating in them. Because the active species in the exchange process are ions, the process can be controlled by the application of electric fields, and there is considerable flexibility available in the choice of the composition of the base glass and of the exchange medium.

Because of the many options available, ion exchange in glass is a difficult process to characterize, and much more research is necessary to define the limits and opportunities of this approach. Until recently, much of the effort on ion-exchanged glass waveguides has been concentrated on multimode waveguides, because ion exchange is just about the only process capable of fabricating integrated optical waveguides that are well matched to conventional multimode fibers.

9.4.4 Semiconductors

At present there is a substantial world-wide effort to develop techniques for epitaxial growth and waveguide and device fabrication in semiconducting materials. The variety of materials and fabrication techniques being investigated is too large, and the field is evolving too rapidly, for us to provide a reasonable summary in this introductory chapter. Typical devices and structures are discussed in Chapter 11, and many of the devices use structures and fabrication techniques similar to those of the semiconductor lasers described in Chapters 13 and 15.

The prospect of integrating waveguide devices on the same chip as lasers and detectors is a significant motivation for work in this area. To date, most semi-conductor work has been carried out in the binary and ternary systems of GaAs and AlGaAs; however, the emphasis has shifted toward devices based on the quatenary compositions of InGaAsP on InP substrates as communication systems have moved to the 1.3–1.5 μm wavelength region. Waveguide devices using the electrooptic and electroabsorption effects have been demonstrated, although the insertion loss when working near the band edge is an important issue. Improved performance using quantum well enhancements are also being actively pursued.

9.5 SUMMARY

In this chapter we have reviewed the basic concepts and techniques of integrated optics. The following two chapters describe the implementation of those concepts and techniques in components for use in telecommunications systems.

REFERENCES

Findakly, T. (1985). Glass wavelengths by ion exchange: a review. *Opt. Eng.* **24**, 244–250.

Granestrand, P., Stoltz, B., Thylen, L., Bergvall, K., Doldissen, W., Heidrich, H., and Hoffman, D. (1986). Strictly nonblocking 8 × 8 integrated-optic switch matrix. *Elect. Lett.* **22**, 816–817.

Hunsperger, R. G. (1984). "Integrated Optics: Theory and Technology." Springer-Verlag, New York.

Hutcheson, L. D., Ed. (1987). "Integrated Optical Circuits and Components." Marcel Dekker, New York.

Iizuka, K. (1983). "Engineering Optics." Springer-Verlag, New York, Chapter 15.

Jackel, J. L., Rice, C. E., and Veselka, J. J. (1982). Proton exchange for high-index waveguides in $LiNbO_3$. *Appl. Phys. Lett.* **41**, 607–608.

Marcatili, E. A. J., and Miller, S. E. (1969). "Improved relations describing directional control in electromagnetic wave guidance," *Bell Syst. Tech. J.* **48**, 2161–2188.

Masuda, M., and Koyama, J. (1977). *Appl. Opt.* **16**, 2994–3000.

Miller, S. E. (1969). Integrated optics: an introduction. *Bell Sys. Tech. J.* **48**, 2059–2069.

Otto, A., and Sohler, W. (1971). "Modification of the total reflection modes in a dielectric film by one metal boundary," *Opt. Commun.* **3**, 254–258.

Ross, L., Lilienhof, H.-J., Holscher, H., Schlaak, H. F., and Brandenberg, A. (1986). Improved substrate glass for planar waveguides by Cs-ion exchange. *Technical Digest Topical Meeting on Integrated and Guided-wave Optics, Atlanta*, paper THBB2.

Tamir, T., Ed. (1979). "Integrated Optics." Springer-Verlag, New York.

Chapter 10

Passive and Low-Speed Active Optical Components for Fiber Systems

W. J. TOMLINSON

Bell Communications Research, Red Bank, New Jersey

10.1 INTRODUCTION

If we define an *optical* component as a device that operates on input streams of photons, and processes them to give output streams of photons, then in the initial applications of optical fibers in telecommunications the fiber itself was about the only optical component. As fiber systems mature, however, and the range of applications for them broadens, there is increasing need for a wide variety of optical components to exploit fully the capacity and capability of the fibers. In this chapter we review the components that have been developed to perform passive optical functions, such as wavelength division multiplexing, and to perform low-speed active optical functions, such as switching. Chapter 11 covers components that can perform high-speed active optical functions, such as modulation and time-division multiplexing. The technologies (primarily integrated optics), that can be used to implement high-speed active functions, can also be used for low-speed and passive functions. To avoid duplication, our coverage of these technologies is quite brief, and focused on features of particular relevance for passive functions. (In addition to the detailed descriptions of active integrated optical components in Chapter 11, the basic concepts and techniques of integrated optics are described in Chapter 9.)

In this chapter we are concerned with components for many different functions, and each function has been implemented in more than one technology. We have chosen to organize the chapter by technology, and to begin in Section

10.2, with microoptic components. The microoptic technology was the first to be developed, and is currently the most fully developed and available. It is also a good starting place because it is capable of implementing all of the functions we will be describing, and it is conceptually simpler than the other component technologies. In Sections 10.3 and 10.4, we describe alternative technologies, which avoid some of the disadvantages of the microoptic technology. These alternative technologies are less versatile, but for certain functions and applications they may be the optimum choice. Section 10.5 briefly describes the rapidly developing field of integrated optical components. In Section 10.6 we summarize our conclusions and attempt to provide some perspective on possible future developments.

It is important to realize that the performance of the components we will be describing is, in general, not limited by fundamental laws of physics, but by the amount and quality of the engineering effort that has been invested in their development. Thus, we cannot predict the maximum practically achievable performance of a particular component; and in most cases, we cannot even explain in detail how the best performance was achieved, because such details are usually treated as proprietary. We have, however, attempted to provide representative performance data for components that have been offered commercially.

In addition to performance, the other key criterion in evaluating components is price. Since none of the components we will be describing has been produced in high volumes, it is almost impossible to predict the price of such components in high-volume competitive production. We have resisted reporting the prices of presently available components, because we do not feel that they are representative of the prices in high-volume manufacture. We have, however, attempted to describe the relative suitabilities of various component technologies for high-volume low-cost manufacture.

Our purpose in this chapter is to provide an overview, which is focused on those components that are already available, or that have the greatest potential for providing improved performance, and we have not attempted to catalog all of the variations that have been proposed. We have tried to provide references to guide the interested reader to more detailed reviews or studies, but we have not attempted to reference all of the technical contributions that have advanced the field to its present state.

10.2 MICROOPTIC COMPONENTS

10.2.1 Introduction

Microoptic components contain conventional, but miniature, optical systems with lenses, mirrors, prisms, filters, gratings, etc. They were originally developed for use with multimode fibers, but as interest in single-mode fibers has grown it has been shown that they can also provide good single-mode performance, although at the expense of tighter alignment tolerances. Except for the lenses and filters, these components do not involve any new technologies, and, as described before, the achieved performances are more a reflection of the amount of detailed engineering development that has been devoted to a particular component, rather than any fundamental limits.

Microoptic components all use lenses (or, in a few cases, concave mirrors) to collimate the beam(s) from the input fiber(s), and to focus the output beam(s) on the output fiber(s). Therefore, before describing individual component designs, we will briefly review the characteristics of the various types of lenses and mirrors that have been used in microoptic components.

10.2.2 Focusing / Collimating Elements

10.2.2.1 GRIN-Rod Lenses. The lenses most widely used in microoptic components are Graded Refractive INdex rod lenses, usually referred to as GRIN-rod lenses (Tomlinson, 1980). These are glass rods with a radially varying refractive index. The index has its maximum on the axis, and decreases approximately as the square of the radius. The index distribution is frequently expressed in the form:

$$n^2(r) = n_a^2 \left[1 - (\alpha r)^2 + \alpha_2(\alpha r)^4 + \alpha_3(\alpha r)^{6+} \cdots \right]. \quad (10.1)$$

Alternative notations in common use are $g = \sqrt{A} = \alpha$, $h_4 = \alpha_2$, and $h_6 = \alpha_3$. In the paraxial approximation it can easily be shown that light rays in such a medium will follow approximately sinusoidal paths, with period $L = 2\pi/\alpha$ (Miller, 1965). This is illustrated in Fig. 10.1, for a source point on axis (a) and off axis (b). From the figure we can see that for an object at one end of a GRIN rod, the rod will form an inverted image, with unity magnification, a distance $L/2$ down the rod, and an erect image after a distance L. At the intermediate points, $L/4$ and $3L/4$, all rays from a given object point are parallel, and thus one has a collimated beam at these points.

The GRIN-rod lenses normally used in fiber components are of length $L/4$ and are usually referred to as quarter-pitch lenses. As is illustrated in Fig. 10.2, such a lens will convert the diverging beam from a fiber at one end of the lens into a collimated beam at the other end of the lens. If the fiber is centered on the axis of the lens, the collimated beam will emerge from the lens parallel to the lens axis. For a fiber centered a distance r off-axis, the collimated beam will emerge at an angle (in air) of $\tan^{-1}(r/f)$, where $f = L/(2\pi n_a)$ is the effective focal length of the lens. Since optical systems are reciprocal, a quarter-pitch GRIN-rod lens will also convert a collimated beam into a converging (focused) beam, which will couple efficiently to an output fiber. The quarter-pitch lens is thus a basic building block for a wide variety of microoptic components.

GRIN-rod lenses have a number of features that make them particularly

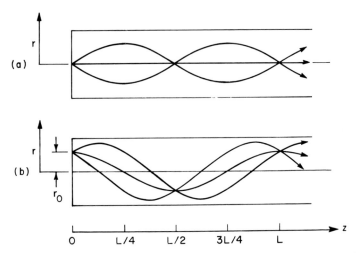

Fig. 10.1 Cross section of GRIN rods, showing ray paths for an object point (a) on-axis and (b) off-axis.

attractive for use in fiber components. (1) The input and output faces are planar, and since the focusing action of the lens results from its internal index gradient (rather than from refraction at the input and output surfaces—as in a conventional lens), the optical elements on either side of it can be glued directly to the lens faces. This reduces reflection losses, eliminates air interfaces that might collect dirt, and gives a rugged, stable, compact, solid assembly. (2) GRIN-rod lenses have aberrations that are several times smaller than an equivalent optimized conventional lens (Tomlinson, 1980a). It has been shown that the aberrations of available lenses will contribute less than a few tenths of a dB of loss for both multimode (Cline and Jander, 1982) and single-mode

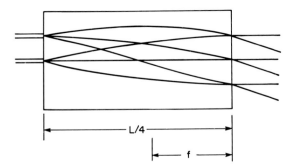

Fig. 10.2 Cross section of a quarter-pitch GRIN-rod lens. The distance marked f is the effective focal length of the lens.

(Wagner and Tomlinson, 1982) components. (3) GRIN-rod lenses are available with focal lengths from about one to a few mm, and numerical apertures of 0.1–0.5, which is just the desired range for most fiber components. (4) The cylindrical shape of GRIN-rod lenses makes them considerably easier to mount and align than conventional lenses. (These features are described in somewhat more detail in Tomlinson, 1980.)

Currently, GRIN-rod lenses are made by preparing rods of a high-index base glass, which are then immersed in a molten salt bath, to undergo an ion exchange process. In this process, which is adaptable to high-volume manufacture, high-polarizability ions from the base glass diffuse into the bath and are replaced by lower-polarizability ions from the bath. Through careful control of the ion species, and the exchange time and temperature, the desired index profile can be obtained. The most successful process uses a Tl-doped base glass, and the problems involved in safely handling this material *in the molten state* have been a deterrent to wide-spread investigations of GRIN-rod lens fabrication technology.

At the present time the only commercial source of GRIN-rod lenses is the Nippon Sheet Glass Company, which markets them under their trademark SELFOC. While SELFOC lenses have been widely used in fiber components, the vast majority of such lenses have been used in lens arrays for compact xerographic copiers. In this high-volume application the price can be as low as a few cents per lens element. While the requirements for lenses for copier applications are rather less stringent than for fiber components, it seems reasonable to assume that if a high-volume market for microoptic components using GRIN-rod lenses were to develop, the lenses would not be a major cost component.

10.2.2.2 Rod Lenses. A possible alternative to GRIN-rod lenses is rod lenses—pairs of glass rods each with a spherical surface on one end, as illustrated in Fig. 10.3. These lenses have several of the features of GRIN-rod lenses. They have planar input and output faces; they have a convenient shape for mounting; and they can have reasonably low aberrations (particularly if made of a high-index glass). On the other hand, they have two air-glass interfaces, which can collect dirt (unless sealed under clean conditions), and will give reflection losses (unless anti-reflection coatings are applied). Rod lenses with

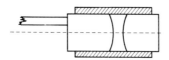

Fig. 10.3 Cross section of a rod lens. The shaded piece is a mechanical sleeve to hold and align the two lens elements.

the focal lengths desirable for fiber components require spherical surfaces with radii of a few mm, and are thus not readily adaptable to high-volume low-cost manufacture using conventional lens grinding techniques. Some of the problems of rod lenses may be overcome by the molding fabrication technology described below, although the molding technology may make other designs even more attractive. To date we are not aware of any major use of rod lenses for fiber components, but there is one report of a wavelength-division multiplexer, using rod lenses (Watanabe *et al.*, 1980). It seems likely that in that case the rod lenses were chosen because the device required a relatively long focal length; a longer focal length than is currently available with GRIN-rod lenses.

10.2.2.3 Ball Lenses. There have been a few reports of components using ball lenses, which are simply transparent spheres. Spheres have the feature that they are easier to fabricate in quantity than conventional or rod lenses. Glass ball lenses 2.5 mm in diameter have been used in a wavelength multiplexer, and provided performance almost equal to GRIN-rod lenses, but required somewhat more complex mounting (Tamura *et al.*, 1986). Ball lenses have been primarily used for laser-fiber coupling, because when made from high-index materials (e.g., semiconductors), they can be used at the high numerical apertures required to collect the light from a laser diode (Lipson *et al.*, 1985b). Ball lenses made from the magnetooptic material yttrium iron garnet (YIG) have the feature that they can also serve as polarization rotators for isolators (Sugie and Saruwatari, 1983).

10.2.2.4 Molded Lenses. Conventional homogeneous thin lenses with ground spherical surfaces have little to offer compared to the GRIN-rod and rod lenses described. However, advances in techniques for precision molding of small lenses offer some attractive alternatives for fiber components. Molding is well established as a practical technique for volume production of moderate-quality lenses. The demand for low-cost high-quality lenses for digital audio disk players, however, has led to considerable advances in lens molding technology. Diffraction-limited single-element lenses with numerical apertures of ∼ 0.4 (using two aspheric surfaces) are now being manufactured in large volume in glass (Izumitami *et al.*, 1985), and in plastic (Kubota, 1985). A major feature of molded lenses is that alignment and mounting fixtures can be integral parts of the lens. On the other hand, since they function by refraction at air-glass (or plastic) interfaces, they will suffer from reflection losses unless provided with anti-reflection coatings. To date, molded lenses have only been used for fiber connectors (described in Chapter 7), but, if a significant demand for microoptic components should develop, molded lenses should be given serious consideration.

10.2.2.5 Mirrors. Concave mirrors are a classical alternative to conventional refractive lenses, and they have found some applications in fiber components.

Because they must almost always be used at least somewhat off-axis, mirrors typically have greater aberrations than the equivalent lens system, but they are free of chromatic aberrations. The advantages and disadvantages of mirrors as the focusing elements in fiber components can be best explained by comparison to components using lenses, and thus we defer detailed consideration of this issue to Section 10.2.4.

10.2.2.6 Coupling Losses. In all microoptic components, the optical system forms images of the input fibers on the ends of the output fibers. To the extent that those images are imperfect, or misaligned, there will be less than perfect coupling of the light to the output fibers, which will add to the insertion loss of the device. The imperfections of optical systems are usually characterized in terms of aberrations, and it has been shown that there is a direct, and simple, relationship between the classical aberrations of an optical system and the coupling loss of a fiber component using that optical system. This relationship has been thoroughly analyzed for single-mode components (Wagner and Tomlinson, 1982). For multimode components the analysis is less precise, because of the practical impossibility of specifying the relative amplitudes and phases of all of the modes of the input fiber, but the available theoretical (Tomlinson *et al.*, 1981) and experimental (Cline and Jander, 1982) results confirm that over the relevant range of parameters there is a linear relationship between aberrations and insertion loss. Since there are well-established techniques for directly measuring the aberrations of an optical system, it is possible to determine the loss resulting from the imperfections of the optical system, independent of the losses resulting from misalignments or other factors. For GRIN-rod lenses, as previously mentioned, there is ample evidence that this loss is typically less than a few tenths of a dB for multimode (Cline and Jander, 1982) or single-mode (Wagner and Tomlinson, 1982) fibers. Since the typical losses reported for components using GRIN-rod lenses are ~ 1 dB or higher, it would appear that in most components, lens aberrations are not the major source of loss. For components using concave mirrors the evidence is less definitive, but it would appear that while mirrors tend to have greater aberrations than GRIN-rod lenses, it is still practical to reduce the aberrations to the point that they do not dominate the component losses.

10.2.3 GRIN-Rod-Lens Components

10.2.3.1 Connectors. Perhaps the simplest fiber component using GRIN-rod lenses is a connector. But since connectors, including lensed connectors, are discussed in great detail in Chapter 7, we simply present a brief description of the basic features of lensed connectors.

In most fiber connectors the fibers are butted directly against each other, and a considerable variety of mechanical schemes has been devised for aligning the

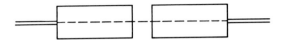

Fig. 10.4 Schematic cross section of a fiber connector using GRIN-rod lenses. The dashed line indicates the path of the center of the optical beam.

fibers and maintaining that alignment. In lensed connectors, each fiber is attached to a lens, and the connection is made in the collimated beam between the lenses, as illustrated schematically in Fig. 10.4. Since the lens transforms the small-diameter diverging beam from the input fiber into a larger-diameter collimated beam, the mechanical system used to align the two parts of the connector must be capable of higher-precision angular alignment, but can achieve a lower-precision lateral alignment than is required for a butt-joint connector. (See Chapter 7 for a full discussion of the advantages and disadvantages of this trade-off.) The major features of lensed connectors, as compared to butt-joint connectors, would seem to be that because the connection is made in an expanded beam, small dust particles will result in much less loss; and because the connection is made in a collimated beam, the lens end faces can be separated by as much as many mm, thus making them less susceptible to being damaged by dust particles while being connected. Thus, lensed connectors may be advantageous for use in dirty environments, provided that the required mechanical alignment can be maintained.

10.2.3.2 Attenuators. The basic design of an attenuator using GRIN-rod lenses is illustrated in Fig. 10.5. The attenuator plates can be fixed, interchangeable, or variable. Because the plates are located in the collimated beam, their positioning is not critical, provided their surfaces are parallel, so that they do not cause angular deviations of the beam. Fixed or interchangeable attenuator plates are typically made of glass, coated with a thin metal film, and oriented with their normal at a small angle with respect to the beam axis, such that reflected light is not coupled back to the input fiber. Attenuations from 3 dB to 30 dB are typically available for multimode or single-mode fibers. Fixed attenuators with attached connectors tend to have outside dimensions of about $15 \times 15 \times 15$ mm (not including the connectors), primarily determined by the

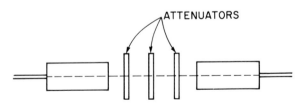

Fig. 10.5 Schematic cross section of an attenuator using GRIN-rod lenses.

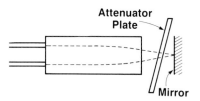

Fig. 10.6   Schematic cross section of a variable attenuator using a single GRIN-rod lens.

connector size. Fixed attenuators with fiber pigtails are typically smaller in diameter (~ 3 mm), but somewhat longer (~ 40 mm), because of the space required for fiber strain relief.

Continuously variable attenuators are needed for characterizing fiber systems and are convenient for use in lightwave receivers to set the input optical signal to the optimum level for the detector. Available components are of the basic design shown in Fig. 10.5, except that the attenuator plate has a spatially varying attenuation and can be translated (approximately normal to the beam direction) to vary the net attenuation of the beam. The spatially varying attenuator plates consist of glass plates with varying thickness metal films; or of a pair of glass wedges, one of which is absorbing. The latter approach is somewhat more difficult to implement, but provides a linear response without calibration. For typical components, the minimum loss is ≥ 1 dB (both multi-mode and single-mode), and the maximum loss is ~ 60 dB.

For use on receiver circuit boards it is convenient to have compact variable attenuators. The basic design of such a component, using a single GRIN-rod lens, is shown in Fig. 10.6, and Fig. 10.7 is a photograph of a commercial device of this design. Note that by using a mirror, only one lens is required and the input and output fibers are attached to the same end of the lens. Other types of compact variable attenuators are described in Section 10.2.5.

10.2.3.3 Isolators.   As the data rates used in optical communications systems increase, there is increasing concern about the effects of light reflected back to the laser source. This is of particular concern for experimental systems using coherent detection, which require extremely stable sources.

All true isolators make use of the Faraday effect, which is a rotation of the plane of polarization of a light beam induced by a magnetic field oriented parallel to the propagation direction. This rotation is in the same direction, independent of whether the light is propagating parallel or antiparallel to the magnetic field direction.

A schematic cross section of an isolator using GRIN-rod lenses is shown in Fig. 10.8. Light traveling from left to right has its polarization rotated 45°, so as to pass through the second polarizer. Light traveling in the opposite direction has its polarization rotated an additional 45° and is blocked by the first polarizer.

Fig. 10.7 Photograph of a commercial variable attenuator of the type illustrated in Fig. 10.6. The attenuator has an overall length of about 38 mm, and transverse dimensions of 24 × 17 mm. (Photo courtesy of JDS Optics Inc.)

Crystals of yttrium iron garnet (YIG) exhibit a very large Faraday rotation such that, with the magnetic fields available from permanent magnets, only a few mm of crystal are required to obtain a 45° rotation. Unfortunately, YIG is strongly absorbing for wavelengths shorter than about 1.1 μm. For the 0.8-μm wavelength region, rotators have been made from Tb-doped glass, which requires a path length of several cm to achieve a 45° rotation. The optical path through the glass rotator can be folded to reduce the overall size of the isolator, but these isolators are still larger and heavier than those for the longer wavelengths. Crystals of bismuth-substituted gadolinium iron garnet have almost the same absorption as YIG, but a considerably larger rotation coefficient (Verdet constant), such that they may make possible compact isolators with acceptable losses for the 0.8-μm region (Shiraishi, 1985). Typical curves of isolation and insertion loss as functions of wavelength are given in Fig. 10.9, for a Tb-glass isolator and for a YIG isolator. Note that peak isolations of greater

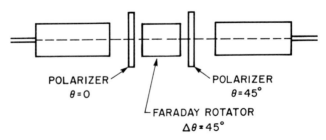

Fig. 10.8 Schematic cross section of an isolator.

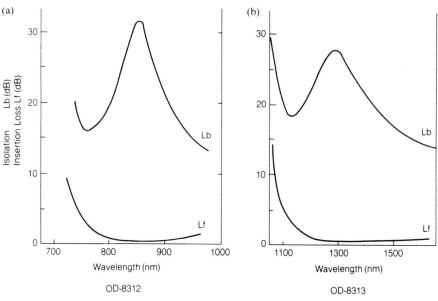

Fig. 10.9 Typical curves of isolation and insertion loss, as functions of wavelength, (a) for an isolator using a Tb-doped glass rotator, and (b) for an isolator using a YIG rotator. (Data courtesy of NEC Electronics Inc.)

than 25 dB are obtainable. (The wavelength dependence of the isolation is a consequence of the wavelength dependence of the polarization rotation.)

To date, isolators have been used primarily in laboratory experiments. If it is determined that future fiber systems need isolators to achieve the required laser stability, the isolator will probably be incorporated into the laser package, along with the optics for coupling the light from the laser to the output fiber. A recent example of this approach, using ball lenses rather than GRIN-rod lenses, is described in (Chikama *et al.*, 1986). (Other designs for isolators, which do not use lenses, are described in Sections 10.3 and 10.4.)

Isolators of the basic design shown in Fig. 10.8 transmit only a single linear polarization, but this is not a problem if the isolator is to be used immediately adjacent to a laser source, which has a well-defined linear polarization. A polarization-independent isolator can be made by separating the two polarization components of the input beam and processing them separately. This approach has been used to make a polarization-independent switch, which is described in Section 10.2.3.7. An isolator of this type has been reported that, instead of GRIN-rod lenses, uses YIG spheres, which serve both as rotators and as coupling lenses (Yokohama *et al.*, 1986). This device combined all-fiber and microoptic techniques, in that the polarizing beamsplitters were of the fused-fiber type, described in Section 10.4.4.

10.2.3.4 Directional Couplers. The directional coupler, or beamsplitter (depending on one's educational antecedents), performs a basic optical function of dividing input beams between pairs of output ports. The conceptually simplest version of a directional coupler, using GRIN-rod lenses, is shown in Fig. 10.10. This design has the features that all lenses are used on-axis and that only a single fiber has to be attached to each lens. On the other hand, it requires four lenses, four separate fiber alignment operations, and has fibers going in four different directions. (The latter two characteristics can be convenient for components using connectors, rather than fiber pigtails.) In addition, since the beamsplitter film is used at an incidence angle of 45°, the performance of the device is generally polarization dependent.

While couplers of the basic design illustrated in Fig. 10.10 have been offered commercially, in both 4-port and 3-port models, most available GRIN-rod-lens couplers are of the design illustrated in Fig. 10.11. In these components one takes advantage of the off-axis properties of GRIN-rod lenses (note Figs. 10.1 and 10.2), so that only two lenses are required with two fibers attached to each lens. In addition, since the beamsplitter film is used at close to normal incidence, the devices are generally insensitive to the polarization of the input signals. Couplers of this design, with fiber pigtails, are available commercially for

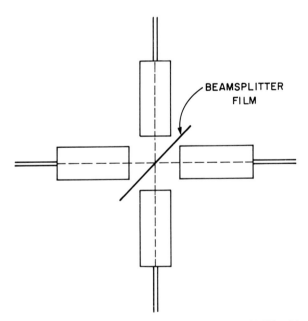

Fig. 10.10 Schematic cross section of a directional coupler using four GRIN-rod lenses. Typically the beamsplitter film is contained in a cube beamsplitter, with the lenses cemented to the faces of the cube so as to give a single solid assembly.

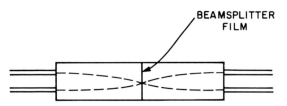

Fig. 10.11 Schematic cross section of a directional coupler using two GRIN-rod lenses.

multimode or single-mode fibers with a variety of splitting ratios and with excess losses typically ≤ 1.0 dB, and crosstalk typically ≤ −30 dB. Typical transverse dimensions are 4–10 mm, and typical longitudinal dimensions are 4–10 cm, primarily determined by the space required for the fiber strain relief.

The greatest use of couplers of the design illustrated in Fig. 10.11 has been as wavelength-division multiplexers, using a beamsplitter film with a wavelength-dependent reflectivity. This application is described in greater detail in the following section.

10.2.3.5 Wavelength-Division Multiplexers Using Filters. Wavelength-division multiplexing (WDM) is a technique for increasing the information capacity of a fiber by transmitting two or more different signals at different wavelengths (Tomlinson, 1977; Winzer, 1984; Ishio *et al.*, 1984). This requires a separate source emitting at each desired wavelength, and multiplexing devices to couple all the signals into the fiber and to separate them out at the other end of the fiber. WDM is currently being used to provide additional capacity in existing multimode systems, because the residual modal dispersion in multimode fibers limits the maximum bit rate in a single channel. Single-mode fibers have much larger bandwidths, and thus far it has generally been more attractive to increase the capacity of single-mode systems by increasing the bit rate, rather than by using WDM. As the bit rate is increased, however, the required high-speed electronic time-division multiplexers becomes more difficult and expensive to produce, and at some undetermined bit rate WDM will probably become attractive again. An innate feature of WDM is that since the channels are completely independent, information with completely different formats and data rates can be transmitted on the same fiber without requiring any synchronization between the channels.

By using a beamsplitter film that is highly reflective at the wavelength of one channel, and highly transmissive at the wavelength of another channel, the directional coupler illustrated in Fig. 10.11 can serve as a two-channel wavelength multiplexer. With recent developments in the technology of fabricating multilayer dielectric filters, it is possible to use wavelengths separated by as little as 20 nm, but larger separations are usually employed to allow for the variations in the wavelengths of the sources. Multimode and single-mode multiplexers of

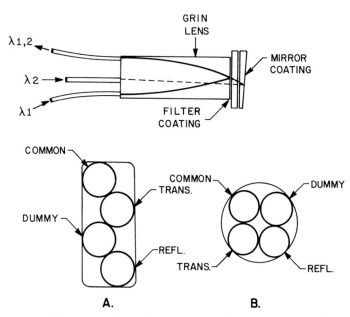

Fig. 10.12 Single-ended wavelength multiplexer using a single GRIN-rod lens. The lower part of the figure illustrates two possible arrangements of the fibers. (Figure courtesy of J. Lipson, AT & T Bell Laboratories.)

this design are available for a variety of wavelength combinations, such as 0.8/1.3 μm (usable with LED sources), 0.80/0.83 μm, 0.86/0.89 μm, 0.80/0.89 μm, 1.2/1.3 μm, 1.3/1.5 μm, etc. Typical insertion losses are 2–3 dB (per pair), and crosstalk levels are < -25 dB.

An alternative design for a two-channel multiplexer is shown in Fig. 10.12. This device uses only a single lens, and has the feature that all the fibers are attached to the same end, which simplifies packaging. Experimental results for a device of this design, using single-mode fibers, are given in Fig. 10.13 (Lipson et al., 1984). Note that for the longer-wavelength channel the insertion loss is only about 0.5 dB.

Wavelength multiplexers using filters can be made for more than two channels, but there are some practical problems. One possible design for a four-channel device is illustrated in Fig. 10.14. Since the filters can be made very highly reflecting at the wavelengths they are supposed to reflect, the insertion loss can be held to acceptable levels, but the large number of individual optical piece parts that must be individually manufactured, aligned, and cemented in place (15 parts for the design in Fig. 10.14) make this a costly approach. A more complicated design for a four-channel multiplexer is shown in Fig. 10.15, along with typical experimental results. With this design the incidence angle on each

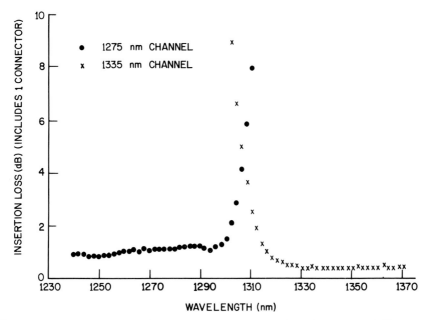

Fig. 10.13 Experimental results for a single-mode multiplexer of the design shown in Fig. 10.12 (Lipson *et al.*, 1984).

filter can be adjusted during manufacture so as to optimize the wavelength response.

Until recently, the minimum channel bandwidths obtainable with filter multiplexers (≥ 20 nm) did not have a major influence on possible systems designs, because the achievable tolerances on laser wavelengths required bandwidths > 20 nm. Recent progress in the development of distributed feedback lasers, however, offers the possibility of much more precise wavelength control, and may make some of the other types of multiplexers, described below, more attractive than the filter devices.

10.2.3.6 Wavelength-Division Multiplexers Using Gratings. For multi-channel wavelength multiplexers, devices using diffraction gratings have some important features. A schematic design for a grating multiplexer, using a GRIN-rod lens, is shown in Fig. 10.16. A multi-wavelength signal input on the upper fiber will be collimated by the lens and strike the grating. The grating has the property that signals at different wavelengths are reflected at different angles. These reflected beams pass back through the lens and are focused on separate output fibers. The key point is that this is a *parallel* device, requiring only a single lens and a single grating, independent of the number of channels. Experimental results for a

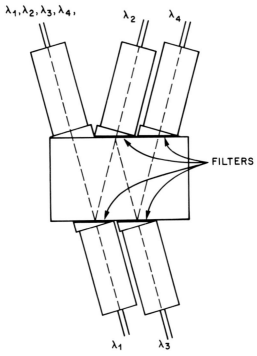

Fig. 10.14 Schematic cross section of a four-channel wavelength multiplexer using GRIN-rod lenses and interference filters.

four-channel multimode device of this type are shown in Fig. 10.17 (Lipson *et al.*, 1985). Note that channel spacings of 18 nm are achieved, with low crosstalk.

The results in Fig. 10.17 also illustrate some of the problems inherent in grating multiplexers. Because the gratings typically have losses of 1–2 dB, the insertion losses are somewhat greater than for filter devices (with modest numbers of channels). In addition, because only a single lens is used, chromatic aberrations of the lens may make it impossible to optimize the focus for all channels, and this is apparently the cause of the higher loss for the 885-nm channel. Of greater importance, however, is the fact that the bandwidths of the individual channels are narrower, relative to the interchannel spacing, than was achieved with the filter multiplexer illustrated in Fig. 10.15. This is a consequence of the angularly-dispersive characteristic of the grating. For a monochromatic input signal, the optical system of the multiplexer will form an image of the end of the input fiber on the plane containing the ends of the output fibers.

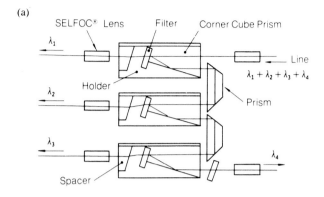

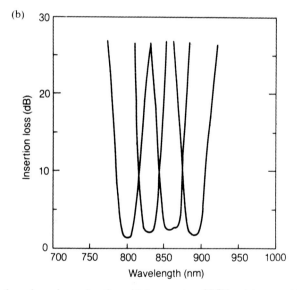

Fig. 10.15 A four-channel wavelength multiplexer, using GRIN-rod lenses, which permits the wavelength response to be optimized during manufacture. (a) Schematic cross section. (b) Typical response curves for a device of this design. (Data courtesy of NEC Electronics Inc.)

Varying the wavelength of that signal will cause the image to translate along the array of output fibers. Thus, the shapes of the channel passbands are the same as would be obtained by translating a fiber along the output array and are completely determined by the characteristics of the fibers (neglecting lens aberrations). The lens and grating determine only the wavelength scale on the response functions, not their shapes.

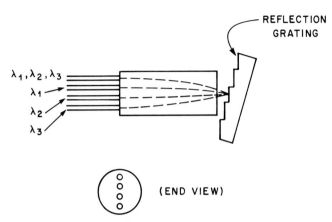

Fig. 10.16 Schematic cross section of a wavelength-division multiplexer using a grating and a GRIN-rod lens.

Despite these problems, it should be clear that grating multiplexers require far fewer parts than multi-channel filter devices (such as those illustrated in Figs. 10.14 and 10.15) and are much simpler to manufacture. Note that since all the fibers are in a single array and attached to the same end of the lens, assembly requires only a single precision alignment operation. In addition, in some cases it is possible to get around the problem of the narrow channel bandwidths.

The data in Fig. 10.17 are for a device that was deliberately made symmetric, with identical input and output fibers, so that it can function equally well as a multiplexer or as a demultiplexer. (Almost all filter devices are symmetric, so up to this point we have not needed to distinguish between multiplexing and demultiplexing.) For a demultiplexer, the output fibers are quite short (typically < 1 m) and go directly to detectors, so it is feasible to use customized fibers for this purpose. By using large-core step-index output fibers, one can substantially broaden the channel bandwidths of a demultiplexer, as shown in the experimental results in Fig. 10.18 (Lipson *et al.*, 1984). This particular device had a single-mode input fiber, also illustrating that grating demultiplexers can be used with single-mode systems. It was designed for use with the single-mode filter multiplexer for which the response is shown in Fig. 10.13, and a grating demultiplexer was chosen to provide better crosstalk rejection than is possible with the filter device.

For a grating multiplexer (as opposed to a demultiplexer), the output fiber must match the transmission fiber chosen for the system, so there is much less flexibility possible in the choice of fibers. For a multimode transmission system,

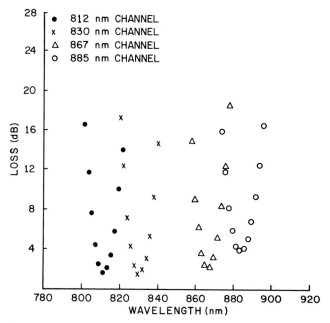

Fig. 10.17 Experimental results for a four-channel grating multiplexer of the general design illustrated in Fig. 10.16 (Lipson *et al.*, 1985). For this device the input and output fibers are all standard 50-μm-core-diam. multimode fibers.

one could obtain some broadening of the channel bandwidths by using single-mode input fibers, but this would require laser sources with single-mode outputs, which is probably not a very attractive option. For a single-mode transmission fiber, even this approach is not available.

One approach for a multi-channel single-mode grating multiplexer is illustrated in Fig. 10.19 (Lipson *et al.*, 1985a). An integrated optical waveguide concentrator (described in more detail in Section 10.5), is used to transfer the beams from the array of fibers, with center-to-center spacings of ~ 250 μm, into a closer-spaced array with ~ 20-μm spacing. Response curves for an experimental device of this design are given in Fig. 10.20, which also includes response curves for a companion demultiplexer using large-core multimode output fibers (Lipson *et al.*, 1985a). For this particular device the design of the concentrator was not fully optimized, and much of the observed loss is a result of mode mismatch between the fibers and the concentrator, with additional loss in the longer-wavelength channels caused by bend losses in the concentrator.

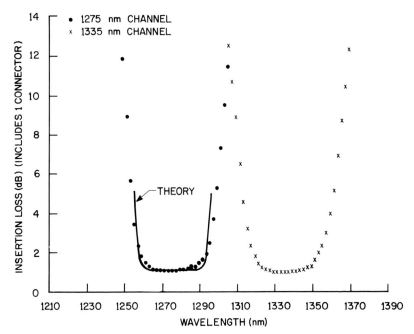

Fig. 10.18 Experimental results for a GRIN-rod grating demultiplexer using custom large-core step-index output fibers, and a single-mode input fiber (Lipson *et al.*, 1984). The output fibers had 47-μm-diam. cores, outside diameters of 57 μm, and were located on 60 μm centers.

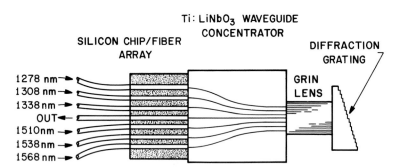

Fig. 10.19 Schematic cross section of a multi-channel single-mode grating multiplexer, using a waveguide concentrator (Lipson *et al.*, 1985a).

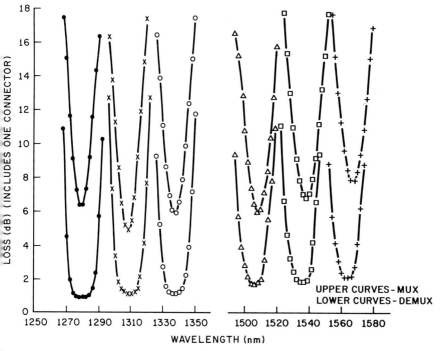

Fig. 10.20 Experimental results for a single-mode multiplexer of the design shown in Fig. 10.19, and for a companion demultiplexer using large-core multimode output fibers (Lipson *et al.*, 1985a).

Another approach has been reported recently (Nishi *et al.*, 1985; Shirasaki *et al.*, 1986). The basic idea is to use the device in double pass, so as to cancel the undesireable effects of the angular dispersion. While the approach is primarily of use for multiplexers, it is somewhat easier to understand its operation as a demultiplexer. If, in the device illustrated in Fig. 10.16, the fibers for the demultiplexed channels are replaced by a mirror on the end of the lens, the spectrally dispersed light incident on the mirror will be reflected back through the device, and all of the wavelengths will retrace their paths and be focused back to an output beam, which is matched to the input fiber. If, instead of a mirror, one uses a set of retroreflecting prisms, each of which intercepts a finite range of wavelengths and directs it back into the device with a unique lateral offset (i.e., perpendicular to the plane of Fig. 10.16), then light in each such wavelength range will be focused to a spot that is a good match to the mode of the input fiber, but is displaced from the input fiber by the lateral offset introduced by its prism. A 9-channel single-mode multiplexer of this type, with a 10-nm channel spacing and channel bandwidths of several nm, has been reported; although absolute insertion losses were not given (Shirasaki *et al.*,

1986). While this approach has demonstrated wider single-mode multiplexer channel bandwidths than the waveguide concentrator approach, the available data are insufficient for an assessment of its practicality.

It is difficult to make GRIN-rod-lens grating multiplexers with much more than about 10 channels, because the available GRIN-rod lenses do not have large enough diameters or focal lengths. Other types of grating multiplexers, suitable for larger numbers of channels, are described in Section 10.2.4.

10.2.3.7 Switches.   The components we have considered thus far are generally considered passive components; although pedants might object to including variable attenuators in this category. Switching is more clearly an active function, but in the present chapter we confine our discussions to low-speed optical switches, by which we mean switches with switching times vastly longer than the reciprocal data rate of the signals that they are likely to transmit. Higher-speed switches and modulators are described in the following chapter.

Mechanically actuated switches.   One of the simplest techniques for controlling the destination of a beam of light is by mechanically moving a mirror or deflector, and GRIN-rod-lens components provide a collimated beam that is convenient for manipulating in this way. Figure 10.21 is a schematic cross section of such a 1 × 2 switch, using a moving prism reflector. Glass parallelograms can be used to interchange two parallel beams to give a 2 × 2 switch element, as illustrated in the detail at the bottom of Fig. 10.22 (Fujii *et al.*, 1979). (The overall figure illustrates the use of six of these 2 × 2 elements to make a nonblocking 4 × 4 matrix switch.) Corner reflecting prisms and parallel-plate deflectors both have the feature that, provided they are perfectly made, small tilts will not result in an angular deviation of the beams passing through them, which eases the requirements on the mechanical actuators used to move them. A number of 1 × 2 and 2 × 2 switches, using the basic designs shown in Figs. 10.21 and 10.22 and variations thereof, are available commercially for multi-

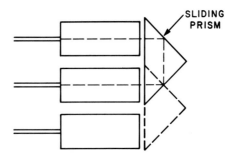

Fig. 10.21   Schematic cross section of a 1 × 2 switch using GRIN-rod lenses and a moving prism.

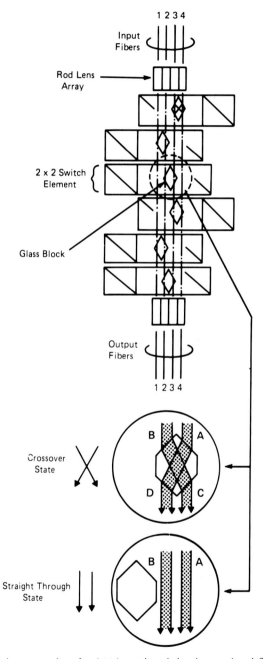

Fig. 10.22 Schematic cross section of a 4 × 4 matrix switch using moving deflector plates (Fujii *et al.*, 1979).

mode or single-mode fibers. Typical insertion losses are ~ 1.5 dB, with < −60 dB crosstalk. For devices with solenoid actuators, the typical switching times are 10–30 msec, and, because of the space required for the actuators, typical package sizes are 40–50 mm square by about 20 mm thick. The solenoid actuators are available in latching models, which require the control voltage to be applied for only a few 10s of msec to cause switching, or in nonlatching models. Nonlatching 2 × 2 switches have been widely considered for use as fail-safe bypass switches in the nodes of local area networks with bus or ring architectures.

Some 1 × n switches, with 3 ≤ n ≤ 100, are also available in manual or electrically-driven models. These presumably use rotating deflector elements, and at present, they are primarily of interest for use in test equipment.

As shown in Fig. 10.22, 2 × 2 switch elements can be combined to provide higher-dimension switches. There is, however, a basic problem with this approach, which will limit the feasible size of such switches. As the switch size is increased, the optical path length through the switch and the variations in that length as a function of the state of the switch, both increase, which necessitates using larger-diameter collimated beams. These larger-diameter beams require increases in the transverse size of the switch, the size of the deflector elements, and the sensitivity of the switch to angular misalignments. It would seem that at some point, not very far beyond a 4 × 4 switch, this approach will become impractical, because of the increased mechanical complexity of providing all the deflection elements and their actuators, and the requirement for increasingly precise fabrication and alignment of all those elements.

In addition to moving a solid reflector, one can also switch a beam by moving a liquid. A schematic cross section of a switch using electrically activated motion of a liquid is given in Fig. 10.23. The switching element, located between GRIN-rod lenses, is a flat capillary tube containing an electrolyte and a drop of mercury (Jackel et al., 1983). When a voltage is applied to electrodes in the ends of the capillary, the electrowetting effect causes the mercury to translate along the capillary. Light incident from fiber 1 is reflected by the mercury into fiber 2, or transmitted to fiber 3, depending on the position of the mercury. Switching times of ~ 20 msec have been reported. Switches of this type have not been offered commercially, but there is a recent report of a related moving-liquid switch that is said to be "in development."

Electrically actuated switches. The operation of all the switches described so far requires the physical transport of a macroscopic mass. Through various electro-optic and magnetooptic effects, it is possible to use electric or magnetic fields to influence an optical beam directly, but these effects are small, and it is difficult to obtain significant deflections with practically-achievable fields. It is, however, possible to change the polarization of a beam, and two different types of

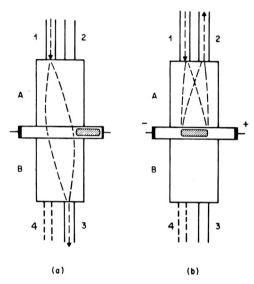

(a)   (b)

Fig. 10.23 Schematic cross section of moving-liquid switch using the electrowetting effect to position a drop of mercury in a flat capillary tube (Jackel *et al.*, 1983).

polarization rotators have been used to make electrically activated switches. In order to make the switches independent of the polarization of the input light, the configuration shown in Fig. 10.24 has been used. The input beam is split into its two polarization components by the polarizing beamsplitter P1. The two polarizations are then operated on separately and recombined at a second polarizing beamsplitter, P2. The switching is accomplished by the polarization rotator, R. If it has a net rotation of zero, as in part (a), all the light will exit at output A. If the rotator has a net rotation of 90°, as in part (b), all the light will exit at output B. (The figure omits the GRIN-rod lenses that would be used on the inputs and outputs.)

One type of polarization rotator, which has been used to demonstrate a switch of the basic design shown in Fig. 10.24, is a liquid crystal twist cell. These cells can be arranged to produce a 90° polarization rotation with no voltage applied. When a voltage is applied through transparent electrodes, the liquid crystal molecules reorient, so that the cell produces a net rotation of zero. An experimental model of such a switch had a switching time of ~ 100 msec, but only required 2.5 volts to operate it (Wagner and Cheng, 1980). The speed was limited by the particular liquid crystal material used, and substantial improvements seem possible.

The other type of polarization rotator, which has been used for optical switching, is based on the Faraday effect, described in Section 10.2.3.3. Because

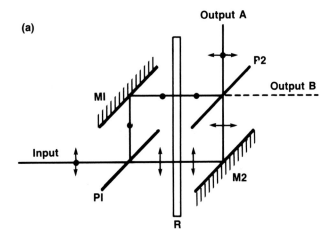

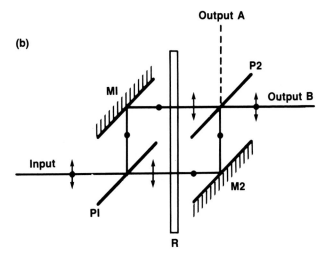

Fig. 10.24 Schematic cross section of a polarization-independent switch using a polarization rotator. The polarizing beamsplitters transmit light polarized in the plane of the figure (indicated by the arrows), and reflect light polarized perpendicular to the plane of the figure (indicated by the dots). (a) With the polarization rotator set for zero net rotation, both polarization components of the input beam are directed to output A. (b) With the rotator set for a 90° rotation, both components are directed to output B.

f the saturation and hysteresis of the magnetization of materials with large araday rotation coefficients (Verdet constants), these rotators have maximum eproducibility when driven to saturation, of either sign. The magnetooptic olarization switches that have been reported all use a passive 45° polarization otator and a magnetooptic plate, which produces a rotation of ±45°. The ombination then produces a rotation of 0° or 90°. This type of switch has been xtensively developed as a single-mode source-sparing switch for undersea cable pplications. Losses as low as 1.2 dB have been reported; and, by using a emi-hard magnetic material to provide the magnetic field, the switch can be nade bistable, such that drive power need only be provided to change the state f the switch (Shiraski *et al.*, 1984).

Because of the non-reciprocal nature of Faraday rotation, switches based on araday rotation can also serve as polarization-independent isolators. A device f the design shown in Fig. 10.24, with a passive +45° rotator and a +45° araday rotator, will direct an input beam to output port B. A reflected beam nput through that port will experience a +45° rotation in the Faraday rotator nd a −45° rotation in the passive rotator, for a net rotation of zero, and will be lirected to the unmarked port below beamsplitter P1. This device is actually a irculator (Emkey, 1983) with the circulation direction determined by the sign of he magnetic field on the Faraday rotator. The full power and versatility of this levice have yet to be exploited.

0.2.4 Concave-Mirror Components

10.2.4.1 Directional Couplers and Filter Multiplexers. The directional couplers nd filter multiplexers previously described can also be implemented using oncave mirrors as the focusing/collimating elements, instead of GRIN-rod enses. A schematic diagram of such a device is shown in Fig. 10.25. Operation s a directional coupler or as a wavelength multiplexer is determined by the vavelength dependence of the coating on the 1st spherical reflector. (Switching an also be obtained by a mechanical motion of a mirror.) Because the mirrors perate in reflection, the device is analogous to the single-ended GRIN-rod-lens nultiplexer illustrated in Fig. 10.12, and the two devices require the same umber of piece parts. The mirror device requires three spherical surfaces, while he lens device requires only plane surfaces and a GRIN-rod lens. Reflectors do ot suffer from chromatic aberration, which can be a problem with GRIN-rod enses. When used with input and output fibers located near the center of urvature of the mirror (as in Fig. 10.25), spherical mirrors are free of spherical berration, but they must be used very close to on-axis to avoid excessive stigmatism. For minimum loss the mirror devices require that the fibers be in a onparallel array (as illustrated in Fig. 10.25), which complicates fabrication, vhile for GRIN-rod-lens devices a parallel array is optimum (Tomlinson, 1980).

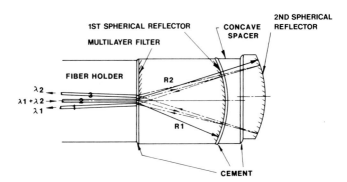

Fig. 10.25 Schematic cross section of a directional coupler or wavelength multiplexer using spherical concave reflectors. (Figure courtesy of N. S. Kapany, Kaptron, Inc.)

(This angular alignment is much more critical for single-mode fibers than for multimode.) Multimode devices of the design illustrated in Fig. 10.25 have been offered commercially with typical insertion losses of ~ 1 dB. Just recently some single-mode devices, which are apparently of the same design, have also been offered. On balance, there do not appear to be any fundamental reasons why either mirror devices or GRIN-rod-lens devices should be superior; thus, any choice between them must be made on the basis of the comparative performance and price of specific components.

There is, however, another type of coupler/multiplexer using concave mirrors, which merits careful attention. Figure 10.26 shows a schematic perspective of a device using molded off-axis parabolic reflectors. The use of parabolic reflectors makes it possible to use them far off-axis, thus gaining flexibility in the overall configuration of the device without suffering the large off-axis aberrations of spherical reflectors. This flexibility makes it possible to use a configuration in which three reflectors, plus alignment surfaces for mounting sources, detectors, a connector, and a beamsplitter, are all molded simultaneously in a single piece. The currently available devices are molded in plastic and have been designed for bidirectional wavelength-multiplexed transmission using LED sources and multimode fiber. Bidirectional transmission is considered attractive for short data links, because it requires half as much fiber and (more importantly) half as many connectors as a pair of unidirectional links. The reported excess loss of the molded device, as compared to a simple unidirectional link, is 3 dB, which is probably quite acceptable for many short-haul data link or local area network applications (Roberts and Rando, 1986). We are not aware of any attempts to use single-piece molded-mirror devices with single-mode fibers; it may be that the current molding technology is not capable of meeting the required tolerances for such complicated pieces. For high-volume applications, however, this approach would seem to be worthy of further investigation.

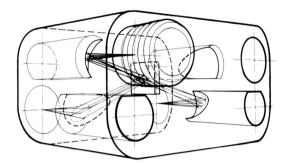

Fig. 10.26 Schematic perspective view of a molded coupler, for bidirectional wavelength-multiplexed transmission, using off-axis parabolic reflectors. The circular object at the front upper left is a fiber connector. Ray paths show input light diverging from the end of the fiber, and being collimated by one of the parabolic reflectors. The collimated beam passes through the filter (the square plate in the center of the device), and is focused on a detector by a second parabolic reflector. The detector is mounted in a recess (not detailed) in the back lower right corner. Light from a source mounted in a similar recess in the back lower left corner is collimated by a third parabolic reflector, reflected by the filter, and focused on the fiber by the first reflector (Figure courtesy of H. Roberts, ADC Fiber Optics Corp.)

10.2.4.2 Plane-Grating Multiplexers. While wavelength multiplexers using gratings have the feature that they are parallel devices and require only a single grating and a single focusing element independent of the number of channels, there are problems in making GRIN-rod-lens devices with large numbers of channels. As described before, this is because GRIN-rod lenses are not available in the desired sizes, and (in some cases) because of the chromatic aberration of the lenses. One can get around these problems by using concave mirrors as the focusing element, as illustrated in Fig. 10.27. In order to use the mirror as close to on-axis as possible, the fiber array enters through a small hole in the grating. When used in this configuration, a concave mirror will exhibit spherical aberration (unless it is parabolized), but this can be canceled by using the configuration shown at the bottom of Fig. 10.27. The detailed shapes of the passbands are determined by the characteristics of the input and output fibers, as described above in Section 10.2.3.6. A wide variety of devices, using multimode and single-mode fibers, with up to 49 channels, has been reported (Laude and Lerner, 1984). Insertion losses ranged from 0.5 dB for a two-channel demultiplexer with single-mode input and multimode output, up to 7 dB for a 42-channel multimode multiplexer. Devices of this configuration are thus very attractive for wavelength multiplexed systems with many channels, and a variety of such devices have been offered commercially.

10.2.4.3 Concave-Grating Multiplexers. In a sense, the simplest grating multiplexer uses a concave grating, such that a single element provides both the

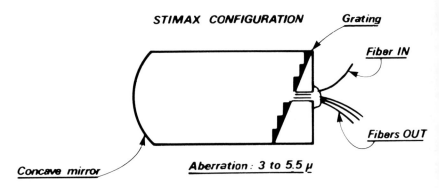

STIMAX CONFIGURATION

Grating

Fiber IN

Fibers OUT

Concave mirror

Aberration : 3 to 5.5 μ

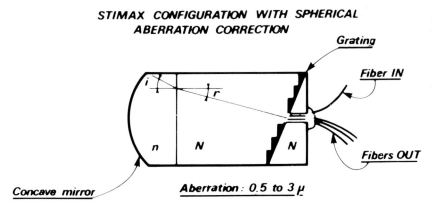

**STIMAX CONFIGURATION WITH SPHERICAL
ABERRATION CORRECTION**

Grating

Fiber IN

Fibers OUT

Concave mirror

Aberration : 0.5 to 3 μ

Fig. 10.27 Schematic cross sections of wavelength multiplexers using plane gratings and concave mirrors. The design at the bottom provides compensation for the spherical aberration of the spherical mirror. (Figure courtesy of Instruments SA/Jobin-Yvon.)

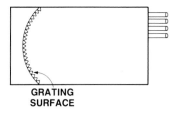

**GRATING
SURFACE**

Fig. 10.28 Schematic cross section of a wavelength multiplexer using a concave grating.

wavelength selectivity and the focusing action, as illustrated in Fig. 10.28. Concave gratings have an important place in the history of atomic spectroscopy, but there are some important problems in using them for fiber multiplexers. To provide the required resolution, concave gratings must be used off-axis, and thus suffer from substantial astigmatism. With a numerically-controlled ruling engine, it is possible to rule aberration-corrected concave gratings, and such a grating was used to make a 10-channel multimode demultiplexer with losses of less than 2.5 dB (Watanabe *et al.*, 1980a). Aberration-corrected concave gratings can also be fabricated holographically and such gratings have been offered commercially (Mikes and Shafer, 1983). Despite the apparent simplicity of concave-grating devices, concave gratings are more difficult to fabricate than plane gratings, and it seems likely (but not certain) that plane-grating devices will be easier to produce.

10.2.5 Lensed-Fiber Components

Techniques for producing hemispherical ends on fibers have been developed to improve laser-fiber coupling efficiencies (Eisenstein and Vitello, 1982; Khoe *et al.*, 1983; Lipson *et al.*, 1985b), and these lensed fibers have been used in two different types of variable attenuators.

Miniature variable attenuators of the basic design illustrated in Fig. 10.5, but with lensed fiber ends rather than GRIN-rod lenses, have been offered with multimode fibers and with single-mode input and multimode output fibers. These have been intended for circuit board mounting, with a package design similar to that illustrated in Fig. 10.7, except that the input and output fibers are on opposite sides of the package.

A quite different attenuator has been made by butting the lensed fiber ends against each other in a V-groove, as illustrated in Fig. 10.29 (Curtis and Young, 1985). Tilting the V-groove varies the angle between the axes of the two fibers, and thus varies the loss. With single-mode fibers the typical minimum loss is ~ 0.5 dB, and the loss can be varied smoothly and reproducibly to at least 70 dB. These attenuators have not been offered commercially, but they have been widely used in laboratory experiments.

10.2.6 Features and Disadvantages

A major feature of microoptic components is that they are available commercially, and most types are available from more than one source. They are also available in versions to perform each of the basic functions, while providing generally good performance with either multimode or single-mode fibers. The major disadvantage of microoptic components is that manufacturing them requires fabrication of multiple individual piece parts, and precision alignment

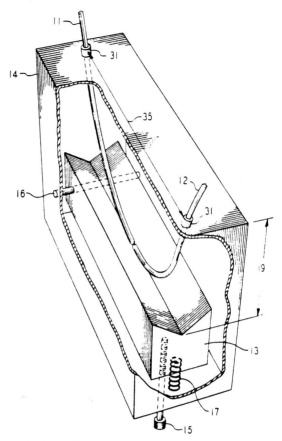

Fig. 10.29 Attenuator using lensed fiber ends (Curtis and Young).

and assembly operations. If a high-volume demand for such components develops, many of the manufacturing processes could be automated, but it seems unlikely that the costs could be reduced to the levels necessary for many applications in local distribution and local area networks. This problem has led to a number of efforts to develop components that are more suitable for high-volume low-cost manufacture, and such components are described in the following three sections. Another disadvantage of microoptic components is that they are not well suited for performing multiple functions. Connecting several components together typically results in an overly complex device, or excessive insertion loss, or both. Solutions to this problem are discussed in Sections 10.4 and 10.5.

10.3 LENSLESS COMPONENTS

10.3.1 Introduction

One approach to designing fiber components, which are simpler to fabricate than microoptic components, is simply to eliminate the lenses. In this section we consider components that do not include lenses (or other focusing elements, such as concave mirrors), but which involve the transfer of light between fibers through cleaved or cut and polished end faces. In Section 10.4 we will consider components in which the desired optical function is accomplished entirely within a single fiber or by evanescent-wave coupling between two fibers.

A pair of fibers with flat end faces can typically be separated by ~ 100 μm (in air) before the loss will exceed 1 dB. This applies for single-mode or multimode fibers, and even greater separations can be tolerated if the space between the fibers is filled with a material with a refractive index greater than unity. Techniques have been developed for performing a variety of different optical functions within this limited space. There is, however, an important limitation to the lensless components. Since there are no lenses to provide collimated beams, the performance of angularly-sensitive elements, such as interference filters, will be limited by the angular spread of the light in the fibers.

10.3.2 Attenuators

Very simple compact fixed attenuators can be made by sandwiching a thin attenuating plate between two fibers or by coating an attenuator film on the end of one of the fibers. If the attenuator plate or film works by reflection (e.g., a thin metal film), the fibers should be cut at a small angle with respect to the fiber axis, such that the reflected light does not couple back into the core of the input fiber. Fixed attenuators of this type are available with either multimode or single-mode fibers, with an overall package only 3 mm in diameter and 38 mm long.

10.3.3 Isolators

An isolator requires three optical elements, a polarization rotator and two polarizers, as illustrated in Fig. 10.8. The required thickness of the rotator depends on its Verdet constant and its saturation magnetization. With some of the new rotator materials this thickness can be as small as 290 μm for a wavelength of 1.3 μm, and 85 μm for a wavelength of 0.78 μm (Shiraishi, 1985). Such rotators have been imbedded in fibers with resulting insertion losses ranging from 6.7 dB for a single-mode fiber at a wavelength of 0.78 μm (dominated by absorption loss), to 2.9 dB for a multimode fiber at a wavelength of 1.3 μm (dominated by diffraction loss), and further refinements have been projected (Shiraishi, 1985). (The high refractive index of the rotator material, $n = 2.4$, has the effect of reducing the diffraction losses.) For an isolator one also needs polarizers, and infrared polarizers as thin as 20–30 μm have been reported (Shiraishi *et al.*, 1986), but it does not appear that they have been used to make complete isolators. It seems clear that reasonably efficient lensless isolators can be developed, but since isolators are most likely to be needed immediately adjacent to laser sources, in-fiber isolators may have fewer advantages than other types of lensless components.

10.3.4 Directional Couplers and Wavelength Multiplexers

A lensless directional coupler can be made by inserting the beamsplitter into the fiber in a lensless version of the coupler illustrated in Fig. 10.10. This approach was implemented by molding triangular pieces on fibers, and then polishing the surfaces so that the fiber end is centered in the apex of a 90° angle. Four such pieces, and a beamsplitter coating, were assembled into four-port directional couplers, as illustrated in Fig. 10.30, and several other related devices were also demonstrated (Wagner, 1979). This approach has the feature that it uses molding, which is generally advantageous for high-volume production, but has the disadvantage that it requires precision polishing operations. No components based on this design have been offered commercially.

There is, however, a lensless directional coupler that has been the subject of extensive development programs, and that has been offered in commercial products. That coupler is a two-channel wavelength multiplexer of the basic

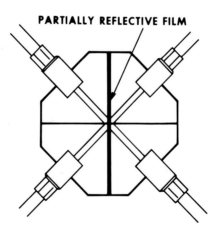

PARTIALLY REFLECTIVE FILM

Fig. 10.30 Lensless four-port directional coupler using molded fiber holders (Wagner, 1979).

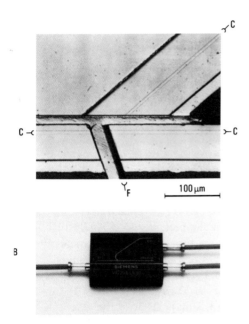

Fig. 10.31 (a) Cross section of a lensless wavelength multiplexer using single-mode fiber (Winzer, 1984). The fiber cores are indicated by C, and the interference filter by F. (b) Photograph of a multiplexer of the design illustrated above (Winzer, 1984).

design illustrated in Fig. 10.31. The beamsplitter coating is inserted in the in-line fiber at an angle chosen to be an optimum between minimizing the polarization dependence of the coupler, and minimizing the coupling loss to the branching fiber. This approach was originally developed for multimode devices, but has been adapted for single-mode fiber by polishing away one side of the in-line fiber, so that the core of the branching fiber can be brought into close proximity to the core of the in-line fiber, as shown in Fig. 10.31. Single-mode devices have been reported with a channel spacing of 40 nm, and insertion losses of ~ 0.5 dB (Winzer, 1984). These are quite impressive results, but it should be pointed out that: because of the angular spread of the light in the input fiber, it will not be possible to achieve substantially smaller channel spacings; and the device is intrinsically a two-channel device, so for multi-channel wavelength multiplexing several devices must be used in series. While these lensless multiplexers require relatively few piece parts, it should be clear from Fig. 10.31 that they do require precision polishing, aligning, and assembly operations, in addition to the problems listed before. The compensating feature of these devices is that multiple devices (perhaps 20) can be fabricated in parallel, such that the cost of the precision operations is shared among many devices, which are then cut apart in the final step of fabrication.

A variety of lensless couplers/splitters/stars has been proposed by Severin and collaborators, using what they call the "fused-head-end" technique (Severin *et al.*, 1986; Severin, 1986). In this technique, fibers with their cladding etched away are fused together, cut (and polished?), and mated with other fibers. The technique is only useful for multimode fibers, but it has been suggested that it could result in low-cost components essential for practical fiber local area networks (Severin, 1986). Relatively little experimental data on such components have been published, and no commercial versions have been offered.

10.3.5 Switches

A considerable variety of lensless switches, which use V-grooves to align a moving fiber with any of two or more fixed fibers, has been demonstrated. An early, but typical, implementation is shown in Fig. 10.32 (Hale and Kompfner, 1976). In this case a glass tube with a square bore provides four V-grooves, although the figure shows only two of them being used. The moving fiber has a small metal sleeve attached to it so that a magnetic field can be used to move it to align with either of the two fixed fibers. An alternative technique for actuating such a switch is simply to bend the glass tube. Switches of this type have been demonstrated with multimode and with single-mode fibers, and reasonably low insertion losses have been reported. They have not been developed into commercial products, however, probably because of concerns about their long-term reliability and the difficulty of fabricating them. The bare unsupported moving fiber seems susceptible to breakage, and rather small particles can prevent the moving fiber from properly aligning in the V-grooves.

A lensless switch, which avoids these problems, is illustrated in Fig. 10.33 (Young and Curtis, 1981). This switch, commonly referred to as a "Si-chip switch," makes use of the precision-grooved Si chips that were developed for multi-fiber connectors. The moving fiber is cemented between two chips, and this assembly is then moved to align with either of the fibers in a fixed array, also held between chips. The outer chips provide alignment surfaces. Since the chips have multiple grooves (typically 12), they can hold multiple fibers and thus make multi-pole switches. Note that the moving fibers are secured within the chip assembly, so there are no bare or unsupported fibers. Also, since the alignment surfaces have macroscopic areas, it is possible to apply sufficient mechanical force to ensure precise and reproducible alignment of the moving fibers. Average insertion losses of 0.2 dB for multimode fibers, and 0.4 dB for single-mode fibers, have been achieved. These low losses make it feasible to cascade switches. For example, a 1×2, a 2×2, and a 4×2 switch have been connected together to obtain a 1×8 switch. A high-reliability Si-chip switch has been developed for possible use as a source-sparing switch for undersea cable applications. A general feature of lensless switches is that they are almost totally

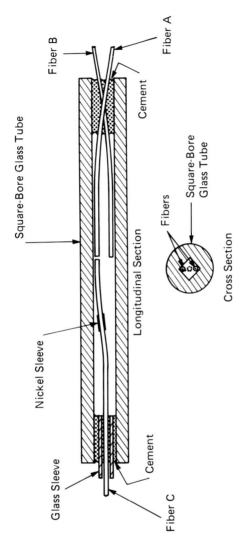

Fig. 10.32 Moving fiber switch (Hale and Kompfner, 1976).

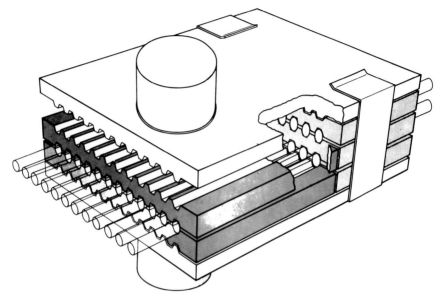

Fig. 10.33   Moving fiber array switch using Si chips to hold and align the fibers (Young and Curtis, 1981).

insensitive to the characteristics of the signals they are transmitting and can switch any signals that can be transmitted by the fibers used in the switch. Despite their excellent performance, Si-chip switches have not been offered as commercial products, probably in part because they are somewhat difficult to assemble and possibly because mechanical switches do not appear to be sufficiently "high tech." None the less, these switches have some important features, and they merit serious attention.

10.3.6   Features and Disadvantages

Lensless components are difficult to summarize. Lensless fixed attenuators are clearly simpler than microoptic versions and offer equivalent performance in a smaller package. Lensless wavelength multiplexers are not much simpler than microoptic versions and are not suitable for closely-spaced channels or for multiple channels, but the possibility of parallel fabrication of multiple devices may give them a decisive advantage for some applications. Lensless switches are more complicated than might be desired, but their reliability and independence of the signals being transmitted may make them a good investment for systems for which it is difficult to predict future demands. Lensless isolators seem effective, but may not be suitable for what is likely to be the greatest need for isolators. Note that each type of lensless component and application requires somewhat different fabrication technology and development; thus, a successful application of one lensless component may have relatively little effect on the attractiveness of other lensless components.

10.4 ALL-FIBER COMPONENTS

10.4.1 Introduction

The next level in the search for simplicity in fiber components is to eliminate not only the lenses, but also fiber ends, by performing the desired optical function entirely within a single fiber or by evanescent-wave coupling between two fibers. In addition to its simplicity, this approach has the feature that since the components are made of fiber, the input and output coupling losses can be made vanishingly small, and for many all-fiber components this is actually the major feature. All-fiber components are being actively developed for a wide variety of applications, particularly for sensors, but we have concentrated on the components that appear to be of greatest importance for telecommunications applications.

10.4.2 Polarization controllers

There is an increasing need for devices to control the polarization of the light in a fiber, particularly for systems using coherent detection, and a number of all-fiber components have been proposed (Okoshi, 1985). Polarization control is primarily an issue for single-mode systems, and we only consider approaches applicable to single-mode fiber.

In general, polarization controllers need at least two independently controllable birefringent elements. These can either have fixed orientations and variable

Fig. 10.34 All-fiber polarization controller. (Figure courtesy of R. A. Bergh, Fibernetics, Inc.)

amplitudes, or fixed amplitudes and variable orientations. All-fiber variable-amplitude birefringent elements have been made by mechanically squeezing a fiber, or by mechanically stretching a birefringent fiber. A fixed-amplitude birefringence can be induced in a fiber by winding it into a coil of the appropriate diameter, and the orientation of the birefringence can be adjusted by rotating the plane of the coil about the axis of the input and output fibers. A polarization controller based on this approach is illustrated in Fig. 10.34. Controllers of this type have been offered as a commercial product and have been widely used in laboratory experiments. None of the available techniques, however, meet all the requirements for systems applications, particularly for coherent detection, and further research in this area is needed (Okoshi, 1985).

10.4.3 Isolators

For an isolator one needs a Faraday rotator (a circular birefringence). Conventional silica fibers have a finite, but small, Verdet constant; thus, it is possible to make a polarization rotator by applying a magnetic field to a fiber. The problem is that fibers tend to have a non-negligible linear birefringence, and this birefringence interacts with the Faraday rotation such that large rotations cannot be obtained.

One solution to this problem is to use a fiber with a large birefringence and to split the magnetic field into many short sections, spaced by the birefringence beat length. The result is that the small rotations produced by each magnet can add to give as large a rotation as is desired. A rotator of this type, made with 14 magnets and a fiber looped back and forth to make 9 passes through the magnets (a total fiber length of 7 m), had a 45° rotation at a wavelength of 0.6328 μm, with negligible excess loss (Turner and Stolen, 1981). Clearly this is not an ideal device for communications systems, but because of its low loss, it has been used within the cavity of a fiber Raman ring laser to give a cavity with a high input coupling efficiency for the pump light and a high Q for the stimulated Raman light.

Another solution involves inducing a large circular birefringence in the fiber. As long as the circular birefringence is much larger than the linear birefringence, the only effect of the linear birefringence is a small increase in the insertion loss, and a periodic magnetic field is not required. A large circular birefringence can be induced in a fiber by twisting it as it is wound into a coil (Ulrich and Simon, 1979). Alternatively, the twist can be built into the fiber by rotating the preform as the fiber is drawn (Hussey et al., 1986). The magnetic field can be provided by a toroidal coil wound on the fiber coil. An isolator of this type, using 140 m of fiber wound in a 9-cm-diam. coil and using a fiber polarizer, has recently been reported, with an isolation of 22 dB at a wavelength of 1.32 μm, and a total insertion loss of 2.7 dB (of which 1.4 dB was from splice losses at the input and output) (Warbrick, 1986).

10.4.4 Directional Couplers

By far the largest use of all-fiber components has been for directional couplers of the type usually referred to as fused biconical taper couplers, or fused-fiber couplers. These are made by twisting or pressing together two (or more) bare fibers, heating them to above their softening temperature so that they fuse, and then stretching them such that the fused region is drawn down to a smaller diameter (Kawasaki *et al.*, 1983). This technique has been successfully applied to multimode and single-mode fibers, but the explanation of their operation is quite different for the two types of fiber.

For a multimode coupler, on the input side the input fiber has a steadily decreasing core diameter, and thus can support fewer and fewer modes. As each mode reaches cutoff, the light in that mode is radiated into the larger multimode guide formed by the fused claddings of the two fibers, until essentially all of the input light is traveling in the cladding. On the output side, as the core diameters increase, light is coupled back into the cores, dividing between the cores approximately in the ratio of their areas. Because of the smooth taper, there is almost no light reflected back to the input fiber and very little excess loss. In addition to the basic 2 × 2 configuration using two fibers, higher-dimension couplers can be made by using more fibers. The fabrication techniques for these couplers have obviously been automated, because they are widely available at relatively low cost. Numerous suppliers quote excess losses of < 1 dB, and port configurations of up to 64 × 64 are available. A major application of these couplers is in multimode local area networks.

For a single-mode fused coupler the operation is somewhat more subtle. Since the input fiber is already single-mode and the fundamental mode of a fiber does not have a cutoff, the light remains in that mode, but the mode size increases. As the two cores get closer together, their modes will begin to overlap, and light will then couple coherently from one core to the other in exactly the same manner as in the integrated optical directional coupler described in the preceding chapter. The light will oscillate between the two cores, with the final split determined by the detailed structure and length of the coupling region. To achieve a particular splitting ratio, it would seem that control of the fabrication process is more critical than for multimode couplers, but these problems appear to have been solved, and single-mode couplers are available from a number of suppliers. Typical excess losses are < 1 dB, with some suppliers offering selected units with < 0.1 dB excess loss. Couplers are available in package sizes as small as 2.8-mm diameter by 4 cm long, as illustrated in Fig. 10.35. Recently, single-mode couplers have been offered that are made from polarization-preserving fiber, and that preserve the polarization of the input signals. With polarization-preserving fiber, it is also possible to make a polarization-sensitive coupler, which functions as a polarizing beamsplitter (Yokohama *et al.*, 1985). Some higher-dimension single-mode couplers have been offered, but with more than three

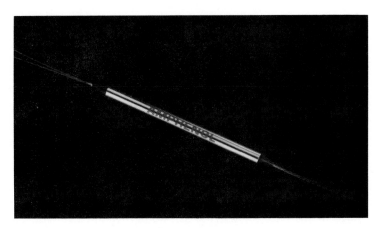

Fig. 10.35 Single-mode fused-fiber directional coupler. (Figure courtesy of Allied Amphenol Products.)

single-mode fibers it is very difficult to control the splitting ratios. For a $1 \times n$ splitter, multiple 1×2 couplers can be connected together in a tree structure, and they can be fabricated with continuous lengths of fiber, so that there are no splices between the couplers.

A somewhat different type of all-fiber single-mode coupler is illustrated in Fig. 10.36. These couplers are made by cementing a fiber in a slightly curved slot in a glass block, and then polishing the block and fiber until the fiber core is very close to the surface. Two such blocks are then placed together with the fiber cores parallel to each other. When the blocks are aligned so that the cores are close together, light couples between the cores, exactly as in the fused coupler (Bergh *et al.*, 1980). The major feature of this design is that the splitting ratio can be adjusted by varying the separation between the cores. Variable couplers of this design have been offered commercially and have been widely used in laboratory experiments. Half of such a coupler can function as a fiber polarizer by placing a birefringent material on the polished surface, so that one polarization sees a refractive index lower than that of the fiber core and remains guided, while the other polarization sees a higher index and is radiated out of the fiber (Bergh *et al.*, 1980a).

10.4.5 Wavelength Multiplexers

In single-mode directional couplers of the fused fiber or polished fiber design, the coupling ratio is a sinusoidal function of the coupling strength, which is approximately inversely proportional to wavelength, so the coupling ratio is approximately a sinusoidal function of wavelength. This wavelength dependence

a.

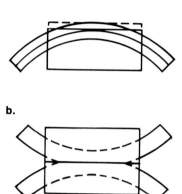

b.

Fig. 10.36 Schematic cross sections showing the construction of an all-fiber directional coupler. (a) A fiber is cemented in a glass block, and the block polished until part of the fiber core is near the surface. (b) Two such blocks are assembled such that light can couple between the fiber cores.

is frequently a problem, but it can be exploited for wavelength-division multiplexing. By making the coupling strength large enough, one can arrange for the coupling ratio to be near unity for one wavelength and near zero for another wavelength. Figure 10.37 shows experimental data for such a fused-fiber coupler designed for multiplexing signals at 1.3 and 1.5 μm (Lawson et al., 1984), and multiplexers of this design have been offered commercially. Fiber wavelength multiplexers have also been made by fabricating a coupler from nonidentical fibers, designed such that their propagation constants are quite different, except over a specified wavelength region (Zengerle and Leminger, 1986). A limitation of these techniques is that they are not well suited for closely-spaced or multiple channels. A narrowband (0.6 nm passband) wavelength multiplexer has been demonstrated by etching a grating into one side of a polished-fiber coupler, so that light at the desired wavelength is reflected by the grating and couples into the other fiber (Whalen et al., 1986).

10.4.6 Features and Disadvantages

The fact that all-fiber components can achieve lower losses and reflectivities than any other type of components makes them the optimum choice for applications for which the overriding consideration is obtaining the lowest possible loss. The prime example of such an application is a fiber ring-resonator gyro. In this application it is also common to take advantage of the possibility of fabricating several components on the same piece of fiber, so as to obtain essentially lossless, reflectionless connections between multiple components. For

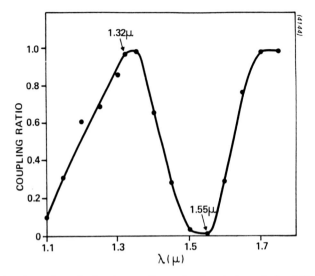

Fig. 10.37 Experimental results for the coupling ratio of a single-mode fused-fiber directional coupler as a function of wavelength (Lawson *et al.*, 1984). This coupler was designed for multiplexing signals at 1.3 and 1.5 μm.

other applications where insertion loss is not quite so critical, however, the complexity of fabricating several devices on the same fiber is unlikely to be cost effective. The fused-fiber directional couplers are clearly very successful devices, and likely to remain so, because they offer very good performance at reasonably low cost. For applications that require using several components connected together, however, the integrated optics techniques described in the following section may eventually be preferable. For polarization control, none of the available approaches is entirely satisfactory, and integrated optics approaches may eventually be the optimum choice.

10.5 INTEGRATED-OPTICAL COMPONENTS

10.5.1 Introduction

It seems likely that integrated optical components will play a major role in future optical communications systems. As described in more detail in the preceding chapter, integrated optical components are based on forming patterns of optical waveguides in the surface of a planar substrate. By arranging for pairs of these guides to come close to each other, and/or by applying electric fields, various optical functions can be performed. The guides also serve to interconnect these elements, so a component involving multiple individual optical elements can be fabricated as a single piece. A major feature is that, since integrated optical circuits are fabricated by the same types of processes as are used for integrated electronic circuits, high-volume low-cost manufacture should be possible. Since the basic concepts and techniques of integrated optics have been covered in the previous chapter, and since the following chapter is devoted to describing active integrated optical devices, in the present chapter we will only consider passive integrated optical components, and our treatment will be quite brief. Readers who do not have some acquaintance with integrated optical components are likely to find Chapter 9 a useful preface to this section.

10.5.2 Ion-Exchanged Glass Components

Ion exchange in glass substrates (see Chapter 9) is an integrated optics fabrication process with good potential for scale-up to high-volume manufacture, but since glass does not exhibit an electrooptic effect, this process is not generally useful for making active components, and it has probably received less attention than it deserves. A particular feature of the process is that it is capable of making multimode waveguides suitable for use with typical multimode fibers.

10.5.2.1 Multimode Components.   The multimode integrated optical component, which has received the most attention, is the $1 \times n$ splitter, made by combining a number of 1×2 Y-branch splitters—most probably intended for application in multimode fiber local area networks. By careful control of the process to give guides with a refractive index distribution closely matching that of multimode fibers, a 1×16 splitter has been made with a fiber-splitter-fiber excess loss of only 2.5 dB, and ± 0.6 dB uniformity between output ports (Cline, 1986). When used with an LED source, such a splitter can accept light from the LED in high-order modes, which cannot be accepted by a fiber, and convert that light to lower-order modes at the output, which can then be accepted by a fiber. The LED-splitter-fiber excess loss was only 0.2 dB (as compared to LED-fiber coupling).

Glass multimode integrated optical waveguides, fabricated by ion exchange, have recently been used in wavelength-division multiplexers, in which interference filters are inserted into thin slots cut across waveguide junctions (Seki *et al*, 1987). Optically these devices are similar to the lensless multiplexer illustrated in Fig. 10.31, and multiple devices can be fabricated in parallel and then cut apart, but the integrated version requires fewer polishing and alignment operations. Multimode waveguide wavelength multiplexers have also been made using waveguides defined by reactive ion etching of a planar SiO_2-TiO_2 waveguide grown on a Si substrate (Kawachi *et al.*, 1985). In addition to defining the waveguides, the etching process is used to form grooves to position the fibers and the multilayer dielectric filters.

10.5.2.2 Single-Mode Components. Single-mode waveguides, which are a good match to typical single-mode fibers, can be made in glass by ion exchange, and thus a variety of single-mode passive components can be fabricated in this way including single-mode versions of the glass wavelength multiplexer described above. Other examples include 1 × *n* splitters similar to the multimode component described above, and waveguide concentrators for grating wavelength multiplexers, as illustrated in Fig. 10.19. The concentrator illustrated in Fig. 10.19 was actually made by Ti indiffusion in lithium niobate, but that process was only chosen because it was more readily available to those working on that particular multiplexer. A glass concentrator could provide equivalent or better performance, and in large volumes would probably be lower in cost. Similarly, the wavelength-selective directional coupler, which has been demonstrated in lithium niobate and is described in detail in Chapter 11, could also be fabricated in glass, possibly at lower cost. Devices like the wavelength-sensitive coupler are quite sensitive to the waveguide characteristics, but it has not yet been shown that the fabrication processes can be sufficiently reproducible for high-yield manufacture. This issue is currently the subject of active research.

Since we have stressed that glass integrated optical components are passive devices, we should give one counter example of a "somewhat active" glass component. It has been shown that a glass Mach Zehnder waveguide interferometer (see Chapter 9) can be tuned by using the power dissipated in an electrode on one of the branches of the interferometer to induce a thermal gradient between the branches, which results in a phase difference because of the temperature dependence of the refractive index of the glass. Devices of this type, illustrated in Fig. 10.38, have exhibited response times of a few msec and have been described as electrically-controllable variable attenuators (Jackel *et al.*, 1985).

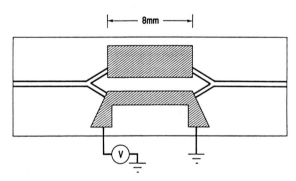

Fig. 10.38 Schematic drawing of a thermally-tuned Mach-Zehnder interferometer, which can be fabricated in glass by ion exchange (Jackel *et al.*, 1985). This device has the feature that it is almost polarization independent.

10.5.2.3 Lithium Niobate Components. As detailed in Chapters 9 and 11, the most extensively developed technology for fabricating integrated optical components is based on the diffusion of Ti into crystals of lithium niobate ($LiNbO_3$). The achievable Ti concentration raises the refractive index sufficiently to make single-mode waveguides with mode sizes similar to those of typical single-mode fiber, and the guides have low propagation losses. The major feature of lithium niobate is that it has a large electrooptic coefficient, and thus can be used for electrically controllable components, such as fast modulators and switches. It is obvious that, in addition to using passive waveguides to interconnect various active elements on the same crystal, lithium niobate can be used to make passive elements, such as splitters and wavelength multiplexers, and it is this capacity for integrating multiple active and passive elements on a single chip that is one of the major motivations for the development of integrated optical techniques. As in the case of the waveguide concentrator described before, lithium niobate has been used to demonstrate completely passive integrated optical components, because the fabrication techniques are more fully developed than other techniques, such as ion exchange in glass. Further development is necessary to insure that the fabrication processes can be sufficiently reproducible for high-yield manufacture of devices, such as wavelength-sensitive couplers, which are particularly sensitive to the waveguide characteristics, but several companies are actively engaged in such development programs. (For experimental investigations of new types of passive integrated optical components, lithium niobate has the important feature that one can add electrodes to a device and use an electric field to trim the performance of the device to compensate for inaccuracies in the fabrication processes.) For passive components that are very sensitive to the waveguide characteristics, the choice between glass and lithium niobate will be largely determined by the reproducibility of the fabrication processes; the jury is still out.

Waveguides can also be made in lithium niobate by ion exchange—usually referred to as proton exchange, because the process is a H-Li exchange (Jackel *et al.*, 1982). This process only increases the extraordinary index, so it is only useful for single-polarization devices, but it is capable of achieving an index difference about an order of magnitude larger than is possible with Ti indiffusion. The potential of the proton-exchange process for integrated optical components has yet to be fully explored, and it is the subject of active research at a number of laboratories.

10.5.2.4 Semiconductor Components. A variety of techniques have been reported for fabricating integrated optical waveguides in semiconducting substrates. There are a number of reasons why it is difficult to make good waveguides in semiconducting materials, but considerable progress has been made in the past few years, and the best waveguides in semiconductors now have losses similar to those for guides in lithium niobate or glass (Kapon and Bhat, 1987) (Deri *et al*, 1987). The motivation for the work on semiconducting materials is not that this is a good way to make a waveguide, but to fabricate monolithically integrated optoelectronic devices with integrated optical components and associated integrated electronic circuits all made on the same chip. This is an attractive and exciting prospect, but since the field is still in an early stage and because it seems unlikely that semiconductors will be attractive for use in completely passive integrated optical components, we will not consider them any further in this chapter.

10.5.2.5 Features and Disadvantages. Integrated optical components have the feature of easily integrating multiple passive and active elements on a single chip, and the promise of low-cost high-volume manufacture. To achieve that goal, however, requires large investments in capital equipment and in process development, and a high-volume application. Several companies are beginning to offer Ti:LiNbO$_3$ modulators and other devices, but these are still essentially samples, and a high-volume application has not developed yet. Assuming that integrated optical chips can be manufactured at low cost, fiber attachment and packaging are likely to be the dominant cost factors. This is an area that requires much more research and development.

10.6 SUMMARY AND CONCLUSIONS

The development of optical components for fiber systems began with microoptic components, which are conceptually simple and are capable of providing good performance, but which seemed less than ideal for high-volume low-cost manufacture. This led to the development of various lensless and all-fiber components, which are somewhat easier to fabricate but which cannot easily provide all of the desired functions. The current research trend is to integrated optical components, which are very difficult and complicated to fabricate, but have the potential (not yet realized) for parallel fabrication of multiple units at very low cost per unit—provided there is a high-volume demand.

Microoptic components seem likely to continue to be useful, particularly in specialized applications requiring low to moderate volumes, but for high-volume applications they are likely to be displaced by other types of components. There is, however, one type of microoptic component that appears to have unique advantages. For wavelength-division multiplexing of multiple closely-spaced channels, only the microoptic grating multiplexers process all of the channels in parallel, and this could make them the optimum choice even for high-volume applications.

Lensless components are somewhat (but only somewhat) easier to fabricate than microoptic components, and it is difficult to assess their future role in telecommunications applications. Lensless switches (particularly of the "Si-chip" design) have the unique feature that they will work equally well for any signal that can be transmitted on fibers of the type used in the switch, but it is not clear what value systems designers will place on this feature.

All-fiber components have the unique feature that they can provide lower loss and reflectivity than any other approach; although this is less critical in telecommunications than in other areas (such as fiber sensors). For applications requiring a single directional coupler, however, the fused-fiber devices appear to be hard to beat.

Integrated optical components have shown great promise, but considerable research and development is still necessary to show if the fabrication technology can be made sufficiently reliable and reproducible to realize that promise.

REFERENCES

Bergh, R. A., Kotler, G., and Shaw, H. J. (1980). Single-mode fibre optic directional coupler. *Elect. Lett.* **16**, 260–261.

Bergh, R. A., Lefevre, H. C., and Shaw, H. J. (1980a). Single-mode fiber-optic polarizer. *Opt. Lett.* **5**, 479–481.

Chikama, T., Watanabe, S., Goto, M., Miura, S., and Touge, T. (1986). Distributed-feedback laser diode module with a novel and compact optical isolator for gigabit optical transmission systems. *1986 Optical Fiber Communication Conference*, Atlanta, GA, February 24-26, 1986, paper ME4.

Cline, T. W., and Jander, R. B., (1982). Wave-front aberration measurements on GRIN-rod lenses. *Appl. Opt.* **21**, 1035–1041.

Cline, T. W. (1986). Diffused channel glass waveguide optical splitters. *Digest of Technical Papers*, Topical Meeting on Integrated and Guided-wave Optics, February 26-28, 1986, Atlanta, GA, paper FDD5.

Curtis, L., and Young, W. C. U.S. Pat. 4,519,671.

Deri, R. J., Kapon, E., and Schiavone, L. M., (1987). Scattering in low-loss GaAs/AlGaAs rib waveguides. *Appl. Phys. Lett.* **51**, 789–791.

Eisenstein, G., and Vitello, D. (1982). Chemically etched conical microlenses for coupling single-mode lasers into single-mode fibers. *Appl. Opt.* **21**, 3470–3474.

Emkey, W. L. (1983). A polarization-independent optical circulator for 1.3 μm. *J. Lightwave Technol.* **LT-1**, 466–469.

Fujii, Y., Minowa, J., Aoyama, T., and Doi, K. (1979). Low-loss 4 × 4 optical matrix switch for fibre-optic communications. *Elect. Lett.* **15**, 427–428.

Hale, P. G., and Kompfner, R. (1976). Mechanical optical-fibre switch. *Electron. Lett.* **12**, 388.

Hussey, C. D., Birch, R. D., and Fujii, Y. (1986). Circularly birefringent single-mode optical fibres. *Elect. Lett.* **22**, 129–130.

Ishio, H., Minowa, J., and Nosu, K. (1984). Review and status of wavelength-division multiplexing technology and its application. *J. Lightwave Technol.* **LT-2**, 448–463.

Izumitami, T., Hirota, S., Ishibai, I., and Kobayashi, R. (1985). Precision molded aspheric lenses for a camera and for a compact disk system. *Proc. SPIE* **554**, 290–294.

Jackel, J. L., Rice, C. E., and Veselka, J. J. (1982). Proton exchange for high-index waveguides in LiNbO₃. *Appl. Phys. Lett.* **41**, 607–608.

Jackel, J. L., Hackwood, S., Veselka, J. J., and Beni, G. (1983). Electrowetting switch for multimode optical fibers. *Appl. Opt.* **22**, 1765–1770.

Jackel, J. L., Veselka, J. J., and Lyman, S. P. (1985). Thermally tuned glass Mach-Zehnder interferometer used as a polarization insensitive attenuator. *Appl. Opt.* **24**, 612–614.

Kapon, E., and Bhat, R., (1987). Low-loss single-mode GaAs/AlGaAs optical waveguides grown by organometalic vapor phase epitaxy. *Appl. Phys. Lett.* **50**, 1628–1630.

Kawachi, M., Yamada, Y., Yasu, M., and Kobayashi, M. (1985). Guided-wave optical wavelength-division multi/demultiplexer using high-silica channel waveguides. *Elect. Lett.* **21**, 314–315.

Kawasaki, B. S., Johnson, D. C., and Hill, K. O. (1983). Configurations, performance, and applications of biconical taper optical fibre coupling structures. *Can. J. Phys.* **61**, 353–360.

Khoe, G. D., Poulissen, J., and deVrieze, H. M. (1983). Efficient coupling of laser diodes to tapered monomode fibres with high-index end. *Electron. Lett.* **19**, 205–207.

Kubota, S. (1985). A lens design for optical disk systems. *Proc. SPIE* **554**, 282–289.

Laude, J.-P., and Lerner, J. M. (1984). Wavelength division multiplexing/demultiplexing (WDM) using diffraction gratings. *Proc. SPIE* **503**, 22–28.

Lawson, C. M., Kopera, P. M., Hsu, T. Y., and Tekippe, V. J. (1984). In-line single-mode wavelength division multiplexer/demultiplexer. *Elect. Lett.* **20**, 963–964.

Lipson, J., Harvey, G., Read, P. H., and Nicol, D. A. (1984). Two-channel single-mode wavelength-division-multiplexing devices. *Conference on Optical Fiber Communication* (OFC-84), New Orleans, LA, 23–25 January, 1984 paper WG2.

Lipson, J., Young, C. A., Yeates, P. D., Masland, J. C., Wartonick, S. A., Harvey, G. T., and Read, P. H. (1985). A four-channel lightwave subsystem using wavelength multiplexing. *J. Lightwave Technol.* **LT-3**, 16–20.

Lipson, J., Minford, W. J., Murphy, E. J., Rice, T. C., Linke, R. A., and Harvey, G. T. (1985a). A six-channel wavelength multiplexer and demultiplexer for single mode systems. *J. Lightwave Technol.* **LT-3**, 1159–1163.

Lipson, J., Ku, R. T., and Scotti, R. E. (1985b). Opto-mechanical considerations for laser-fiber coupling and packaging. *Proc. SPIE* **554**, 308–312.

Mikes, T. L., and Shafer, R. A. (1983). Wavelength division multiplexing systems using concave holographic diffraction gratings. *Proc. SPIE* **417**, 61–66.

Miller, S. E. (1965). Light propagation in generalized lens-like media. *Bell Sys. Tech. J.* **44**, 2017–2064.

Nishi, I., Oguchi, T., and Kato, K. (1985). Broad-passband-width optical filter for multi-demulti-plexer using a diffraction grating and a retroreflecting prism. *Elect. Lett.* **21**, 423–424.

Okoshi, T. (1985). Polarization-state control schemes for hetrodyne or homodyne optical fiber communications. *J. Lightwave Technol.* **LT-3**, 1232–1237.

Roberts, H., and Rando, J. (1986). WDM active coupler facilitates bidirectional transmission. *Laser Focus/Electro Optics* **22**, (4) 98–108.

Seki, M., Sugawara, R., Hanada, Y., Okuda, E., Wada, H., Yamasaki, T. (1987). High-performance guided-wave multi/demultiplexer based on novel design using embedded gradient-index waveguides in glass. *Electron. Lett.* **23**, 948–949.

Severin, P. J., Severijns, A. P., and van Bommel, C. H. (1986). Passive components for multimode fiber-optic networks. *J. Lightwave Technol.* **LT-4**, 490–496.

Severin, P. J. (1986). On the relation between passive optical fiber component properties and the network configuration. *J. Lightwave Technol.* **LT-4**, 1425–1433.

Shiraishi, K. (1985). Fiber-embedded micro-Faraday rotator for the infrared. *Appl. Opt.* **24**, 951–954.

Shiraishi, K., Sugaya, S., Baba, K., and Kawakami, S. (1986). Microisolator. *Appl. Opt.* **25**, 311–314.

Shirasaki, M., Nakajimi, H., Fukushima, N., and Asama, K. (1986). Broadening of bandwidths in grating multiplexer by original dispersion-dividing prism. *Elect. Lett.* **22**, 764–765.

Shiraski, M., Wada, F., Takamatsu, H., Nakajima, H., and Asama, K. (1984). Magnetooptical 2 × 2 switch for single-mode fibers. *Appl. Opt.* **23**, 3271–3276.

Sugie, T., and Saruwatari, M. (1983). An effective nonreciprocal circuit for semiconductor laser-to-optical-fibre coupling using a YIG sphere. *J. Lightwave Technol.* **LT-1**, 121–130.

Tamura, Y., Maeda, H., Satoh, N., and Katoh, K. (1986). Single-mode fiber WDM in the 1.2/1.3 μm wavelength region. *J. Lightwave Technol.* **LT-4**, 841–845.

Tomlinson, W. J. (1977). Wavelength multiplexing in multimode optical fibers. *Appl. Opt.* **16**, 2180–2194.

Tomlinson, W. J. (1980). Applications of GRIN-rod lenses in optical fiber communications systems. *Appl. Opt.* **19**, 1127–1138.

Tomlinson, W. J. (1980a). Aberrations of GRIN-rod lenses in multimode optical fiber devices. *Appl. Opt.* **19**, 1117–1126.

Tomlinson, W. J., Wagner, R. E., Stakelon, T. S., Cline, T. W., and Jander, R. B. (1981). Coupling efficiency of the optics in single-mode and multimode fiber components. *3rd International Conference on Integrated Optics and Optical Communications* (IOOC-81), San Francisco, CA, April 27–29, 1981 Paper TuL2.

Turner, E. H., and Stolen, R. H. (1981). Fiber Faraday circulator or isolator. *Opt. Lett.* **6**, 322–323.

Ulrich, R., and Simon, A. (1979). Polarization optics of twisted single-mode fibers. *Appl. Opt.* **18**, 2241–2251.

Wagner, R. E. (1979). Multimode optical fiber access port. *Digest of Technical Papers*, Topical Meeting on Optical Fiber Communications, March 5–9, 1979 Washington, DC, paper WG3.

Wagner, R. E., and Cheng, J. (1980). Electrically controlled optical switch for multimode fiber applications. *Appl. Opt.* **19**, 2921–2925.

Wagner, R. E., and Tomlinson, W. J. (1987). Coupling efficiency of optics in single-mode fiber components. *Appl. Opt.* **21**, 2671–2688.

Warbrick, K. (1986). Single-mode optical isolator at 1.3 μm using all-fiber components. *Elect. Lett.* **22**, 711–712.

Watanabe, R., Nosu, K., and Fujii, Y. (1980). Optical grating multiplexer in the 1.1–1.5 μm wavelength region. *Elect. Lett.* **16**, 108–109.

Watanabe, R., Nosu, K., Harada, T., and Kita, T. (1980a). Optical demultiplexer using concave grating in 0.7–0.9 μm wavelength region. *Elect. Lett.* **16**, 106–108.

Whalen, M. S., Divino, M. D., and Alferness, R. C. (1986). Demonstration of a narrowband Bragg-reflection filter in a single-mode fibre directional coupler. *Elect. Lett.* **22**, 681–682.

Winzer, G. (1984). Wavelength multiplexing components—A review of single-mode devices and their applications. *J. Lightwave Technol.* **LT-2**, 369–378.

Yokohama, I., Okamoto, K., and Noda, J. (1985). Fibre-optic polarizing beam splitter employing birefringent-fibre coupler. *Elect. Lett.* **21**, 415–416.

Yokohama, I., Okamoto, K., and Noda, J. (1986). Polarization-independent optical circulator consisting of two fibre-optic polarizing beamsplitters and two YIG spherical lenses. *Elect. Lett.* **22**, 370–372.

Young, W. C., and Curtis, L. (1981). Cascaded multipole switches for single-mode and multimode optical fibres. *Electron. Lett.* **17**, 571–573.

Zengerle, R., and Leminger, O. G. (1986). Wavelength-selective directional coupler made of nonidentical single-mode fibers. *J. Lightwave Technol.* **LT-4**, 823–827.

Chapter 11

Waveguide Electrooptic Devices for Optical Fiber Communication

STEVEN K. KOROTKY

AT & T Bell Laboratories, Holmdel, New Jersey

ROD C. ALFERNESS

AT & T Bell Laboratories, Holmdel, New Jersey

11.1 INTRODUCTION

11.1.1 Optical Communications and Integrated Optics

The rapid growth of single-mode optical fiber transmission since the printing of the first edition of this book has been nothing short of explosive. This trend toward single-mode technology to fulfill the need for increasing channel capacity has stimulated vigorous research and exploratory development in the field of integrated optics, which offers the prospect of manipulating and processing information signals at the optical level with compact *photonic* circuits. In this chapter we examine a selected set from the variety of waveguide devices that have been demonstrated with this purpose in mind. Here the emphasis is on those devices that are designed to play a temporally active, as opposed to passive, role within the system. The perspective is from the standpoints of their functional operation, performance, and system use with the intention of conveying the state of the art.

Communication with light is accomplished by modulating (changing) the quality (level, strength, state) of a characteristic of light according to a predetermined coding scheme. It can be as straightforward as shuttering the beam of a flashlight (intensity modulation) using the Morse code. Integrated optics aims to provide compact devices compatible with optical fiber that can modulate and control all of the characteristics of light. These characteristics include the *optical* frequency (free-space wavelength), amplitude and intensity, phase, polarization, and direction of propagation. Corresponding to these characteristics are devices such as frequency shifters, tunable wavelength filters, amplitude and intensity modulators, polarization controllers and combiners, and spatial switches.

11.1.2 Scope and Format of Chapter

In this chapter we describe integrated-optic implementations of each of the types of devices mentioned. The basic principles of operation are reviewed and the performance of representative devices that have been realized is presented. Where these devices have been used in research fiber-optic communication systems, we have made an effort to illustrate their application.

There are a variety of mechanisms that may be used to modify the value of the characteristics of light. In nearly all of the devices described here, the control is via the linear electrooptic effect, or Pockels effect, which permits fast control of the optical signals and has proved quite versatile. We begin, therefore, with a brief and elementary review of the electrooptic effect. Other techniques for modulation and control include electro-reflection, electro-absorption, magneto-optic, and higher order nonlinear electrooptic effects.

The waveguide fabrication technologies have been outlined in the preceding chapter and are dealt with in more detail in other places (see, for example, Hutcheson, 1987). A more rigorous treatment of the electromagnetic theory of optical guided waves in crystals and the electrooptic interaction are found in several excellent texts (Tamir, 1979; Yariv and Yeh, 1984; Marcuse, 1974; Kaminow, 1974). Reviews of recent advances in waveguide devices may also be found in the technical literature (for example, see Alferness, 1981). The device designer should note that a large body of knowledge pertaining to the physical and material considerations and issues relevant to the understanding, design and performance of waveguide devices may be found in the pioneering research on bulk devices (see, for example, Kaminow, 1974).

11.2 WAVEGUIDE ELECTROOPTIC DEVICES

11.2.1 The Electrooptic Effect

Crystals exhibiting the linear electrooptic effect, by definition, experience a change in the index of refraction (more precisely a change of the dielectric tensor) in proportion to an applied electric field. The response time of the

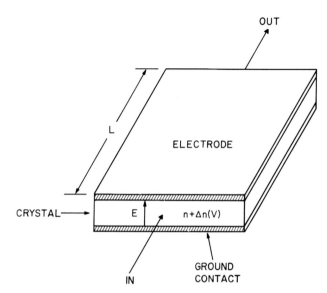

Fig. 11.1   Bulk electrooptic phase modulator.

material to changes in the applied electric field can be extremely fast ($\sim 10^{-13}$ − 10^{-12}s; Kaminow *et al.*, 1970), thereby permitting the possibility of very high-speed devices. To illustrate the concepts, we first consider a predecessor of waveguide electrooptic devices, namely the bulk-optic electrooptic phase modulator. As shown in Fig. 11.1, it consists of an electrooptic crystal with electrodes on the upper and lower surfaces. A collimated beam of light enters through one end face and exits through the opposite face. If the lateral width is large compared to the thickness of the sandwich, then the combination of electrodes and dielectric may be viewed as an ideal parallel plate capacitor—the electric field lines of an applied voltage being normal to the electrode surface within the crystal.

In the absence of an applied voltage, the optical phase of the emerging light beam relative to the input phase is determined by the optical path length traversed by the beam and the optical frequency. The optical path length being the product of the physical length, L, and the index of refraction n. With the application of a voltage V, the index is changed by a small amount, $\Delta n_{eo} \propto V$, as is the optical path length. Consequently, the optical phase, ϕ, at the output of the crystal is also modified and the change is given by

$$\Delta\phi = \frac{2\pi}{\lambda} \Delta n_{eo} L, \qquad (11.1)$$

where λ is the free-space wavelength. In a digital coding scheme, such as two-level phase-shift-keying (PSK), the voltage may be chosen to introduce a 180° (π) phase shift. Then, when referenced to an unmodulated beam, the light passing through the modulator with no voltage applied may represent a logical "0" and light passing through with voltage applied may represent a logical "1."

If PSK were the only optical function possible using the electrooptic effect, there would be little use for waveguide electrooptics except for modulators in "coherent" communication systems. As we shall see later, phase modulation provides the basis for implementing many of the desired functions for controlling light, such as amplitude and intensity modulation, spatial switching, and filter tuning. In addition to changing the index of refraction, the electrooptic effect also permits optical polarization conversion within the crystal. This is accomplished by changing the spatial orientation of the axes defining the polarization eigenmodes within the crystal by application of the electric field. This phenomenon has been used to fabricate polarization controllers and combiners, wavelength filters, and frequency shifters.

These electrooptic phenomena are conveniently and formally described as changes of the dielectric tensor, which relates the polarization and electric field vectors. We note that the dielectric tensor is a second rank tensor implying that it may be represented as a 3×3 matrix. Although the properties of the crystal are not dependent on its orientation, the effect of the crystal on an optical wave passing through it does depend on the relative orientations of the optical polarization, applied electric field, and crystal axes. Consequently, the values of the matrix elements of the dielectric tensor are a function of the coordinate system chosen to define the optical fields and the orientation of the crystal. There is, however, always one coordinate system in which the dielectric tensor is diagonal; it is referred to as the principal axis system. Values of the dielectric constants are tabulated for the principal axis system as are the elements of the third-rank tensor specifying the nature and strength of the linear electrooptic effect.

The change, $\overline{\Delta\epsilon}$, in the dielectric tensor induced via the linear electrooptic effect may be written in the principal axis system as

$$\Delta\epsilon_{ij} = -\frac{n_i^2 n_j^2}{\epsilon_o} r_{ij,k} E_k^a, \qquad (11.2)$$

where n_i is the index for light polarized along the i-axis, ϵ_o is the free-space permittivity, $r_{ij,k}$ are the coefficients of the linear electrooptic tensor, $\vec{E}^a$ is the applied electric field, and the summation over the Cartesian component index k is implied. The notation for the paired indices, ij, is often simplified according to the convention: $(11) \to 1$, $(22) \to 2$, $(33) \to 3$, $(23, 32) \to 4$, $(13, 31) \to 5$, and $(12, 21) \to 6$. The values of several of the r-coefficients are often interrelated or constrained to be identically zero by nature of the crystalline symmetry.

We show how the dielectric tensors of $LiNbO_3$, having a rhombohedral crystalline structure, and GaAs, having a cubic crystalline structure, are modified through the electrooptic effect. For LiNbO$_3$ we have:

$$\Delta\epsilon_{ij} \sim \begin{pmatrix} -r_{22}E_y^a + r_{13}E_z^a & -r_{22}E_x^a & r_{51}E_x^a \\ -r_{22}E_x^a & r_{22}E_y^a + r_{13}E_z^a & r_{51}E_y^a \\ r_{51}E_x^a & r_{51}E_y^a & r_{33}E_z^a \end{pmatrix}. \tag{11.3}$$

For the binary, ternary, and quatenary III-V semiconductor compounds we have:

$$\Delta\epsilon_{ij} \sim \begin{pmatrix} 0 & r_{41}E_z^a & r_{41}E_y^a \\ r_{41}E_z^a & 0 & r_{41}E_x^a \\ r_{41}E_y^a & r_{41}E_x^a & 0 \end{pmatrix}. \tag{11.4}$$

It is important to remember that these matrices have been specified in the principal axis system. Although it would seem that pure phase modulation is not possible for the III-V systems, it can be achieved by proper selection of the applied field direction, direction of the optical polarization and direction of propagation as we shall see later. Finally, we note that the electrooptically induced index change for the idealized bulk phase modulator can be expressed as

$$\Delta n_{eo} = -\frac{n^3 r}{2}\frac{V}{d}, \tag{11.5}$$

where d is the gap between the electrode plates.

Values of the linear electrooptic coefficients for many crystals have been tabulated (Kaminow, 1974; Kaminow, 1987). The largest of the r-coefficients of LiNbO$_3$ is r_{33} with a value of $\sim 31 \times 10^{-12}$ m/V. The r_{41} coefficient of the III-V semiconductor compounds is $\sim 1.3 \times 10^{-12}$ m/V. At first glance these values would seem to indicate that LiNbO$_3$ has an enormous advantage over the semiconductors. The difference between the two is reduced, however, when full consideration of device design is given. First, a relevant figure of merit summarizing the strength of the electrooptic effect is $n^3 r$, as suggested by Eq. 11.5. Near 1.3 μm the index of refraction for LiNbO$_3$ is $n \sim 2.1$ and for InGaAsP it is $n \sim 3.1$. Second, semiconductor devices generally have smaller lateral dimensions owing to the larger index change used to form the waveguide. Typically, the semiconductor waveguide is ~ 5 times smaller for the same wavelength and the electric field strength is correspondingly larger for the same applied voltage. Finally, for high-speed operation, a relevant figure of merit is bandwidth per unit power. As the semiconductors have a lower permittivity at rf frequencies, they require less power to develop the same voltage. Thus, it would seem reasonable to expect comparable electrooptic performance in ideal devices of both material systems.

At present, and taken as a group, the device technology based on the selective diffusion of titanium into LiNbO$_3$ has demonstrated the highest level of performance. This is reflected in the bias of the descriptions that follow. This is not intended to imply that semiconductor devices designed to perform a similar function have not been demonstrated or that similar overall performance might not be achieved in the future. Where semiconductor devices have also demonstrated system level performance, we have made an effort to include them.

11.2.2 Waveguide Devices

Waveguide electrooptic devices differ from their bulk optic counterparts by virtue of the optical wave being confined to small transverse dimensions as it propagates along the waveguide. Intrinsic to the waveguiding action is the elimination of restrictions that diffraction imposes in the case of beam propagation. There, the diameter of the beam cannot be made arbitrarily small for arbitrarily long propagation distances, but the beam waist must be traded off against the collimation length. Thus, the voltage-length product and absolute voltage required to accomplish a function are significantly lower for waveguide devices. The small device sizes and the ability to fabricate and interconnect many devices on a chip are other attractive features in considerations of manufacture and use.

In Fig. 11.2 we illustrate the construction of two common implementations of basic waveguide electrooptic phase modulators. Myriad other variations are certainly possible. The device in Fig. 11.2a consists of an embedded-type waveguide fabricated in a crystal either of a dielectric material, such as LiNbO$_3$, or a semi-insulating material, such as Fe-doped InGaAsP/InP. The device in Fig. 11.2b consists of a rib waveguide arrangement in a semiconducting material, such as the III-V compound semiconductors. To attain an electric field in the waveguide volume of the semiconducting material, Fig. 11.2b, a p-n, p-i-n, or Schottky junction is employed. In the examples considered here, the modulator design makes use of an applied electric field oriented along the crystallographic z-axis, which is oriented perpendicular to the direction of the optical propagation and normal to the upper surface of the crystal. In LiNbO$_3$ this axis is often referred to as the c-axis. It is often defined as the [001] direction in the cubic semiconductor compounds.

To achieve pure phase modulation the input linear optical polarization would be oriented along the c-axis of LiNbO$_3$ to take advantage of the r_{33} electrooptic coefficient. This corresponds to the TM mode of the waveguide for the orientation chosen. The direction of optical propagation is typically chosen as the crystallographic y-direction of the LiNbO$_3$ substrate. In the cubic semiconductors, the input linear polarization would be oriented normal to the applied

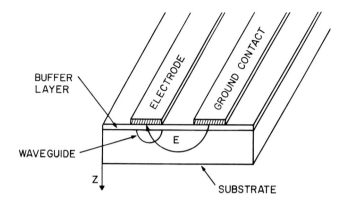

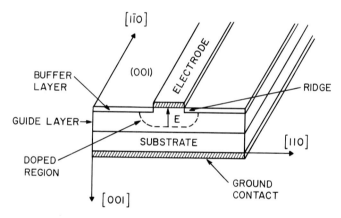

Fig. 11.2 Waveguide phase modulators.

electric field in the [110] direction with the wave propagating in the [1̄10] direction. This polarization orientation corresponds to the TE mode of the waveguide.

The details of the electrodes, their dimensions, configuration, and connections, in general depend on the type of device as will become evident. Rarely, however, is the spatially uniform electric field of an ideal parallel plate capacitor achieved. Because the electrooptic index change is small, its effect on the optical phase velocity depends on the overlap integral of the optical and electrical fields. The consequence of these non-uniform fields is summarized by a factor, Γ, which represents the efficiency of the electrooptic interaction relative to an idealized parallel plate capacitor of the same electrode gap. The electrooptic index change is then written as

$$\Delta n_{eo} = -\frac{n^3 r}{2} \frac{V}{d} \Gamma.$$ (11.6)

The optical connections to and from the device waveguide are usually made with optical fibers. The total insertion loss of the device then consists of a waveguide-fiber coupling loss as well as optical propagation loss. By careful choice of fabrication and diffusion parameters, single-mode Ti:LiNbO$_3$ waveguide devices have been fabricated with mode spot sizes very well matched to that of typical single-mode fibers. Coupling losses in the range of 0.5–1.0 dB/interface are now common for Ti:LiNbO$_3$ devices (Alferness et al., 1982; Veselka and Korotky, 1986). For the same sets of parameters, propagation losses are less than 0.2 dB/cm and excess bend loss may be maintained below 0.1 dB/bend. Thus, once on chip, the excess loss for incorporating a device is in principle less than 0.1 dB. Consequently, sophisticated optical circuits fabricated on a single chip with the number of inter-connected elements numbering in the tens have been fabricated with a total insertion loss nearly identical to that of a straight waveguide (Granestrand et al., 1986).

In general, semiconductor waveguide devices have larger fiber-coupling losses because of their tendency toward small mode sizes. The semiconductors, however, offer the prospect of integrating waveguide devices on the same substrate as the injection lasers and detectors, thereby significantly reducing coupling losses between chips and providing added functionality, such as optical amplification. Recently, propagation losses of ~ 1 dB/cm have been attained when operating in the region of the bandgap, and losses of ~ 0.2 dB/cm have been observed for operating wavelengths far below the bandgap (Hiruma, 1985; Kapon and Bhat, 1987). Significant progress toward integration is being made (see, for example, Reinhart and Logan, 1974; Sakano et al., 1986); however, much work is still required to achieve the flexibility in crystalline growth that is necessary to optimize devices with and without gain on the same substrate.

11.3 HIGH-SPEED PHASE MODULATION

11.3.1 Phase Modulation

In addition to optical insertion loss and operating voltage, the speed with which the desired optical function can be carried out is an important concern. High-speed phase modulators, for example, play a key role in systems employing phase-shift-keying and heterodyne detection since the direct current modulation of injection lasers is not easily adapted to phase modulation. As many other optical operations are implemented through clever use of phase modulation, it is convenient to discuss the issues pertaining to high-speed performance in the

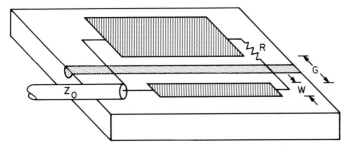

Fig. 11.3 Integrated-optic traveling-wave phase modulator.

framework of phase modulators. This also simplifies the discussion somewhat, as phase modulators based on the linear electrooptic effect are inherently linear devices.

The electrodes used to introduce the applied electric fields within the waveguides can be classified according to whether they behave as lumped circuit elements or transmissions lines. The latter are often referred to as traveling-wave electrodes. By definition, the electrodes may be considered a lump element if the electrical transit time along the interaction length is negligible compared to the period of the highest frequency of interest and if the resistance of the conductor is low, so there is little electrical attenuation along the active length. If these conditions are met, the electrode behaves as an ideal capacitor. If they are not met, then the details of the electrode structure, excitation and termination must be carefully considered to design an efficient traveling-wave arrangement. A schematic of a traveling-wave phase modulator is shown in Fig. 11.3.

Normally, regardless of the electrode type, the source electrical signal is provided by a transmission line, such as a coaxial cable or microstrip line, having a characteristic impedance, Z_s. To achieve the flattest frequency response and optimal power transfer, the source line should be terminated with a resistor of the same impedance. In the case of the lumped electrode criterion, the modulation bandwidth is limited by the RC time constant of the source/electrode combination with the *electrical 3 dB bandwidth* given by $\nu_{3\,dB} = 1/\pi RC$. Typical electrode geometries result in a capacitance per unit length of ~ 3 pF/cm (Alferness, 1982). Thus, the bandwidth-length product for a 50 Ω source impedance is ~ 2 GHz-cm. The performance can be improved if a reduced driver impedance can be tailored to the device design.

The lumped electrode limitation is circumvented for the same electrode length, and hence same drive voltage and power, by configuring the electrode and termination in a traveling-wave arrangement. There, the bandwidth is ultimately limited by any difference in the electrical and optical velocities and/or frequency dependent electrical attenuation (Korotky and Alferness,

1983). The bandwidth limit (3 dB electrical) resulting from electrical/optical walkoff is $\nu_{3\,dB} = a/\pi T_w$, where T_w is a walkoff time, and $a \simeq \sqrt{2}$ is the solution of $\sin a/a = 1/\sqrt{2}$. The walkoff time is given by $T_w = (L/c)(N_m - N_o)$. Here, L is the electrode length, c is the speed of light, N_m is the effective microwave index of refraction, and N_o is the optical index of refraction. In $LiNbO_3$, the effective microwave index for the typical planar transmission lines is $N_m \sim 4.2$ and the optical index is $N_o \sim 2.2$. This results in a small-signal bandwidth-length product of approximately 6.5 GHz-cm. In the semiconductors, the microwave and optical velocities are more nearly equal and the bandwidth-length product is approximately 45 GHz-cm.

Electrical attenuation reduces the attainable bandwidth from the above values. In $LiNbO_3$ and semi-insulating materials, the rf attenuation in the microscopic structures is dominated by conductor loss originating in the skin depth effect. Empirically, attenuations for both material systems are in the range of $1-1.5$ dB/cm-$\sqrt{GHz}$ for gold electrodes of $3-4$ μm thickness and lateral dimensions of ~ 10 μm. This limits bandwidths to approximately 5 GHz for a 1 cm active length. For ultra high-speed traveling-wave devices ($L \sim 1$ mm) walkoff is the limiting factor for $LiNbO_3$ based devices, whereas electrical attenuation is the limiting factor for semi-insulating III-V based devices.

11.3.2 Performance and Applications

Multi-gigahertz bandwidth phase modulators have been demonstrated in both $Ti:LiNbO_3$ and semiconductor technologies. In $LiNbO_3$, devices designed for 1.55 μm wavelength have been demonstrated with total insertion losses of 1.8 dB and modulation voltage, V_π, of 8 V for an active length of 1 cm (Alferness et al., 1986a). Broadband traveling-wave phase modulators in the GaAs system have been recently reported with a drive voltage of 20 V at 1.3 μm wavelength (Wang et al., 1987).

Several optical fiber transmission systems experiments using PSK and DPSK (differential PSK) modulation with heterodyne detection have been carried out at multi-gigabit/sec rates. The highest data rate achieved so far has been 2 Gb/s using a $Ti:LiNbO_3$ phase modulator with a transmission distance of 170 km and receiver sensitivity of -35 dBm (Gnauck et al., 1987). The longest distance attained using a waveguide electrooptic phase modulator is currently 260 km at a data rate of 400 Mb/s (Linke, 1987). System level details of these experiments are discussed in Chapter 20.

11.4 AMPLITUDE AND INTENSITY MODULATION

Although phase modulation is perhaps the most basic operation from the viewpoint of device operation, fiber communication systems based on optical intensity modulation are more easily implemented. There, information is encoded by transmitting or not transmitting light during predetermined time slots. This method of communication is by far the most pervasive as can be judged from the discussions in this book and its predecessor. One reason for this is that the components that generate light, such as LEDs and injection lasers, easily lend themselves to ON/OFF modulation.

Given that semiconductor lasers can be intensity modulated directly by controlling the injection current, it is logical to ask why separate waveguide intensity modulators are necessary at all. There are several device issues that are involved, but all address the optimization of the overall system performance often measured by the data rate, repeater span, and minimum channel spacing. At very high data rates (> 2 Gb/s), it becomes increasingly more difficult to modulate lasers directly without significant wavelength broadening. The broadening may be divided into two categories. For cleaved facet injection lasers, the dominant broadening occurs via an increase in the number of longitudinal Fabry-Perot modes that are excited and effectively compete for the optical power. In addition to this behavior, the frequency (wavelength) of each mode is varied, or chirped, during the transitions between on and off states because of the change of the complex index of refraction as a function of the charge carrier density. This phenomenon is the dominant source of spectral broadening in so-called single-frequency lasers, such as ideal distributed feedback lasers.

When the optical pulses are transmitted through the fiber, the pulses are lengthened in temporal duration depending on the fiber dispersion characteristics and the spectral width transmitted. At the wavelength (1.55 μm) of the loss minimum ($\sim$ 0.2 dB/km) for silica based fibers the dispersion is roughly 17 ps/km-nm. This degradation causes inter-symbol interference and, for a given fiber and spectral width, places a limit on the distance over which the signal can be sent and reliably recovered. To minimize the amount of spectral broadening, the directly modulated laser is usually not turned fully off to encode the low light level state. Typical ON/OFF ratios are 3–4 : 1, which introduces power penalties for low modulation depth and excess receiver noise (Gnauck *et al.*, 1985). In addition to reducing the distance between repeaters, spectral broadening also reduces the number of independent wavelength channels that may be transmitted over the fiber.

The problem of spectral broadening at high data rates is effectively eliminated using waveguide electrooptic modulators external to the laser. An optical transmitter consisting of a single-frequency laser and an external electrooptic intensity modulator emits an optical signal with a spectral width that is essentially determined by the frequency content of the driving information signal (for a theoretical discussion see, for example, Koyama and Iga, 1987). System experiments at 4 and 8 Gb/s have indeed shown that the significantly narrower spectral width and higher modulation depth possible with external modulation more than offsets the modulator insertion loss ($\sim$ 3 dB) (Korotky *et al.*, 1985; Gnauck *et al.*, 1986). Because of the potential advantages, several groups are pursuing external modulation as an alternative to direct modulation (Leboutet and Sorel, 1985; Auracher *et al.*, 1984). Before discussing the results of such experiments in further detail, we briefly review the mechanism for attaining intensity and amplitude modulation and the performance that has been achieved.

It is also worth noting that the same waveguide electrooptic devices that serve as intensity modulators for direct detection are intrinsically amplitude modulators and can therefore serve as modulators for amplitude-shift-keyed (ASK) systems using heterodyne detection (Park *et al.*, 1987). In the latter systems the spectral characteristics of the transmitter are even more stringent than for direct detection, and external waveguide modulators have provided the only means of attaining the required level of spectral purity. As we will see later, the directional coupler switch not only provides for amplitude and intensity modulation, but permits high-speed spatial switching as well, making it a versatile component for lightwave communication systems.

11.4.1 Amplitude Modulators

There are several techniques to accomplish amplitude modulation and recent work in both LiNbO$_3$ and semiconductor materials is reviewed in the literature (Alferness, 1982). Two of the basic device structures for implementing amplitude/intensity modulation—the Y-branch interferometric modulator and the directional coupler switch—have been briefly introduced in Chapter 9. Here we will expand on their operational characteristics. For completeness, we also introduce the Mach-Zehnder switch, which is shown with the other two devices in Fig. 11.4. Although we treat these devices as distinct, as has traditionally been done, fundamentally the mechanisms of operation can be placed in a unifying framework (Korotky, 1986).

The devices illustrated in Fig. 11.4 operate as amplitude modulators by providing a means to adjust the relative weighting, or power distribution, of two orthogonal states of optical propagation and a filter that can selectively isolate the optical wave propagating in one of the states. These two states are distinguished by their spatial characteristics. In the cases considered here, the method of changing the relative weighting of the light propagating in the states is accomplished with phase modulation. The function of projecting out (filtering) the component of interest is accomplished in a region physically distinct from the region of phase modulation. The phase modulation is carried out in the center regions of these devices, which may be viewed as the arms of a generalized interferometer. The filtering is accomplished in the outgoing Y-junction of the Y-branch modulator, in the outgoing 50/50 coupler of the Mach-Zehnder switch, and in the isolated outgoing waveguides of the directional coupler.

The functional dependence of the output amplitude as a function of the applied phase modulation differs between the Y-branch and Mach-Zehnder

**PHASE
MODULATOR**

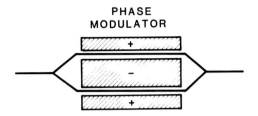

Y-BRANCH MODULATOR

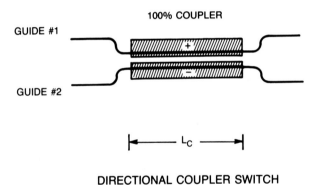

DIRECTIONAL COUPLER SWITCH

Fig. 11.4   Single-mode waveguide modulators and switches.

interferometers and the switched directional coupler. As discussed in Chapter 9, the amplitude output from the former devices depends on the cosine of the accumulated phase difference between the two waveguides. Denoting the local phase velocity difference, or mismatch, as $\Delta\beta = (2\pi/\lambda)\,\Delta n_{eo}$, the amplitude response of the two interferometric modulators is

$$\cos\left(\frac{\Delta\beta L}{2}\right). \tag{11.7}$$

We note that the value of $\Delta\beta$ may be increased by a factor of two for the same applied voltage by configuring the electrodes so as to induce index changes of

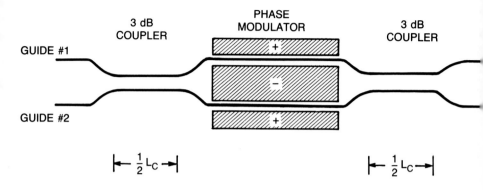

MACH-ZEHNDER SWITCH

Fig. 11.4 Continued

opposite sign in the two waveguides. Such an arrangement also avoids the occurrence of a phase term in the amplitude response (Eq. 11.7). (Note that the electrode configuration necessary to implement the *push-pull* effect for the interferometer also results in an increase in the electrode capacitance.) The intensity output of these modulators is thus a periodic function of the applied voltage ($\Delta\beta$) as shown in Fig. 11.5b.

The amplitude output of the switched directional coupler differs in functional form because the induced mismatch can cause a change in the relative amount of optical power carried on the two sides of the symmetry plane between the two waveguides in the interaction region, which does not occur in the interferometric modulators. The theoretical description of the switched directional coupler is described by the coupled mode equations (Miller, 1954). Qualitatively, the switched coupler's behavior is as follows: The coupler is fabricated from *identical* waveguides with an interaction length corresponding to one coupling length (see Chapter 9). Without application of a voltage, light entering one of the input waveguides then exits the opposite waveguide. This situation is referred to as the crossover, or cross state. When voltage is applied the index in one waveguide is slightly increased while in the other it is slightly decreased via the electrooptic effect. The phase synchronism between the two waveguides is thus destroyed by application of a voltage, and the resonant transfer of power from one waveguide to the other is quenched. For specific values of the phase-mismatch, light entering the input waveguide exits entirely from that waveguide. This situation

DC MODULATION RESPONSE

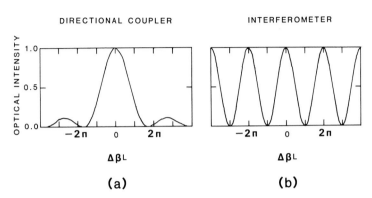

Fig. 11.5 Intensity response of modulators.

is referred to as the straight-through, or bar state. If one now considers the output from the crossover waveguide, the output intensity is given by

$$\frac{\sin^2\left(\kappa L \sqrt{1 + (\delta/\kappa)^2} \right)}{(1 + \delta/\kappa)^2},\tag{11.8}$$

where $\delta = \Delta\beta/2$ and κ is the interwaveguide-coupling coefficient.

The crossover efficiency versus applied phase-mismatch, i.e., the switching curve, for the directional coupler is plotted in Fig. 11.5a. We see that unlike the interferometric modulators, the on state (cross state) is achieved for only one voltage. We also note that the value of the phase-mismatch times length product required to switch between the cross and bar states for a single-coupling-length device is $\Delta\beta L = \sqrt{3}\,\pi$, which is $\sqrt{3}$ times larger than for the interferometric modulators. Naively, one might then assume it is always more advantageous to use an interferometer if the device length is predetermined. However, the large-signal bandwidth of the traveling-wave directional coupler modulator is larger than that of the interferometers having identical electrode length for the same reason its switching voltage is higher, namely its effective interaction length is shorter because of a center weighting effect of the coupler (Korotky and Alferness, 1983).

11.4.2 Performance

As mentioned, the more relevant characteristics of modulators are 1) bandwidth, 2) required drive voltage, and 3) optical insertion loss. These performance figures are not entirely independent, however. For example, both lumped and

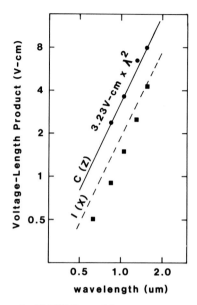

Fig. 11.6 Switching voltages for Ti:LiNbO$_3$ modulators.

traveling-wave modulators exhibit bandwidth-length products as noted. Consequently, to increase the operating bandwidth requires that the active length be decreased proportionately. These devices, however, also require a specific phase shift change for full modulation and are, therefore, constrained by a voltage-length product as well. The voltage necessary may be minimized by proper design of the electrodes so as to maximize the overlap factor, Γ. Much work has been carried out to optimize the electrooptic efficiency for Ti:LiNbO$_3$ devices. Best values for voltage-length products needed to switch are plotted in Fig. 11.6 for both directional coupler and interferometric modulators as a function of design wavelength. The data represent nearly a decade of work at laboratories around the world. As can be seen, the voltage-length product for directional couplers is described by $VL = 3.2$ V-cm $\times \lambda^2(\mu\text{m})$, and for interferometers it is approximated by $VL = 2$ V-cm $\times \lambda^2(\mu\text{m})$. From these data and the electrode gaps, which scale as wavelength, it is estimated that the overlap factor of the Ti:LiNbO$_3$ devices is $\Gamma \sim 0.3$. In the 1.3–1.5 μm wavelength region, the low voltage-length products have been attained at the same time the total fiber-to-fiber insertion losses have been kept low, i.e., 2–3 dB. Similar overall performance has been reported for GaAs directional couplers operated at 1.3 μm (Ishida *et al.*, 1985).

In Fig. 11.7 we illustrate the ability of these modulators to work at very high frequencies. Plotted is the electrical-to-optical-to-electrical power transfer

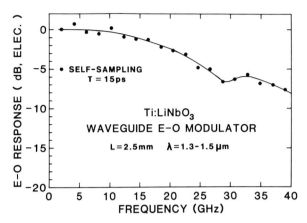

Fig. 11.7 Frequency response of waveguide modulator.

efficiency as a function of drive frequency for a 2.5 mm-long $Ti:LiNbO_3$ modulator. This transfer efficiency is the relevant measurement for communications systems based on square-law optical detectors. (The reader is cautioned that both optical and electrical -3 dB bandwidths are found quoted in the literature and that the former value corresponds to the frequency at which the electrical power transfer is -6 dB.) The experimental device exhibits a small signal electrical -3 dB bandwidth of approximately 22 GHz and only an 8 dB rolloff at 40 GHz (Korotky et al., 1987). These combinations of characteristics have made $Ti:LiNbO_3$ integrated-optic devices competitive for application in both digital and analog (Blauvelt et al., 1984; Bulmer and Burns, 1984; Stephens and Joseph, 1987) high-speed fiber-optic systems. Recently, there has also been significant progress on discrete broadband traveling-wave semiconductor optical modulators (Wang et al., 1987).

11.4.3 System Application / Demonstration

Directional coupler amplitude/intensity modulators have been used in a variety of laboratory direct and heterodyne detection transmission experiments. Figure 11.8 depicts a point-to-point fiber-optic transmission system based on direct detection and external modulation as an example. The system consists of an injection laser, a directional coupler intensity modulator, a long length of low-loss single-mode optical fiber, an APD-FET photo-receiver front end, drive and receiver amplifiers, and the digital bit-error-rate (BER) test set. The laser is operated cw and oscillates in a single longitudinal mode and serves as the light source. Light from the laser is coupled into a fiber and routed to the modulator, in this case a $Ti:LiNbO_3$ directional coupler switch. The BER test set provides a

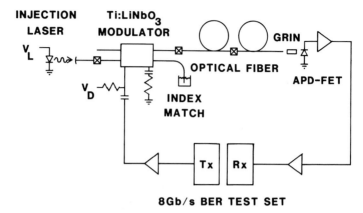

Fig. 11.8 Direct detection fiber-optic transmission system.

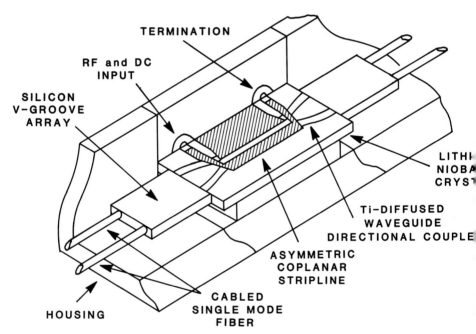

Fig. 11.9 Schematic diagram of the waveguide amplitude modulator.

pseudo-random digital non-return-to-zero (NRZ) electrical signal that drives the modulator. The output of the modulator is taken from the crossover waveguide and is connected to the transmission fiber. Depending on the voltage level from the driver, the light level is turned on or off at the modulator output, thereby replicating the data on the optical wave. The light signal travels through the single-mode fiber and impinges on the photodetector. There, the signal is converted back to an electrical signal and is returned to the BER test set to compare the received data stream with the transmitted stream to assess the number of errors.

The structure of the Ti:LiNbO$_3$ integrated-optic modulator is shown in Fig. 11.9. Visible are the optical waveguide directional coupler and traveling-wave electrode, as well as the fiber pigtails. For the experiments described, the active length of the modulators was 1 cm long. A voltage to switch of 8 V and total insertion loss of 3 dB were achieved at 1.56 μm. In addition, these devices incorporated anti-reflection coatings at the LiNbO$_3$/fiber interface to reduce the

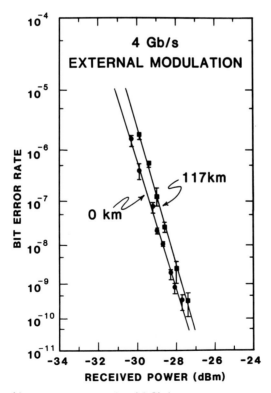

Fig. 11.10 System bit-error-rate curves at 4 and 8 Gb/s.

SYSTEM BIT-ERROR-RATE CURVES

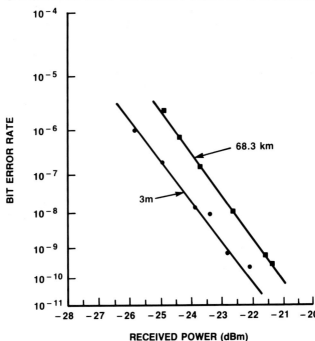

Fig. 11.10   Continued

reflection back toward the laser to ~ -35 dB. Using this type of device, data have been transmitted and recovered with a BER below 10^{-9} at data rates of 4 and 8 Gb/s over fiber lengths of 117 km and 68 km, respectively. In both cases only very small dispersion penalties were incurred. This is demonstrated by the BER versus received power curves shown in Fig. 11.10. Plotted are curves for transmission through the long length of the system fiber and also through a very short length (< 3 m). At 4 Gb/s the dispersion penalty is ~ 0.4 dB and at 8 Gb/s it is ~ 1 dB. These penalties are just those expected for a spectral width corresponding to the bandwidth of the electrical signal representing the digital information combined with the dispersion characteristic of the fiber at 1.5 μm. This result is to be compared with dispersion penalties of several dB for direct current modulation of the laser at similar rates.

11.5 OPTICAL SWITCHING

Optics is already having a profound effect on long-distance, ground-based communications. Lightwave transmission systems based on very low loss, wide bandwidth optical fiber and high-speed opto-electronic repeaters are penetrating throughout worldwide communication networks. The virtual elimination of electrical coaxial long haul systems seems inevitable. As the optical routes extend closer and closer to end users, optical technologies are beginning to set their sights on switching within the network. This is an area where electronics, with its ability to treat data and control on an equal level, has a strong foothold. Because control and data are carried and can interact in the same electronic medium, it is unlikely that electronic switching will be completely displaced until *and unless* digital optical processing becomes practical. This is likely to be some time off, and present approaches to optical switching attempt to take advantage of the best characteristics of both optical and electronic technologies. Not surprisingly then, opto-electronic and electro-optic approaches to optical switching are among the more promising near-term possibilities to provide connectivity among tens of multi-gigabit/sec fiber-optic lines.

11.5.1 Space-Division Switching

The 4-port (two inputs + two outputs) waveguide directional coupler and Mach-Zehnder interferometer described in the section on modulators both may also be used as optical crosspoints. The former is more common and is often preferred because it requires fewer waveguide bends and thus may be made shorter for the same active length and large-signal bandwidth-to-power ratio. Both devices have the important characteristic that they are virtually data-rate transparent. Thus, they are ideally suited as circuit switches for broadband facility switching, time-multiplexed switching, and protection switching applications. In addition, they are capable of being reconfigured at very high speeds if desired; therefore, they are also useful for time-division multiplexing/demultiplexing, add/drop switching, and time-slot interchange environments.

A key difference between the use of these devices as modulators and as switches is that the latter application requires low-crosstalk performance in addition to low switching voltage and insertion loss. The possibility of crosstalk arises, for example, when two different signals are entering the two input ports (referred to as both inputs active) and these signals are to be routed to separate outputs. The degree to which the isolation between the two signals is degraded is measured as the ratio of the undesired signal power to the desired signal power and is typically specified in dB.

In the discussion of space-division switching, we refer to the directional coupler switch, although the general principles and techniques are applicable to the Mach-Zehnder switch as well. It is convenient to consider the situation where an optical signal is launched into one of the input ports of the switchable waveguide directional coupler. The two possible states of the switch corresponding to the signal exiting one or the other of the output ports are referred to as the straight-through state and the crossover state. In the former state, the light exits from the output port of the same waveguide in which it had entered. In the crossover state the optical signal crosses over to the second waveguide and exits. As described in the section on intensity modulators, the optical switch is typically formed by fabricating a pair of identical waveguides in an electrooptic crystal such as LiNbO$_3$ or GaAs. In one implementation the strength of the inter-wave-guide coupling coefficient, κ, and the total interaction length, L, are chosen so that the product of the two, κL, corresponds to one coupling length, i.e., $\kappa L = \pi/2$. In this way the crossover state is achieved passively.

To attain the straight-through state, electrodes may be placed over the waveguides as illustrated in Fig. 11.4b. When a voltage is applied to the electrodes, the electric fields penetrating the crystal cause a differential change of the index of refraction in the neighborhood of the two waveguides via the electrooptic effect. This corresponds to introducing a phase velocity mismatch, $\Delta\beta$, between the two guides and the switch structure is referred to as a uniform-$\Delta\beta$ directional coupler switch. If the proper amount of mismatch (in this case $\Delta\beta L = \sqrt{3}\,\pi$) is introduced, then the crosspoint may be switched from the cross state into the straight-through state as indicated by Eq. 11.8 and shown in Fig. 11.5a. Switches of this type have been fabricated with crosstalks less than -30 dB for both switch states. Mass producing such devices, however, would require very tight fabrication tolerances since the cross state is not under electrical control. For this reason the concept of the $\Delta\beta$-reversal switch was invented (Kogelnik and Schmidt, 1976).

The $\Delta\beta$-reversal switch is illustrated schematically in Fig. 11.11. The technique requires that the total interaction length be longer than one coupling length; typically it is arranged to be between one and two coupling lengths. The electrodes are placed over the electrodes as before with the difference that the polarity of the applied voltage, and hence the phase mismatch, is reversed at the midpoint of the coupler. It can be shown that both the cross state and the straight-through state can then be attained by application of two appropriate voltages, without precise control of the coupling length. The minimum voltage swing to switch between states is attained for this two-section device when the

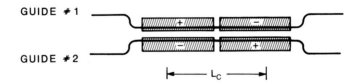

$\Delta\beta$-REVERSAL COUPLER

Fig. 11.11 $\Delta\beta$-reversal directional coupler switch.

4×4 OPTICAL CROSSBAR SWITCH ARCHITECTURE

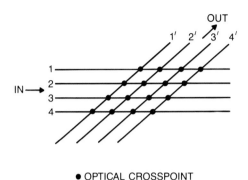

● OPTICAL CROSSPOINT

Fig. 11.12   Crossbar inter-connection architecture.

coupler is two coupling lengths long. If the coupler is near one coupling length long, then it is usual to use $\Delta\beta$-reversal to obtain the cross state and to use uniform-$\Delta\beta$ to obtain the straight-through, or bar state. With these methods crosstalks of -30 dB or below are routinely achieved (see, for example, Bogert *et al.*, 1986).

For communications networks it is desirable to increase the cross-connect capability from the 2×2 (one-way) capacity of the directional coupler crosspoint. One scheme for interconnecting crosspoints is the crossbar configuration, or crossbar architecture. The logical interconnect scheme for the crossbar architecture is illustrated in Fig. 11.12. In the crossbar architecture a crosspoint is placed at each of the N^2 points formed by the intersection of N incoming channels and N outgoing (orthogonal) channels. With this arrangement a connection can be made from any one of the incoming channels to any one of the outgoing channels in a strictly non-blocking fashion, i.e., so long as the desired input and output are not busy, then the connection can be made without rearrangement of the existing traffic. The physical implementation of a waveguide crossbar switch based on directional coupler crosspoints is depicted in Fig. 11.13. In typical implementations the crossover state of the coupler is used as the off state of the inter-connect. Then to provide a connection from the ith input line to the jth output line it is only necessary to switch the i, j crosspoint in the array from the crossover state to the straight-through state. It is easy to see that the overall crosstalk performance of the crossbar switch described depends on the quality of the crossover state, whereas it is relatively insensitive to the quality of the straight-through state. For this reason the $\Delta\beta$-reversal configuration is normally used.

Besides the crossbar architecture, there are a host of other cross-connect architectures that may be considered. Some are drawn directly from electrical

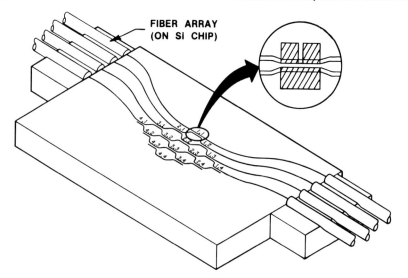

Fig. 11.13 4 × 4 integrated-optic crossbar switch array.

switching systems, while others have been specifically designed to make best use of the advantages of optics and to avoid any possible shortcomings. It is beyond the scope of this chapter to describe even the more relevant architectures in any detail. In addition to the crossbar architecture, the following architectures are among those being investigated for optical switching circuits: Active-Splitter/Active-Combiner (Spanke, 1986), Benes (Benes, 1965), Clos (Clos, 1970), Dilated (Padmanabhan and Netravali, 1987), Duobanyon (Milbrodt, 1987), Folded (Suzuki *et al.*, 1987), Reflective (Duthie *et al.*, 1987), Richards (Richards and Hwang, 1984), and Shuffle networks. For a given cross-connect size these architectures differ in the number of couplers required, the number of stages required, the manner in which crosstalk builds from the crosspoint value, their blocking/nonblocking characteristics, and the degree to which the network can be broken down into modular components without significantly increasing the optical insertion loss (see, for example, Spanke, 1987). In addition, some architectures (for example, the Duobanyan) take advantage of the fact that optical waveguides can be made to cross (intersect) with negligible crosstalk or loss (Bogert, 1987; Agrawal *et al.*, 1987).

Factors that influence the size of the switch that can be constructed on a single substrate are the length required for the active section to ensure a given switching voltage, the length required to make bends in waveguides to inter-connect couplers with negligible insertion loss, the propagation loss, the waveguide fiber coupling loss, and the practical crystal size. At present the Ti:LiNbO$_3$ technology can attain a voltage-length product for switching directional couplers at 1.3 μm of < 7 V-cm (see Fig. 11.6). Waveguide bends that permit the parallel waveguides of a coupler to be separated from the very small inter-waveguide gap used in the coupler (~ 5 μm) to the larger separations necessary for connection to optical fibers (~ 250 μm) in a 5 mm transition length and with negligible loss (< 0.1 dB) are routine (Minford et al., 1982). Total waveguide propagation loss is now commonly at or below 0.2 dB/cm at 1.3–1.5 μm wavelength and can be as low as 0.05 dB/cm (Suche et al., 1985). Single-mode waveguide to single-mode fiber coupling has been demonstrated with losses as low as a few tenths of a dB per interface, with 1 dB being commonplace (Alferness et al., 1982). Finally, wafer sizes of 3 and 4 inch diameter are standard, with chip sizes of 1 cm × 6 cm not being unusual. Based on these loss figures, the typical optical circuit in LiNbO$_3$ can be fabricated with a path-independent fiber-to-fiber loss of ~ 4 dB. Use of standard fiber connectors typically increases the total insertion loss by ~ 1 dB.

To date both 4 × 4 and 8 × 8 crossbar arrays have been demonstrated by several laboratories around the world (McCaughan and Bogert, 1985; Bogert et al., 1986; Granestrand et al., 1986; Neyer et al., 1986; Hoffmann et al., 1986; Suzuki et al., 1987). Typical insertion losses reported are ~ 5–6 dB with crosspoint crosstalk less than − 30 dB. Experimental demonstrations of switching of digital video signals have been carried out as one potential application (Erickson et al., 1987a). It is anticipated that nonblocking networks as large as 32 × 32 (512 × 512) may be built by interconnecting chips in such a way that the total insertion loss is less than 20 dB (40 dB)— loss figures within the power budget of multi-gigabit/sec (several hundred megabit/sec) optical repeaters—and the overall crosstalk level is below − 20 dB. The reconfiguration speed of the optical switches can be in the range of 100 ps—10 ns depending on the size of the array and the density of the crosspoints. From the characteristics mentioned, it is clear that the forte of such optical switches is to provide complete connectivity among moderate numbers (10–100) of very high speed (1–10 Gb/s) lines. Switches of this capacity may be useful to provide routing capability in both centralized hubs and remote terminals. It is important to remember, however, that this class of optical switch requires an electrical intelligence to orchestrate and drive the switching elements.

In some applications (see, for example, Erickson et al., 1987b) the real-time connectivity between two end points capable of a spatial switch may not be required. Rather, effectively continuous and distinct connections from any one end point with each of the other end points may be desirable. Such a virtual connectivity can be accomplished by operating a spatial switch to make connections among the end points in a cyclic, i.e., time-multiplexed, fashion. In such a deterministic interconnection environment the number of crosspoints that are sufficient to provide full connectivity is smaller than required for both strictly non-blocking and rearrangeably non-blocking architectures. This can translate into a smaller circuit size and a lower per line cost. It should be noted, however, that architectures employing fewer crosspoints generally require better crosstalk performance of each crosspoint than architectures with more combinatoric capability. In addition, for the time-multiplexed mode of operation to be efficient, the switch reconfiguration time should be small compared to the hold time and not very large compared to the data bit period. Thus, the optical crosspoints for this application should be relatively high-speed devices. Finally, the synchronization of the data channels and switch drive need to be considered in the overall design of such systems. Experimental switching systems to examine these issues are necessary before any broadly applicable conclusions can be drawn.

11.5.2 Time-Division Switching

11.5.2.1 Time-Slot Interchange. An alternative to space-division switching for interconnecting a number, N, of logical users is switching via time-slot interchange, which is widely used in digital electronic switches. In this approach, data on the incoming fiber channel are assigned to a recurring pattern of N time slots referred to as a frame. The switch permits connections among the logical users by permuting the data within a frame according to the desired interconnection scheme. To accomplish this the switch must include some form of memory and a means for writing and reading the memory at speeds corresponding to the aggregate data rate. The memory element may be as basic as a fiber delay line (Kondo *et al.*, 1983) or may include nonlinear devices such as bistable laser diodes (Suzuki *et al.*, 1986). In several implementations it is common to use fast $1 \times N$ space switches as the write and read gating elements of the N optical memories. Using a combination of Ti:LiNbO$_3$ space switches and bistable laser diodes, researchers have demonstrated switching among four 64 Mb/s data streams multiplexed on a 256 Mb/s optical highway (Susuki *et al.*, 1986).

11.5.2.2 Optical Time-Division Multiplexing. As the data rate is increased it becomes increasingly more difficult for electronically multiplexed systems to meet the bandwidth requirements. Two approaches to increase the use of the capacity of the optical fiber channel without increasing the bandwidth of the electronics, modulators, and receivers are wavelength division multiplexing (WDM) and optical time division multiplexing (OTDM). In general, WDM may be used as a final step to increase the throughput over the channel, and devices for implementing such systems have been described in the preceding chapter and are also discussed later in this chapter. Here we consider the use of high-speed directional coupler switches in an optical time-division multiplexing environment.

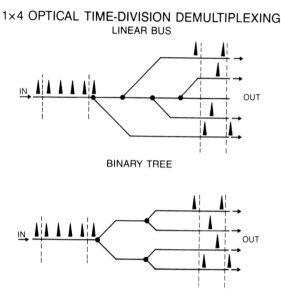

Fig. 11.14 Optical time-division multiplexing architectures.

Two architectures that immediately come to mind for carrying out optical time-division multiplexing/demultiplexing are a linear bus arrangement and a binary hierarchical tree. The basic operation of these two schemes is illustrated in Fig. 11.14, where the demultiplexing function is depicted. In the linear system, data on the incoming bus are demultiplexed by a serial sequence of switches. Each of the switches is configured to be normally in the straight-through state and is activated by a pulse with a $1/N$ duty cycle to demultiplex every N^{th} bit from the bus. The main advantage of such an approach is that only the crosstalk when in the straight-through is of concern and this may be made very low without special techniques. The disadvantages are that the drive electronics and switches must respond over a wide frequency range to accommodate the pulsed operation. In addition, the total loss of the switches scales basically as N, the number of multiplexed channels. The advantage of the binary tree for multiplexing/demultiplexing is that the switches can be operated with narrow-band sinusoidal drives and that the loss scales as $\log N$. The tradeoff required is that now both switch states must be achieved with low crosstalk. This is possible using traveling-wave versions of the $\Delta\beta$-reversal switch (see, for example, Veselka *et al.*, 1988). The switchable directional coupler has the convenient characteristic that its response under large signal conditions is nonlinear, and thus provides an intrinsic "squaring up" of the sinusoidal drive signal.

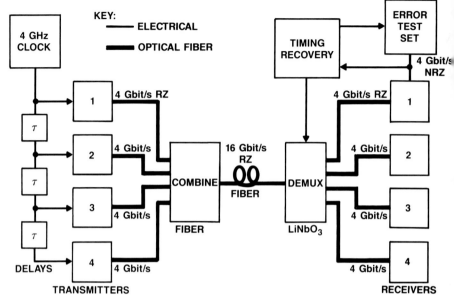

Fig. 11.15   Four channel OTDM system experiment.

In Fig. 11.15 we illustrate a recently demonstrated optical time-division multiplexed system experiment based on Ti:LiNbO$_3$ switches and modulators and integrated-optic mode-locked lasers (Tucker *et al.*, 1987). Four fiber-extended-cavity semiconductor lasers are mode-locked to provide optical pulses at a common repetition rate of 4 GHz. The optical output from each laser is modulated using an external Ti:LiNbO$_3$ waveguide modulator. A useful aspect of this arrangement is that the action of mode-locking inherently provides pulses in a return-to-zero (RZ) format. Thus, the modulators may be driven with a non-return-to-zero (NRZ) electronic data stream as is used for non-optically-multiplexed experiments to minimize the required electronic bandwidth. The optical output of the modulators is at the same time an RZ-encoded data stream with the necessary spacing between optical bits to allow the four incoming channels to be interleaved.

Directional coupler switches are used at the tail end of the system to demultiplex the higher data rate stream back down to the initial data rate. The drive to the demultiplexer is synchronized using a local timing-recovery circuit. In the OTDM experiments, data at the aggregate rate of 16 Gb/s were transmitted over 8 km of fiber with a bit-error-rate below 1×10^{-10}.

11.6 POLARIZATION CONTROL

In sophisticated fiber optic systems, where devices process as well as transport light, the optical polarization state can become an important parameter. For example, devices based on $LiNbO_3$, which is a birefringent crystal, are normally polarization sensitive. Thus, care must be taken to ensure the proper polarization mode of the waveguide is excited. Even in situations where there are no explicitly polarization-sensitive devices, the orthogonality of polarization states can be a concern. Such a situation occurs in coherent detection schemes, where the received signal is mixed with a strong cw beam from a frequency-locked local oscillator. To generate the largest beat signal the waves being mixed must have identical polarizations. Standard single-mode fiber does not preserve the state or orientation of the polarization, and the received polarization over a length of such fiber changes slowly with time. The degree of polarization, however, is maintained.

For these reasons components that, under feedback control, can convert the polarization state of an input beam from one value to any other, are required. Generally they must be able to handle an arbitrary input state and to convert it to any other output polarization state. It is frequently sufficient, however, for the output to be linearly polarized. In addition, devices that can separate orthogonal linear polarizations are often useful.

To understand the operation of integrated optic polarization controllers, we first clarify the description of the polarization state. The polarization state of purely polarized light can be written as the Jones vector (see, for example, Yariv and Yeh, 1984)

$$\begin{pmatrix} E_x \\ E_y \end{pmatrix} = \begin{pmatrix} \cos\theta \\ \sin\theta e^{j\phi} \end{pmatrix}, \tag{11.9}$$

where θ is a measure of the relative $TE(E_x)$ and $TM(E_y)$ components and ϕ is the phase between them. For example, $\theta = 0$ represents TE polarized light while $\theta = 45°$ and $\phi = \pi/2$ represents circularly polarized light. Furthermore, for $\phi = 0$, θ is the polarization angle of linearly polarized light. The most comprehensive polarization transformer must be capable of converting arbitrary values of (θ, ϕ) to arbitrary output values.

11.6.1 Electrooptic TE ↔ TM Mode-Convertor

The basic element for polarization transformations is the electrooptic TE ↔ TM mode converter. This device is designed to use an off-diagonal coefficient of the electro-optic tensor to induce coupling between the TE and TM polarization modes. (Eqs. 11–3 and 4). In birefringent crystals, such as lithium niobate, the TE and TM polarizations may sense different indices depending on the direction

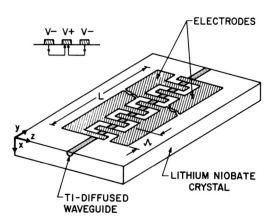

Fig. 11.16 TE ↔ TM mode-converter in LiNbO₃.

of propagation through the crystal. In such cases it is necessary that the electrode be designed to provide a periodic interaction so as to achieve phase-matched coupling between these asynchronous-modes.

An example of an electro-optic mode converter using Ti:LiNbO₃ waveguides is shown in Fig. 11.16. Efficient phase matching via periodic coupling is achieved only for the wavelength, λ_0, that satisfies the condition

$$\Lambda |N_{TM} - N_{TE}| = \lambda_0, \tag{11.10}$$

where Λ is the electrode period and N_{TM} and N_{TE} are the effective indices of the TM and TE polarization, respectively.

In this case for TE(TM) polarized light input to the TE ↔ TM converter in Fig. 11.16, the voltage-induced conversion efficiency to the TM(TE) polarization is given by

$$\eta = \sin^2(\kappa_c L), \tag{11.11}$$

where $\kappa_c = \dfrac{\pi}{\lambda} r_{51} n^3 \dfrac{V}{G} \Gamma$. TE ↔ TM mode converters have been fabricated in LiNbO₃ (Alferness and Buhl, 1981), GaAs (McKenna and Reinhart, 1976), and InP (Schlak *et al.*, 1986).

11.6.2 Polarization Transformer

To allow transformation of arbitrary input polarization, not just TE or TM, it is necessary to also control the relative phase between the TE and TM components. The most basic polarization transformer circuit is shown as implemented in lithium niobate in Fig. 11.17. Its operation may be understood as follows. The

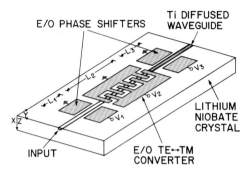

Fig. 11.17 Polarization transformer.

first element is used to adjust the relative phase between the TE and TM component to be $\pm \pi/2$. When this condition is satisfied it can be shown that the phase-matched mode converter (MC) can be operated as a linear polarization rotator such that the change in the polarization angle after the MC is given by

$$\theta_{\text{out}} - \theta_{\text{in}} = \kappa_c L. \tag{11.12}$$

This type of three section polarization controller has been demonstrated in LiNbO$_3$ (Alferness and Buhl, 1981) and is feasible in III-V semiconductor based systems (Schlak *et al.*, 1986). In some applications a linear output polarization and, in fact, TE or TM is sufficient. In this case the second phase shifter is not needed. For general elliptical output polarization, however, the output value of ϕ can be adjusted by the output phase shifter.

The electro-optic polarization controller (PC) described has an important feature for active feedback control, i.e., the values of the phase shift and mode converter voltages can be independently optimized. Suppose an input signal must be converted to TE polarization. The state of polarization can be sensed with the polarization splitter described next. As the input polarization to the PC changes, the voltage V_1 is adjusted to find the value that minimizes the error signal. Voltage V_1 is then fixed at that value and V_2 is swept to further reduce the error signal. This done, there is no need to further optimize V_1 until the input polarization changes.

11.6.3 Polarization Splitters, Combiners, and Filters

Polarization selective devices may be required for several applications including polarization sensing, as mentioned before, and polarization multiplexing. These devices can be classified as either polarizers or polarization splitter/combiners.

The simplest linear polarizer is a metal clad waveguide for which induced currents in the metallic overlay result in as much as a 10 dB/cm loss to the TM polarized component, while producing negligible loss increase to the TE mode (Buhl, 1983). This effect can be enhanced by placing a thin dielectric layer between the waveguide and electrode to resonantly enhance coupling of the TM mode to the metallic overlay, thereby increasing its loss (Kaminow et al., 1974). The differential TM/TE loss coefficient in this case can be as high as 20 dB/mm (Thyagarajan et al., 1985).

In another approach, polarizers have taken advantage of polarization dependent index changes with waveguide fabrication processes such as proton exchange or lithium out-diffusion in lithium niobate. For example, the proton exchange (PE) process increases the extraordinary index but decreases the ordinary index of $LiNbO_3$. Therefore, proton exchange waveguides in z-cut lithium niobate, for example, guide only the TM mode resulting in a TM-pass linear polarizer. By integrating a short section of PE waveguide between titanium diffused waveguides, a high extinction ratio polarizer can be fabricated on a substrate that otherwise guides both polarizations. Polarizers with 40 dB extinction and very low excess loss have been demonstrated by this technique (Findakly and Chen, 1985; Veselka and Bogert, 1986).

Another versatile device is the linear polarization splitter capable of spatially separating the light carried in the TE and TM waveguide modes. This function has been achieved with specially designed Y-branch splitters (Burns et al., 1976), intersecting waveguides (Nakajima et al., 1982), and directional couplers (Alferness and Buhl, 1984). As an example, we consider the polarization selective directional coupler, shown schematically in Fig. 11.18. A polarization selective inter-waveguide power transfer can be achieved by making either the inter-waveguide coupling coefficient, κ, or phase-mismatch, $\Delta\beta$, strongly polarization dependent. In the former case one makes use of the fact that the substrate-waveguide refractive index difference, Δn, is generally different for the TE and TM modes of titanium diffused lithium niobate waveguides. Therefore, the lateral waveguide mode width and κ also depend upon polarization. This

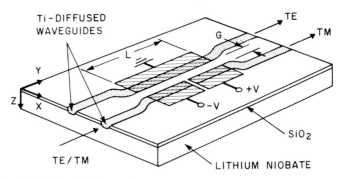

Fig. 11.18 Polarization selective coupler.

difference can be enhanced or reduced by choice of the diffusion parameters. Therefore by appropriate selection of fabrication parameters and the interaction length L, one can approximately make $\kappa_{TE}L = \pi$ and $\kappa_{TM}L = \pi/2$. Generally, the limited value of $|\kappa_{TE} - \kappa_{TM}|$ demands a relatively long device, ~ 1 cm. In this case, for light incident in waveguide 1, the TE component exits waveguide 1 while the TM exits waveguide 2. Such a polarization splitter is a passive component and its performance is limited only by fabrication tolerances.

The second approach to attain polarization splitting makes use of a polarization dependence of $\Delta\beta$. In one implementation this polarization dependence was achieved by a direct metallic electrode on one waveguide and a buffer layer and electrode on the other. The metallic overlay loads the TM mode and changes its propagation constant but has little effect on the TE mode. Therefore $\Delta\beta_{TM}$ can be made finite while $\Delta\beta_{TE} \approx 0$. By choosing L such that $\kappa_{TE}L = \pi/2$, and $\Delta\beta_{TM}L \simeq \sqrt{3}\,\pi$ for light incident in the unloaded waveguide (this is necessary to avoid loss to the TM component), the TE component will couple to the second waveguide while the TM component stays in the incident waveguide. In practice it can be somewhat difficult to satisfy both conditions. A polarization splitter of this type in Ti:LiNbO$_3$ waveguides, however, has been achieved with a polarization crosstalk of ~ -20 dB (Mikami, 1980).

To reduce tolerances and to allow input in either waveguide, one can use a weak polarization dependence of κ and the strong polarization dependence of the electro-optic induced $\Delta\beta$ in lithium niobate. These effects and the use of the reversed $\Delta\beta$ electrode have resulted in a polarization splitter with crosstalk below -20 dB for both outputs and values as low as -27 dB when optimized for one polarization (Alferness and Buhl, 1984).

11.7 WAVELENGTH AND FREQUENCY CONTROL

11.7.1 Interferometric and Directional Coupler Filters

Several types of waveguide wavelength selective devices that can be generically classed as coupled mode or interferometric have been demonstrated. While several implementations of interferometric filters are possible, for purposes of explanation, the Y-branch interferometer with unequal arms shown in Fig. 11.19 is a good example. The Y-branch splitter and combiner perform the same

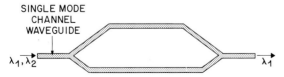

Fig. 11.19 Interferometric wavelength filter.

function as in the interferometric modulator. However, as a result of the physical path length difference ΔL, the optical phase difference at the combiner is strongly wavelength dependent

$$\Delta\phi = \frac{2\pi}{\lambda}N\Delta L. \qquad (11.13)$$

As a result, the intensity transmittance is periodic in wavelength

$$I_{out} = I_i\cos^2\left(\frac{2\pi N\Delta L}{\lambda}\right), \qquad (11.14)$$

yielding a null-to-null bandwidth of

$$\Delta\lambda = \frac{\lambda^2}{2\,\Delta L}. \qquad (11.15)$$

As an example, for a path length difference of $100\dfrac{\lambda}{N_\circ}$ (~ 100 μm in glass for 1.5 μm wavelength) the periodic response peaks at 150 Å wavelength intervals.

Unfortunately, building an integrated-optic interferometer with different arm lengths is limited by technological issues, particularly bend loss. An alternate approach is the interferometer whose arms have equal physical length but unequal optical mode path lengths. One example is shown in Fig. 11.20. Light from a single mode waveguide is injected into a double mode waveguide. After length L, the light is split into two different single mode waveguides. Assuming 3 dB modal splitter and combiners and no inter-mode coupling in the straight waveguide section, for input in the lower guide, the efficiency of light exiting in the upper waveguide is identical to Eq. 11.13 with

$$\Delta\phi = \frac{2\pi}{\lambda}(N_2 - N_1)L = \frac{2\pi}{\lambda}\,\Delta NL, \qquad (11.16)$$

where N_i are the effective refractive indices of each mode. Due to the different mode distributions, application of an electric field changes the difference in

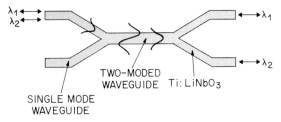

Fig. 11.20 Double moded waveguide filter.

DIRECTIONAL COUPLER FILTER

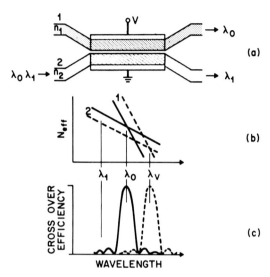

Fig. 11.21 Directional coupler filter.

effective indices of the two modes resulting in tuning of the filter center wavelength. Using 1.2 cm long waveguides, a periodic filter response at $\lambda_0 = 1.5$ μm with 350 Å separation between wavelength peaks was demonstrated. The achieved tuning rate was 25 Å/volt (Neyer, 1984).

The periodic filter response of interferometric devices has somewhat limited utility, although it is well suited to separate two appropriately chosen wavelength channels. To be more useful, periodic filters should exhibit a select/reject wavelength range ratio much less than one. This can be achieved by cascading filters with values of ΔL that are power-of-2 multiples to provide effectively increased finesse.

Alternatively wavelength division multiplexing can be obtained with filters that have a passband response. This is a generic advantage of coupled mode filters over interferometric ones. In these devices the wavelength dependence results from the distributed coupling between two waveguide modes that generally have different propagation indices. Finite power transfer at the design wavelength results from achieving effective phasematching at that wavelength. Two general techniques have been used. The first, shown in Fig. 11.21a, is to design the two mode indices versus wavelength functions to be generally different but such that they intersect at the desired filter center wavelength (Taylor, 1973). This can be achieved using different waveguide dimensions with compensatingly different refractive index differences. At the phasematch wavelength the coupler is phasematched and coupling between the two waveguides is possible. For wavelengths increasingly removed from λ_0 the mismatch increases with a resulting decrease in coupling efficiency. The filter response is functionally the same as the switched directional coupler (Eq. 11.18), with the mismatch depending linearly upon $\lambda - \lambda_0$ rather than voltage. The filter bandwidth depends upon how rapidly the effective index difference changes with $\Delta\lambda = \lambda - \lambda_0$ and upon the interaction length. The fractional bandwidth can be written (Alferness and Schmidt, 1978)

$$\frac{\Delta\lambda}{\lambda} = \frac{1}{L} \left[\frac{d}{d\lambda} [N_2 - N_1] \right]_{\lambda=\lambda_0}. \qquad (11.17)$$

The ability to make $\Delta\lambda$ a strong function of λ (a large intersection angle of the effective index versus λ curves in Fig. 11.21b) to obtain narrow bandwidth without requiring excessive length, depends upon the ability to make waveguides with large values of waveguide substrate index difference. Therefore, for a given length, narrower filter bandwidths can be achieved in the semiconductor material systems than with Ti:LiNbO$_3$ waveguides.

The directional coupler filter can be readily tuned by increasing the effective index of one waveguide while decreasingt that of the other using the push-pull electrodes arrangement shown in Fig. 11.21a. This moves the phasematch to $\lambda = \lambda_T$ where

$$\frac{\lambda - \lambda_T}{\lambda_0} = \frac{2\,\Delta N(V)}{\lambda_0} \left[\frac{d}{d\lambda} [N_2 - N_1] \right]_{\lambda=\lambda_0}. \qquad (11.18)$$

Directional coupler filters have been demonstrated with both Ti:LiNbO$_3$ and InGaAsP/InP waveguides, although tuning has been demonstrated only with the former. In Ti:LiNbO$_3$ for operation near $\lambda_0 = 1.5\ \mu$m, a 700 Å bandwidth was demonstrated with a tunability of ~ 100 Å/volt and essentially complete transfer efficiency at the filter center wavelength. Because this device is, in fact, a tunable wavelength-dependent switch, it can be used to multiplex together two wavelengths while simultaneously modulating one of them (Alferness and Veselka, 1985) as shown in Fig. 11.22. An InGaAsP/InP vertical directional coupler filter has been demonstrated in which the two guiding regions are epitaxially grown planar waveguides (Lindgren et al., 1985). A 170 Å bandwidth was achieved. This vertical wavelength selective coupling is also essential to the operation of InGaAsP/InP semiconductor lasers which integrate, via evanescent coupling, an active gain region with low-loss passive waveguide reflector sections (Sakakibara et al., 1980).

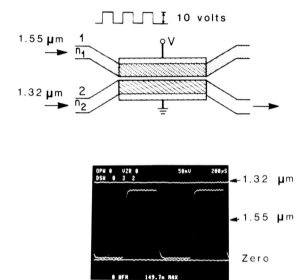

Fig. 11.22 Multiplexing and modulating with a waveguide filter.

11.7.2 Mode-Convertor and Grating Filters

Phasematching for a particular wavelength can also be achieved in coupled mode filters via periodic coupling. In this case the two modes are designed to have different effective indices for all wavelengths. As a result, no net power transfer occurs even though means of local coupling exists between the two modes. By making the coupling strength between the two modes periodic along the propagation direction, however, complete coupling can occur at a wavelength λ_0, which satisfies the phasematch condition

$$\frac{2\pi}{\lambda_0}(N_2 - N_1) = \frac{2\pi}{\Lambda}, \tag{11.19}$$

where Λ is the coupling period. The filter response is again the familiar coupled mode response (Eq. 11.8) with the phase mismatch as a function of $\Delta\lambda = \lambda - \lambda_0$

given by

$$\phi(\Delta\lambda) = \frac{\pi \Delta N \Delta\lambda}{\lambda_0^2} L. \qquad (11.20)$$

The 3 dB fractional bandwidth is given approximately by

$$\frac{\Delta\lambda}{\lambda} \approx \frac{\Lambda}{L}. \qquad (11.21)$$

An example of a wavelength filter based upon periodic coupling in Ti:LiNbO$_3$ makes use of the rather strong birefringence of lithium niobate to achieve a large difference in effective indices of the TE and TM modes in a single waveguide. Coupling is achieved electro-optically with a periodic electrode (Alferness and Buhl, 1980). The filter center wavelength can be tuned by electro-optically changing the birefringence (Alferness and Buhl, 1982). A schematic diagram of the device is shown in Fig. 11.23 together with the measured response with several tuning voltages. As shown the filter bandwidth is about 12 Å. The filter exhibits about seven resolvable channels (Heismann et al., 1987a). This device requires input light of fixed polarization and a polarization selective element as described earlier.

The second example of a periodically coupled mode device is the reflection filter shown in Fig. 11.24. In this case phase-matched coupling between the counter propagating ($N_2 = -N_1$ in Eq. 11.19) modes of a single waveguide is achieved via a fine period grating. Because of the large change in effective index and resulting fine period for this contra-direction interaction, the filter bandwidth, given approximately by Eq. 11.21, can be quite small with modest interaction lengths. The required fine grating is typically written as a photoresist mask using two interferring ultraviolet beams and then chemically or physically etched. In glass waveguides, a filter bandwidth as narrow as 0.1 Å (centered around 6000 Å) was achieved for a 1 cm long grating filter (Schmidt et al., 1974). In InGaAsP/InP waveguides filter bandwidths as narrow as 6 Å have been reported (Alferness et al., 1984). Because of the modest substrate-waveguide index difference and difficulty to etch lithium niobate, reflection filter results for Ti:LiNbO$_3$ waveguides have been very limited.

To convert the wavelength dependent retro-reflection into a useful filter device a separate coupler can be used to separate the input beam from the reflected (filtered) one. This, however, results in an additional 6 dB splitting/combining loss. A better approach is to place the etched grating in the interwaveguide region of a mismatched directional coupler as has been done with a fiber device (Whalen et al., 1986).

Finally, very narrow bandwidth has been achieved in a compact waveguide structure by using the two waveguide grating reflectors to form the effective

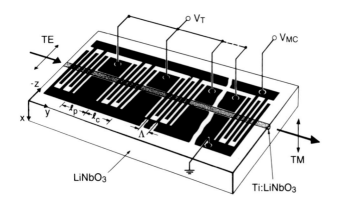

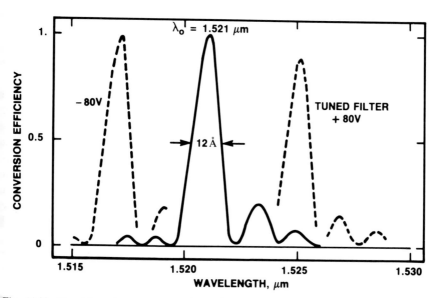

Fig. 11.23 Tunable mode-converter wavelength filter.

mirrors of a resonator. A filter bandwidth of 1 Å near 1.5 μm was achieved using a grating resonator made with InGaAsP/InP waveguides (Alferness *et al.*, 1986b). The device was only 250 μm long.

11.7.3 Frequency Shifter

Devices that can change the optical frequency of a signal are useful for many systems that can be envisioned using coherent optical techniques. To change the

InGaAsP/InP WAVEGUIDE GRATING FILTER

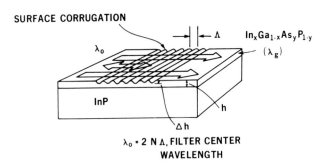

Fig. 11.24 Grating filter.

optical frequency, it is necessary to impart or extract energy, as well as change the momentum, of individual photons constituting the signal. One technique to accomplish this makes use of electrooptic mode conversion. If the region at which the conversion takes place can be made to move relative to the direction of propagation of the original optical wave, then the source of the converted signal appears to be moving for a stationary observer and the light is consequently Doppler shifted. In practice, to generate the effect of a moving coupling grating, the mode-converter is divided into several sections and each is driven with a properly phase-shifted sinusoidal signal having a frequency equal to the desired up/down frequency conversion. Such a frequency shifter has been demonstrated in Ti:LiNbO$_3$ waveguides (Heismann and Ulrich, 1984). In addition, frequency translators based on amplitude modulation have been integrated into sophisticated optical circuits for optical heterodyne receivers (Stallard et al., 1985).

11.7.4 Tunable Extended Cavity Laser

In all of the discussions so far, the optical waveguide devices have acted on optical signals external to the lightwave source. If the devices are inserted internal to laser cavities then these basic elements may be used to implement additional functions. For example, if a waveguide electrooptic phase-shifter similar to those described for phase modulation is placed within the laser cavity, then that device can be used to adjust the optical path length of the resonator and serve as a fine frequency tuning element (Korotky et al., 1986). If a tunable filter, such as the tunable TE ↔ TM mode-converter filter described before, is placed with the laser cavity, then frequency tuning over a wide range is possible.

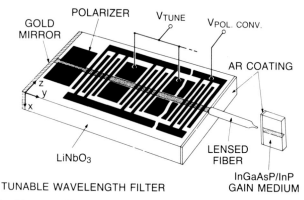

Fig. 11.25 Tunable waveguide extended cavity laser.

Using the arrangement shown in Fig. 11.25, incremental tuning over a range of 70 Å has been demonstrated (Heismann *et al.*, 1987b).

11.8 SUMMARY

In conclusion, in this chapter we have discussed the more basic waveguide electrooptic devices that may potentially find use in fiber optic systems. Their performance in laboratory systems has been described. It is expected that vigorous research in the area of integrated optical circuits and opto-electronic integration in semiconductors will continue to address the growing need for high-performance, low cost devices necessary to realize the full potential of fiber optic communication.

REFERENCES

Agrawal, N., McCaughan, L., and Bogert, G. A. (1987). Novel physical effects in intersecting waveguides. *Tech. Digest of Conf. on Lasers and Electro-Optics*, Baltimore, paper ThU9.

Alferness, R. C. (1981). Guided-wave devices for optical communication. *IEEE J. Quantum Electron.* **QE-17**, 946–959.

Alferness, R. C. (1982). Waveguide electrooptic modulators. *IEEE Tran. Microwave Theory and Tech.* **MTT-30**, 1121–1137.

Alferness, R. C., and Buhl, L. L. (1980). Electro-optic waveguide TE ↔ TM mode converter with low drive voltage. *Opt. Lett.* **5**, 473–475.

Alferness, R. C., and Buhl, L. L. (1981). Waveguide electrooptic polarization transformer, *Appl. Phys. Lett.* **38**, 655–657.

Alferness, R. C., and Buhl, L. L. (1982). Tunable electro-optic waveguide TE ↔ TM converter/wavelength filter, *Appl. Phys. Lett.* **40**, 861–862.

Alferness, R. C., and Buhl, L. L. (1984). Low-crosstalk waveguide polarization multiplexer/demultiplexer for λ = 1.32 μm. *Opt. Lett.* **10**, 140–142.

Alferness, R. C., and Schmidt, R. V. (1978). Tunable optical waveguide directional coupler filter. *Appl. Phys. Lett.* **33**, 161–163.

Alferness, R. C., and Veselka, J. J. (1985). Simultaneous modulation and wavelength multiplexing with a tunable Ti:LiNbO₃ directional coupler filter. *Electron Lett.* **21**, 466–467.

Alferness, R. C., Ramaswamy, V. R., Korotky, S. K., Divino, M. D., and Buhl, L. L. (1982). Efficient single-mode fiber to titanium diffused lithium niobate waveguide coupling for $\lambda = 1.32$ μm. *IEEE J. Quantum Electron.* **QE-18**, 1807–1813.

Alferness, R. C., Joyner, C. H., Divino, M. D., and Buhl, L. L. (1984). InGaAsP/InP waveguide grating filters for $\lambda = 1.5$ μm. *Appl. Phys. Lett.* **45**, 1278–1280.

Alferness, R. C., Buhl, L. L., Divino, M. D., Korotky, S. K., and Stulz, L. W. (1986a). Low-loss, broadband Ti:LiNbO₃ waveguide phase modulators for coherent systems. *Electron. Lett.* **22**, 309–310.

Alferness, R. C., Joyner, C. H., Divino, M. D., Martyak, M. J. R., and Buhl, L. L. (1986) Narrowband grating resonator filters in InGaAsP/InP waveguides. *Appl. Phys. Lett.* **49**, 125–127.

Auracher, F., Schicketanz, D., and Zeitler, K.-H. (1984). High-speed $\Delta\beta$-reversal directional coupler modulator with low insertion loss for 1.3 μm in LiNbO₃. *J. Opt. Comm.* **5**, 7–9.

Benes, V. E. (1965). "Mathematical Theory of Connecting Networks and Telephone Traffic." Academic Press, New York.

Blauvelt, H., Parsons, J., Lewis, D., and Yen, H. (1984). High-speed GaAs Schottky barrier photodetectors for microwave fiber-optic links. *Tech. Digest of Conf. on Opt. Fiber Comm.*, New Orleans, paper TUH3.

Bogert, G. A. (1987). Ti:LiNbO₃ intersecting waveguides. *Electron Lett.* **23**, 72–73.

Bogert, G. A., Murphy, E. J., and Ku, R. T. (1986). Low crosstalk 4 × 4 Ti:LiNbO₃ optical switch with permanently attached polarization maintaining fiber array. *J. Lightwave Technol.* **LT-4**, 1542–1545.

Buhl, L. L. (1983). Optical losses in metal/SiO₂ clad Ti:LiNbO₃ waveguides. *Electron. Lett.* **19**, 659–660.

Bulmer, C. H. and Burns, W. K. (1984). Linear interferometric modulators in Ti:LiNbO₃. *J. Lightwave Technol.* **LT-2**, 512–521.

Burns, W. K., Lee, A. B., and Milton, A. F. (1976). Active branching waveguide modulator. *Appl. Phys. Lett.* **29**, 790–792.

Clos, C. (1953). A study of non-blocking switch networks. *Bell. Sys. Tech. J.*, **32**, 406–424.

Duthie, P. J., Wale, M. J., and Bennion, I. (1987). New architecture for large integrated optical switch arrays. *Tech. Digest of Top. Meet. on Photonic Switching*, Incline Village, NV, paper ThD4.

Erickson, J. R., Hseih, C. G., Huisman, R. F., Larson, M. L., Tokar, J. V., Bogert, G. A., Murphy, E. J., and Ku, R. T. (1987). Photonic switching demonstration display. *Tech. Digest of Top. Meet. on Photonic Switching*, Incline Village, NV, paper ThA5.

Erickson, J. R., Nordin, R. A., Payne, W. A., and Ratajack, M. P. (1987). A 1.7 Gigabit-per-second, time-multiplexed photonic switching experiment. *IEEE Communications Magazine* **25**, 56–58.

Findakly, T., and Chen, B. (1984). Single-mode transmission selective integrated-optical polarizers in LiNbO₃. *Electron. Lett.* **20**, 128–129.

Gnauck, A. H., Kasper, B. L., Dawson, R. W., Linke, R. A., Koch, T. L., Bridges, T. J., Burkhardt, E. G., Yen, R. T., Wilt, D. P., Ciemiecki Nelson, K., and Campbell, J. C. (1985). 4 Gb/s transmission over 103 km of optical fiber using a novel electronic multiplexer/demultiplexer. *Tech. Digest of Conf. on Opt. Fiber Comm.*, San Diego, paper PDP2.

Gnauck, A. H., Korotky, S. K., Kasper, B. L., Campbell, J. C., Talman, J. R., Veselka, J. J., and McCormick, A. R. (1986). Information-bandwidth-limited transmission at 8 Gb/s over 68.3 km of single-mode optical fiber. *Tech. Digest of Conf. on Opt. Fiber Comm.*, Atlanta, paper PDP9.

Gnauck, A. H. (1987). Coherent lightwave transmission at 2 Gbit/sec over 170 km of optical fiber using phase modulation. *Electron. Lett.* **23**, 286–287.

Granestrand, P., Stolz, B., Thylén, L., Bergvall, K., Döldissen, W., Heinrich, H., and Hoffmann, D. (1986). Strictly nonblocking 8 × 8 integrated optical switch matrix. *Electron. Lett.* **22**, 816–818.

Heismann, F., and Ulrich, R. (1984). Integrated-optical frequency translator with strip waveguide. *Appl. Phys. Lett.* **45**, 490–492.

Heismann, F., Buhl, L. L., and Alferness, R. C. (1987). Electro-optically tunable, narrowband Ti:LiNbO$_3$ wavelength filter. *Electron. Lett.* **23**, 572–574.

Heismann, F., Alferness, R. C., Buhl, L. L., Eisenstein, G., Korotky, S. K., Veselka, J. J., Stulz, L. W., and Burrus, C. A. (1987). Narrow-linewidth, electro-optically tunable InGaAsP-Ti:LiNbO$_3$ extended cavity laser. *Appl. Phys. Lett.* **51**, 164–166.

Hiruma, K., Inoue, H., Ishida, K., and Matsumura, H. (1985). Low loss GaAs optical waveguides grown by metal-organic vapor deposition. *Appl. Phys. Lett.* **47**, 186–188.

Hoffmann, D., Heidrich, H., Döldissen, W., Ahlers, H., Kleinwächter, A., and Schmidt, R. has been published. (1986). Low-loss rearrangeably nonblocking 4 × 4 switch matrix module. *Tech. Digest of Eur. Conf. on Opt. Comm.*, Barcelona, 167–170.

L. D. Hutcheson, Ed. (1987). "Integrated Optical Circuits and Components." Marcel Dekker, New York.

Ishida, K., Hirma, K., Inoue, H., Asai, T., Hirao, M., and Matsumura, H. (1985). Low loss GaAs optical switches. *Tech. Digest of Eur. Conf. on Integ. Opt.*, Berlin, post-deadline.

Kaminow, I. P. (1974). "An Introduction to Electro-Optic Devices." Academic Press, New York.

Kaminow, I. P. (1987). "CRC Handbook of Laser Science and Technology, Vol. IV." edited by M. J. Weber, CRC Press, Inc., Boca Raton, FL.

Kaminow, I. P., Bridges, T. J., and Pollack, M. A. (1970). A 964-GHz traveling-wave electro-optic light modulator. *Appl. Phys. Lett.* **16**, 416–418.

Kaminow, I. P., Mammel, W. L., and Weber, H. P. (1974). Metal clad optical waveguides: analytical and experimental study. *Appl. Opt.* **13**, 396–405.

Kapon, E., and Bhat, R. (1987). Low-loss GaAs/AlGaAs ridge waveguides grown by organometallic vapor-phase epitaxy. *Tech. Digest of Conf. on Lasers and Electrooptics*, Baltimore, paper WQ3.

Kogelnik, H., and Schmidt, R. V. (1976). Switched directional couplers with alternating Δβ. *IEEE J. Quantum Electron.* **QE-12**, 396–401.

Kondo, M., Ohta, Y., Fujiwara, M., and Sakaguchi, M. (1982). Integrated optical switch matrix for single-mode fiber networks. *IEEE J. Quantum Electron.* **QE-18**, 1759–1765.

Kondo, M., Komatsu, K., Ohta, Y., Suzuki, S., Nagashima, K., and Goto, H. (1983). High-speed optical time switch with integrated optical 1 × 4 switches and single-polarization fiber delays lines. *Technical Digest of 4th Int. Conf. on Integrated Optics and Opt. Fiber Commun.*, Tokyo, paper 29D3-7.

Korotky, S. K. (1986). Three-space representation of phase-mismatch switching in coupled two-state optical systems. *IEEE J. Quantum Electron.* **QE-22**, 952–958.

Korotky, S. K., and Alferness, R. C. (1983). Time- and frequency-domain response of directional-coupler traveling-wave optical modulators. *J. Lightwave Technol.* **LT-1**, pp. 244–251.

Korotky, S. K., Eisenstein, G., Gnauck, A. H., Kasper, B. L., Veselka, J. J., Alferness, R. C., Buhl, L. L., Burrus, C. A., Huo, T. C. D., Stulz, L. W., Ciemiecki Nelson, K., Cohen, L. G., Dawson, R. W., and Campbell, J. C. (1985). 4 Gb/s transmission experiment over 117 km of optical fibver using a Ti: LiNbO$_3$ external modulator. *J. Lightwave Technol.* **LT-3**, 1027–1031.

Korotky, S. K., Marcatili, E. A. J., Eisenstein, G., Veselka, J. J., Heismann, F., and Alferness, R. C. (1986). Integrated-optic, narrow-linewidth laser. *Appl. Phys. Lett.* **49**, 10–12.

Korotky, S. K., Eisenstein, G., Tucker, R. S., Veselka, J. J., and Raybon, G. (1987). Optical intensity modulation to 40 GHz using a waveguide electrooptic switch. *Appl. Phys. Lett.* **50**, 1631–1633.

Koyama, F., and Iga, K. (1987). Frequency chirping in some types of external modulator. *Tech. Digest of Int'l. Conf. on Integrated Optics and Opt. Fiber Comm.*, Reno, Nevada, paper WO4.

Leboutet, A., and Sorel, Y. (1985). 1.7 Gbit/s direct and external modulation of lasers: A comparison in the second and third windows over 40 km of installed link. *Tech. Digest of the Int'l. Conf. on Integ. Opt. and Opt. Comm.*, Venice, 757–760.

Lindgren, S., Broberg, B., Oberg, M., and Jiang, H. (1985). Integrated optics wavelength filter in InGaAsP-InP. *Tech. Digest of Int'l. Conf. on Integrated Optics and Opt. Fiber Comm.*, Venice, 175–178.

Linke, R. A. (1987). Beyond gigabit-per-second transmission rates. *Tech. Digest of Conf. on Opt. Fiber Comm.*, Reno, NV, paper WO3.

Marcuse, D. (1974). "Theory of Dielectric Optical Waveguides." Academic Press, New York.

McCaughan, L., and Bogert, G. A. (1985). 4×4 Ti:LiNbO$_3$ integrated-optical crossbar switch array. *Appl. Phys. Lett.* **47**, 348–350.

McKenna, J. C., and Reinhart, F. K. (1976). Double-heterostructure GaAs-Al$_x$Ga$_{1-x}$As [110] p-n-junction-diode modulator. *J. Appl. Phys.* **47**, 2069–2078.

Mikami, O. (1980). LiNbO$_3$ coupled-waveguide TE/TM mode splitter. *Appl. Phys. Lett.* **36**, 491–493.

Milbrodt, M. A., Veselka, J. J., Bahadori, K., Chen, Y. C., Bogert, G. A., Coult, D. G., Holmes, R. J., Erickson, J. R., and Payne, W. A. (1988). A tree-structured 4×4 switch array in lithium niobate with attached fibers and proton-exchange polarizers. *Tech. Digest on Conf. on Integ. and Guided-Wave Opt.*, Sante Fe, paper MF9.

Miller, S. E. (1954). Coupled-wave theory and waveguide applications. *Bell Syst. Tech. J.* **33**, 661–719.

Minford, W. J., Korotky, S. K., and Alferness, R. C. (1982). Low-loss Ti:LiNbO$_3$ waveguide bends at $\lambda = 1.3$ μm. *IEEE J. Quantum Electron.* **QE-18**, 1802–1806.

Nakajima, H., Horimatsu, T., Seino, M., and Sawaki, I. (1982). Crosstalk characteristics of Ti:LiNbO$_3$ intersecting waveguides and their application as TE/TM mode splitters. *IEEE J. Quantum Electron.* **QE-18**, 771–776.

Neyer, A. (1984). Integrated-optical multi-channel wavelength multiplexer for monomode systems. *Elect. Lett.* **20**, 744–746.

Neyer, A., Mevenkamp, W., and Kretzschmann, B. (1986). Nonblocking 4×4 switch array with sixteen X-switches in Ti:LiNbO$_3$. *Tech. Digest on Conf. on Integ. and Guided-Wave Opt.*, Atlanta, paper WAA2.

Padmanabhan, K., and Netravali, A. (1987). Dilated networks for photonic switching. *Tech. Digest of Top. Meet. on Photonic Switching*, Incline Village, NV, paper ThB3.

Park, Y. K., Berstein, S. S., Tench, R. E., Smith, R. W., Korotky, S. K., and Burns, K. J. (1987). Crosstalk and prefiltering in a two-channel ASK heterodyne detection system without the effect of laser phase noise. *Tech. Digest of Conf. on Opt. Fiber Comm.*, Reno, NV, paper PDP13.

Reinhart, F. K., and Logan, R. A. (1974). Monolithically integrated AlGaAs double heterostructure optical components. *Appl. Phys. Lett.* **25**, 622–624.

Richards, G. W., and Hwang, F. K. (1985). A two stage rearrangeable broadcast switching network. *IEEE Trans. on Comm.* **33**, 1025–1035.

Sakakibara, Y., Furuya, K., Utaka, K., and Suematsu, Y. (1980). Single-mode oscillation under high speed direct modulation in GaInAsP/InP integrated twin-guide lasers with distributed Bragg reflector. *Electron. Lett.* **16**, 456–458.

Sakano, S., Inoue, H., Nakamura, H., Katuyama, T., and Matsumura, H. (1986). InGaAsP/InP monolithic integrated circuit with lasers and an optical switch. *Electron. Lett.* **22**, 594–596.

Schlak, M., Nolting, H. P., Albrecht, P., Döldissen, W., Franke, D., Niggebrugge, U., and Schmitt, F. (1986). Integrated-optic polarization converter on (001)-InP substrate. *Electron. Lett.* **22**, 883–885.

Schmidt, R. V., Flanders, D. C., Shank, C. V., and Standby, R. D. (1974). Narrow-band grating filters for thin-film optical waveguides. *Appl. Phys. Letts.*, 25, 651–652.

Spanke, R. A. (1986). Architectures for large nonblocking optical space switches. *IEEE J. Quantum Electron.* QE-22, 964–967.

Spanke, R. A. (1987). Architectures for guided-wave optical space switching systems. *IEEE Comm.*, 25, 42–48.

Stallard, W. A., Hodgkinson, T. G., Preston, K. R., and Booth, R. C. (1985). Novel LiNbO₃ integrated-optic component for coherent optical heterodyne detection. *Electron Lett.* 21, 1077–1079.

Stephens, W. E., and Joseph, T. R. (1987). System characteristics of directly modulated and externally modulated rf fiber-optic links. *J. Lightwave Technol.* LT-5, 380–387.

Suche, H., Hampel, B., Seibert, H., and Sohler, W. (1985). Parametric fluorescence, amplification, and oscillation in Ti:LiNbO₃ optical waveguides. *Proc. Conf. on Integ. Opt. Cir. Engr. II, SPIE* 578, Boston, 156–161.

Suzuki, S., Terakado, T., Komatsu, K., Nagashima, K., Suziki, A., and Kondo, M. (1986). An experiment on high-speed optical time-division switching. *J. Lightwave Technol.* LT-4, 894–899.

Suzuki, S., Kondo, M., Kagashima, K., Mitsuhashi, M., Komatsu, K., and Miyakawa, T. (1987). Thirty-two-line optical space-division switching system. *Tech. Digest. of Int 'l. Conf. on Integ. Opt. and Opt. Fiber Comm.*, Reno, NV, paper WB4.

Tamir, T., Ed. (1979). "Integrated Optics," 2nd ed., Springer-Verlag, New York.

Taylor, H. F. (1973). Frequency-selective coupling in parallel dielectric waveguides. *Opt. Comm.* 8, 421–425.

Thyagarajan, K., Bourbin, Y., Enard, A., Vatoux, S. and Papuchon, M. (1985). Experimental demonstration of TM mode-attenuation resonance in planar metal-clad optical waveguides. *Opt. Lett.* 10, 288–290.

Tucker, R. S., Eisenstein, G., Korotky, S. K., Buhl, L. L., Veselka, J. J., Kasper, B. L., and Alferness, R. C. (1987). 16 Gbit/s fiber transmission experiment using optical time-division multiplexing. *Electron. Lett.* 23, 1270–1271.

Veselka, J. J., and Bogert, G. A. (1987). Low-loss proton exchange channel waveguides and TM-pass polarizers in z-cut LiNbO₃. *Tech. Digest of Conf. on Integ. Opt. and Opt. Fiber Comm.*, Reno, Nevada, paper TUH3.

Veselka, J. J., and Korotky, S. K. (1986). Optimization of Ti:LiNbO₃ optical waveguides and directional coupler switches for 1.56 μm wavelength. *IEEE J. Quantum Electron.* QE-22, 933–938.

Veselka, J. J., Herr, D. A., Murphy, T. O., Buhl, L. L., and Korotky, S. K. (1988). Integrated high-speed Ti:LiNbO₃ Δβ-reversal switching circuits. *Tech. Digest on Conf. on Integ. and Guided-Wave Opt.*, Santa Fe, paper WD2.

Wang, S. Y., Lin, S. H., and Houng, Y. M. (1987). GaAs traveling-wave electrooptic waveguide modulator with bandwidth > 20 GHz at 1.3 μm. *Tech. Digest of Int 'l. Conf. on Integrated Opt. and Opt. Fiber Comm.*, Reno, Nevada, paper WK3.

Whalen, M. S., Divino, M. D., Alferness, R. C. (1986). Demonstration of a narrowband Bragg-reflection filter in a single-mode fiber directional coupler. *Electron. Lett.* 22, 681–682.

Yariv, A., and Yeh, P. (1984). "Optical Waves in Crystals." Wiley, New York.

Chapter 12

Light-Emitting Diodes for Telecommunications

TIEN PEI LEE

Bell Communications Research, Inc., Red Bank, New Jersey

C. A. BURRUS, JR.

AT&T Bell Laboratories, Inc., Holmdel, New Jersey

R. H. SAUL

AT&T Bell Laboratories, Inc., Murray Hill, New Jersey

12.1 INTRODUCTION

Under proper but easy-to-achieve conditions, forward-biased p-n junctions of many semiconductors, notably those composed of elements from group III and V of the periodic table, can emit external spontaneous radiation in the visible or infrared regions of the spectrum. Such devices are called light-emitting diodes or LEDs. In operation, the normally empty conduction band of the semiconductor is populated with electrons injected into it by a forward current through the junction, and light is generated when these electrons recombine with holes in the valence band and emit a photon. The energy of the emitted photon is approximately that of the energy gap between the conduction and valence bands of the particular semiconductor. This spontaneous emission, which usually occurs physically in the p-layer close to the p-n junction, is referred to as recombination radiation. Unfortunately, the recombination of the injected electrons also can occur by processes that do not emit photons, the so-called nonradiative processes; thus, the internal quantum efficiency of an LED is not 100%. One aim in the fabrication of LEDs is to maximize this internal conversion, i.e., minimize the crystalline imperfections and impurities that lead to nonradiative traps for the injected electrons. In practice the internal quantum efficiency can be quite high, certainly exceeding 50% in simple homostructure LEDs (Hill, 1965; Archer and Kreps, 1967). In the double-heterostructure (DH) LEDs to be the principal topic of discussion here, measurements of recombination lifetime suggest that internal quantum efficiencies of 60–80% are being achieved (Lee and Dentai, 1978), with the losses primarily ascribed to nonradiative recombination at semiconductor surfaces and interfaces. Another aim is to produce the radiation in a geometry from which it can be collected and, thus, to maximize the useful external power efficiency. A third aim is to produce diodes in which the light output can be directly current-modulated at high rates with information-carrying signals. A last aim is to produce geometries from which heat can be extracted efficiently, since the output of an LED drops by 2–3 dB if the junction temperature rises 100°C.

It was recognized early that, if significant power were to be coupled from an incoherent LED into a small fiber, the source would have to exhibit very high radiance. For light-emitting diodes, this meant that it would be necessary to use direct energy gap semiconductors and structures that could be driven at high current densities would have to be devised. It also was recognized that such a diode probably would take the geometrical form of either a very small-area emitter (Kibler et al., 1964), in which the light from the surface of a small junction would be collected perpendicular to the junction plane through a thin or transparent layer of semiconductor above the junction, or of an edge-emitter (Zargar'yants et al., 1971) in which the light would be emitted directly from the exposed edge of the junction. Both configurations have been made and applied to optical fiber uses.

Optical fiber systems using LED sources can provide reliable, inexpensive alternatives to laser-based systems. LEDs utilize relatively simple driving circuits without need for feedback to control output power, and they are capable of operating over a wide range of temperatures with projected device lifetimes one to two orders of magnitude longer than those of laser diodes made of the same material. First generation LED systems used GaAlAs sources emitting in the 0.85–0.9 μm wavelength region, but because of the relatively high attenuation and chromatic dispersion of silica-based fibers in this wavelength region, transmission distances are limited to short hauls ($\leq$ 4 Km) and data rates are moderate ($\leq$ 50 Mb/s); thus such systems are primarily useful for inexpensive data links. Second generation systems utilize long-wavelength InGaAsP LEDs emitting near 1.3-μm wavelength, where silica fibers exhibit low attenuation and minimum dispersion. Repeater spacings of tens of kilometers and data rates of a few hundred Mb/s are feasible (Gloge et al., 1980). Here the primary applications are for high speed data links, intra-city trunks, and loop feeders. In fact, many LED-based large telecommunication systems, such as the SLC 96(™) are in use today (Bohn *et al.*, 1984). While long-distance optical fiber trunk lines are in rapid deployment at present, applications of lightwave in the subscriber loop are also emerging. However, the enormous number of devices ($\sim 10^8$) required for the subscriber loop demands a high degree of device reliability, at least an order of magnitude higher than that exhibited by present laser diodes. This requirement favors the choice of LEDs rather than lasers in this application, until laser reliability can be improved. Recent experiments have shown that 1.3-μm LEDs can launch sufficient power into single-mode fibers for subscriber loop applications at speeds as high as 560 Mb/s [Gimlett *et al.*, 1985b].

This chapter begins with a description of various device structures for high radiance LEDs, both surface emitting and edge emitting, suitable for lightwave communications using optical fibers at both 0.9-μm and 1.3-μm wavelengths, with discussions of the optimization of output power and modulation speed. Next, the spectral and temperature characteristics of practical devices are described, followed by discussions regarding the coupling of power to both multimode and single-mode fibers. Whenever appropriate, comparisons of the performance of surface emitters and edge emitters are made. Finally, the chapter closes with descriptions of some LED-based multimode system experiments and various LED-based single-mode system experiments aimed at future subscriber loop applications.

Several earlier review papers on small-area high-radiance LEDs (Lee, 1980; Lee, 1982; Saul, 1983; Saul *et al.*, 1985) may be used as supplementary reading material.

12.2 DEVICE STRUCTURES

12.2.1 Surface Emitters

Historically, homojunction surface emitters designed for optical fiber communication were made in GaAs emitting at 0.9-μm wavelength (C. A. Burrus, unpublished, 1969; Burrus and Dawson, 1970; Goodfellow and Mabbit, 1976). The principle of these designs, still largely unchanged, utilized the fact that a small-area *p-n* junction will benefit from rapid spreading of heat into a large heat sink, and can be driven at relatively high current densities before overheating, thus providing a light source with high radiance. The emitting area of the junction is confined in various ways to a small dot, usually 15 to 100 μm diameter, which is smaller or equal to the core diameter of the fiber to achieve maximum coupling efficiency. The semiconductor through which the emission must be collected is made transparent and/or very thin, 10–15 μm, to minimize absorption and allow the end of the fiber to be very close to the emitting surface. Such devices may be operated at current densities of a few kiloamperes per square centimeter for the larger sizes to several tens of kiloamperes per square centimeter for the smallest.

Considerable advantage can accrue from the use of somewhat more complicated semiconductor structures, as illustrated in the example of Fig. 12.1. These structures (Burrus and Miller, 1971; King and SpringThorpe, 1975), employing GaAlAs/GaAs double heterostructures, have several advantages compared to the single-material or homostructure devices. These advantages are (1) increased efficiency resulting from the electron confinement provided by the layers of higher energy gap semiconductor surrounding the recombination regions near the junction; (2) increased transmission of the emitted radiation to the outside, since the higher energy gap confining layers do not absorb radiation from the lower energy gap emitting regions; and (3), not limited to heterostructures but especially easy to obtain with them, the emitted wavelength may be varied readily by controlling the composition of the semiconductor layers containing the junction. In the particular example shown in Fig. 12.1, the small emitting area was made by restricting the current flow in the contact area, for example, by using a SiO_2 layer as an insulator. The GaAs substrate material above the emitting area was removed by chemical etching, forming a well, to reduce absorption loss and, simultaneously, to allow the end of the fiber to be very close to the emitting surface for maximum light coupling. Alternatively, planar devices have been fabricated by totally removing the substrate (Berkstresser et al., 1980; Abe et al., 1977). Other methods for confining the current in a small emitting area include proton bombardment (Dyment et al., 1977) of regions surrounding the light-emitting area, using a reverse-biased p-n

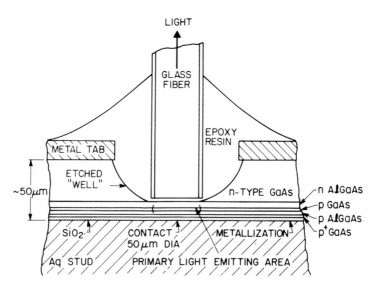

Fig. 12.1 GaAlAs/GaAs Surface-emitting LED structures. [After Burrus and Miller (1971)]

junction for isolation (T. P. Lee, unpublished 1976) and employing Schottky barriers in place of an oxide-confined p-contact (Chin et al., 1981).

The double-heterostructure wafer, usually grown by liquid phase epitaxy (LPE), consists of four layers (from the substrate): an n-type (10^{18} cm^{-3}) $Ga_{1-x}Al_xAs$ confining layer, a p-type (10^{17} to 10^{19} cm^{-3}) GaAs or $Ga_{1-y}Al_yAs$ (y < x) active layer, another p-type (10^{19} cm^{-3}) $Ga_{1-x}Al_xAs$ confining layer, and a top p-type (> 10^{19} cm^{-3}) GaAs contacting layer to assure low p-contact resistance. Sn or Te is used for the n-type dopant and Ge as the p-type dopant.

The structure for the long-wavelength (usually 1.3 μm, to match the minimum dispersion wavelength of silica-based fibers) LED is very similar, except that the material system is based on InGaAsP/InP, as shown in Fig. 12.2. The first 2–5 μm thick n-InP buffer layer is generally Sn-doped ($\sim$ 2 $\times$ 10^{18} cm^{-3}), followed by an n-type ($\sim$ 5 $\times$ 10^{17} cm^{-3}) InGaAsP active layer (0.5–1.5 μm thick) whose composition is chosen to provide a bandgap energy corresponding to an emission wavelength of 1.3 μm and, simultaneously, a crystal-lattice match to InP. Subsequent layers consist of a 1- to 2-μm p-InP confining layer doped with Zn or Cd (0.5 $\sim$ 5 $\times$ 10^{18} cm^{-3}) and a 0.5 μm thick "cap" layer of InGaAsP heavily doped with Zn ($\sim$ 1 $\times$ 10^{19} cm^{-3}) to reduce contact resistance. Etching of a well may be omitted since the InP substrate is transparent to 1.3-μm radiation. If desired, the active layer composition in this semiconductor system can be varied to provide output wavelengths from about 1 μm to nearly 1.6 μm.

In the planar structure for lateral current confinement described, significant current spreading occurs for contact diameters smaller than 25 μm. This current spreading results in a reduced current density and in an effective emitting area substantially larger than the contact area. In order to reduce the current spreading in very small devices, mesa-structure surface emitting LEDS, as shown

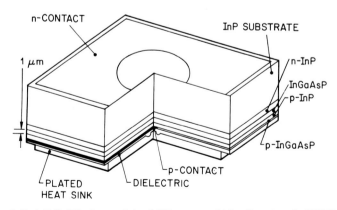

Fig. 12.2 InGaAsP/InP Surface-emitting LED structure. [After Dentai et al., (1977)]

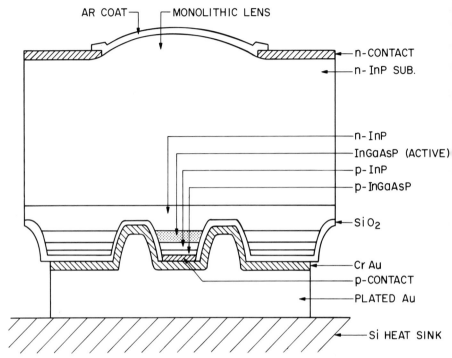

Fig. 12.3 Small-area mesa etched InGaAsP/InP Surface-emitting LED structure. [After Uji and Hayashi (1985)]

in Fig. 12.3, have been fabricated (Uji and Hayashi, 1985). Mesas 20–25 μm in diameter at the active layer were formed by chemical etching. An integral lens was also formed at the exit face of the InP substrate to improve coupling efficiency to fibers (more detailed discussion on coupling using lenses will be given in Section 4.) The advantages of the mesa LEDs are (1) improvement in power launched into fibers due to the elimination of lateral current spreading, and (2) higher modulation speed due to a reduction in parasitic capacitance. The disadvantages are increased complication in fabrication and an exposed junction that may degrade the reliability of the device.

12.2.2 Edge Emitters

An efficient edge emitter LED, as illustrated in Fig. 12.4 (Kressel and Ettenberg, 1975; Horikoshi *et al.*, 1976; Ettenberg *et al.*, 1976; Pearsall, *et al.*, 1976), emits part of its radiation in a relatively directed beam and thus has the advantage of improved efficiency in coupling light into a fiber. This is particularly important in coupling to a single-mode fiber having a small acceptance angle. The

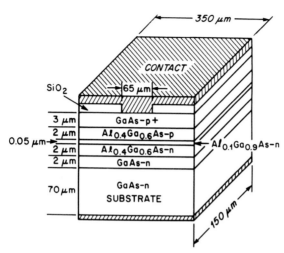

Fig. 12.4 GaAlAs stripe geometry edge emitter. [After Ettenberg *et al.*, (1976)]

decrease in emission angle for this configuration is in the plane perpendicular to the junction, and it results from waveguiding effects of the heterostructure (Dumke, 1975; Wittke, 1975; Seki, 1976). In the structures of Fig. 12.4 the active layer is extremely thin, ~ 500 Å, so that light generated there is not totally contained but leaks into the surrounding waveguide layers where it is strongly coupled to the lowest-order guided mode perpendicular to the junction. Although absorption in the active layer itself is high, the absorption in the surrounding Al-containing waveguide layers is low, so that most of the light propagating in this mode is transmitted to the end faces and emitted with a beamwidth determined by the waveguide parameters. A reflector at one end face and an antireflection coating at the other assures that most of the propagating light is emitted at one end face. The emitted beam then is Lambertian with a half-power width of 120° in the plane of the junction, where there is no waveguide effect, but it has been made as small as 25–35° in the plane perpendicular to the junction (Ettenberg *et al.*, 1976) by properly proportioning the waveguide.

A second advantage of this structure is that, as a result of the channeling of light to a very small end face by the waveguide, the effective radiance at this face can be very high. The value achieved to date is 1000–1500 W/cm²-sr in an emitting area of $2 - 4 \times 10^{-6}$ cm² (Horikoshi *et al.*, 1976; Ettenberg *et al.*, 1976). This value is several times the 300 W/cm²-sr achieved in air without antireflection coating with a 15-μm diameter DH surface emitter (C. A. Burrus, unpublished, 1974), and an order of magnitude greater than that reported for 50-μm diameter DH devices operated at one-half to two-thirds saturation

(Burrus and Miller, 1971). With the aid of a lens at the end of the fiber (effective because the emitter area was smaller than the fiber cross section), 0.8 mW of optical power has been coupled into a 0.14-NA 90-μm core fiber from one of these edge-emitters operating near power saturation at 5.1 kA/cm^2 (Ettenberg et al., 1976). This is a coupling improvement of 7 dB compared to that expected from a purely Lambertian source.

On the basis of idealized assumptions, an analysis (Marcuse, 1977) has shown that an edge-emitter LED with a guiding region should be capable of coupling 7.5 times more power into a fiber than can a surface emitter of equal intrinsic radiance and active layer thickness. When more realistic assumptions are made, including, for example, the more difficult heat-sinking geometry of the edge emitter, this same analysis shows that the expected advantage for multimode fibers drops toward a factor of two for practical devices. Because of the narrow beam width, however, an edge-emitter can couple significantly higher power into a single-mode fiber than can a surface emitting LED.

12.2.3 Superluminescent LED

A third device geometry that may eventually offer advantages of (1) increased power, (2) a well-directed output beam, and (3) reduced spectral bandwidth is an elongated version of the stripe-geometry edge emitter. It has been called a superluminescent diode, or SLD, and one form of its construction is illustrated in Fig. 12.5 (Lee et al., 1973). While optimization for each application may lead to some variations in the configuration of the semiconductor layers, the edge emitter, the SLD and the laser are strikingly similar in construction.

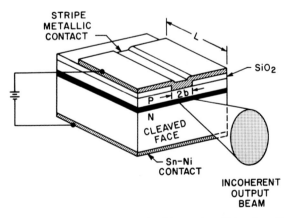

Fig. 12.5 Superluminescent diode (SLD) with absorbing region. [After Lee, Burrus and Miller (1973)]

Similar to a laser diode, the SLD structure requires a *p-n* junction in the form of a long rectangular stripe. One end of the stripe is made optically lossy to prevent reflections and thus suppress lasing, and the output is from the opposite end. In operation, the injection current is increased until stimulated emission (the first step in the onset of lasing) and amplification occurs, but because of high loss at one end of the stripe, no feedback exists and no oscillation builds up. Therefore, in the current region of stimulated emission, there is gain and the output increases rapidly with current due to the single-pass amplification. At the same time, the spectral width of the output decreases to less than 100 Å. Devices have been made and operated in a pulsed mode to provide a peak output of 60 mW in an optical bandwidth of 60–80 Å at a center wavelength near 0.87 μm (Lee *et al.*, 1973), and CW devices have delivered 25 mW (Iwamoto *et al.*, 1976). Other methods for suppressing Fabry-Perot resonance utilize an oblique output facet (Kurbatov *et al.*, 1971) or, more recently, antireflection coatings on the cleaved facets (Wang *et al.*, 1982; Kaminow *et al.*, 1983; Fye *et al.*, 1986).

The disadvantage of these devices is that, to produce power comparable to that of a laser, the required current density is about three times that of a laser and the absolute current is high due to the long length (large area); thus recent improvements in laser diodes, resulting in lowered thresholds and higher output per watt input, have made the superluminescent diode more practical. An additional penalty is the enhanced temperature dependence of output power (See Section 12.3.3).

12.3 DEVICE CHARACTERISTICS

12.3.1 Power Efficiency

In considering the operating efficiency of an LED, the total external quantum efficiency and similar measures of junction performance often are of little practical use. It is more convenient to think of a "useful power efficiency" as the ratio of the optical power that can be collected in a useful way to the electrical power applied at the diode terminals. The latter is easy to determine unambiguously, but the collectible optical power depends upon the application. It may vary from the total forward emission collected by a large detector near the top of the well of the device of Fig. 12.1, which emits into a unobstructed forward angle of about 130°, to that collected by a 0.14-NA fiber with an acceptance angle of only 16°, or to that collected by a single-mode fiber with an even smaller acceptance angle. In the first case, the total power collected by a large detector can be a useful measure of the total available power from the device, which must be multiplied by the coupling efficiency to fibers. The coupling efficiency ranges from 5–10% for multimode fibers to 0.1–1% for single-mode fibers (see Section 12.4 for detailed discussions of coupling power to fibers).

The total available power from a 75-µm diameter GaAlAs/GaAs DH LED of the type illustrated in Fig. 12.1 at 150 mA (one-half diode saturation) and 2V can be 15–18 mW for air interfaces and no antireflection coatings (22–24 mW at saturation) and results in a useful external power efficiency of 6%. Even the simple addition of a drop of epoxy resin in the well to decrease the index mismatch and thus increase the internal emission cone at the semiconductor surface can increase this number to 8–10%; more sophisticated lenses, mentioned later, are now commonly used. This is not the total optical power emitted by the junction, but the total useful power for a particular application. Each application then leads to a different value.

For 1.3 µm InGaAsP LEDs, the total available power is lower than that of 0.9 µm LEDs for a similar emitting area because of the lower internal quantum efficiency due to the Auger non-radiation recombination. At high driving current, power output becomes saturated due to in-plane superluminescence (Goodfellow et al., 1981). More discussions of power saturation will be given in Section 12.3.3.

Although the total available power from an edge emitter is much lower than that of a surface emitter, the higher coupling efficiency to optical fiber compensates for it. In most applications, the edge emitter can provide higher launched power into fibers than can surface emitters.

12.3.2 Power and Modulation Bandwidth

12.3.2.1 Non-Radiative and Radiative Recombination Lifetimes. The power of light generated in LED devices (Lee and Dentai, 1978) is related directly to the effective electron density in the active region. This density in turn is a function of the active layer thickness, the carrier diffusion length and the surface recombination velocity, as well as self-absorption in the active layer, which is a function of the doping density. The speed at which the junction can be directly current modulated with an information-carrying signal is fundamentally limited by the recombination lifetime of the carriers. The injected electrons recombine with holes in the active region via both radiative and non-radiative processes. The total carrier lifetime τ, is,

$$\frac{1}{\tau} = \frac{1}{\tau_r} + \frac{1}{\tau_{nr}}, \tag{12.1}$$

where τ_r is the radiative recombination time and τ_{nr} is the nonradiative recombination time of carriers. For a thin (a few micron) active region, the non-radiative recombination is dominated by surface recombination at the heterostructure interfaces. Equation (12.1) can be rewritten as

$$\frac{1}{\tau} = \frac{1}{\tau_r} + \frac{2s}{w}, \tag{12.2}$$

where w is the thickness of the active layer, and s is the surface recombination velocity, which is related to the strain due to lattice mismatch at the heterostructure interface (Kressel *et al.*, 1980),

$$s \approx (2 \times 10^7)(\Delta a/a), \qquad (12.3)$$

where a is the lattice constant and Δa is the amount of the mismatch in lattice constants.

The internal quantum efficiency is defined as

$$\eta_{\text{int}} = \tau/\tau_r. \qquad (12.4)$$

For η_{int} to be higher than 50 percent, $\Delta a/a$ must be less than 10^{-3}. For GaAlAs/GaAs DH structures, an interfacial recombination velocity of 2×10^3 cm/sec has been achieved (Lee and Dentai, 1978) and a value as low as 500 cm/sec has been reported (Nelson and Sobers, 1978). A higher value of 10^4 cm/sec was measured for InGaAsP/InP (Sakai *et al.*, 1980).

The spontaneous radiative recombination rate of a band-to-band transition is given by

$$R_{sp} = Bnp, \qquad (12.5)$$

where $B(\text{cm}^3/\text{s})$ is the radiative recombination probability, a characteristic of band structure depending on material. Under the nonequilibrium condition in which excess electrons, Δn, are generated (for example, by the injection of carriers in a forward biased junction) in the conduction band and excess holes Δp are created in the valence band, the net carriers are

$$n = n_o + \Delta n \qquad (12.6)$$

and

$$p = p_o + \Delta p, \qquad (12.7)$$

where n_o and p_o are electron and hole concentrations at thermal equilibrium. Substituting (12.6) and (12.7) into (12.5), and using $\Delta p = \Delta n$, Eq. 12.5 becomes

$$R_{sp} = Bn_o P_o + B\Delta n(n_o + p_o + \Delta n). \qquad (12.8)$$

The first term $Bn_o p_o$ represents the thermal-equilibrium radiative recombination rate, which is usually negligible. Thus, the radiative recombination rate can be written as

$$R_{sp} = B\Delta n(n_o + p_o + \Delta n). \qquad (12.9)$$

The radiative recombination lifetime is defined as

$$\tau_r = \frac{\Delta n}{R_{sp}}. \qquad (12.10)$$

Then, with the help of Eq. (12.9) we obtain

$$\tau_r = \frac{1}{B(n_o + p_o + \Delta n)} . \tag{12.11}$$

For p-type material $p_o \simeq N_A, n_o \ll p_o$ and if the excess carrier density is less than the doping (acceptor) concentration, i.e., $\Delta n \ll N_A$ (the case at low injection levels), the carrier lifetime is inversely proportional to the doping concentration. That is

$$\tau_r = \frac{1}{BN_A} . \tag{12.12}$$

Similarly, for n-type material N_A in Eq. (12.12) is replaced by the donor concentration, N_D. Thus, the value of B can be determined by the measurement of τ_r vs. N_A. For GaAs, B is $0.64-1.3 \times 10^{-10}$ cm^3/sec (Casey and Stern, 1976; Acket, Nijam, and Lam, 1974). The value of B ranges from 0.35×10^{-10} cm^3/sec (Wada $et\ al.$, 1979) to 0.98×10^{-10} cm^3/sec (Wada $et\ al.$, 1981) for InGaAsP.

At high injection levels such that Δn is much larger than the background doping concentration, that is $\Delta n > N_A$, (or $\Delta n > N_D$), Eq. (12.11) becomes

$$\tau_r = \frac{1}{B\Delta n} . \tag{12.13}$$

Since the charge neutrality condition must be maintained, an equal number of holes must be injected into the active region, and this can be achieved in a double heterostructure only by means of hole injection at the heterojunction. This condition is called bi-molecular recombination. The injected carrier density in a 50-μm diameter LED usually ranges from 1×10^{18} cm^{-3} to 5×10^{18} cm^{-3} under ordinary operating conditions, and bi-molecular recombination is dominant only for active layers lightly doped (below $\sim 1 \times 10^{18}$ cm^{-3}). Shortening of the radiative lifetime through heavy doping, according to Eq. (12.12), however, does not necessarily improve the internal quantum efficiency because heavy doping may also introduce nonradiative recombination through an Auger process. In fact, heavy doping is used to increase the speed of an LED at the expense of output power, as we shall describe in detail later.

12.3.2.2 Modulation Bandwidth. Intensity modulation of LED light output can be accomplished by direct modulation of the injection current. If the current is modulated at an angular frequency ω, the intensity of the light output will vary with ω as (Namizaki $et\ al.$, 1974; Liu and Smith, 1975)

$$|I(\omega)| = \frac{I(O)}{\sqrt{1 + (\omega\tau)^2}} , \tag{12.14}$$

where $I(0)$ is the intensity (optical power) at zero modulation frequency and τ is the total carrier lifetime. The modulation bandwidth Δf (or cutoff frequency f_c) is defined as the frequency at which the detected electrical power ($P_{elec} \sim I^2(\omega)$) is one half that at zero modulation frequency, that is

$$\Delta f = \frac{\Delta \omega}{2\pi} = \frac{1}{2\pi\tau}, \qquad (12.15)$$

where τ is the total carrier lifetime discussed in the previous section.

When bi-molecular injection prevails in the lightly doped active region, the injected carrier density is proportional to the current density and inversely proportional to the thickness of the active layer, i.e.,

$$\Delta n = \frac{J\tau_r}{e\mathrm{w}}, \qquad (12.16)$$

where J is the current density. Combining (12.16) and (12.13), and solving for τ_r, we obtain

$$\tau_r = \left(\frac{e\mathrm{w}}{BJ}\right)^{1/2}. \qquad (12.17)$$

Using the fact that $\tau \approx \tau_r$, and substituting τ in (12.15) by (12.17), the modulation bandwidth becomes

$$\Delta f = \frac{1}{2\pi}\left(\frac{BJ}{e\mathrm{w}}\right)^{1/2}. \qquad (12.18)$$

Thus, the bandwidth of a DH LED having a lightly doped (mid 10^{17} cm^{-3}) active layer and operated in the bi-molecular recombination regime increases with $(J/\mathrm{w})^{1/2}$. Fig. 12.6 and Fig. 12.7 confirm this relationship in practical GaAlAs LEDs (Lee and Dentai, 1978) and in InGaAsP LEDs (Wada et al., 1981), respectively.

For doping densities higher than 10^{18} cm^{-3} in the active layer, the modulation bandwidth becomes independent of the injected carrier density (upper curves in Fig. 12.6); rather, it is inversely proportional to the doping concentration in the active layer as given by Eq. (12.12). Modulation bandwidths greater than 1 GHz have been achieved with $N_A \approx 1.5 \times 10^{19}$ cm^{-3} (Heinen et al., 1976). However, the increased bandwidth is obtained at the expense of the output power as shown in Fig. 12.8.

For InGaAsP LEDs with an undoped (residual n-type concentration varying from $\sim 10^{17}$ cm^{-3} to $\sim 5 \times 10^{17}$ cm^{-3}) active layer, the nominal bandwidth of such device ranges from 50 MHz to 100 MHz (Dentai et al., 1977; Umebu et al., 1978; Goodfellow et al., 1979; Wada et al., 1981; Temkin et al., 1983).

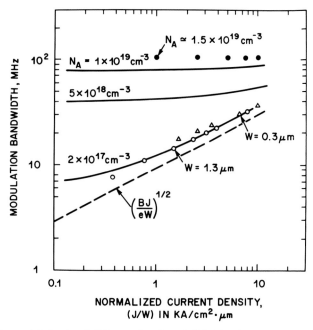

Fig. 12.6 The modulation bandwidth as a function of (J/w) for GaAlAs [After Lee and Dentai (1978)].

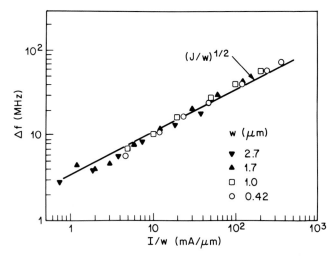

Fig. 12.7 The modulation bandwidth as a function of (J/w) for InGaAsP LEDs [After Wada *et al.*, (1981)].

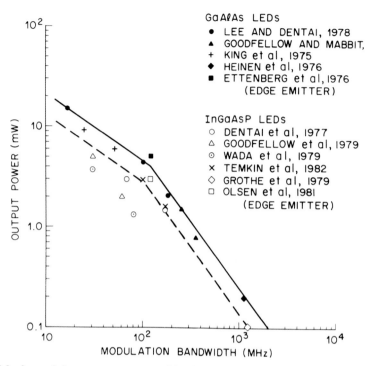

Fig. 12.8 State of the art output power and bandwidth of 1.3-μm LEDs. [After Lee (1982) and Saul (1983)].

With Zn-doping at $2 - 4 \times 10^{18}$ cm^{-3} in the active layer, the bandwidth increases to 170 MHz (Gloge *et al.*, 1980; Goodfellow *et al.*, 1979). Using a relatively thin (0.4 μm) active layer doped to 5×10^{18} cm^{-3}, bandwidth of 690 MHz was achieved (King *et al.*, 1985). Bandwidths as high as 1.2 GHz have been obtained for diodes with a Mg-doped active layer (Grothe *et al.*, 1979), or with a Zn-doped (1.3×10^{19} cm^{-3}) layer (Suzuki *et al.*, 1984).

In practical devices, parasitic capacitance can impose a limitation on the device modulation bandwidth. For the planar structure shown in Fig. 12.1, the capacitance C is comprised of a space charge capacitance associated with the *p-n* junction area (entire diode chip) and a diffusion capacitance, which is related to the carrier lifetime in the small light-emitting area. As the diode is turned on with a step current pulse, the initial injection level is low and the current spreads in the *p*-layers so that the full area of the diode junction contributes to the device capacitance that, for a typical diode chip, amounts to 150 to 250 pF. As the current is increased, the high spreading resistance of the *p*-layer "crowds" the current so that the portion passing through the emitting region (above the p-contact) becomes a larger fraction of the total current. The space charge capacitance associated with the emitting region C_c is charged rapidly relative to the "perimeter" capacitance of the surrounding p-n junction, C_p. Thus, when a step current pulse is applied, the resulting light pulse is delayed slightly due to C_c. The delay time decreases from $\sim$ 3–4 nsec at 30 mA drive to 1 nsec at 100 mA drive current (Hino and Iwamoto, 1979). The initial rise-time (10–90%) is primarily determined by the carrier recombination lifetime as previously discussed. Because of the large RC time constant of C_p, however, the steady state may be reached only slowly. Various improvements in the slow portion of the rise time can be achieved by applying a small dc prebias (Hino and Iwamoto, 1979), or by rapid injection of a current spike (Lee, 1975; Zucker, 1978). On the other hand, the fall time can be shortened by using a reverse bias pulse to accelerate the removal of carriers (Dawson, 1980). Alternatively, the perimeter capacitance can be completely eliminated by isolation of the emitting areas by proton bombardment (Heinen et al., 1976), by etching mesas (Harth et al., 1976; Uji and Hayashi, 1985) and by use of buried heterostructures (Heinen, 1982a). Thus, by using a variety of driver techniques or structural designs, the parasitic effects can be made so small that the LED transient response is determined largely by the carrier lifetime discussed in the previous section. Detailed transient analyses can be found in previous publications (Lee, 1975; Hino and Iwamoto, 1979).

12.3.2.3 Practical Device Performance. As described in previous sections, the carrier recombination lifetime is primarily a function of the doping concentration, the number of electrons injected into the active region, the surface recombination velocity (the rate at which electrons recombine without radiation at crystalline defects at nearby semiconductor interfaces), and the physical thickness of the active layer. All of these parameters controlling the efficiency and the modulation bandwidth, many of them interdependent, are adjustable, within limits, in present-day technology. In practice, then, the device fabricator seeks to optimize both the radiative output and the modulation capabilities. The desirable characteristics for the materials are large injection rates, proper (not necessarily highest) doping levels, thin active layers, and highly perfect crystalline layers and interfaces. For GaAlAs/GaAs diodes at 0.9 μm wavelength, maximum output power has been obtained experimentally with w between 2 and 2.5 μm and with 2×10^{17} cm^{-3} p-type (Ge) doping (Lee and Dentai, 1978). For InGaAsP/InP 1.3-μm LEDs, maximum power was achieved with w between 1 and 1.5 μm and an undoped ($\sim 10^{17}$ cm^{-3}) n-type layer (Wada et al., 1981). In the latter case, junction displacement in the active layer was found to be an additional factor affecting the output power. The output power decreases as the p-n junction moves into the InGaAsP active layer from the

InP/InGaAsP heterointerface and as the doping in the active layer changes from predominantly n-type to predominantly p-type (Temkin *et al.*, 1982). This suggests that the radiative recombination efficiency of p-type InGaAsP must be lower than that of n-type material. The junction displacement can result from contamination of the active layer growth solution by Zn vapor from the neighboring p-confining-layer growth solution. This contamination can be eliminated by covering the melt wells and adjusting the flow and direction of the carrier gas, or by use of Cd as the dopant (Wada *et al.*, 1981), since Cd has a smaller diffusion coefficient than Zn (Kundukhov *et al.*, 1967).

State-of-the-art results are shown in Fig. 12.8, where the reciprocal relationship between modulation bandwidth and output power is apparent. For GaAlAs surface-emitting LEDs, the highest power reported for a 50-μm diameter emitting area has been 15 mW with a 3-dB bandwidth of 17 MHz (Lee and Dentai, 1978), whereas the largest bandwidth of 1.1 GHz was achieved at a much lower output power of 0.2 mW (Heinen *et al.*, 1976). Experimental results such as those shown in Fig. 12.8 indicate that output power P can be related to bandwidth Δf by $P \sim \Delta f^{-\nu}$. For bandwidth $\Delta f < 100$ MHz, $\nu \approx 2/3$ and for $\Delta f > 100$ MHz, $\nu \approx 4/3$. For LEDs with small bandwidth (20–30 MHz) the output power can be maximized by using a thick active layer (2–2.5 μm) with low doping densities ($< 5 \times 10^{17}$ cm^{-3}). Diodes in the intermediate bandwidth region (50–100 MHz) require a thinner active layer (1–1.5 μm) and moderate doping densities (0.5–1 $\times 10^{18}$ cm^{-3}). Increase in the modulation bandwidth beyond the 100 to 200 MHz range can be achieved only by doping the active layer in excess of $\sim 5 \times 10^{18}$ cm^{-3}. This heavy doping evidently introduces nonradiative centers and, as a result, the value of ν is greater than unity and the output power is significantly reduced.

Similar power vs. bandwidth characteristics are observed for InGaAsP 1.3-μm LEDs. The solid line and the dotted line in Fig. 12.8 are contours of best reported results for GaAlAs and InGaAsP LEDs, respectively. The output power of GaAlAs LEDs is a factor of two higher than that of InGaAsP LEDs at all bandwidths. Because the photon energy at 1.3-μm wavelength is smaller by a factor of 1.53 than that at 0.85 μm, the output power of InGaAsP LEDs is reduced by the same factor compared with GaAlAs LEDs of the same quantum efficiency. When this factor is taken into account, the best performance of 1.3-μm LEDs is only slightly below that of 0.85-μm LEDs, probably due both to the fact that the technology is more advanced for GaAlAs than for InGaAsP and to the enhanced radiance saturation in the longer wavelength material.

The best performance of both GaAlAs and InGaAsP edge emitters is also included in Fig. 12.8 for comparison. These devices utilize thin ($\lesssim 0.2$ μm), low-doped active layers, which promotes bi-molecular recombination. Consequently, edge emitters typically have high bandwidth, generally in the 100–150

MHz range. The thin active layer also narrows the output beam divergence, which enhances coupling to fibers, as to be discussed in Section 12.4.

12.3.3 Output Power Limitations and Temperature Effect

The output power of LEDs as a function of current is observed to increase approximately linearly at first, then to saturate gradually, and with further increase of current, to diminish light output. At the point of maximum output the diode is said to be saturated. It is of interest to see what limits the power output of these devices.

First, it is an accepted but not completely understood fact that the internal quantum efficiency of p-n junction light-emitting devices decreases exponentially with increasing temperature. Second, the thermal conductivity of the III-V semiconductor compounds is an exponentially decreasing function of increasing temperature (Carlson *et al.*, 1965). Thus the junction temperature at maximum power output (the diode saturates) can be calculated for a given semiconductor, and it is found to be about $360°K$ for GaAlAs and slightly higher, about $400°K$, for GaAs (Lee and Dentai, 1978). The maximum output is set by the junction temperature regardless of the size of the junction, but the temperature rise due to resistive heating from the driving current is very much dependent upon the structure.

Since in the structures of Fig. 12.1 the oxide layer, used to define the contact and to confine the current flow, has very low heat conductivity, the heat flow in the emitting region can be approximated by a simple model assuming several stacked layers and one-dimensional heat flow. With such a model, and separation of the heat conduction into temperature-dependent (semiconductor) and temperature-independent (metallic) parts, the temperature rise expected from a bias current applied to several emitter diameters, semiconductor compositions and layer configurations has been calculated (Lee and Dentai, 1978). The results show that, since power saturation occurs at a constant junction temperature and smaller devices run cooler than larger ones for a given current density, small-area units can operate at relatively higher current densities (and hence higher radiance) than can larger area diodes. Similarly, smaller devices can be driven at higher current densities before the critical junction temperature is reached, so that again the smaller devices provide higher radiance. The results of this work provide a remarkably close fit to experimental observations on the small-area surface emitters.

Another mechanism known as in-plane superluminescence (Goodfellow *et al.*, 1981), which results from stimulated emission in the junction plane at high injection levels, has been postulated to account for the observed excessive radiance saturation of the surface emission. Observations of the light-current characteristics of faceted (to reflect in-plane luminescence) surface emitters

(SpringThorpe *et al.*, 1982) and comparison of the facet (single pass gain) and side (unpumped) emission of edge-emitting LEDs (Dutta and Nelson, 1982) are consistent with this mechanism. On the basis of this mechanism alone, saturation is expected to be strongly temperature dependent.

A third factor limiting the power output was revealed by the measurement of diode current (Uji *et al.*, 1981) and carrier lifetimes (Dutta and Nelson, 1982; Sermaze *et al.*, 1983) as a function of injection level, which indicated that at high carrier density ($\geq 2 \times 10^{18}$ cm^{-3}) the nonradiative component increases with the square of injection level, consistent with an Auger-process contribution to the power saturation characteristic. Nonradiative lifetimes decrease rapidly with increasing free carrier concentration, also consistent with the Auger effect (Henry *et al.*, 1983b). Furthermore, the enhanced saturation with increased injected carrier density (observed when the active layer thickness is decreased and the diameter of the emitting region is reduced) has been successfully modeled assuming an nonradiative Auger-type component that depends on injection level (Uji *et al.*, 1982; Su *et al.*, 1982a).

For the more common unintentionally-doped (mid-10^{17} cm^{-3}) active layers, however, the saturation characteristic is consistent (Su *et al.*, 1982b) with a drift-leakage model (Anthony and Schumaker, 1980). Others have pointed to the role of injected carriers that are not confined by the heterobarriers cladding the active layers (Yano *et al.*, 1981). The existence of leakage of injected electrons over the *p*-cladding interface has been verified (Yamakoshi *et al.*, 1982) in LEDs specially prepared to have a thin active layer, and direct measurement of carrier leakage in laser-bipolar-transistor structures have indicated that it is an appreciable fraction (15–30%) of the total current (Chen *et al.*, 1983). Carrier leakage may be enhanced when the temperature of the injected carriers exceeds the lattice temperature (Shah *et al.*, 1981). The rise in carrier temperature at 1.5-μm wavelength was determined to be larger by a factor of ~ 3 than at 1.3-μm wavelength and has been correlated to the enhanced saturation observed at the longer wavelengths (Wada *et al.*, 1982c). Measurements of spontaneous emission and carrier lifetimes in 1.58-μm InGaAsP devices has led some workers (Asada and Suematsu, 1982) to conclude that the major contribution to the nonradiative lifetime is carrier leakage, with a more minor contribution from Auger recombination.

While there are conflicting data and diverse interpretations, it appears likely that several interrelated mechanisms are operative, the dominant ones being dependent on the details of the device structure and excitation conditions. For example, hot carriers [if present (Henry *et al.*, 1983b; Manning *et al.*, 1983)] may be the result of Auger recombination, especially in heavily doped active layers and can lead to the observed carrier leakage. At sufficiently high injected carrier densities ($\geq 3 \times 10^{18}$ cm^{-3}) stimulated emission leads to in-plane super-luminescence that then becomes an important loss mechanism in surface emitters.

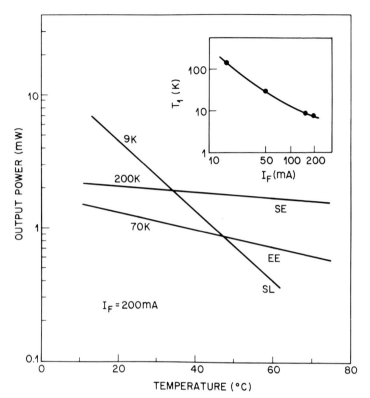

Fig. 12.9 Temperature dependence of light output for various 1.3-μm LED structures. The inset shows for an SL LED the variation of the characteristic temperature T_1 with drive current. [After Dutta *et al.*, (1983).]

Additional mechanisms, including interface recombination (Yano *et al.*, 1980) and intervalence band absorption (Adams *et al.*, 1980), have been proposed to explain the temperature dependence of threshold on lasers, but at present their role in determining the spontaneous emission characteristics of LEDs is unclear.

For comparison, Fig. 12.9 illustrates the light-temperature characteristics for the various LED structures. The light output is characterized by the empirical expression

$$L = L_0 \exp[-T/T_1], \qquad (12.19)$$

where T is in Kelvin and T_1 ranges from 180–220K and 300–350K for InGaAsP and GaAlAs LEDs, respectively (Temkin *et al.*, 1983), and is independent of drive current (Temkin *et al.*, 1981a). For edge emitters, however, T_1 is smaller, typically in the range 70–80K and 100–120K, respectively (Kressel *et al.*, 1980). Further, with increased stimulated emission, as in superluminescent LEDs, the effective value of T_1 can be strongly dependent on drive current (Fig. 12.9 inset). Thus, to utilize the high power potential of these edge-emitting devices at elevated temperatures, thermoelectric coolers may be required. Some care must be exercised, however, since there is always the possibility that these devices will lase at low temperatures. In principle, lasing can be avoided by judicious device design or by use of a thermoelectric heater.

12.3.4 Output Spectrum

The peak electroluminescence wavelength λ_p is determined primarily by the bandgap (composition) of the active layer. Figure 12.10 shows typical spectra for InGaAsP surface- and edge-emitting LEDs. Modulation on the emission spectra is commonly observed in GaAlAs surface emitting LEDs (but rarely in GaAsP devices), apparently due to interference effects resulting from reflections between the emitting surface and the back contact. Under pulsed current operation that eliminates Joule heating, increasing the peak current shifts λ_p toward shorter wavelengths due to band-filling (Nelson *et al.*, 1963). Under CW drive conditions where junction heating occurs, however, λ_p shifts toward longer wavelengths with increasing current since the bandgap decreases at higher junction temperatures. For the same reason, λ_p increases with ambient temperature, shifting by 0.35 and 0.6 nm/°C for GaAls (Burrus and Miller, 1971) and InGaAsP (Temkin *et al.*, 1983), respectively. Increased doping of the active layer shifts λ_p to higher values as a result of impurity banding (Casey and Stern, 1976).

The spectral width, taken here to be the full width at half maximum intensity (FWHM), varies roughly as λ_p^2 for undoped active layers (Fukui and Horikoshi, 1979); this corresponds to a constant energy width of ~ 2 kT. While this is

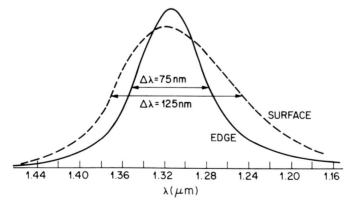

Fig. 12.10 Spectral output curve for InGaAsP surface and edge-emitting LEDs.

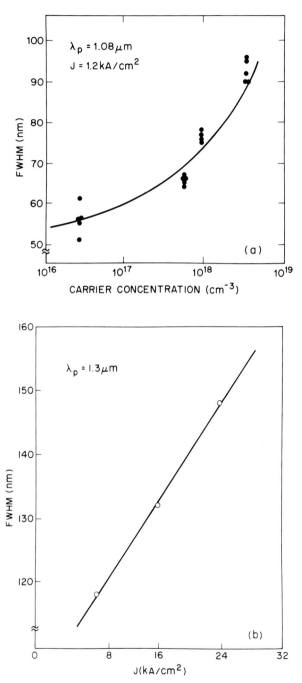

Fig. 12.11 Dependence of FWHM spectral width on (a) doping of the active layer [After Wright *et al.*, (1975)] and (b) current density [After Burton, (1983)].

borne out in $Ga_{1-x}Al_xAs$ and in $In_{1-x}Ga_xAs_yP_{1-y}$ near the InP and InGaAs ends of the composition range, FWHM is nearly 3 kT at the composition corresponding to 1.3-μm emission, a fact attributed to compositional grading in the active layer (Temkin *et al.*, 1981b). GaAlAs surface-emitting LEDs operating at 0.85 μm have a spectral width of 40 nm compared with ~ 110 nm for InGaAsP LEDs emitting at 1.3 μm. As a result of self absorption along the length of the active layer, FWHM of edge emitters is ~ 1.6 times smaller than for surface emitters, as indicated in Fig. 12.10. For superluminescent LEDs, the spectra is further narrowed by the onset of optical gain, and FWHM can be much smaller than for surface emitters. The shift toward broader FWHM with increased temperature (~ 0.3 nm/°C) and increased pulsed drive current (Escher *et al.*, 1982) is attributed to band-filling, while the shift with increased doping of the active layer (Wright *et al.*, 1975) is attributed to impurity banding. These effects are illustrated in Figs. 12.11(a) and 12.11(b). Thus, a variety of factors contributes to spectral broadening. In general, high speed LEDs (with doped active layers) driven at high current density (to achieve maximum radiance) will have wide spectra; for example, up to 160 nm for 1.3-μm LEDs.

12.4 COUPLING TO OPTICAL FIBERS

12.4.1 Direct (Butt) Coupling

The output of surface emitting devices is approximately Lambertian; that is, the surface radiance is constant in all directions, but maximum emission is perpendicular to the junction and falls off to the sides in proportion to the cosine of the viewing angle because the apparent area varies with this angle. That is

$$B(r, \theta) = B(r)\cos \theta, \qquad (18)$$

where $B(r)$ accounts for radial variation in radiant intensity within the emitting plane. The farfield pattern is symmetric, and the full beam angle to half intensity is ~ 120°.

The expected coupling of this output to the fiber depends upon the exact geometry and conditions, such as the LED emission pattern and radiance, fiber size and refractive index grading, effective NA of the fiber, relative area of the fiber core and LED emitting area, distance and alignment between the two, the medium between them, etc. Thus, coupling in an individual case is complex (Colvin, 1974; Yang and Kingsley, 1975; DiVita and Vannucci, 1975) and probably can be determined accurately only by measurement. Here, we describe some principles applicable to LED-fiber coupling.

The power emitted into air by a surface emitting LED can be expressed as

$$P_{LED} = \pi B A_{LED}, \qquad (12.20)$$

where B is the average radiance $(w/sr/cm^2)$ of the source (Barnoski, 1981; Plihal, 1982; Hudson, 1974). The total power accepted by a step index fiber with a numerical aperture NA is

$$P_{\text{Fiber}} = \pi(1 - R)BA_{LED}(NA)^2$$

$$= TP_{LED}(NA)^2, \qquad (12.21)$$

where R accounts for Fresnel reflection at the fiber end and $T = (1 - R)$ is the transmissivity. Equation (12.21) is valid for the case where the fiber core is larger than the emitting area of the LED, since use of a lens cannot increase the power coupled from a large-area incoherent source into a small-area fiber. Thus, the coupling efficiency to a step index fiber is simply

$$\eta_s = T(NA)^2. \qquad (12.22)$$

In a graded index fiber, the index of refraction varies parabolically, which results in an effective NA that is smaller than that of the step index fiber. In this case, equation (12.22) must be modified to give

$$\eta_g = T(NA)^2\left(1 - \frac{1}{2D^2}\right), \qquad (12.23)$$

where the parameter $D(= d_f/d_s \geq 1)$ is defined as the ratio of the core diameter of the fiber, d_f, to the diameter of the source d_s. We note that when $d = 1$, $\eta_g = \eta_s/2$. In practice, the optimum direct coupling for a graded index fiber requires that the source diameter be about one-half the fiber core diameter.

It is generally observed, however, that the experimentally determined value of η_g for small-core fibers with low NA is appreciably less than that predicted by Eq. (12.23). Measurements of the near field pattern of oxide-confined (current confined) surface emitters reveal that the light intensity is not constant (Wada *et al.*, 1982a), as assumed for Eq. (12.23); rather, the light intensity is uniform only over a small central area (due to current "spreading" and carrier drift) and then decreases with a Gaussian-like tail (Wada *et al.*, 1982a). This light intensity profile reflects the current distribution across the emitting area: the current is constant directly above the p-contact and spreads laterally according to the conductivity of the epitaxial layers. Using numerical integration to account for the measured variation in $B(r)$, good agreement between the experimental and the theoretical value of η_g is obtained for small-core fibers (Borusk, 1983). The effect of the Gaussian tail on η_g is illustrated in Fig. 12.12, where the abscissa is the standard deviation σ of the Gaussian function. Note that with typical current spreading, η_g is reduced by $\sim 30\%$ compared to a perfectly confined, uniform light spot ($\sigma = o$). Restricting the junction area by mesa etching, which eliminates current spreading, has been shown to improve the coupling efficiency (Uji and Hayashi, 1985).

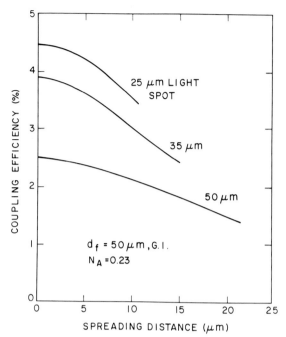

Fig. 12.12 Butt coupling efficiency as a function of spreading of the light spot beyond the uniform central portion defined by the contact size. [After Borsuk (1983).]

A factor that further complicates the accurate prediction of the results to be expected in LED-fiber coupling is the transmission characteristics of the so-called leaky modes or large-angle off-axis rays. Much of the light from an incoherent source is initially coupled into these large-angle rays, which fall within the acceptance angle of the fiber but which have much higher attenuation than do the meridional or on-axis rays; energy from these rays tunnels or "leaks" into the cladding and is lost. The result is that much of the light coupled into a multimode fiber from an LED is lost within meters to hundreds of meters, and the power apparently coupled into a shorter fiber significantly exceeds that coupled effectively into a longer fiber. A concise review is available in the literature (Snyder, 1974).

Modeling the coupling of edge emitters is more complex because of the asymmetric far-field pattern, but it has been attempted by several authors (Yang and Kingsley, 1975; Marcuse and Kaminow, 1981; Wittke $et\ al.$, 1976). In general, it is found that for low NA (small θ_c fibers), the narrower beam divergence of the edge emitter ($\theta_\perp = 30°$) in the plane perpendicular to the junction results in appreciably higher coupling efficiencies than is possible with surface emitters, but this advantage becomes less important for high NA (large

θ_c fibers). The coupling of edge emitters to fibers can be further enhanced by using devices with real-index guiding or gain guiding structures in the junction plane to narrow $\theta_{\parallel}$. Coupling efficiencies into a 50 μm core, 0.2 NA graded index fiber as high as 26% have been realized with super-radiant 1.3 μm LEDs (Dutta *et al.*, 1983).

12.4.2 Coupling with Lenses

If the fiber core can be significantly larger than the emitter, a lens may be used between the two to decrease the angular divergence of the source. The coupling improvement relative to butt coupling has been calculated using conventional optics theory for a variety of situations: lenses placed between the LED and fiber (Barnoski, 1981; Plihal, 1982), fibers with integral lenses (Abe *et al.*, 1977), truncated lenses cemented to the LED surface (Abram *et al.*, 1975), and lenses that are integral with the LED structure (Hasegawa and Namazu, 1980). Ray tracing techniques also have been used for calculation of coupling gain for a variety of lens shapes (Ackenhusen, 1978).

For a simple spherical lens, the coupling gain can be expressed as

$$g_s = T_L D^2 \qquad (12.24)$$

for a step index fiber where T_L is the transmissivity of the lens, and $D = d_f/d_s$; and

$$g_g = T_L \frac{D^2}{2} \left(\frac{1}{1 - \dfrac{1}{2D^2}} \right) \qquad (12.25)$$

for a graded-index fiber. Thus, for $D = 2$, a gain of 3 dB (for graded index fiber) and 6 dB (for step index fibers) can be expected and, in fact, much greater gains (11 dB) have been achieved with a large (85 μm/18 μm) core-to-emitter diameter ratio using a complex truncated lens to optimize the coupling (Goodfellow *et al.*, 1979). Coupling efficiency also has been enhanced using monolithic or integral lenses formed on the emitting surface of the LED, by epitaxial regrowth (King and SpringThorpe, 1975), ion-beam milling (Wada *et al.*, 1981), photoelectrochemical (PEC) etching (Ostermayer *et al.*, 1982), and masked chemical etching (Heinen, 1982b).

The above general principles also apply to edge emitters. In practice, lens-ended fibers or tapered fibers are widely used to increase coupling. Coupling gains as high as ~ 4 have been realized using lensed 50-μm core, 0.2-NA fibers; launched power as high as 225 μW was achieved (Frahm, 1982). Ure *et al.*, (1982) reported super-radiant edge emitters with a truncated spherical lens glued onto the emitting facet. A coupling gain of ~ 5 was obtained; coupled power was 350 μW at 250 mA. Using a ridge-loaded, lateral waveguide (to reduce $\theta_\perp$) in a 1.3-μm superluminescent LED, Kaminow et al. (1982) achieved a 0.5 mW of power launched into a lensed 50-μm core, 0.2-NA fiber. In gain-guided GaAlAs superluminescent LEDs, launched power as high as 2.5 mW ($\sim$ 60% coupling efficiency) was realized (Davies et al., 1981).

The power coupled to a single-mode fiber is significantly reduced compared to that coupled to a multimode fiber. It ranges from 0.5–1 μW from a butt-coupled standard surface emitter to $\sim$ 10 μW from an edge-emitter (Reith and Shumate, 1986, unpublished; Saul et al., 1985). Because of the small core (9 μm) of the single-mode fiber, no coupling gain has been obtained for surface emitter but 5 dB gain for edge-emitters has been realized using a tapered fiber (Saul et al., 1985).

The use of lenses, although generally improving the coupling efficiency, increases the complexity of the transmitter package, and the following effects must be taken into consideration when using lenses:

(1) For both surface and edge emitters, achievable coupling efficiency increases as spot size is decreased. In general, however, as spot size is reduced output power decreases because of enhanced device heating. For surface emitters, radiance saturation compounds the problem. Consequently, the maximum coupled power will not, in general, correspond to the maximum coupling efficiency. For a given fiber and operating current, spot size and magnification must be optimized to give the highest coupled power (Wada et al., 1982b).

(2) Increased coupling efficiency is obtained at the expense of reduced mechanical tolerance to alignment (Ackenhusen, 1978; Berg et al., 1981; Escher et al., 1982, Johnson et al., 1980). While use of microlenses provides larger tolerance in lateral misalignment, very close placement of the fiber end to the lens is required. The use of macrolenses, on the other hand, relaxes the axial separation requirement but suffers from tight lateral alignment tolerances and reduced coupling efficiency. Figure 12.13a and b illustrate this point. Thus, the optimum type of lens is dependent upon the application.

(3) In most practical surface emitters, the near-field intensity is maximum at the center of the emitting area and reduces radially toward the perimeter of the light spot due to carrier diffusion, which implies a fall-off of the carrier density, $J(r)$. Since the rise time, t_r is proportional to $\sqrt{J(r)}$, it follows that the modulation bandwidth is higher for the center emission than for the perimeter emission (Wada et al., 1982a). Consequently, the effective speed depends on the portion of the spot from which light is collected. Thus, the common practice of measuring t_r (or bandwidth) by collecting LED light

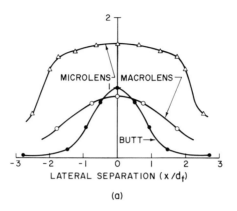

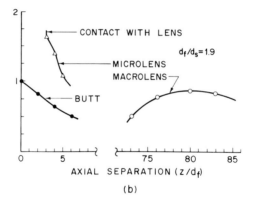

Fig. 12.13 Coupling gain for various lensing schemes as (a) lateral separation and (b) axial separation of LED and fiber. [After Ackenhusen (1978).]

using an objective lens with a large NA can lead to a larger value of t_r (lower bandwidth) than that with maximum coupling to a fiber using microlenses. Optimizing the lens magnification usually yields the highest power-bandwidth product.

12.5 DUAL-WAVELENGTH LEDs

The very wide transmission window from 0.8 to 1.6 μm wavelength in present low-loss optical fibers offers opportunities for use of wavelength-division multiplexing (WDM) to increase the transmission capacity of a single optical fiber. The conventional way to accomplish WDM is to combine the signal from each single-wavelength source by means of a passive optical multiplexer, which may be either a grating, filter or fiber device. A simpler way, however, is to provide a source that emits two (or more) wavelength bands simultaneously from a single

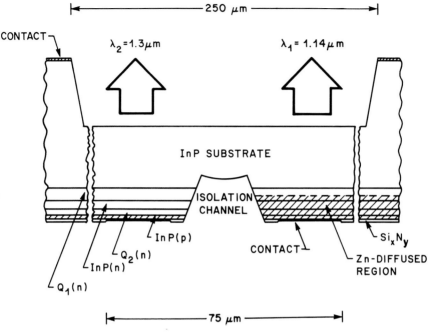

Fig. 12.14   Schematic illustration of a dual wavelength LED. [After Lee *et al.*, (1980).]

chip, where each wavelength band can be modulated independently. Figure 12.14 shows a schematic cross-section of a dual-wavelength surface-emitting LED (Lee *et al.*, 1980). Basically, the LED wafer consisted of five epitaxial layers grown by LPE on an InP substrate. A buffer layer of InP was first grown, followed by a 2-μm thick $In_{0.77}Ga_{0.23}As_{0.5}P_{0.5}$ layer (denoted by Q_1), and InP barrier layer, a 1.5-μm thick $In_{0.66}Ga_{0.34}As_{0.75}P_{0.25}$ layer (denoted by Q_2), and finally, a top InP (p) layer. The Q_1 layer had a bandgap energy of 1.09 eV, corresponding to 1.14-μm emission wavelength, whereas the Q_2 layer had a bandgap energy of 0.95 eV, corresponding to 1.3-μm wavelength. The *p-n* junction in the Q_1 layer was formed by Zn diffusion, and the *p-n* junction in the Q_2 layer was a grown junction. The contact windows were 25 × 75-μm separated by a 25-μm channel chemically etched through the grown layer to provide electrical isolation; each emitter then was 25 × 25 μm.

The emission spectrum of each junction, superimposed on one graph, is shown in Fig. 12.15. The measured total cross-talk was −22 to −26 dB. The maximum output of the diode at each wavelength was 1.0 mW when driven at 50 mA d.c., with a modulation bandwidth of 25 MHz. A dual-wavelength LED of this type was used in conjunction with a dual-wavelength demultiplexing photodetector (Campbell *et al.*, 1980) in a WDM experiment operated at 33

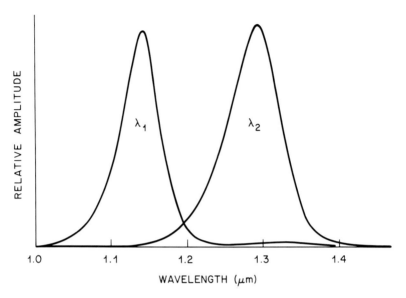

Fig. 12.15   Emission spectra of dual wavelength LED. [After Lee *et al.*, (1980).]

Mbit/s (Ogawa *et al.*, 1981). Approximately 60μW average optical power for each channel was coupled into a 0.36NA, 100μm core-silicone cladded fiber used in the experiment. A receiver sensitivity of −38.5 dBm and −39.7 dBm was achieved for the two wavelength channels respectively. Recently, Chin *et al.* (1985) demonstrated a dual wavelength edge-emitting LED, which is fabricated by attaching discrete diodes so that the emission spots are in close proximity.

12.6 SYSTEM EXPERIMENTS AND PERFORMANCE

12.6.1 LED-Based Multimode Fiber System Experiments

With presently available low-loss (< 1 dB/km at 1.3 μm) multimode fiber, an LED-based lightwave system can achieve repeaterless transmission distance up to 20 km and information rates up to a few hundred megabits per second. Applications of such a system include interoffice trunks, subscriber loops, local-area networks and on-premises data links.

Chromatic-dispersion-limited transmission in multimode fibers was first demonstrated in an experiment on a 137-Mb/s data link operating at both 0.9-μm and 1.23-μm wavelengths using an AlGaAs and an InGaAsP LED respectively [Muska *et al.*, 1977]. In a subsequent experiment [Gloge *et al.*, 1980a], 1.3-μm surface-emitting LEDs, which coupled −20 dBm power into a 50-μm core multimode fiber, were used as the transmitters. The receivers consisted of back-illuminated InGaAs pin photo-diodes (Lee *et al.*, 1980a; Lee *et al.*, 1981a) and GaAs FET preamplifiers (Ogawa and Chinnock, 1979) had measured sensitivities of -46.7 dBm and -36 dBm at 44.7 Mb/s and 274 Mb/s data rates, respectively. Repeaterless transmission distances of 23 km at the low rate and 7 km at the high rate were achieved (Gloge *et al.*, 1980). The overall bandwidth of the fiber used was approximately 800 MHz-km. These initial experiments demonstrated the feasibility of high bit-rate lightwave systems using 1.3-μm LEDs.

Similar laboratory experiments at both 32 Mb/s and 100 Mb/s were conducted in Japan (Kunita *et al.*, 1979; Shikata *et al.*, 1982). In the 100 Mb/s experiment a small surface-emitting LED with a 23-μm diameter light-emitting area and a 140-nm spectral width (FWHM) centered at the wavelength of 1.25 μm was used. The output power was coupled to a 50-μm core, 0.2-NA, graded-index fiber by means of a graded-index-rod lens. A -13-dB coupling efficiency resulted in a launched power of -17.9 dBm into the fiber. The measured fiber bandwidth was 320 MHz-km. Using a Ge-APD for the photodiode, an average received signal power of -44.3 dBm was obtained for 10^{-9} error probability. The power penalty due to fiber dispersion (15 km) was 2.8 dB. Allowing for sufficient system margins, repeater spacings up to 10 km were considered feasible.

The first long-wavelength LED-based system to carry commercial telephone services was installed between two central offices in Sacramento, California in June 1980. It operated at 12.6 Mb/s with interoffice spacing of 6 km. More recently, a 34-Mb/s lightwave system was installed for long-haul telephone service (Davis *et al.*, 1982). This system, between London and Birmingham in Great Britain, operated with 1.3-μm LEDs over a total distance of 205 km, within which the maximum repeater spacing was 10.5 km. The surface-emitting LEDs, equipped with a truncated spherical lens for improved coupling, produced peak launched powers of -17 dBm into the fiber. InGaAs pin photodiodes and GaAs FETs were used for the front-end of the receiver. A sensitivity of -52 dBm was realized. The mean bandwidth of the fiber was 270 MHz-km and the mean fiber loss was 1.32 dB/km. Using edge emitting LEDs, this system can be up-graded to 140 Mb/s. (Ure *et al.*, 1982).

Similar application opportunities exist for LED-based lightwave systems in the feeder portion of the subscriber loop, where digital systems become increasingly attractive with the introduction of digital end offices. An example is the Fiber SLC^{TM} carrier system (Bohn *et al.*, 1984) which first went into commercial service in Chester Heights, Pennsylvania in November 1982. This was followed by a second installation in early 1983 in Kernersville, North Carolina with a rapid build-up leading to approximately 500 Fiber SLC^{TM} systems by the end of 1983 [Olsen and Schepis, 1984]. The fibers, specifically designed for LED-based system applications, have a numerical aperture of 0.29, a core diameter of

62.5 μm and a cladding diameter of 125 μm. The installed fiber cable loss is less than 1 dB/km (at λ = 1.3 μm) and the bandwidth is capable of supporting system bit rates up to 100 Mb/s for a maximum feeder length of 20 km. The surface-emitting LEDs with monolithic integral lens are capable of launching more than 100 μW into the fiber. The receivers employed planar InGaAs pin photodiodes and GaAs FETs (Bohn et al., 1984). The initial transmission rate in the field trial was at 12.6 Mb/s with capabilities of upgrading to 44.5 Mb/s, and ultimately to 90 Mb/s, as economic and growth conditions become favorable.

The use of digital carrier in the loop plant naturally places many remote feeder terminals in a hostile, unattended environment. Reliability considerations favor the choice of LED/pin-photodiode-based systems over laser-based systems. The system margin of 25 dB, achieved at 44.5 Mb/s, takes into account a local component temperature range of $-40\,^\circ$C to $+85\,^\circ$C, degradation of LED power due to aging and other system tolerances. A total of 7 dB is allocated to system margin and 18 dB to fiber loss. Device mean operating life times (to failure) of these LEDs are expected (from the accelerated aging results) to be greater than 10^9 hours at $60\,^\circ$C [Saul, 1983]. The projected mean life (to failure) for pin photodiodes is estimated more than 10^9 hours at $60\,^\circ$C, with a worst estimate of about 10^7 hours [Saul et al., 1984].

The relentless drive to increase the speed of operation of on-premises equipment such as computers, multiplexers and digital switches has served as a stimulus to push higher the transmission speed of short-distance LED-based data links. An example of recent work addressing this need is the demonstration of a 4.4-km link using a high-speed InGaAsP LED operating at 560 Mb/s [Groth et al., 1983]. The specially-designed high-speed LED has a modulation bandwidth of 900 MHz and, with a ball-lens, is able to couple 7.3 μW into a 0.2 NA graded-index fiber at 100 mA drive current. A total link attenuation of 5.3 dB at a bit error rate less than 10^{-9} has been achieved. A similar experiment with a 2.7-km link at 500 Mb/s has also been demonstrated [Uji et al., 1984]. King and Olsson (1986) transmitted data at 1 Gb/s over 3 km of fiber achieving the highest bit-rate-distance product with a multimode fiber. Very high modulation rates for surface-emitting LEDs (1.6 Gb/s) has been achieved [Kobayashi and Nomura, 1984; Suzuki et al., 1984].

12.6.2 LED-Based Single-Mode-Fiber System Experiments

Because of the tremendous transmission bandwidth, single-mode fiber has been considered a favorable choice over multimode fiber for subscriber loops as well as other short-haul installations. What has prohibited the rapid deployment of single-mode fiber in the subscriber loop is the high cost and insufficient reliability of present-day 1.3-μm laser diodes, which until recently were assumed to be the only source for single-mode systems. When compared with laser diodes, LEDs offer the advantages of high reliability (see Chapter 17), reduced temperature sensitivity, immunity to optical feedback, simple driving circuits, and low cost due to high yield and simpler packaging technology.

It has been recognized recently that LEDs can couple useful power into single-mode fibers (Pophillat, 1984; Arnold and Krumpholz, 1985a; Shumate et al., 1985). Initial experiments (Shumate et al., 1985) used a small-area (20-μm dia) surface-emitting LED and a commercial edge-emitting LED, both operated at 140 Mb/s. The average (50% duty cycle) launched power into a single-mode fiber (butt-coupled) was -37.6 dBm and -28.1 dBm for the surface emitting and edge-emitting devices, respectively. With the -40.4-dBm sensitivity of a commercial pin-FET receiver, transmission distances of 4.5 km using the surface emitter and 22.5 km using the edge-emitter were realized.

Improvement in the device design and coupling method (Uji and Hayashi, 1985; Olshansky et al., Saul et al., 1985; Plastow et al., 1985, Fye et al., 1985; Arnold and Krumpholz, 1985a) has increased both the bit rates and the distances in single-mode transmission experiments using LEDs. For example, a small-area (20-μm dia) surface emitting mesa LED has been designed (Uji and Hayashi, 1985); with lensed coupling into a 10-μm core single-mode fiber, this device launched a maximum of -26.7 dBm (2.1 μW) near the power saturation point. Transmission distances of 25 km at 140 Mbit/s and 5 km at 565 Mbit/s have been reported (Uji et al., 1985). In another experiment, using commercial edge-emitting LEDs, transmission distances of 15 km at 140 Mbit/s and 35 km at 560 mbit/s were achieved, as compared to 25 km and 50 km using a superluminescent LED at these two bit-rates, respectively (Gimlett et al., 1985b; Ulbricht et al., 1985). Bidirectional transmission on 15-km long single-mode fiber using 1.3-μm and 1.5-μm edge-emitting LEDs, operated at 140 Mbit/s and 45 Mbit/s respectively, was also demonstrated (Stern et al., 1985, Hall et al., 1985). Because of the absence of mode-partition noise in LEDs, the power penalties observed in these experiments were due entirely to the inter-symbol interference (ISI) arising from the pulse broadening, which increased at higher bit-rates (Gimlett et al., 1985a). For minimizing the dispersion penalties and fiber losses, temperature effects must be considered (Shumate, 1986). The optimum location of the center emission wavelength lies at the shorter wavelength side of the zero dispersion wavelength of the silica fiber.

12.7 CONCLUSIONS

Both practical high-radiance surface-emitting and edge-emitting LEDs have demonstrated high power and large bandwidth capabilities for use as optical sources in lightwave communications systems. There is an inverse relationship, however, between the output power and the modulation bandwidth of LEDs. This relationship dictates the maximum power that can be obtained for a given system data rate. The best power-times-bandwidth product can be achieved by the optimization of the thickness and doping concentration of the active layer as well as the bandgap energy and doping concentrations of the confining layers in the double-heterostructure wafer. Edge-emitting LEDs and superluminescent diodes have narrower spectral width and higher coupling efficiency to optical fibers, but the output power of these devices decreases more rapidly with increasing temperature than does that of surface-emitting LEDs. A variety of lensing techniques have been developed to increase coupling efficiency to fibers, especially for surface emitters. In general, higher coupled power comes at the expense of greater complexity in device packaging (external lens) or chip fabrication (integral lens).

Present LED materials have been grown primarily by liquid-phase epitaxy (LPE), which is somewhat limited for large volume production. Other epitaxial growth methods, more suitable for volume production, which include vapor-phase epitaxy (VPE), metal-organic vapor-phase epitaxy (MOVPE), and molecular beam epitaxy (MBE), have been developed. Devices made by both VPE and MBE have shown performance (Olsen et al., 1981; Lee et al., 1978a; Lee et al., 1981) comparable to that of devices from LPE materials.

It is evident that LED-based lightwave systems operating at the wavelength of minimal dispersion (near 1.3 μm) of doped-silica graded-index fiber can have many applications that offer advantages of simplicity, reliability and economy over laser-based systems. Data rates up to a few hundred megabits per second and repeater spans over tens of kilometers are achievable with practical LEDs and pin photodetectors for both single-mode and multimode fiber systems. Using simple wavelength-division-multiplexing, channel capacity can be increased severalfold and cost savings on the complex electronic multiplexing equipment may be realizable. Present commercial systems are deployed as on-premises data links and for subscriber-loop carrier transmission. Future growth may extend the application of these LED-based systems into the areas of local area networks and wideband services, including voice, data and video transmission in the subscriber loop.

The ultimate limitations of LED-based multimode fiber systems are set by the available power of LEDs at high modulation rates (especially for single-mode fibers) and by the chromatic-dispersion-limited bandwidth of multimode fibers. In the area of long-haul, high-capacity transmission, laser-based single-mode lightwave systems are the natural choice. The market niche for LED-based fiber systems in short-distance, moderate-bandwidth applications is not at all insignificant, however, and favorable future growth, as new applications evolve, may further expand this market sector.

REFERENCES

Abe, M., Umebu, I., Hasegawa, O., Yamakoshi, S., Yamaoka, T., Kotami, T., Okada, H., and Takanashi, H. (1977). "High Efficiency Long Lived GaAlAs LEDs for Fiber-Optical Communications," *IEEE Trans. Electron Devices* **ED-24**, 990.

Abram, R. A., Allan, R. W., and Goodfellow, R. C. (1975). "The Coupling of Light-Emitting Diodes to Optical Fibers Using Spherical Lenses," *J. Appl. Phys.* **46**, 3468.

Ackenhusen, J. (1978). "Micro Lenses for Improved LED to Fiber Optical Alignment Tolerance," *Appl. Optics* **18**, 3694.

Acket, G. A., Nijam, W., and Lam, H. (1974). "Electron Lifetime and Diffusion Constant in Germanium-Doped Gallium Arsenide," *J. Appl. Phys.* **45**, 3033.

Alavi, K., Pearsall, T. P., Forrest, S. R., and Co, A-Y. (1983). "$Ga_{0.47}In_{0.53}As/Al_{0.48}In_{0.52}As$ Multiquantum-Well LEDs Emitting at 1.6 µm," *Electron Lett.* **19**, 227.

Adams, A. R., Asada, M., Suematsu, Y., and Arai, S. (1980). "The Temperature Dependence of the Efficiency and Threshold Current of $In_{1-x}Ga_xAs_yP_{1-y}$ Lasers Related to Intervalence Band Absorption," *Jap. J. Appl. Phys.* **19**, L621.

Anthony, P. J., and Schumaker, N. E. (1980). "Temperature Dependence of Lasing Threshold Current of DH Injection Lasers Due to Drift current Loss," *J. Appl. Phys.* **51**, 5038–5040.

Archer, R. J., and Kreps, D. (1967). "The Quantum Efficiency of Electro-Luminescence in Gallium Arsenide Diodes," *Gallium Arsenide Symposium*, Inst. Phys. and Phys. Soc., p. 103.

Arnold, G., and Krumpholz, O. (1985a). *Technical Digest*, Conference on Optical Fiber Communication, San Diego, February 11–14, p. 48.

Arnold, G., and Krumpholz, O. (1985b). "Optical Transmission with Single-Mode Fibres and Edge-Emitting Diodes," *Electron Lett.* **21**, 9, 390.

Asada, M., and Suematsu, Y. (1982). "Measurement of Spontaneous Emission Efficiency and Nonradiative Recombinations in 1.58 µm Wavelength GaInAsP/InP Crystals," *Appl. Phys. Lett.* **41**, 353.

Barnoski, M. K. (1982). "Coupling Components for Optical Fiber Waveguides," Chap. 2 *in* "Fundamentals of Optical Fiber Communications," (M. K. Barnoski, ed.), Academic Press, pp. 147–185.

Berg, H. M., Lewis, G. L., and Mitchell, C. W. (1981). "A High Performance Connectorized LED Package for Fiber Optics," *Proceedings 31st Electronic Components Conference*, 372–379.

Berkstresser, G. W., Keramidas, V. G., and Zipfel, C. L. (1980). "Planar, Fast, Reliable Single-Heterojunction Light-Emitting Diode for Optical Links," *Bell Syst. Tech. J.* **59**, 1549.

Bohn, P. P., Brackett, C. A., Buckler, M. J., T. N., and Saul, R. H. (1984). "The Fiber SLC Carrier System," *AT & T Bell Laboratories Tech. J.* **63**, 2389.

Borsuk, J. A. (1983). "Light Intensity Profiles of Surface-Emitting InGaAsP LEDs: Impact on Coupling Efficiency," *IEEE Trans. Electron Devices* **ED-30**, 296.

Burrus, C. A. (1969). Unpublished Work.

Burrus, C. A., and Dawson, R. W. (1970). "Small-Area High-Current-Density GaAs Electroluminescent Diodes and a Method of Operation for Improved Degradation Characteristics," *Appl. Phys. Lett.* **17**, 97.

Burrus, C. A., and Miller, B. J. (1971). "Small-Area, Double Heterostructure AlGaAs Electroluminescent Diode Sources for Optical-Fiber Transmission Lines," *Opt. Commun.* **4**, 307–309.

Burton, R. H. (1983). Unpublished, Bell Laboratories.

Campbell, J. C., Dentai, A. G., Lee, T. P., and Burrus, C. A. (1980). "Improved Two-Wavelength Demultiplexing InGaAsP Photodetector," *IEEE J. Quantum Electron.* **16**, 601.

Carlson, R. O., Slack, G. A., and Silverman, S. T. (1965). "Thermal Conductivity of GaAs and GaAsP Laser Semiconductors," *J. Appl. Phys.* **36**, 505.

Casey, H. C., Jr., and Stern, F. (1976). "Concentration-Dependent Absorption and Spontaneous Emission of Heavily Doped GaAs," *J. Appl. Phys.* 47, 631.

Chen, T. R., Chiu, L. C., Hasson, A., Koren, U., Margalit, S., and Yariv, A. (1983). "Direct Measurement of the Carrier Leakage in an InGaAsP/InP Laser," *Appl. Phys. Lett.* 42, 1000.

Chin, A. K., Chin, B. H., Camlibel, I., Zipfel, C. L., and Minneci, G., (1985). "Practical Dual-Wavelength Light-Emitting Double Diode," *J. Appl. Phys.* 57, 5519.

Chin, A. K., Zipfel, C. L., Dutt, B. V., DiGiuseppe, M. A., Bauers, K. B., and Roccaseccs, D. D. (1981). "New Restricted Contact LEDs Using a Schottky Barrier," *Japan J. Appl. Phys.* 20, 1487.

Colvin, J. (1974). "Coupling (Launching) Efficiency for a Light-Emitting Diode, Optical Fiber Termination," *Opto-Electronics* 6, 387.

Davies, I. G. A., Goodwin, A. R., and Plumb, R. G. (1981). *Conf. on Lasers and Electrooptics*, Washington, D.C..

DiVita, P., and Vannucci, R. (1975). "Geometrical Theory of Coupling Errors in Dielectric Optical Waveguides," *Opt. Commun.* 14, 139.

Dawson, R. W. (1980). "LED Bandwidth Improvement by Bipolar Pulsing," *IEEE J. Quantum Electron.* 16, 697.

Dentai, A. G., Lee, T. P., and Burrus, C. A. (1977). "Small-Area High Radiance CW InGaAsP LEDs Emitting at 1.2 to 1.3 Micrometers," *Electron. Lett.* 13, 484.

Dumke, W. P. (1975). "The Angular Beam Divergence in Double-Heterojunction Lasers with Very Thin Active Regions," *IEEE J. Quantum Electron.* 11, 400.

Dutta, N. K., and Nelson, R. J. (1981). "Temperature Dependence of Threshold of InGaAsP DH Lasers and Auger Recombination," *Appl. Phys. Lett.* 38, 407.

Dutta, N. K., and Nelson, R. J. (1982). "Light Saturation of InGaAsP-InP LEDs," *IEEE J. Quantum Electron* 18, 375.

Dutta, N. K., Nelson, R. J., Wright, P. D., Besomi, P., and Wilson, R. B. (1983). "Optical Properties of a 1.3 μm InGaAsP Superluminescent Diode," *IEEE Trans. Electron Devices* ED-30, 360.

Dyment, J. C., SpringThorpe, A. J., King, F. D., and Straus, J. (1977). "Proton Bombarded Double-Heterostructure LEDs," *J. Electron. Mater.* 6, 173.

Escher, J. S., Berg, H. M., Lewis, G. L., Moyer, C. D., Robertson, T. V., and Wey, H. A. (1982). "Junction-Current-Confinement Planar Light-Emitting Diodes and Optical Coupling into Large-Core Diameter Fibers Using Lenses," *IEEE Trans. Electron Devices* ED-19, 1463.

Ettenberg, M., Kressel, H., and Wittke, J. P. (1976). "Very High Radiance Edge-Emitting LED," *IEEE J. Quantum Electron.* 12, 360.

Frahm, R. E. (1982). Unpublished, Bell Laboratories.

Fukui, T., and Horikoshi, Y. (1979). "Anomalous Luminescence Near the InGaAsP-InP Heterojunction Interface," *Japan J. Appl. Phys.* 18, 961.

Fye, D. M., Olshansky, R., LaCourse, J., Powaznik, W., and Lauer, R. B. (1986). "Low-Current 1.3 μm Edge-Emitting LED for Single-Mode-Fiber Subscriber Loop Applications," *Electron. Lett.* 22, 2, 87.

Gimlett, J. L., Stern, M., Curtis, L., Young, W. C., Cheung, and Shumate, P. W. (1985a). "Dispersion Penalties for Single-Mode-Fibre Transmission Using 1.3 and 1.5 μm LEDs," *Electron. Lett.* 21, 16, 668.

Gimlett, J. L., Stern, M., Vodhanel, R. S., Cheung, N. K., Chang, G. K., Leblanc, H. P., Shumate, P. W., and Suzuki, A. (1985b). "Transmission Experiments at 560 Mbit/s and 140 Mbit/s Using Single-Mode Fibre and 1300 nm LEDs," *Electron Lett.* 21, 1198.

Gloge, D., Albanes, A., Burrus, C. A., Chinnock, E. L., Copeland, J. A., Dentai, A. G., Lee, T. P., Li, T., and Ogawa, K. (1980). "High-Speed Digital Lightwave Communication Using LEDs and PIN Photodiodes at 1.3 μm," *Bell System Tech. J.* 59, 1365.

Goodfellow, R. C., and Mabbitt, A. W. (1976). "Wide-Bandwidth High-Radiance Gallium-Arsenide Light-Emitting Diodes for Fibre-Optic Communication," *Electron. Lett.* **12**, 50.

Goodfellow, R. C., Carter, A. C., Griffith, I., and Bradely, R. R. (1979). "GaInAsP/InP Fast, High Radiance 1.05–1.13 μm Wavelength LEDs with Efficient Lens Coupling to Small Numerical Aperture Silica Optical Fibers," *IEEE Trans. Electron Devices* **ED-26**, 1215.

Goodfellow, R. C., Carter, A. C., Rees, G. J., and Davis, R. (1981). "Radiance Saturation in Small-Area GaInAsP/InP and GaAlAs/GaAs LEDs," *IEEE Trans. Electron Devices* **ED-28**, 365.

Grothe, H., Proebster, W., and Harth, W. (1983). "Mg-Doped InGaAsP/InP LEDs for High Bit Rate Optical Communication Systems," *Electron Lett.* **19**, 910.

Hall, R. D., Betts, and Moss, J. P. (1985). "Bidirectional Transmission over 11 km of Single-Mode Optical Fibre at 34 Mbit/s Using 1.3 μm LEDs and Directional Couplers," *Electron. Lett.* **21**, 628.

Harth, W., Huber, W., and Heinen, J. (1976). "Frequency Response of GaAlAs Light-Emitting Diodes," *IEEE Trans. Electron Devices* **ED-23**, 478.

Hasegawa, O., and Namazu, R. (1980). "Coupling of Spherical-Surfaced LED and Spherical-Ended Fiber," *J. Appl. Phys.* **51**, 30.

Heinen, J., Huber, W., and Harth, W. (1976). "Light-Emitting Diodes with a Modulation Band-Width of More than 1 GHz," *Electron. Lett.* **12**, 533.

Heinen, J. (1982a). "1.3 μm InGaAsP/InP Light Emitting Diodes with Internally Defined Emission Area Prepared by Single-Step LPE Technique," *Electron Lett.* **18**, 23.

Heinen, J. (1982b). "Preparation and Properties of Monolithically Integrated Lenses on InGaAsP/InP Light Emitting Diodes," *Electron Lett.* **18**, 831.

Henry, C. H., Logan, R. A., and Merritt, F. R. (1977). "Origin of $n = 2$ Injection Current in Al$_x$Ga$_{1-x}$As Heterojunctions," *Appl. Phys. Lett.* **31**, 454.

Henry, C. H., Logan, R. A., Temkin, H., and Merritt, F. R. (1983a). "Absorption, Emission and Gain Spectra of 1.3 μm InGaAsP Quaternary Lasers," *IEEE J. Quant. Electronics* **19**, 941.

Henry, C. H., Levine, B. F., Logan, R. A., and Bethea, C. G. (1983b). "Minority Carrier Lifetime and Luminescence Efficiency of 1.3 μm InGaAsP-InP DH Layers," *IEEE J. Quant. Electronics* **19**, 905.

Hill, D. E. (1965). "Internal Quantum Efficiency of GaAs Electroluminescent Diodes," *J. Appl. Phys.* **36**, 3405.

Hino, I., and Iwamoto, K. (1979). "LED Pulse Response Analysis Considering the Distributed CR Constant in the Peripheral Junction," *IEEE Trans. Electron Devices* **ED-26**, 1238.

Horikoshi, Y., Takanashi, Y., and Iwane, G. (1976). "High-Radiance Light-Emitting Diodes," *Japan J. Appl. Phys.* **15**, 485.

Hudson, M. C. (1974). "Calculation of the Maximum Optical Coupling Efficiency into Multimode Optical Waveguides," *Appl. Optics* **13**, 1029.

Iwamoto, K., Hino, T., Matsumoto, S., and Inoue, K. (1976). "Room Temperature CW Operated Superluminescent Diodes for Optical Pumping of Nd: YAG Laser," *Japan J. Appl. Phys.* **15**, 2191.

Johnson, B. H., Ackenhusen, J. G., and Lorimor, O. G. (1980). "Connectorized Optical Link Package Incorporating a Microlens," *Proceedings 30th Electronic Components Conference*, 279.

Kaminow, I. P., Eisenstein, G., Stulz, L.W., and Dentai, A. G. (1982). *IEEE Specialist Conference on Light Emitting Diodes and Photodetectors*, Ottawa-Hull, Canada, p. 212.

Kaminow, I. P., Eisenstein, G., Stulz, L. W., and Dentai, A. G. (1983). "Lateral Confinement InGaAsP Superluminescent Diode at 1.3 μm," *IEEE J. Quantum Electron* **19**, 78.

Kibler, L. U., Burrus, C. A., and Trambarulo, R. (1964). "Light-Emitting, Formed-Point-Contact Gallium Arsenide and Gallium Arsenide-Phosphide Diodes," *Proc. IEEE* **52**, 1260.

King, F. D., and SpringThorpe, A. J. (1975). "The Integrated Lens Coupled LED," *J. Electron. Mat.* **4**, 243.

King, W. C., Chin, B. H., Camlibel, I., and Zipfel, E. L. (1985). "High-Speed High-Power 1.3 μm InGaAsP/InP Surface Emitting LEDs for Short-Haul Wide-Bandwidth Optical Fiber Communications," *IEEE Electron Device Lett.* **6**, 335.

King, W. C., and Olson, N. A. (1986). "Gb/s Transmission Experiments with a 1.3 μm High-Speed Surface-Emitting LED and Multimode Graded-Index Fiber," *Electron Lett.* **22**, 761.

Kobayashi, K., and Nomura, H. (1984). *Technical Digest*, Opt. Fiber Commun. New Orleans, MJ2.

Kressel, H., and Ettenberg, M. (1975). "A New-Edge-Emitting (AlGa)As Heterojunction LED for Fiber-Optic Communications," *Proc. IEEE* **63**, 1360.

Kressel, H., Ettenberg, M., Wittke, J. P., and Ladany, I. (1980). *In* "Semiconductor Devices for Optical Comm.," Springer-Verlag, New York, Kressel, H., ed. Ch. 2.

Kunita, M., Tonge, T., and Fujimeto, N. (1979). *Technical Digest*. Topic Meeting Optical Fiber Commun, Washington, DC, TUD1.

Kurbatov, L. N., Shakhidzhanov, S. S., Bystrova, L. V., Krapuhkin, V. V., and Kolonenkova, S. I. (1971). "Investigation of Superluminescence Emitted by a Gallium Arsenide Diode," *Sov. Phys.—Semiconductor* **4**, 1739.

Kundukhov, R. M., Metreveli, S. G., and Siukaev, N. V. (1967). "Diffusion of Cadmium and Zinc in Indium Phosphide," *Sov. Phys.—Semiconductor* **1**, 765.

Lee, T. P., Burrus, C. A., and Miller, B. I. (1973). "A Stripe-Geometry Double-Heterostructure Amplified-Spontaneous-Emission (Superluminescent) Diode," *IEEE J. Quantum Electron* **9**, 820.

Lee, T. P. (1975). "Effect of Junction Capacitance on the Rise Time of LEDs and on the Turn-On Delay of Injection Lasers," *Bell Syst. Tech. J.* **54**, 53.

Lee, T. P., and Dentai, A. G. (1978). "Power and Modulation Bandwidth of GaAs-AlGaAs High Radiance LEDs for Optical Communication Systems," *IEEE J. Quantum Electron.* **14**, 150.

Lee, T. P., Holden, W. S., and Cho, A. Y. (1978a). "AlGaAs-GaAs Double-Heterostructure Small-Area Light-Emitting Diode by Molecular Beam Epitaxy," *Appl. Phys. Lett.* **32**, 415.

Lee, T. P. (1980). "Recent Development in LEDs for Optical Fiber Communications Systems," *Proc. Society of Photo-Optical Instrument Engineers, (SPIE)*, Paper 224-16.

Lee, T. P., Burrus, C. A., and Dentai, A. G. (1980). "Dual Wavelength Surface Emitting InGaAsP LEDs," *Electron. Lett.* **16**, 845.

Lee, T. P., Burrus, C. A., Dentai, A. G., and Ogawa, K. (1980a) "Small-area InGaAs/InP p-i-n photodiodes: Fabrication, Characteristics and Performance of Devices in 274 Mbls and 45 Mbls Lightwave Receivers at 1.31 μm Wavelength," *Electron. Lett.* **16**, 155.

Lee, T. P., Holden, W. S., and Cho, A. Y. (1981). "Improved Molecular Beam Epitaxial Growth of $Al_xGa_{1-x}As$/GaAs High Radiance LEDs for Optical Communications," *IEEE J. Quantum Electron.* **17**, 387.

Lee, T. P., Burrus, C. A., and Dentai, A. G. (1981a). "InGaAs/InP p-i-n Photodiodes for Lightwave Communications at 0.95–1.65 μm Wavelength," *IEEE J. Wuantum Electron.*, QE-17, 232.

Lee, T. P. (1982). "LEDs and Photodetectors for Wavelength-Division-Multiplexed Light-Wave Systems," *Opt. and Laser Tech.* **14**, 15.

Liu, Y. S., and Smith, D. A. (1975). "The Frequency Response of an Amplitude-Modulated GaAs Luminescence Diode," *Proc. IEEE* (lett.) **63**, 542.

Manning, J., Olshansky, R., Su, C. B., and Powazinik, W. (1983). "Measurement of Carrier and Lattice Heating in 1.3 μm InGaAsP Light-Emitting Diodes," *Appl. Phys. Letters* **43**, 134.

Marcuse, D. (1977). "LED Fundamentals: Comparison of Front and Edge Emitting Diodes," *IEEE J. Quantum Electron.* **13**, 819.

Marcuse, D., and Kaminow, I. P. (1981). "Computer Modal at a Superluminescent LED with Lateral Confinement," *IEEE J. Quant. Electr.* **17**, 1234.

Muska, W. M., Li, Tingye, Lee, T. P., and Dentai, A. G. (1977). "Material-Dispersion-Limited Operation of High-Bit Rate Optical-Fibre Data Links Using LEDs," *Elect. Lett.* **13**, 605.

Namizaki, H., Nagano, M., and Nakahara, S. (1974). "Frequency Response of GaAlAs Light Emitting Diodes," *IEEE Trans. Electron Devices* **21**, 688.

Nelson, D. F., Gershenzon, M., Ashkin, A., D'Asaro, L. A., and Saraca, J. C. (1963). "Band-Filling Model for GaAs Injection Luminescence," *Appl. Phys. Lett.* **2**, 182.

Nelson, R. J., and Sobers, R. G. (1978). "Interfacial Recombination Velocity in GaAlAs/GaAs Heterostructures," *Appl. Phys. Lett.* **32**, 761.

Ogawa, K., and Chinnock, E. L. (1979). "GaAs F.E.T. Transimpedance Front-End Design for a Wideband Optical Receiver," *Electron. Lett.* **15**, 650.

Ogawa, K., Lee, T. P., Burrus, C. A., Campbell, J. C., and Dentai, A. G. (1981). "Wavelength Division Multiplexing Experiment Employing Dual-Wavelength LEDs and Photodetectors," *Electron Lett.* **17**, 857.

Olson, J. W., and Schepis, A. J. (1984). "Optical Fiber Transmission in the Subscriber Loop Plant," *Technical Digest*, Opt. Fiber Commun. New Orleans, MD1, 6.

Olshansky, R., Fye, D. M., Manning, J., Stern, M., Meland, E., Powazinik, W., Ulbricht, L., and Lauer, R. (1985). "High-Power InGaAsP Edge-Emitting LEDs for Single-Mode Optical Communication Systems," *Electron. Lett.* **21**, 17, 730.

Olsen, G., Hawrylo, F., Channin, D. J., Botez, D., and Ettenberg, M. (1981). "1.3 μm LPE and CVD-Grown InGaAsP Edge-Emitting LEDs," *IEEE J. Quant. Electron.* **17**, 2130.

Ostermayer, F. W., Jr., Kohl, P. A., and Burton, R. H. (1982). "Photoelectrochemical Formation of Integral Lenses on InP/InGaAsP LEDS," *IEEE Specialist Conf. on LEDs and Photodetectors*, Ottawa-Hull, Canada, Unpublished, Bell Laboratories.

Pearsall, T. P., Miller, B. I., Capik, R. J., and Backmann, K. J. (1976). "Efficient Lattice-Matched Double-Heterostructure LED's at $1 - 1$ μm from $Ga_x I_{1-x} As_y P_{1-y}$," *Appl. Phys. Lett.*, **28**, 499.

Plastow, R., Monham, K. L., Carter, A. C., Ritter, J. E., Croft, T. D., and Gibson, M. (1985). "Transmission over 107 km of Dispersion-Shifted Fibre at 16 Mbit/s Using a 1.55 μm Edge-Emitting Source," *Electron. Lett.* **21**, 369.

Plihal, M. (1982). "Improvement of Launching Efficiency of High-Radiance Surface-Emitting IREDS with Hybrid or Integrated Spherical Lenses into Step-Index and Graded-Index Fibers," *Siemens Forsch. & Entwicklungsber.* **11**, 221.

Pophillat, L. (1984). "Video Transmission Using a 1.3 m LED and Monomode Fibre," *10th European Conference on Optical Communications*, Stuttgart, Sept. 3–6, p. 238.

Rode, D. L. (1974). "How Much Al in the AlGaAs-GaAs Laser?" *J. Appl. Phys.* **45**, 3887.

Saul, R. H. (1983). "Recent Advances in the Performance and Reliability of InGaAsP LEDs for Lightwave Communication Systems," *IEEE Trans. Electron Devices* **ED-30**, 285.

Saul, R. H., King, W. C., Olsson, N. A., Zipfel, C. L., Chin, B. H., Chin, A. K., Camlibel, I., and Minneci, G. (1985). "180 Mbit/s, 35 km Transmission over single-Mode Fibre Using 1.3 μm Edge-Emitting LEDs," *Electron. Lett.* **21**, 17, 773.

Saul, R. H., Lee, T. P., and Burrus, C. A. (1985). "Light-Emitting Diode Device Design," *Semiconductors and Semimetals* **22**, Part C, 193.

Sakai, S., Umeno, M., and Amemiya, Y. (1980). "Measurement of Diffusion Coefficient and Surface Recombination Velocity for p-InGaAsP Grown on InP," *Japan. J. Appl. Phys.* **19**, 109.

Seki, Y. (1976). "Light Extraction Efficiency of the LED with Guide Layers," *Japan. J. Appl. Phys.* **15**, 327.

Sermaze, B., Eichler, H. J., Heritage, J. P., Nelson, R. J., and Dutta, N. K. (1983). "Photoexcited Carrier Lifetime and Auger Recombination in 1.3 μm InGaAsP," *Appl. Phys. Lett.* **42**, 259.

Shah, J., Laheny, R. F., Nahory, R. E., and Temkin, H. (1981). "Hot Carrier Effects in 1.3 μm InGaAsP Light Emitting Diodes," *Appl. Phys. Lett.* **39**, 618.

Shikata, M., Nomura, H., Suzuki, A., Minenura, K., and Sugimoto, S. (1982). *Technical Digest*, 5th Topic Meeting Opt. Fiber Commun., Phoenix, AZ, TUDD3.

Shumate, P. W., Gimlett, J. L., Stern, M., Romeiser, M. B., and Cheung, N. K. (1985). "Transmission of 140 Mbit/s Signals over Single-Mode Fibre Using Surface and Edge-Emitting 1.3 μm LEDs," *Electron. Lett.* **21**, 12, 522.

Shumate, P. W. (1986). "Temperature Effects on Dispersion and Loss for High-Bit-Rate LED Based Light Wave Systems," *Electron. Lett.* **22**, 1.

SpringThorpe, A. J., Look, C. M., and Emmerstorfer, B. F. (1982). "High Radiance Burrus LEDs with Integral 45° Mirrors," *IEEE Trans. Electron Devices* **ED-29**, 876.

Stern, M., Gimlett, J. L., Curtis, L., Cheung, N. K., Romeiser, M. B., Young, W. C., and Shumate, P. W. (1985). "Bidirectional LED Transmission on Single-Mode Fibre in the 1300 and 1500 nm Wavelength Regions," *Electron. Lett.* **21**, 20, 928.

Su, C. B., Schlafer, J., Manning, J., and Olshansky, R. (1982a). "Measurement of Radiative and Auger Recombination Rates in p-Type InGaAsP Diode Lasers," *Electronics Letters* **18**, 595.

Su, C. B., Schlafer, J., Manning, J., and Olshansky, R. (1982b). "Measurement of Radiative Recombination Coefficient and Carrier Leakage in 1.3 μm InGaAsP Lasers with Lightly Doped Active Layers," *Electronics Letters* **18**, 1108.

Snyder, A. W. (1974). "Leaky-Ray Theory of Optical Wavelengths of Circular Cross Section," *Appl. Phys.* **4**, 273.

Suzuki, A., Inomoto, Y., Hayashi, J., Isoda, Y., Uji, T., and Nomura, H. (1984). "Gbit/s Modulation of Heavily Zn-Doped Surface-Emitting InGaAsP/InP DH LED," *Electron. Lett.* **20**, 274.

Temkin, H., Chin, A. K., and DiGiuseppe, M. A. (1981a). "Light-Current Characteristics of InGaAsP LEDs," *Appl. Phys. Lett.* **39**, 405.

Temkin, H., Keramidas, V. G., Pollack, M. A., and Wagner, W. R. (1981b). "Temperature Dependence of Photoluminescence of n-InGaAsP," *J. Appl. Phys.* **52**, 1574.

Temkin, H., Joyce, W. B., Chin, A. K., DiGiuseppe, M. A., and Ermanis, F. (1982). "Effect of p-n Junction Position on the Performance of InGaAsP Light Emitting Diodes," *Appl. Phys. Lett.* **41**, 745.

Temkin, H., Zipfel, C. L., DiGiuseppe, M. A., Chin, A. K., Keramidas, V. G., and Saul, R. H. (1983). "InGaAsP LEDs for 1.3 μm Optical Transmission," *Bell System Technical Journal* **62**, 1.

Uji, T., Iwamoto, K., and Lang, R. (1981). "Nonradiative Recombination in GaAsP/InP Light Sources Causing Light Emitting Diode Saturation and Strong Laser-Threshold-Current Temperature Sensitivity," *Appl. Phys. Lett.* **38**, 193.

Uji, T., Onabe, K., Hayashi, J., Isoda, Y., Morihisa, Y., Iwamoto, K., and Sakuma, J. (1982). "Dependence of Optical Output on Diameter of Light Emitting Region and Active Layer Thickness in InGaAsP/InP DH LED," *Electronic Communications Society National Conference*, Japan.

Uji, T., Isoda, Y., Inomoto, Y., Suzuki, A., Hayashi, J., and Nomura, H. (1984). "Highly Reliable Zn-Doped InGaAsP 1.3 μm Surface Emitting LEDs for High Speed Optical Communication Systems," *Technical Digest*, Opt. Fiber Commun., New Orleans, MJ3.

Uji, T., and Hayashi, J. (1985). "High-Power Single-Mode Optical-Fibre Coupling to InGaAsP 1.3 μm Mesa-Structure Surface-Emitting LEDs," *Electron. Lett.* **21**, 10, 418.

Uji, T., Shikada, M., Fujita, S., Hayashi, J., and Isoda, Y. (1985). "565 Mb/s-5 km and 140 Mb/s-25 km Single-Mode Fiber Transmission Using 1.3 μm Mesa-Structure Surface Emitting LEDs," *Technical Digest*, Post Deadline Papers, IOOC/ECOC'85, Venice, Italy, 57.

Ulbricht, L. W., Teare, M. J., Olshansky, R., and Lauer, R. B. (1985). "Loss-Limited Transmission at 140 Mbit/s over 30 km of Single-Mode Fibre Using a 1.3 μm LED," *Electron. Lett.* **21**, 19, 860.

Umebu, I., Hasegawa, O., and Akita, K. (1978). "InGaAsP/InP DH LEDs for Fiber-Optical Communication," *Electron. Lett.* **14**, 499.

Ure, J., Carter, A. C., Goodfellow, R. C., and Harding, M. (1982). "High Power Lens Coupled 1.3 μm Edge Emitting LED for Long Haul 140 Mb/s Fiber Optics Systems," *IEEE Specialist Conference on Light Emitting Diodes and Photodetectors*, Ottawa-Hull, Canada, paper 20, 204.

Wada, O., Yamakoshi, S., Abe, M., Akita, K., and Toyama, Y. (1979). "A New Type InGaAsP/InP DH LED for Fiber Optical Communication System at 1.2-1.3 μm," *Proceedings of the Optical Communication Systems Conference*, Amsterdam, 4.6.

Wada, O., Yamakoshi, S., Abe, M., Yishitoni, Y., and Sakwai, T. (1981). "High Radiance InGaAsP Lensed LEDs for Optical Communication Systems at 1.2-1.3 μm," *IEEE J. Quantum Electron.* **QE-17**, 174.

Wada, O., Hamaguchi, H., Nishitani, Y., and Sakurai, T. (1982a). "High Speed Response InGaAsP/InP DH LEDs in the 1.3 μm Wavelength Region," *IEEE Electron Device Lett.* **EDL-3**, 129.

Wada, O., Hamaguchi, H., Nishitani, Y., and Sakurai, T. (1982b). "Optimized Design and Fabrication of High Speed and High Radiance InGaAsP/InP DH LED in the 1.3 μm Wavelength Range," *IEEE Trans. Electron Devices* **ED-29**, 1454.

Wada, O., Yamakoshi, S., and Sakurai, T. (1982c). "Band-Gap Enhanced Carrier Heating in InGaAsP/InP DH Light-Emitting Diodes," *Appl. Phys. Lett.* **41**, 981.

Wang, C. S., Cheng, W. H., and Hwang, C. J. (1982). "High-Power Low-Divergence Superradiance Diode," *Appl. Phys. Lett.* **41**, 587.

Wittke, J. P., Ettenberg, M., and Kressel, H. (1976). "High Radiance LED for Single-Fiber Optical Links," *RCA Rev.* **37**, 159.

Wittke, J. P. (1975). "Spontaneous-Emission-Rate Alteration by Dielectric and Other Waveguiding Structures," *RCA Rev.* **36**, 655.

Wright, P. D., Chai, Y. G., and Antypas, G. A. (1975). "InGaAs-InP Double-Heterojunction High Radiance LEDs," *IEEE Trans. Electron Devices* **ED-26**, 1220.

Yamakoshi, S., Sanada, T., Wada, O., Umebu, I., and Sakurai, T. (1982). "Direct Observation of Electron Leakage in InGaAsP/InP Double Heterostructure," *Appl. Phys. Lett.* **40**, 144.

Yang, K. H., and Kingsley, J. D. (1975). "Calculation of Coupling Losses Between Light Emitting Diodes and Low-Loss Optical Fibers," *Appl. Optics* **14**, 288.

Yano, M., Nishi, H., and Takusagawa, M. (1980). "Influences of Interfacial Recombination Oscillation Characteristics of InGaAsP/InP DH Lasers," *IEEE J. Quantum Electron.* **QE-16**, 661.

Yano, M., Imai, H., and Takusagawa, M. (1981). "Analysis of Threshold Temperature Characteristics for InGaAsP/InP DH Lasers," *J. Appl. Phys.* **52**, 3172.

Zipfel, C. L., Saul, R. H., Chin, A. K., and Keramidas, V. G. (1982). "Competing Processes in Long-Term Accelerated Aging of DH GaAlAs Light Emitting Diodes," *J. Appl. Phys.* **53**, 1781.

Zargar'yants, M. N., Mezin, Yu. S., and Kolonenkova, S. I. (1971). "Electroluminescent Diode with a Flat Surface Emitting Continuously 25 W/cm²-sr at 300°K," *Sov. Phys.-Semicond.* (Engl. Transl.) **4**, 1371.

Zucker, J. (1978). "Closed-Form Calculation of the Transient Behavior of (Al,Ga) As Double-Heterojunction LEDs," *J. Appl. Phys.* **49** 2543.

Chapter 13

Semiconductor Lasers for Telecommunications

J. E. BOWERS

AT&T Bell Laboratories, Inc., Holmdel, New Jersey

M. A. POLLACK

AT&T Bell Laboratories, Inc., Holmdel, New Jersey

13.1 INTRODUCTION

When the first edition of this book went to press in 1979, semiconductor lasers were just beginning to move from the laboratory into commercial systems. Great progress has been made in the intervening years, and we now seek to present an updated picture of semiconductor laser technology as it applies to modern lightwave communications. We will describe how a semiconductor laser works, how it behaves in a lightwave system, and which new laboratory devices might move into the marketplace within the next decade. We also aim to provide those readers more acquainted with semiconductor lasers with a general reference to areas outside of their own expertise.

13.1.1 Organization of the Chapter

The semiconductor laser is such a well accepted part of optical fiber telecommunications that we often fail to appreciate the enormous research and development effort that has taken place since the early 1960s. In the remainder of this section, we lead the reader on a brief journey through the 25 years or so of semiconductor laser history. We next review, in Section 13.2, the physics underlying the operation of semiconductor lasers. In Section 13.3, we discuss the principal lateral and longitudinal device structures, as well as lasers incorporating some degree of integration. Laser modulation and noise are covered in Section 13.4, and Section 13.5 describes mid-infrared lasers compatible with fluoride-based optical fibers. Finally, Section 13.6 offers a view of future prospects for semiconductor lasers.

13.1.2 Historical Perspective

Although the use of semiconductors in lasers had been suggested earlier, the first achievement of laser action in p-n junction diodes was reported in 1962 by three groups, Hall et al. (1962) at GE, Nathan et al. (1962) at IBM, and Quist et al. (1962) at Lincoln Labs. These GaAs lasers operated at a temperature of 77 K, at wavelengths near 0.85 μm. Holonyak and Bevacqua (1962) at GE reported 77 K operation of shorter wavelength GaAsP laser diodes, and laser emission at other wavelengths followed rapidly. All of these early devices were homojunction lasers, which relied on carrier injection at junctions diffused into bulk single crystals. At room temperature, they had threshold current densities of the order of 10^5 A/cm^2 or more; thus, the required current was far too high to permit continuous (cw) lasing.

A step of crucial importance to the room temperature cw semiconductor laser was the lattice-matched growth by liquid phase epitaxy of $Al_xGa_{1-x}As$ (hereafter abbreviated AlGaAs) on GaAs (Woodall et al., 1967). This work stimulated several groups to immediately use the new technology of heteroepitaxy. The "double heterostructure," in which a narrow band-gap gain region is sandwiched between two regions of wider band-gap material to provide carrier confinement and more efficient recombination, had been suggested earlier by both Kroemer (1963) and Alferov and Kazarinov (1963). In the case of GaAs/AlGaAs, the narrow band-gap GaAs region also has a larger refractive index, thus forming an optical waveguiding structure. The first heterojunction lasers were GaAs/AlGaAs single, rather than double, heterostructures; their room temperature threshold current densities were reduced an order of magnitude below those obtained in homojunction devices (Hayashi et al., 1969; Kressel and Nelson, 1969). Lasers based on AlGaAs/GaAs/AlGaAs double heterostructures (DH) came shortly thereafter, and the reduced threshold current densities permitted the first cw room temperature operation of an injection laser (Hayashi et al., 1970; Alferov et al., 1970).

The second decade of the semiconductor laser, the 1970s, saw major improvements in GaAs/AlGaAs DH laser performance and reliability. The first devices lasted only minutes at room temperature, and only after enormous effort were projected lifetimes extended to years. This period saw the introduction of a wide variety of lateral index guiding structures, and the achievement of well-controlled lateral laser modes. At the same time, advances were being made in bringing cooled semiconductor lasers to new infrared wavelengths using a variety of materials. An update of this work for the mid-infrared region is the subject of Section 13.5.

A major milestone in the development of semiconductor lasers for telecommunications applications was the demonstration of cw room temperature devices operating at wavelengths beyond 1 μm, where silica fibers have reduced losses. The first of these DH lasers was made from the $GaAs_{1-x}Sb_x/Al_yGa_{1-y}As_{1-x}Sb_x$ mixed-crystal alloy system and grown on a GaAs substrate (Nahory $et\ al.$, 1976). It was followed within a few months by lasers in the $Ga_xIn_{1-x}As_yP_{1-y}/InP$ system, which has the important advantage of being lattice-matched to InP substrates (Hsieh $et\ al.$, 1976). The GaInAsP/InP DH laser now covers the entire 1.1 to 1.65 μm wavelength range and has overtaken the GaAs/AlGaAs laser in many aspects of performance and reliability.

A significant achievement of the 1980s has been the realization of high performance single frequency lasers. The distributed feedback (DFB) laser concept, originally demonstrated in dyes by Kogelnik and Shank (1971) and Kaminow $et\ al.$ (1971) was applied soon after to injection lasers. The first room temperature semiconductor lasers with distributed Bragg reflectors (Reinhart $et\ al.$, 1975) or distributed feedback (Casey $et\ al.$, 1975; Aiki $et\ al.$, 1975) were demonstrated in GaAs/AlGaAs devices, and these concepts have been extended to GaInAsP/InP lasers at 1.3 μm and 1.55 μm by many workers. The focus of recent research has been on single frequency lasers with narrow linewidths under continuous operation, and low levels of frequency chirp under intensity modulation. Other goals have been the attainment of ever-higher output powers and modulation frequencies, and the achievement of wavelength tunability. The following sections are intended to bring the reader up to date in these important areas.

13.2 PHYSICS OF SEMICONDUCTOR LASERS

This chapter is concerned primarily with the communications applications of semiconductor lasers. As discussed in Chapter 2, the lowest optical fiber transmission loss in silica fiber is at a wavelength of about 1.55 μm. Most long distance communication systems, therefore, use GaInAsP sources emitting either at 1.55 μm or at 1.3 μm, the wavelength of zero dispersion of conventional silica fiber. Lasers based on GaAs are important for optical storage and printing applications and may be important for short distance communications applications. As GaAs lasers are already comprehensively discussed in many books (Kressel and Butler, 1977; Casey and Panish, 1978; Thompson, 1980), GaInAsP lasers will be the focus of this chapter. Lasers at even longer wavelengths may become important if the loss of one of several alternative fibers (Tran $et\ al.$, 1986) becomes sufficiently low; longer wavelength lasers are the subject of Section 13.5. This section describes band structures and band offsets, gain and

512   J. E. Bowers and M. A. Pollack

loss in semiconductors, the condition for inversion, heterostructures, the rate
equation for threshold, the temperature dependence of threshold, and the
differential quantum efficiency.

13.2.1 Band Structure

The band-gap and lattice constant dependence on composition of several
important compound semiconductors are shown in Fig. 13.1. For
$Ga_xIn_{1-x}As_yP_{1-y}$ compositions that are lattice-matched to InP ($y \approx 2.2x$), the
band-gap in eV varies as (Nahory *et al.*, 1978):

$$E_g = 1.35 - 0.72y + 0.12y^2. \qquad (13.1)$$

Band-gap wavelengths from 0.92 to 1.65 μm are covered by this material
system. The band structure of GaInAsP is shown in Fig. 13.2a. All GaInAsP
compositions lattice matched to InP are direct band-gap. The split-off valence
band and the conduction band are separated by approximately the same energy
from the valence band maximum at the Γ-valley in 1.55 μm GaInAsP, unlike
the situation in InP or GaAs. Consequently, absorption due to valence band
transitions can cause significant optical loss.

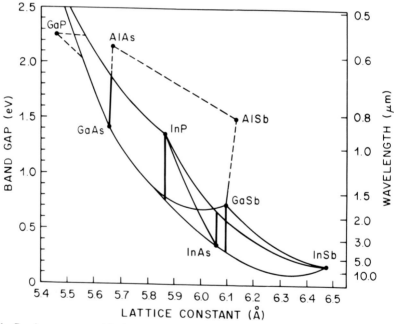

Fig. 13.1 Bandgap energy and lattice constant of several binary and ternary compound semiconductors. The heavy lines indicate three important sets of compositions: $Ga_xIn_{1-x}As_yP_{1-y}$ (hereafter
referred to as GaInAsP) lattice-matched to InP, GaInAsSb lattice-matched to GaSb, and AlGaAs
nearly lattice-matched to GaAs.

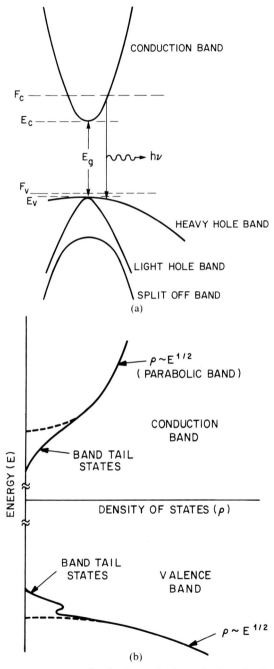

Fig. 13.2 (a) Energy versus wavenumber for the conduction and valence bands near the Γ valley. (b) Energy dependence of the density of states. (After Dutta and Nelson, 1982)

13.2.2 Heterostructures

Most semiconductor lasers use the double heterostructure design, in which the active gain region is sandwiched between layers of larger band-gap material. This design serves to confine the electrons and holes to a narrow region and achieves a high carrier density. In some semiconductor alloy systems such as AlGaAs and GaInAsP, the narrower band-gap material also has a higher refractive index, and the double heterostructure also serves as an optical waveguide. Figure 13.3 shows the wavelength dependence of the refractive index for several GaInAsP alloys lattice matched to InP. For a 1.3 μm-band-gap-GaInAsP/InP heterojunction, the index difference is about 0.3, giving a strongly guided structure. The band-gap difference is 0.4 eV or 15 kT at room temperature, which gives good carrier confinement, although there is some leakage across the heterojunction at high temperatures or at high bias currents. Active layer thicknesses of 0.2 μm or less are desirable for low thresholds, but cause a rapid beam divergence of the radiated lightwave in the transverse direction.

13.2.3 Gain

The simplified band structure around the Γ valley in Fig. 13.2a shows the energy definitions and quasi-Fermi-levels, F_c, F_v, for our discussion of transition rates. There are two primary techniques used to calculate laser gain. The approach described here requires evaluation of the matrix element for transitions between an upper and lower state, with a density of states model that includes bandtails. The other technique uses a density matrix calculation that includes a relaxation broadening model (Yamada and Suematsu, 1979). Both techniques model the absorption within and above the band-gap.

To begin our discussion of gain, we consider a forward-biased laser, with electrons and holes injected into the conduction and valence bands, respectively. We assume that the carriers within a band are in equilibrium with each other; and therefore, the probability that a state in the conduction band of energy E is occupied by an electron is given by Fermi-Dirac statistics:

$$f_c(E) = \frac{1}{e^{(E-F_c)/kT} + 1}. \tag{13.2}$$

A similar expression describes the occupation probability in the valence band. The conduction band density of states per unit energy (Fig. 13.2b) is approximated, ignoring bandtails, by the parabolic dependence:

$$\rho_c(E) = 4\pi\left(\frac{2m_e}{h^2}\right)^{3/2}(E - E_c)^{1/2} \qquad E > E_c, \tag{13.3}$$

where m_e is the effective electron mass and h is Planck's constant. Similar

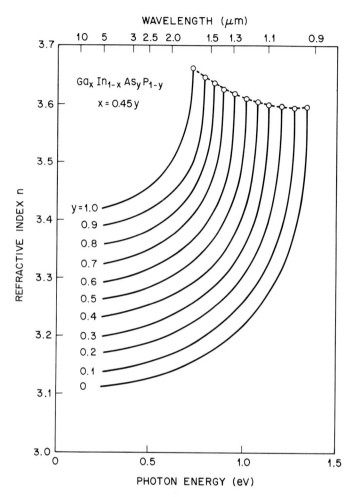

WAVELENGTH (μm)

$Ga_x In_{1-x} As_y P_{1-y}$

$x = 0.45 y$

REFRACTIVE INDEX n

PHOTON ENERGY (eV)

Fig. 13.3 Calculated wavelength dependence of the real part of the optical index in $Ga_x In_{1-x} As_y P_{1-y}$ (after Adachi, 1982).

expressions are used for the heavy hole (hh) and light hole (lh) valence bands. The electron density is:

$$N = \int_{E_c}^{\infty} f_c(E) \rho_c(E) \, dE. \qquad (13.4)$$

Given a particular injection level, these equations can be used to calculate the quasi Fermi levels. For typical values of masses ($m_e = 0.06 m_o$, $m_{hh} = 0.45 m_o$, $m_{lh} = 0.08 m_o$, where m_o is the free electron mass) the heavy hole band has the largest density of states (Fig. 13.2b). The electron quasi Fermi level generally penetrates farther into the conduction band because of the smaller electron mass and smaller density of states.

The stimulated emission and absorption *probabilities* are equal and related to the spontaneous emission probability (Einstein, 1917). The net transition rate (i.e., absorption rate minus stimulated emission rate) at photon energy $h\nu$ depends on the number of occupied conduction band states, the number of vacant valence band states and the dipole matrix element M between a conduction band state of energy $(E + h\nu)$ and a valence band state of energy E (Stern, 1971):

$$\alpha(h\nu) = \frac{q^2}{2m_o^2\epsilon_o n_g c\nu} \int_{-\infty}^{\infty} \rho_c(E + h\nu)\rho_v(E)|M(E + h\nu, E)|^2$$

$$\times [f_v(E) - f_c(E + h\nu)] \, dE, \qquad (13.5)$$

where ϵ_o is the vacuum permittivity, c is the velocity of light and n_g is the group index.

Ignoring, for the moment, the calculation of the matrix element M, we note that gain occurs (i.e., $\alpha(h\nu) < 0$) when $f_c(E + h\nu) > f_v(E)$ for the whole range of the integration. This condition is satisfied (Eq. 13.2) if:

$$F_c - F_v > h\nu, \qquad (13.6)$$

which was first pointed out by Bernard and Duraffourg (1961). Thus, gain occurs for all transitions with energies less than the separation of the quasi Fermi levels.

Evaluation of the matrix element is discussed extensively in the literature (Stern, 1976; Casey and Panish, 1978) and depends on the modeling of the bandtail impurity states (Kane, 1963; Halperin and Lax, 1966; Huang, 1970) and on the assumptions about k-selection rules. A common assumption is that recombination takes place between a parabolic conduction band and an acceptor band.

The wave function for a bound carrier can be written as the product of Bloch functions associated with the lattice and an envelope function associated with the properties of the hydrogenic impurity. The matrix element is then the product (Casey and Panish, 1978):

$$|M|^2 = |M_{bb}|^2|M_{env}|^2 \qquad (13.7)$$

of terms due to the Bloch waves (Kane, 1957):

$$|M_{bb}|^2 = \frac{m_o^2 E_g(E_g + \Delta)}{12m_e(E_g + 2\Delta/3)} \qquad (13.8)$$

and due to the envelope wave functions (Eagles, 1960; Dumke, 1963):

$$|M_{env}|^2 = \frac{64\pi a^{*3}}{\left(1 + a^{*2}k_b^2\right)V}, \qquad (13.9)$$

where Δ is the spin-orbit splitting, a^* is the effective Bohr radius of the localized state, $k_b^2 = 8\pi^2 m_c(E - E_c)/h^2$, and V is the active volume.

A more complicated envelope matrix element due to Stern (1971) with a concentration dependent density of states was used by Dutta (1980) to calculate the gain in 1.3 μm GaInAsP (Fig. 13.4). For higher injection levels, the peak gain moves to higher energies, because at higher injection levels, states farther from the band edge become inverted and the density of states increases with

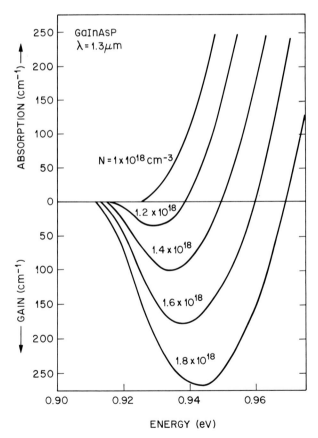

Fig. 13.4 Calculated gain as a function of photon energy for $\lambda = 1.3$ μm GaInAsP at various injected carrier densities (after Dutta, 1980).

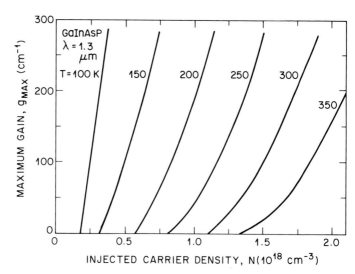

Fig. 13.5 Maximum gain as a function of injected carrier density for undoped $\lambda = 1.3$ μm GaInAsP at different temperatures (after Dutta and Nelson, 1982).

energy. The maximum gain as a function of injected carrier density is shown in Fig. 13.5 for different temperatures. Note that the differential gain generally can be approximated as

$$g = \hat{g}(N - N_t), \tag{13.10}$$

where N_t is the carrier density for transparency. For a temperature of 300 K, $\hat{g} = 3.7 \times 10^{-16}$ cm^2 and $N_t = 1.2 \times 10^{18}$ cm^{-3}.

We have just discussed the change in gain (Δg) with carrier injection. In addition to the change in the imaginary part of the refractive index $\Delta n''$, there is of course a corresponding change in the *real* part of the index $\Delta n'$, which may be calculated using the Kramers-Kronig dispersion relations (Henry *et al.*, 1981; Asada, 1985). Figure 13.6 shows the wavelength dependence of these changes in index for a GaAs laser. For such laser properties as chirping and linewidth, it is the ratio of these changes, commonly called the linewidth enhancement factor, $\alpha \equiv (\partial n'/\partial N)/(\partial n''/\partial N) = \Delta n'/\Delta n''$, that is important. From Fig. 13.6 we see that this ratio is strongly wavelength dependent. Figure 13.7 shows measurements of the wavelength dependence of α for a $\lambda = 1.5$ μm GaInAsP laser. Operating a laser on the short wavelength side of the gain peak, where α is smaller, is desirable to reduce chirping and minimize the linewidth.

13.2.4 Laser Oscillation

Optical feedback is necessary to sustain lasing and is usually provided by 1) a cleaved or etched semiconductor surface, 2) an integrated grating or 3) an

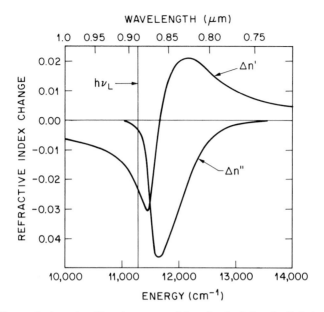

Fig. 13.6 Changes in the real and imaginary parts of the refractive index of a GaAs laser when it is excited from low current up to the threshold current (after Henry, 1981).

external mirror or grating. We consider here a double heterostructure laser with cleaved facets; the alternatives are discussed in Section 13.3.2.

If the gain, g, and waveguide loss, α_i, are constant over the length, L, of the laser, then the optical wave grows exponentially along the length and lasing occurs if the total loop gain is unity:

$$R_1 R_2 e^{2(\Gamma g_{th} - \alpha_i)L} = 1, \qquad (13.11)$$

where R_1 and R_2 are the fractions of optical power reflected at each facet. Equivalently, the gain at threshold equals the total loss:

$$\Gamma g_{th} = \alpha_i + \alpha_m, \qquad (13.12)$$

where Γ is the optical confinement factor, the fraction of optical power in the active layer. The waveguide loss depends on α_a, the loss in the active layer and on α_c, the loss in the confinement layers: $\alpha_i = \Gamma\alpha_a + (1 - \Gamma)\alpha_c$. From Eq. 13.11, the mirror loss is $\alpha_m = (1/2L)\ln(1/R_1 R_2)$. Typically, the reflectivity of an uncoated facet is around 30%, so $\alpha_m = 48$ cm^{-1} for a 250 μm long laser. The gain and internal loss have a weak wavelength dependence (Fig. 13.4), with a spectral width of the order of 500 Å. The mirror loss is minimized at the Fabry-Perot modes, i.e., the wavelengths for which the cavity length is an integral number of half wavelengths. These resonant wavelengths are separated

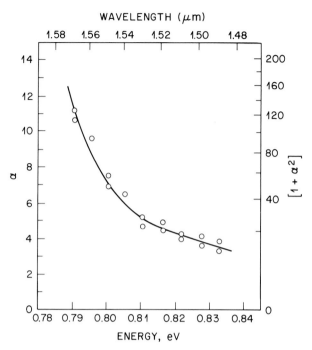

Fig. 13.7 Measured wavelength dependence of the linewidth enhancement factor for $\lambda \sim 1.5\ \mu m$ GaInAsP (after Westbrook, 1985).

in wavelength by

$$\Delta\lambda = \frac{\lambda^2}{2n_g L}.\tag{13.13}$$

At low gain, the finesse of the cavity is low and the ratio of the peaks to the troughs in the light output, $P(h\nu)$, is not large, while at higher gain, the width of a given mode can be one thousandth of the mode spacing and $P(h\nu)$ can be greater than 10^4. The peak to trough ratio below threshold can be used to calculate the net gain (Hakki and Paoli, 1975):

$$\Gamma g(h\nu) - \alpha_i = \frac{1}{L}\ln\left[\frac{\sqrt{P(h\nu)} - 1}{R\left(\sqrt{P(h\nu)} + 1\right)}\right].\tag{13.14}$$

where $R_1 = R_2 = R$.

13.2.5 Rate Equations (Steady State)

We now consider a simple set of rate equations for the electron (N) and photon (S_k) densities in the active layer of a laser with a family of modes having different constants g_k and N_{tk}. The electron density increases due to the injection of a current I into a volume V, and decreases due to stimulated and spontaneous emission:

$$\frac{dN}{dt} = -\sum_k g_k(N - N_{tk})S_k + \frac{I}{qV} - \frac{N}{\tau_n(N)}, \qquad (13.15)$$

where $\tau_n(N)$ is the spontaneous emission electron lifetime, and the differential gain, g_k, can be approximated by a parabolic dependence (see Fig. 13.4). Note that $g_k(N - N_{tk})$ is a recombination rate and is obtained from Fig. 13.5 by dividing by the group velocity. Similarly the photon density in the mode k is increased by stimulated and spontaneous emission and decreased by internal and mirror losses with a photon lifetime $\tau_p = [v_g(\alpha_i + \alpha_m)]^{-1}$:

$$\frac{dS_k}{dt} = \Gamma g_k(N - N_{tk})S_k - \frac{S_k}{\tau_p} + \frac{\Gamma \beta N}{\tau_n(N)}, \qquad (13.16)$$

where β is the fraction of spontaneous emission coupled into the active layer. For steady-state single-mode operation ($g_k = g_o$, $N_{tk} = N_t$), the solution to Eqs. (13.15)–(13.16) is

$$S = \frac{\Gamma \tau_p I}{qV} - \frac{(1 - \beta)}{g_o \tau_n}\left(\frac{1 + \Gamma g_o N_t \tau_p}{1 + \beta/Sg_o \tau_n}\right) \qquad (13.17)$$

$$N = \frac{\dfrac{1}{\Gamma g_o \tau_p} + N_t}{1 + \dfrac{\beta}{g_o \tau_n S}}. \qquad (13.18)$$

These results are plotted in Fig. 13.8 for single-mode lasers. The light output gradually increases up to threshold. The laser below threshold is a superluminescent light-emitting diode (LED) with a power typically below 100 μW, and a large linewidth. The light output above threshold can be approximated as:

$$S = \frac{\Gamma \tau_p}{qV}(I - I_{th}), \qquad (13.19)$$

where the threshold current is:

$$I_{th} = \frac{N_{th}qV}{\tau_n}, \qquad (13.20)$$

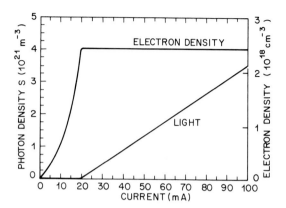

Fig. 13.8 Calculated dependence of photon and carrier densities on injected current.

and the threshold carrier density is

$$N_{th} = \left(N_t + \frac{1}{\Gamma g_o \tau_p} \right). \tag{13.21}$$

In Fig. 13.8, we see that the carrier density is essentially clamped at threshold and increases in input current produce additional light output. The light output in the non-lasing modes is essentially clamped at the level attained at threshold, since the carrier density and the gain do not further increase at higher current levels. In practice though, semiconductor lasers are not single mode without additional filtering of the gain or loss (Section 13.3.2), and also the carrier density is not as tightly clamped as indicated in Fig. 13.8. These two results are due to effects such as photon-density-dependent gain (spectral hole-burning) or loss (two-photon absorption), dynamic effects such as mode beating, longitudinal or lateral spatial hole-burning, or spontaneous emission effects. In particular, Manning *et al.* (1985) have shown that calculated spectra only agree with the results of an equation like Eq. (13.16) if a significant amount of symmetric and asymmetric hole-burning exists.

The threshold current density, J_{th}, of a typical GaInAsP laser of area A is expected from Eq. (13.20) to be $J_{th} = I_{th}/A \sim 1500$ A/cm^2. The temperature dependence of J_{th} is of particular importance. Empirically, threshold current density is usually found to vary with temperature as:

$$J_{th} = J_o e^{T/T_o}, \tag{13.22}$$

where the "characteristic" temperature, T_o, may also have a weak temperature dependence (Fig. 13.9).

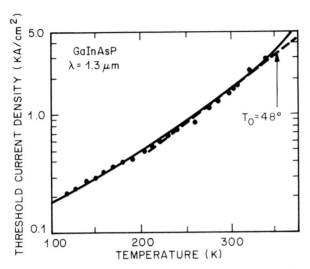

Fig. 13.9 Typical dependence of threshold current density on temperature for a $\lambda = 1.3$ μm GaInAsP laser.

From Fig. 13.5, we see that N_t and g_o vary with a T_o greater than 100 K at a gain of 100 cm^{-1}. However, the non-radiative recombination represented in τ_n is strongly temperature dependent due to Auger recombination (Dutta and Nelson, 1981), and experimental values of T_o for 1.3 μm GaInAsP lasers are typically 60 to 80 K. For 1.5 μm GaInAsP lasers, the valence band splitting is approximately equal to the band-gap energy, so that inter-valence-band absorption (Asada *et al.*, 1981; Henry *et al.*, 1981) becomes important.

The external differential quantum efficiency, η, is the internal differential quantum efficiency, η_i, reduced by the ratio of the mirror loss to the total loss:

$$\eta = \eta_i \frac{\alpha_m}{\alpha_m + \alpha_i}. \tag{13.23}$$

If η_i and α_i are not carrier density dependent, then a plot of:

$$\frac{1}{\eta} = \frac{1}{\eta_i}\left(1 + L\frac{\alpha_i}{\ln 1/R}\right) \tag{13.24}$$

versus cavity length, L, provides a means of measuring η_i ($L = 0$ intercept) and α_i (slope). Typically, α_i is higher in GaInAsP lasers (25 cm^{-1}) than in GaAs lasers while η_i is lower (0.7) than is usually seen in GaAs lasers. The lower value of η_i could be due to leakage across the heterojunction (Yano *et al.*, 1983) or leakage around the active layer. It is probably not due to Auger recombination

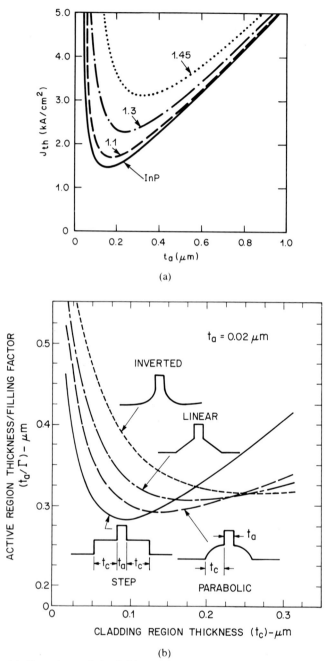

Fig. 13.10 (a) Dependence of threshold current density on active layer thickness for 1.5 μm GaInAsP double heterostructures with cladding compositions of either InP or GaInAsP of bandgap wavelength 1.1, 1.3 and 1.45 μm (after Bowers and Wilt, 1984), and (b) dependence of threshold current density on confinement layer thickness for several types of GaAs separate confinement heterostructures (after Streifer *et al.*, 1983).

because in an ideal laser the carrier density and thus the Auger recombination current are approximately clamped at threshold.

The dependence of threshold current on laser design parameters is given in Eqs. (13.20) and (13.21). The dependence on the active layer thickness, d, primarily enters through the active layer volume and the confinement factor. It also enters through a dependence of the nonradiative component of the current density due to Auger and other effects (Dutta and Nelson, 1982). Figure 13.10a shows the calculated dependence of threshold current on active layer thickness for a 1.55 μm GaInAsP laser clad with several different composition cladding layers. The predominant dependence is as $d/\Gamma(d)$.

For large active layer thicknesses, $\Gamma \approx 1$ and the threshold is linearly dependent on the active layer thickness (Fig. 13.10a) due to the dependence of gain on carrier *density*, not total number of carriers. For active layers thinner than ~ 0.1 μm and an index step of ~ 0.3, the threshold rapidly increases because the confinement factor decreases as d^2. The solution to this tradeoff (Thompson and Kirkby, 1973; Panish *et al.*, 1973) is to provide a separate means of confining the carriers and the photons. Several variations in separate confinement heterostructures (SCH) are shown in Fig. 13.10b and use changes in the refractive index in the form of parabolically graded GRIN-SCH (Tsang, 1981), inverse parabolically graded profiles, or step linear graded profiles (Hersee *et al.*, 1982). They can all be designed to maintain a constant confinement factor Γ, while decreasing the active layer thickness, increasing the carrier density, and decreasing the laser threshold. Minimum thresholds of 175 A/cm^2 have been obtained in AlGaAs GRIN-SCH lasers (Fuji *et al.*, 1984).

13.2.6 Quantum Size Effects

Thin active layers, such as described before, give rise to quantum size effects, which in the case of AlGaAs multi-quantum-well (MQW) lasers have produced superior characteristics such as less temperature dependence, narrower gain spectrum, higher modulation frequency, lower linewidth enhancement factor, and lower threshold. In the longer wavelength region, GaInAsP MQW lasers should also have some of these superior qualities, although there are significant limitations such as the Auger effect, intervalence band absorption and carrier leakage.

In a quantum well structure (Fig. 13.11), the confinement of electrons in one dimension causes a quantization in the allowed energy levels, and the formation of subbands of energy

$$E_{ln} = \frac{l^2 h^2}{8 m_e L_z^2},$$ (13.25)

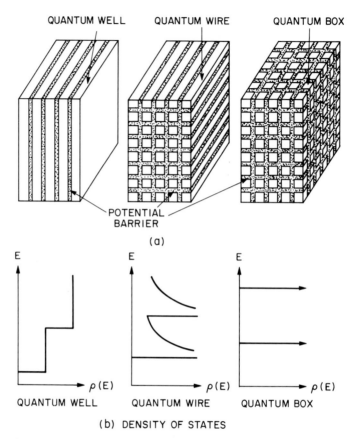

Fig. 13.11 Schematic diagram and density of states for quantum well, quantum wire and quantum box lasers (after Arakawa and Yariv, 1986).

where L_z is the thickness of the quantum well. A similar expression holds for the valence band. The density of states changes from the parabolic dependence, Eq. (13.3), to a steplike structure (Fig. 13.11):

$$\rho_c(E) = \sum_{l=1}^{\infty} \frac{4\pi m_e}{h^2 L_z} H(E - E_{ln}), \tag{13.26}$$

where H is the Heaviside function. Since the density of states is constant over a band of energies, rather than gradually increasing from zero density, there is a group of electrons of nearly the same energy available to recombine with a group of holes with nearly the same energy, and gain can be larger than in

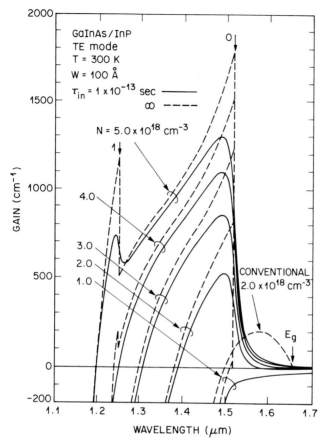

Fig. 13.12 Linear gain of GaInAs/InP multiquantum well lasers with intraband relaxation included. The result for a conventional double heterostructure is shown with a long dashed curve (after Asada *et al.*, 1984).

double heterostructures. Calculations of the gain for multi-quantum wells (MQW), using an equation of the form of Eq. (13.5) with the modified density of states, Eq. (13.26), and allowing for intraband relaxation, are shown in Fig. 13.12. The maximum gain is much larger than for conventional double hetero-structures, although the exact level depends on the assumed value of the intraband relaxation time. It is this effect that reduces the threshold, increases the resonance frequency and decreases the linewidth enhancement factor in MQW lasers.

If the electrons are confined in two dimensions (quantum wire) or even three dimensions (quantum box), the peak density of states becomes even larger

(Arakawa and Yariv, 1986):

$$\rho_c^{\text{wire}}(E) = \left(\frac{2m_e}{\hbar^2} \right)^{1/2} \sum_{l, m} \frac{1}{\sqrt{E - E_l - E_m}} \tag{13.27}$$

$$\rho_c^{\text{box}}(E) = \sum_{l, m, k} \delta(E - E_l - E_m - E_k), \tag{13.28}$$

as shown in Fig. 13.11. The effects of quantum confinement on the modulation response of the laser are discussed in Section 13.4. Due to the peaked structure of the density of states, it has been suggested that the threshold current of quantum wire and quantum box lasers would have a reduced temperature dependence (Arakawa and Sakaki, 1982), just as predicted earlier (Chin *et al.*, 1979) that MQW lasers would have less threshold temperature dependence than conventional double heterostructure lasers. Further experimental and theoretical work is needed to determine if this will indeed be true for GaInAsP devices. In particular, calculations on GaInAsP MQW devices show that intervalence band absorption severely reduces the potential increase in T_o, particularly in the $0°C$ to $100°C$ temperature range (Asada *et al.*, 1984).

13.3 STRUCTURES

In the previous section, the double heterostructure laser was modeled as a simple slab waveguide resonator of unspecified width, having cleaved crystal-facet end mirrors. In this section we will outline the evolution of practical three-dimensional laser structures and describe some of their important characteristics.

13.3.1 Lateral Mode Control and Confinement

The simplest laser configuration, and the earliest to have been demonstrated, is the broad-area laser formed by dicing or sawing a two-dimensional slab waveguide resonator into chips a few hundred micrometers wide. The longitudinal (Fabry-Perot) modes of the laser (Eq. (5.13)) are specified by the resonator length, L, which is usually in the range of 100 to 400 μm. The active region thickness d determines the optical and electrical confinement, as well as the beam divergence in the direction transverse to the junction plane. In a typical laser, d is small enough (< 0.5 μm) that only the lowest order transverse mode can oscillate. The lateral modes, and the field pattern in the plane of the junction, are controlled by the lateral structure of the laser. (Many authors fail to distinguish between the lateral and transverse modes and refer to both simply as the transverse modes of the laser.) In a broad-area laser, the lateral field pattern is poorly controlled, and several lateral modes can oscillate simultaneously. Only a portion of the lateral dimension of the chip may be active, resulting in filamentary emission. The beam pattern in the plane of the junction and the output spectrum are complex, and both may be unstable in time. In addition, the current needed to drive a broad-area laser to threshold is very large, because of the large area over which the required current density must be maintained. While broad-area lasers may find use in some applications, such as the pumping of solid state lasers, virtually all semiconductor lasers used for telecommunications are designed with lateral control of carriers, current, and optical mode.

13.3.1.1 Gain-Guided Laser Structures. The lateral extent of the laser gain distribution can be controlled with the addition of a narrow current-confining structure. Laser oscillation is then limited to one or a small number of lateral modes, and the optical source spot is confined to facilitate efficient coupling to an optical fiber. The threshold current is typically reduced by at least an order of magnitude, as the drive current is restricted to flow through a region only a few micrometers, rather than a few hundred micrometers wide.

Gain-guided planar stripe structures, used in the earliest homojunction lasers for current confinement (Dyment and D'Asaro, 1967), remain available for some multimode fiber applications. In a common configuration, shown in cross-section in Fig. 13.13a, one current contact is limited to a conducting stripe opened along the longitudinal direction in a current-blocking layer of dielectric material (SiO$_2$ or Si$_3$N$_4$). The dielectric layer can also be replaced by a Schottky-barrier contact. In either case, the current always spreads laterally from the stripe region, as shown schematically by the dashed lines in the figure. Current spreading may be eliminated by making the side regions non-conduc-

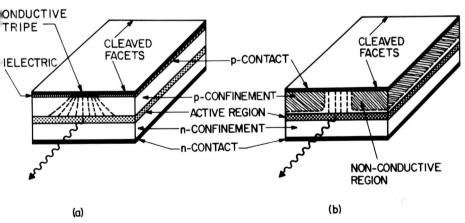

Fig. 13.13 Planar stripe geometry double heterostructure lasers. (a) dielectric stripe (SiO$_2$ or Si$_3$N$_4$) (b) proton-bombarded stripe.

tive, for example by using proton bombardment to create ion implantation damage (Fig. 13.13b). This provides more effective current injection, resulting in lower threshold currents and smaller lateral mode widths. With typical stripe widths of $W = 10$ μm, planar stripe lasers can be coupled with high efficiency to multimode fibers, and with somewhat less efficiency to the small diameter cores of single-mode fibers.

In general, gain-guided lasers suffer from "kinks," or non-linear light vs. current (L-I) characteristics, which are caused by jumps between sets of lateral modes. As the drive current is increased, the carrier density at the center of the laser stripe increases more slowly than the density near the edges, because of the higher optical field and consequently higher carrier recombination. The resulting gain distribution then favors the next higher order lateral mode, and the laser power typically dips on the mode change (Kirkby et al., 1977). Self-pulsations and restricted high temperature operation are also problems with gain-guided structures.

13.3.1.2 Index-Guided Laser Structures. The problems of kinks, astigmatism, unstable far-field patterns and self-pulsations have been greatly reduced by introducing some real refractive index variation into the lateral structure of the laser. In some structures with weak index guiding, the *active* region waveguide thickness is varied by growing it over a ridge or a channel in the substrate. Alternatively, loading of a uniformly thick, planar active waveguide can be obtained by means of lateral variations in the *confinement* layer thickness or refractive index. The channeled-substrate planar (Aiki et al., 1978) or inverted-rib waveguide (Turley et al., 1981) lasers are examples of this design. In the GaInAsP/InP version shown in Fig. 13.14a, the lower quaternary guide layer is chosen to have a refractive index smaller than that of the 1.3 μm wavelength active layer. Room temperature cw thresholds of 70–90 mA, output powers to 20 mW, and operation to 90°C have been reported for 1.3 μm lasers (Turley et al., 1981).

In the ridge waveguide laser (Kawaguchi and Kawakami, 1977; Kaminow et al., 1983), the ridge not only provides the loading for weak index guiding, but also acts as a narrow current confining stripe. Single lateral mode ridge lasers have been made at several wavelengths, and one (Kaminow et al., 1979) was among the three nearly simultaneous reports of a first 1.55 μm cw room temperature laser. In this structure (Fig. 13.14b), the $\lambda_g = 1.2$ μm anti-melt-back layer also serves as an etch-stop in the fabrication. Room temperature cw thresholds of 18 mA, outputs of 25 mW and operation to 90°C have been reported for 1.5 μm devices (Armistead et al., 1986).

In general, index guiding and current confinement are provided by separate parts of the structure in weakly-guided lasers, and appreciable lateral current spreading can take place. This usually leads to higher threshold currents than

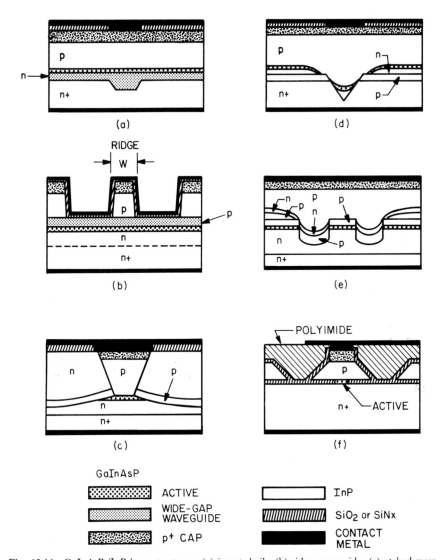

GaInAsP

▨ ACTIVE

▨ WIDE-GAP WAVEGUIDE

▨ p⁺ CAP

☐ InP

▨ SiO₂ or SiNx

■ CONTACT METAL

Fig. 13.14 GaInAsP/InP laser structures: (a) inverted rib; (b) ridge-waveguide; (c) etched-mesa buried heterostructure; (d) channeled substrate buried heterostructure; (e) double channel planar buried heterostructure; (f) constricted mesa.

are possible in buried heterostructure lasers, in which the active volume is completely buried in a material of wider band-gap and lower refractive index. In buried-heterostructure (BH) lasers, the optical field is well confined not only in the transverse but also in the lateral direction, providing *strong* index guiding of the optical mode. Excellent carrier confinement also results. The wide band-gap, low refractive index confinement material is AlGaAs in 0.8–0.9 μm GaAs lasers and is InP in 1.3–1.6 μm GaInAsP lasers. Note that for some heterostructure combinations, the wider band-gap material does *not* provide a smaller refractive index (see Section 13.5). A variety of buried-heterostructure lasers have found their way into commercial use, and still others are under laboratory investigation. Next we discuss some of the more common GaInAsP/InP BH laser designs.

Fabrication of the "etched-mesa" GaInAsP/InP BH laser of Fig. 13.14c (Hirao *et al.*, 1980) starts with a planar double heterostructure wafer prepared by any of the common growth techniques, usually topped by an GaInAsP contact layer to reduce the device resistance. Then a narrow mesa stripe is formed by anisotropically etching away the side regions through to the InP buffer layer below the GaInAsP active layer. A typical stripe width for single lateral mode operation is 1.5 μm or less. An etch mask protects the mesa stripe during the etching and subsequent regrowth. The side regions are back-filled by growth of InP current-blocking layers. As shown in Fig. 13.14c, the blocking layers form a *p-n* junction, which is reverse-biased when the main laser junction is under normal forward bias. The proper alignment of the blocking layers with respect to the laser junction is critical to the achievement of low leakage current, high performance devices. The resulting etched-mesa BH laser wafer may not be planar, but can be made so by the growth of a *p*-layer over the entire wafer surface to form a "buried-mesa" laser.

Excellent performance characteristics have been demonstrated for both GaAs/AlGaAs and 1.3 μm or 1.5 μm GaInAsP/InP etched-mesa BH lasers. For example, for 1.3 μm etched-mesa BH lasers with active region widths of 1–2 μm, room temperature threshold currents of 20 mA, cw output powers of 30 mW per facet and operation to 100°C are typical (Hirao *et al.*, 1980). The performance limits of BH lasers are reached both at high current drive levels and at elevated temperatures, where excessive shunt-path leakage currents flow through the blocking layers. These layers form a *p-n-p-n* thyristor that switches on at high current, shunting the desired current path and dropping the laser output toward zero at sufficiently high currents (Dutta *et al.*, 1984).

The V-groove laser (Ishikawa *et al.*, 1982), also called a channeled-substrate BH laser, is fabricated by the LPE growth of a non-planar active region in a channel or groove etched in the initial substrate (Fig. 13.14d). This design requires no regrowths, and the tapered active layer provides discrimination against higher order lateral modes. A restricted current path can be formed by etching the channel into the n-type substrate through a surface layer of semi-insulating material, or through a grown or ion-implanted p-layer as shown in the figure. The performance characteristics are generally similar to those of etched-mesa BH lasers. Room temperature thresholds of 15–25 mA at 1.3 μm, and 25–35 mA at 1.55 μm are typical, as are maximum cw operating temperatures of 90°C (Dutta et al., 1985).

A modification of the etched-mesa BH first developed for the GaInAsP/InP system is the double-channel planar buried-heterostructure (DC-PBH) laser (Mito et al., 1983; Kobayashi and Mito, 1985) shown in Fig. 13.14e. In this laser structure, channels are etched on both sides of the mesa stripe after the initial wafer growth. In the subsequent regrowth, a p-n-p-n current confinement structure is automatically aligned to the active region stripe through a unique property of the liquid phase epitaxial growth process. The existence of the narrow band-gap GaInAsP layer outside of the channel region reduces the parasitic (thyristor) leakage current. Because of the reduced leakage, very high power operation has been achieved, and lasing at temperatures up to 130°C has been reported. Room temperature threshold currents of 15–20 mA are typical of both 1.3 μm and 1.5 μm devices (Dutta et al., 1985), as are cw output powers of 40 mW or more (Kobayashi and Mito, 1985).

The high-speed modulation capabilities of buried-heterostructure lasers are affected adversely by the parasitic capacitance added with the use of reverse-biased current-blocking layers. To overcome this problem, lasers can be fabricated with current blocking provided by either the regrowth of semi-insulating material (Miller et al., 1986) or the deposition of a dielectric material (Bowers et al., 1985). An example is the constricted-mesa laser, shown in Fig. 13.14f (Bowers et al., 1986). Modulation speeds in excess of 20 GHz, limited by the properties of the active region rather than by parasitic elements, have been achieved and are discussed in detail in section 13.4.

13.3.2 Longitudinal Mode Control—Short and Coupled Resonators

The structures just described provide control of the laser's lateral modes. As discussed in Section 13.2.3, several longitudinal modes (resulting from the Fabry-Perot resonances of the cavity formed by the cleaved laser mirrors) generally have sufficient gain to reach threshold and oscillate simultaneously. Although some lasers show a tendency toward single longitudinal mode oscillation under cw excitation, under high speed current modulation most Fabry-Perot lasers are prone to multimode operation. This leads to pulse spreading during propagation through dispersive fiber and to partition noise arising from fluctuations in the modal distribution from pulse to pulse. To avoid these effects, much effort has gone into the development of single frequency, or dynamic single-mode (DSM) lasers, which are stable under high speed modulation.

The key to achieving single longitudinal mode operation is to provide adequate gain or loss discrimination between the desired mode and all of the unwanted modes of the laser resonator. For a Fabry-Perot cavity, all the longitudinal modes have nearly equal losses and for a length $L = 250$ μm are spaced by about 1 nm at a wavelength of 1.3 μm (Eq. 5.13)). The broad gain spectrum of a 1.3 μm laser operated well above threshold can support oscillation over a wavelength range of several tens of nanometers, which leads to the highly multimode operation.

Conceptually, the simplest way to increase mode discrimination is to shorten the cavity. By reducing L from 250 to 25 μm, the mode spacing can be increased from 1 to 10 nm. Centering the desired mode on the peak of the gain curve by current or temperature tuning results in nearest neighbor modes with insufficient gain to oscillate. Conventional Fabry-Perot lasers with very short cavities have been demonstrated with side-mode powers suppressed by one or two orders of magnitude, but they have been limited to output powers of only a few milliwatts (Lee *et al.*, 1983). While the conventional cleaved mirror structures are difficult to fabricate at lengths below 50 μm, some success has also been achieved using etched (Iga *et al.*, 1980; Coldren *et al.*, 1982) or "microcleaved" (Blauvelt *et al.*, 1982) mirrors. Resonators as short as 20 μm now have been reproducibly made using reactive ion etching (Yamada *et al.*, 1986). Another short-cavity configuration is the surface-emitting laser in which a resonator less than 10 μm long is formed in the direction normal to the active region. Such an AlGaAs/GaAs "microcavity" laser (Fig. 13.15) has been operated with a room temperature threshold current as low as 6 mA (Iga *et al.*, 1986) and GaInAsP/InP versions are under development.

Multiple-element resonators, or resonators with distributed reflectors, provide a loss function with a frequency dependence strong enough to result in single frequency oscillation under most operating conditions. A three-mirror resonator geometry is formed when an external, highly reflecting mirror is coupled to a cleaved-facet Fabry-Perot laser. In one version of the three-mirror resonator, a graded-refractive-index (GRIN) rod lens is used to enhance the coupling. The result, shown in Fig. 13.16a, is a GRIN-rod-external-coupled-cavity (GRECC) laser (Liou *et al.*, 1984). The external mirror approach has the advantage that a standard laser chip of presumably proven reliability is used in the composite cavity. The composite-cavity modes of the resulting double cavity are widely spaced in frequency when the optical path lengths of the two sections are unequal. In order to assure that one of these modes is close to the gain peak, the current and temperature of the laser must be carefully controlled.

A four-mirror resonator consisting of two active laser sections separated by a gap of the order of the wavelength provides increased flexibility and improved performance. In this structure, shown schematically in Fig. 13.16b, the gap is formed either by etching part way into the laser chip (Coldren *et al.*, 1981;

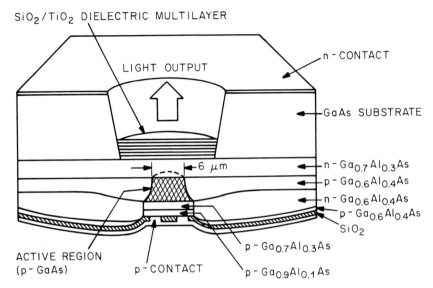

Fig. 13.15   Surface-emitting laser (after Iga *et al.*, 1986).

Coldren *et al.*, 1982) or by recleaving a finished chip into two partially-attached segments to yield a cleaved-coupled-cavity (C^3) laser (Tsang *et al.*, 1983; Tsang, 1985). The two sections are driven separately, and either one or both can be operated with sufficient gain to reach threshold. With amplification in both sections, gap lengths that are multiples of $\lambda/2$ provide the optimum selectivity. Unlike a three-mirror cavity, the effective reflectivity shows a narrow high-Q resonance rather than a narrow notch (Henry and Kazarinov, 1984). Dynamic

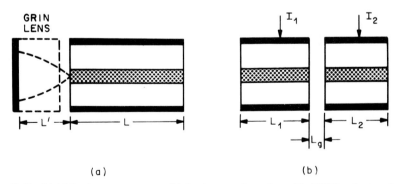

Fig. 13.16   Multiple element resonators. (a) 3-mirror configuration with GRIN-rod-external-coupled cavity (b) 4-mirror configuration, in which the gap is formed by etching, or by cleaving to form a cleaved-coupled cavity (C^3) laser.

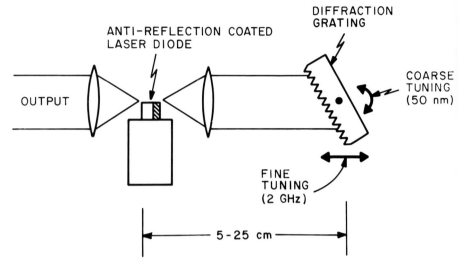

Fig. 13.17 Tunable single frequency laser with external diffraction grating.

single-mode operation of C^3 lasers with side-mode suppression ratios of several thousand under high speed modulation has been achieved by controlling the magnitudes and relative phases of the two injection currents, and the temperature (Coldren *et al.*, 1985). A further advantage of the C^3 laser is that its frequency can be stepwise tuned. If only one section is driven above threshold, the other can be used as a current-controlled tuning element, which makes use of the free carrier dependence of the refractive index and hence the effective resonator length.

The coupled-cavity laser designs just discussed have a loss that repeats periodically in wavelength. This effect provides discrimination against nearby modes, but can allow widely separated modes to oscillate simultaneously. The use of an external diffraction grating as part of the resonator eliminates this difficulty, and also allows tunability. In a very useful laboratory approach to tunable single frequency lasers, the laser output is collimated with a lens and the external mirror is replaced by a diffraction grating (Fig. 13.17) to provide strong wavelength discrimination to the cavity feedback (Wyatt and Devlin, 1983).

13.3.3 Longitudinal Mode Control—Distributed Resonators

An elegant approach to single-frequency operation is the integration of wavelength selectivity directly into the semiconductor laser structure using a distributed Bragg grating. In a distributed-feedback (DFB) laser, the grating region is built into the pumped part of the gain region. In a distributed-Bragg-reflector

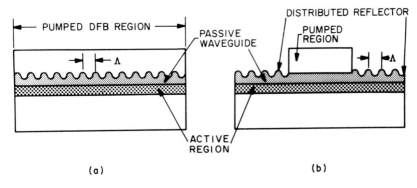

Fig. 13.18 Schematic representation of (a) DFB laser, (b) DBR laser.

(DBR) laser, an unpumped Bragg grating coupled to a low loss waveguide is used to replace the usual cleaved mirror on one or both ends of the resonator. Figure 13.18 illustrates schematically the DFB and DBR structures. In both cases, the grating is shown in a passive waveguide layer adjacent to the active gain region. The grating can be produced with a periodic variation of either gain, phase, or both along the structure.

The operation of lasers with distributed resonators can be understood in terms of the distributed Bragg phase grating reflector and already has been treated in considerable detail (Kogelnik and Shank, 1972; Streifer *et al.*, 1975). In effect, a phase grating is a region of periodically varying refractive index that serves to couple two counter-propagating traveling waves. The coupling is a maximum for wavelengths close to the Bragg wavelength, λ_B, which is related to the grating spatial period Λ by

$$\lambda_B = 2n_e\Lambda/l, \tag{13.29}$$

where n_e is the effective refractive index of the mode and l is the integer order of the grating. First order gratings provide the strongest coupling, but second order gratings are sometimes used because they are easier to fabricate, with their larger spatial period.

In a DFB laser without facet or other extraneous reflections, and with an ideal grating, longitudinal modes are spaced symmetrically around λ_B at wavelengths given by:

$$\lambda = \lambda_B \pm \left[(m + 1/2)\lambda_B^2/2nL_e\right], \tag{13.30}$$

where m is the mode index and L_e is the effective grating length. These modes exist outside of the transmission stop-band centered on λ_B. There are two equivalent lowest order modes ($m = 0$) and oscillation on at least two frequencies is expected. Because of grating imperfections and end-facet reflections, actual DFB lasers are not perfectly symmetric and single frequency operation on one of the modes is often the result. Single frequency operation with increased power output can also be achieved by intentionally using DFB lasers with cleaved end facets, one coated for high reflectivity and the other (output) coated for low reflectivity. In several experimental variations of the DFB laser, the grating is modified at some point to introduce an additional optical phase shift, typically a quarter-wave or less. The result is oscillation on a single mode near the Bragg wavelength (Utaka et al., 1986). Buried-heterostructure DFB lasers have been developed by many laboratories; their threshold currents and output powers are now comparable to those of Fabry-Perot lasers of similar BH geometries (Kitamura et al., 1984).

In the distributed-Bragg-reflector (DBR) laser, the longitudinal modes are determined by the effective length of the overall optical cavity as well as by the details of the grating reflectors (Suematsu et al., 1983). Generally, the mode closest to the Bragg wavelength has the highest gain and oscillates. The coupling of the different regions of the DBR laser is difficult, whereas the coincidence of the gain and grating regions in the DFB laser leads to only minor degradation in performance. The Bragg wavelength can be tuned electronically in a DBR laser by separately injecting carriers to change the effective refractive index (Tohmori et al., 1986). At this point, DBR lasers are less well developed than their DFB counterparts, and their performance characteristics require further improvement. The aging characteristics of both laser types remain to be studied in detail, particularly with regard to spectral characteristics.

13.3.4 Line-Narrowed Lasers

The single-frequency lasers described have been developed to minimize the transmission penalties due to fiber dispersion in high-bit-rate digital systems. In those systems using direct detection of the modulated optical intensity, neither the laser linewidth nor the absolute stability of the laser frequency is important. Stability and linewidth are critical factors, however, in coherent lightwave applications involving heterodyne detection. While the precise requirements are discussed in Chapter 20, heterodyne systems possibly will require linewidths in the 1 MHz range or below. This is about two orders of magnitude smaller than the 100 MHz linewidths typical of lasers without special linewidth control.

The finite linewidth of a laser is a fundamental consequence of the spontaneous emission process and is related directly to fluctuations in the phase of the optical field. The phase fluctuations arise both from 1) phase noise contributed directly by spontaneous photon emission, and 2) conversion of spontaneous emission amplitude noise to phase noise through coupling of the photon and carrier densities. This latter process generates phase noise because the carrier density fluctuations induced by the field amplitude fluctuations in turn produce refractive index, and therefore, frequency/phase fluctuations in the optical field. Analysis of the fluctuations leads to a relationship for the linewidth of a semiconductor laser (Henry, 1982) in terms of its output power P_o:

$$\Delta f = v_g^2 \frac{h\nu}{8\pi} \frac{n_{sp}}{P_o} \alpha_m (\alpha_i + \alpha_m)(1 + \alpha^2), \qquad (13.31)$$

where v_g is the group velocity and n_{sp} ($\sim 2\text{-}3$) is the spontaneous emission factor. The mirror loss, α_m, and the internal waveguide loss, α_i, (both per unit length) have been defined in Section 13.2.4. The $(1 + \alpha^2)$ term arises from the contributions to the linewidth of the two fluctuation effects. The linewidth enhancement factor α, already discussed in Section 13.2.3, is a measure of the amplitude-to-phase fluctuation conversion, and can range from 2 to 16, depending on the material composition and wavelength (Fig. 13.7). As the laser power increases, the linewidth decreases because spontaneous emission becomes relatively less important at high photon densities. The linewidth also decreases with increasing laser length, as the effective mirror loss α_m per unit length decreases; the exact dependence is a function of the details of the laser structure. Typically, 250 μm long Fabry-Perot or DFB lasers emitting a few milliwatts at 1.3 or 1.55 μm exhibit the 100 MHz linewidths mentioned above.

As the output power of a laser cannot be made arbitrarily large, a more practical way to reduce the linewidth is to increase the laser cavity length. The laser chip itself may be made longer, or the cavity may be extended with a passive medium such as air, an optical fiber, or a suitable semiconductor integrated passive waveguide. Use of a passive extended waveguide has been shown to reduce the amplitude fluctuations of the laser as well as the linewidth (Miller, 1986). When the chip facet coupling to the extended cavity is antireflection coated, the structure acts as a single long resonator. Because this long resonator has very closely spaced longitudinal modes, additional wavelength selectivity must be provided. A diffraction grating can be included at the end of the extended cavity (Fig. 13.17) or in the laser chip in the form of a DFB or DBR structure. Linewidths in the tens of kilohertz have been obtained in some experimental structures. The external grating approach has the added advantage of providing a mechanically tunable laser, while the DBR approach allows the possibility of electronic tuning, although over a smaller range.

13.4 MODULATION CHARACTERISTICS

13.4.1 Rate Equation Analysis

Most applications of semiconductor lasers involve modulating either the intensity or frequency of the laser output. The simplest way to do this is to modulate the current driving the laser, which changes the carrier density. This changes the gain of the laser cavity and thus the power output, and it also changes the optical cavity length and thus the lasing frequency. The interplay between current, electron and photon densities was described by rate Eqs. (13.15)–(13.16) for the electrons and photons. A linear dependence of optical gain on electron density was assumed in these equations (see Fig. 13.5 for a more exact dependence). Additionally, there is an optical intensity dependence due to spectral hole-burning or dynamic carrier heating $g = g_o/(1 + \epsilon S)$, where ϵ characterizes the nonlinear gain[1]. This effect has been measured directly in experiments on carrier dynamics (Stix *et al.*, 1986) and inferred from intensity (Bowers *et al.*, 1986) and frequency (Koch and Bowers, 1985) modulation experiments.

The rate equations of Section 13.2 were presented for multimode lasers. Now we shall treat the simpler and very important case of single longitudinal mode lasers. The results for multimode lasers are similar but are not nearly as amenable to analytic solution. We will describe two methods of analyzing the coupled differential Eqs. (13.15) and (13.16). The first is a small signal, frequency domain analysis, and the second is a large signal approach in which we solve the rate equations in a number of regions and join the solutions together.

13.4.2 Small Signal Characteristics

We assume sinusoidal modulation ($I = I_o + ie^{j2\pi ft}$, $S = S_o + se^{j2\pi ft}$, $N = N_o + ne^{j2\pi ft}$), ignore the small product terms ns and s^2, which give rise to second and higher order generation as well as intermodulation products, and solve for the transfer function relating intensity modulation and current (Ikegami and Suematsu, 1968):

$$\frac{s(f)}{i(f)} = \frac{s(o)}{i(o)} \frac{f_o^2}{f_o^2 - f^2 + jff_d}, \qquad (13.32)$$

where f_o is given approximately by (Lau *et al.*, 1981):

$$f_o \approx \frac{1}{2\pi} \left[\frac{g_o S_o}{\tau_p(1 + \epsilon S_o)} \right]^{1/2}, \qquad (13.33)$$

---

[1]The reduction in gain can also be modeled as $g = g_o(1 - \epsilon S)$, (Tucker, 1985), which is the same to first order in ϵS, but becomes negative (unphysical) for large S.

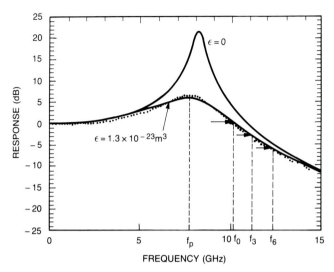

Fig. 13.19 Typical experimental modulation characteristic (dotted line) and calculated responses using Eq. 13.32 (solid lines).

and f_d is approximated by:

$$f_d \approx \frac{\epsilon S}{2\pi\tau_p}. \qquad (13.34)$$

This transfer function is plotted in Fig. 13.19 and has several interesting characteristics. The response is peaked at f_p, the resonance or relaxation oscillation frequency ($f_p^2 = f_o^2 - f_d^2/4$). We have written f_o as a function of internal photon density (S_o) rather than input current, and this form shows that to obtain a higher resonance frequency, one should (1) increase the differential gain coefficient by quantum confinement or by cooling the laser, (2) increase the photon density by making narrower index-guided waveguides with strong guiding and (3) decrease the photon lifetime by making shorter cavity lasers. The photon density is linearly related to the output optical power, $P = SVh\nu\alpha_m/2\Gamma\tau_p(\alpha_m + \alpha_i)$, and so we expect a linear dependence of resonance frequency on the square root of optical power (Fig. 13.20). The slope of the line is typically 3–5 GHz/mW$^{1/2}$ for GaInAsP lasers, and is lower (1–3 GHz/mW$^{1/2}$) for GaAs lasers, as expected from the band-gap energy dependence in the power to photon density conversion. The amount of damping is given by f_d. The roll-off beyond the resonance frequency goes as f^2. Simple expressions for the frequency at which the response has fallen to its dc value

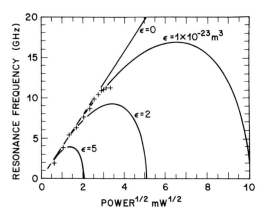

Fig. 13.20 Dependence of resonance frequency on output optical power. The curves are calculated using full expressions for f_o and f_d in Eq. (13.32). The experimental points were measured on a 1.3 μm GaInAsP constricted-mesa laser.

($f_{0\ dB}$), the 3 dB electrical frequency[2] ($f_{3\ dB}$) and the 6 dB electrical (3 dB optical) frequency ($f_{6\ dB}$) are (Bowers *et al.*, 1986):

$$f_{0\ dB} = \sqrt{2}\, f_p \tag{13.35}$$

$$f_{3\ dB} \approx \sqrt{1 + \sqrt{2}}\ f_p \tag{13.36}$$

$$f_{6\ dB} \approx \sqrt{3}\, f_p. \tag{13.37}$$

The first expression is valid for any value of ϵ or S and does not require the approximations used in Eqs. (13.33–13.34). The second two expressions neglect terms of order $(\epsilon S)^2$.

As the current to the laser is increased, the photon density increases, the stimulated electron lifetime decreases, and the resonance frequency increases, producing a larger modulation bandwidth. In addition, the damping of the response increases, which is a generally desirable effect. There is, however, a power level beyond which the response becomes critically damped, no resonance is observed, and the bandwidth of the laser decreases for increasing current. The maximum resonance frequency is given by (Bowers, 1987):

$$f_0^{\max} = \frac{1}{2\pi\sqrt{2}} \frac{g_o}{\epsilon + g_o\tau_p}. \tag{13.38}$$

This discussion has related the intensity modulation to the current in the lasing region. In practice, the laser structure and package may considerably

---

[2]$s(f)/i(f)$ down by $\sqrt{2}$, electrical detected power down by 2.

affect the current actually passing through the active layer. For the analysis of parasitics, we assume that the laser chip is connected to a microstrip line of impedance R_S without significant discontinuities. The first significant parasitic element is the bond wire inductance (L). For a 1 mm long wire, $L \approx 1$ nH. The second is typically the bonding pad capacitance (C). For a typical chip size ($500 \times 250 \ \mu m^2$) with a 1000 Å SiO_2 layer under the bonding pad, this capacitance is 40 pF. The next significant parasitic element is usually the resistance of the p-contact layer, typically several ohms. Under a forward bias current of 50 mA, the p-n junction itself has an impedance of less than 1 Ω. The n^+-substrate (or n^+ cap-layer on p^+-substrate devices) typically has a negligible resistance, as do properly alloyed p- and n-contacts to highly-doped layers. Other additional elements may be significant, such as parallel leakage paths or the diffusion and depletion capacitance associated with adjacent p-n junctions used for current confinement.

With the simple circuit model described here, the ratio of current getting to the active layer, $i(f)$, to the source current, i_s, is (Lau and Yariv, 1985):

$$\frac{i(f)}{i_s} = \frac{1}{1 + \dfrac{jf}{f_l Q} + \dfrac{(jf)^2}{f_l^2}}, \tag{13.39}$$

where

$$f_l = \frac{1}{2\pi} \left(\frac{R_S + R}{LRC} \right)^{1/2} \tag{13.40}$$

and

$$Q = \frac{\left(LRC(R_S + R) \right)^{1/2}}{L + RR_S C}. \tag{13.41}$$

The 3 dB bandwidth is (Bowers, 1987):

$$f_{3\,dB} = f_l \left(1 - \frac{1}{2Q^2} + \left(2 - \frac{1}{Q^2} + \frac{1}{4Q^4} \right)^{1/2} \right)^{1/2}. \tag{13.42}$$

Figure 13.21 shows contours of constant parasitic-limited bandwidth for a laser with $R = 4 \ \Omega$ operating from a source resistance of 50 Ω. The laser described before ($C = 40$ pF, $L = 1$ nH) would be expected to have a bandwidth of 1 GHz, which could be increased slightly with proper biasing to place the resonance frequency at the proper point. To design a laser for a bandwidth

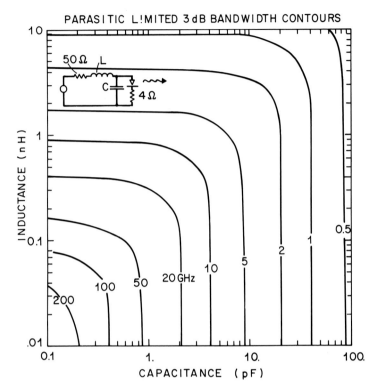

Fig. 13.21 Bandwidth contours (3 dB) in the capacitance-inductance plane calculated from Eq. 13.42, using $R_s = 50$ Ω and $R = 4$ Ω with a circuit model consisting of a series bond-wire inductance and a parallel capacitance (C) and resistance (R) (after Bowers *et al.*, 1986).

of 30 GHz, the bonding pad capacitance[3] must be kept small (< 2 pF) by limiting the bonding pad area and using a thick, low-dielectric-constant layer under it, and by reducing the area of current confining *p-n* junctions. A schematic diagram of one such structure, a constricted mesa laser, which implements these two guidelines, was shown in Fig. 13.14f. The capacitance of this laser can be 1 pF or less. The other requirement for 30 GHz operation is to reduce package effects and reduce L to less than 0.2 nH. Bowers *et al.* (1986) achieved such a low inductance by using a short length (< 0.3 mm) of Au wire or mesh to connect the laser to a 2.5 mm coaxial connector, overmoded at 46 GHz.

[3] The bonding pad and *p-n* junction capacitances are proportional to cavity length, and the chip resistance is inversely proportional to cavity length, so the product RC is independent of cavity length. However, the ratios R/R_s and L/R are length dependent, so a slight overall dependence of parasitic roll-off on cavity length is observed.

13.4.3 Large Signal Characteristics

The small signal modulation characteristics depend strongly on the bias current. Consequently, the large-signal impulse response of the laser is not simply the Fourier transform of the small-signal frequency response. We will analyze the impulse response of the laser, a regime commonly called gain switching, where optical pulses are emitted, which are usually shorter than the electrical drive pulse. This behavior is important for high speed, time division multiplexed systems, for return-to-zero (RZ) transmission where a "one" may be a single relaxation oscillation, and for physics experiments and characterization systems where a compact, tunable source of short ($\sim$ 15 ps), high power ($\sim$ 20 mW) pulses is needed.

Short electrical pulses for gain switching can be obtained from comb generators (sinusoidally-driven step-recovery diodes), sinusoidal microwave modulation (possibly combined with a negative bias voltage), and photoconductive pulse generators. In each case, the electrical pulse causes the electron density to rise to a maximum, which is maintained during a turn-on delay until a large photon density builds up and depletes the carriers. This behavior is easily seen from the rate equations. Typically, the spontaneous electron lifetime τ_n is long ($\sim$ 1 ns) compared to the other time scales, and the spontaneous emission component of the photon density is not significant. We also assume for the purposes of analysis that the laser is initially below threshold, so the photon density S is initially small. Then, during the period when the photon density is small ($t < t_{on}$),

$$\frac{dN}{dt} \approx \frac{I(t)}{qV} \tag{13.43}$$

or

$$N(t) = N_o + \frac{1}{qV} \int_0^t I(t)\, dt. \tag{13.44}$$

For a short electrical pulse arriving at $t = 0$, this is just:

$$N(t) = \begin{cases} N_o & t \leq 0 \\ N_o + \dfrac{Q_t}{qV} & 0 < t < t_{on} \end{cases} \tag{13.45}$$

where Q_t is the total charge in the electrical pulse. The shape of the electrical pulse is unimportant—it must simply be short. From the photon rate equation

(13.16), we have

$$\frac{dS}{S} \approx \left(g\Gamma(N(t) - N_t) - \frac{1}{\tau_p} \right) dt = g\Gamma(N(t) - N_{th}) \, dt \quad (13.46)$$

$$S = \begin{cases} S_o & t < 0 \\ S_o e^{t/\tau_r} & 0 < t < t_{on} \end{cases} \quad (13.47)$$

where

$$\frac{1}{\tau_r} = g\Gamma\left(\frac{Q_t}{qV} + N_o - N_t \right) - \frac{1}{\tau_p}.$$

For a laser biased at threshold

$$N_o = N_t + \frac{1}{g\Gamma\tau_p}, \quad \text{and} \quad \tau_r = \frac{qV}{g\Gamma Q_t}.$$

The rise time is inversely proportional to the charge in the pulse. Once the photon density is high, the electron density is rapidly depleted, and the optical pulse ends. Note that the fall time is not given by the photon lifetime ($\sim$ 1 ps), but rather by a time constant determined by how far the electron density has been driven below threshold.

This behavior has been verified in a recent experiment (Downey et al., 1987), in which a photoconductive pulse generator was used to produce large (7.5 volt), short (FWHM = 17 ps) electrical pulses. The impulse response of the semiconductor laser was measured accurately by mixing its output with a subpicosecond pulse from the same laser that drove the photoconductive switch (i.e., no pulse jitter) in a nonlinear crystal. Figure 13.22 shows the measured current waveform, the calculated electron density (i.e., the integral of the current), and the measured photon level (output power) plotted on linear and log scales. The electrical pulse is over before the optical pulse begins, verifying the delta function electrical pulse assumption in Eq. (13.45). It can be seen from the logarithmic plot of photon output that the optical pulse rises exponentially with a time constant of 3.0 ps and falls exponentially with a time constant of 8.2 ps. The dependence of the rise and fall times and the FWHM on the bias current are shown in Fig. 13.23. The shortest pulse is obtained for a bias at threshold. Biases below threshold produce longer pulses because the peak electron density is lower. Biases above threshold produce longer pulses because the presence of initially high photon density (lasing) does not allow the electron density to rise to as high a value. Consequently, the peak photon density is smaller, and the fall time is larger. This second conclusion is partly the result of having finite electrical pulse widths (17 ps) and is less true for shorter electrical pulses.

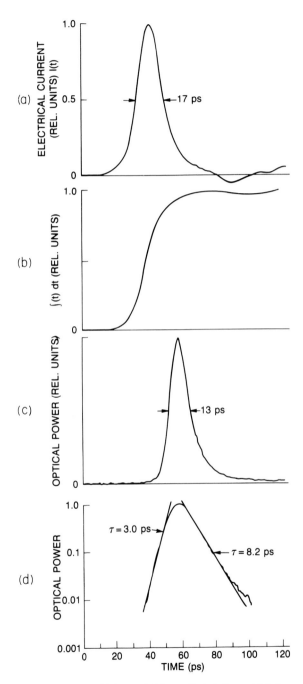

Fig. 13.22 Impulse response of a high speed 1.3 μm GaInAsP constricted-mesa semiconductor laser biased at threshold. (a) Measured electrical pulse from the photoconductive pulse generator. (b) Integral of electrical pulse in (a), approximately the electron density in the active layer in the absence of light emission. (c) Measured optical output plotted on a linear scale showing a FWHM of 13 ps. (d) Measured optical output plotted on a log scale showing the exponential rise and fall (after Downey *et al.*, 1987).

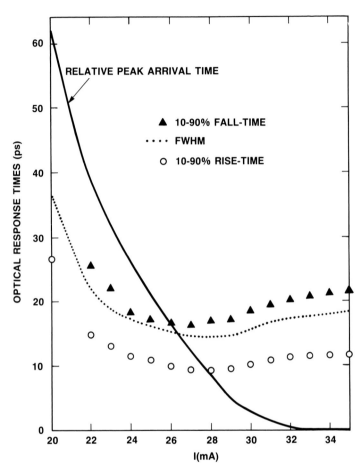

Fig. 13.23 Dependence of the rise and fall times and FWHM on the bias current to the laser used in Fig. 13.22 (after Downey *et al.*, 1987).

In a laser with significant parasitic capacitance, the electrical pulse reaching the laser is broadened, and the capacitance provides a source of current during the time when the photon density is high. Consequently, the laser output may consist of two or more pulses as the electron density is repetitively built up and extinguished (relaxation oscillations).

13.4.4 Pulse Code Modulation

Most optical communications systems transmit information using pulse code modulation of the laser intensity, by directly modulating the current to the laser. The two most common formats are non-return-to-zero (NRZ) modulation,

DIGITAL MODULATION OF A CONSTRICTED MESA LASER (50 ps/div, 100 mA DC BIAS)

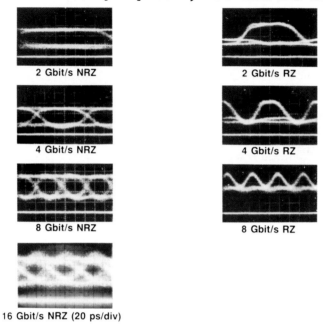

2 Gbit/s NRZ	2 Gbit/s RZ
4 Gbit/s NRZ	4 Gbit/s RZ
8 Gbit/s NRZ	8 Gbit/s RZ

16 Gbit/s NRZ (20 ps/div)

Fig. 13.24 Pseudorandom NRZ and RZ modulation of a high speed 1.3 μm GaInAsP constricted mesa laser at bit rates of 2 Gbit/s, 4 Gbit/s, 8 Gbit/s, and 16 Gbit/s.

where the current stays on during multiple ones, and return-to-zero (RZ) modulation where a "one" transmission consists of a transition from off to on to off again. The disadvantage of RZ transmission is that the bandwidth of the driver must be twice as large (a "one" is only half as long in RZ modulation), but the advantages are easier clock recovery and the possibility of gain switching the laser, i.e., biasing the laser at threshold and using a single relaxation oscillation as a "one" (Tucker et al., 1987). The responses of a constricted mesa laser to large signal NRZ and RZ modulation at bit rates from 2 Gbit/s to 16 Gbit/s are shown in Fig. 13.24. The ringing that can be seen during a series of ones is at the bias-dependent resonance frequency and is a major source of eye closure. Note that the eye closure at 4 and 8 Gbit/s is not a bandwidth limitation; it is due to the inherent nonlinearity of the laser that causes ringing, timing jitter and zero distortion. Consequently, the NRZ eye at 16 Gbit/s is not much worse than at 8 Gbit/s. If large extinction ratios are desired, Fig. 13.24 shows the advantage of RZ format at the higher bit rates because a one is essentially a single relaxation oscillation.

13.4.5 Mode-Locking

While gain switching has produced pulses as short as 13 ps, even shorter pulses are obtainable from mode-locked lasers. Such repetitive strings of pulses are useful for physics experiments, optical device characterization and for time division multiplexed systems using external modulation. The shortest pulses have been obtained from passively mode-locked external cavity structures with antireflection-coated semiconductor lasers and a saturable absorber formed either by proton bombardment, by an external element such as a multi-quantum-well, or by damaging the laser (such as operating a GaAs laser at high output and inducing mirror damage). Active mode-locking, where the current and thus gain of the cavity are modulated, has been used to obtain short pulses even at 20 GHz rates (Eisenstein *et al.*, 1986).

One approach to understanding mode-locked lasers with composite cavities is to examine their operation in the frequency domain. The laser has a number of longitudinal modes at frequencies ν_k, spaced by (ignoring dispersion): $\Delta = c/2\sum_i^N n_i l_i$, where n_i and l_i are the index and length of each of the elements of the composite cavity and N is the number of elements in the cavity. For most experiments using an anti-reflection coated laser and an external cavity, N is 2. For experiments using an external saturable absorber, N is 3. A nonlinearity such as saturable absorption, or a time-varying gain at a frequency Δ, induces modulation sidebands at the mode spacing. This produces energy exchange between the modes and produces a fixed, coherent phase relationship between them. The phases are such that short pulses result from the interference between the standing wave cavity modes. If M modes are locked with a spectral width $\gamma = M\Delta$, then the pulse width τ is the transform limit of the locked modes; $\tau = 0.441/\gamma$ for Gaussian pulses and $\tau = 0.315/\gamma$ for hyperbolic secant pulses (Ippen and Shank, 1977). There have been extensive theoretical and experimental studies of mode-locked lasers, which have been summarized by Haus (1981) and van der Ziel (1985).

Active mode-locking using high speed semiconductor lasers and a microwave drive signal, often at a harmonic of the cavity round trip time, has resulted in very high repetition rates. The variations of the current, carrier and photon densities for active mode-locking are shown in Fig. 13.25a. The dependence of pulse width on drive frequency for an actively mode-locked GaInAsP laser is shown in Fig. 13.25b. At the highest modulation frequencies, the gain modulation is decreased, resulting in broadened laser pulses. Modulation at rates of 40 GHz should be possible using newer lasers, which have bandwidths three times as large as the lasers used for the data of Fig. 13.25. Using such lasers, pulsewidths as short as 560 fs have been demonstrated (Corzine and Bowers, 1987).

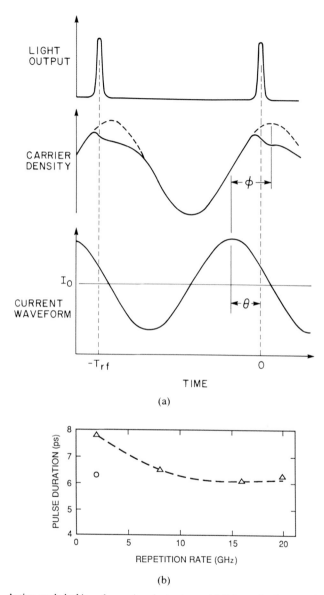

(a)

(b)

Fig. 13.25 Active mode locking of a semiconductor laser: (a) Schematic diagram of the sinusoidal current drive, and resultant carrier and photon densities (after van der Ziel, 1981); (b) dependence of pulse width on drive frequency for an active mode-locked GaInAsP laser in an external fiber cavity (after Eisenstein, 1986).

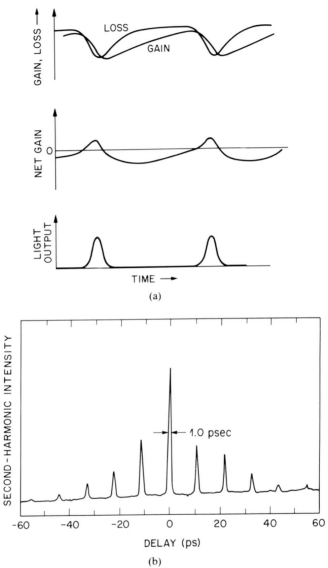

Fig. 13.26 Passive mode locking: (a) Gain, loss and photon levels (after van der Ziel, 1985); (b) autocorrelation trace of the output of a passively mode locked AlGaAs laser (after van der Ziel, 1981).

The time dependent gain and loss in a passively mode-locked laser with a slow[4] saturable absorber are shown in Fig. 13.26a. Two conditions must be satisfied by the saturable absorber material for such short pulse operation: (1) the loss must saturate faster than the gain, and (2) the recovery time of the loss must be faster than the recovery time of the gain. If these conditions are met, then the leading edge of the optical pulses experiences loss because the loss is initially larger than the gain. As the pulse intensity increases, however, the loss saturates faster than the gain, and the peak of the pulse is amplified. When the gain becomes depleted below the threshold level, the trailing edge experiences a loss that will continue until the pulse returns, provided condition (2) is met.

Subpicosecond pulses in a mode-locked semiconductor laser were first observed by van der Ziel *et al.* (1981) using a proton-bombarded facet as the saturable absorber. The autocorrelation trace (Fig. 13.26b) consists of a series of peaks spaced by 11 ps due to the nonideal antireflection coating and multiple transits within the semiconductor laser cavity itself. The laser pulses are generally chirped due to coupling between the real and imaginary parts of the index of the active layer (see next section) and further reduction in pulse width is possible by pulse compression with an external cavity (Silberberg and Smith, 1986).

13.4.6 Frequency Modulation

The previous section described the modulation of the intensity of the laser by current modulation. Current modulation causes electron density modulation, which modulates the gain *and* refractive index of the active layer. Consequently, the frequency of the laser is also modulated. This frequency modulation (FM) can be used to advantage in frequency shift keyed (FSK) coherent communication systems, but it is a problem in intensity modulated single longitudinal mode transmission systems because the resulting spectral width and the fiber dispersion combine to limit the maximum possible transmission distance. Most of the applications where FM is of interest involve single longitudinal mode lasers, and this assumption will be made in the analysis.

The ratio, α, between the change in the real and imaginary parts of the active layer index with a carrier density change was discussed in Section 13.2.3. Using the photon rate equation (13.16) and the relation between a frequency shift and an index shift $\Delta\nu/\nu = -\Delta n'/n'$, the frequency shift and power of the laser can

---

[4]A slow saturable absorber is one whose recovery time is long compared to the mode-locked pulse duration.

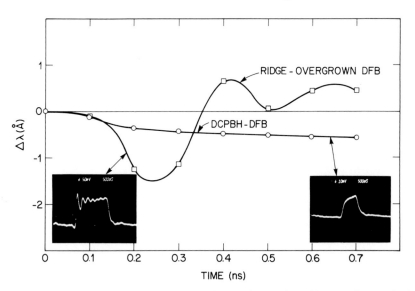

Fig. 13.27 Time variation of (a) wavelength and (b) optical power for a ridge type (heteroepitaxial ridge overgrown) and a DCPBH laser in response to a step current pulse applied at $t = 0$. The lasers were biased with an optical extinction ratio of $4 : 1$ in both cases (after Koch and Linke, 1986).

be related (Koch and Bowers, 1985):

$$\Delta\nu = \frac{-\alpha}{4\pi}\left(\frac{\dfrac{dP}{dt}}{P} + \kappa P\right), \qquad (13.48)$$

where $\kappa = 2\Gamma\epsilon(\alpha_m + \alpha_i)/(L\alpha_m)$. The first term is independent of the laser structure and leads to severe chirping during relaxation oscillations. The second term leads to a wavelength chirp between the high and low points in the optical waveform. These differences are readily apparent in actual lasers (Fig. 13.27). In lasers with small active layer volumes, such as DC-PBH structures, κ is large. The second term in Eq. (13.48) is thus large, but the laser is heavily damped (see Eq. (13.34)), so the relaxation oscillations are small and "transient" chirping is insignificant. In weakly-guided lasers, such as ridge waveguide structures, κ is smaller, and the laser intensity and frequency oscillate, but there is an insignificant shift in frequency between the on and off levels. Note that it is the transient chirp that causes eye closure and severely limits the transmission distance in a dispersive fiber, while the frequency shift between the on and off levels causes a much smaller system penalty, particularly when the power in the off level is insignificant.

13.4.7 Noise

We have just described the intensity and frequency modulation of a laser. The intensity and frequency noise spectra determine the signal-to-noise ratios in most applications. Noise characteristics can be studied from the rate equations by adding Langevin noise sources to the electron and photon rate equations. Other approaches include using the Fokker-Planck equation for the photon amplitude probability density, the classical van der Pol equation with a noise driving source (Haug and Haken, 1967) or a density matrix analysis (Scully and Lamb, 1967). Yamamoto (1983) compared these approaches, applied them to calculate the intensity noise spectrum, and compared the results to experimental data (Fig. 13.28). The general shape of the intensity noise is similar to the modulation response shown earlier (Fig. 13.20). The frequency dependence of the intensity noise is (Haug, 1969):

$$M(f) = \left(\frac{f_o^2}{a} \right) \frac{1 + b^2 f^2}{f_o^2 - f^2 + jf f_d}, \qquad (13.49)$$

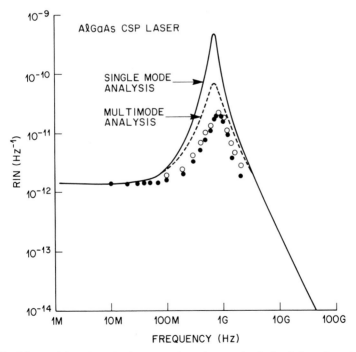

Fig. 13.28 Measured intensity noise frequency dependence and calculated dependence based on single-mode and multimode analyses (after Yamamoto *et al.*, 1983).

where a and b depend on the Langevin noise sources. For narrow band microwave signal transmission (Bowers, 1986), the frequency dependence of the signal to noise ratio is:

$$SNR(f) = \frac{a}{1 + bf^2}.$$ (13.50)

The signal to noise ratio decreases monotonically at high frequencies and is not reduced at the relaxation oscillation frequency where the intensity noise is a maximum.

The curves in Fig. 13.28 are representative of a single-mode laser and of a multimode laser if the noise power in all the modes is measured. However, the power in a particular longitudinal mode fluctuates (mode partition noise) and the relative intensity noise (RIN) of one of the modes may be tens of dBs higher than the RIN of the entire laser output. Mode partition noise is manifested when the mode distribution is filtered or dispersed, as in an optical transmission system. (Ogawa, 1982).

13.5 MID-INFRARED LASERS

The laser structures discussed to this point have emission wavelengths in the 0.8–0.9 μm (GaAs) or 1.2–1.6 μm (GaInAsP) wavelength ranges. Laser sources for wavelengths beyond 2 μm have found use in such nontelecommunications applications as high resolution spectroscopy, materials processing, and remote monitoring. As a result of recent progress in potentially ultra-low-loss fluoride fibers, new applications in mid-infrared (2–4 μm) lightwave communications now are receiving serious consideration as well (Lines, 1984). The loss minimum in these fibers, just below 1 dB/km at present, falls near 2.5 μm (Tran et al., 1986). Some low temperature injection lasers, as well as gas and solid-state lasers, are available for the mid-infrared region and may prove acceptable for very specialized applications, but most practical communications systems will contain semiconductor lasers capable of operation at or close to room temperature.

Semiconductor materials with direct band-gaps covering 2–4 μm include many II-VI, IV-VI, and III-V alloys. For lasers in this wavelength range, carrier losses due to Auger recombination are even more important than they are in GaInAsP lasers, and optical losses due to free-carrier absorption are also larger because of their dependence on λ^2. These effects play a major role in determining laser threshold and efficiency and limit the maximum operating temperatures of 2–4 μm wavelength injection lasers.

The total current required at threshold is larger than the value attributable solely to radiative recombination, $I_{th,r}$, by the addition of an Auger current,

$I_{th,\,a}$. The internal quantum efficiency of the laser at threshold is reduced to:

$$\eta_i^{th} = \frac{I_{th,\,r}}{I_{th,\,r} + I_{th,\,a}} = \frac{\tau_a}{\tau_a + \tau_r}, \qquad (13.51)$$

where τ_a and τ_r are the Auger and radiative lifetimes, respectively. The radiative current is proportional to the ratio of the carrier density, N_{th}, to the spontaneous lifetime, τ_n, at threshold (Eq. 13.20). The radiative threshold current increases slowly with temperature and decreases for materials with longer band-gap wavelengths through a decrease in N_{th}/τ_n. The Auger current depends on the exact electronic band structure of the material, and consists of contributions from several different Auger transitions. The largest contribution for some of the III-V alloys, for example, involves the transition of a conduction band electron to the heavy hole band along with the excitation of another conduction electron to a higher energy in the band (the "CHCC" process). The resulting Auger current increases exponentially with temperature and is especially large for materials with longer band-gap wavelengths.

Based on these considerations, calculations of thresholds and internal quantum efficiencies for several long-wavelength alloys lead to some interesting conclusions. Figure 13.29 compares the highest predicted oscillation temperatures of pulsed DH lasers from various materials as a function of wavelength, based on estimates of the temperature at which the internal quantum efficiency at threshold falls to 2.5% (Horikoshi, 1983; Horikoshi, 1985). The calculations predict a general limit to room temperature laser oscillation at a little beyond 2

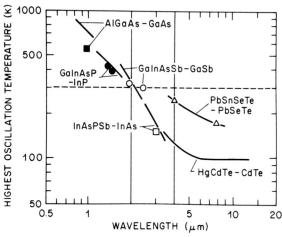

Fig. 13.29 Maximum temperature of pulsed laser operation vs. wavelength for several material systems (after Horikoshi, 1985).

μm for any of the alloy materials considered. The data points in the figure are experimental observations, which are now discussed for several particular laser systems.

The HgCdTe material system, from which excellent infrared detectors have been fabricated, is representative of the II-VI alloy semiconductors covering the 2–4 μm wavelength range. Both LEDs and optically-pumped lasers have been demonstrated in these alloys, but no injection laser sources have been reported so far.

Injection lasers based on IV-VI lead-salt alloys are now used for many high-resolution spectroscopic applications. Lasers fabricated from quaternary PbSnSeTe and related ternary materials emit at wavelengths longer than about 4 μm. Auger effects are calculated (Sugimura, 1982; Horikoshi, 1985) to be smaller in some of these alloys than in the III-V materials, which could result in lower thresholds and higher maximum operating temperatures. The replacement of Sn with Ge, Cd, or Eu shifts the band-gap to shorter wavelengths, and PbEuSeTe/PbTe DH lasers covering 2.7 to 6.6 μm have been reported (Partin, 1984). The highest operating temperature for a pulsed 4 μm DH laser made from this alloy was 241 K, in approximate agreement with the calculated curves in Fig. 13.29.

Among the III-V alloys with band-gap wavelengths beyond 2 μm (Fig. 13.1), GaInAsSb and InAsPSb, lattice-matched to either GaSb or InAs substrates, are the most promising for lasers. Only the injection-pumped laser results are discussed here, but electron-beam-pumped and optically-pumped semiconductor lasers based on both of these materials also have been demonstrated.

The band-gap wavelength of the quaternary GaInAsSb, grown lattice-matched to a GaSb substrate, can be varied from 1.7 μm (corresponding to GaSb) to 4.3 μm by appropriate choice of composition. In practice, a "miscibility gap" in the center of this range prevents the formation of alloys from about 2.4 to 4 μm, at least by an equilibrium growth technique such as LPE (DeWinter et al., 1985). Growth by MOVPE (Cherng et al., 1986) or MBE (Tsang et al., 1985) may provide device quality material over a wider range. Another difficulty with this system is that the refractive indices of GaInAsSb compositions lattice-matched to GaSb decrease with increasing band-gap wavelength (the index of InAs is smaller than that of GaSb), thereby limiting the effectiveness of optical waveguiding at the longer wavelengths.

Early work on GaInAsSb/AlGaAsSb DH lasers (Dolginov et al., 1978) led to room temperature operation at 1.8 μm (Kobayashi et al., 1980). At this wavelength, calculations predict that the Auger current becomes comparable in magnitude to the radiative current at room temperature (Sugimura, 1982). As the wavelength is increased, the Auger current increases sharply. The room temperature internal quantum efficiency is estimated to fall below 5% at 2.2 μm, and the normalized threshold current density is calculated to be about 20 kA/cm^2 μm. This threshold prediction is pessimistic in light of the most recent experimental results discussed next.

Room temperature pulsed operation of GaInAsSb/AlGaAsSb DH injection lasers has now been obtained at wavelengths as long as 2.2–2.4 μm (Bochkarev et al., 1985; Caneau et al., 1985). The lowest reported room temperature threshold current density is 1.7 kA/cm^2, corresponding to a normalized value of about 4 kA/cm^2 μm (Pollack, 1987; Caneau, et al., 1987); observed T_o values are 60 to 80 K and cw operation has been extended to 235 K (Caneau et al., 1986). Further improvements could lead to room temperature cw operation. Nevertheless, the calculations indicate that near room temperature operation of lasers involving any of the III-V alloys will not be possible at much longer wavelengths. At 77 K, for comparison, laser oscillation is predicted to extend out to 4.4 μm because of the temperature dependence of the Auger current.

The InAsPSb alloy, grown lattice-matched to InAs substrates, potentially covers the 2 to 3.5 μm range. The calculations predict a dependence of the maximum operating temperature on wavelength similar to that of GaInAsSb (Fig. 13.29). So far, however, only cooled DH InAsPSb/InAs injection lasers with emission wavelengths around 3 μm have been reported (Kobayashi and Horikoshi, 1980). The maximum operating temperature of these lasers was 145 K, where the threshold current density reached 57 kA/cm^2, and T_o was 23 K.

13.6 FUTURE PROSPECTS

Semiconductor lasers are enjoying great success in many fields: telecommunications, optical memories, printing, and as pumping sources for other lasers. They have emerged from "flash bulbs" to high performance devices with projected 20-year lifetimes. In many respects, the field has matured, and future research will largely involve integrating lasers with other optical and electronic devices. However, the demands for higher performance (e.g., higher power, higher temperature operation, higher speed) require new inventions and new laser structures. One example of a potentially important advance is the suggestion (Tsang and Logan, 1986) of using impurity doping to modify the gain curve to possibly allow dynamic single frequency operation without the need for frequency dependent reflectors (DFB, DBR, etc.).

This section examines some of the inherent limits in laser performance and asks questions such as: How low can thresholds become? How much power could a single laser or an array produce? How fast can a laser be modulated?

How low can laser thresholds become? This question is particularly important for integrated optoelectronics where many devices are integrated together, and the heat generated is an important problem. Presently, the lowest threshold current reported (Tsang et al., 1982) is 2.5 mA, for GRIN-SCH GaAs lasers. The equation for threshold current in terms of N_{th} was given in Eq. (13.20), and the expression for N_{th} was given in Eq. (13.21). For the lowest threshold, one wants to decrease the mirror loss (by high reflectivity coating of the facets) to make τ_p long, so that $1/(\Gamma g_o \tau_p)$ is small compared with N_t. Then

$$I_{th} \approx \frac{qVN_t}{\tau_n}. \tag{13.52}$$

To minimize the threshold, the volume must be small. Using etched facets, cavity lengths can be very short, and lasers with 20 μm cavities have already been demonstrated without high reflectivity coatings. Cavities as short as 5 μm are certainly possible provided the facets are coated to keep the mirror loss small. Waveguide widths of 0.5 μm have been demonstrated; this is probably a lower limit since narrower waveguides greatly increase the scattering loss and decrease the carrier lifetime τ_n below the typical value of 1 ns. The density for transparency, N_t, is $\sim 1 \times 10^{18}$ cm^{-3} so that for a 200 Å single quantum well, the threshold might be as low as 0.1 mA. Modulation doping the active layer p-type lowers the transparency level, and additional quantum confinement to quantum wires or boxes may allow threshold currents as low as 10 μA. Such low thresholds could allow the integration of thousands of lasers on a single chip for printing, optical interconnect or integrated optic applications.

How much power could be obtained from a single laser? This is a complicated, multifaceted problem that depends largely on the system requirements in terms of lifetime, beam divergence and other parameters, but a few guidelines can be given. For GaAs lasers, an important limit is facet damage (Henry et al. 1979), and a power density of $2 - 10 \times 10^6$ W/cm^2 is a limit for uncoated facets, while somewhat higher limits exist for catastrophic damage to coated facets. Another way to increase the power in a GaAs laser is to use a window laser where the current injection and optical waveguide do not extend to the facet, thus reducing carrier recombination at the facet and allowing the beam to diverge before reaching the facet. For any of these lasers, the power limit for long-term degradation may be much lower (Hakki and Nash, 1974). Assuming the maximum mode size for single transverse mode operation is of the order of 4 μm in diameter, this translates to a maximum power level of 1 W for a large output spot. Presently, 250 mW has been obtained for a single stripe. For 100-μm wide arrays of AlGaAs lasers, cw powers of 2.1 W have been obtained (Thornton et al., 1987) and pulsed (50 ns pulses) powers of 20 W have been obtained (Welch et al., 1987). This is the highest power per unit width, but higher powers have been obtained in wider arrays and in two-dimensional arrays. For GaInAsP lasers, the threshold for catastrophic mirror damage is much higher and is usually not the limitation. Rather, the problem is heating, which causes increased leakage around and across the active layer and causes decreased power output because of the relatively low T_o of quaternary lasers. For 1.3 μm single guide lasers, cw powers of 250 mW have been demonstrated, and even higher powers are certainly possible.

How fast can a laser be modulated? We have already indicated (Eq. 13.38) that there is an inherent limit caused by nonlinear gain effects. At very high photon densities, the modulation response is critically damped, and a maximum 3 dB bandwidth is obtained. For single-longitudinal-mode double heterostructure semiconductor lasers, this limit is of the order of 40 GHz. As discussed earlier, however, the use of quantum size effects, and the additional use of modulation doping (Uomi *et al.*, 1986) greatly increase the differential gain coefficient, and this limit increases to 80 GHz or more. Presently the maximum 3 dB bandwidth obtained is 31 GHz, and the maximum resonance frequency is around 20 GHz for both AlGaAs lasers (Yuasa *et al.*, 1986) and GaInAsP lasers (Bowers *et al.*, 1986).

How narrow can the linewidth of a single-mode laser be? This question is particularly important for coherent transmission systems. The expression for linewidth was given in Eq. (13.31); the limitation is due to phase noise generated by spontaneous emission. While linewidths of 10 MHz are common, by the use of long, integrated passive or possibly active cavities (Miller, 1986), as well as by operation on the short wavelength side of the gain peak to reduce the linewidth enhancement factor α, linewidths of 3 MHz have been demonstrated (Kobayashi and Mito, 1987) and linewidths of 1 kHz seem possible in an integrated laser ($P_o = 100$ mW, $L = 1$ cm, $\alpha_i = 20$ cm^{-1}, $\alpha = 2$ in Eq. 13.31). With external cavities, either bulk or fiber optic, linewidths of 50 kHz have been demonstrated (Wyatt and Devlin, 1983), and 1 kHz linewidths seem possible, limited by the effects of coherence collapse in long cavity lasers (Henry and Kazarinov, 1986). The challenge is to make a manufacturable device that has a narrow linewidth and is tunable and perhaps capable of broadband frequency modulation. Various multiterminal DFB and DBR lasers with tuning and phase control regions have been proposed for this purpose and these goals should soon be achieved.

There are many other performance issues where the spread is large between what has been achieved and what is achievable. Many new inventions are needed to fully utilize the semiconductor laser in future systems. There are many challenges to integrating lasers with other electronic and optical elements to provide cost effective solutions. One important example is integrating the local oscillator laser, directional coupler, polarization controller, and photodetectors needed for low-cost high-performance coherent receivers.

REFERENCES

Adachi, S. (1982). Refractive indices of III-V compounds: key properties of InGaAsP relevant to device design. *J. Appl. Phys.* **53**, 5863–5869.

Aiki, K., Nakamura, M., Umeda, J., Yariv, A., Katzir, A., and Yen, H. W. (1975). GaAs-GaAlAs distributed-feedback diode lasers with separate optical and carrier confinement. *Appl. Phys. Lett.* **27**, 145–146.

Aiki, K., Nakamura, M., Kuroda, T., Umeda, J., Ito, R., Chinnone, N., and Maeda, M. (1978). Transverse mode stabilized $Al_xGa_{1-x}As$ injection lasers with channeled-substrate-planar structure. *IEEE J. Quantum Electron.* **QE-14**, 89–94.

Alferov, Zh. I., and Kazarinov, R. F. (1963). Authors Certificate 28448 (U.S.S.R.) (as cited in Casey and Panish, 1978).

Alferov, Zh. I., Andreev, V. M., Garbuzov, D. Z., Zhilyaev, Yu. V., Morozov, E. P., Portnoi, E. L., and Trofim, V. G. (1970). Investigation of the influence of the AlAs-GaAs heterostructure parameters on the laser threshold current and the realization of continuous emission at room temperature. *Fiz. Tekh. Poluprovodn.* **4**, 1826–1829 [translated in: *Sov. Phys. Semicond.* **4**, 1573–1575 (1971).].

Arakawa, Y., and Sakaki, H. (1982). Multiquantum well laser and its temperature dependence of the threshold current. *Appl. Phys. Lett.* **40**, 939–941.

Arakawa, Y., and Yariv, A. (1986). Quantum well lasers—gain, spectra, dynamics. *J. Quantum Electron.* **QE-22**, 1887–1899.

Armistead, C. J., Wheeler, S. A., Plumb, R. G., and Musk, R. W. (1986). Low-threshold ridge waveguide lasers at $\lambda = 1.5$ μm. *Electron. Lett.* **22**, 1145–1147.

Asada, M. (1985). Theoretical linewidth enhancement factor α of GaInAsP/InP lasers. *Trans. IECE Japan.* **68**, 518–520.

Asada, M., Adams, A. R., Stabkjaer, K. E., Suematsu, Y., Itaya, Y., and Asai, S. (1981). The temperature dependence of the threshold current of GaInAsP/InP DH lasers. *IEEE J. Quantum Electron.* **QE-17**, 611–619.

Asada, M., Kameyama, A., and Suematsu, Y. (1984). Gain and intervalence band absorption in quantum well lasers. *J. Quantum Electron.* **QE-20**, 745–753.

Bernard, M. G. A., and Duraffourg, G. (1961). Laser conditions in semiconductors. *Phys. Status Solidi.* **1**, 699–703.

Blauvelt, H., Bar-Chaim, N., Fekete, D., Margalet, S., and Yariv, A. (1982). AlGaAs lasers with micro-cleaved mirrors suitable for monolithic integration. *Appl. Phys. Lett.* **40**, 289–290.

Bochkarev, A. E., Dolginov, L. M., Drakin, A. E., Druzhinina, L. V., Eliseev, P. G., and Sverdlov, B. N. (1985). Injection InGaAsSb lasers emitting radiation of wavelengths 1.9–2.3 μ at room temperature. *Sov. J. Quantum Electron.* **15**, 869–870.

Bowers, J. E. (1986). Optical transmission using PSK modulated subcarriers at frequencies to 16 GHz. *Electron Lett.* **22**, 1119–1121.

Bowers, J. E. (1987). High speed semiconductor laser design and performance. *Solid State Electron.* **30**, 1–11.

Bowers, J. E., and Wilt, D. P. (1984). Optimum design of 1.55 μm double heterostructures and ridge-waveguide lasers. *Optics Lett.* **9**, 330–332.

Bowers, J. E., Hemenway, B. R., Gnauck, A. H., Bridges, T. J., and Burkhardt, E. G. (1985). High-frequency constricted mesa lasers. *Appl. Phys. Lett.* **47**, 78–80.

Bowers, J. E., Hemenway, B. R., Gnauck, A. H., and Wilt, D. P. (1986). High-speed InGaAsP constricted-mesa lasers. *J. Quantum Electron.* **QE-22**, 833–844.

Caneau, C., Srivastava, A. K., Dentai, A. G., Zyskind, J. L., and Pollack, M. A. (1985). Room temperature GaInAsSb/AlGaAsSb DH injection lasers at 2.2 μm. *Electron. Lett.* **21**, 815–817.

Caneau, C., Srivastava, A. K., Dentai, A. G., Zyskind, J. L., Burrus, C. A., and Pollack, M. A. (1986). Reduction of the threshold current density of 2.2 μm GaInAsSb/AlGaAsSb injection lasers. *Electron. Lett.* **22**, 992–993.

Caneau, C., Zyskind, J. L., Sulhoff, J. W., Glover, T. E., Centanni, J., Burrus, C. A., Dentai, A. G., and Pollack, M. A. (1987). 2.2 μm GaInAsSb/AlGaAsSb injection lasers with low threshold current density. *Appl. Phys. Lett.* **51**, 764–766.

Casey, H. C., Jr., and Panish, M. B. (1978). Heterostructure lasers. Parts A and B. Academic Press, New York.

Casey, H. C., Jr., Somekh, S., and Illgems, M. (1975). Room-temperature operation of low-threshold separate-confinement heterostructure injection laser with distributed feedback. *Appl. Phys. Lett.* **27**, 142–144.

Cherng, M. J., Stringfellow, G. B., Kisker, D. W., Srivastava, A. K., and Zyskind, J. L. (1986). GaInAsSb metastable alloys grown by organometallic vapor phase epitaxy. *Appl. Phys. Lett.* **48**, 419–421.

Chin, R., Holonyak, Jr., N., Bojak, B. A., Hess, K., and Dupuis, R. D. (1979). Temperature dependence of threshold current for quantum well $Al_xGa_{1-x}As/GaAs$ heterostructure laser diodes. *Appl. Phys. Lett.* **36**, 19–21.

Coldren, L. A., Miller, B. I., Iga, K., and Rentschler, J. A. (1981). Monolithic two-section GaInAsP/InP active-optical-resonator devices formed by reactive ion etching. *Appl. Phys. Lett.* **38**, 315–317.

Coldren, L. A., Furuya, K., Miller, B. I., and Rentschler, J. A. (1982). Etched mirror and groove-coupled GaInAsP/InP laser devices for integrated optics. *IEEE J. Quantum Electron.* **QE-18**, 1679–1688.

Coldren, L. A., Boyd, G. D., Bowers, J. E., and Burrus, C. A. (1985). Reduced dynamic linewidth in three-terminal two-section diode lasers. *Appl. Phys. Lett.* **46**, 125–127.

Corzine, S. and Bowers, J. E., (1987). Unpublished.

DeWinter, J. C., Pollack, M. A., Srivastava, A. K., and Zyskind, J. L. (1985). Liquid phase epitaxial $Ga_{1-x}In_xAs_ySb_{1-y}$ lattice-matched to (100) GaSb over the 1.71 to 2.33 μm wavelength range. *J. Electron. Mat.* **14**, 729–747.

Dolginov, L. M., Druzhinina, L. V., Eliseev, P. G., Lapshin, A. N., Mil'vidskii, M. G., and Sverdlow, B. N. (1978). Injection heterolaser based on InGaAsSb four-component solid solution. *Sov. J. Quantum Electron.* **8**, 416.

Downey, P., Bowers, J. E., Tucker, R. S., and Agyekum, E. (1987). Picosecond dynamics of a gain-switched InGaAsP laser. *J. Quantum Electron.* **QE-23**, 1039–1047.

Dumke, W. P. (1963) Optical transitions involving impurities in semiconductors. *Phys. Rev.* **132**, 1998–2002.

Dutta, N. K. (1980). Calculated emission, absorption and gain in $In_{0.72}Ga_{0.28}As_{0.6}P_{0.4}$. *J. Appl. Phys.* **51**, 6095–6100.

Dutta, N. K., and Nelson, R. J. (1981). Temperature dependence of threshold of InGaAsP/InP double heterostructure lasers and auger recombination. *Appl. Phys. Lett.* **98**, 407–409.

Dutta, N. K., and Nelson, R. J. (1982). The case for auger recombination in $In_{1-x}Ga_xAs_yP_{1-y}$. *J. Appl. Phys.* **53**, 74–91.

Dutta, N. K., Wilt, D. P., and Nelson, R. J. (1984). Analysis of leakage currents in 1.3 μm InGaAsP real-index-guided lasers. *J. Lightwave Technol.* **LT-2**, 201–208.

Dutta, N. K., Wilson, R. B., Wilt, D. P., Besomi, P., Brown, R. L., Nelson, R. J., and Dixon, R. W. (1985). Performance comparison of InGaAsP lasers emitting at 1.3 and 1.55 μm for lightwave system applications. *AT & T Technical Journal* **64**, 1857–1884.

Dyment, J. C., and D'Asara, L. A. (1967). Continuous operation of GaAs junction lasers on diamond heat sinks at 200°K. *Appl. Phys. Lett.* **11**, 292–294.

Eagles, D. M. (1960). Optical absorption and recombination radiation in semiconductors due to transitions between hydrogen-like acceptor impurity levels and the conduction band. *J. Phys. Chem. Solids.* **16**, 76–83.

Einstein, A. (1917). Zur quantentheorie der strahlung. *Phys. Z.* **18**, 121–128.

Eisenstein, G., Tucker, R. S., Koren, J., and Korotky, S. K. (1986). Active mode-locking characteristics of InGaAsP-single mode fiber composite cavity lasers. *J. Quantum Electron.* **QE-22**, 142–148.

Fuji, T., Yamakoshi, S., Nanbu, K., Waldo, O., and Hiyamiza, S. (1984). MBE growth of extremely high quality GaAs-AlGaAs GRIN-SCH lasers with a superlattice buffer layer. *J. Vac. Sci. and Tech.* **2**, 259–261.

Hakki, B. W., and Nash, F. P. (1974). Catastrophic failure in GaAs double heterostructure lasers. *J. Appl. Phys.* **45**, 3907–3912.

Hakki, B. W., and Paoli, T. L. (1975). Gain spectra in GaAs double heterostructure injection lasers. *J. Appl. Phys.* **46**, 1299–1306.

Hall, R. N., Fenner, G. E., Kingsley, J. D., Soltys, T. J., and Carlson, R. O. (1962). Coherent light emission from GaAs junctions. *Phys. Rev. Lett.* **9**, 366–368.

Halperin, B. I., and Lax, M. (1966). Impurity band tails in the high density limit. I. minimum counting methods. *Phys. Rev.* **148**, 722–740.

Haug, H. (1969). Quantum mechanical rate equations for semiconductor lasers. *Phys. Rev.* **184**, 338–348.

Haug, H., and Haken, H. (1967). Theory of noise in semiconductor laser emission. *Z. Phys.* **204**, 262–275.

Haus, H. A. (1981). Modelocking of semiconductor laser diodes. *Jpn. J. Appl. Phys.* **20**, 1007–1020.

Hayashi, I., Panish, M. B., and Foy, P. W. (1969). A low-threshold room temperature injection laser. *IEEE J. Quantum Electron.* **QE-5**, 211–212.

Hayashi, I., Panish, M. B., Foy, P. W., and Sumski, S. (1970). Junction lasers which operate continuously at room temperature. *Appl. Phys. Lett.* **17**, 109–111.

Henry, C. H. (1982). Theory of the linewidth of semiconductor lasers. *IEEE J. Quantum Electron.* **QE-18**, 259–264.

Henry, C. H., and Kazarinov, R. F. (1984). Stabilization of single frequency operation of coupled-cavity lasers. *IEEE J. Quantum Electron.* **QE-20**, 733–744.

Henry, C. H., and Kazarinov, R. F. (1986). Instability of semiconductor lasers due to optical feedback from distant reflectors. *J. Quantum Electron.* **QE-22**, 294–301.

Henry, C. H., Petroff, P. M., Logan, R. A., and Merritt, F. R. (1979). Catastrophic damage of $Al_xGa_{1-x}As$ double heterostructure laser material. *J. Appl. Phys.* **50**, 3721–3732.

Henry, C. H., Logan, R. A., and Bertness, K. A. (1981). Measurement of spectrum, bias dependence and intensity of spontaneous emission in GaAs lasers. *J. Appl. Phys.* **52**, 4453–4456.

Henry, C. H., Logan, R. A., Merritt, F. R., and Luongo, J. P. (1983). The effect of intervalence band absorption on the thermal behavior of InGaAsP lasers. *J. Quantum Electron.* **QE-19**, 947–952.

Hersee, S., Baldy, M., Assenat, P., DeCremoux, B., and Duchemin, J. P. (1982). Low threshold GRIN-SCH GaAs-GaAlAs laser structure grown by OM VPE. *Electron. Lett.* **18**, 618–620.

Hirao, M., Dori, A., Tsuji, S., Nakamura, M., and Aiki, K. (1980). Fabrication and characterization of narrow stripe InGaAsP/InP buried heterostructure lasers. *J. Appl. Phys.* **51**, 4539–4540.

Holonyak, N., Jr., and Bevacqua, S. F. (1962). Coherent (visible) light emission from $Ga(As_{1-x}P_x)$ junctions. *Appl. Phys. Lett.* **1**, 82–83.

Horikoshi, Y. (1983). Lasing characteristics of PbSnSeTe-PbSeTe lattice-matched double heterostructure laser diodes. *Proc. SPIE* **438**, 29–36.

Horikoshi, Y. (1985). Semiconductor lasers with wavelengths exceeding 2 μm. In "Semiconductors and semimetals: lightwave communication technology," (R. K. Willardson and A. C. Beer, eds. Vol. 22C, W. T. Tsang, Vol. ed.), Academic Press, London, 93–151.

Hsieh, J. J., Rossi, J. A., and Donnelly, J. P. (1976). Room-temperature CW operations of GaInAsP/InP double-heterostructure diode lasers emitting at 1.1 μm. *Appl. Phys. Lett.* **28**, 709–711.

Huang, C. J. (1970). Properties of stimulated and spontaneous emission in GaAs junction lasers II: temperature dependence of threshold current and excitation dependence of superradiant spectra. *Phys. Rev.* **B2**, 4126–4134.

Iga, K., Pollack, M. A., Miller, B. I., and Martin, R. J. (1980). GaInAsP/InP DH lasers with a chemically etched facet. *IEEE J. Quantum Electron.* **QE-16**, 1044–1046.

Iga, K., Kinoshita, S., and Koyama, F. (1986). Microcavity GaAlAs/GaAs surface emitting laser with I_{th} = 6 mA. Paper PD-4, Tenth IEEE International Semiconductor Laser Conference, Kanazawa, Japan, October 14–17, 1986.

Ikegami, T., and Suematsu, Y. (1968). Direct modulation of semiconductor junction laser. *Electron. Comm. Japan* **51-B**, 51–58.

Ippen, E., and Shank, C. V. (1977). In "Ultrashort Light Pulses," Vol. 18 (S. L. Shapiro, ed.), Springer-Verlag, New York, p. F3.

Ishikawa, H., Imai, H., Tanahashi, T., Hori, K., and Takahei, K. (1982). V-grooved substrate buried heterostructure InGaAsP/InP laser emitting at 1.3 μm wavelength. *IEEE J. Quantum Electron.* **QE-18**, 1704–1711.

Kaminow, I. P., Weber, H. P., and Chandross, E. A. (1971). Poly(Methyl Methacrylate) dye laser with internal diffraction grating resonator. *Appl. Phys. Lett.* **18**, 497–499.

Kaminow, I. P., Nahory, R. E., Pollack, M. A., Stulz, L. W., and DeWinter, J. C. (1979). Single-mode CW ridge-waveguide laser emitting at 1.55 μm. *Electron. Lett.* **15**, 763–765.

Kaminow, I. P., Stulz, L. W., Ko, J. S., Dentai, A. G., Nahory, R. E., DeWinter, J. C., and Hartman, R. L. (1983). Low-threshold InGaAsP ridge waveguide lasers at 1.3 μm. *IEEE J. Quantum Electron.* **QE-19**, 1312–1319.

Kane, E. O. (1957). Band structure of indium antimonide. *J. Phys. Chem. Solids* **1**, 249.

Kane, E. O. (1963). Thomas-Fermi approach to impure semiconductor band structure. *Phys. Rev.* **131**, 79–88.

Kawaguchi, H., and Kawakami, T. (1977). Transverse-mode control in an injection laser by a strip-loaded waveguide. *IEEE J. Quantum Electron.* **QE-13**, 556–560.

Kirkby, P. A., Goodwin, A. R., Thompson, G. H. B., and Selway, P. R. (1977). Observations of self-focusing in stripe geometry semiconductor lasers and the development of a comprehensive model of their operation. *IEEE J. Quantum Electron.* **QE-13**, 705–719.

Kitamura, M., Yamaguchi, M., Murata, S., Mito, I., and Kobayashi, K. (1984). High-performance single-longitudinal-mode operation of InGaAsP/InP DFB-DC-PBH LD's. *J. Lightwave Technol.* **LT-2**, 363–369.

Kobayashi, N., and Horikoshi, Y. (1980). DH lasers fabricated by new III-V semiconductor material InAsPSb. *Jpn. J. Appl. Phys.* **19**, L641–L644.

Kobayashi, K., and Mito, I. (1985). High light output-power single-longitudinal-mode semiconductor laser diodes. *J. Lightwave Technol.* **LT-3**, 1202–1210.

Kobayashi, K., and Mito, I. (1987). Progress in narrow-linewidth tunable laser sources. Paper WC1, in *Conference on Optical Fiber Communication/International Conference on Integrated Optics and Optical Fiber Communication Technical Digest Series 1987*, Vol. 3, (Optical Society of America, Washington, DC 1987), 150.

Kobayashi, N., Horikoshi, Y., and Uemura, C. (1980). Room temperature operation of the InGaAsSb/AlGaAsSb DH laser at 1.8 μm wavelength. *Jpn. J. Appl. Phys.* **19**, L30–L32.

Koch, T. L., and Bowers, J. E. (1985). Nature of wavelength chirping in directly modulated semiconductor lasers. *Electron. Lett.* **20**, 1038–1040.

Koch, T. L., and Linke, R. A. (1986). Effect of nonlinear gain reduction on laser wavelength chirping. *Appl. Phys. Lett.* **48**, 613–615.

Kogelnik, H., and Shank, C. V. (1971). Stimulated emission in a periodic structure. *Appl. Phys. Lett.* **18**, 152–154.

Kogelnik, H., and Shank, C. V. (1972). Coupled-wave theory of distributed feedback lasers. *J. Appl. Phys.* **43**, 2327–2335.

Kressel, H., and Butler, J. K. (1977). "Semiconductor lasers and heterojunction LEDs." Academic Press, New York.

Kressel, H., and Nelson, H. (1969). Close confinement gallium arsenide PN junction lasers with reduced optical loss at room temperature. *RCA Rev.* **30**, 106–113.

Kroemer, H. (1963). A proposed class of heterojunction injection lasers. *Proc. IEEE* **51**, 1782–1783.

Lau, K. Y., and Yariv, A. (1985). High frequency current modulation of semiconductor injection lasers. In "Semiconductors and semimetals: lightwave communication technology." (R. K. Willardson and A. C. Beer, eds., Vol. 22B, W. T. Tsang, Vol. ed.), Academic Press, London, 69–152.

Lau, K. Y., Harder, C., and Yariv, A. (1981). Ultimate frequency response of GaAs injection lasers. *Optics Comm.* **36**, 472–474.

Lee, T. P., Burrus, C. A., Linke, R. A., and Nelson, R. J. (1983). Short-cavity single-frequency InGaAsP buried heterostructure lasers. *Electron. Lett.* **19**, 82–84.

Lines, M. E. (1984). The search for very low loss fiber-optic materials. *Science* **226**, 663–668.

Liou, K. Y., Burrus, C. A., Linke, R. A., Kaminow, I. P., Granlund, S. W., Swan, C. B., and Besomi, P. (1984). Single-longitudinal-mode stabilized graded-index-rod external coupled-cavity laser. *Appl. Phys. Lett.* **45**, 729–731.

Manning, J., Olshansky, R., Fye, D. M., and Powazinik, W. (1985). Strong influence of nonlinear gain on spectral and dynamic characteristics of InGaAsP lasers. *Electron. Lett.* **21**, 496–497.

Miller, S. E. (1986). Integrated low-noise lasers. *Electron. Lett.* **22**, 256–257.

Miller, B. I., Koren, U., and Capik, R. J. (1986). Planar buried heterostructure InP/GaInAs lasers grown entirely by OMVPE. *Electron. Lett.* **22**, 947–949.

Mito, I., Kitamura, M., Kobayashi, K., Murata, S., Seki, M., Odagiri, Y., Nishimoto, H., Yamaguchi, M., and Kobayashi, K. (1983). InGaAsP double-channel-planar-buried-heterostructure laser diode (DC-PBH LD) with effective current confinement. *J. Lightwave Technol.* **LT-1**, 195–202.

Nahory, R. E., Pollack, M. A., Beebe, E. D., DeWinter, J. C., and Dixon, R. W. (1976). Continuous operation of 1.0 μm wavelength $GaAs_{1-x}Sb_x/Al_yGa_{1-y}As_{1-y}Sb_x$ double-heterostructure injection lasers at room temperature. *Appl. Phys. Lett.* **28**, 19–21.

Nahory, R. E., Pollack, M. A., Johnston, W. D., and Barnes, R. L. (1978). Bandgap versus composition and demonstration of Vegard's Law for $In_{1-x}Ga_xAs_yP_{1-y}$ lattice matched to InP. *Appl. Phys. Lett.* **33**, 659–661.

Nathan, M. I., Dumke, W. P., Burns, G., Dill, F. H., Jr., and Lasher, G. (1962). Stimulated emission of radiation from GaAs p-n junctions. *Appl. Phys. Lett.* **1**, 62–64.

Ogawa, K. (1982). Analysis of mode partitioning noise for laser diode system. *IEEE J. Quantum Electron.* **QE-17**, 849–855.

Panish, M. B., Casey, Jr., H. C. Sumski, S., and Foy, P. W. (1973). Reduction of threshold current density in $GaAs-Al_xGa_{1-x}As$ heterostructure lasers by separate optical and carrier confinement. *Appl. Phys. Lett.* **22**, 590–591.

Partin, D. L., and Thrush, C. M. (1984). Wavelength coverage of lead-europium-selenide-telluride diode lasers. *Appl. Phys. Lett.* **45**, 193–195.

Phelan, R. J., Calawa, A. R., Rediker, R. H., Keyes, R. J., and Lax, B. (1963). Infrared InSb laser diode in high magnetic fields. *Appl. Phys. Lett.* **3**, 143–145.

Pollack, M. A. (1987). Devices and device issues for wavelengths beyond 2 μm. Paper WH2, *Technical Digest of Optical Fiber Communication Conference*, Reno, Jan. 19–22, 1987.

Quist, T. M., Rediker, R. H., Keyes, R. J., Krag, W. E., Lax, B., McWhorter, A. L., and Zeigler, H. J. (1962). Semiconductor maser of GaAs. *Appl. Phys. Lett.* **1**, 91–92.

Reinhart, F. K., Logan, R. A., and Shank, C. V. (1975). $GaAs-Al_xGa_{1-x}As$ injection lasers with distributed bragg reflectors. *Appl. Phys. Lett.* **27**, 45–48.

Scully, M. D., and Lamb, Jr., W. E. (1967). Quantum theory of an optical maser: I general theory. *Phys. Rev.* **159**, 208–226.

Silberberg, Y., and Smith, P. W. (1986). Subpicosecond pulses from a mode-locked semiconductor laser. *J. Quantum Electron.* **QE-22**, 759–761.

Stern, F. (1971). Band tail model for optical absorption and the mobility edge in amorphous silicon. *Phys. Rev.* **B3**, 2636–2645.

Stern, F. (1976). Calculated spectral dependence of gain in excited GaAs. *J. Appl. Phys.* **47**, 5382–5386.

Streifer, W., Burnham, R. D., and Scifres, D. R. (1975). Effect of external reflectors on longitudinal modes of distributed feedback lasers. *IEEE J. Quantum Electron.* **QE-11**, 154–161.

Streifer, W., Burnham, R. D., and Scifries, D. R. (1983). Modal analysis of separate-confinement heterojunction lasers with inhomogeneous cladding layers. *Optics Lett.* **8**, 283–285.

Stix, M., Kesler, M. P., and Ippen, E. (1986). Observation of subpicosecond dynamics in GaAlAs laser diodes. *Appl. Phys. Lett.* **48**, 1722–1724.

Suematsu, Y., Arai, S., and Kishino, K. (1983). Dynamic single-mode semiconductor lasers with a distributed reflector. *J. Lightwave Technol.* **LT-1**, 161–176.

Sugimura, A. (1982). Band-to-band auger effect in long wavelength multinary III-V alloy semiconductor lasers. *IEEE J. Quantum Electron.* **QE-18**, 352–363.

Thompson, G. H. B. (1980). "Physics of semiconductor laser devices." Wiley, New York.

Thompson, G. H. B., and Kirkby, P. A. (1973). (GaAl)As lasers with an heterostructure for optical confinement and additional heterojunctions for extreme carrier confinement. *IEEE J. Quantum Electron.* **QE-9**, 311–318.

Thornton, R. L., Welch, D. F., Burnham, R. D., Paoli, T. L., and Cross, P. S. (1986). High power (2.1W) 10-stripe AlGaAs laser arrays with Si disordered facet windows. *Appl. Phys. Lett.* **49**, 1572–1574.

Tohmori, Y., Oohashi, H., Kato, T., Arai, S., Komori, K., and Suematsu, Y. (1986). Wavelength stabilization of 1.5 μm GaInAsP/InP bundle-integrated-guide distributed-bragg-reflector (BIG-DBR) lasers integrated with wavelength tuning region. *Electron. Lett.* **22**, 138–140.

Tran, D. C., Levin, K. H., Burk, M. J., Fischer, C. F., and Brower, D. (1986). Preparation and properties of high optical quality IR transmitting glasses and fibers based on metal fluorides. *Proceedings of SPIE* **618**, 48–50.

Tsang, W. T. (1981). A graded-index waveguide separate-confinement laser with very low threshold and a narrow Gaussian beam. *Appl. Phys. Lett.* **39**, 134–137.

Tsang, W. T. (1985). The cleaved-coupled-cavity (C^3) laser. In "Semiconductors and semimetals: lightwave communications technology." (R. K. Willardson and A. C. Beer, Eds., Vol. 22B, W. T. Tsang, Vol. ed.), Academic Press, London, 257–373.

Tsang, W. T., and Logan, R. A. (1986). Observation of enhanced single-longitudinal mode operation in 1.5 μm GaInAsP erbium-doped semiconductor injection lasers. *Appl. Phys. Lett.* **49**, 1686–1688.

Tsang, W. T., Logan, R. A., and Ditzenberger, J. A. (1982). Ultra low threshold, graded index waveguide, separate confinement, CW buried heterostructure lasers. *Electron. Lett.* **18**, 845–847.

Tsang, W. T., Olsson, N. A., and Logan, R. A. (1983). High-speed direct single-frequency modulation with large tuning rate and frequency excursion in cleaved-coupled-cavity semiconductor lasers. *Appl. Phys. Lett.* **42**, 650–652.

Tsang, W. T., Chiu, T. H., Kisker, D. W., and Ditzenberger, J. A. (1985). Molecular beam epitaxial growth of $In_{1-x}Ga_xAs_{1-y}Sb_y$ lattice matched to GaSb. *Appl. Phys. Lett.* **46**, 283–285.

Tucker, R. S. (1985). High speed modulation of semiconductor lasers. *J. Lightwave Technol.* **3**, 1180–1192.

Tucker, R. S., Gnauck, A. H., Wiesenfield, J. M., and Bowers, J. E. (1987). 8 Gb/s return-to-zero modulation of a semiconductor laser by gain switching. Paper WK4, in *Conference on Optical Fiber Communication/International Conference on Integrated Optics and Optical Fiber Communication Technical Digest Series 1987*, Vol. 3, (Optical Society of America, Washington, DC 1987), 178.

Turley, S. E. H., Henshall, G. D., Greene, P. D., Knight, V. P., Moule, D. M., and Wheeler, S. A. (1981). Properties of inverted rib-waveguide lasers operating at 1.3 μm wavelength. *Electron. Lett.* **17**, 868–870.

Uomi, K., Ohtoshi, T., and Chinone, N. (1986). Ultra high relaxation oscillation frequency (50 GHz) in modulation doped multiquantum well (MD-MQW) lasers: theoretical analysis. Paper M-6, Tenth IEEE International Semiconductor Laser Conference, Kanazawa, Japan, October 14–17, 1986.

Utaka, K., Akiba, S., Sakai, K., and Matsushima, Y. (1986). λ/4-Shifted InGaAsP/InP DFB lasers. *IEEE J. Quantum Electron.* **QE-22**, 1042–1051.

van der Ziel, J. P. (1981). Active mode locking of double heterostructure lasers in an external cavity. *J. Appl. Phys.* **52**, 4435–4446.

van der Ziel, J. P. (1985). Mode locking of semiconductor lasers. In "Semiconductors and semimetals: lightwave communications technology." (R. K. Willardson and A. C. Beer, eds., Vol. 22B, W. T. Tsang, Vol. ed.), Academic Press, London, 1–68.

Welch, D. F., Cross, P. S., Thornton, R. L., Burnham, R. D., and Paoli, T. L. (1987). High power diode laser arrays with silicon impurity induced disordered nonabsorbing mirrors. Paper ME3, in *Conference on Optical Fiber Communication/International Conference on Integrated Optics and Optic Fiber Communication Technical Digest Series 1987*, Vol. 3, (Optical Society of America, Washington, DC 1987), 18.

Westbrook, L. D. (1985). Dispersion of linewidth-broadening factor in 1.5 μm laser diodes. *Electron. Lett.* **21**, 1018–1019.

Woodall, J. M., Rupprecht, H., and Pettit, G. D. (1967). Efficient electroluminescence from epitaxially grown $Ga_{1-x}Al_xAs$ p-n junctions. *IEEE Trans. Electron. Devices* **ED-14**, 630.

Wyatt, R., and Devlin, W. J. (1983). 10 kHz linewidth 1.5 μm InGaAsP external cavity laser with 5.5 nm tuning range. *Electron. Lett.* **19**, 110–112.

Yamada, M., and Suematsu, Y. (1979). A condition of single longitudinal mode operation in injection lasers with index-guiding structure. *IEEE J. Quantum Electron.* **QE-15**, 743–749.

Yamada, T., Yuasu, T., Uchida, M., Asakawa, K., Sugata, S., Takado, N., Kamon, K., Shimizu, M., and Ishii, M. (1986). Fabrication and characteristics of dry-etched-cavity GaAs/AlGaAs MQW laser. Paper PD-4, Tenth IEEE International Semiconductor Laser Conference, Kanazawa, Japan, October 14–17, 1986.

Yamamoto, Y. (1983). AM and FM quantum noise in semiconductor lasers—Part I: theoretical analysis. *J. Quantum Electron.* **QE-19**, 34–46.

Yamamoto, Y., Saito, S., and Mukai, T. (1983). AM and FM quantum noise in semiconductor lasers —Part II: comparison of theoretical results for GaAlAs lasers. *J. Quantum Electron.* **QE-19**, 47–58.

Yano, M., Nishitani, Y., Hori, K., and Takusagawa, M. (1983). Temperature characteristics of double-carrier-confinement (DCC) heterojunction InGaAsP (λ = 1.3 μm)/InP lasers. *IEEE J. Quantum Electron.* **QE-19**, 1319–1327.

Yuasa, T., Yamada, T., Uchida, M., Asakawa, K., and Ishii, M. (1986). Observation of 24 GHz relaxation oscillation frequency in short cavity GaAs/AlGaAs multiquantum well lasers using streak camera. Paper PD-1, Tenth IEEE International Semiconductor Laser Conference, Kanazawa, Japan, October 14–17, 1986.

Chapter 14

Optical Detectors
for Lightwave Communication

S. R. FORREST

University of Southern California, Los Angeles, California

14.1 INTRODUCTION

During the past several years, enormous strides have been made in developing optical detectors for lightwave communications systems (Stillman *et al.*, 1982; Forrest, 1986). In general, the research efforts have been directed toward fabricating devices that, when installed in receiver amplifiers, have a high sensitivity to weak optical signals, and also can be operated at very high bandwidths such that the detecting element is rarely the bandwidth "bottleneck" in a particular system. Secondary areas in which significant progress has also been made is in fabricating low-cost devices, as well as devices (such as arrays) that can be advantageously employed in high fiber density networks such as long haul trunks, local area networks and photonic system interconnects. In this chapter we are concerned with photodetectors used for detecting digital data streams transmitted in optical communication systems. At present, these systems operate in two transmission "windows" determined by the low loss of silica fibers. The short wavelength window lies between 0.8 to 0.9 μm, and uses Si, or less commonly GaAs, as the light detecting material. Typically, the fiber loss in this region is relatively high at 2–3 dB/km; therefore, short wavelength communication systems are optimally employed for high density local area network and computer "data linking" applications. These cost-sensitive applications thus rely on the comparative low cost and availability of high quality GaAs substrates for their success. On the other hand, very long-haul, high capacity systems operate at 1.3 μm and 1.55 μm where the fiber loss is between only 0.5 and 0.25 dB/km, and the dispersion can be at or near zero. Such systems employ $In_{0.53}Ga_{0.47}As$ as the detecting semiconductor and have been demonstrated to operate with 200 km between repeaters in prototype system experiments.

We will discuss the three major categories of photodetectors considered for use in optical communication systems. These are photoconductors, p-i-n photodetectors and avalanche photodetectors (APDs). The demands placed on these detectors are several and include the need for very high sensitivity, bandwidth, and reliability, and they must be manufacturable at relatively low cost. To date, most short wavelength systems employ p-i-n photodiodes and APDs, whereas at long-wavelengths only the former detector has found widespread use due to the extreme difficulties met in fabricating APDs, and the inherent shortcomings of photoconductors for these applications. In the sections that follow, we will discuss each of these various devices in terms of their noise (and hence their sensitivity) and bandwidth characteristics. We will also discuss photodetector structures that are being developed for use in future generation optical communication and signal processing systems.

14.2 PHOTOCONDUCTORS

After having been ignored for many years for lightwave communication applications, there has recently been a renewed interest in photoconductors, particularly for use at long-wavelengths due to the suitability of some III-V semiconductors for photoconductor applications (Chen *et al.*, 1984a, 1984b; Forrest, 1985). In addition, these particular devices demonstrate in a simple manner the relationship between gain and bandwidth, which presents a fundamental limitation to the performance of all photoconductor structures.

Figure 14.1 shows a prototypical photoconductor for use in the long-wavelength transmission region. The conducting channel consists of a thin layer of $In_{0.53}Ga_{0.47}As$, which can absorb radiation with wavelengths as long as 1.65 μm. The conducting layer is approximately 1–2 μm thick, allowing for the absorption of a significant fraction of the incident light. The channel can be either p-type, semi-insulating, or n-type. As the best sensitivities at reasonably high bandwidths have been reported using n-type channel layers, in this discussion we will henceforth assume that the photoconductive channel is in fact very lightly doped ($< 5 \times 10^{14}$ cm^{-3}) n-type $In_{0.53}Ga_{0.47}As$. As will be shown next, however, the thickness or the doping of the layer must not be too great, thereby resulting in low channel resistance, and hence high noise. Since the absorption process is exponential, the internal quantum efficiency of a photoconductor, i.e., the ratio of carrier pairs generated per incident photon, is

$$\eta = \left(1 - \exp(-\alpha d)\right) \cdot 100\%. \tag{14.1}$$

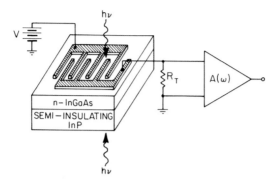

Fig. 14.1 Prototypical photoconductor receiver for use in long-wavelength (1.3 μm and 1.55 μm) communication systems. Here we show the bias circuit in conjunction with a receiver amplifier with a frequency (ω) dependent transfer function, $A(\omega)$.

Here, α is the absorption coefficient of the particular semiconductor used, and d is the channel thickness. In $In_{0.53}Ga_{0.47}As$ and other direct band gap semiconductors, the absorption coefficient is $\alpha > 10^4 \, cm^{-1}$—thus, a 2 μm thick channel results in $\eta = 88\%$. Note that the particular composition of $In_{0.53}Ga_{0.47}As$ chosen is lattice-matched to the semi-insulating InP substrate to avoid the formation of dislocations and other crystalline imperfections in the epitaxial layer. There are several techniques commonly employed to grow this compound, the most common being liquid phase epitaxy (LPE), molecular beam epitaxy (MBE), and the various forms of vapor phase epitaxy (VPE). Of these methods, VPE has been shown to produce the most uniform layers of $In_{0.53}Ga_{0.47}As$ with the highest purity (and therefore highest resistivity). In fact, free carrier concentrations as low as $3 \times 10^{14} \, cm^{-3}$ have been reported for chloride transport-grown VPE $In_{0.53}Ga_{0.47}As$ (Cox et al., 1985; Brown et al., 1986); in contrast to typical doping densities of greater than $8 \times 10^{14} \, cm^{-3}$ obtained via LPE, and $3 - 5 \times 10^{15} \, cm^{-3}$ typical of MBE, hydride transport VPE and metallographic VPE. Thus, chloride-transport VPE appears to be the most ideally suited means of growing simple photodetector structures where high purity and uniformity over large substrate areas are critically important.

To fabricate the photoconductor, low resistance contact is made to the conducting layer by the use of interdigitated cathodes and anodes that are designed to maximize coupling of the light into the absorbing region by obstructing as little of the active area as possible, while minimizing the distance that photogenerated carriers must travel before being collected at one of the electrodes. Furthermore, optical coupling efficiency can be maximized by applying an antireflection coating to the surface facing the optical input. One important advantage of using $In_{0.53}Ga_{0.47}As$ as the absorbing material is that it is grown on wide band gap InP substrates, which are transparent to long-wavelength (> 0.95 μm) radiation. Thus, light can be coupled into the photoconductive channel via the substrate, thereby avoiding excess loss of top-coupled light, which is incident on the cathode and anode metallization. It should be noted, however, that even light incident via the substrate in the vicinity of the electrodes will be less efficiently coupled than light incident between the electrodes, since the electric field in the former region is at or near zero. Thus, carriers generated in the low field regions cannot be separated prior to recombining, thereby resulting in a loss of detector efficiency.

In the dark, the conductivity of the device is determined by the number of charges present between cathode and anode. When the channel is illuminated, some of the light incident on the channel region is absorbed, thereby generating electron-hole pairs. The photogenerated carrier densities also contribute to, and thus increase, the channel conductivity. This results in an increase in current in the receiver circuit. Thus, an optical receiver must be sensitive to small changes in resistance induced by an incident light beam. For example, Fig. 14.1 illustrates an optical receiver consisting of a photoconductor placed in series with a bias resistor. This voltage division network is then connected to the input of a voltage preamplifier followed by a postamplifier, and an equalizer for the re-establishment of a square wave signal which might have undergone integration at the photoconductor and preamplifier stages. The equalizer is then followed by a filter to limit the noise bandwidth of the receiver (Smith and Personick, 1982), and the last stage is a decision circuit for recovering the input waveform and increasing its signal-to-noise ratio. Note that although the circuit in Fig. 14.1 is applied to the case of a photoconductor receiver, in practice, a similar generic receiver is used regardless of the details of the photodetector.

Thus far, the discussion has been confined to the steady-state performance of the device. Of considerably greater interest for optical communications is the response of the photoconductor to a very high frequency stream of optical pulses representing information transmitted in a digital format of "ones" (light on) and "zeroes" (light off). Clearly, the faster the device can respond to such a bit stream, the higher the density of the transmitted data, thereby increasing the information density and capacity of the transmission system. To understand the frequency response of the photoconductor, we must examine what happens to a single electron-hole pair optically generated in the region between the cathode and anode. When the photoconductor is placed in the receiver circuit, there is a small voltage present between cathode and anode due to the existence of the battery. Once the carriers have been generated, the electron is swept toward the anode, while the hole moves toward the cathode. This movement of carriers results in the detected photocurrent and will persist until both carriers are collected at the electrodes, or until they recombine in the bulk of the semiconductor. Thus, the *minimum* time for detection of the onset of the photogenerated current (i.e., the photoconductor rise time) is limited by the transit time between electrodes of the fastest charge carrier. In general, the mobility (μ_p) of holes is considerably smaller than that of electrons (μ_n), with the difference in carrier mobilities being most pronounced in such III-V materials as GaAs and $In_{0.53}Ga_{0.47}As$. Thus, the rise time of a photoconductor can be no shorter than the electron transit time (t), given by $t = L^2/\mu_n V$, where L is the distance between cathode and anode, and V is the applied voltage. From this expression, it is evident that the highest speed response is obtained by minimizing the distance that the carriers must travel between electrodes.

While the electron is collected almost immediately at the anode, the slow holes continue to proceed across the channel. The absence of the fast carrier results in a net positive charge of the channel region due to the continued presence of the photogenerated hole. This excess charge is immediately compensated by the injection of a second electron from the cathode into the channel region. In short, two electrons have now circulated through the circuit, although only a single photon has been absorbed. This addition, or excess current, is photocurrent gain. The fast carriers are drawn into the channel until the excess hole either recombines with an electron or is collected at the cathode. Thus, the photoconductive gain (G) is simply the ratio of the slow carrier lifetime (or transit time, depending upon which is the shortest), τ, to the fast carrier transit time, t. Thus,

$$G = \frac{\tau}{t}. \qquad (14.2)$$

Since the current response persists a time τ after the end of the optical pulse, the maximum bandwidth of the device is $B = 1/2\pi\tau$.

Now, the basic relationship that expresses the speed of response of a photodetector is the gain-bandwidth product, given by

$$G \cdot B = 1/(2\pi\tau). \qquad (14.3)$$

Equation (14.3) implies that in a given device, where t is fixed, photoconductive gain can only be obtained at the expense of response speed. That is, one cannot design a device that has arbitrarily large gain (and therefore sensitivity) along with an arbitrarily large frequency response. Rather, these two properties are physically linked and form the fundamental sensitivity-frequency response tradeoff present in all photodetectors. Thus, although very large gains of 1000 or more can be observed in Si photoconductors, contrasted with gains of 50–100 for $In_{0.53}Ga_{0.47}As$ devices, this gain is accompanied by a concomitant reduction of bandwidth. Typically, the response time ranges between 1 μs and 1 ms for Si photoconductors. On the other hand, response times of 1–2 ns can be achieved with $In_{0.53}Ga_{0.47}As$, making these devices considerably more attractive for moderate bit rate (< 500 Mb/s) applications. The higher speed performance of this latter material is due to its very high room-temperature mobility (10,000 cm^2/V-s) and peak velocity (with $v_p = 2.5 - 3 \times 10^7$ cm/s) as compared with Si (where $\mu = 1000$ cm^2/V-s and $v_p = 1 \times 10^7$ cm/s).

Finally, in considering the usefulness of a particular device for a given application, we must consider the sensitivity, or signal-to-noise performance of the device. It has been shown (Forrest, 1985) that there are numerous sources contributing to the noise generated in photoconductors. The dominant source, however, arises from the finite dark conductivity of the device generating a randomly fluctuating background dark current. This so-called time-averaged mean square Johnson noise current is given by $\langle i_n^2 \rangle = 4kTB_{eff}/R_T$, where k is Boltzmann's constant, T is the absolute temperature, B_{eff} is the effective receiver bandwidth, and R_T is the total input resistance of the amplifier.

In considering the noise performance of the preamplifier-photoconductor combination, it is useful to model the input as a simple voltage divider network consisting of the photoconductor (with dark resistance, R_D) in series with a bias resistor, R_F (see Fig. 14.1). Then, the signal (or output) voltage of this network induced by a reduction of photoconductor resistance, dR_D, due to the generation of excess carriers is simply:

$$\frac{dv_o}{dR_D} = \frac{V_{DD}R_F}{\left(R_F + R_D\right)^2} \tag{14.4}$$

Here $dR_D \ll R_D$ has been assumed, and V_{DD} is the supply voltage. Now, taking R_T as the parallel combination of R_D and R_F, and recognizing that the signal current is $di_s = dv_o/R_F$, it can be shown that the signal-to-noise ratio, S/N is maximized when $R_F = 1/4R_D$. Note that this condition is considerably different than that usually employed in photoconductive receiver experiments, where, generally, R_F is allowed to be much greater than R_D. In our analysis, we will assume the receiver has a "matched" input load, i.e., $R_F = R_D$ and calculate the total noise accordingly. The reason for employing this simplification is that it closely resembles what has been done in laboratory experiments. Later we will compare our calculation to actual data.

Now, the signal current (di_s) of the photoconductor is simply equal to the gain times the primary photogenerated current (i_o) determined by the time average power ($\bar{P}_o$) of the incident optical signal. Thus, the signal-to-noise ratio is

$$S/N = \frac{\left(\dfrac{q\eta\lambda G\bar{P}_o}{hc}\right)^2}{\langle i_n^2 \rangle + \langle i_a^2 \rangle}, \tag{14.5}$$

where the noise current due to the receiver preamplifier ($\langle i_a^2 \rangle$) has been included. In this equation, we have let $i_o = q\eta\bar{P}_o\lambda/hc$, where λ is the wavelength of the incident photon, q is the electronic charge, and hc is Planck's constant (h) times the speed of light in vacuum (c). Note that the signal-to-noise ratio of the photoconductor receiver increases with increasing channel resistance and gain. Thus, to increase the sensitivity of the receiver, we must increase the gain of the channel at the expense of frequency response, as discussed previously. These material-dependent tradeoffs inherent in the photoconductor necessarily limit its ultimate usefulness to low and moderate bit rate applications. Although photoconductors are characterized by relatively large gains, the gain is usually insufficient to fully offset the dark noise ($\langle i_n^2 \rangle$) to the point where a significant improvement can be achieved over the sensitivity of reverse-biased detectors such as the p-i-n photodiode and the APD. We will have more to say on this topic in the section on sensitivity comparison between detector types.

14.3 p-i-n PHOTODETECTORS

The most commonly employed photodetector in long-wavelength optical communication systems is the p-i-n photodiode. This reverse-biased device is characterized by the relative ease with which it can be fabricated, its extremely high reliability, low noise, and compatibility with low voltage amplifier circuits. In addition, the p-i-n photodiode is an extremely high bandwidth device.

In contrast to the photoconductor, the p-i-n detector has a p-n junction which is operated under reverse bias. Due to the presence of the moderate electric fields (roughly 1×10^5 V/cm) in the vicinity of the p-n junction, the region is depleted of free carriers; thus, no large dark currents flow in the absence of light. Nevertheless, very small, randomly varying currents generated thermally or by tunneling breakdown of the material give rise to a background, or noise current to be discussed below. A photon incident on the depletion region can be absorbed, thereby generating an electron and a hole, which are rapidly drawn to the electrodes where they are collected and appear as current in the external circuit. The separation of the carrier pair occurs on a time scale very short compared with its recombination time, and therefore, the detection process is intrinsically efficient and fast.

The ultimate bandwidth of the device is limited by the time (t) it takes for the charges to be collected. Assuming a depletion region of width W and a saturated drift velocity of v_{sat}, the bandwidth is:

$$B = \frac{1}{2\pi t} = \frac{v_{sat}}{2\pi W}. \tag{14.6}$$

Furthermore, the maximum possible efficiency of this device is unity, since there is no gain mechanism present. Thus, the gain-bandwidth product of the p-i-n detector is simply equal to its bandwidth as given in Eq. (14.6).

Whereas the transit time of the p-i-n photodetector is generally limited by the high-field, *saturated* carrier velocity of the *slowest* carrier, the transit time of the photoconductor (optimally operated at low voltage and therefore low electric field) is determined by the *peak* velocity of the *fastest* carrier. This subtle distinction is unimportant in Si photodetectors, where the saturated and peak velocities of both electrons and holes are equal. For III-V materials such as GaAs and $In_{0.53}Ga_{0.47}As$, however, the peak velocity of electrons is much larger than the velocity of holes (by a factor of 2 to 3), and the peak velocity of electrons obtained at low electric fields is considerably larger than their high electric field saturated velocities. Thus, the gain-bandwidth product of photoconductors made from these important III-V materials can be significantly higher than for p-i-n detectors. Nevertheless, the absence of gain in the latter device generally results in much higher operating bandwidths. Indeed, the bandwidth of efficient $In_{0.53}Ga_{0.47}As$ p-i-n photodiodes has been found to be greater than 20 GHz (Bowers, 1985) as compared to 100MHz typical of fast photoconductors, and is usually limited in the p-i-n detector by such extrinsic effects as the RC time constant of the detector and receiver circuit. If the i-region is allowed to be narrow ($\sim$ 0.5 μm), frequency responses as high as 67 GHz have been reported (Tucker, 1986); although at these layer thicknesses a significant loss of quantum efficiency is expected.

Now, R is the effective resistance of the external bias circuit, and C is the photodiode capacitance due to the removal of carriers from the depletion region of width W. Thus, for a photodiode of area, A, the capacitance is

$$C = \frac{\kappa_s A}{W}, \tag{14.7}$$

where κ_s is the permittivity of the semiconductor. Now, W is a function of applied voltage, V, and the concentration of free carriers (or net doping density, $|N_D - N_A|$) via:

$$W^2 = -\frac{2\kappa_s(V + V_{bi})}{q|N_D - N_A|}, \tag{14.8}$$

where V_{bi} is the junction built-in voltage. Equation (14.8) is valid for a p^+-i-n, abrupt junction photodiode such that both V_{bi} and V are negative under reverse bias operation.

Comparing Eqs. (14.7) and (14.8), it is clear that the capacitance decreases as the reverse bias (and hence the depletion region width) increases. Thus, a trade-off exists in very high bandwidth devices whereby the transit time (Eq. 14.6) is balanced against device capacitance (Eq. 14.7) to optimize high speed operation.

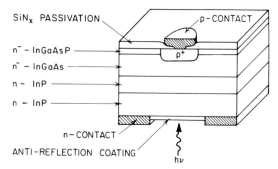

Fig. 14.2 Schematic cross-sectional view of a back-illuminated, planar p-i-n photodiode for use in the wavelength range 0.9 μm $< \lambda < 1.65$ μm.

In Fig. 14.2, we show a schematic cross-section of a typical *p-i-n* photodiode fabricated using $In_{0.53}Ga_{0.47}As$ as the light absorbing material. As in the case of the photoconductor, the absorbing region is thick (typically 4–5 μm) and is generally an *n*-type material grown on an *n*-type InP substrate. Sometimes a thin layer of a lattice-matched composition of wider band gap InGaAsP is grown on top of the narrow band gap $In_{0.53}Ga_{0.47}As$ to reduce dark currents generated on the surface of the latter material (Kim *et al.* 1985). Next, in this planar structure, the top surface is coated by a thin ($\sim$ 1000 Å) insulating layer. After a hole is etched into the insulator, an acceptor atom such as Cd or Zn is diffused at high temperature into the absorbing region, thereby forming the *p*-type region and the *p-n* junction. Thus, the insulating layer acts as a diffusion mask to confine the extent of the *p-n* junction. In addition, it has been found that when SiN_x is used for this layer, it also "passivates" the *p-n* junction region where it intersects the diode surface. In this way, the high electric field junction region is protected from environmental variations, rendering the detector highly stable and reliable over extended periods of time (Saul *et al.*, 1985).

Finally, ohmic metal contacts are deposited on the front and rear surfaces of the device. Light is generally coupled into the absorbing region through a window in the transparent InP substrate contact metallization. Substrate, or rear-illumination, has two advantages over illumination incident via the *p*-region: (1) The device can be made smaller since the junction area can be matched to the size of the incident light beam. For top-illuminated structures, the top contact must have a hole to admit light, and thus must necessarily be larger than the light source; and (2) The coupling efficiency of the light to the diode can be near unity, i.e., nearly one electron-hole pair is collected for each incident photon. This results in photodiodes where the depletion region extends all the way to the $In_{0.53}Ga_{0.47}As/InP$ interface under reverse-biased operation, and thus all photocarriers are generated far from any surfaces where they might recombine. In contrast, the diffused p-region contains a very high density of free holes, which are not depleted by the applied field. Since no electric field exists in the p-region, photogenerated carriers are not separated and collected; thus they recombine leading to a loss in detector efficiency.

It is apparent from the discussion that high efficiency, high bandwidth operation is achieved by a combination of illumination through the substrate, and complete depletion of a thick absorbing region. This latter requirement implies that the background carrier concentration ($|N_D - N_A|$) be as small as possible in the absorbing region such that it can be fully depleted of free carriers at the low voltages ($< 5\text{V}$) available in digital receiver circuits. Furthermore, low-voltage operation is essential if the background current arising from electric-field-generated processes such as tunneling are to be minimized. Using Eq. (14.1), we find that approximately 98% of the incident radiation is absorbed in 4 μm of material. Thus, for high-efficiency, approximately 4 μm of depleted material must exist between the p-region and the $In_{0.53}Ga_{0.47}As/InP$ interface.

Typical dark current characteristics for an $In_{0.53}Ga_{0.47}As$ p-i-n photodiode are shown in Fig. 14.3. Data are indicated by closed circles. At low voltage, in the region where the detector is generally operated, the dark current increases only gradually with reverse bias, characteristic of thermally generated current.

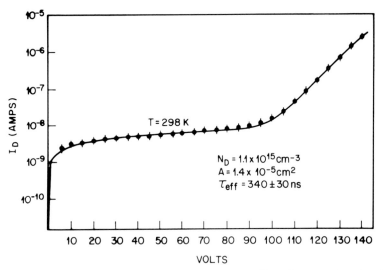

Fig. 14.3 Typical measured dark current characteristics of an $In_{0.53}Ga_{0.47}As$ p-i-n photodiode (closed circles). The solid line indicates a theoretical fit to the data where, at low voltage, the dark current is assumed to be due to thermally generated recombination, and at high voltage it is due to tunneling (from Forrest et al., 1980).

The thermal generation can occur both at the diode surface or in the bulk active region. As voltage is increased, the dark current then begins a nearly exponential increase characteristic of carrier tunneling between states in the valence and conduction bands. This region of reverse bias is the tunneling breakdown region. Due to the high dark currents, the noise of the photodiode is excessive at high voltage and is therefore not useful for high sensitivity detection applications. A theoretical fit to the data assuming low voltage thermal current generation followed by high voltage tunneling is indicated by the solid line in the figure (Forrest *et al.*, 1980).

The interrelationship between voltage, capacitance, quantum efficiency and the free carrier concentration is shown for $In_{0.53}Ga_{0.47}As$ *p-i-n* photodiodes in Fig. 14.4. Also shown by the diagonal line is the breakdown voltage (V_B), which is seen to decrease with increasing carrier concentration. The shaded region indicates the voltages at which tunneling leakage becomes dominant, thereby rendering the device useless. It is apparent that the requirements placed on the device by the several performance parameters restricts the net doping concentration of usable materials to values less than 3×10^{15} cm^{-3}. In practice, $In_{0.53}Ga_{0.47}As$ can be obtained with doping densities of 5×10^{14} cm^{-3}, provided

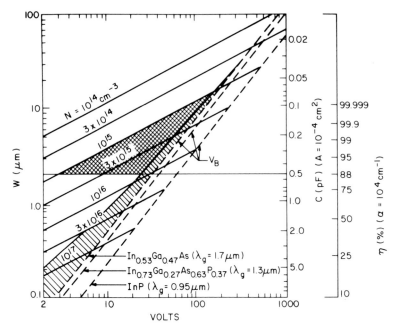

Fig. 14.4 The dependence of depletion region width, breakdown voltage, diode capacitance and internal quantum efficiency on applied voltage. Free carrier concentration is a parameter.

great care is used in the growth process. Thus, very high performance p-i-n photodiodes are routinely fabricated using this material system, making the p-i-n photodiode the most frequently employed device for long wavelength communication systems.

Silicon p-i-n detectors are similar in structure to that shown in Fig. 14.2, although light must be incident via the thin diffused layer on the top surface of the device. In addition, since the absorption coefficient of Si at 0.8 μm is roughly 10 times less than for III-V materials (Melchior, 1972), wider depletion layers and lower dopings are required in the active, absorption region of the detector.

The noise performance of the p-i-n detector is limited at low bit rates by shot noise in the reverse-biased junction region. In this case, the time-averaged mean-square noise current of the detector is: $\langle i_n^2 \rangle = 2qI_D B_{eff}$, where I_D is the reverse-biased dark current. Note that this shot noise is generally several orders of magnitude smaller than Johnson noise obtained in photoconductors. Thus, although the p-i-n detector signal is not enhanced by gain, the device is more sensitive due to better coupling with the light beam and a lower background noise contribution.

At higher bit rates (> 200Mb/s), it can be shown that the noise is dominated in FET amplifiers commonly employed in optical receivers by noise generated in the channel of the front-end transistor. This noise current is given by (Smith and Personick, 1982):

$$\langle i_n^2 \rangle \simeq 4kT\Gamma \left(\frac{4\pi^2 C_T^2}{g_m} \right) B_{eff}^3 \qquad (14.9)$$

Here Γ is the noise figure of the FET ($\simeq 1.1$–1.5 for GaAs), C_T is the total front end capacitance consisting of the parallel combination of the photodiode, front-end FET and other parasitic capacitance, and g_m is the transconductance of the front-end FET. Clearly, the channel noise current increases rapidly at high bit rates (due to its dependence on B_{eff}^3) reducing the sensitivity of p-i-n receivers at these frequencies.

The general reduction of receiver sensitivity with increased bit rate can be readily understood if we recognize that most direct detection receivers require roughly 1000 photons/bit to achieve a BER of 10^{-9}. Thus, as bit rate is increased, the bit time slot must be decreased. To maintain a constant 1000 photons/bit, the bit amplitude must be correspondingly increased with bit rate. In this manner, the mean minimum detectable power, $\eta \bar{P}_o$, is increased, thereby degrading receiver sensitivity, as observed.

For the p-i-n detector, the S/N ratio is given by Eq. (14.5) used for photoconductors, where we employ the expressions above for the noise current, and recall that $G = 1$ for the p-i-n photodiode.

14.4 AVALANCHE PHOTODETECTORS

Avalanche photodetectors (APDs) are similar to p-i-n photodiodes in that they are operated under reverse bias, and therefore in the absence of large background dark currents. Unlike p-i-n photodiodes, however, APDs are operated at sufficiently high reverse voltages such that photocurrent gain due to impact ionization of carriers with the lattice atoms occurs. An illustration of the avalanche process is illustrated in Fig. 14.5. Here, a single electron-hole pair is shown to be generated at one edge of the depletion region. The hole is immediately collected at one electrode of the device, while the electron is accelerated in the electric field toward the counter electrode. For avalanche, or photocurrent gain to occur, the field must be high enough such that the energy gained by the electron in a given distance must be greater than that lost due to the emission of optical phonons, which constitute the principal mechanism of energy loss. At this point, the electron has gained sufficient energy such that it undergoes an ionizing collision with a lattice atom, thereby generating a secondary electron-hole pair. These secondary carriers, along with the initial primary carrier, can then undergo additional ionizing collisions such that large numbers of charges are collected at the electrodes of the device. In effect, the APD is similar to a photomultiplier, where the photogeneration of a single charge carrier in a high electric field region results in a shower, or avalanche of secondary carriers.

The process of ionization is exponentially dependent on the magnitude of the electric field, E. Thus, the ionization coefficient for electrons, which is defined as the inverse of the mean distance between ionizing collisions, is given by: $\alpha' = A \exp(-[b/E]^m)$ where A and b are constants, and m is greater than or equal to 1. There is a similar expression for the ionization coefficient for holes (β'), where the values of A, b and m are in general different from those of electrons. Another feature of the avalanche process is, therefore, that it is asymmetrical, i.e., the probability of initiating an avalanche is different for holes than it is for electrons. This asymmetry is expressed by the ionization coefficient ratio, k, whereby:

$$k = \frac{\alpha'}{\beta'} \qquad (14.10)$$

Depending on structure, the actual k-value, or "effective k-value", k_{eff} can differ significantly from that obtained via Eq. (14.10). For example, in InP-based devices, where holes are more ionizing than electrons (i.e., $\beta' > \alpha'$), and in devices designed such that avalanche is initiated by photogenerated holes, the "effective" k-value is defined as $k_{eff} = \beta'/\alpha' = 1/k$. In fact, in properly de-

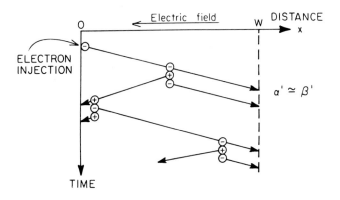

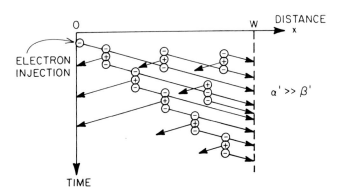

Fig. 14.5 Schematic time evolution of the avalanche process in (a) materials where the electron (α') and hole (β') ionization coefficients are roughly equal and (b) where $\alpha' \gg \beta'$. Here W indicates the depletion region width. The top diagram is typical of avalanche occurring in Ge and GaAs, whereas the bottom corresponds to the case of Si. In InP, $\beta' \simeq 2\alpha'$ (after G. E. Stillman and C. M. Wolfe, 1977).

signed APDs where avalanche is initiated by the most ionizing carriers, the effective k-value can always be taken to be less than or equal to 1.

The gain process is inherently probablistic. That is, the probability that an ionization occurs within a given distance, l, is exponentially dependent on the product $\alpha'l$ or $\beta'l$, depending on the carrier type. As in any statistical process, there are random fluctuations in the actual distance between successive ionizing collisions. This gives rise to fluctuations in the total number of secondary carriers generated per primary photogenerated carrier injected into the gain region; therefore, there is randomness, or noise, in the total signal current. The magnitude of the fluctuations depends on the mean avalanche gain, $\langle M \rangle$, and is minimized in structures in which avalanche is initiated by the carrier with the highest ionization rate. In effect, the noise associated with ionization is shot noise due to the signal and dark currents, increased by the excess noise factor, $F(M)$. Now $F(M)$ is a monotonically increasing function of the gain (McIntyre, 1966), and is largest for k-values close to unity. The shot noise current in an APD receiver is:

$$\langle i_n^2 \rangle = 2q \langle M^2 \rangle F(M) \left(I_{dm} + \frac{q \eta \lambda \bar{P}_0}{hc} \right) B_{eff}, \qquad (14.11)$$

where I_{dm} is the primary, unmultiplied bulk dark current of the APD. Thus, the total signal-to-noise ratio of the APD is:

$$S/N = \frac{\left(\dfrac{q \eta \lambda \bar{P}_0 \langle M \rangle}{hc} \right)^2}{\langle i_n^2 \rangle + \langle i_a^2 \rangle}. \qquad (14.12)$$

From Eqs. (14.11) and (14.12), it is apparent that the square signal current, which is proportional to $\langle M \rangle^2$, increases less rapidly than the mean square noise current, proportional to $\langle M \rangle^2 F(M) \simeq \langle M \rangle^x$, where $x \geq 2$. The gain process is useful in increasing the S/N of the optical receiver only because of the finite noise of the receiver amplifier, $\langle i_a^2 \rangle$. Thus, S/N for the APD receiver is maximized at the value of gain where $\langle i_n^2 \rangle$ is equal to the amplifier noise. The mean optimum gain, M_{opt} at which this occurs depends, of course, on how rapidly $F(M)$ increases with M.

It is apparent that the fluctuations in photocurrent must be larger in materials where the probability of ionization is approximately the same for both electrons and holes (Fig. 14.5a). In this case, where $F(M) \sim M$, which is typical of GaAs, InP and Ge, a large shower of carriers of both types is generated during an avalanche, since both secondary holes as well as electrons can generate further secondary carriers. Thus, a large fluctuation in the magnitude of the photocurrent can occur depending on whether or not a secondary carrier is generated during the traverse of a primary carrier across a given distance in the depletion region. On the other hand, in some materials such as Si, electrons have a much higher probability (by a factor of $\simeq 50$) than holes of undergoing an ionizing collision (Fig. 14.5b). In this case, although large numbers of secondary carriers can be induced by a single photogenerated electron, the avalanche is less regenerative, and thus, statistical fluctuations in the collected current are considerably smaller than in cases where $k_{eff} \simeq 1$. Since the excess noise factor is smaller for materials with smaller k_{eff}-values, such devices, can be used at higher values of gain than detectors made using large k_{eff}-value materials.

One further consideration of avalanche gain is its effect on receiver bandwidth. It can be shown that the bandwidth of an avalanche detector is ultimately limited by the effective transit time, $\tau_1 = \gamma k_{eff} t$, where γ is a constant of proportionality, and t is the carrier transit time across the depletion region determined in a manner similar to that used in the case of p-i-n detectors (Kaneda et al., 1976). The bandwidth of the device is determined by the avalanche buildup time (τ) and is given by: $B = 1/2\pi\tau = 1/2\pi\tau_1 \langle M \rangle$ for $\langle M \rangle > 1/k_{eff}$.

The signal persists after the initiation of avalanche gain until some fluctuation in the carrier density in the depletion region terminates the process. In the case where there is only a single ionizing carrier type ($k_{eff} = 0$), the effective transit time vanishes, and the response time is limited by the duration of a single transit of the carrier across the depletion region. The worst case occurs when the ionization ratio approaches unity, resulting in a continuous recycling of secondary carriers through the depletion region, thereby inducing a very long response time accompanied by high gain. Now from the above, the gain-bandwidth product of an APD is given by:

$$\langle M \rangle \cdot B = 1/(2\pi\tau_1) \tag{14.13}$$

and is clearly maximized in devices and materials where the effective transit time is smallest. As discussed above, this is accomplished in structures where k is minimized.

A typical Si APD used in short wavelength communication systems is shown in Fig. 14.6. Here, the light is absorbed in p-type material, thus generating electrons that traverse the device while undergoing avalanche multiplication. Since the ratio of ionization coefficients of holes to electrons is $k_{eff} \simeq 0.02$, very low noise, high gain operation is achieved by initiating avalanche with the most ionizing (electron) carrier. Note also the presence of guard-rings at the edge of the p-n junction. These non-planar extensions of the n-type region reduce the electric field at the curved edges of the junction, thereby reducing edge-breakdown and the accompanying, highly noisy, non-uniform gain characteristic of simple junction profiles.

It is considerably more difficult to fabricate a high quality APD for use at long-wavelengths. Much of the difficulty arises from the need to use narrow band gap materials such as Ge or $In_{0.53}Ga_{0.47}As$ to absorb the relatively low energy, long-wavelength radiation. It has been found that many such narrow band gap semiconductors undergo tunneling breakdown at only moderate values of electric field ($< 2 \times 10^5$ V/cm) below the threshold for carrier

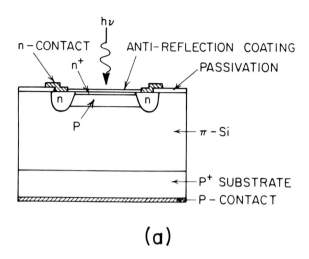

(a)

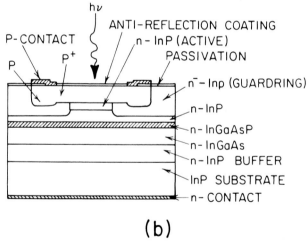

(b)

Fig. 14.6 Schematic cross-sectional views of (a) a planar Si avalanche photodiode (APD) for $\lambda \leq 0.9\ \mu$m and (b) a Separate Absorption and Multiplication region APD (SAM-APD) for use at $0.95\ \mu$m $< \lambda < 1.65\ \mu$m. Note that the multiplication region of the Si diode is p-π type, whereas it is n-type for the long-wavelength detector to optimize low noise performance.

multiplication (see Fig. 14.3). Thus, when using $In_{0.53}Ga_{0.47}As$ as the absorbing material, it has been necessary to fabricate heterostructure APDs of the type shown in Fig. 14.6b, where light is absorbed in a narrow band gap region consisting of $In_{0.53}Ga_{0.47}As$ in which the electric field never exceeds values that would induce significant amounts of tunneling leakage. Once generated, the holes are swept into the large band gap, InP layer. The p-n junction, and therefore the highest electric fields ($> 5 \times 10^5$ V/cm) are present in this latter region, and are sufficient for avalanche multiplication to occur. The structure is known as the SAM-APD denoting *s*eparate *a*bsorption and *m*ultiplication region APD. In InP, holes are more highly ionizing than electrons, and therefore hole injection is optimized by fabricating the device using n-type material.

It has been found that the large differences in band gaps between InP (with $\epsilon_g = 1.35$ eV) and $In_{0.53}Ga_{0.47}As$ (with $\epsilon_g = 0.75$ eV) results in an energy step of 0.4 eV at the valence band edge in the region of the interface between these materials (Forrest and Kim, 1981). Thus, holes photogenerated in the $In_{0.53}Ga_{0.47}As$ become trapped at the heterointerface where they either recombine, resulting in a loss of detector efficiency, or are thermionically emitted over the barrier into the multiplication region, where they undergo the normal gain process. This latter phenomenon results in extremely slow detector response (Forrest *et al.*, 1982), making a simple SAM-APD useless for moderate to high bit rate applications. The problem can be eliminated, however, by inserting a thin layer of InGaAsP whose band gap is intermediate between InP and $In_{0.53}Ga_{0.47}As$, as shown in Fig. 14.6b. Alternatively, it has been suggested that grading the composition of InGaAsP, and hence the bandgap, between that of the $In_{0.53}Ga_{0.47}As$ absorbing region and the InP gain region over a distance of a few hundred angstroms will have the same effect of reducing the effective energy barrier to hole emission, and hence speeding up the diode response (Forrest *et al.*, 1982).

One further technique that has been demonstrated to "speed up" the APD response is to interpose a thin multiple quantum well (MQW) structure between the narrow and wide bandgap layer (Capasso, 1985). This MQW region (Fig. 14.7) consists of several alternating layers of InP and $In_{0.53}Ga_{0.47}As$, where the ratio of widths of the InP to $In_{0.53}Ga_{0.47}As$ layers varies from near infinity at the InP multiplication region boundary, to zero at the $In_{0.53}Ga_{0.47}As$ side. In this way, a so-called "pseudo-quaternary" compound whose "effective bandgap" is determined by the local ratio of layer thicknesses, can be graded from that of InP to $In_{0.53}Ga_{0.47}As$ as required for high bit rate performance.

We note that of all the above techniques, the simplest to employ, and hence the most frequently demonstrated, is the interposition of only one or two layers of InGaAsP whose compositions lie intermediate between $In_{0.53}Ga_{0.47}As$ and InP (Matsushima *et al.*, 1982). Indeed, it has been found that devices with only two such layers can exhibit ideal gain-bandwidth limited response with no evidence for hole trapping.

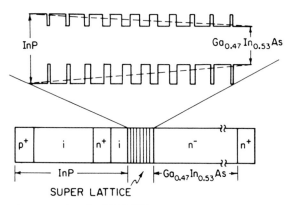

Fig. 14.7 Schematic cross-section of a SAM-APD with a "pseudo-quaternary" graded-gap super-lattice interposed between the multiplication and absorption regions (F. Capasso, 1985).

Typically, the gain-bandwidth product of these modified SAM-APDs is approximately 20 GHz, making them useful for receivers operating at roughly 2 GHz or less (Forrest, 1984). More recently, however, gain-bandwidth products as high as 60 GHz have been demonstrated (Holden et al., 1986) suggesting that such devices might find use at 5 GHz and higher. Caution should be used, however, in inferring that devices optimized for bandwidth are useful in practical applications. There are usually performance tradeoffs which must be made whenever a device is optimized for a single parameter. Thus, it remains undetermined whether the noise or optical sensitivity of such high bandwidth devices are acceptable for very high bit rate system applications.

Although high speed SAM-APD structures with guard rings similar to the one shown have been fabricated, they are extremely high-tolerance structures where slight deviations in layer thickness, uniformity, and carrier concentration can result in disasterously high leakage currents or small detection efficiencies (Kobayashi et al. 1984; Shirai et al., 1981). Thus, such devices are not yet generally available, nor have they been deployed to any significant extent in optical communication systems. In addition, there is little information regarding the reliability of these devices where high fields extend to the insulator/semiconductor interface at the surface. In general, such large surface fields increase dark current along with the potential for catastrophic breakdown of the device. Thus, to date, the most sensitive APDs for use at long-wavelengths have been mesa structures of questionable reliability.

One alternative means of fabricating long wavelength detectors is to use Ge as the absorbing material. This avenue has been heavily pursued with considerable success in Japan. Indeed, the $p^+ n n^-$ Ge APD has been shown to have very high quantum efficiency along with a high bandwidth (Yamada et al., 1982a). Nevertheless, Ge APD sensitivities are limited by large dark currents (approximately 10 to 100 times that of $In_{0.53}Ga_{0.47}As/InP$ APDs) and low detection efficiencies at $\lambda \geq 1.55$ μm. Thus, it is not expected that the very highest receiver sensitivities can be reached using Ge APDs. Considerable work has yet to be done before long-wavelength APDs can be extensively employed in practical communication systems.

14.5 COMPARISON OF PHOTODETECTOR SENSITIVITIES

To gain an appreciation of the relative sensitivities of the three basic photodetector types, it is useful to consider devices used in the same application. Thus, in this section we will compare the sensitivity, or the minimum detectable time-average signal power, of $In_{0.53}Ga_{0.47}As$ photoconductors, *p-i-n* photodiodes and APDs used in the same, hypothetical optical receiver operating at a wavelength of $\lambda = 1.3$ μm. Wherever possible, experimental and theoretical results will be compared.

The mean detectable coupled power, $\eta \bar{P}_o$, can be obtained from our various signal-to-noise expressions derived in the above treatment. Assuming Gaussian noise statistics, it can be shown that mistakenly identifying a "one" as a "zero," or vice versa, will occur once in the transmission of 10^9 bits of data if we set $S/N = 6$ (Smith and Personick, 1982). In general, this so-called bit error rate of $BER = 10^{-9}$ is sufficiently low for most system applications.

Using practical values for the various terms in the S/N expressions above, we then plot the theoretical sensitivities of the various detector types as a function of bit rate in Fig. 14.8. In generating the photoconductor curve, a channel and feedback resistance of 400Ω, a bandwidth of 80 MHz, a gain of 40, and a 50% optical filling factor (which relates the ratio of the area of the contact electrodes to the total area of the channel) are assumed. For the *p-i-n* photodetector, a total receiver front-end capacitance of 0.5 pF, and a dark current of 50 nA was used. Finally, the curve for the APD assumes $k = 0.5$ and is taken in the zero dark current ($I_{dm} = 0$) limit. Note that the APD line is only extended to 4 Gb/s, which is considered to be the practical range of response of gain-bandwidth limited SAM-APD structures. Experimental data for these various detectors are shown as points in the figure.

It is immediately apparent that the APD can give between 5 and 7 dB more sensitivity than either a *p-i-n* or a photoconductor if used within the appropriate bit rate range. Note that the reduction in sensitivity with increasing bit rate is due to the need for wider bandwidth (and therefore higher noise) receivers at the higher bit rates considered. On the other hand, the sensitivity difference between the *p-i-n* and photoconductor is not nearly as apparent except at low bit rates, where the high Johnson noise of the photoconductor makes it significantly less attractive than the *p-i-n* photodetector for applications where high sensitivity is

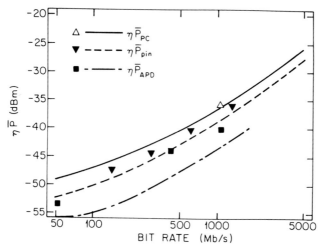

Fig. 14.8 Minimum time-average coupled optical power $(\eta \bar{P})$ versus bit rate calculated for the several detector types considered. Here, triangles and squares correspond to recent results reported in the literature. In these calculations, we assume $\lambda = 1.3$ μm and BER $= 10^{-9}$.

essential. At higher bit rates, the detectors appear to be nearly comparable, although extreme caution should be used in considering the operation of the two devices in this region. Recall that the bandwidth of the photoconductor is only 80 MHz, and its use at high bit rates can only be achieved by equalization of the amplified signal. Such techniques, although possible, greatly reduce other performance characteristics of the receiver such as dynamic range, and are therefore not desirable in many applications. Thus, the best combination of bandwidth and sensitivity for very high and low bit rates is accomplished with the *p-i-n* photodiode, while the APD is useful in the moderate bit rate range of 100 Mb/s < B < 2 Gb/s.

Perhaps a more useful comparison of these three detectors can be made if we consider how they would influence the design of an optical transmission system. One figure of merit by which to compare the devices is to calculate the distance between repeaters in a hypothetical long-wavelength system. Thus, in Fig. 14.9 we plot repeater spacing versus bit rate for the three detectors used in Fig. 14.8. In this plot, we assume that single-mode fiber is employed using a laser light source coupling -3 dBm of power into the fiber. Furthermore, a system margin of -15 dB is assumed, with the remainder of the available power being lost in the fiber. On the left-hand ordinate, we plot the repeater spacing for fiber with 0.5 dB/km loss, and on the right-hand ordinate a loss of 0.25 dB/km is assumed. These values bracket what can be achieved with state-of-the-art single-mode fiber. Once again, the relative sensitivities of the various detectors

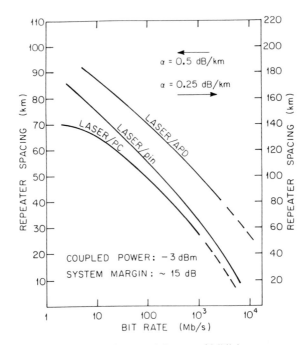

Fig. 14.9 Maximum repeater spacings for several "comparable" lightwave systems using different detector types. Here, two different fiber losses (α) are assumed.

are readily apparent, with a 15 km increase in repeater spacing achieved using an APD rather than a p-i-n, assuming a fiber loss of 0.5 dB/km and a bit rate of < 2 Gb/s. It is also apparent that the fractional increase in repeater spacing using the APD is greatest at the highest possible bit rates.

This plot can also be used to understand the improvement obtained in an existing transmission system, where the repeater spacing was determined at the time that the optical cables were installed. Thus, for example, a transmission system originally installed using a p-i-n photodetector operating at 100 Mb/s has a repeater spacing of 55 km (with a fiber loss of 0.5 dB/km). By replacing the p-i-n photodetectors with APDs as they become available, it is possible to increase the bit rate (and therefore the system capacity) by more than an order of magnitude without affecting the repeater spacing. Considerations such as these provide strong motivation for many industrial laboratories worldwide to pursue the development of photodetectors with the characteristics of ever increasing bandwidth and sensitivity.

Table 14.1 gives a compilation of the best receiver sensitivities experimentally obtained for the various detector types reported to date.

TABLE 14.1
Optical Receiver Sensitivities Measured for Various Long-Wavelength Photodetector Types

Type	Bit Rate (Mb/s)	Wavelength (μm)	Signal SFormat[a]	Normalized Receiver Sensitivity (Approx.) (dBm)	Transmission Distance[b] (km)	Reference
Photoconductor						
InGaAs	2000	1.51	NRZ	− 30.3	—	Chen et al. (1985)
	1000	1.3	NRZ	− 35.9	—	Chen et al. (1984b)
p-i-n						
InGaAs	1200	1.53	NRZ	− 36.6	—	Chidgey et al. (1984)
				− 33.6	113.7	"
	420	1.55	NRZ	− 35.6	119	Tsang et al. (1983)
	140	1.55	NRZ	− 46.5	—	Malyon and McDonna (1982)
				− 45.0	102	"
	565	1.3	NRZ	− 40.2	—	Smith et al. (1982)
	280	1.3	NRZ	− 44.7	—	"
	140	1.3	NRZ	− 47.7	—	"
	420	1.3	NRZ	− 35.0	84	Boenke et al. (1982)
	274	1.3	NRZ	− 38.9	101	"
	447	1.3	NRZ	− 48.8	23.3	Ogawa et al. (1981)
APD						
Ge	2000	1.55	RZ	− 33.6	51.5	Yamada et al. (1982b)
				− 34.2	—	"
	800	1.55	RZ	− 34.1	20	Yamada et al. (1980)
	420	1.55	NRZ	− 40.4	108	Tsang et al. (1983)
	4000	1.3	RZ	− 22.0	—	Takano et al. (1985)
APD						
Ge	2240	1.3	RZ	− 24.9	21	Albrecht et al. (1982)
				− 28.2	—	
	2000	1.3	RZ	− 31.6	44.3	Yamada and Kimura (1982)
				− 34.1	—	"
	1200	1.3	NRZ	− 31.7	22.7	Yamada et al. (1979)
	800	1.3	NRZ	− 35.5	11	Yamada et al. (1978)
	400	1.3	NRZ	− 38.8	11	"
	100	1.3	NRZ	− 42.1	11	"
InGaAs/InP						
SAM-APD	4000	1.51	NRZ	− 32.7	—	Kasper et al. (1985)
	2000	1.51	NRZ	− 38.1	—	"
	1000	1.55	NRZ	− 40.0	—	Campbell et al. (1983)
	1000	1.3	NRZ	− 39.4	—	"
	420	1.55	NRZ	− 44.9	—	"
	420	1.3	NRZ	− 42.9	—	"
	45	1.3	NRZ	− 53.2	—	Forrest et al. (1981b)

TABLE 14.1 *(Continued.)*
Optical Receiver Sensitivities Measured for Various Long-Wavelength Photodetector Types

Type	Bit Rate (Mb/s)	Wavelength (μm)	Signal SFormat[a]	Normalized Receiver Sensitivity (Approx.) (dBm)	Transmission Distance[b] (km)	Reference
Integrated Detectors						
InGaAs *p-i-n*/InP						
MISFET Receiver	100	1.3	NRZ	-36.0	—	Kasahara *et al.* (1
	295	1.54	NRZ	-29.5	—	Tell *et al.* (1985)
	90	1.54	NRZ	-34.0	—	"
1 × 12 InGaAs p-i-n array	45	1.3	NRZ	-40.5	—	Kaplan *et al.* (198

[a] NRZ = non-return-to-zero; RZ = Return-to-zero.
[b] Maximum distance measure between transmitter and receiver with BER = 10^{-9}.

14.6 PHOTODETECTORS FOR THE FUTURE

Thus far, we have considered the three most important detector types for use in optical communication systems. Although considerable work needs to be done to improve the sensitivities and bandwidths of APDs and photoconductors, the approaches to the solutions to these problems are generally agreed upon and are being energetically pursued in laboratories worldwide. What have we to look forward to in the future for second-generation photodiodes and beyond?

Essentially, the advanced photodiode research has one of two goals: (1) to develop photodiodes with improved sensitivities for operation at very high bandwidths, and (2) to develop structures with improved functionality at low cost. Efforts to develop improved sensitivity detectors have recently focused on the staircase and multiquantum well (MQW) structures (Chin *et al.*, 1980; Capasso *et al.*, 1982; Williams *et al.*, 1982). These complex structures are APDs where the gain region consists of multiquantum wells formed by alternately growing thin layers of wide and narrow band-gap materials such as AlGaAs and GaAs, respectively (Fig. 14.10a). Due to asymmetries of the band offsets at the valence and conduction band edges, the holes and electrons can have markedly different ionization coefficients for the two carrier types. This situation should therefore lead to enhancements in the noise performance of APDs made using such III-V materials, where the ionization coefficients of holes and electrons are approximately equal. Note that although the reduction of the *k*-value of these devices is generally regarded as essential for low excess noise performance, perhaps the greatest advantage of using MQW APDs will arise from their expanded bandwidth due to the concomitant reduction in avalanche build-up time (Eq. 14.13).

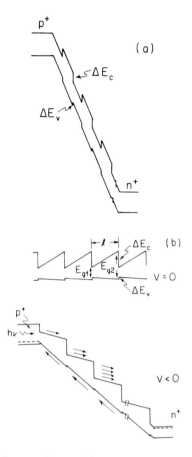

Fig. 14.10 Energy-band diagram of an (a) AlGaAs/GaAs MQW APD under reverse bias and (b) a staircase APD for use at longer wavelengths under zero bias (top) and reverse bias (bottom).

The most convenient means of growth for a MQW is molecular beam epitaxy (MBE). Thus, MQW devices have been grown using AlGaAs and GaAs; although these are of only limited interest since Si APDs with very low k-values are already widely available for use at short wavelengths. Unfortunately, growth of such structures for use at long-wavelengths is extremely difficult since MBE growth of InP-based compounds is still at an immature stage of development. In addition, the band edge asymmetries in this particular system are not as favorable for k-value reduction as they are in the GaAs-based material system, and problems with tunneling leakage in the high-field regions containing narrow band-gap layers will destroy any potential sensitivity improvement afforded by the MQW (Forrest *et al.*, 1981a; Forrest and Kim, 1980). For this reason, more complex schemes such as staircase APDs have been devised (Fig. 14.10b). This device employs a narrow bandgap region that is compositionally graded over a distance of ~ 100 Å–200 Å into a material with at least double the bandgap at the narrow end of the step. The composition is then abruptly changed to attain, once gain, the narrow bandgap as a second step is formed. The advantage of the staircase device is that carrier multiplication, which is induced by the transition of the carrier from the wide to narrow gap material, can occur at much lower electric fields than is possible with the MQW APD. As in the case of the MQW device, this structure favors the multiplication of one carrier type over another due to band edge asymmetries, thereby resulting in lower k_{eff}. Thus, in principle, carrier multiplication in the absence of tunneling in the narrow bandgap materials can be achieved. Unfortunately this structure presents severe difficulties in growth, and therefore has not yet been demonstrated.

On the other hand, integrated circuits containing photodetectors are now attracting considerable interest due to their potential for providing increased functionality and sensitivity at higher bandwidths and lower costs. Perhaps the most interesting such structure is an optical receiver consisting of a monolithically integrated photodiode and preamplifier circuit. The sensitivity increase is obtained by minimizing the capacitance of the front-end by replacing a discrete photodiode wire-bonded to the input transistor by a very small area photodiode integrally interconnected with an optimized FET. Thus, very low capacitance, small area (and therefore small dark current) amplifiers capable of very high bandwidth operation can be fabricated. The integrated structure will possibly also result in a cost reduction due to the reduced assembly costs associated with hybrid component technologies.

The difficulty in fabricating these devices arises from the conflicting requirements that must be reconciled in order that both the photodiode and the transistor circuit be independently optimized. As discussed with regard to p-i-n photodiode design, the absorbing region of the detector must be very low doped, fully depleted material of necessarily high purity. On the other hand, a high gain FET channel is generally very thin (2000 Å to 5000 Å thick) and has a high carrier concentration unsuitable for photodetector applications. Furthermore, optimized performance is achieved by separately biasing the photodetector and the FET, placing further constraints on the design of the integrated structure. The most sophisticated structure demonstrating many of these attributes for use at short wavelengths (Miura et al., 1985) is shown in Fig. 14.11. Here, the GaAs MESFETs are made on layers grown separately from the thick, light-absorbing p-i-n layers and are interconnected over a gentle slope "machined" into the substrate using Argon ion etching. As shown from the accompanying schematic diagram, the complete circuit consists of the photodiode and receiver preamplifier. Although the sensitivity reported for this device was considerably inferior to that obtained using hybrid components, the techniques employed exhibit great potential for improving receiver performance in the future.

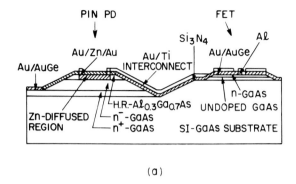

(a)

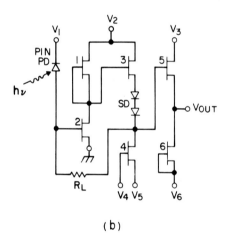

(b)

Fig. 14.11 A monolithically integrated receiver consisting of a p-i-n photodiode and transimpedance amplifier for use at short wavelengths (0.8 μm < λ < 0.9 μm) (after Miura *et al.*, 1985).

Considerably less sophisticated integrated receivers have been demonstrated for operation at long-wavelengths. Due to the relative immaturity of the InP-based material technologies, there is as yet no generally accepted means of fabricating transistors as there is in the GaAs-based system. However, the motivation for developing such a transistor technology goes beyond the desirability of optoelectronic integration. It has been found that both InP and $In_{0.53}Ga_{0.47}As$ have effective electron drift and peak velocities roughly 2.5 times higher than GaAs. Furthermore, the ternary material has a room temperature electron mobility of 10,000 cm^2/V-s: also approximately 2.5 to 3 times that of

GaAs. These characteristics imply that transistors made from such materials will have correspondingly higher bandwidths than those achievable using GaAs; thus, they are attractive for both very high bandwidth electronic as well as optoelectronic applications. With the intense effort and excellent results now being achieved in the development of transistors (Cheng *et al.*, 1986; Wake *et al.*, 1984) and photonic devices in this long-wavelength material system, it is probable that highly complex and high-performance integrated devices will be demonstrated in the very near future. Indeed, these may eventually go on to supplant GaAs-based optoelectronic components in order that a single material technology will be used to serve in all optoelectronic applications.

Finally, a different type of second-generation detector is the array. Such a device recently demonstrated for use at long-wavelengths (Brown *et al.*, 1984) is shown in Fig. 14.12. This particular device is a linear array of 12 photodiodes, where each array element is individually addressable and is coupled to an independent optical fiber. Such a device is of interest due to its use in applications where a large number of optical fibers converge into a single location. Examples of such system architectures are active star networks and local loops where each subscriber is connected to the central office with an optical fiber. In addition, these arrays are of interest in parallel data bus communication between computers, or between different areas in a single computer. Such applications require that the array operate at high bandwidths such that the "effective" bandwidth of the device is the product of the bit rate times the number of elements. Thus, the particular device shown consists of 12 elements each operating at 50 Mb/s, giving a total available bandwidth of 600 Mb/s.

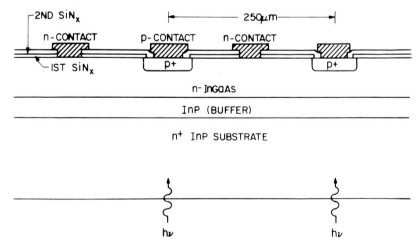

Fig. 14.12 Cross-sectional schematic view of a linear array of planar p-i-n photodiodes for use at 0.95 μm $< \lambda < 1.65$ μm (after Brown *et al.*, 1984).

Clearly, extremely high bandwidths can be achieved by this technique. The limiting factor in arrays is interchannel cross-talk induced by capacitive coupling between closely spaced elements (Kaplan et al., 1986). This problem can be somewhat alleviated by integrating the entire receiver front-end with each photodiode element—a level of sophistication that is still just beyond present state-of-the-art materials and fabrication capabilities.

One variation on the array concept is the wavelength demultiplexing photodiode (Campbell, 1980; Sakai, 1981). This is usually a "vertically" integrated device whereby photodetecting layers consisting of different bandgap materials are epitaxially stacked in the device in a direction perpendicular to the incoming optical signal. These layers can consist, for example, of different compositions of InGaAsP grown lattice-matched to an InP substrate, where the widest bandgap p-i-n is placed nearest to the optical input. In this way, the highest energy light is absorbed by the wide bandgap material, which is transparent to the low energy signal (this assumes a two-wavelength multiplexed input signal). The longer wavelength signal is thus absorbed in the underlying p-i-n consisting of the narrow bandgap composition of InGaAsP. Although such devices have been demonstrated to have good sensitivity and low optical cross-talk between wavelength channels, they have not yet been investigated for their reactive cross-talk performance at high frequencies. Nevertheless, the cross-talk limitations to such a device are expected to be comparable to, or even worse than encountered in linear photodiode arrays due to the extremely close proximity of neighboring detectors.

In spite of such problems, the wavelength division demultiplexing photodiode presents new capabilities and functions that can be used to expand the possibilities for photonic system applications. Probably the most important aspect of this and other integrated device technologies is that they press our current capabilities for device fabrication and materials growth to their limits. It is anticipated that any such exercise will lead to significant advances in photodiode technology.

REFERENCES

Albrecht, W., Elze, G., Enning, B., Walf, G., and Wenke, G. (1982). *Electron. Lett.* **18**, 746.

Boenke, M. M., Wagner, R. E., and Will, D. J. (1982). *Electron. Lett.* **18**, 898.

Bowers, J. E., Burrus, C. A., and McCoy, R. J. (1985). *Electron. Lett.* **21**, 812.

Brown, M. G., Forrest, S. R., Hu, P. H-S., Kaplan, D. R., Koza, M., Ota, Y., Potopowicz, J. R., Seabury, C. W., and Washington, M. A. (1984). *IEDM Tech. Digest*, Paper 31.5, 727.

Campbell, J. C., Dentai, A. G., Lee, T. P., and Burrus, C. A. (1980). *IEEE J. Quant. Electron.* **QE-16**, 601.

Campbell, J. C., Dentai, A. G., Holden, W. S., and Kasper, B. L. (1983). *Int. Electron Devices Meet. Tech. Digest*, 464.

Capasso, F., Tsang, W. T., Hutchinson, A. L., and Williams, G. F. (1982). *Appl. Phys. Let.* **40**, 38.

Capasso, F. (1985) in Lightwave Communications Technology (W. T. Tsang, Ed.) "Semiconductors and Semimetals." Vol. 22D, Ch. 1, 166, Academic Press, New York.

Chen, C. Y., Pang, Y. M., Alavi, K., Cho, A. Y., and Garbinski, P. A. (1984a). *Appl. Phys. Lett.* **44**, 99.

Chen, C. Y., Kasper, B. L., and Cox, H. M. (1984b). Seventh Topical Meet. on Integrated and Guided-Wave Optics, Orlando, FL.

Chen, C. Y., Kasper, B. L., Cox, H. M., and Plourde, J. K. (1985). *Appl. Phys. Lett.* **46**, 379.

Cheng, J., Guth, G., Washington, M., Forrest, S. R., and Wunder, R. (1986). *IEEE Electron Device Lett.* **EDL-7**, 225.

Chidgey, P. J., White, B. R., Brain, M. C., Hooper, R. C., Smith, D. R., Smyth, P. P., Fiddyment, P. J., Nelson, A. W., and Westbrook, L. D. (1984). *Electron. Lett.* **20**, 707.

Chin, R., Holonyak, N., Jr., Stillman, G. E., Tang, G. E., and Hess, K. (1980). *Electron. Lett.* **16**, 467.

Cox, H. M., Humme, S. G., and Keramidas, V. G. (1985). Electronic Materials Conf., Boulder, CO.

Forrest, S. R., Leheny, R. F., Nahory, R. E., and Pollack, M. A. (1980). *Appl. Phys. Lett.* **37**, 322.

Forrest, S. R., and Kim, O. K. (1981a). *J. Appl. Phys.* **52**, 5838.

Forrest, S. R., Williams, G. F., Kim, O. K., and Smith, R. G. (1981b). *Electron. Lett.* **17**, 917.

Forrest, S. R., Kim, O. K., and Smith, R. G. (1982). *Appl. Phys. Lett.* **41**, 95.

Forrest, S. R. (1984). *IEEE J. Lightwave Technol.* **LT-2**, 34.

Forrest, S. R. (1985). *IEEE J. Lightwave Technol.* **LT-3**, 347.

Forrest, S. R. (1986). *IEEE Spectrum* **23**, 76.

Holden, W. S., Campbell, J. C., Ferguson, J. F., Dentai, A. G., and Jhee, Y. K. (1986). *Optical Fiber Communication Conf. Tech. Digest*, paper WCC3, 98, Atlanta, GA.

Kaneda, T., Takanashi, H., Matsumoto, H., and Yamaoka, T. (1976). *J. Appl. Phys.* **47**, 4960.

Kaplan, D. R., Forrest, S. R., and Johnson, J. G. (1986). *Optical Fiber Communication Conf. Tech. Digest*, Paper WCC4, 100, Atlanta, GA.

Kasahara, K., Hayashi, J., Makita, K., Taguchi, K., Suzuki, A., Nomura, H., and Matsushita, S. (1984). *Electron. Lett.* **20**, 314.

Kim, O. K., Dutt, B. V., McCoy, R. J., and Zuber, J. R. (1985). *IEEE J. Quant. Electron.* **QE-21**, 138.

Kobayashi, M., Yamazaki, S., and Kaneda, T. (1984). *Appl. Phys. Lett.* **45**, 759.

Malyon, D. J., and McDonna, A. P. (1982). *Electron. Lett.* **18**, 445.

Matsushima, Y., Akiba, S., Sakai, K., Kushiro, Y., Noda, Y., and Utaka, K. (1982). *Electron. Lett.* **18**, 945.

McIntyre, R. J. (1966). *IEEE Trans. Electron Devices* **ED-13**, 164.

Melchior, H. (1972) in "Laser Handbook." (F. T. Arecchi and E. O. Schulz-Dubois, Eds.) Vol. 1, Ch. C7, 725. North-Holland, Amsterdam.

Miura, S., Machida, H., Wada, O., Nakai, K., and Sakurai, T. (1985). *Appl. Phys. Lett.* **46**, 389.

Ogawa, K., Chinnock, E. L., Gloge, D., Kaiser, P., Nagel, S. R., Jang, S. J. (1981). *Electron. Lett.* **17**, 71.

Sakai, S., Aoki, T., Tobe, M., and Umeno, M. (1981). *Japan J. Appl. Phys.* **20**, L205.

Saul, R. H., Chen, F. S., and Shumate, P. W., Jr. (1985). *AT & T Tech. J.* **64**, 861.

Shirai, T., Osaka, F., Yamasaki, S., Nakajima, K., and Kaneda, T. (1981). *Electron. Lett.* **17**, 826.

Smith, D. R., Hooper, R. C., Smyth, P. P., and Wake, D. (1982). *Electron. Lett.* **18**, 453.

Smith, R. G., and Personick, S. D. (1982). In "Semiconductor Devices for Optical Communication." (H. Kressel, Ed.) Ch. 4, 89, Springer-Verlag, Berlin.

Stillman, G. E., and Wolfe, C. M. (1977), in "Infrared Detectors II" (R. K. Willardson and A. C. Beer, Eds.), "Semiconductors and Semimetals," Vol. 12, p. 291, Academic Press, NY.

Stillman, G. E., Cook, L. W., Bulman, G. E., Tabatabaie, N., Chin, R., and Dapkus, P. D. (1982). *IEEE Trans. on Electron Devices* **ED-29**, 1355.

Takano, T., Iwakami, T., Mito, I., and Tashiro, Y. (1985). *Optical Fiber Commun. Conf. Tech. Digest*, Paper WB5, 88.

Tell, B., Liao, A. S., Brown-Goebeler, K. F., Bridges, T. J., Burkhardt, G., Chang, T. Y., and Bergano, N. S. (1985). *IEEE Trans. on Electron Devices* **ED-32**, 2319.

Tsang, W. T., Logan, R. A., Olsson, N. A., Temkin, H., Van der Ziel, J. P., Kaminow, I. P., Kasper, B. L., Linke, R. A., Mazurczyk, V. J., Miller, B. I., and Wagner, R. E. (1983). *Optical Fiber Communication Conf. Tech. Digest*, Paper PD-9, New Orleans, LA.

Tucker, R. S., Taylor, A. J., Burrus, C. A., Eisenstein, G., and Westfield, J. M. (1986). *Electron. Lett.* **22**, 917.

Wake, D., Livingstone, A. W., Andrews, D. A., and Davies, G. J. (1984). *IEEE Electron Device Lett.* **EDL-5**, 285.

Williams, G. F., Capasso, F., and Tsang, W. T. (1982). *IEEE Electron Device Lett.* **EDL-3**, 71.

Yamada, J.-I., Saruwatari, M., Asatani, K., Tsuchiva, H., Kawana, A., Sugiyama, K., and Kimura, T. (1978). *IEEE J. Quantum Electron.* **QE-14**, 791.

Yamada, J.-I., Machida, S., Kimura, T., and Takata, H. (1979). *Electron. Lett.* **15**, 278.

Yamada, J.-I., Susumu, M., Mukai, T., and Kimura, T. (1980). *Electron. Lett.* **16**, 115.

Yamada, J.-I., and Kimura, T. (1982). *IEEE J. Quantum Electron.* **QE-18**, 718.

Yamada, J.-I., Kawana, A., Miya, T., Nagai, H., and Kimura, T. (1982a). *IEEE J. Quant. Electron.* **QE-18**, 1537.

Yamada, J.-I., Kawana, A., Nagai, H., Kimura, T., and Miya, T. (1982b). *Electron. Lett.* **18**, 98.

Chapter 15

Integrated Optical and Electronic Devices

KOHROH KOBAYASHI

Optoelectronics Research Laboratories, NEC Corporation, Miyamae-ku, Kawasaki-city, Japan

15.1 INTRODUCTION

In recent years, remarkable progress has been made worldwide in the practical use of optoelectronics technologies for communications. This progress has been supported by the realization of high performance discrete optical devices, such as semiconductor laser diodes (LD), light emitting diodes (LED), PIN photo-diodes (PD) and avalanche photo-diodes (APD). In order to further expand the applications area of optical fiber communications from the trunk transmission lines to subscriber loops, local area networks and data links, optical devices with higher performance or with new functions are urgently needed. Optoelectronic integrated circuits (OEICs) or integrated optoelectronic circuits (IOECs) are expected to be key elements in these systems.

OEICs were first proposed by S. Somekh and A. Yariv (1972), where a new concept was proposed for monolithic integration of optical devices with electronic devices on the same semiconductor substrate. One item of the background for this idea was the fact that optical semiconductor devices are always used in connection with electronics circuits. The other is that both optical and electronic devices can be fabricated from the same semiconductor material systems. The first experimental trial for OEICs was demonstrated by Lee *et al.* (1978) in the integration of a semiconductor laser and a Gunn diode on a semi-insulating GaAs substrate. Since then, a variety of effort has been devoted to monolithic opto-electronic integration (Bar-Chaim *et al.*, 1982; Forrest, 1985; Matsueda *et al.*, 1983; Wada *et al.*, 1986; Yariv, 1984). Another concept has been introduced to guide the progress of OEICs, that is, the signal transmission function is carried out mainly by optics and signal processing is carried out mainly by electronics, respectively (Hayashi, 1983; 1986). These ideas lead to OEICs with optical devices at the front portion or at the end portion of the OEIC chips. Almost all the OEICs devised so far will be included in this category. As an advanced version of OEICs, a more complicated OEIC has been proposed, where optical components, electronic components or optoelectronically integrated gates are distributed within an LSI (large-scale integrated circuit) chip (Goodwin *et al.*, 1984).

An aim of this chapter is to provide an overview of OEICs and also to show the present status of OEIC research and development. Basic concepts on which OEIC development has been and will be performed are described in Section 15.2. Components for OEICs are described briefly in Section 15.3. Design principles, including structural and material problems and fabrication procedures specific to OEICs, are discussed in Section 15.4. The state-of-the-art in OEIC devices is summarized in Section 15.5. Experimental results on optical fiber transmission using OEIC transmitter and receivers are summarized in Section 15.6.

15.2 OEIC PURPOSES

The potential advantages inherent in monolithic integration of optical and electronic components include size, ruggedness, cost and reliability, as has been well established in the field of electronic integration using both Si and GaAs. In addition, speed and noise performance improvement will be expected by parasitic reactance reduction.

Based on these potential advantages, three directions can be considered in which OEICs will progress, as shown in Fig. 15.1. They are (1) to improve optical device performances, (2) to improve electronics device performances and (3) to create new functions. Devices in category 1 can be called "optical-device-oriented OEICs," where electronic device incorporation into optical devices is expected to improve the total performance of the optical devices. The main target for this direction will be transmitter/receiver (TX/RX) chips for optical fiber communications. They are segregated into two categories: (A) moderate performance optical transmitters/receivers and (B) those with high speed and high sensitivity capability. In the first category, the speed or the noise performance is not expected to be improved by optoelectronic integration. Instead, reliability increase and cost reduction will be gained by reducing the optical device chip mounting and wire bonding processing, which will be necessary to the so-called hybrid combination of optical devices and electronic devices/circuits. Size reduction by optoelectronic integration will attract much interest in

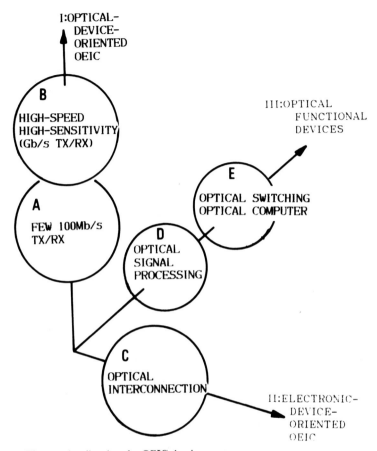

Fig. 15.1 Three major directions for OEIC development.

some applications such as parallel signal transmission. Transmitters/receivers for less than a few hundred Mb/s will be included in this category.

In category 1-B, higher performance in speed or sensitivity characteristics will be achieved in OEIC schemes than in hybrid schemes. Feasibility studies using computer simulation with SPICE indicated the superiority of OEIC integration over hybrid integration with respect to high speed modulation characteristics in a laser diode (LD)-MESFET combination (Nakamura *et al.*, 1986). Even a fairly small amount of inductance of the interconnection wire induces direct modulation response degradation at high frequencies. According to the simulation, for example, 0.3 nH inductance, which corresponds to a 35 μm diameter 0.5-mm-long bonding wire, will greatly degrade modulated light waveforms at 6.3 Gb/s. OEIC integration reduces the interconnection inductance, resulting in substantial elimination of the waveform distortion. In a photodiode (PD)-FET combination, receiver sensitivity depends strongly on the total front-end capacitance consisting of contributions from both the PD and the front-end FET gate capacitance (Forrest, 1985). In the PD capacitance, a fairly large fraction is due to the bonding pad, which is necessary to connect the detector and the FET in a hybrid receiver. Therefore, integrated receivers have the potential for higher sensitivity than hybrid receivers.

Category 2 aims to remove performance limits in electronic circuits, due to signal delay time and crosstalk inherent in closely packed high speed electronic integrated circuits, by introducing optical interconnection (Goodwin *et al.*, 1984). In electronic LSIs, the gate delay time consists of charging or discharging time for the logic gates and the interconnecting wires. With the increase in the gate number within an LSI chip, power consumption per gate must be reduced, resulting in an increase in the signal delay time. This signal delay, as well as crosstalk, may become a bottleneck for high speed LSI. Optical interconnections between LSI chips or even within LSI chips are considered pertinent to solving interconnect and communication problems in integrated circuits and systems (Hayashi, 1983; 1986). A conceptual drawing of OEICs for this category is shown in Fig. 15.2. Optical signal transmission will be done through optical waveguides as well as through free space. Signal delay time in optical interconnection consists of a transmission time through the optical waveguide and a time required to convert from the electronic to optical, or vice versa, in the light sources and the light detectors. The transmission delay does not depend on the power consumption, and the OE/EO conversion time has only a weak depen-

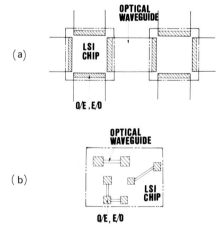

Fig. 15.2 Conceptual drawing for OEICs in direction II. Optical interconnections between electronic LSI chips or inside an LSI chip will solve communication problems in electronic LSIs.

dence on the power injected into each light source. According to theoretical estimation, optical interconnection will be advantageous over electronic interconnection when, for example, the gate number per chip exceeds a few thousand to ten thousands (Nii, 1984). Many problems, however, must be solved to gain the advantage. Among them are development of complex optical wiring using optical waveguides, light sources and detectors operable at ultra-high speed under very low power, optical isolation between optical components, etc. Fundamental research efforts have been started both theoretically and experimentally (Goodwin *et al.*, 1984).

Other intense interest has been shown to optical integrated circuits as an effective way to create new functional devices (Category 3). It still seems difficult to describe a clear image for this direction. Optical signal processing, utilizing the parallelism and high speed capability inherent in optics, appears to be a basis for this category. Key components for future optical computing, optical switching and other optical signal processing may be provided from this area. For example, optical interconnections, which exchange signals with very high speed and without any fanout limitation, will make parallel pipeline processors as the first realization of optical computers. Non-linear optical-optical interaction in materials will be fundamental to realize optical integrated functional devices/circuits. OEIC technologies may not occupy a principal part for this direction, but will support their realization by providing electronic control means and optical interconnection means.

15.3 COMPONENTS FOR OEICS

Important components for OEICs are light sources such as laser diodes (LDs), light emitting diodes (LEDs), light detectors such as photodiodes (PDs), and electronic components such as FETs and bipolar transistors. For the optical components, please refer to Chapters 11–14. In this section, component features specific to OEIC incorporation will be described.

15.3.1 Optical Component

Requirements for OEIC light sources (especially LD case) are (1) low threshold current and high efficiency, yielding low power consumption; (2) facetless or having facets made by methods other than by cleaving; (3) low device height to reduce the surface step between the light sources and the electronic circuits; (4) high-speed capability; and (5) fabrication ease. In addition to these requirements, fundamental characteristics commonly required for discrete LDs are also necessary, including transverse mode control and high reliability.

Many approaches have been tried to achieve these requirements. Quantum well structures have been introduced into the active region of LDs, resulting in a marked reduction in the threshold current density. The best record for the threshold current density is 190 A/cm^2 (Fujii *et al.*, 1984), obtained with an AlGaAs/GaAs quantum well LD. To reduce the total threshold current, buried heterostructures and many variations have been achieved. The minimum threshold current reported so far is 2.5 mA (Tsang *et al.*, 1982) achieved by AlGaAs/GaAs graded-index separate-confinement heterostructure (GRIN-SCH) LDs with a buried heterostructure (BH). These low threshold-current densities and absolute current values are most attractive from the thermal design viewpoints in OEICs. In the future, LDs, which can be operated with a few mW or less than a mW, will be realized based on these achievements.

Problems concerning the device processing, i.e., facetless structure (2) and low device height (3), are described in Section 15.4.2 in this chapter. Various schemes to achieve high-speed direct modulation will be described in a separate chapter.

For OEIC photo-detectors, a simple structure, fabrication ease and a low driving voltage are most desirable for monolithic integration. In addition to a PIN-photodiode (PD), a metal-semiconductor-metal (MSM) PD, a photoconductor and an FET have been used. Photo-detectors often require a bulky active region to fully absorb the light for higher efficiency. In order to incorporate a substantially planar PD structure suitable for monolithic integration with electronic components, a buried structure (Dawe *et al.*, 1986; Miura *et al.*, 1986) and a lateral version have been applied as well as the ordinary diffusion or ion implantation isolated PD structures. The operating voltage is relatively low (a few to ten volts) and comparable to the applied voltage for electronic circuits.

15.3.2 Electronic Components

FETs and bipolar transistors, especially heterobipolar transistors (HBTs), are the main electronic components for OEICs. Their fundamental structures are depicted in Fig. 15.3. In GaAs/AlGaAs material systems, metal-semiconductor (MES) FETs are now being established as a high-speed electronic circuit component. In InP/InGaAsP material systems, metal-insulator-semiconductor (MIS) FETs and junction (J) FETs have been investigated for use in OEICs. Since the Schottky barrier height is small in InP material systems, MESFETs are hard to realize. InP MISFETs are relatively simple to fabricate, but have problems involving current drift and low-frequency noise. The current drift problem will be solved, if the surface states at the semiconductor-insulator interface are controlled. JFETs seem to be suited for OEIC electronic circuits. Short gate length and self-alignment gate contact processing have realized short gate length JFETs with a high gain and high transition frequency.

Heterobipolar transistors have a multilayered structure, where electrons and holes pass vertically through the layers. The layer structure and the carrier flowing behavior seem to be similar to those for laser diodes. HBTs have an

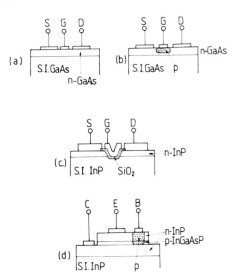

Fig. 15.3 Fundamental structures for electronic transistors for OEICs. (a) MESFET; Metal-semiconductor FET. (b) JFET; Junction FET. (c) MISFET; Metal-insulator-semiconductor FET. (d) HBT; Hetero-bipolar transistor.

ability for higher current, because a high bandgap emitter and a low bandgap base are applied, resulting in a higher current gain. This feature is attractive for high frequency LD direct modulation. In HBTs, shown in Fig. 15.3d, an f_T as high as 7.5 GHz, 100–200 h_{FE}, and 200 mA saturation collector current have been reported (Inomoto *et al.*, 1986).

15.4 OPTOELECTRONIC INTEGRATION TECHNOLOGIES

15.4.1 Design Principles

15.4.1.1 Structural Considerations. Two schemes for integration have been devised, from the view-point of relative optical and electronic component position with respect to the substrate. They are (1) vertical integration and (2) horizontal integration (Matsueda *et al.*, 1983). The fundamental concept is shown in Fig. 15.4.

In vertical integration, optical and electronic components are stacked in the substrate thickness direction on a conductive substrate. The current flows from the electronic components to the optical components, or vice versa, in series. The interconnection between the optical and the electronic components can be made as small as the intermediate layer thickness, possibly less than 1 μm. This would be advantageous for high-speed operation. The component layout freedom is

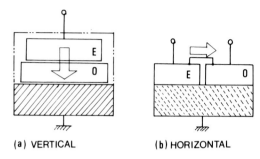

(a) VERTICAL   **(b) HORIZONTAL**

Fig. 15.4 Two fundamental OEIC integration schemes (Matsueda *et al.*, 1983). (a) Vertical integration. (b) Horizontal integration.

fairly limited, however, resulting in a very limited number of integrable constituents.

In horizontal integration, both components are placed side-by-side on a semi-insulating substrate. Interconnections are made by conductors over the epitaxial layers. The integration size is not fundamentally limited. For electrical isolation between components, a semi-insulating substrate or layers are effectively used. As described in Section 15.5 (State-of-the Art OEIC), almost all examples of the OEICs devised so far are included in this structure.

In designing OEICs, the difference in optical and electronic components should be fully considered. The most significant difference to be noted is that electronic components are basically surface devices, while optical components are bulk devices. Optical components require a certain amount of bulk crystal for the active region because of the optical field spreading. Among the electronic components are bipolar transistors, which have bulky active layers similar to the laser diodes. Structure and processing compatibility between optical and electronic components is a key point for realizing high performance OEICs.

15.4.1.2 Material Considerations.   Important materials, used so far for optical and electronic components, are GaAs/AlGaAs, InGaAsP/InP, Si and Ge. Table 15.1 summarizes features and present status of these materials. Si is the most thoroughly established material for use in electronic components, including up to large-scale integrated circuits. From the optical component reliability and fabrication viewpoints, InGaAsP/InP material systems seem to be most attractive, while they have rather few results on electronic components. GaAs/AlGaAs material systems have the most balanced results in regard to both optical and electronic components.

Besides the materials described, research has begun to realize new material systems by combining these materials. For example, GaAs/AlGaAs epitaxial layers have been stacked on a Si substrate. Pulse operation at room temperature

TABLE 15.1
Materials for OEIC

	Optical Components	Electronic Components
AlGaAs/GaAs	•Enough experience on 0.8 μm light sources (LD, LED) •Limited results on detectors •Clear demonstration of quantum size effect in quantum wells.	•Small scale integration started •MESFET 2DEG* Effect for high speed electronics (HEMT; etc.)
InGaAsP/InP or InGaAs/InP	•Highly reliable light sources (1.3/1.5 μm LD, LED) •Sensitive detectors (APD, PIN-PD) •Quantum size effect is not clear yet	•R & D started •MISFET, JFET, HBT. •Difficulty in surface control
Si	•Highly sensitive detectors (< 0.8 μm) •Light sources are not possible	•Large scale integration has been well established (LSI, VLSI)

*Two-dimensional electron gas

has been realized with GaAs/AlGaAs LDs made on Si (Windhorn et al., 1984). The new hybrid material systems will play an important role in future OEIC integration, where optimized material systems can be utilized for both optical and electronic components. Moreover, the huge achievements in Si microelectronics can be effectively used in future OEICs.

15.4.2 Fabrication Procedures

15.4.2.1 Epitaxial Growth. The most important key technology for optical device fabrication, at least so far, has been epitaxial crystal growth. This arises partly from the fact that optical devices, in general, need a bulk active region and partly from the fact that heterointerfaces are effectively utilized. The situation is the same in OEIC fabrication. As will be described in separate chapters, a variety of epitaxial growth technologies for III-V compound semiconductor materials have been developed, including liquid phase epitaxy (LPE), vapor phase epitaxy (VPE), metal organic vapor phase epitaxy (MOVPE), molecular beam epitaxy (MBE) and their combinations or variations. Higher uniformity in the layer thickness and the doping concentration over a larger wafer and finer control in the doping concentration or background impurity concentration, compatible with both the optical and electronic components, will be required for OEIC fabrication than for discrete device fabrication. Spatially selective epitaxial growth technology will also be very important for OEIC fabrication.

15.4.2.2 Device Processing. One of the most important problems in OEIC fabrication processing is related to surface steps between the optical and electronic components. This problem arises from the fact that, in principle, most optical components use a bulk material and most electronic components use the area near a crystal surface, as described in Section 15.4.1.1. For example, optical components, such as LDs, LEDs and PDs, require at least a total thickness of 5–10 μm including the active layer, cladding layers, and contact layer. Electronic components, such as FETs, need only about 1 μm thickness. These geometrical steps make high resolution photo-lithography difficult for high-gain/high-speed FETs. Device yield will be degraded due to the wiring imperfections near the step edges.

Several methods have been investigated as means to avoid this step problem. Among these are using (a) optical components-in-a-well (Carney *et al.*, 1982), (b) slowly varying, slanting surfaces (Miura *et al.*, 1985), (c) thick wiring lines (Inomoto *et al.*, 1986), (d) thick electronic components, such as bipolar transistor (Inomoto *et al.*, 1986; Shibata *et al.*, 1984), and (e) vertical integration schemes (see 15.4.1.1) and their combinations. The first method uses epitaxial material for optical components in a well to assure a planar surface or to reduce the steps, as shown in Fig. 15.5. Selective LPE, VPE, or whole surface epitaxy followed by

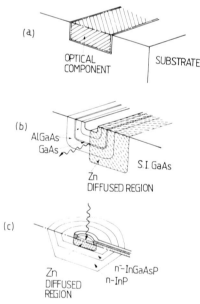

Fig. 15.5 Optical components-in-a-well method for a planar surface in OEICs. (a) Fundamental concept. (b) TJS-LD in-a-well (Carney *et al.*, 1982). (c) Planar embedded PIN photodiode (Miura *et al.*, 1986).

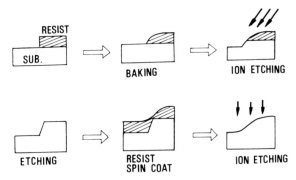

Fig. 15.6   Gentle slope formation method by ion etching with a thick photoresist layer (Miura *et al.*, 1985).

etching was applied to fill the well with multilayered epitaxial materials. The second and the third methods were intended to improve wiring yield. Gentle slopes in the second method were made by applying an ion etching process with a thick photoresist, as shown in Fig. 15.6. Thick wiring contact layers were made at the side wall of a hetero-bipolar transistor by selective gold plating through SiO_2 masks (Inomoto *et al.*, 1986). Substantially flat surfaces can be realized by integrating bipolar transistors, which have nearly the same device thickness as the laser diode (Shibata *et al.*, 1984).

15.4.2.3 Facet Formation.   For laser diode integration into OEICs, LD cavity facet formation is indispensible. Faceting processings for LDs developed so far are (a) conventional cleaving, (b) micro-cleaving, (c) wet chemical etching, and (d) dry ion beam etching. Optical feedback by corrugation is a highly effective and efficient alternative to optical feedback by discrete mirrors.

In case of conventional cleaving facet formation, the length is determined by the LD cavity length, i.e., about 200–500 μm, which imposes a severe limitation on the circuit layout. In order to meet the requirement for the circuit size increase, OEIC chip length must be increased, resulting in a very elongated stripe chip. Micro-cleaving techniques (Blauvelt *et al.*, 1982, Wada *et al.*, 1982a) have revealed the possibility of making a cleaved facet inside the wafer. Micro-cleaving uses ultrasonic vibration to break a narrow cantilevered portion, extending from the LD stripe, to make a cleavage facet.

Etching by wet or dry processes appears to be effective not only for LD mirror formation, but also for optical component integration, such as PDs for monitoring. Recently, etched mirrors, whose quality was almost equal to that for cleaved mirrors, have been realized by wet etching as well as by reactive ion etching (Asakawa *et al.*, 1985; Shimizu *et al.*, 1985; Uchida *et al.*, 1986). The reactive ion beam etching was carried out under chlorine plasma pressure excited by electron cyclotron resonance.

In a discrete LD, optical feedback by corrugation has been introduced to achieve single frequency light sources for high bit-rate, long-distance optical fiber transmission systems. The corrugation optical feedback eliminates the necessity for optical feedback facets. The first experiments on corrugation feedback in OEICs, instead of on mirrors, have been reported (Kasahara *et al.*, 1985; 1986), where an InGaAsP DFB-LD was integrated with a PD and MISFETs on a semi-insulating InP substrate. That will be described in the next section, with other OEICs with cleaved or etched mirrors.

15.5 STATE-OF-THE-ART OEICs

Since the first demonstration of an OEIC in 1978 by Lee *et al.*, optoelectronic integration efforts have been continuing steadily. In Fig. 15.7, the total number of integrated components, optical as well as electronic, of typical OEICs reported so far are plotted as a function of a year. Before 1982, most OEICs consisted of only one optical component and only one electronic component, which is the basic OEIC form. In recent years, the number of constituents has been rapidly increased; although the number is still limited to about ten, with one exception (Carney *et al.*, 1982). In this section, various OEICs (mainly for optical communications use) are reviewed. Structures, constituents and features of experimental OEICs are summarized in Tables 15.2 to 15.6; short wavelength transmitter OEICs in Table 15.2, short wavelength receiver OEICs in Table

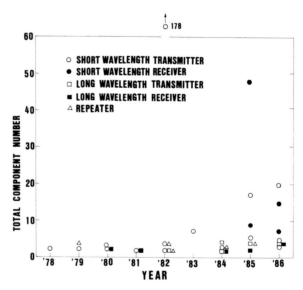

Fig. 15.7 Progress in integration scale of OEIC. Total number of integrated components is plotted with respect to year.

TABLE 15.2
Short wavelength transmitter OEICs

HV Optical component	Electronic component	Substrate	Epitaxy	Year	Reference
H DH-LD	Gunn Diode	S.I.GaAs	LPE	1978	Lee *et al.* (1978)
H Mesa-LD	MESFET	S.I.GaAs	LPE	1979	Ury *et al.* (1979)
V DH-LD	2S.G.-FET	n-GaAs	LPE	1980	Fukuzawa *et al.* (1980)
H Be-I/I LD	MESFET	S.I.GaAs	LPE	1980	Wilt *et al.* (1980)
H Be-I/I LD	HBT	n-GaAs	LPE	1980	Katz *et al.* (1980)
V MCS-LD	MESFET	n-GaAs	LPE	1980	Matsueda *et al.* (1981)
H BH-LD	MESFET	S.I.GaAs	LPE	1982	Ury *et al.* (1982)
H TJS-LD	MESFET	S.I.GaAs	LPE	1982	Carney *et al.* (1982)
H LED	3MESFET	n-GaAs	MBE	1982	Wada *et al.* (1982b)
H TS-LD, PD	5MESFET	S.I.GaAs	LPE	1983	Matsueda *et al.* (1983)
H MQW-RW-LD	2MESFET	S.I.GaAs	MBE	1984	Sanada *et al.* (1984)
H MQW-LD, PD	12MESFET, 3R	S.I.GaAs	MOVPE	1985	Matsueda *et al.* (1985)
H GRIN-SCH-LD	4MESFET	S.I.GaAs	MBE	1985	Sanada *et al.* (1985)
H TJS-LD	2MESFET	S.I.GaAs	MBE	1986	Ohta *et al.* (1986)
H LD, monitor PD	3MESFET	S.I.GaAs	MBE	1986	Sanada *et al.* (1986)

H: Horizontal integration, V: Vertical integration, Be-I/I: Be-ion implantation, MCS: Modified channeled substrate, BH: Buried heterostructure, TJS: Transverse junction stripe, TS: Terraced substrate, MQW: Multi-quantum well, RW: Ridge waveguide, GRIN: Graded-index, SCH: Separate confinement heterostructure, S.G.: Schottky gate, R: Resistor, S. I.: Semi-insulating.

15.3, long wavelength tᵣ nsmitter OEICs in Table 15.4, long wavelength receiver OEICs in Table 15.5 and others in Table 15.6.

15.5.1 Short Wavelength Optical Transmitter / Receiver OEICs

Monolithic integration of light sources or light detectors and electronic components on a GaAs substrate has been most extensively studied, as indicated in Table 15.2 and 15.3. LD structures have been improved to stabilize the transverse mode and to reduce the threshold current, yielding a low power

TABLE 15.3
Short wavelength receiver OEICs

HV Optical component	Electronic component	Substrate	Epitaxy	Year	Reference
H PIN-PD	6MESFET, 2Di	S.I.GaAs	LPE	1985	Wada *et al.* (1985a)
H MSM-PD	6MESFET, 4Di	S.I.GaAs	MOVPE	1985	Wada *et al.* (1985b)
H PIN-PD	3MESFET, 5Di,	S.I.GaAs	LPE	1985	Matsueda *et al.* (1985)
H PD	7MESFET, 7DI	S.I.GaAs	—	1986	Lee *et al.* (1986)

MSM: Metal-semiconductor-metal
Di: Diode

TABLE 15.4
Long wavelength transmitter OEICs

HV	Optical component	Electronic component	Substrate	Epitaxy	Year	Reference
H	Groove-LD	MISFET	S.I.InP	LPE	1982	Koren et al. (1982)
–	BH-LD	3HBT	n-Inp	LPE	1984	Shibata et al. (1984)
H	LD	2MISFET	S.I.InP	LPE	1984	Kasahara et al. (1984)
H	DFB-LD, PD	2MISFET	S.I.InP	LPE	1985	Kasahara et al. (1985)
H	DC-PBH-LD	3HBT	S.I.InP	LPE	1986	Inomoto et al. (1986)

DFB: Distributed feedback
DC-PBH: Double channel planar buried heterostructure

TABLE 15.5
Long wavelength receiver OEICs

HV	Optical component	Electronic component	Substrate	Epitaxy	Year	Reference
H	PIN-PD	JFET	S.I.InP	LPE	1980	Leheny et al. (1980)
H	PC	MESFET	InP	MBE	1981	Bernard et al. (1981)
H	PIN-PD	FET	S.I.InP	LPE	1984	Hata et al. (1984)
H	PIN-PD	MISFET	S.I.InP	LPE	1984	Kasahara et al. (1984)
H	PIN-PD	JFET	S.I.InP	LPE	1985	Ohnaka et al. (1985)
H	PIN-PD	3JFET	S.I.InP	LPE	1986	Suzuki et al. (1986)
V	PD	JFET	S.I.InP	MBE	1986	Wake et al. (1986)
H	PD	MISFET	S.I.InP	LPE	1986	Ohtsuka et al. (1986)

PC: Photo-conductor

TABLE 15.6
Repeater OEICs

HV	Optical component	Electronic component	Substrate	Epitaxy	Year	Reference
H	DH-LD	3MESFET	S.I.GaAs	LPE	1979	Yust et al. (1979)
–	BH-LD, HPT	HPT	n-GaAs	LPE	1982	Bar-Chaim et al. (1982)
H	LED, PD	MESFET, R	S.I.GaAs	LPE	1982	Carter et al. (1982)
H	BH-LD, PIN-PD	MESFET	S.I.GaAs	LPE	1984	Bar-Chaim et al. (1984)
H	BH-LD, PIN-PD	2 C-FET	S.I.InP	LPE	1985	Hata et al. (1985)
H	HPT	2HBT, 4R	S.I.GaAs	MBE	1986	Wang and Ankri (1986)

HPT: Hetero photo-transistor
C-FET: Column gate FET

consumption suitable for OEIC integration. Very simple double heterostructure (DH) or current confined structures in the early stage of OEIC development have been replaced by buried heterostructure (BH) (Ury *et al.*, 1982), transverse junction stripe (TJS) (Carney *et al.*, 1982), terraced substrate (TS) (Matsueda *et al.*, 1983) structures, etc. The threshold current has been reduced to about one tenth of that of oxide stripe lasers. To furthers reduce the threshold current density, quantum well structures have been applied by using MBE and MOVPE epitaxial techniques.

Although there have been attempts to integrate various kinds of electronic components, such as Gunn diodes, Schottky gate (S. G.) FETs, heterobipolar transistors and metal-semiconductor (MES) FETs, recent efforts have been focused on MESFETs. GaAs MESFETs have been well established as key elements for GaAs electronic integrated circuits. They have a simple structure and can be easily fabricated on a semi-insulating GaAs substrate.

Figure 15.8 shows external views and circuit diagrams for (a) vertical (Fukuzawa *et al.*, 1980) and (b) horizontal integration OEICs (Ury *et al.*, 1982). As indicated in the "HV" column in Table 15.2 by notation "H" the most recent approach tends to be horizontal integration. The example for horizontal integration, shown in Fig. 15.8b, has a BH LD and a MESFET on a semi-insulating GaAs substrate. The BH LD has a low threshold current, 20 mA, and a high external quantum efficiency, 55%. This device exhibits a direct modulation bandwidth exceeding 4 GHz. This OEIC is the first one that has successfully demonstrated the feasibility of high-frequency capability in OEICs. Such characteristics have been expected from the beginning of OEIC development.

An example for advanced GaAs transmitter OEIC versions is shown in Fig. 15.9 (Matsueda *et al.*, 1985; Nakano *et al.*, 1986). It includes a self-aligned structure multi-quantum well (SAS-MQW) LD, a monitoring PD, ten MESFETs and three resistors for the LD driver, two additional MESFETs and a resistor for the monitoring circuit. Epitaxial layers for the MQW LDs were made using an MOVPE technique on a semi-insulating GaAs substrate with selectively etched wells. The LD threshold current is 40 mA and the differential quantum efficiency is 19%. The rear end-facet of the LD, facing to the monitoring PD, was made by reactive ion beam etching using Cl_2 as the major constituent gas.

Another example for the transmitter OEIC is depicted in Fig. 15.10 (Carney *et al.*, 1982). A transverse junction stripe (TJS) LD was introduced into a small-scale GaAs integrated circuit, which has a 36 gate 4:1 multiplexer and a large FET driver, fabricated using direct ion implantation MESFET technology. The TJS LD was fabricated by liquid phase epitaxy in a well. The LD epi-material was grown below the surface of the substrate (in-a-well) yielding a planar surface for high resolution photolithography. Since GaAs IC technology is now being established, this integration method seems to be an effective approach to achieving large-scale OEICs.

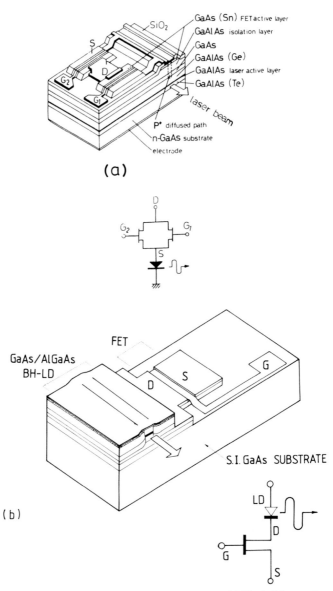

Fig. 15.8 Examples of two fundamental integration schemes. (a) Vertical integration. (Matsueda *et al.*, 1981). (b) Horizontal integration (Ury *et al.*, 1982).

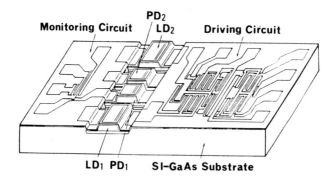

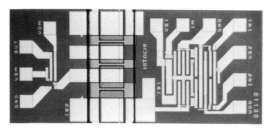

Fig. 15.9 An OEIC transmitter for a short wavelength region consisting of an MQW laser diode, a monitoring photodiode, twelve MESFETs and four resistors (Nakano *et al.*, 1986). (a) Overall structure. (b) Chip photograph. (c) Circuit diagram. (Copyright © 1986 IEEE)

Short wavelength receiver OEIC fabrication trials, based on a GaAs substrate, have been rather limited so far, as shown in Table 15.3. This would be partly due to the existence of Si as an excellent detector material at short wavelengths. Among the experimental receiver OEICs in the 0.8 μm wavelength region is a four channel photodiode/amplifier array (Wada *et al.*, 1985a). The cross-section and circuit diagram for the single channel are shown in Fig. 15.11. The single channel circuit consists of a simple metal-semiconductor-metal (MSM) PD, six MESFETs and four diodes. The MSM PD was made by forming interdigital metal contacts on a low-doped GaAs light absorption layer. The epi-layers were made using MOVPE. The MSM PD responsivity was 0.2 A/W. The rise time was estimated to be 300 ps. Four channel operation was demonstrated with fairly good results regarding the crosstalk and the output voltage uniformity. Multi-channel function will be a significant target for OEICs, because a hybrid combination of optical devices and electronic circuits would encounter severe problems with regard to crosstalk, as well as to size, when a multi-channel function is required.

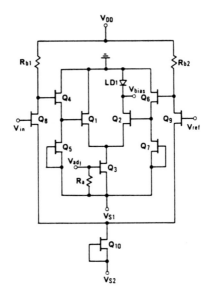

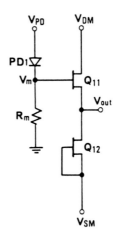

Fig. 15.9 (*Continued.*)

15.5.2 Long Wavelength Optical Transmitter / Receiver OEICs

Transmitters or receivers of complexity comparable to those discussed in the previous section have not yet been demonstrated using the InP/InGaAsP material system, which is most important for wavelengths ranging from 1.1 μm to 1.6 μm. In contrast to the mature electronic component technology for the

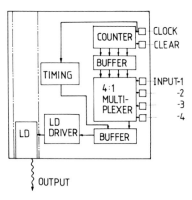

Fig. 15.10 Structure diagram of an OEIC transmitter incorporating a TJS laser diode and a multiplexer circuit on a semi-insulating GaAs substrate (Carney *et al.*, 1982).

GaAs material system, that for the InP material system is still in a preliminary stage. This makes complex optoelectronic integration difficult. Basic combinations, however, have already been investigated, as summarized in Table 15.4 and 15.5. Metal-insulator-semiconductor (MIS) FETs, Junction (J) FETs and heterobipolar transistors have been utilized in OEICs for the long wavelength regions, instead of MESFETs used for the short wavelength region.

The first trial for a long wavelength transmitter OEIC was a monolithic integration of an InGaAsP grooved-LD with 1.3 μm emission wavelength and a MISFET on a semi-insulating InP (Koren *et al.*, 1982). Following that, HBT integration was studied, as shown in Figs. 15.12 and 15.13. Figure 15.12 shows the cross-sectional view and the equivalent circuit diagram for an LD-HBT OEIC (Shibata *et al.*, 1984). A buried heterostructure (BH) LD and three HBTs were fabricated on a conductive (*n*-type) InP substrate by two-step liquid phase epitaxy. The HBT was chosen from the viewpoints of high current and high speed capability. The HBTs were fabricated using layers regrown as burying layers for BH structures. For the electrical isolation between the HBT and the LD, etched grooves, embedded with polymide, were utilized. A differential circuit is formed for LD high-speed switching. The LD threshold current was 20-30 mA. Single-transverse-mode operation was maintained, up to 10 mW light output power. External quantum efficiency was 20%. HBT current gain h_{FE} was typically 400 at 60 mA collector current. High-speed direct modulation was demonstrated, up to 1.6 GHz.

Another example of an LD-HBT OEIC is shown in Fig. 15.13 (Inomoto *et al.*, 1986). A 1.3 μm double channel planar buried heterostructure (DC-PBH) LD and three HBTs were fabricated on a semi-insulating InP substrate doped with Fe. The semi-insulating substrate leads to easier component electrical isolation, as well as to less parasitic capacitance, yielding higher speed operation capabil-

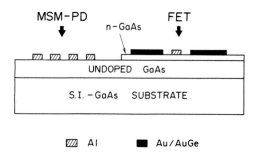

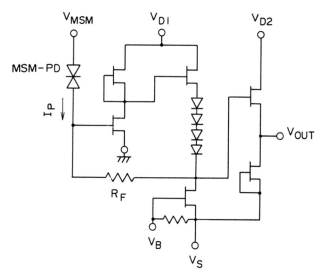

Fig. 15.11 Cross section (a) and circuit diagram (b) of a short wavelength OEIC receiver monolithically integrating a metal-semiconductor-metal photodiode, six MESFETs, four diodes, and a resistor on a semi-insulating GaAs substrate (Wada *et al.*, 1985).

ity. Device processing, however, may become more complicated compared with that for a conductive substrate. For example, a three-step LPE process is required for this specific OEIC. The first step is for DH crystal growth, the second step is for DC-PBH formation, and the third step is for HBT crystal growth, respectively. For high-speed operation, narrow mesa structures were applied, both for the LD and the HBT. Thick gold plating wiring line processing was applied for reproducible interconnection. The driver circuit consists of a current-mode-logic circuit and an input buffer amplifier. LD threshold current was 23 mA and differential quantum efficiency was 38%. HBT

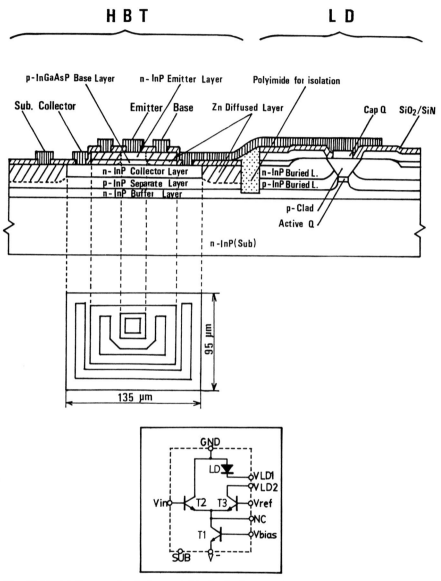

Fig. 15.12 (a) Cross-sectional view and (b) circuit diagram of a long wavelength OEIC transmitter, where a buried heterostructure and three heterobipolar transistors are monolithically integrated on a conducting InP substrate (Shibata *et al.*, 1984).

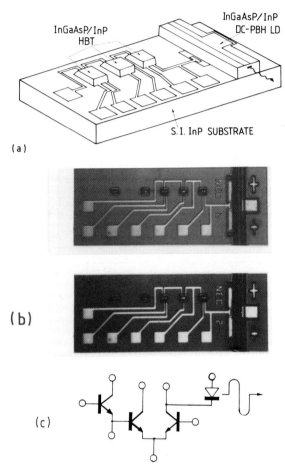

Fig. 15.13 (a) Structure diagram, (b) chip photograph and (c) circuit diagram of a long wavelength OEIC transmitter, where a DC-PBH laser diode, three heterobipolar transistors are monolithically integrated on a semi-insulating InP substrate (Inomoto *et al.*, 1986).

current gain h_{FE} was 100–200 and the collector saturation current was larger than 200 mA. High-speed co-operative operation was achieved up to 2 Gb/s (NRZ) pulse rate.

Progress to be noted in long wavelength transmitter OEIC research and development is the adoption of the DFB LD as the OEIC light source. A 1.3 μm DFB LD, a monitor PD and two MISFET were monolithically integrated on a semi-insulating InP, as shown in Fig. 15.14 (Kasahara *et al.*, 1985; 1986). Facetless LDs, such as DFB LDs, increase the freedom in the electronic circuit layout and will allow future large-scale integration. With a mesa stripe DFB-

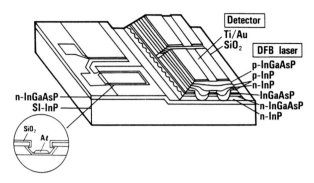

Fig. 15.14 External view of a DFB-DC-PBH laser diode/monitor-PD/MISFET transmitter OEIC (Kasahara *et al.*, 1986). This would be the first trial for DFB laser diode incorporation into OEICs. (Copyright © 1986 IEEE)

DC-PBH LD, a 4 GHz small signal direct modulation bandwidth was achieved. Open eye patterns were observed for RZ random pulse modulation, up to 2 Gb/s.

Integrated optical receivers for the long wavelength region have attracted intense interest since the first trial by Leheny *et al.* (1980) due to their potential for receiver sensitivity improvement, compared with conventional hybrid schemes, as for the short wavelength region. Since high performance photo-detectors have been developed as discrete devices based on InGaAs on an InP substrate, the idea of InGaAs-PD integration with InP-based electronic components appears to be rather straightforward, compared with the method used for the GaAs case.

Figure 15.15 shows an external view, a chip photograph and the circuit diagram of a receiver OEIC developed recently (Suzuki *et al.*, 1986). It consists of a PIN-PD and three junction FETs. The active region of the photodiode is non-doped InGaAsP with 1.5 μm composition wavelength, prepared by LPE. The JFET was made using self-aligned gate contact processing to realize a short gate length for high-speed and low noise characteristics. The electronic circuit consists of the source-ground amplifier with the first FET and the buffer amplifier with the second and third FETs. JFET f_T was observed to be 3 GHz and the transconductance was 65 mS/mm. Total input capacitance was measured to be about 3 pF. Receiver sensitivities were evaluated; for 10^{-9} bit-error-rate, the mean required detector input powers were -21.5 dBm and -14.2 dBm for 565 Mb/s and 1.2 G/s NRZ signals, respectively. A fairly large amount still remains to be gained in receiver sensitivity. There have been no reports to date, where the OEIC receiver sensitivity is improved, compared with the hybrid scheme, not only at long wavelengths but also in the short wavelength region.

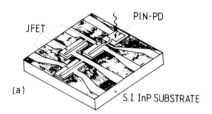

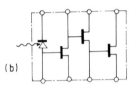

Fig. 15.15 An integrated InGaAsP/InP optical receiver consisting of a PIN photodiode, three JFETs (Suzuki *et al.*, 1986). (a) External view. (b) Circuit diagram.

15.5.3 Optical Repeater OEICs

Integration of a light source, a light detector, and an electronic circuit will lead, in principle, to an optical repeater. Table 15.6 summarizes repeater OEICs. Up to now their configurations were very simple and they acted as an optical amplifier. That means the input optical signal is converted into an electrical signal by the light detector, amplified electronically by circuit components and again converted into an optical signal by the light source. No timing regeneration or signal decision function has been incorporated in OEIC repeaters.

Three MESFETs were fabricated with an LD on an S.I. GaAs substrate, as shown in Fig. 15.16 (Yust *et al.*, 1979). One MESFET among them acts as an optical detector, resulting in a device acting as an optical repeater. Overall optical power gain of 10 dB was observed in this specific OEIC with short wavelength light from an external GaAlAs LD. In the long wavelength region, a BH-LD, two column gate FETs and a PD were integrated on an S.I. InP substrate (Hata *et al.*, 1985). Overall repeater gain was observed as 5.5 dB.

15.6 OEIC APPLICATIONS TO OPTICAL FIBER COMMUNICATIONS

The feasibility of OEICs has been fairly well demonstrated by optical fiber transmission experiments using the OEIC transmitter and receiver. Typical results reported are summarized in Table 15.7. The OEIC receiver sensitivities are plotted with respect to the signal bit rate, as shown in Fig. 15.17. In the figure, the receiver sensitivity results achieved by a combination of a discrete optical detector and an electronic circuit are also shown as references. It is clear

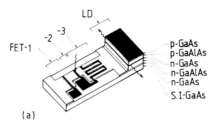

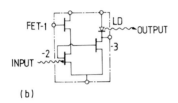

Fig. 15.16 An integrated AlGaAs/GaAs optical repeater consisting of a DH laser diode and three MESFETs, one of which (FET-2) acts as a photodiode also (Bar-Chaim *et al.*, 1982) (a) External view. (b) Circuit diagram.

that, at the present, OEIC integration does not induce any advantageous effect on the receiver sensitivity. Low quantum efficiency and high dark current in the optical detector as well as problems with the front end electronic components may be the causes for the sensitivity degradation. At short wavelengths, 4 km and 2 km, multimode fiber transmission were carried out at 400 Mb/s and 800 Mb/s, respectively, with transmitter/receiver modules in which an OEIC chip was installed (Iwama *et al.*, 1986). A higher-bit-rate signal transmission experiment was reported, using single-mode fiber in the 0.8 μm wavelength region. A one Gb/s optical signal was transmitted through 1 km of single-mode fiber (Minami *et al.*, 1986). At 1.3 μm, in the long wavelength region, 22 km and 12 km transmission were achieved for signal bit rates of 565 Mb/s and 1.2 Gb/s

TABLE 15.7
Transmission experiments using transmitter / receiver OEICs

Wave-length	Bit-rate	Distance	Receiver Sensitivity	Transmitter electronics	Receiver electronics	Reference
μm		km	dBm			
1.3	565Mb/s	22	−21.5	3HBT	3JFET	Suzuki *et al.* (1986)
1.3	1.2Gb/s	12	−14.2	3HBT	3JFET	Suzuki *et al.* (1986)
0.8	1Gb/s	1	−19.5	12MESFET	3MESFET + 5Di	Minai *et al.* (1986)
0.8	400Mb/s	4	−20.6	4MESFET + R	6MESFET + 2DiR	Iwama *et al.* (1986)
0.8	800Mb/s	2	−18.5	4MESFET + R	6MESFET + 2DiR	Iwama *et al.* (1986)

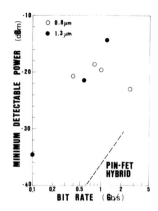

Fig. 15.17 OEIC receiver sensitivity as a function of signal bitrate (Suzuki *et al.*, 1986; Minai *et al.*, 1986; Iwama *et al.*, 1986; Lee *et al.*, 1986).

with level margins of 9.9 dB and 7.7 dB, respectively (Suzuki *et al.*, 1986). The transmission lengths have been extended five to ten times in the 1.3 μm wavelength region compared with those in 0.8 μm wavelength region, because of the low attenuation loss and the low dispersion in the 1.3 μm region. These results have successfully indicated good prospects for OEICs as future key elements in optical fiber communications, although they have not yet fully realized their potential ability.

15.7 CONCLUDING REMARKS

A number of OEICs have been demonstrated experimentally, based on GaAs and InP material systems, using a variety of material epitaxial methods and wafer fabrication procedures. Although their fundamental attractive characteristics have been demonstrated, they are still in a relatively preliminary stage in the integration scale as well as the performance scale. There are many problems to be solved in achieving their expected potentials, including (i) full compatibility in the epitaxy and the fabrication processing between the optical and electronic components; (ii) further improvements in the structure, epitaxy, and processing for both discrete optical and electronic components; and (iii) optical, electronic and thermal isolation between optical components and electronic components in OEICs. These problems will become more severe as the integration size increases. The isolation problems may finally limit the integration size or even the possible function of OEICs. It would be most serious in OEICs for optical interconnections within LSI chips (category 2 in Sec. 15.2), or for new functions for optical switching or optical computing (category 3 in Sec. 15.2), where a large number of optical and electronic components will be integrated in a mixed manner on the same substrate. Compact and miniaturized optical isolators, which can be fabricated by wafer processing compatible with that for optical and electronic components, are eagerly anticipated.

GaAs and InP have been widely used as the substrate semiconductor materials for OEICs. In the future, Si may play an important role. Recent demonstrations of high quality GaAs on Si (Akiyama *et al.*, 1984; Windhorn *et al.*, 1984) will open new Si-based OEIC areas, where mature Si-technologies will be effectively combined with III-V compound semiconductor technologies for optical as well as electronic components. This scheme will be most suitable for OEICs with a few optical components and very many electronic components or large size circuits. Furthermore, if III-V compound semiconductor material systems with quality high enough for optical device fabrication are made on dielectric materials or glass, a new prospect will be provided for future optical integrated circuits with new functions.

For OEICs to become practical devices, discrete device technologies should be significantly improved. For example, discrete device yield close to 100% will be necessary. Each discrete device reliability should be improved to such a level that a screening test for each device, which has been done for almost all optical devices, is not necessary. Without achieving such a high technological level in the discrete device characteristics and reliability, it will be impossible to fabricate high performance and reliable OEICs, in which many discrete devices are included as the key components and at reasonable cost. Therefore, when we consider this from the opposite point of view, OEIC research and development may accelerate discrete optical device fabrication technology improvements. It will expand further the applications area of optoelectronics, for example, from signal transmission to signal switching and processing. Following the discrete device technologies, OEIC technologies will also play an important role in future fields.

REFERENCES

Akiyama, M., Kawarada, Y., and Kaminishi, K. (1984). Growth of single domain GaAs layer on (100)-oriented Si substrate by MOCVD. *Japan. J. Appl. Phys.* **23**, L843–845.

Asakawa, K., and Sugeta, S. (1985). GaAs and AlGaAs anisotropic fine pattern etching using a new reactive ion beam etching system. *J. Vac. Soc. and Tech.* **B3**, 402–405.

Bar-Chaim, N., Margalit, S., Yariv, A., and Ury, I. (1982a). GaAs integrated optoelectronics. *IEEE Tasns. Electron Devices* **ED-29**, 1372–1381.

Bar-Chaim, N., Harder, Ch., Katz, J., Margalit, S., and Yariv, A. (1984). Monolithic integration of a GaAlAs buried-heterostructure laser and a bipolar phototransistor. *Appl. Phys. Lett.* **40**, 556–557.

Bar-Chaim, N., Lau, K. Y., Ury, I., and Yariv, I. (1984). Monolithic optoelectronic integration of a GaAlAs laser, a field-effect transistor, and a photodiode. *Appl. Phys. Lett.* **44**, 941–943.

Bernard, J., Ohno, H., Wood, E. C., and Eastman, L. F. (1981). Integrated double heterostructure GaInAs photoreceiver with automatic gain control. *IEEE Electron Device Lett.* **EDL-2**, 7–9.

Blauvelt, H., Bar-Chaim, N., Fekete, D., Margalit, S., and Yariv, A. (1982). AlGaAs lasers with micro-cleaved mirrors suitable for monolithic integration. *Appl. Phys. Lett.* **40**, 289–291.

Carney, J. K., Helix, M. J., Kolbas, R. M., Jamison, S. A., and Ray, S. (1982). Monolithic optoelectronic/electronic circuits. GaAs IC symposium, 38–41.

Carter, A. C., Forbes, N., and Goodfellow, R. C. (1982). Monolithic integration of optoelectronic electronic and passive components in GaAlAs/GaAs multilayers. *Electron. Lett.* **18**, 72–74.

Dawe, P. J. G., Spear, D. A. H., and Thompson, G. H. B. (1986). Planar embedded GaInAs photodiode on semi-insulating InP substrate for monolithic integration. *Electron Lett.* **22**, 722–724.

Forrest, S. R. (1985). Monolithic optoelectronic integration: A new component technology for lightwave communications. *J. Lightwave Tech.* **LT-3**, 1248–1263.

Fujii, T., Yamakoshi, S., Wada, O., and Hiyamizu, S. (1984). Extremely high-quality GaAs-AlGaAs GRIN-SCH lasers with a superlattice buffer layer by MBE for OEIC applications. Extended Abstract. 16th Conf. Solid State Devices and Materials, Kobe 1984, 145–148.

Fukuzawa, T., Nakamura, M., Hirao, N., Kuroda, T., and Umeda, J. (1980). Monolithic integration of a GaAlAs injection laser with a Schottky-gate field effect transistor. *Appl. Phys. Lett.* **36**, 181–183.

Goodwin, J. W., Leonberger, F. J., Jung, S., and Athale, R. A. (1984). Optical interconnections for VLSI systems. *IEEE*, **72**, 850–866.

Hayashi, I. (1983). OEIC: Its concepts and prospects. *Tech. Digest*, IOOC'83, Tokyo 1983, 170–174.

Hayashi, I. (1986). Future direction of optoelectronics: Basic concepts of optoelectronic integration. *OPTOELECTRONICS-Devices and Technologies* **1**, 1–10.

Hata, S., Ikeda, M., Amanao, T., Motosugi, G., and Kurumada, K. (1984). Planar InGaAs/InP p-i-n-FET fabricated by Be ion implantation. *Electron. Lett.* **20**, 247–248.

Hata, S., Ikeda, M., Kondo, S., and Noguchi, Y. (1985). PIN-FET-LD integrated device in long wavelength region. *Tech. Digest*, Domestic Conf, IECE Japan, 58 (in Japanese).

Inomoto, Y., Terakado, T., and Suzuki, A. (1986). High performance InGaAsP/InP LD-HBTs light source OEIC. *Tech. Digest*, The first Optoelectronics Conference (OEC'86) Tokyo, A6–4.

Iwama, T., Oikawa, Y., Horimatsu, T., Makiuchi, M., Yamaguchi, K., and Touge, T. (1986). Design and fabrication of OEIC modules and application to optical LANs. *Tech. Digest*, Optical and Quantum Electron. Meeting, IECE Japan, **OQE86-13**, 93–100 (in Japanese).

Kasahara, K., Hayashi, J., Makita, K., Taguchi, K., Suzuki, A., Noumra, H., and Matsushita, S. (1984a). Monolithically integrated InGaAs/InP-MISFET photoreceiver. *Electron. Lett.* **20**, 314–315.

Kasahara, K., Hayashi, J., and Nomura, H. (1984b). Gigabit per second operation by monolithically integrated InGaAsP/InP LD-FET. *Electron. Lett.* **20**, 618–619.

Kasahara, K., Terakado, T., Suzuki, A., and Murata, S. (1985). Monolithically integrated high speed light source using 1.3 μm wavelength DFB-DC-PBH lasers. *Tech. Digest* IOOC-ECOC'85, Venetzia 1985, 295–297.

Kasahara, K., Terakado, T., Suzuki, A., and Murata, S. (1986). Monolithically integrated high-speed light source using 1.3-μm wavelength DFB-DC-PBH laser. *J. Lightwave Tech.*, **LT-4**, 908–912.

Katz, J., Bar-Chaim, N., Chen, P. C., Margalit, S., Ury, I., Wilt, D., Yust, M., and Yariv, A. (1980). A monolithic integration of GaAs/GaAlAs bipolar transistor and heterostructure laser. *Appl. Phys. Lett.* **37**, 211–213.

Koren, Y., Yu, K. L., Chen, T. R., Bar-Chaim, N., Margalit, S., and Yariv, A. (1982). Monolithic integration of a very low threshold GaInAsP laser and metal-insulator-semiconductor field-effect transistor on semi-insulating InP. *Appl. Phys. Lett.* **40**, 643–645.

Lee, C. P., Margalit, S., Ury, I., and Yariv, A. (1978). Integration of an injection laser with a Gunn oscillator on a semi-insulating GaAs substrate. *Appl. Phys. Lett.* **32**, 806–807.

Lee, W. S., Adams, G. R., Mun, J., and Smith, J. (1986). Monolithic GaAs photoreceiver for high-speed signal processing applications. *Electron. Lett.* **22**, 147–148.

Leheny, R. F., Nahory, R. E., Pollack, M. A., Ballman, A. A., Beebe, E. D., DeWinter, J. C., and Martin, R. J. (1980). Integrated InGaAs p-i-n F.E.T. photoreceiver. *Electron. Lett.* **16**, 353–355.

Matsueda, H., Fukuzawa, T., Kuroda, T., and Nakamura, M. (1981). Integration of a laser diode and a twin FET. *Japan. J. Appl. Phys.* **20**, suppl. 20-1, 193–197.

Matsueda, H., Sasaki, S., and Nakamura, M. (1983). GaAs optoelectronic integrated light sources. *J. Lightwave Tech.* **LT-1**, 261–269.

Matsueda, H., Hirao, N., Tanaka, T., Kodera, H., and Nakamura, H. (1985). Integration of optical devices with electronic circuits for high speed communications. Inst. Phys. Conf. Ser. No. 79: Chapter 12, Int. Symp. GaAs and Related Compounds, Kaurizawa, Japan.

Minai, Y., Uehara, H., Motegi, Y., Maeda, M., and Kodera, H. (1986). A 1Gb/s optical transmission experiment with opto-electronic integrated circuits. *Tech. Digest*, Communication System Meeting, IECE Japan. **CS85-130**, 67–74 (in Japanese).

Miura, S., Machida, H., Wada, O., Nakai, K., and Sakurai, T. (1985). Monolithic integration of a pin photodiode and a field-effect transistor using a new fabrication technique-graded step process. *Appl. Phys. Lett.* **46**, 389–391.

Miura, S., Kuwatsuka, H., Mikawa, T., and Wada, O. (1986). Low capacitance, high speed InP/GaInAs PIN photodiode with a planar, embedded structure. *Tech. Digest*, The first Optoelectronic Conference (OEC'86) Tokyo, A1-1.

Nakamura, M., Suzuki, N., and Ozeki, T. (1986). The superiority of optoelectronic integration for high-speed laser diode modulation. *IEEE J. Quantum Electron.* **QE-22**, 822–826.

Nakano, H., Yamashita, S., Tanaka, T. P., Hirao, M., and Maeda, M. (1986). Monolithic integration of laser diodes, photo-monitors, and laser driving and monitoring circuits on semi-insulating GaAs. *J. Lightwave Tech.* **LT-4**, 574–582.

Nii, M. (1984). Problems of LSI technology and possibility for highly integrated OEIC. Investigation Report on Materials for Optical Information Processing, 59-M-214, 5–10 (in Japanese).

Nobuhara, H., Kuno, M., Makiuchi, M., Fujii, T., and Wada, O. (1985). Optoelectronic integrated transmitter with a microcleaved facet AlGaAs/GaAs quantum well laser. *Tech. Digest* IEDM'85, 650–653.

Ohnaka, K., Inoue, K., Uno, T., Hasegawa, K., Hase, N., and Serizawa, H. (1985). A planar InGaAs PIN/JFET fiber-optic detector. *IEEE J. Quantum Electron.* **QE-21**, 1236–1239.

Ohta, J., Kuroda, K., Mitsunaga, K., Kyuma, K., Hamanaka, K., and Nakayama, T. (1986). Monolithic integration of a transverse junction stripe laser and metal semiconductor field effect transistors on a semi-insulating GaAs substrate. *Tech. Digest*, The first Optoelectronics Conference (OEC'86) Tokyo, A7-2.

Ohtsuka, K., Sugimoto, H., Abe, Y., and Matsui, T. (1986). Monolithic integration of InGaAs/InP PIN-PD with MISFET on stepless substrate. *Tech. Digest*, The first Optoelectronics Conference (OEC'86) Tokyo,

Sanada, T., Yamakoshi, S., Wada, O., Fujii, T., Sakurai, T., and Sasaki, M. (1984). Monolithic integration of an AlGaAs/GaAs multiquantum well laser and GaAs metal-semiconductor field-effect transistors on a semi-insulating GaAs substrate by molecular beam sepitaxy. *Appl. Phys. Lett.* **44**, 325–327.

Sanada, T., Yamakoshi, S., Hamaguchi, H., Wada, O., Fujii, T., Horimatsu, T., and Sakurai, T. (1985). Monolithic integration of a low threshold current quantum well laser and a driver circuit on a GaAs substrate. *Appl. Phys. Lett.* **46**, 226–228.

Sanada, T., Kuno, M., Nobuhara, H., Makiuchi, M., Fujii, T., Wada, O., and Sakurai, T. (1986). Fabrication of multi-channel optical transmitter. *Tech. Digest*, Optical and Quantum Electronics Meeting, **OQE86-12**, 85–92 (in Japanese).

Shibata, J., Nakao, I., Sasai, Y., Kimura, S., Hase, N., and Serizawa, H. (1984). Monolithic integration of an InGaAsP/InP laser diode with heterojunction bipolar transistors. *Appl. Phys. Lett.* **45**, 191–193.

Shimizu, H., Wada, M., Hamada, K., Kume, K., Shibutani, T., Yoshikawa, N., Itoh, K., Kano, G., and Teramoto, I. (1985). A new monolithic composite-cavity (GaAl)As laser. *Tech. Digest*, IOOC-ECOC'85, Venetzia, 111–114.

Somekh, S., Yariv, A. (1972). Fiber optic communications. *Proc. Conf. International Telemetry*, Los Angeles, 407–418.

Suzuki, A., Kasahara, K., Fujita, S., Inomoto, Y., Terakado, T., and Shikada, M. (1986). Long wavelength high speed transmitter-receiver OEIC. *Tech. Digest*, Optical and Quantum Electronics Meeting, **OQE86-14**, 101–108 (in Japanese).

Tsang, W. T., Logan, R. A., and Ditzenberger, J. A. (1982). Ultra-low threshold graded-index waveguide separate confinement cw buried-heterostructure lasers. *Electron. Lett.* **18**, 845–847.

Uchida, M., Matsumoto, S., Asakawa, K., and Kawano, H. (1986). Integrated AlGaAs two-beam LD-PD array fabricated by reactive ion beam etching. *Electron. Lett.* **22**, 585–587.

Ury, I., Margalit, S., Yust, M., and Yariv, A. (1979). Monolithic integration of an injection laser and a metal semiconductor field effect transistor. *Appl. Phys. Lett.* **34**, 430–431.

Ury, I., Lau, Y., Bar-Chaim, N., and Yariv, A. (1982). Very high frequency GaAlAs laser field-effect transistor monolithic integrated circuit. *Appl. Phys. Lett.* **41**, 126–128.

Wada, O., Yamakoshi, S., Fujii, T., Hiyamizu, S., and Sakurai, T. (1982a). AlGaAs/GaAs microleaved facet (MCF) laser monolithically integrated with photodiode. *Electron. Lett.* **18**, 189–190.

Wada, O., Sanada, T., and Sakurai, T. (1982b). Monolithic integration of an AlGaAs/GAas DH LED with a GaAs FET driver. *IEEE Electron Device Lett.* **EDL-3**, 305–307.

Wada, O., Hamaguchi, H., Miura, S., Makiuchi, M., Nakai, K., Horimatsu, H., and Sakurai, T. (1985a). AlGaAs/GaAs p-i-n photodiode/preamplifier monolithic photoreceiver integrated on a semi-insulating GaAs substrate. *Appl. Phys. Lett.* **46**, 981–983.

Wada, O., Hamaguchi, H., Nakauchi, M., Kumai, T., Ito, M., Nakai, K., and Sakurai, T. (1985b). Monolithic four channel photodiode/amplifier array integrated on a GaAs substrate. *Tech. Digest*, IOOC-ECOC'85, Venetzia, 303–306.

Wada, O., Sakurai, T., and Nakagami, T. (1986). Recent progress in optoelectronic integrated circuits (OEIC's). *IEEE J. Quantum Electron.* **QE-22**, 850–821.

Wake, D., Scott, E. G., and Henning, I. D. (1986). Monolithically integrated InGaAs/InP PIN-JFET photoreceiver. *Electron. Lett.* **22**, 719–721.

Wang, H., and Ankri, D. (1986). Monolithic integrated photo-receiver implemented with GaAs/GaAlAs heterojunction bipolar phototransistor and transistors. *Electron. Lett.* **22**, 391–393.

Wilt, D., Bar-Chaim, N., Margalit, S., Ury, I., Yust, M., and Yariv, A. (1980). Low threshold Be implanted (GaAl)As laser on semi-insulating substrate. *IEEE J. Quantum Electron.* **QE-16**, 390–391.

Windhorn, T. H., Metze, G. M., Tsaur, B.-Y., and Fan, J. C. (1984). AlGaAs double-heterostructure diode lasers fabricated on a monolithic GaAs/Si substrate. *App. Phys. Lett.* **45**, 309–311.

Yariv, A. (1984). The beginning of integrated optoelectronic circuits. *IEEE Trans. Electron. Devices* **ED-31**, 1656–1661.

Yust, M., Bar-Chaim, N., Izadphanah, S. H., Margalit, S., Ury, I., Wilt, D., and Yariv, A. (1979). A monolithically integrated optical repeater. *Appl. Phys. Lett.* **35**, 795–797.

Chapter 16

Epitaxial Growth Methods for Lightwave Devices

J. A. LONG

AT&T Bell Laboratories, Inc., Murray Hill, New Jersey

R. A. LOGAN

AT&T Bell Laboratories, Inc., Murray Hill, New Jersey

R. F. KARLICEK, JR.

AT&T Bell Laboratories, Inc., Murray Hill, New Jersey

16.1 INTRODUCTION

Epitaxial crystal growth techniques are essential for the development of optoelectronic devices based on compound semiconductor materials. Generally speaking such crystal growth techniques can be divided into four categories: Liquid phase epitaxy (LPE), vapor phase epitaxy (VPE), metalorganic chemical vapor deposition (MOCVD), and molecular beam epitaxy (MBE). This chapter reviews, in sequence, applications of these epitaxial growth techniques to the fabrication of optoelectronic devices based on III-V compound semiconductors used primarily in optical fiber-based communication systems.

The selection of an epitaxial growth technique, besides being influenced by the device requirements, is also governed by considerations of system cost, production throughput, the purity and toxicity of the source chemicals, and the degree of versatility required (e.g., different compounds to be grown). While the bulk of current epitaxial crystal growth research is aimed at extending the usefulness of VPE, MOCVD, and MBE growth techniques, most commercially available optoelectronic devices are grown by LPE. Each growth technique has advantages and limitations that are reviewed, along with examples of growth properties and device applications, in the discussion that follows.

16.2 LIQUID PHASE EPITAXY

16.2.1 Growth Method

This growth technique is the principal method for constructing many optical communication devices (Laudise, 1983; Li, 1983) due to the simplicity of the growth process. Layer growth is achieved by placing a substrate crystal under an appropriate saturated solution melt that is being slowly cooled, with layer growth occurring as the solubility in the melt adjusts to the reduced temperature. Typical carbon growth boats are shown in Fig. 16.1.

Figure 16.1a shows a simple boat design for AlGaAs growth (Logan and Reinhart, 1975). The boat is machined from dense carbon and is contained in a pure H_2 ambient in a quartz furnace liner. The layers are grown by successively sliding a (100) oriented rectangular substrate with (110) edges, under saturated melts, while cooling the furnace from $850°C$ at a rate of $0.2°C/min$. After a preliminary bake-out of the Ga melts at $800°C$ for 16 hrs., dopants and excess GaAs are added through a porthole in the quartz furnace liner to prevent the baked-out Ga from being reoxidized. Saturation at the bottom of the melt is ensured by use of a saturation seed, which precedes the sample to the growth region. This greatly improves the layer morphology, especially in thin layers (≤ 2000 Å). Although random growth occurs on the saturation seed, this does not adversely affect the growth on the substrate since AlGaAs is always lattice matched to GaAs. A similar saturation seed is not used in the InGaAsP system because quaternary layer growth must occur at preselected temperatures with melts of specified composition. For example, growth of a quaternary layer on the saturation seed during growth of a thick buffer layer on the substrate would lead to lattice mismatched growth on the seed, with ensuing poor melt wiping and degradation of the layer growth sequence.

The boat design of Fig. 16.1 differs from conventional ones in that each melt is contained in thin-walled (0.5 mm) modules that fit snugly into recesses in the main boat. This design permits great flexibility in layer growth since individual modules can be modified to grow layers of varying complexity without disturbing the growth conditions of adjacent layers. For example, module 2 of Fig. 16.1b contains a thin partition that has permitted growth of a single bicompositional layer from two different adjacent melts in the module. The use of a thin insulating mask at the bottom of module 3 of Fig. 16.1b causes layer growth to occur in two predetermined areas, with layer segments terminating in smooth tapers, decreasing to zero thickness in distances of ~ 100 μm.

In the InGaAsP system where lattice matched layer growth must occur at a specified temperature and melt composition, a typical, simple growth boat (Nordland *et al.*, 1984) is shown in Fig. 16.1c. The design is similar to the AlGaAs boat of Fig. 16.1a, but the slider is modified to permit saturation of each

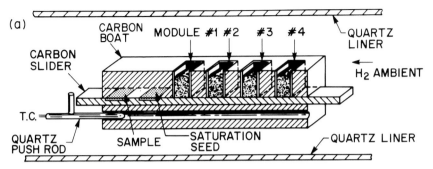

Fig. 16.1a Schematic cross section of the growth boat and furnace lining, used for GaAlAs growth where a "saturation" seed precedes the substrate under each growth melt.

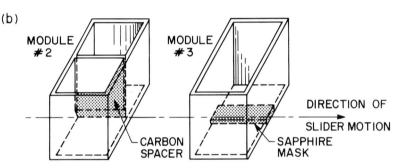

Fig. 16.1b Module inserts for obtaining growth from two separate melts onto the same substrate and also a module for holding a sapphire mask to grow layer segments with peripheries tapering smoothly to zero thickness over a distance of ~ 100 μm. Typical substrate dimensions are $1 \times 1.5 \times 0.02$ scm^3.

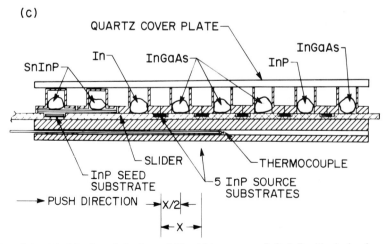

Fig. 16.1c Modification of the boat of Fig. 16.1a to grow InGaAsP with single phase growth, supersaturation and a Sn-InP source of P to preserve the substrate from dissociation prior to growth.

melt with P from an InP source seed on the slider, after dissolving the InAs and GaAs sources in the melt at the equilibration temperature. This procedure overcomes the observed interaction of P and As in the melt, where the presence of P inhibits dissolution of the GaAs and InAs constituents. After the melt is saturated with P at the equilibration T, the InP source is removed by sliding the source wafers to locations between the melts, and the furnace temperature is lowered to achieve the desired melt supersaturation (typically 3–9°C) prior to growth on the substrate. The boat of Fig. 16.1c can also be used for conventional two-phase growth by placing the InP sources on top of each melt after dissolving the InAs and GaAs constituents. In two-phase growth, the melt composition and supersaturation is less accurately specified than in single-phase growth, as described, since growth occurs on the saturation seeds that float on the melt and locally relieve the supersaturation.

Another modification required in using InP is the inclusion of a source of P over the substrate to prevent decomposition due to P evaporation during the heat treatment prior to epitaxy. The "basket" type containers at the left end of the boat in Fig. 16.1c contain the undersaturated Sn-InP source (Antypas, 1980).

While the InGaAsP quaternary composition is accurately and uniformly specified by this procedure, the large distribution coefficient (the ratio of solid to liquid solute concentration) of Al, coupled with its high temperature dependence (Panish & Ilegems, 1978), limit the uniformity of composition of AlGaAs that can be achieved. With layer growth at 850°C, the addition of only 1 mg Al per gram of Ga correlates with the solid composition $Al_{0.3}Ga_{0.7}As$. This limit on uniformity is not a severe restriction in most device formation; for example, AlGaAs is used in laser cladding layers where small compositional fluctuations have an insignificant effect on the device characteristics.

16.2.2 Dopants and Growth Morphology

The basic dopant solubilities and composition dependences in the AlGaAs system are well documented in the literature and summarized by Casey and Panish (1978). Similar data are also available for the InGaAsP system (Astles et al., 1973; Kuphal, 1984; Nakajima et al., 1980). The common dopants, such as Sn donors and Ge acceptors, are both nonvolatile and easily handled. Two dopants require special precautions. The acceptor Zn is sufficiently volatile that the doping correlation with melt composition depends on the effectiveness with which the melt is capped. Thus, during the growth cycle sufficient Zn may evaporate to have a significant effect on the ensuing doping of the grown layer. Hence, for each boat configuration and growth cycle, the Zn doping level must be calibrated by Hall measurements or other similar procedures.

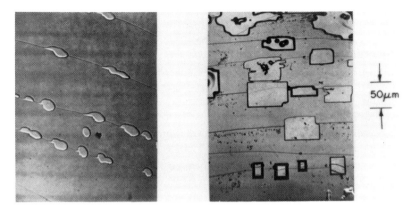

Fig. 16.2 Nomarski phase contrast photomicrograph of meniscus lines on the surface of a AlGaAs layer after growth of a second layer for about 1 second. The second layer nucleates on the meniscus lines and the initial growth platelets are either with featureless contours or with edges parallel to (110) directions. The meniscus lines lie approximately perpendicular to the direction of slider motion.

Similar precautions must be used with the volatile donor Te (or Se), which may be used to achieve high dopant concentrations ($\sim 10^{19}$ cm^{-3}). For growth at 850°C, 17 mg of Sn per gram of Ga gives a donor concentration of 10^{17} cm^{-3} in GaAs, whereas only 10 μg of Te per gram of Ga would give an equal doping level. Contamination by Te vapor in the growth apparatus leads to serious nucleation problems (Logan *et al.*, 1979), especially at the higher growth temperature (~ 850°C).

Nucleation of growth occurs on meniscus lines. Meniscus lines (Small *et al.*, 1975) are formed on the wafer surface at the trailing edge of the melt as it slides from the surface in a series of steps rather than a continuous wiping motion. These lines run approximately normal to the direction of sliding. They are S-shaped in cross section with peaks and valleys deviating from the crystal surface by a few hundred angstroms and, although not all of equal prominence, they can be readily observed in phase contrast microscopy. The detailed mechanism of meniscus line formation is not understood, but the important role played by such lines in layer growth is demonstrated in Fig. 16.2. After growth of several microns of GaAs, the seed was slid right through a second melt to permit very limited growth. As evident in the two cases shown, nucleation occurs on the meniscus lines formed by removing the melt from the first layer and growth initiates as a series of pancake-like platelets. These platelets may have featureless contours or have edges parallel to {110} planes. Growth occurs by enlarging and overlapping of these platelets. In contrast to this behavior, similar nucleation for [110] oriented wafers also originates on meniscus lines but, instead of platelets, growth occurs in narrow stripes, which extend in a direction parallel to a 90° (110) cleavage plane. The occurrence of meniscus lines obviously contributes to the inability to grow a succession of very thin layers (quantum wells), but the mechanism of growth nucleation discussed also inhibits formation of thin layers (< 1000 Å) with precise thickness control. Growth at low temperature ($< 650°$C) has permitted formation (Kelting et al., 1986) of single quantum wells (70–240 Å thick).

The presence of Te vapor in the growth apparatus completely inhibits the nucleation mechanism described before so that growth originates between the meniscus lines (Logan et al., 1979). The meniscus lines are the last place to fill in and do not fill in if the layer is $\lesssim 0.2$ μm thick, as in the active layer of a laser. These breaks in the active layer have been termed "rake" lines. While the detailed nature of this contamination upon growth nucleation is not understood, the problem can be easily eliminated by dismantling and acid cleaning or replacing the quartz furnace liner. The growth of thick layers (> 0.5 μm) in the presence of similar levels of Te contamination leads to a severe terracing of the layer surface (Logan et al., 1979). Growth of degenerately doped n-layers with Te doping in the InGaAsP system at $\sim 650°$C leads to similar rake line and terracing effects in this system also. The presence of oxygen (air leaks) in the growth chamber causes poor layer growth, characterized by pitted surfaces and carry-over of melt when transferring the substrate from well to well (or a final wipe-off after layer growth).

High resistivity semi-insulating layers have been grown in the InGaAsP system by doping with Co (Rezek et al., 1983). The success of this procedure requires concentration on crystal growth procedures to achieve a doping level of $N_D - N_A \sim 10^{15}$ cm^{-3} in undoped layers. This requires careful bake-out, sample cleaning and handling procedures to exclude airborne contaminants (e.g., S) from the growth apparatus and cleaning procedures. In the AlGaAs system, high resistivity, compensated layers of $Al_{0.65}Ga_{0.35}As$ were found ideal for current confinement in buried heterostructure (BH) lasers (Henry et al., 1981).

16.2.3 Regrowth

Layer growth in the InGaAsP system is somewhat more difficult than in the AlGaAs system because (1) the solubility of P in In is about an order of magnitude lower than that of As in Ga, and (2) the high volatility of P dictates that layer growth occurs at $\leq 650°$C compared to $\sim 850°$C in the AlGaAs system so that the growth rates are reduced. Moreover, comparison of layer growth with the superior wetting $In_{1-x}Sn_x$ melts (Logan and Temkin, 1986) demonstrates the inherently poor wetting encountered with In melts, so that supersaturation is needed to initiate growth in the InGaAsP system. In the AlGaAs system, however, regrowth around etched mesas used in forming BH lasers could be preceded by a slight meltback (caused by placing a substrate under a saturated melt and raising the furnace temperature 1 °C) and excellent regrowth achieved without use of melt supersaturations (Henry et al., 1981).

In the AlGaAs system, layer segments with smoothly tapered peripheries were grown (Reinhart and Logan, 1975) where the tapered sections efficiently coupled light out of the active layer of a laser into an adjacent waveguide. The taper was formed by the melt wetting underneath a sapphire mask, which blocked the melt from a region of the substrate, as shown in Fig. 16.1b. Solute replenishment limited crystal growth under the mask, giving rise to the smooth taper in layer thickness over a distance of ~ 100 μm. In the InGaAsP system, as a result of the impaired wetting, similar procedures generally produce abruptly discontinuous layer edges that do not appear useful as optical waveguide couplers.

In the AlGaAs system, no regrowth difficulties were encountered in forming BH lasers and similar devices, where regrowth over an etched substrate was needed (Henry et al., 1981). An exception to this is the difficulty of wetting $Al_xGa_{1-x}As$ with x > 0.1, which may simply reflect the stability of the oxidized, exposed Al at high concentration.

In the InGaAsP system regrowth of buried heterostructure devices is more difficult and the chief obstacle is the difficulty of producing an interface on the {111}A plane that is free from nonradiative defects. The feasibility of obtaining low threshold buried heterostructure lasers was demonstrated by Hirao et al. (1980) with lasers emitting at 1.3 μm. The mesa structure was formed by means of a Br-methanol etching solution using SiO_2 stripes as masks. A mask such as SiO_2 must be used since photoresist is soluble in this etchant, which is one of the limited number of isotropic etchants in the InGaAsP system. The stripe masks are again oriented to form narrow-waisted mesas (1–2 μm wide) with $\langle 111 \rangle$A stop-etch sidewalls.

Studies of regrowth used to form this laser (Logan et al., 1983; 1984) have shown that the regrowth around the mesa exhibits a nonradiative interface at the {111}A planes when the usual cleaning and regrowth procedures described before are used. This is demonstrated in Fig. 16.3 by means of the regrowth around a mesa of InP containing {111}A planes, similar to that used to form the BH laser. Recombination at the laser active layer thus does not obscure the observation of the photoluminescence of this regrown structure at a cleaved mirror. The figure shows that the regrown {111}A face is highly nonradiative. The facet luminescence was excited with weakly focused Ar ion laser light and observed with an infrared microscope equipped with an X100 objective and an S-1 image converter tube. The only nonradiative interface is {111}A, with the (100) interface nearly as bright as the regrown or first grown InP. To simulate the growth situation on the {111}A mesa faces homoepitaxial growth on {111}A faces of InP substrates was performed; and, to further probe this growth

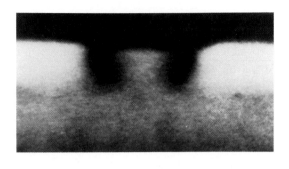

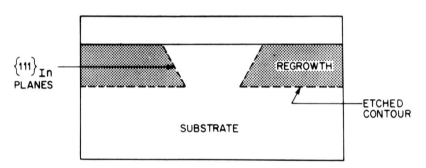

Fig. 16.3 Shown schematically in the lower area, a 10 μm wide mesa with {111}A side walls is formed on a (100) InP wafer by masking with SiO₂ and etching in a 1% solution of bromine-methanol and regrowing InP between the mesas. The photoluminescence at a cleaved cross section is shown in the upper area. The dark regions at the {111}A regrown interfaces indicate the nonradiative behavior.

anomaly, different substrate preparations were used to examine the cause of the difficulty in obtaining good epitaxy on the {111}A surface.

The substrates were obtained from 500 μm thick slices of ⟨111⟩ oriented InP and were chemically-mechanically polished in Br₂-methanol etchant on both surfaces to ∼ 300 μm thickness. To test the effectiveness of an etchant as a surface preparation prior to regrowth a further 25 μm was removed in a stagnant solution of the etchant.

Even though the {111}A plane etches very slowly, the surface produced always exhibited a coarse orange peel finish. This roughness is attributed to the inability to etch sufficient material from this surface to get uniformly beyond the saw damage present on the original wafer.

The layer growth characteristics encountered in growth on {111}A planes are summarized in Fig. 16.4. Figure 16.4d shows the uniform wetting achieved after two anodizations (Studna and Gualtieri, 1981) to 100 V with successive dissolution of the native oxide in buffered HF. By measurement of the step height of

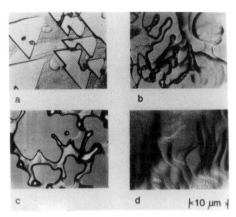

Fig. 16.4 The {111}A plane of InP after etching in Br$_2$-methanol solution and regrowth of InP by liquid phase epitaxy with a final cleaning step prior to epitaxy of (a) anodization to a 100 V and dissolution of the native oxide in buffered HF; (b) etching in a 1% solution of Br$_2$-methanol; (c) etching in a 10-1-1 by volume of H$_2$SO$_4$—H$_2$O—H$_2$O$_2$; and (d) anodization 2 times to 100 V with oxide dissolution in buffered HF.

such anodization adjacent to a surface protected by SiO$_2$, it was observed that 860 Å was anodically etched. Etching with more than two anodizations gives similar growth, but a single anodization produced the local wetting shown in Fig. 16.4a. Figures 16.4a and b show the two types of surfaces encountered after etching in Br$_2$-methanol, largely independent of the Br$_2$ concentration in the range 0.01 to 10%. The locally wetted surface with irregular growth and melt entrapment are both encountered. The differences between Fig. 16.4a and 16.4b in the amount of surface coverage by platelet growth are not understood but may relate to the degree of saw damage introduced in the wafer preparation. The commonly used etchant 10 : 1 : 1 volume ratio of H$_2$SO$_4$: H$_2$O : H$_2$O$_2$ was used as a final preparation for the growth shown in Fig. 16.4c. These results as well as those obtained with HCl or HBr-based etchants are qualitatively similar to Fig. 16.4b. An additional final deoxidation in HF with H$_2$O rinse did not affect the subsequent layer growth. Photoluminescence at cleaved cross-sections similar to the data of Fig. 16.3 show that indeed the cleanest regrowth of Fig. 16.4d has the most radiative interface.

Studies (Logan and Temkin, 1986) of regrowth with In$_{1-x}$Sn$_x$ melts show that with $x > 0.5$ clean regrowth occurs on the {111}A plane if a phosphorus source is used to protect the substrate and a supersaturation ~ 10°C is used with regrowth at 650°C. While the regrown layer is heavily doped, since Sn is a donor, the growth of a thin wetting layer preceding the normal regrowth can be successfully incorporated into buried heterostructure formation and other device applications.

16.3 VAPOR PHASE EPITAXY

The expression vapor phase epitaxy (VPE) generically refers to any non-solution, moderate pressure crystal growth technique involving a vapor-solid interface. In the context of III-V compound semiconductor growth, however, VPE commonly describes two similar growth technologies based on deposition of epitaxial films via the reaction of a group III chloride (InCl, GaCl) with group V species (As_2, P_2, As_4, P_4). Since these species are volatile only at elevated temperatures, VPE is a hot wall technique conducted in a quartz reaction vessel heated inside a furnace. Growth of III-V materials, which contain Al, is not readily achieved by VPE because quartz reacts with volatile aluminum compounds at temperatures required for crystal growth.

The two variants of VPE, termed hydride and trichloride VPE, differ in the way in which the reactive species are prepared, as shown schematically in Fig. 16.5. The hydride approach supplies group V species via the pyrolysis of PH_3 and AsH_3, and the group III chlorides are generated via reaction of HCl with liquid metal sources contained in graphite or quartz boats. A wide variety of reactor designs used in hydride epitaxy have been reviewed by Olsen (1982), and most optoelectronic device results obtained from materials grown by VPE are based on the hydride technique.

Trichloride VPE supplies both the group V species, as well as the HCl required for group III transport, by the pyrolysis of $AsCl_3$ or PCl_3 in the presence of H_2. The technique employs solid binary III-V sources and has been used primarily for the growth of high purity GaAs used to fabricate microwave devices (Jain and Purohit, 1984). The use of solid sources in the trichloride technique means that the reactor source region must be at a higher temperature than the growth region, since this temperature difference determines the degree of gas phase supersaturation in the growth region. In the growth of InGaAsP alloys, four separate binary sources (e.g., InP, InAs, GaAs, GaP) are required for the growth of the entire range of alloy compositions, which lattice match InP (Seki and Koukitu, 1985). Applications of the trichloride approach to the growth of InGaAsP films for optoelectronic devices are relatively recent (Vohl, 1981; Cox et al., 1985; 1986).

Of these two VPE techniques, the hydride technique is better suited for the flexible growth of all alloy compositions in a single reactor system at moderate levels of purity ($N_D - N_A \geq 10^{15}$ cm^{-3}). The trichloride system, however, excels at the growth of epitaxial films of consistently high purity owing to the high purity of commercially available In, Ga, $AsCl_3$, and PCl_3. Both techniques offer the advantage of large area growth, high throughput, good layer thickness uniformity, and excellent surface morphology. A recent review discussing both techniques has been published by Beuchet (1985).

16.3.1 Thermodynamics and Kinetics

The chemistry of the crystal growth process for both forms of VPE can be expressed by the following set of chemical equilibria:

$$InCl + 1/4\, P_4 + 1/2\, H_2 \rightleftharpoons InP_{(ss)} + HCl \qquad (16.1)$$

$$InCl + 1/4\, As_4 + 1/2\, H_2 \rightleftharpoons InAs_{(ss)} + HCl \qquad (16.2)$$

$$GaCl + 1/4\, As_4 + 1/2\, H_2 \rightleftharpoons GaAs_{(ss)} + HCl \qquad (16.3)$$

$$GaCl + 1/4\, P_4 + 1/2\, H_2 \rightleftharpoons GaP_{(ss)} + HCl \qquad (16.4)$$

$$P_2 \rightleftharpoons 1/2\, P_4 \qquad (16.5)$$

$$As_2 \rightleftharpoons 1/2\, As_4 \qquad (16.6)$$

$$PH_3 \rightleftharpoons 1/2\, P_2 + 3/2\, H_2. \qquad (16.7)$$

This basic set of chemical reactions is somewhat incomplete in that higher chlorides of In and Ga (of the form MCl_3, MCl_2 and M_2Cl_4) are omitted since their concentrations are negligible under typical VPE conditions (Ban, 1971; Shaw, 1975). Interpnictide compounds of the form AsP and $As_n P_m$ ($n + m = 4$) are known to be present at significant concentrations (Ban, 1971) but are omitted because thermodynamic data for these species are unavailable.

Several workers, under the assumption of thermodynamic equilibrium between the vapor phase and the deposition alloy, have developed models useful in estimating the alloy composition grown at a given temperature and set of input gas reactant concentrations for both the hydride (Nagai, 1980; Koukitu and Seki, 1980) and halide (Seki and Koukitu, 1985) systems. These models use published thermodynamic data for the gas phase species and semiempirical estimates for the activity coefficients of the deposited binary constituents (e.g., InP, InAs, GaAs, and GaP) in solid solution (Stringfellow, 1974; Onabe, 1982). While the models do not provide a precise prediction of the alloy composition for a given set of input gas concentrations, the agreement with available experimental data is reasonable, as shown in Fig. 16.6. These models provide a useful starting point for alloy growth in a particular reactor system and can also be used to evaluate the sensitivity of the alloy stoichiometry to instrumental inaccuracy for various (e.g., hydride, trichloride, etc.) VPE techniques (Yoshida and Watanabe, 1985).

In addition to thermodynamics, kinetics play an important role in the growth of InGaAsP alloys. While the kinetics of the VPE source reactors, such as the reaction of HCl with the In and Ga sources and the pyrolysis of PCl_3, $AsCl_3$, AsH_3, and PH_3 can be important in determining reactor performance, these

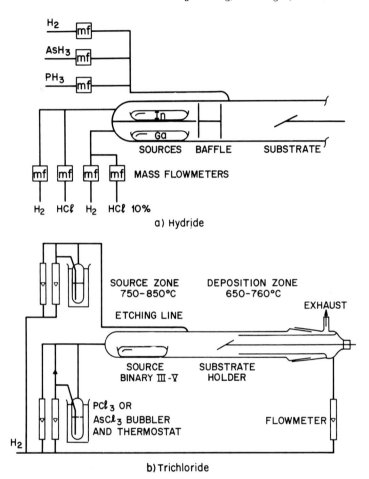

Fig. 16.5 Schematics of typical hydride (a) and trichloride (b) VPE systems illustrating basic differences in source preparation for the two techniques.

reactions can be made to approach thermodynamic equilibrium with careful selection of a reactor source region design. Kinetic factors involved in the growth of the epitaxial films can influence the growth rate and the stoichiometry of the grown films and depend upon the substrate orientation, the growth temperature, and the gas phase saturation. The most detailed understanding of kinetic factors and growth mechanisms for VPE is based on studies of trichloride GaAs VPE (Jain & Purohit, 1984; Theeten et al., 1977). Evidence for the importance of kinetic factors in VPE of InGaAsP alloys is based on the substrate orientation dependence of the growth rate and stoichiometry of ternary InGaAs films

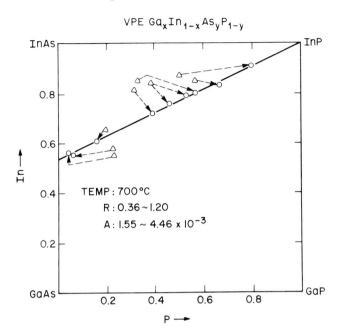

Fig. 16.6 Comparison of experimental alloy growth results (0) and thermodynamic calculations (Δ) (from Olsen, 1982).

(Kanbe *et al.*, 1979; Olsen *et al.*, 1982). Since the binary constituents of InGaAsP films have different free energies of formation at a given growth temperature, conventional kinetic studies such as the variation of growth rate with temperature and gas phase saturation are not easily used to determine the kinetics of alloy growth.

Most hydride and halide VPE systems use purified H_2 as a carrier gas, but some workers have studied the effects of using inert (N_2, He, Ar) gas as a replacement. Generally speaking, the primary effect of inert gas substitution is a significant reduction in the growth rate (Mizuno and Watanabe, 1975). In the case of InP, total replacement of H_2 with an inert carrier renders growth impossible unless PH_3 (as a hydrogen source) is added to the gas flow in the reactor growth region (Giles *et al.*, 1983). The dependence of the growth rate on the H_2 partial pressure stems from its presence as a reactant in the deposition process (Eqs. 16.1–4). In the absence of H_2, deposition must proceed via the formation of some chlorine-containing compound (possibly $GaCl_3$, $InCl_3$, or Cl_2), much less stable than HCl. The effect of H_2 on the growth of III-V compounds has been cleverly applied to the trichloride growth of GaAs, where the gas phase supersaturation required for growth was achieved in a flat furnace temperature profile by adding H_2 to the reactor downstream of the reactor source region in which H_2 was absent (Koukitu *et al.*, 1976). Pulsed introduction of H_2 in a similar reactor has been used to induce pulsed growth of GaAs, with growth as thin as 40 Å/pulse (Koukitu *et al.*, 1985).

16.3.2 Reactor Design

VPE reactor systems used for the growth of epitaxial films for optoelectronic devices require reproducible control of alloy composition, good control of heterointerface abruptness and the ability to grow thin (typically 0.01 μm to 10 μm) uniform layers. Gas input flow rates are selected using electronic mass flow controllers usually under computer control, and input concentrations are adjusted by diluting high concentration mixtures of AsH_3, PH_3, and HCl stored in compressed gas cylinders with H_2 purified by palladium diffusion. Since these gases are highly toxic, laboratories housing VPE reactor systems should contain toxic gas monitoring systems as well as provisions for treating the reactor exhaust effluent safely. Trichloride VPE uses $AsCl_3$ and PCl_3 and poses a somewhat reduced toxic gas hazard but requires increased care in handling since these liquid sources are shipped, stored, and operated in glass (or quartz) containers.

Crystal growth requirements for optoelectronic devices outlined above, coupled with the need to prevent thermal decomposition of the InP (or alloy) during heat-up or in between layer growths, have resulted in two general types of VPE reactor designs: single barrel and multibarrel. In the single barrel design, gas inputs for a particular alloy composition are selected, and the substrate is stored in a separate, heated portion of the reactor until the input flows are stabilized and/or the substrate is at the growth temperature (typically 650° to 725°C). In this separate chamber, the substrate is protected against thermal decomposition by providing a slight phosphorus overpressure obtained by adding PH_3 to the H_2 flow in that portion of the reactor. Multibarrel reactor systems (Beuchet, 1985; Olsen, 1982) are actually multiple single barrel reactors contained within a single quartz shell. Input gas flows for successive layers (or PH_3 preservation flows) can be set in one barrel, while growth is proceeding in another. These reactors also include some provision for rotating or translating the sample between barrels for the growth of different layers. Both reactor types have provisions for adding HCl directly to the growth region for cleaning between growth runs as well as for suppressing extraneous reactor wall deposition during growth (Mizutani and Watanabe, 1982).

Single barrel systems are much simpler in terms of quartzware and gas mixing manifold designs and are best suited for growth of single epitaxial layers or heterostructures, which do not require extremely abrupt (~ several lattice spacings) interfaces. Difficulties relating to interface abruptness stem from what may be called memory effects, such as outgassing of GaCl for the $Ga_{(l)}$ source during growth of an InP layer and following the growth of an InGaAsP layer

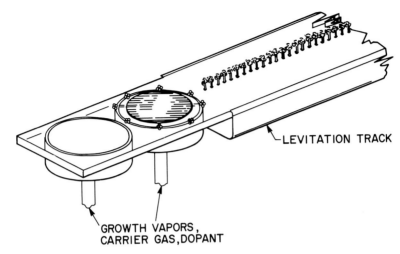

LEVITATION TRACK

GROWTH VAPORS,
CARRIER GAS,DOPANT

Fig. 16.7 Illustration of the levitation concept for growth in a two frit vapor levitation epitaxial (VLE) reactor (from Cox, 1984).

(Karlicek, 1985), or from surface deterioration during wafer storage between layers. Multibarrel reactors are well suited for the growth of abrupt heterointerfaces and multiple quantum well structures, but they are more complicated to construct and operate. Limitations in the size of multibarrel reactors imposed by quartzware fabrication and handling difficulties usually restrict the substrate size. For a more extensive discussion of various published reactor designs the reader is referred to reviews by Olsen (1982) and Beuchet (1985).

A recent, novel approach to VPE reactor design has been discussed by Cox (1984). Termed vapor levitation epitaxy (VLE), this technique uses the input gas flows to float the wafer above a quartz frit. By placing two frits connected to different reactor source regions adjacent to each other (Fig. 16.7), the sample size limitation inherent in a conventional multibarrel reactor can be avoided. Used to date as a trichloride reactor, the technique has been applied to the growth of InGaAsP and InGaAs-containing heterostructures (Cox et al., 1985).

16.3.3 Device Applications

Both hydride and trichloride VPE techniques produce high quality InGaAsP alloys useful in the fabrication of optoelectronic devices. Although trichloride VPE lacks flexibility for quaternary alloy growth, the high purity of the technique makes it well suited for the growth of heterostructures used for detectors (PINs and APDs), which require the growth of InP and In$_{.53}$Ga$_{.47}$As. Cox et al. (1985) have reported the use of a single barrel trichloride reactor for the growth of high purity InGaAs (low 10^{14} cm^{-3}) and InP (mid 10^{14} cm^{-3}) for the fabrication of PIN photodiodes (90 μm diameter, 28 nA leakage and 27 pF capacitance at -5 V).

Hydride VPE has been more widely used for sources (LEDs and lasers) where flexible control of the alloy stoichiometry is important. Yanase *et al.* (1983a) have reported the growth of 1.3 μm DCPBH lasers (LPE used for confining layer regrowth) with a 19.5 mA threshold current and CW operation to 120°C, as well as low threshold MQW 1.3 μm lasers (1.2 kA/cm^2) (Yanase *et al.*, 1983b). Hydride VPE has also been used to grow visible (671 nm emission) InGaAsP/InGaP lasers capable of continuous operation at room temperature (Usui *et al.*, 1985). These three lasers structures were grown in two-barrel hydride VPE reactor designs.

16.4 METALORGANIC CHEMICAL VAPOR DEPOSITION (MOCVD)

16.4.1 Basic Principles and System Design

In the MOCVD of III-V materials a vapor phase mixture of the desired Group III and Group V sources is introduced into the reaction chamber and transported to the heated growth zone, where pyrolysis and subsequent deposition occur. The technique offers the advantage of large-area uniform growth coupled with the ability to produce abrupt interfaces; it may be used to grow virtually all III-V compounds. Several recent reviews have appeared (Stringfellow, 1985; Razeghi, 1985; Ludowise, 1985; Ziko, 1986) to which the reader is referred for a more complete discussion. MOCVD is known by a variety of names, including MOVPE (metalorganic vapor phase epitaxy) and OMVPE (organometallic VPE) derived from the use of metalorganic sources in the process. In particular, Group III alkyls, such as trimethyl indium (TMIn), are used in combination with Group V hydrides, for example, PH$_3$. Since the initial studies, Group V alkyls have also been employed, in addition to Lewis acid–Lewis base adducts of the Group III and V alkyls. The alkyls are air-sensitive, in many cases pyrophoric, liquids, with the notable exception of TMIn, which is a solid. They are contained in stainless steel bubblers or sublimers through which a carrier gas is passed to transport the alkyl vapor into the reactor. The overall reaction occurring is exemplified by the following for InP:

$$Me_3In + PH_3 \rightarrow InP + 3CH_4.$$

The substrate rests on a susceptor (generally silicon-carbide coated graphite) heated, in most cases, by r.f. induction heating or infrared radiation from quartz-halogen lamps. Thus, MOCVD differs from hydride and chloride VPE in that it is a cold-wall process, where the susceptor alone is intentionally heated.

Two types of reaction chamber are most common, horizontal and vertical. A typical horizontal system is shown in Fig. 16.8. In this configuration, gas flow is

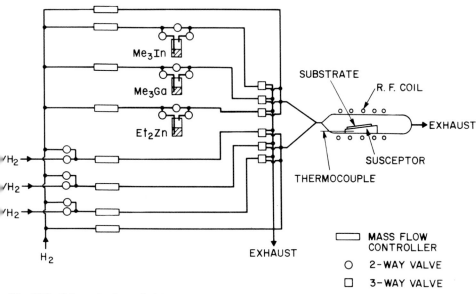

Fig. 16.8 Schematic of a typical horizontal MOCVD system.

approximately parallel to the substrate, the susceptor being angled a few degrees into the flow to prevent reactant depletion occurring toward the downstream end. In contrast, for the vertical design, gas flow is introduced from the top of the chamber, perpendicular to the substrate. A number of variations have been applied to the basic reactor design, including the use of a chimney reactor (Leys et al., 1984), where the substrates are held vertical, substrate transfer mechanisms (Moss and Spurdens, 1984a) and the multichamber approach (Shealy, 1986). The large majority of reactors are operated either at atmospheric or "low" pressure ($\sim$ .1 atmospheres). In contrast, the technique of vacuum chemical epitaxy or vacuum MOCVD (Fraas et al., 1986), developed for the growth of solar cell material, employs a hot-wall reactor (thereby leading to more efficient use of AsH_3), operated at a few mTorr pressure. In most reactors the susceptor can hold one or two wafers, although barrel reactors have been designed to accommodate much larger numbers of wafers. Simultaneous growth on as many 60 GaAs wafers, each measuring 2×4 cm, has been reported (Komeno et al., 1985).

16.4.2 Growth Mechanism and Chemistry

MOCVD is a non-equilibrium process, since the input reactant concentrations are in excess of those that would be in equilibrium with the substrate at the growth temperature. The flow dynamics of the reactor result in formation of a boundary layer, close to the surface of the substrate, within which the gas flow is approximately laminar. A steep thermal gradient exists within this layer, through which the Group III and V reactants diffuse to reach the substrate. As yet there is no detailed understanding of the mechanism by which the growth process takes place and a certain amount of controversy exists as to whether decomposition of the reactants occurs in the boundary layer or on the substrate surface itself. Several groups have established that, for GaAs growth, decomposition of AsH_3 is slow unless catalyzed by the presence of a GaAs substrate (Schlyer and Ring, 1976; Leys and Veenvliet, 1981; Nishizawa and Kurabayashi, 1983). IR and atomic absorption spectroscopy have been employed to monitor the gas-phase Group III species present in the boundary layer (Leys and Veenvliet, 1981; Nishizawa and Kurabayashi, 1983). For GaAs growth, only CH_4 was observed, suggesting that trimethyl gallium, TMGa, is completely decomposed in the gas phase. In the case of InP (Haigh, 1985), however, no gas-phase In atoms were detected, from which the authors concluded that the final stage of TMIn decomposition must occur at the growth interface. Clearly, further work is required before a definitive mechanism can be established.

Prior to decomposition in the heated growth zone, gas-phase reactions may occur between the Group III alkyls and Group V hydrides. These generally result in undesirable adduct or polymer formation, leading to non-reproducible growth and depletion of reactants. The growth of In-containing compounds, in particular, has been plagued by such parasitic reactions. Triethyl indium, TEIn, and, to a lesser extent, TMIn undergo elimination reactions with AsH_3 and PH_3 followed by polymer formation. For example:

$$nR_3In + nPH_3 \rightarrow nR_3In \cdot PH_3 \xrightarrow{-2nRH} \left(\underset{\text{polymer}}{-RIn-PH-} \right)_n$$

$$R = Me, \quad Et.$$

Several strategies have been employed to avoid these unwanted reactions ranging from mixing the reactants as close to the growth region as possible (Fukui and Horikoshi, 1980), through operation of the reactor at low pressure (Duchemin *et al.*, 1981) to use of adducts, such as $Me_3In \cdot PEt_3$, formed from Group III and Group V alkyls (Moss, 1984). More recently, there have been a number of reports of the use of TMIn with PH_3 and AsH_3 at atmospheric pressure with no apparent side reactions occurring (Carey, 1985; Kuo *et al.*, 1985a). It has been suggested that previous difficulties resulted from impurities in the TMIn, which have since been eliminated. Careful attention to reactor design no doubt also has a role to play, since small changes in reactor geometry can greatly affect the growth parameters.

The growth of InGaAs is inherently more difficult than that of InP. In addition to the possibility of side reactions occurring, it is necessary to rapidly change from a PH_3 to an AsH_3 ambient and establish the correct In- and Ga-alkyl flows, without degrading the InP surface or producing a defective interface. TEM studies (Hockley and White, 1984) of InGaAs/InP interfaces demonstrated that considerable improvements in X-ray linewidth and reduction in the number of threading dislocations at the interface could be achieved by increasing the speed at which gas flows were switched. A great deal of attention has been devoted to developing methods for fast switching of gas compositions and elimination of pressure transients. Several different approaches can be used: low-pressure growth (Duchemin *et al.*, 1981), use of vent-run systems coupled with fast-switching manifolds and pressure balancing (Griffiths *et al.*, 1983; Roberts *et al.*, 1984), and substrate transfer techniques (Moss and Spurdens, 1984a), where the substrate is moved between pre-established gas flows.

The two most important variables in MOCVD are growth temperature and input reactant mole fractions, which can be used to determine growth rate and alloy composition. For III-V semiconductors, growth temperatures ranging from ~ 550°C–750°C are normally used, and studies of the dependence of growth rate on growth temperature are helpful in understanding the kinetics of the MOCVD process. As shown in Fig. 16.9 for GaAs, three growth regions are evident (Reep and Ghandi, 1983). In the middle region, where growth usually takes place, the growth rate is almost independent of temperature, indicating

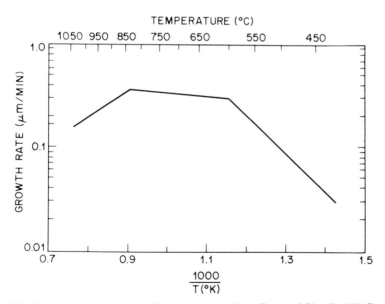

Fig. 16.9 Temperature dependence of GaAs growth rate (from Reep and Ghandi, 1983. Reprinted by permission of the publisher, the Electrochemical Society, Inc.).

that diffusion of reactants through the boundary layer to the growth interface is the rate-limiting step. Here, the growth rate is directly proportional to the Group III input mole fraction and independent of the AsH_3 concentration, provided the latter is in excess. At temperatures above and below this region the growth rate is highly temperature dependent. At low temperatures the growth rate is limited by AsH_3 pyrolysis, although the Group III decomposition rate may also be reduced sufficiently to affect the growth rate (Tsang, 1984). In the high temperature regime, the decrease in growth rate is probably due to desorption of Ga or As from the GaAs surface or to deposition on the reactor walls, resulting from premature decomposition of the reactants in the gas phase at the high temperatures involved. Similar results have been obtained for InP (Duchemin et al., 1981; Razeghi et al., 1983).

The Group V hydrides, AsH_3 and particularly, PH_3, possess significantly greater thermal stability than the Group III alkyls. This necessitates the use of high V/III ratios in MOCVD, especially for low pressure growth, where, in the case of InP, V/III ratios in excess of 100 are commonly employed (Razeghi

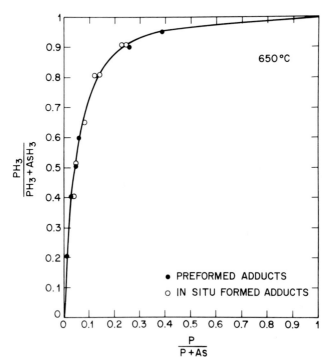

Fig. 16.10 Relationship between the solid and gas compositions for the Group V components in InGaAsP (from Moss and Spurdens, 1984b).

et al., 1983). For growth of materials containing both As and P, the P/As incorporation ratio depends principally on the pyrolysis rates of the PH_3 and AsH_3. As a result of the greater stability of PH_3, the gas-solid composition relationship is highly nonlinear, as shown in Fig. 16.10 for InGaAsP (Moss and Spurdens, 1984b). In addition, mass spectrometric studies (Ban, 1972) indicate that, although PH_3 is more stable than AsH_3, the rate of PH_3 decomposition increases more rapidly with temperature than that of AsH_3. Thus, the P/As incorporation also varies with temperature for a fixed PH_3/AsH_3 input ratio. This places stringent requirements on control of both the susceptor temperature and the input AsH_3 and PH_3 mole fractions, since slight variations in either can significantly affect the alloy composition.

16.4.3 Materials' Properties

For lightwave devices, the III-V compounds of greatest importance are InP, InGaAs and InGaAsP, although InAlAs has recently generated considerable interest. Rapid progress has been made in the past two to three years in the growth of these materials by MOCVD, and it is now clear that the high purity and abrupt interfaces required by current device technology can be achieved using this technique. A more detailed discussion of the properties of these materials and other important III-V compounds grown by MOCVD can be found in the reviews by Stringfellow (1985) and Ludowise (1985).

16.4.3.1 InP. The major factor influencing the background doping levels and mobility of the InP is source purity, particularly that of the In alkyls, and large variations in purity have been observed from different TEIn and PH_3 sources (Razeghi, 1985). However, the purity of the available In alkyls has steadily improved, and 77 K mobilities in excess of 130,000 $cm^3V^{-1}s^{-1}$ have been reported for InP grown using TEIn (De-Forte-Poisson, 1985) and TMIn (Zhu *et al.*, 1985a).

Dopant sources for InP are typical of those used for most of the III-V compounds. The *n*-type dopants are obtained from gaseous sources, H_2S, H_2Se and SiH_4 being the most popular, whereas organometallics such as diethyl zinc, dimethyl cadmium and biscyclopentadienyl magnesium, are normally favored for *p*-doping. Growth of semi-insulating Fe-doped InP has also been reported (Long *et al.*, 1984), with resistivities as high as 2×10^8 Ω-cm obtained. This material has a particularly important application in providing current confinement in buried heterostructure lasers (Dutta *et al.*, 1986; Wilt *et al.*, 1986) and may be used for isolation layers in FETs (Cheng *et al.*, 1985).

16.4.3.2 InGaAs. Until recently, InGaAs of sufficient quality for photodetector and FET applications was obtained routinely only by low-pressure MOCVD. With improvements in growth techniques as described, before, and also in source purity, however, InGaAs with room temperature mobility greater than 10,000 $cm^2V^{-1}s^{-1}$ has been reported by a number of workers using atmospheric (Carey, 1985; Kuo *et al.*, 1985a; Chan *et al.*, 1985) and low-pressure (Razeghi *et al.*, 1983) growth. In addition, enhanced low temperature mobilities have been measured ($> 90,000$ $cm^2V^{-1}s^{-1}$ at 4.2 K) (Zhu *et al.*, 1985b) indicative of a two-dimensional electron gas at the InGaAs/InP interface, confirmed by Shubnikov-de Haas and cyclotron resonance experiments.

Attention has also been turned toward the growth of InGaAs/InP quantum wells. Such structures are of particular interest, not only because of their novel optical and electrical properties, but also because they have the potential for improving the device performance of lasers and detectors. Razeghi *et al.* (1984) have reported growth of multiquantum well structures by low-pressure growth. Well widths as narrow as 8 Å were obtained, having strong photoluminescence emission intensities and narrow linewidths. Although the low-pressure technique offers the advantage of lower growth rate, thereby leading to greater control of thin-layer growth, several groups have reported atmospheric pressure growth of both single- and multiquantum well structures (Kuo *et al.*, 1985b; Miller *et al.*, 1986; Scholz *et al.*, 1986; Skolnick *et al.*, 1986). Photoluminescence linewidths and energy shifts comparable to those obtained by low-pressure have now been measured (Miller *et al.*, 1986).

16.4.3.3 InGaAsP. As in the case of InGaAs, growth of InGaAsP by atmospheric pressure MOCVD lagged behind its low-pressure counterpart, which for several years has produced double heterostructure lasers with excellent operating characteristics (Razeghi, 1985). Recent results, however, indicate that the atmospheric pressure technique is also suitable for production of lightwave sources (Nelson *et al.*, 1985).

A remarkable degree of compositional uniformity has been obtained for InGaAsP grown at low pressure (Razeghi *et al.*, 1983; Razeghi, 1985). For epitaxial layers 10 cm^2 in area the variation in lattice mismatch and photoluminescence wavelength is less than 2%. Lattice mismatch, $\Delta a/a$, $\leq 4 \times 10^{-4}$ has been obtained repeatedly.

16.4.4 Selective Area Epitaxy and Regrowth

Selective area epitaxy and regrowth techniques have important roles to play in the fabrication of double heterostructure lasers and in integrated optoelectronics. To date, studies of selective growth have concentrated on GaAs-based materials (Gale *et al.*, 1982; Nakai and Ozeki, 1984; Ghosh and Layman, 1984; Kamon *et al.*, 1985; Okamoto and Yamaguchi, 1986). With the exception of one report of growth on ion-implanted substrates (Favennec *et al.*, 1983), results of selective epitaxy of InP have only recently begun to appear in the literature (Oishi and Kuroiwa, 1985; Blaauw *et al.*, 1986; Clawson *et al.*, 1986). Much of the work has involved growth on masked substrates, and some controversy has arisen concerning the degree of selectivity obtained. Some authors observed essentially complete selectivity (Kamon *et al.*, 1985), with growth occurring only on the exposed substrate, whereas others reported extensive polycrystalline growth on the mask (Gale *et al.*, 1982). These apparently conflicting observations almost certainly result from differences in the growth conditions employed, since factors such as growth temperature and pressure, mask orientation and material, degree of mask undercut and the ratio of exposed substrate to mask area significantly influence the nature of the selective growth.

Selectivity is dependent on the ability of the reactant Group III species to diffuse across the mask to the adjoining exposed substrate surface. Thus, by reducing the growth pressure (Kamon *et al.*, 1985) and increasing the growth temperature (Ghosh & Layman, 1984), thereby increasing reactant migration, selectivity is improved. However, selectivity is also a function of the concentration of the Group III species in the vicinity of the mask. Migration of these species across the mask leads to an increase in their concentration at the mask edge, resulting in an accelerated growth rate and ridge formation (Nakai and Ozeki, 1984; Okamoto and Yamaguchi, 1986). As the mask width and area increase, the Group III reactant concentration at the mask surface can build up sufficiently to cause nucleation and subsequent polycrystalline deposition on the mask (Oishi and Kuroiwa, 1985).

Overgrowth at the mask edge can be suppressed by mask undercutting (Zilko *et al.*, 1986), as shown in Fig. 16.11a for a mesa etched InP substrate. The effects of increasing the mask area are apparent, however, from the channel etched substrate in Fig. 16.11b. Although the oxide overhang again prevents overgrowth at the mask edge, spherical nodules of InP are nucleated on the SiO_2 (actually through pinholes in the oxide) due to the greatly increased In concentration at the mask surface.

At both atmospheric and low pressure, faceted growth occurs at the mask edge (Nakai and Ozeki, 1984; Kamon *et al.*, 1985; Okamoto and Yamaguchi, 1986). When the mask is aligned in the [110] direction, the epitaxial layer is bounded by the {111}B planes, whereas when alignment is in the [$\bar{1}$10] direction, the layer is bounded by the {111}A and {111}B planes (Kamon *et al.*, 1985). Results obtained using low pressure growth suggest that while growth occurs on the {111}A planes, no growth occurs on the {111}B planes.

Using a judicious choice of the above variables, selective MOCVD regrowth has been achieved successfully for a number of laser structures. In the InP/InGaAsP materials' system, these have typically involved MOCVD regrowth over LPE base structures (Dutta *et al.*, 1986). However, Nelson *et al.*, (1985) have now reported a BH laser grown entirely by MOCVD.

Another important device structure requiring regrowth is that of the DFB laser. Several problems exist in DFB laser fabrication, not least of which is the

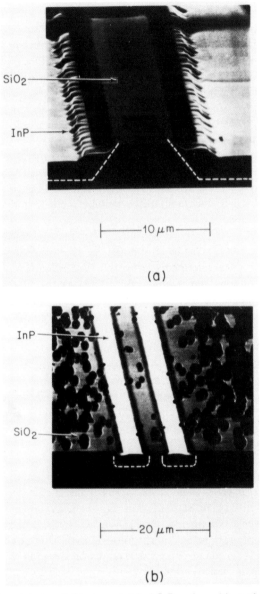

Fig. 16.11 (a) Mesa-etched and (b) channel-etched InP wafers with mask undercutting (with thanks to J. L. Zilko).

surface deformation occurring during the conventional LPE overgrowth of the diffraction gratings. By careful control of the pre-growth conditions, deformation-free overgrowth of the gratings has been achieved by MOCVD (Nelson *et al.*, 1983), and hybrid LPE-MOCVD DFB lasers fabricated (Westbrook *et al.*, 1984). More recently, low-pressure MOCVD has been used to grow the complete DFB laser structure (Razeghi *et al.*, 1984).

16.5 MOLECULAR BEAM EPITAXY

16.5.1 Basic Principles and System Design

MBE, unlike the other epitaxial growth techniques, takes place in ultra-high vacuum (UHV). The basic process for III-V material deposition involves generation of Group III and Group V molecular beams which then impinge on a heated substrate where epitaxial growth occurs. A typical system (Cho and Cheng, 1981) is shown in Fig. 16.12. A variety of pumping systems can be employed to maintain the UHV, but cryopumps or turbomolecular pumps are favored when elements of high volatility, such as phosphorus, are present.

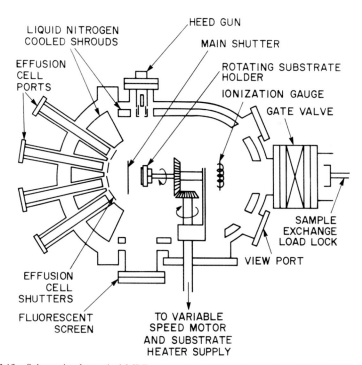

Fig. 16.12   Schematic of a typical MBE system viewed from the top (from Cho and Cheng, 1981).

Extensive use is made of cryopanels both to reduce background contamination from water vapor and carbon-containing gases and also to reduce the chamber pressure after growth. A load-lock allows the UHV to be maintained in the growth chamber during insertion and removal of samples.

Knudsen effusion cells, normally made from pyrolytic boron nitride, are used to generate the molecular beams. The cells are equipped with externally controlled mechanical shutters, which can be rapidly opened or closed, thereby allowing abrupt transitions between grown layers. The beam flux intensity is controlled by the effusion cell temperature and by the aperture of the cell.

In traditional MBE solid, elemental sources are normally employed, although polycrystalline GaAs and InP have been used for the Group V sources. More recently, the development of gas-source MBE and chemical beam epitaxy, as described later in this chapter, has led to the introduction of gas and metal alkyl sources. Elemental arsenic and phosphorus generate the tetramers As_4 and P_4 respectively, while the compound sources evaporate to give As_2 or P_2 dimers. Use of the dimers is generally preferred because they have greater sticking coefficients than the tetramers and have been observed to improve growth quality (Neave et al., 1980; Tsang et al., 1983). Thus, when arsenic or phosphorus sources are employed, a high temperature cracker is sometimes incorporated at the end of the effusion cell to thermally dissociate the tetramers into dimers.

Elemental sources are also used for the n- and p-type dopants in MBE. Si and Sn are the most widely used n-type sources although Sn tends to accumulate at the surface during growth. Ge is amphoteric with the type of doping obtained dependent on the growth conditions employed (Cho and Hayashi, 1971). In contrast to the other epitaxial growth techniques, Zn is not a preferred p-dopant because, as is the case for Cd, it has a negligible incorporation rate under normal growth conditions. Thus Be, being generally well behaved, is by far the most common p-type dopant.

The UHV present in the MBE growth chamber permits use of *in-situ* surface analytical techniques to monitor surface structures both prior to and during growth. Perhaps the most important of these is reflection high electron energy diffraction (RHEED), which provides information on surface structure, smoothness and cleanliness. It is particularly useful in monitoring the substrate surface prior to growth. For example, as a GaAs substrate is heated the surface oxide desorbs and, in the presence of an As beam, a streaked RHEED pattern, characteristic of a clean, smooth, single-crystal surface is observed. Information concerning the type of surface reconstruction occurring is also obtained. Auger electron spectroscopy (AES) is another frequently employed technique for surface analysis, again with particular application in characterizing the initial substrate surface. A more detailed discussion of *in-situ* diagnostic techniques can be found in a recent review by Tsang (1985a).

16.5.2 Growth Chemistry

In MBE growth, the atomic Group III species impinge on the heated substrate and migrate into appropriate lattice sites, epitaxial growth subsequently occurring in the presence of the impinging Group V beam. The growth process has been studied in most detail for GaAs, but the results are generally applicable to InP and other III-V semiconductors.

Neither As_2 nor As_4 stick to the GaAs substrate surface in the absence of a Ga beam, but occupy As surface sites to form an As-stabilized surface (Foxon and Joyce, 1975, 1977). Ga, on the other hand, sticks well to the substrate surface with a sticking coefficient of unity below $\sim 480°C$ (Arthur, 1968). The Ga flux density thus controls the growth rate and, in the presence of an excess As flux, growth is obtained over a wide temperature range. However, it has been found that high quality growth is obtained only at temperatures in excess of the congruent sublimation temperature (Wood, 1982).

For GaAs, the growth rate for a fixed Ga flux decreases above $\sim 640°C$ due to the decrease in sticking coefficient at elevated temperatures (Fischer et al., 1983). For growth of III-V compounds containing only one Group V element, it is sufficient to have the Group V flux in excess of that of the Group III. In contrast, when two Group V elements are involved, notably for growth of InGaAsP, precise control of the Group V fluxes is required as a result of the marked difference in Group V sticking coefficients. This is particularly difficult to achieve since it also requires precise temperature control of the large effusion cells needed for the Group V sources. In contrast, materials containing two or more Group III elements, such as InGaAs or InGaAlAs, do not suffer from this problem since Group III sticking coefficients are all close to unity.

16.5.3 Growth and Properties of InP-Based Materials

Growth of materials lattice-matched to InP is complicated by the fact that the InP surface oxide is removed thermally only above $\sim 500°C$, that is, above the decomposition temperature of InP. By heating the substrate in the presence of either a phosphorus (Roberts et al., 1981) or an arsenic (Davies et al., 1980) flux, however, the surface is stabilized against noncongruent decomposition. Furthermore, growth of P-containing compounds presents a particular problem because the low sticking coefficient of phosphorus, coupled with its highly volatility, places great demands on the MBE pumping system, even when P_2 rather than P_4 is used. Background doping levels for InP are usually in the 10^{16} cm^{-3} range, although somewhat lower values were reported by Tsang et al. (1982) using a specially designed system. Recently, it has been suggested that sulfur, from the red-phosphorus source, is the principal residual donor in MBE-grown InP

(Martin *et al.*, 1985). Use of vacuum-packed phosphorus may also be beneficial by reducing the concentration of donor impurities (Roberts *et al.*, 1986).

Given the difficulties associated with the growth of phosphorus-containing compounds, particularly InGaAsP, attention has been focused on the $In_xGa_yAl_{1-x-y}As$ system, lattice-matched to InP. Here the bandgap can be adjusted from .76 eV (1.65 μm, InGaAs) to 1.47 eV (.87 μm, InAlAs). Growth of high-purity InGaAs has been difficult to achieve by MBE, background levels being similar to those reported for InP. With careful attention to growth conditions, however, InGaAs with room temperature mobility > 11,000 $cm^2V^{-1}s^{-1}$ and background doping levels in the low 10^{15} cm^{-3} range has now been obtained (Mizutani and Hirose, 1985; Lee and Fonstad, 1986). The electrical properties of undoped InAlAs, on the other hand, are determined by the growth temperature, semi-insulating material being obtained below ~ 500°C and *n*-type material resulting at higher temperatures (Davies *et al.*, 1984).

InGaAs/InAlAs heterostructures have recently generated a considerable amount of interest for devices operating in the 1.3–1.55 μm range. One particular application is that of avalanche photodetectors (Capasso *et al.*, 1984). The smaller valence band discontinuity of the InGaAs/InAlAs heterojunction compared to that for InGaAs/InP (0.2 eV vs. 0.4 eV) (People *et al.*, 1983) should reduce accumulation of photogenerated holes at the heterointerface, thereby producing faster devices.

More recently InGaAs/InAlAs superlattices have been described where layer thicknesses of 105 Å were chosen to provide 1.55 μm emission (Stolz *et al.*, 1985). Even for relatively low substrate temperatures (500°C), the high quality of the superlattices was evident from the narrow X-ray diffraction linewidths (20–25″) of the satellite peaks. The optical properties of the layers, however, as evaluated by 77K photoluminescence, were observed to improve considerably as the growth temperature was increased from 500 to 590°C.

16.6 GAS-SOURCE MBE

16.6.1 Growth Technique

Several important advantages can be gained by employing the gases AsH_3 and PH_3 as the Group V sources for MBE. The need to replenish the Group V sources inside the growth chamber is eliminated, a more stable beam intensity is obtained and, most significantly, for growth of InGaAsP, precise control of the As/P flux is possible. Early work by Panish (1980) demonstrated the feasibility of GaAs and InP growth by gas-source MBE and further studies by Calawa (1981) indicated that high quality GaAs could be obtained using this technique. Two types of gas source have been employed, low-pressure and high-pressure. In the low-pressure system a leak valve is placed between the gas handling

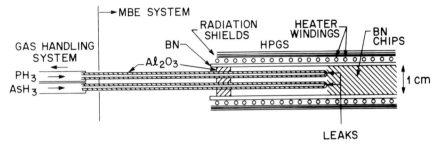

Fig. 16.13 High pressure gas source (from Panish and Hamm, 1986).

equipment and the thermal cracker, where decomposition of the Group V hydrides occurs, facilitated by means of a Ta catalyst. Several groups have used this technique for growth of both GaAs and InP (Calawa, 1981; Chow and Chai, 1983; Huet *et al.*, 1985; Panish *et al.*, 1985). The high-pressure gas source has been developed by Panish (Panish *et al.*, 1985; Panish and Hamm, 1986). In the current set-up, shown in Fig. 16.13, AsH_3 and PH_3 are introduced separately into the alumina tubes, which have small leaks at the sealed ends. The heated region is generally maintained between 900° and 1000°C. The initial decomposition products, As_4, P_4 and H_2 leak into a low-pressure region, packed with boron nitride chips, where further cracking occurs, yielding principally As_2, P_2 and H_2.

16.6.2 Materials' Properties

The success of gas-source MBE in growth of abrupt interfaces and ultra-thin layers is readily apparent from the results obtained for quantum-well structures (Panish *et al.*, 1986). The structures assessed consisted of a thick (600 Å) layer and four quantum wells (30, 15, 7.5 and 5 Å) of either InGaAs or InGaAsP, separated by InP barriers. Strong photoluminescence emission intensities were observed and large energy shifts due to quantum confinement. In particular, for the 5 Å InGaAs well, a shift of 534 meV was measured, ~ 85% of the bandgap discontinuity. Figure 16.14 shows a high resolution lattice image transmission electron micrograph of the InGaAs quantum-well structure, indicating the measured well thicknesses to be in very good agreement with those predicted from extrapolation of the growth parameters. Furthermore, it can be seen that the interfaces are abrupt on the order of one monolayer.

InGaAs/InP superlattices have also been grown, in the form of PIN photodetectors (Panish *et al.*, 1986). The high quality of the structures is attested to by the ability to resolve the $n = 1$ heavy and light hole excitons at room temperature, even for a structure consisting of 100 periods with 20 Å wells.

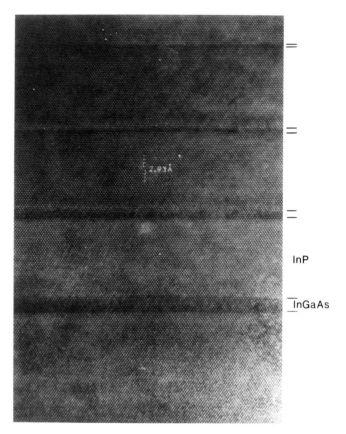

Fig. 16.14 High resolution lattice image TEM of a structure containing four quantum wells of InGaAs, each separated by 150 Å of InP (from Panish *et al.*, 1986).

16.7 CHEMICAL BEAM EPITAXY (CBE)

16.7.1 Growth Technique

CBE is a newly developed technology, designed to incorporate the advantages of MOCVD and MBE in one growth method. Gaseous sources are used that, in the original implementation of the technique, consisted of Group III and Group V alkyls (Tsang, 1984). However, the relatively low purity of the Group V alkyls led to their replacement by the Group V hydrides AsH_3 and PH_3, despite the toxicity of the latter species.

As in MOCVD, H_2 is used to transport the metal alkyl vapors and mass flow controllers control the gas flow rates. Low-pressure thermal crackers, upstream of the flow controllers, decompose the AsH_3 and PH_3 into elemental arsenic and phosphorus. The Group III flows are combined and a single effusion oven used to provide good compositional uniformity during growth of materials containing mixed Group III elements. In CBE the Group III alkyl molecular beam impinges along a line-of-sight onto the heated substrate where dissociation or reevaporation may occur, depending on the substrate temperature and the beam flux. Thus, at a temperature sufficiently high to decompose all the impinging Group III species, the growth rate is determined by their arrival rate. At lower temperatures dissociation of the alkyls is the limiting step. The temperature at which complete dissociation occurs is found to be dependent on the H_2 flow rate through the alkyls, decreasing as the flow rate is reduced.

16.7.2 Materials' Properties

Extremely impressive results have been achieved by CBE, despite its recent development. InGaAs with room-temperature mobilities of 10,000–12,000 $cm^2V^{-1}s^{-1}$ and X-ray linewidths as small as 24 arc seconds have been obtained (Tsang et al., 1986a). In addition, high intensity, low temperature photoluminescence peaks have been observed due to excitonic transitions, with a linewidth as narrow as 1.2 meV. The presence of a two-dimensional electron gas at the CBE InGaAs/InP interface is apparent from the enhanced low temperature mobilities measured (as high as 130,000 $cm^2V^{-1}s^{-1}$ at 4.2K) (Tsang et al., 1986b).

An assessment of quantum-well structure grown by CBE is a further indication of the success of the technique for growth of abrupt interfaces and ultra-thin

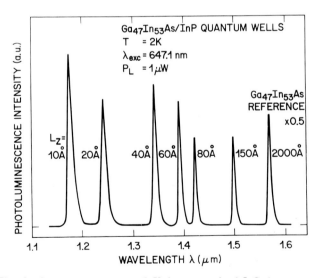

Fig. 16.15 Photoluminescence spectrum at 2 K from a stack of InGaAs quantum wells with different thicknesses, separated by 700 Å InP barriers (from Tsang and Schubert, 1986).

layers (Tsang and Schubert, 1986). InGaAs/InP quantum wells with thicknesses as small as 6 Å have been evaluated by low-temperature photoluminescence. The measured energy shifts were in good agreement with theoretical predictions and sharp peaks due to excitonic transitions were observed as shown in Fig. 16.15 for a multiquantum well structure of varying well-width.

16.7.3 Selective Area Growth

Relatively few studies exist relating to MBE growth on masked or etched substrates. Early work (Cho and Ballamy, 1975) involved growth of GaAs on SiO_2-masked substrates, where polycrystalline growth was observed on the SiO_2 and single crystal growth on the exposed substrate. There have also been reports concerning selective growth by MBE and lately, CBE involving use of shadow masks, Si in particular (Tsang and Ilegems, 1977, 1979; Tsang and Cho, 1978; Tsang, 1985b). The mask is generally pressed against the substrate with the aid of Ta clamps attached to the heating block, although in one experiment a GaAs substrate was moved parallel to the mask during MBE growth in order to achieve epitaxial writing (Tsang and Cho, 1978). The masked substrate area remains clean during growth; thus, further depositions may be made on the same substrate without additional surface cleaning. Multilevel masking techniques have been used to fabricate GaAs two-dimensional optical waveguide tapered couplers (Tsang and Ilegems, 1979), where the first mask defined the stripes and the second mask produced the optical tapers. Patterned growth of InGaAs by CBE has now been reported using the same type of shadow masks, thereby demonstrating the beam nature of the CBE process (Tsang, 1985b). A screen of 100 μm Mo wire was held 2 μm above an InP substrate during InGaAs growth and sharp replication of the screen image in the InGaAs observed.

16.8 CONCLUSION

A variety of crystal growth techniques are available for the epitaxial growth of semiconductor films for optoelectronic device applications. The choice of a particular technique must include consideration of the semiconductor material requirements, safety and cost. LPE, the simplest and least expensive growth technique, has received widespread use in industry for fabrication of commercially available devices. However, it will almost certainly give way to VPE, MOCVD or MBE as these growth techniques mature. This is already the case for short wavelength emitters ($Al_{1-x}Ga_xAs$), where MOCVD is the growth technique of choice for commercial applications.

For the long wavelength (0.9–1.6 μm) material system based on InGaAsP, VPE, MOCVD and MBE are all potential candidates for the preferred growth technique. MBE is currently unsurpassed in its ability to engineer epitaxial films atomically, layer by layer, although it is the most expensive process and has limited application for long wavelength devices due to the difficulty in handling phosphorus. This problem has been solved with the development of gas-source MBE and CBE, but at present only a limited number of results are available for devices fabricated using these techniques. Furthermore, use of AsH_3 and PH_3 is involved, which raises safety concerns typically voiced only for hydride VPE and MOCVD.

Next to LPE, hydride VPE has received the most attention for the growth of InGaAsP films used in device fabrication, primarily for lasers and LEDs, where purity issues are unimportant. The use of trichloride VPE for alloy growth is relatively recent, but the technique is capable of providing extremely high purity binary and ternary films necessary for sensitive detectors. Efforts to improve the purity of the hydride process or to increase the flexibility of the trichloride process may ultimately produce one VPE technique suitable for both emitters and detectors.

MOCVD, like MBE and its variants, has been used to grow a wide variety of structures with excellent material properties and device results. At present, however, only a limited data base exists for devices fabricated by MOCVD, particularly for 1.3 μm and 1.55 μm sources. Unlike VPE and MBE, MOCVD has demonstrated the routine growth of semi-insulating InP, which is important in several InGaAsP laser structures and will become increasingly important in the area of monolithic integrated optics.

In summary, existing epitaxial growth techniques for optoelectronic device fabrication vary considerably in cost, complexity (both in terms of hardware and growth chemistry), and suitability for various device applications. As commercial demand for optoelectronics increases, more emphasis on high volume growth techniques (VPE, MOCVD, and to a lesser degree MBE) will develop. Furthermore, as the level of complexity in integrated optoelectronics also increases, the ability of the selected epitaxial growth technique to produce films suitable for detectors (high purity), sources, and device isolation (semi-insulating films) in a single reactor will become important. While a single growth technique with all the necessary capabilities does not, at present, exist, future research in epitaxial crystal growth will be directed toward realizing this goal.

REFERENCES

Antypas, G. A. (1980). Prevention of InP surface decomposition in liquid phase epitaxial growth. *Phys. Lett.* **37**, 64.

Arthur, J. R. (1968). Interaction of Ga and As_2 molecular beams with GaAs surfaces. *J. Appl. Phys.* **39**, 4032.

Astles, M. G., Smith, F. G. H., and Williams, E. W. (1973). Indium phosphide. II. Liquid epitaxial growth. *J. Electrochem. Soc.* **120**, 1750.

Ban, V. S. (1971). Mass spectrometric studies of vapor phase crystal growth. I. $GaAs_xP_{1-x}$ system. *J. Electrochem. Soc.* **118**, 1473.

Ban, V. S. (1972). Mass spectrometric and thermodynamic studies of some III-V compounds. *J. Cryst. Growth*. **17**, 19.

Beuchet, G. (1985). Halide and Chloride Transport Vapor-Phase Deposition of InGaAsP and GaAs. "Semiconductors and Semimetals," Vol. 22A, ed. W. T. Tsang, Academic Press, New York, pg. 261.

Blaauw, C., Szaplonczay, A., Fox, K., and Emmerstorfer, B. (1986). MOCVD of InP and mass transport on structured InP substrates. *J. Cryst. Growth*. **77**, 326.

Calawa, A. R. (1981). On the use of AsH_3 in the molecular beam epitaxial growth of GaAs. *Appl. Phys. Lett*. **38**, 701.

Capasso, F., Kasper, B., Alavi, K., Cho, A. Y., and Parsey, J. M. (1984). New low dark current, high speed $Al_{0.48}In_{0.52}As/Ga_{0.47}In_{0.53}As$ avalanche photodiode by molecular beam epitaxy for long wavelength fiber optic communication systems. *Appl. Phys. Lett*. **44**, 1027.

Carey, K. W. (1985). Organometallic vapor phase epitaxial growth and characterization of high purity GaInAs on InP. *Appl. Phys. Lett*. **46**, 89.

Casey, Jr., H. C., and Panish, M. B. (1978). Heterostructure lasers. Part A: Fundamental principles. "Heterostructure Lasers." Academic Press, New York, Chapt. 6.

Chan, K. T., Zhu, L. D., and Ballantyne, J. M. (1985). Growth of high quality GaInAs on InP buffer layers by metalorganic chemical vapor deposition. *Appl. Phys. Lett*. **47**, 44.

Cheng, J., Stall, R., Forrest, S. R., Long, J. A., Cheng, C. L., Guth, G., Wunder, R., and Riggs, V. G. (1985). Self-aligned $In_{0.53}Ga_{0.47}As$/semi-insulating InP/n^+InP junction field-effect transistors. *IEEE Electron Dev. Lett*. **EDL-6**, 384.

Cho, A. Y., and Ballamy, W. C. (1975). GaAs planar technology by molecular beam epitaxy (MBE). *J. Appl. Phys*. **46**, 783.

Cho, A. Y., and Cheng, K. Y. (1981). P-N junction formation during molecular beam epitaxy of Ge-doped GaAs. (Photoluminescence). *Appl. Phys. Lett*. **38**, 360.

Cho, A. Y., and Hayashi, I. (1971). Growth of extremely uniform layers by rotating substrate holder with molecular beam epitaxy for applications to electro-optic and microwave devices. *J. Appl. Phys*. **42**, 4422.

Chow, R., and Chai, R. G. (1983). A PH_3 cracking furnace for molecular beam epitaxy. *J. Vac. Sci. Technol*. **A1**, 49.

Clawson, A. R., Hanson, C. M., and Vu, T. T. (1986). MOVPE growth of SiO_2-masked InP structures at reduced pressures. *J. Cryst. Growth* **77**, 334.

Cox, H. M. (1984). Vapor levitation epitaxy: a new concept in epitaxial crystal growth. *J. Cryst. Growth* **69**, 641.

Cox, H. M., Hummel, S. G., Keramidas, V. G., and Temkin, H. *In* "Proc. 12th Symp. on GaAs and Related Cmpds." Karuizawa, 1985, Vapor Levitation Epitaxial Growth of InGaAsP Alloys Using Trichloride Sources AIP Conf. Ser. 79, p. 735.

Cox, H. M., Koza, M. A., Keramidas, V. G., and Young, M. S. (1985). Vapor phase epitaxial growth of high purity InGaAs, InP and InGaAs/InP multilayer structures. *J. Cryst. Growth* **73**, 523.

Davies, G. J., Heckingbottom, R., Ohno, H., Wood, C. E. C., and Calawa, A. R. (1980). Arsenic stabilization of InP substrates for growth of $Ga_xIn_{1-x}As$ layers by molecular beam epitaxy. *Appl. Phys. Lett*. **37**, 290.

Davies, G. J., Kerr, T., Andrews, D. A., Wakefield, B., and Tuppen, C. G. (1984). The growth and characterization of nominally undoped $Al_{1-x}In_xAs$ grown by molecular beam epitaxy. *J. Vac. Sci. Technol*. **B2**, 219.

Di Forte-Poisson, M. A., Brylinski, C., and Duchemin, J. P. (1985). Growth of ultrapure and Si-doped InP by lower pressure metalorganic chemical vapor deposition. *Appl. Phys. Lett*. **46**, 476.

Duchemin, J. P., Hirtz, J. P., Razeghi, M., Bonnet, M., and Hersee, S. D. (1981). GaInAs and GaInAsP materials grown by low pressure MOCVD for microwave and optoelectronic applications. *J. Cryst. Growth* **55**, 64.

Dutta, N. K., Zilko, J. L., Cella, T., Ackerman, D. A., Shen, T. M., and Napholtz. G. (1986). InGaAsP laser with semi-insulating current confining layers. *Appl. Phys. Lett.* **48**, 1572.

Favennec, P. N., Salvi, M., Di Forte-Poisson, M. A., and Duchemin, J. P. (1983). Selected area growth of InP by lower pressure metalorganic chemical vapor deposition on ion implanted InP substrates. *Appl. Phys. Lett.* **43**, 771.

Fischer, R., Klem, J., Drummond, T. J., Thorne, R. E., Kopp, W., Morkoc, H., and Cho, A. Y. (1983). Incorporation rates of Gallium and Aluminum on GaAs during Molecular Beam Epitaxy at high substrate temperatures. *J. Appl. Phys.* **54**, 2508.

Foxon, C. T., and Joyce, B. A. (1975). Interaction Kinetics of As_4 and Ga on (100) GaAs surfaces using a modulated molecular beam technique. *Surface Sci.* **50**, 434.

Foxon, C. T., and Joyce, B. A. (1977). Interaction kinetics of As_2 and Ga on (100) GaAs surfaces. *Surface Sci.* **64**, 293.

Fraas, L. M., McLeod, P. S., Partain, L. D., and Cape, J. A. (1986). Epitaxial growth from organometallic sources in high vacuum. *J. Vac. Sci. Technol.* **B4**, 22.

Fukui, T., and Horikoshi, Y. (1980). Properties of InP films grown by organometallic VPE method. *Jpn. J. Appl. Phys.* **19**, L395.

Gale, R. P., McClelland, R. W., Fan, J. C. C., and Bozler, C. O. (1982). Lateral epitaxial overgrowth of GaAs by organo-metallic chemical vapor-deposition. *Appl. Phys. Lett.* **41**, 545.

Ghosh, C., and Layman, R. L. (1984). Selective area growth of gallium arsenide by metalorganic vapor phase epitaxy. *Appl. Phys. Lett.* **45**, 1229.

Giles, P. L., Davies, P., and Hasdell, N. B. (1983). The growth of epitaxial InP by the chloride process in nitrogen and in the presence of phosphine. *J. Cryst. Growth* **61**, 695.

Griffiths, R. J. M., Chew, N. G., Cullis, A. G., and Joyce, G. C. (1983). Structure of GaAs-$Ga_{1-x}Al_xAs$ superlattices grown by metal-organic chemical vapor deposition. *Electron. Lett.* **19**, 989.

Haigh, J. (1985). Mechanisms of metallo-organic vapor phase epitaxy and routes to a ultraviolet-assisted process. *J. Vac. Sci. Technol.* **B3**, 1456.

Henry, C. H., Logan, R. A., and Merritt, F. R. (1981). Single mode operation of buried heterostructure lasers by loss stabilization. *IEEE J. Quant. Elect.* **QE17**, 2196.

Hirao, M., Doi, Tsuji, S., Nakamura, M., and Aiki, K. (1980). Fabrication and characterization of narrow stripe InGaAsP/InP buried heterostructure lasers. *J. Appl. Phys.* **51**, 4539.

Hockly, M., and White, E. A. D. (1984). TEM studies of MOVPE (Ga, In)As interfaces with InP substrates. *J. Cryst. Growth*, **68**, 334.

Huet, D., Lambert, M., Bonnevie, D., and Dufresne, D. (1985). Molecular beam epitaxy of $In_{0.53}Ga_{0.47}As$ and InP on InP by using cracker cells and gas cells. *J. Vac. Sci. Technol.* **B3**, 823.

Jain, B. P., and Purohit, R. K. (1984). Physics and technology of vapor phase epitaxial growth of GaAs-A review. *Prog. Crystal Growth and Charact.* **89**, 51.

Kamon, K., Takagishi, S., and Mori, H. (1985). Selective epitaxial growth of GaAs by low-pressure MOVPE. *J. Cryst. Growth* **73**, 73.

Kanbe, H., Yamauchi, Y., and Susa, N. (1979). Vapor-phase epitaxial $In_xGa_{1-x}As$ on (100), (111)A, and (111)B InP substrates. *Appl. Phys. Lett.* **35**, 603.

Karlicek, Jr., R. F., and Bloemeke, A. (1985). Remote optical monitoring of reactants in a vapor phase epitaxial reactor. *J. Cryst. Growth* **73**, 364.

Kelting, K., Kochler, K., and Zwicknagle, P. (1986). Luminescence of $Ga_{1-x}Al_xAs/GaAs$ single quantum wells grown by liquid phase. *Appl. Phase Lett.* **48**, 157.

Komeno, J., Tanaka, H., Itoh, H., Ohori, T., Kasai, K., and Shibatomi, A. (1985). Multi-wafer growth of extremely uniform GaAs layers by organometallic vapor phase epitaxy. Electronic Materials Conference, Boulder, Colorado.

Koukitu, A., Seki, H., and Fujimoto, M. (1976). Vapor growth of GaAs by H_2 introduction into an inert carrier gas stream. *Japan J. Appl. Phys.* **15**, 1591.

Koukitu, A., and Seki, H. (1980). Thermodynamic analysis for InGaAsP epitaxial growth by the chloride-CVD process. *J. Cryst. Growth* **49**, 325.

Koukitu, A., Suzuki, T., and Seki, H. (1985). Vapor phase epitaxy of GaAs by the pulsed introduction of H_2. *J. Cryst. Growth* **71**, 450.

Kuo, C. P., Cohen, R. M., Fry, K. L., and Stringfellow, G. B. (1985a). Characterization of $Ga_xIn_{1-x}As$ grown with TMIn. *J. Electron. Mater.* **14**, 231.

Kuo, C. P., Fry, K. L., and Stringfellow, G. B. (1985b). GaInAs/InP quantum wells grown by organometallic vapor phase epitaxy. *Appl. Phys. Lett.* **47**, 855.

Kuphal, E. (1984). Preparation and characterization of LPE InP. *J. Cryst. Growth* **54**, 117.

Laudise, R. A. (1983). Crystal growth progress in response to the needs for optical communications. *J. Cryst. Growth* **63**, 3.

Lee, W., and Fonstad, C. G. (1986). The growth of high mobility InGaAs and InAlAs layers by molecular-beam epitaxy. *J. Vac. Sci. Technol.* **B4**, 536.

Leys, M. R., and Veenvliet, H. (1981). A study of the growth mechanism of epitaxial GaAs as grown by the technique of metalorganic vapor phase epitaxy. *J. Cryst. Growth* **55**, 141.

Leys, M. R., van Opdorp, C., Viegers, M. P. A., and Talen-van der Mhen, H. J. (1984). Growth of multiple thin layer structures in the GaAs-AlAs system using a novel VPE reactor. *J. Cryst. Growth* **68**, 431.

Li, Tingye (1983). Advances in Optical Fiber Communications: An Historical Perspective. *IEEE J. Selected Areas Commun.* **SAC1**, 356.

Logan, R. A., and Reinhart, F. K. (1975). Integrated GaAs-$Al_xGa_{1-x}As$ double-heterostructure laser with independently controlled optical output divergence. *IEEE J. Quant. Elect.* **QE11**, 461.

Logan, R. A., and Temkin, H. (1986). Liquid phase epitaxial growth on InP using $In_{1-x}Sn_x$ melts. *J. Cryst. Growth* **76**, 17.

Logan, R. A., Schumaker, N. E., Henry, C. H., and Merritt, F. R. (1979). Doping effects on rake-line formation in LPE growth of $Al_xGa_{1-x}As$ DH lasers. *J. Appl. Phys.* **50**, 5970.

Logan, R. A., Henry, C. H., Merritt, F. R., and Mahajan, S. (1983). Growth on Indium (III) Planes of Indium Phosphide. *J. Appl. Phys.* **54**, 5462.

Logan, R. A., Merritt, F. R., and Mahajan, S. (1984). GaInAsP/InP buried heterostructure formation by liquid phase epitaxy. *Appl. Phys. Lett.* **45**, 1275.

Long, J. A., Riggs, V. G., and Johnston, W. D. (1984). Growth of Fe-doped semi-insulating InP by MOCVD. *J. Cryst. Growth* **69**, 10.

Ludowise, M. J. (1985). Metalorganic chemical vapor deposition of III-V semiconductors. *J. Appl. Phys.* **58**, R31.

Martin, T., Stanley, C. R., Iliadis, A., Whitehouse, C. R., and Sykes, D. E. (1985). Identification of the major residual donor in unintentionally doped InP grown by molecular beam epitaxy. *Appl. Phys. Lett.* **46**, 994.

Miller, B. I., Schubert, E. F., Koren, U., Ourmazd, A., Dayem, A. H., and Capik, R. J. (1986). High quality narrow GaInAs/InP quantum wells grown by atmospheric organometallic vapor phase epitaxy. *Appl. Phys. Lett.* **49**, 1384.

Mizuno, O., and Watanabe, H. (1975). Vapor growth kinetics of III-V compounds in a hydrogen-inert gas mixed carrier system. *J. Cryst. Growth* **30**, 240.

Mizutani, T., and Hirose, K. (1985). High mobility GaInAs thin layers grown by molecular beam epitaxy. *Jpn. J. Appl. Phys.* **24**, L119.

Mizutani, T. M., and Watanabe, H. (1982). Suppression of extraneous wall deposition by HCl injection in hydride vapor phase epitaxy of III-V semiconductors. *J. Cryst. Growth* **59**, 507.

Moss, R. H. (1984). Adducts in the growth of III-V compounds. *J. Cryst. Growth* **68**, 78.

Moss, R. H., and Spurdens, P. C. (1984b). New approach to growth of abrupt heterojunctions by MOVPE. *Electron. Lett.* **20**, 978.

Moss, R. H., and Spurdens, P. C. (1984b). Growth of $Ga_x In_{1-x} As$ and $Ga_x In_{1-x} As_y P_{1-y}$ using preformed adducts. *J. Cryst. Growth* **68**, 96.

Nagai, H. (1980). Thermodynamic analysis of $Ga_x In_{1-x} As_y P_{1-y}$ CVD: Ga-In-As-P-H-Cl system. *J. Cryst. Growth* **48**, 359.

Nakai, K., and Ozeki, M. (1984). Selective metalorganic chemical vapor deposition for GaAs planar technology. *J. Cryst. Growth* **68**, 200.

Nakajima, K., Kusunoki, T., and Akita, K. (1980). InGaAsP phase diagram and LPE growth conditions for lattice matching in InP. *Fujitsu Sci. Techn. J.* **16**, 59.

Neave, J. H., Blood, P., and Joyce, B. A. (1980). A correlation between electron traps and growth processes in n-GaAs prepared by molecular beam epitaxy. *Appl. Phys. Lett.* **36**, 311.

Nelson, A. W., Westbrook, L. D., and Evans, J. S. (1983). Deformation-free overgrowth of InGaAsP DFB corrugations. *Electron. Lett.* **19**, 34.

Nelson, A. W., Devlin, W. J., Hobbs, R. E., Lenton, C. G. D., and Wong, S. (1985). High-power, low-threshold BH lasers operating at 1.52 μm grown entirely by MOVPE. *Electron. Lett.* **21**, 888.

Nishizawa, J., and Kurabayashi, T. (1983). On the reaction mechanism of GaAs MOCVD. *J. Electrochem. Soc.* **130**, 413.

Nordland, W. A., Kazarinov, R. F., Merritt, F. R., and Savage, A. (1984). Modified single-phase LPE technique for $In_{1-x} Ga_x As_{1-y} P_y$ laser structures. *Electron. Lett.* **20**, 806.

Oishi, M., and Kuroiwa, K. (1985). Low pressure metalorganic vapor phase epitaxy of InP in a vertical reactor. *J. Electrochem. Soc.* **132**, 1209.

Okamoto, K., and Yamaguchi, K. (1986). Selectively buried epitaxial growth of GaAs by metalorganic chemical vapor deposition. *Appl. Phys. Lett.* **48**, 849.

Olsen, G. H. (1982). Vapor-Phase Epitaxy of GaInAsP. *In* "GaInAsP Alloy Semiconductors," ed. T. P. Pearsall, John Wiley and Son, Ltd., New York, p. 11.

Olsen, G. H., Zamerowski, T. J., and Hawrylo, F. Z. (1982). Vapor growth of InGaAs and InP on (100), (110), (111), (311), and (511) InP substrates. *J. Cryst. Growth.* **59**, 654.

Onabe, K. (1982). Thermodynamics of type $A_{1-x} B_x C_{1-y} D_y$ III-V quaternary solid solutions. *J. Phys. Chem. Solids* **43**, 1071.

Panish, M. B. (1980). Molecular beam epitaxy of GaAs and InP with gas sources for As and P. *J. Electrochem. Soc.* **127**, 2729.

Panish, M. B., and Hamm, R. A. (1986). A mass spectrometric study of AsH_3 and PH_3 gas sources for molecular beam epitaxy. *J. Cryst. Growth*, in press.

Panish, M. B., and Ilegems, M. (1972). Phase Equilibria in Ternary III-V Systems. *In* "Progress in Solid State Chemistry." Vol. 1, H. Reiss and W. O. McCaldin, Ed., Pergamon, N.Y., pp. 39–83.

Panish, M. B., Temkin, H., and Sumski, S. (1985). Gas source MBE of InP and $Ga_x In_{1-x} P_y As_{1-y}$: materials properties and heterostructure lasers. *J. Vac. Sci. Technol.* **B3**, 657.

Panish, M. B., Temkin, H., Hamm, R. A., and Chu, S. N. G. (1986). Optical properties of very thin GaInAs(P)/InP quantum wells grown by gas source molecular beam epitaxy. *In* "Proceedings of the TMS/MRS Northeast Regional Meeting," Murray Hill, NJ, May 1986, to be published.

People, R., Wecht, K. W., Alavi, K., and Cho, A. Y. (1983). Measurement of the conduction-band discontinuity of molecular beam epitaxial grown $In_{0.52} Al_{0.48} As/In_{0.53} Ga_{0.47} As$, n-n heterojunction by C-V profiling. *Appl. Phys. Lett.* **43**, 118.

Razeghi, M. (1985). Low-Pressure Metallo-Organic Chemical Vapor Deposition of GaInAsP Alloys. *In* "Semiconductors and Semimetals," Vol. 22A, ed. W. T. Tsang, Academic Press, p. 299.

Razeghi, M., and Duchemin, J. P. (1984). Recent advances in MOCVD growth of $In_xGa_{1-x}As_yP_{1-y}$ alloys. *J. Cryst. Growth* **70**, 145.

Razeghi, M., Poisson, M. A., Larivain, J. P., and Duchemin, J. P. (1983). Low pressure metalorganic chemical vapor deposition on InP and related compounds. *J. Electron. Mater.* **12**, 371.

Razeghi, M., Blondeau, R., Kazmierski, K., Krakowski, M., de Cremoux, B., and Duchemin, J. P. (1984). CW operation of 1.57 μm $Ga_xIn_{1-x}As_yP_{1-y}$/InP distributed feedback lasers grown by low-pressure metalorganic chemical vapor deposition. *Appl. Phys. Lett.* **45**, 784.

Reep, D. H., and Ghandhi, S. K. (1983). Deposition of GaAs epitaxial layers by organometallic CVD. Temperature and orientation dependence. *J. Electrochem. Soc.* **130**, 675.

Reinhart, F. K., and Logan, R. A. (1975). GaAs-AlGaAs double heterostructure lasers with taper-coupled passive waveguides. *Appl. Phys. Lett.* **26**, 516.

Rezek, E. A., Zinkiewicz, L. M., and Law, H. D. (1983). High-resistivity ($> 10^5$ Ωcm) InP layers by liquid phase epitaxy. *Appl. Phys. Lett.* **43**, 378.

Roberts, J. S., Dawson, P., and Scott, G. B. (1981). Homoepitaxial molecular beam growth of InP on thermally cleaned (100) oriented substrates. *Appl. Phys. Lett.* **38**, 905.

Roberts, J., Mason, N. J., and Robinson, M. (1984). Factors influencing doping control and abrupt metallurgical transitions during atmospheric pressure MOVPE growth of AlGaAs and GaAs. *J. Cryst. Growth* **68**, 422.

Roberts, J. S., Claxton, P. A., David, J. P. R., and Marsh, J. H. (1986). Improved molecular beam epitaxial growth of InP using solid sources. *Electron. Lett.* **22**, 506.

Schlyer, D. J., and Ring, M. A. (1976). Examination of product-catalyzed reaction of trimethylgallium with arsine. *J. Organomet. Chem.* **114**, 9.

Scholz, F., Weidmann, P., Benz, K. W., Tränkle, G., Lach, E., Forchel, A., Laube, G., and Weidlein, J. (1986). GaInAs-InP multiquantum well structures grown by metalorganic gas phase epitaxy with adducts. *Appl. Phys. Lett.* **48**, 911.

Seki, H., and Koukitu, A. (1985). Thermodynamic calculation of the VPE growth of $In_{1-x}Ga_xAs_yP_{1-y}$ by the trichloride method. *Jpn. J. Appl. Physics* **24**, 458.

Shaw, D. W. (1975). A comparative thermodynamic analysis of InP and GaAs deposition. *J. Chem. Phys. Solids* **36**, 111.

Shealy, J. R. (1986). Device quality AlGaAs/GaAs heterostructures grown in a multichamber organometallic vapor phase epitaxial apparatus. *Appl. Phys. Lett.* **48**, 925.

Skolnick, M. S., Tapster, P. R., Bass, S. J., Apsley, N., Pitt, A. D., Chew, N. G., Cullis, A. G., Aldred, S. P., and Warwick, C. A. (1986). Optical properties of InGaAs-InP single quantum wells grown by atmospheric pressure metalorganic chemical vapor deposition. *Appl. Phys. Lett.* **48**, 1455.

Small, M. B., Blakeslee, A. E., Shih, K. K., and Potamski, R. M. (1975). A phenomenological study of meniscus lines on the surfaces of GaAs layers grown by LPE. *J. Cryst. Growth* **30**, 257.

Stolz, W., Tapfer, L., Breitschwerdt, A., and Ploog, K. (1985). Optical and structural properties of molecular-beam epitaxially grown $Ga_{0.47}In_{0.53}As/Al_{0.48}In_{0.52}As$ superlattices, emitting at 1.55 μm at room temperature. *Appl. Phys. A.* **38**, 97.

Stringfellow, G. B. (1974). Calculation of ternary and quatenary III-V phase diagrams. *J. Cryst. Growth* **27**, 21.

Stringfellow, G. B. (1985). Organometallic Vapor-Phase Epitaxial Growth of III-V Semiconductors. *In* "Semiconductors and Semimetals," Vol. 22A, ed. W. T. Tsang, Academic Press, New York, p. 209.

Studna, A. A., and Gualtieri, G. J. (1981). Optical properties and water absorption of anodically grown native oxides on InP. *Appl. Phys. Lett.* **39**, 965.

Theeten, J. B., Hollan, L., and Cadoret, R. (1976). Growth Mechanisms in CVD of GaAs. *In* "1976 Crystal Growth and Materials," E. Kaldis and H. J. Scheel, eds., North-Holland Publishing Co., Amsterdam, 1977, p. 196.

Tsang, W. T. (1984). Chemical beam epitaxy of InP and GaAs. *Appl. Phys. Lett.* **45**, 1234.

Tsang, W. T. (1985a). Molecular Beam Epitaxy for III-V Compound Semiconductors. *In* "Semiconductors and Semimetals," Vol. 22A, ed. W. T. Tsang, Academic Press, New York, p. 95.

Tsang, W. T. (1985b). Selective area growth of GaAs and $In_{0.53}Ga_{0.47}As$ epilayer structures by chemical beam epitaxy using silicon shadow masks: a demonstration of the beam nature. *Appl. Phys. Lett.* **46**, 742.

Tsang, W. T., and Cho, A. Y. (1978). Molecular beam epitaxial writing of patterned GaAs epilayer structures (for integrated optics). *Appl. Phys. Lett.* **32**, 491.

Tsang, W. T., and Ilegems, M. (1977). Selective area growth of GaAs $Al_xGa_{1-x}As$ multilayer structures with molecular beam epitaxy using Si shadow masks (for integrated optics). *Appl. Phys. Lett.* **31**, 301.

Tsang, W. T., and Ilegems, M. (1979). The preparation of GaAs thin-film optical components by molecular beam epitaxy using Si shadow masking technique. *Appl. Phys. Lett.* **35**, 792.

Tsang, W. T., and Schubert, E. F. (1986). Extremely high quality $Ga_{0.47}In_{0.53}As/InP$ quantum wells grown by chemical beam epitaxy. *Appl. Phys. Lett.* **49**, 220.

Tsang, W. T., Miller, R. C., Capasso, F., and Bonner, W. A. (1982). High quality InP grown by molecular beam epitaxy. *Appl. Phys. Lett.* **41**, 467.

Tsang, W. T., Ditzenberger, J. A., and Olsson, N. A. (1983). Improvement of photoluminescence of molecular beam epitaxially grown $Ga_xAl_{1-x-y}As$ by using an As_2 molecular beam. *IEEE Electron Dev. Lett.* **EDL-4**, 275.

Tsang, W. T., Dayem, A. H., Chiu, T. H., Cunningham, J. E., Schubert, E. F., Ditzenberger, J. A., Shah, J., Zyskind, J. L., and Tabatabaie, N. (1986a). Chemical beam epitaxial growth of extremely high quality InGaAs on InP. *Appl. Phys. Lett.* **49**, 170.

Tsang, W. T., Chang, A. M., Ditzenberger, J. A., and Tabatabaie, N. (1986b). Two-dimensional electron gas in a $Ga_{0.47}In_{0.53}As/InP$ heterojunction grown by chemical beam epitaxy. *Appl. Phys. Lett.* **49**, 960.

Usui, A., Matsumoto, T., Inai, M., Mito, I., Kobayashi, K., and Watanabe, H. (1985). Room temperature CW operation of visible InGaAsP double heterostructure laser at 671 nm grown by hydride VPE. *Jpn. J. Appl. Phys.* **24**, L163.

Vohl, P. (1981). Vapor-phase epitaxy of GaInAsP and InP. *J. Crystal Growth* **54**, 101.

Westbrook, L. D., Nelson, A. W., Fiddyment, P. J., and Evans, J. S. (1984). Continuous-wave operation of 1.5 μm distributed-feedback ridge-waveguide lasers. *Electron. Lett.* **20**, 225.

Wilt, D. P., Long, J. A., Dautremont-Smith, W. C., Focht, M. W., Shen, T. M., and Hartman, R. L. (1986). Channeled-substrate buried-heterostructure InGaAsP/InP laser with semi-insulating OMVPE base structure and LPE regrowth. *Electron. Lett.* **22**, 869.

Wood, C. E. C. (1983). III-V alloy growth by molecular-beam epitaxy. *In* "GaInAsP Alloy Semiconductors," ed. T. P. Pearsall. John Wiley and Sons Ltd., New York, p. 87.

Yanase, T., Kato, Y., Mito, I., Kobayashi, K., Nishimoto, H., Usui, A., and Kobayashi, K. (1983a). VPE-Grown 1.3 μm InGaAsP/InP double-channel planar buried-heterostructure laser diode with LPE-burying layers (English). *Jpn. J. Appl. Phys.* **22**, L415.

Yanase, T., Kato, Y., Mito, I., Yamaguchi, M., Nishi, K., Kobayashi, K., and Lang, R. (1983b). 1.3 μm InGaAsP/InP multiquantum-well lasers grown by vapor-phase epitaxy. *Electron. Lett.* **19**, 700.

Yoshida, M., and Watanabe, H. (1985). Thermodynamic comparison of InGaAsP vapor phase epitaxy by chloride, hydride, and metalorganic-chloride methods. *J. Electrochem. Soc.* **132**, 1733.

Zhu, L. D., Chan, K. T., and Ballantyne, J. M. (1985a). MOCVD growth and characterization of high quality InP. *J. Cryst. Growth* **73**, 83.

Zhu, L. D., Sulewski, P. E., Chan, K. T., Muro, K., and Ballantyne, J. M. (1985b). Two-dimensional electron gas in $In_{0.53}Ga_{0.47}As/InP$ heterojunctions grown by atmospheric pressure metalorganic chemical-vapor deposition. *J. Appl. Phys.* **58**, 3145.

Zilko, J. L. (1986). Metal-organic chemical vapor deposition technology and equipment. *In* "Handbook of Deposition Technology and Equipment," ed. K. Schuegraf, Noyes Publication, to be published.

Zilko, J. L., Koren, U., Dutta, N. K., and Napholtz, S. G. (1986). Unpublished.

Chapter 17

Reliability of Lasers and LEDs

N. K. DUTTA

AT&T Bell Laboratories, Inc., Murray Hill, New Jersey

C. L. ZIPFEL

AT&T Bell Laboratories, Inc., Murray Hill, New Jersey

17.1 INTRODUCTION

The performance characteristics of optical components used in lightwave systems such as injection lasers, light emitting diodes (LEDs) and photodiodes can degrade during their operation. The degradation is generally characterized by a change in the operational characteristics of these devices and is often associated with the formation and/or multiplication of defects in the active region. For lasers, the degradation is usually characterized by an increase in the threshold current, which is often accompanied by a decrease in the external differential quantum efficiency (Nash *et al.*, 1985). For LEDs the degradation is characterized by a decrease in the light output at a given operating current (Zipfel, 1985).

The dominant mechanism responsible for the degradation is determined by any or all of the several fabrication processes including epitaxial growth, wafer quality, device processing and bonding. In addition, the degradation rate of devices processed from a given wafer depends on the operating conditions, i.e., the operating temperature and the injection current. Although many of the degradation mechanisms are not fully understood, extensive amounts of empirical observations exist in the literature, which have allowed the fabrication of InGaAsP laser diodes with extrapolated median lifetimes in excess of 25 years at an operating temperature of 10°C (Nash *et al.*, 1985), LEDs with lifetimes in excess of 100 years at 70°C (Zipfel, 1985).

The detailed studies of degradation mechanisms of optical components used in lightwave systems have been motivated by the desire to have a reasonably accurate estimate of the operating lifetime before they are used in practical systems. Since for many applications, the components are expected to operate reliably over a period in excess of 10 years, an appropriate reliability assurance procedure becomes necessary, especially for applications such as an undersea lightwave transmission system where the replacement cost is very high. The reliability assurance is usually carried out by operating the devices under a high stress (e.g., high temperature), which enhances the degradation rate so that a measurable value can be obtained in an operating time of a few hundred hours. The degradation rate under normal operating conditions can then be obtained from the measured high temperature degradation rate using the concept of an activation energy.

The purpose of this chapter is twofold: (i) to discuss the various degradation mechanisms and (ii) to discuss the reliability assurance strategies. The specific reliability considerations related to lasers and LEDs are described in Sections 17.2 and 17.3 respectively.

17.2 LASER RELIABILITY

Aside from an unpredictable damage, which is a sudden failure mechanism caused by unforeseen events (such as power supply failure), the degradation mechanisms of injection lasers can be separated into three categories. They are

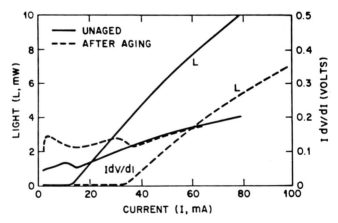

Fig. 17.1 Optical (light L vs. current I) and electrical (IdV/dI vs. I) characteristics of a buried heterostructure laser before and after aging. (Courtesy of E. J. Flynn)

(1) defect formation in the active region, (2) degradation of current confining junctions and (3) catastrophic degradation. It is important to mention at the outset that although reliable semiconductor lasers have been fabricated using both the AlGaAs and InGaAsP alloy system, the study of degradation modes has been descriptive, and the results in many cases are tentatively interpretative.

For the purpose of illustration, the light output vs. current characteristics of a laser before and after stress aging is shown in Fig. 17.1. Note the increase in threshold current and the decrease in external differential quantum efficiency following the stress aging. Also shown in Figure 17.1 is the electrical characteristics IdV/dI vs. I (where V is the voltage across the laser and I is the current) of the laser before and after aging. A change in IdV/dI vs. I such as a "bump" shown in Figure 17.1 is usually characteristic of breakdown of a current confining structure internal to the laser.

17.2.1 Defect Formation in the Active Region

The high density of recombining electrons and holes and a possible presence of strain and thermal gradients can promote defect formation in the active region of the laser. The defect structures that are generally observed are the dark spot defect (DSD) and the dark line defect (DLD).

The "dark line defect" (DLD) as the name suggests, is a region of greatly reduced radiative efficiency of roughly linear form. The DLD was first reported by DeLoach et al., 1973, in the active region of an aged AlGaAs proton stripe double heterostructure laser. The DLD appears as a linear dark feature crossing the luminescent stripe at 45° in degraded lasers. Since the active stripe is oriented along the ⟨110⟩ direction, the DLDs are oriented along the ⟨100⟩ direction.

The observation of nonluminescent regions or regions of low radiative efficiency such as a DLD inside the active region of a semiconductor laser requires the fabrication of a special laser structure. This laser structure commonly known as "the window laser" is shown in Fig. 17.2. A "window" typically 20–25 μm wide is formed using photolithographic techniques on the substrate side. Since the InGaAsP laser emits at energies smaller than the band gap in the n-InP substrate, the spontaneous emission from the active region can be directly observed through the window. The window-laser structure can allow continuous monitoring of luminescence efficiency of the active region and is also compatible with the normal "p-down" bonding configuration. Very often, degraded lasers show dark regions in the active stripe luminescence when observed through the window. DLDs and DSDs have been observed in both AlGaAs and InGaAsP lasers and LEDs. Detailed study of these defect structures requires the use of techniques such as transmission electron microscopy (TEM) (Mahajan et al., 1975), deep level transient spectroscopy (DLTS) (Petroff and Lang, 1977), scanning electron microscopy (SEM), electron beam induced current (EBIC)

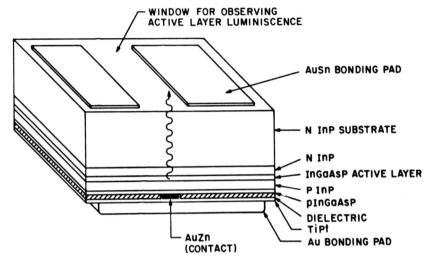

Fig. 17.2   Schematic of a "window" laser structure.

mode of SEM, cathodoluminescence (Mahajan *et al.*, 1985), and scanning photoluminescence (Johnston *et al.*, 1978).

Accelerated aging techniques, which include high temperature and high power operation, are generally used to estimate the usable lifetime of injection lasers under normal operating conditions. The generation rates of DSDs and DLDs are enhanced under accelerated aging. When the accelerated aging is done at high temperature, an activation energy can be defined for the generation rate of defects in the active region. The activation energy is defined using

$$t_d = t_0 \exp(E_a/kT), \tag{17.1}$$

where t_d is the generation time for the first defect, t_0 is a constant, E_a is the activation energy, k is the Boltzmann constant and T is the operating temperature. Fukuda *et al.* (1983) have measured the generation time of DSDs and DLDs in InGaAsP gain guided lasers operating at 1.3 μm and 1.55 μm wavelengths. The generation of DSDs and DLDs was observed by electroluminescence using a window laser configuration. The lasers that did not exhibit DSDs and DLDs operated for a long time without degradation. The measured pulsed threshold current normalized to the initial value is shown in Fig. 17.3. The increase in threshold current was associated with an increase in the number of DSDs and ⟨100⟩ DLDs. GaAs rich regions in the active layer were correlated with the location of the DSDs. Fukuda *et al.* (1983) observed a saturation in the number of DSDs and DLDs in about 50 hrs. of aging at 250°C, beyond that time the increase in threshold current caused by further aging also showed a

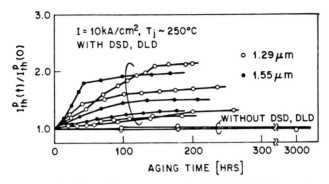

Fig. 17.3 Change in pulsed threshold current normalized by the initial value as a function of aging time (after Fukuda *et al.*, 1983).

tendency to saturate. This observation is very important for the purpose of reliability assurance, because it shows a possible presence of a saturable mode of degradation, i.e., after a certain period of aging the degradation rate may be very small.

17.2.2 Degradation of Current Confining Junctions

Many index guided laser structures utilize current restriction layers so that most of the injected current will flow through the active region. Effective current injection to the active region is necessary in order to obtain low threshold current and high output powers. The current flowing outside the active region in buried heterostructure (BH) type lasers is called the leakage current. A mode of degradation that is associated with BH lasers is an increase in the leakage current (which increases the device threshold and decreases the external differential quantum efficiency) under accelerated aging. An increase in leakage current not only increases the device threshold but also generally decreases the external differential quantum efficiency. Increased leakage current usually appears as a "soft turn on" in the I-V characteristics of the laser. Electron beam induced current (EBIC) observation of laser facets is an useful technique in detecting defects at current confining junctions (Mizuishi *et al.*, 1983).

17.2.3 Catastrophic Degradation

Catastrophic degradation is defined here as the sudden degradation in the performance characteristics of laser diodes associated with the application of a large current pulse. For AlGaAs lasers, the degradation occurs due to strong absorption of the stimulated emission at the facet, which causes a local heating and subsequent melting of the material near the facet. The degraded facet region typically exhibits dislocation networks and multiple dislocation loops, which are thought to be generated during the cooling of the molten region.

Temkin *et al.* (1982) have studied catastrophic degradation in InGaAsP/InP double heterostructure material under intense optical excitation. The threshold power for catastrophic damage is about an order of magnitude larger in InGaAsP double heterostructure (DH) material than in AlGaAs DH material.

Ueda *et al.* (1984) have reported the observation of catastrophic degradation of InGaAsP lasers following a large current pulse. Generally, no facet degradation characteristic of melting at the active region (similar to AlGaAs lasers) is observed. This may be partly due to smaller surface recombination rate in InGaAsP compared to that of AlGaAs. In cases where facet degradation is seen, it extended from the top contact to the p-InP cladding layer, implying that heating was caused by a large amount of current passing near the facet. They suggest catastrophic degradation may be caused by local degradation of some internal current confining region, which can result in large localized leakage current and subsequent melting.

17.2.4 Reliability Assurance

In some semiconductor laser applications, the system design lifetimes are long (~ 20–25 years) and the replacement of components (e.g., lasers) can be too expensive. An example of such an application is provided by the repeaters of an undersea lightwave transmission system where the failure of just a few lasers can cripple the system. Thus, it is important to have a strategy by means of which it is possible to establish the expected operating lifetime of a laser.

Some lasers exhibit an initial rapid degradation after which the operating characteristics of the lasers are very stable. Given a population of lasers, it is possible to quickly identify the "stable" lasers by a high stress test (also known as the purge test) (Gordon *et al.*, 1983). The stress test implies that operating the laser under a set of high stress conditions (e.g., high current, high temperature, high power) would cause the weak lasers to fail and stabilize the possible winners. Observations on the operating current after stress aging have been reported by Nash *et al.* (1985). Figure 17.4 shows their measured data. The operating current required for an output power of 3 mW/facet at 60°C is plotted as a function stress aging time. The stress conditions required the CW operation of the laser at 100°C with a 250 mA current. The lasers were of the buried heterostructure type. Some lasers exhibit an increase in operating current before they stabilize similar to the observations of Fukuda *et al.* (1983), while others exhibit stable characteristics without a significant increase in the operating current. Figures 17.3 and 17.4 show that the high stress test can be used as a screening procedure to identify robust devices. It is important to point out that the determination of the duration and the specific conditions for stress aging are critical to the success of this screening procedure.

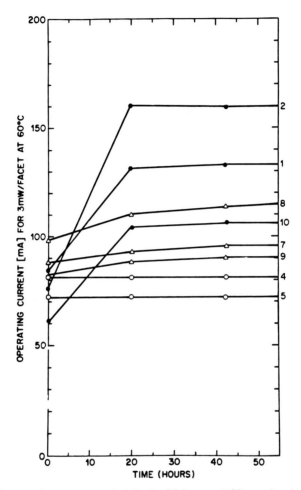

Fig. 17.4 The operating current required for 3 mW/facet at 60°C as a function of stress aging time. The stress conditions were CW operating at 100°C with 250 mA current. (After Nash *et al.*, 1985).

Central to the determination of the expected operating lifetime is the concept of thermally accelerated aging, the validity of which for the AlGaAs injection laser was shown by Hartman and Dixon, 1975. The lifetime (τ) at a temperature T varies as

$$\tau = \tau_0 \exp(-E_a/kT), \tag{17.2}$$

where E_a is the activation energy and k is the Boltzmann constant. Measurements of activation energy of InGaAsP-InP BH lasers have been reported by Mizujishi *et al.* (1982) and Hakki *et al.* (1985) by measuring the degradation rate

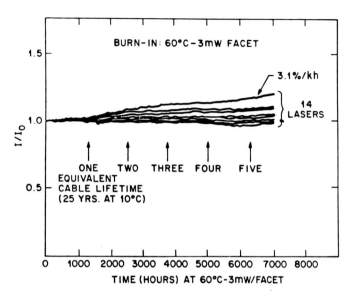

Fig. 17.5 Accelerated aging rate at 60°C, 3 mW/facet of a group of 14 prescreened lasers (after Nash *et al.*, 1985).

at various temperatures. Mizujishi *et al.* (1982) obtained $E_a = 0.9$ eV and Hakki *et al.* (1985) obtained $E_a = 1 \pm 0.13$ eV from their measurements.

The data of Nash *et al.* (1985) on the reliability of BH lasers are shown in Fig. 17.5. The normalized operating currents (I/I_0) for 14 lasers aged for ~ 7000 hrs. at 60°C—3 mW operating conditions are shown. The maximum degradation rate observed is 3.1% per 1000 hrs. The measured degradation rate can be used to obtain an expected operating life using a 50% change in I/I_0 as a failure criterion. The extrapolated life at an operating temperature of 10°C (which is the ocean bottom temperature) is then obtained using Eq. (17.2) and an activation energy of 0.9 eV. 1200 hrs. of aging time at 60°C—3 mW/facet is then equivalent to 25 years of operating time at 10°C. The arrows in Fig. 17.5 represent 25 years of equivalent operating life at 10°C, which is the expected cable lifetime of the transatlantic lightwave system (TAT8) that is currently being deployed.

17.3 LED RELIABILITY

Chapter 12 (T. P. Lee, C. A. Burrus and R. H. Saul, "LEDs for Telecommunications") discusses in detail how light emitting diodes (LEDs) can be the source of choice in lightwave transmission systems such as interoffice trunks, subscriber loops, local-area networks and on-premises data links. InGaAsP/InP surface emitters operating at 1.3 μm have been most commonly used, although edge emitters have recently been proposed for single-mode systems. There also remains considerable interest in GaAlAs/GaAs 0.87 μm surface emitters for short-haul applications.

17.3.1 General Reliability Considerations

The wide temperature range of operation, simplicity of drive circuitry and high manufacturing yield can make LEDs economical alternatives to lasers (Saul, 1983). It is often stated that the high reliability of these non-threshold devices is another factor in their favor. Certainly many of the factors complicating laser reliability such as changes in threshold current, degradation of current confining junctions and catastrophic facet damage are missing in LEDs. The only important degrading parameter in LEDs is the launched power. However, the strategies employed to maximize the coupling efficiency of light into the fiber (e.g., small contacts with high current densities or complex geometries) can negatively affect the reliability. Additionally, LEDs are typically used without thermoelectric coolers and are therefore expected to operate at the high ambient temperatures found in racks or remote terminals. Expensive certification procedures for individual devices such as those discussed in Section 17.2.4 for submarine cable lasers are not practical for the kinds of systems in which LEDs are used.

Although the device structures for GaAs and InP-based LEDs are similar, the relevant short-term failure mechanisms are very different (Yamakoshi *et al.*, 1979, 1981). Dislocation-like defects (dark line defects or DLDs), which propagate by climb, are the dominant cause for early failure in GaAs-based LEDs. In contrast, precipitate-like defects (dark spot defects or DSDs) and glide of existing misfit dislocations predominate in InP-based LEDs. Defects in GaAs-based devices have been extensively studied and many excellent reviews have been published (Newman and Ritchie, 1981). Although great progress has been made since the early days when large numbers of LEDs and lasers degraded rapidly due to climb phenomena, there is still an infant failure population due to DLDs in virtually all types of GaAs devices. In contrast, although the precipitate-like defects appear in most InP-based devices, they have been seldom studied because they do not seem to compromise the reliability.

Unlike lasers, LEDs can be lifetested at arbitrarily high temperatures consistent only with the maximum temperature of the die bond. Virtually all reports of LED reliability have been at constant DC current. LEDs are brought to room temperature for power measurements during life testing at elevated temperatures, and typically only the broad area light power is measured.

LED failures can be classified into three populations: infant, freak and main distributions. Infant failures are due to random, and not always preventable fabrication defects, which can typically be screened out in a burn in. Freak failures pass burn in, but fail earlier in the service life than the main population. The main populations of both GaAs-based and InP-based LEDs degrade gradually in light output with a time dependence, which is well-defined mathematically (Zipfel, 1985).

LED failures have been dealt with by using the log-normal statistics developed for semiconductor devices in general (Newman and Ritchie, 1981). Unfortunately, reports of LED reliability typically give only MTTF or median life (ML) and seldom mention the spread in the distribution. It is clear that a long MTTF means nothing if the spread in the distribution is such that there will be a high percentage of failures before the service life is over. Now that LED systems are going into wide usage, field data on such failures will begin to accumulate.

Since repeater spacings in LED-based transmission systems are limited by LED launched power, a trade-off can be made between the power margin allotted for end-of-life and reliability (Saul, 1983). Because LED degradation is gradual with no specific time at which the LED can be said to have failed, there is no intrinsic reason why LED end-of-life must be given as a 50% drop in power. The amount allotted can be smaller, e.g., 20%. The result will be greater repeater spacings but reduced mean-time-to-failure (MTTF).

In the sections ahead, the published reports of the various LED structures will be presented followed by a discussion of the failure mechanisms that must be considered in reliability assurance.

17.3.2 GaAs / GaAlAs LEDs

17.3.2.1 Results of Life Testing. Early reports of the aging behavior of Burrus LEDs (i.e., LEDs with a well etched in the substrate as shown in Fig. 12.1 of Chapter 12) demonstrated the feasibility of these devices for transmission systems (Burrus and Miller, 1971). Most reports suggest, however, that the proportion of infant DLD failures and freak failures in these devices is not well controlled (King *et al.*, 1975); Dyment *et al.*, 1977; Zucker and Lauer, 1978). The inconsistencies in the reports are most likely due to problems inherent in the fabrication of the well. The epitaxial material left when the well is etched away is very thin, with nothing to support or to protect it, and the active area can easily be damaged during assembly. To further promote DLD growth, there are high stresses at the edges of the well where the substrate has been removed.

Other device structures aimed at increasing the launched power have tended to limit reliability because of excess stress and dislocations. For example, devices with a reflector ring etched into the p surface have shown a high percentage of infant failures (SpringThorpe *et al.*, 1982), probably because of stress and damage resulting from the complicated geometry. An LED structure with current confinement achieved by a regrown buried heterostructure has shown a comparatively low ML at 25°C (3 × 10E5 hr) and a large spread in the failure distribution (Hawkins, 1984). Defects and dislocations at the regrown surfaces may well limit the life of this type of structure.

Much more uniform reliability has been reported for a planar double hetero-structure LED, which has current confinement achieved by thin layers and small diameter p contact (Abe *et al.*, 1978; Yamakoshi *et al.*, 1978). In this device, described in Chapter 12, the substrate is etched off to expose the window layer and to allow close placement of the fiber to the emitting area. As expected, these devices exhibit a certain percentage of infant failures, the percentage being directly related to the dislocation density in the substrate. This percentage can be kept as low as 5% by the use of low dislocation substrates. DLDs propagating from grown-in dislocations are strongly current activated, but only weakly temperature activated, if at all. The inevitable infant failures are screened out after a 100-hr. room temperature burn-in at enhanced current, followed by visual inspection of the electroluminescent spot for DLDs with an IR microscope (Yamakoshi *et al.*, 1977; Zipfel *et al.*, 1981).

Planar DH LEDs that remain clear after burn-in have proven to be excep-tionally reliable. Yamakoshi *et al.* (1978) find that L/Lo, the light output L normalized to its initial value, decreases exponentially with a time constant whose activation energy is 0.56 eV and with a MTTF at 25°C of 4×10^7 hr. Zipfel *et al.* (1981) find similar results; L/Lo varies as $\exp[-(t/\tau)^{1/2}]$, where τ is a degradation rate constant with an activation energy of 0.65 eV as shown in Fig. 17.6. The ML for a 50% drop in light output is 9×10^7 hrs. at 25°C. The closeness of these results suggests that gradual degradation is a well-controlled process in the planar device structure. The enhanced reliability of this device over the etched-well and buried-heterostructure LEDs is not surprising since the active area is entirely isolated from any exposed surface and the structure is relatively stress-free. The main source of stress in the structure is the edge of the opening in the dielectric. This can be relieved to some extent by applying the dielectric over the p-contact and then opening a hole in the dielectric (Chin *et al.*, 1980). The price to be paid for the high reliability of this structure is the difficulty of growing the thick graded Al n-layer and the problems with processing a wafer only about 50 μm thick (Chin *et al.*, 1981).

17.3.2.2 Degradation Mechanisms. If the burn-in and visual inspection have been properly performed, all DLDs from dislocations in the high current area will have been screened out (Chin *et al.*, 1980). Reports exist in the literature, however, that suggest that there is a small but finite percentage of subsequent freak failures that eludes a short-term screen. Therefore, since MTTF for the main population of planar DH LEDs is very long, the ultimate reliability will depend on reducing the size of the freak population to its minimum. Several

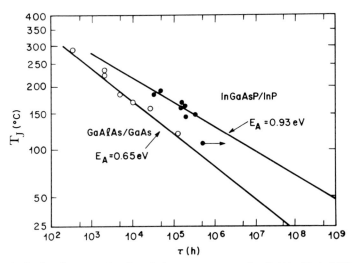

Fig. 17.6 Arrhenius plot comparing degradation rate constant τ for GaAlAs/GaAs LEDs (Zipfel et al., 1982) and InGaAsP/InP LEDs (King et al., 1985).

causes for these freak DLDs have been identified:

(a) Both $\langle 110 \rangle$ and $\langle 100 \rangle$ DLDs can begin to grow slowly at dislocations or damage far from the contact area where the current density is very low. When they reach the contact area where current density is high, they grow rapidly and the device fails, but in times much longer than any reasonable burn-in. Other than keeping substrate dislocation density at a minimum and avoiding processing damage, these DLDs are inevitable. (Chin et al., 1981).
(b) At some threshold of mechanical or thermal stress, $\langle 100 \rangle$ DLDs can form, for example at the dielectric-metal interface (Ueda et al., 1980b).
(c) The ultimate limit to reliability may be gold migration from the contact to the junction. Dark patches formed at long times have been attributed to this effect. The gold-rich precipitates formed may contribute to DLD formation in the long term (Hersee, 1977). This effect puts a limit on the operating current that can be used.

17.3.3 InGaAsP / InP LEDs

17.3.3.1 Results of Life Testing. The important features of the reliability of surface-emitting InGaAsP/InP LEDs were described by Yamakoshi et al. (1979; 1981). In contrast to GaAs-based devices, no DLDs were observed after 100 hrs. of operation even though high dislocation density substrates (1×10^5 cm^{-2}) were used, leading the authors to conclude that dislocations in this material are not strong regions of nonradiative recombination. Unlike GaAlAs/GaAs double heterostructures, lattice mismatch often appears in the InGaAsP/InP system when crystal growth is faulty. Yamakoshi observed that devices with lattice mismatch developed $\langle 110 \rangle$ lines during aging. These lines were shown by TEM to be misfit dislocations. The most surprising observation was that even though small DSDs appeared during aging, little decrease in light output was seen. The activation energy was found to be between 0.9 and 1.0 eV and the extrapolated MTTF at 60°C and 8 kA/cm^2 was 10^9 hr.

King *et al.* (1985) studying highly doped, high current density devices found that degradation in light output follows the exponential root-time law discussed in Section 17.3.2.1. An Arrhenius plot for τ using their data is shown in Fig. 17.6. Their activation energy is similar to that of Yamakoshi. Suzuki *et al.* (1985) report an activation energy of 1 eV and extrapolated half-power lifetimes in excess of 10^8 hrs. at 60°C for very high speed (> 200 Mb/s) LEDs operated at current densities up to 128 kA/cm^2.

17.3.4 Degradation Mechanisms

The notable feature of all reports of InGaAsP LED reliability so far has been the extraordinarily long values of MTTF. It has been unclear whether there is a natural infant failure population that has to be eliminated in a burn-in as there is for GaAlAs/GaAs devices. There are, however, three types of defects that have been reported to cause freak failures and that can be screened out either at the chip or wafer level: indium inclusions, DLDs from stress and misfit dislocations. A fourth type of defect, DSDs, appears almost universally in this material although is not clear how severe its effect is on reliability. These four types of defects, as they relate to device reliability, will be discussed next.

1. In inclusions. Both anomalous $\langle 110 \rangle$ DLD growth and instabilities in I-V characteristics can result from In inclusions. Temkin *et al.* (1981) have shown that freak $\langle 110 \rangle$ DLD growth can result from grown-in inclusions. Indium inclusions originate at either the meltback or wipe-off stages of the LPE cycle or by thermal decomposition of the substrate (Temkin *et al.*, 1983). When a buffer layer is not completely wiped off, In inclusions left at the buffer-layer surface locally prevent growth from the InGaAsP melt and thus produce holes in the active layer. An InGa-rich phase can be expected to have a lower melting point. Such inclusions might be expected to move under thermal gradients through the crystal lattice (Johnston, 1983) changing the I-V characteristic and, in extreme cases, the light output. Such inclusions can be readily screened out by looking for changes in the I-V before and after a room temperature burn-in.

2. DLDs from stress. Ueda *et al.* (1982a) studied several types of dark defects generated during crystal growth or processing (misfit dislocations, stacking faults, precipitates, mechanical damage, and alloyed regions produced by penetration of the electrode metals). It was found that recombination-enhanced climb and glide occur only with difficulty in InGaAsP devices. By contrast, these processes occur readily in GaAs-based devices. In a rare observation, however, Temkin *et al.* (1983) did report that ⟨110⟩ DLDs can be produced by dielectric stress over the entire area of the p-InP confining layer, causing rapid degradation in LED light output. The SiO_2 film must therefore be sufficiently thick to prevent pinholes, yet as thin as possible to minimize stress on the semiconductor.

(3) Misfit dislocations. Attention must be paid to the reproducible growth of quaternary active layers with a composition corresponding to the 1.3 μm emission wavelength, while also being lattice matched to the InP substrate (Temkin *et al.*, 1983). The lattice mismatch must be in the range of 0 to -0.08% at room temperature to make the area without misfit dislocations as wide as possible (Yamazaki *et al.*, 1982). This criterion depends on the confining layer thickness. At sufficiently high current densities (> 8kA/cm2) the growth of misfit dislocations from the p-InP confining layer into the active layer becomes an important failure mode (Chin *et al.*, 1983a).

(4) Dark Spot Defects. Although, as discussed in Section 17.2.1, DSDs appear in both LEDs and lasers, they apparently do not lead to rapid degradation under most circumstances. Ueda *et al.* (1980a; 1981) used TEM to study DSDs in LEDs and concluded that they are formed by precipitation of host atoms during operation. There has been a lack of consistency among the reports of various groups, suggesting that DSDs may encompass more than one phenomenon.

Chin *et al.* (1982; 1983b) found that during operation DSDs appear to propagate from the contact into the active layer as shown in Fig. 17.7. From this series of experiments they proposed a model in which DSDs form by electrother-momigration of gold from the p contact. Camlibel *et al.* (1982) found that the gold contact interacts strongly and nonuniformly with the semiconductor. The conclusion is that the gold-semiconductor interaction and details of alloying may be the most important factor in LED reliability.

17.4. CONCLUSIONS

Many of the studies of laser and LED degradation have been reviewed in this chapter with the purpose of presenting any information that might be useful in reliability assurance, either in design or in use. The procedure for certifying individual lasers for high reliability applications such a submarine cables is

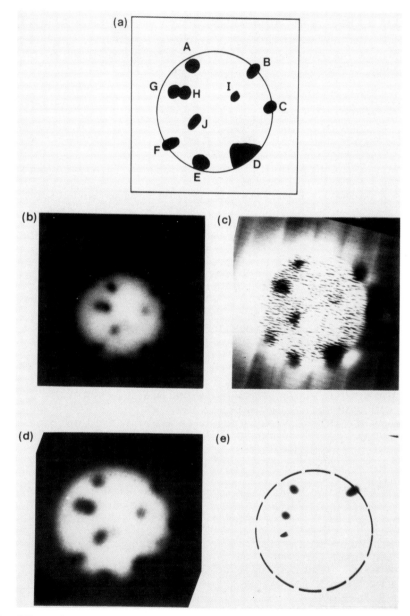

Fig. 17.7 (a) Schematic of DSDs in degraded InP/InGaAsP LED. (b) EL image of light-emitting spot. (c) EBIC image of p contact region with all four epilayers intact. (d) EBIC image of p contact region after removal of contact layer. (e) EBIC image of p contact region after removal of the p-InP confining layer. (After Chin *et al.*, 1983c).

discussed. LEDs typically are not subjected to extensive tests such as this. LED reliability depends on device design that minimizes such factors as stress and contact metal migration and that allows for consistently good crystal growth and processing.

REFERENCES

Abe, M., Hasegawa, O., Komatsu, Y., Toyama, Y., and Yamaoka, T. (1978). *IEEE Trans. Electron Devices* **ED-25**, 1344.

Burrus, C. A., and Miller, B. I. (1971). *Opt. Commun.* **4**, 307.

Camlibel, I., Chin, A. K., Ermanis, F., DiGiuseppe, M. A., Lourenco, J. A., and Bonner, W. A. (1982). *J. Electrochem. Soc.* **129**, 2585.

Chin, A. K., Keramidas, V. G., Johnston, W. D., Jr., Mahajan, S., and Roccasecca, D. D. (1980). *J. Appl. Phys.* **51**, 978.

Chin, A. K., Temkin, H., and Mahajan, S. (1981). *Bell Syst. Tech. J.* **60**, 2187.

Chin, A. K., Zipfel, C. L., Mahajan, S., Ermanis, F., and DiGiuseppe, M. A. (1982). *Appl. Phys. Lett.* **41**, 555.

Chin, A. K., Zipfel, C. L., Chin, B. H., and DiGiuseppe, M. A. (1983a). *Appl. Phys. Lett.* **42**, 1031.

Chin, A. K., Zipfel, C. L., Ermanis, F., Marchut, L., Camlibel, I., DiGiuseppe, M. A., and Chin, B. H. (1983b). *IEEE Trans. Electron Devices* **ED-30**, 304.

DeLoach, B. C., Jr., Hakki, B. W., Hartman, R. L., and D'Asaro, L. A. (1973). *Proc. IEEE* **61**, 1042.

Dyment, J. C., Spring Thorpe, A. J., King, F. D., and Straus, J. (1977). *J. Electron. Mater.* **6**, 173.

Fukuda, M., Wakita, K., and Iwane, G., (1983). *J. Appl. Phys.* **54**, 1246.

Gordon, E. I., Nash, F. R., and Hartman, R. L. (1983). *IEEE Electron Dev. Lett.* **EDL-4**, 465.

Hakki, B. W., Fraley, P. E., and Eltringham, T. (1985). *AT & T Tech. J.* **64**, 771.

Hartman, R. L., and Dixon, R. W. (1975). *Appl. Phys. Lett.* **26**, 239.

Hawkins, B. M. (1984). *Proc. Electron. Component Conf. 34th*, New Orleans, 1984, p. 239.

Hersee, S. D. (1977). *Tech Dig.—Int. Electron Devices Meet.* p. 567.

Johnston, W. D., Jr., Epps, G. Y., Nahory, R. E., and Pollack, M. A. (1978). *Appl. Phys. Lett.* **33**, 992.

Johnston, W. D., Jr. (1983). *Mater. Res. Soc. Symp. Proc.* **14**, 453.

King, F. D., SpringThorpe, A. J., and Szentesi, O. I. (1975). *Tech Dig.—Int. Electron Devices Meet.*, p. 480.

Kingm W. C., B. H. Chin, I. Camlibel, and Zipfel, C. L. (1985). *IEEE Electron Device Lett.* **EDL-6**, 335.

Mahajan, S., Johnston, W. D., Jr., Pollack, M. A., and Nahory, R. E. (1979). *Appl. Phys. Lett.* **34**, 717.

Mahajan, S., Chin, A. K., Zipfel, C. L., Brasen, D., Chin, B. H., Tung, R. T., and Nakahara, S., (1984). *Materials Letts.* **2**, 184.

Mizujishi, K., Hirao, M., Tsuji, S., Sata, H., and Nakamura, M. (1982). *Jpn J. Appl. Phys.* **21**, 359.

Mizujishi, K., Wakita, K., and Nakamura, M. (1983). *IEEE J. Quantum Electron* **QE-19**, 1294.

Nash, F. R., Sundberg, W. J., Hartman, R. L., Pawlik, R. L., Ackerman, D. A., Dutta, N. K., and Dixon, R. W. (1985). *AT & T Tech. J.* **64**, 809 and references therein.

Newman, D. H. and Ritchie, S. (1981). In "Reliability and Degradation." (M. J. Howes and D. V. Morgan, eds.), Chapter 6. Wiley, Chichester.

Petroff, P. M., and Lang, D. V. (1977). *Appl. Phys. Lett.* **31**, 60.

Runge, P. K., and Trischitta, P. R. (1984). *IEEE J. Lightwave Tech.* **LT-2**, 744.

Saul, R. H. (1983). *IEEE Trans. Electron Devices* **ED-30**, 285.

Saul, R. H., Chen, F. S., and Shumate, P. W. (1985). *AT & T Tech. J.* **64**, 861.

SpringThorpe, A. J., Look, C. M., and Emmerstorfer, B. F. (1982). *IEEE Trans. Electron Devices* **ED-29**, 876.

Suzoki, A., Uji, T., Inomoto, Y., Hayashi, J., Isoda, Y., and Nomura, H. (1985). *J. Lightwave Technology*, **LT-3**, 1217.

Temkin, H., Zipfel, C. L., and Keramidas, V. G. (1981). *J. Appl. Phys.* **52**, 5377.

Temkin, H., Mahajan, S., DiGiuseppe, M. A., and Dentai, A. G. (1982). *Appl. Phys. Lett.* **40**, 562.

Temkin, H., Zipfel, C. L., DiGiuseppe, M. A., Chin, A. K., Keramidas, V. G., and Saul, R. H. (1983). *Bell Syst. Tech. J.* **62**, 1.

Ueda, O., Yamakoshi, S., Komiya, S., Akita, K., and Yamaoka, T. (1989b). *Appl. Phys. Lett.* **36**, 300.

Ueda, O., Imai, H., Fujiwara, T., Yamakoshi, S., Sugawara, T., and Yamaoka, T. (1980b). *J. Appl. Phys.* **51**, 5316.

Ueda, O., Komiya, S., Yamakoshi, S., and Kotani, T. (1981). *Jpn. J. Appl. Phys.* **20**, 1201.

Ueda, O., Umebu, I., Yamakoshi, S., and Kotani, T. (1982). *J. Appl. Phys.* **53**, 2991.

Ueda, O., Imai, H., Yamaguchi, A., Komiya, S., Umebu, I., and Kotani, T. (1984). *J. Appl. Phys.* **55**, 665.

Yamakoshi, S., Hasegawa, O., Hamaguchi, H., Abe, M., and Yamaoka, T. (1977). *Appl. Phys. Lett.* **31**, 627.

Yamakoshi, S., Sugahara, T., Hasegawa, O., Toyama, Y., and Takanashi, H. (1978). *Tech. Dig.—Int. Electron Devices Meet.* p. 642.

Yamakoshi, S., Abe, M., Komiya, S., and Toyama, Y. (1979). *Tech. Dig.—Int. Electron Devices Meet.* p. 122.

Yamakoshi, S., Abe, M., Wada, O., Komiya, S., and Sakurai, T. (1981). *IEEE J. Quantum Electron.* **QE-17**, 167.

Yamazaki, S., Kishi, Y., Nakajima, K., Yamaguchi, A., and Akita, K. (1982). *J. Appl. Phys.* **53**, 4761.

Zipfel, C. L., Chin, A. K., Keramidas, V. G., and Saul, R. H. (1981). *Proc. 19th Ann. IEEE Int. Reliab. Phys. Symp.* p. 124.

Zipfel, C. L., (1985) In "Semiconductors and Semimetals." Vol. 22, Part C, ed. by W. T. Tsang, Academic Press, New York.

Zucker, J., and Lauer, R. B. (1978). *IEEE J. Solid-State Circuits* **SC-13**, 119.

Chapter 18

Receiver Design

BRYON L. KASPER

AT&T Bell Laboratories, Inc., Holmdel, New Jersey

18.1 INTRODUCTION

The purpose of an optical receiver is to convert a modulated optical signal to an electrical signal and to recover from the electrical signal whatever information had been impressed on the optical carrier. The information may be digital or analog, and the optical carrier may be modulated in a variety of ways, including amplitude modulation (AM), frequency modulation, and phase modulation. Because simple optical detectors such as PIN diodes or avalanche photodiodes (APDs) are ideal AM envelope detectors, but are insensitive to phase or to small changes in wavelength, it is amplitude modulation that is used almost exclusively in present optical communications systems. Frequency and phase modulation are being investigated in many laboratories for use in coherent optical transmission, and may become important in future systems. This chapter will be concerned with the conventional approach of amplitude-modulated optical signals and the design of appropriate direct-detection receivers.

Analog transmission is not commonly used in optical systems because of the nonlinear light-versus-current characteristics of optical sources (lasers and LEDs). Digital modulation (on-off keying), on the other hand, is easily accomplished with these sources by modulation of bias current. Simplicity of modulation plus the extremely wide bandwidth of optical fibers have made binary digital transmission the method of choice for the majority of optical communications systems. Thus, this chapter will concentrate on receivers intended for the direct detection of digital data consisting of a stream of light pulses, where the presence of a pulse corresponds to the transmission of a binary "one," and the absence of light corresponds to the transmission of a binary "zero."

18.2 BASIC RECEIVER STRUCTURE

A block diagram of a basic telecommunications receiver is shown in Fig. 18.1. The photodiode acts as a square-law detector and has an electrical output that reproduces the envelope of the received optical signal. The electrical signal is amplified by a low-noise preamplifier, which is followed by the receiver linear-channel section. The linear channel consists of an equalizer to correct for roll-off in the front end (not always necessary), a high-gain postamplifier (usually with some form of automatic gain control), and a low-pass filter to limit the bandwidth to the minimum required to pass the signal.

For optimum performance, it is necessary to extract timing information from the received signal in order to synchronize the process of making decisions on the noisy linear-channel output. For this purpose, a portion of the linear-channel output is fed to a timing recovery or clock recovery circuit that generates a clock signal at the baud rate synchronized with transitions in the received data (Maione et al., 1978). With return-to-zero (RZ) data format, a spectral component at the baud rate is present in the received signal, and clock recovery can be accomplished with a narrow bandpass filter such as a phase-locked loop or surface-acoustic-wave filter. With non-return-to-zero (NRZ) data format, a nonlinear operation such as squaring must be performed on the received signal to generate a spectral line at the baud rate.

The final receiver stage is the decision circuit. This stage consists of a comparator with its threshold level set to the center of the received eye pattern to give an equal probability of error for decisions on both "ones" and "zeroes." The comparator is followed by a D flip-flop clocked by the recovered timing signal. The clock phase is timed to store the best estimate of the transmitted data by triggering the flip-flop at the center of each bit when the eye opening is the greatest.

Low-cost receivers such as data links often use asynchronous detection without clock recovery in which the comparator output is used directly as the data output. The receiver bandwidth must be much greater in this case in order to have fast rise and fall times for clearly defined pulse edges; hence, less noise is filtered from the receiver output and sensitivity is reduced. All sensitivity calculations in this chapter assume clock recovery and synchronous detection.

18.3 FUNDAMENTALS

The following sections discuss the current-source nature of photodiodes, the theoretical limit on the sensitivity of optical receivers imposed by the randomness of photon detection, how amplifier noise prevents this quantum limit from

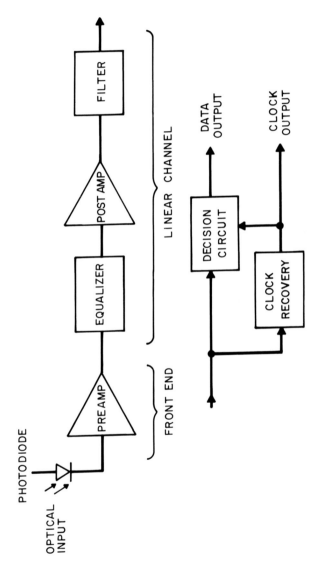

Fig. 18.1 Block diagram of a basic digital lightwave receiver.

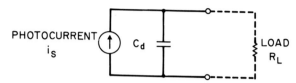

Fig. 18.2 AC equivalent circuit of photodiode.

being reached, and how avalanche photodiodes can provide improved sensitivity over that attainable with PIN photodiodes.

18.3.1 Detector Characteristics

A variety of devices can serve as photodetectors, including PIN photodiodes, avalanche photodiodes, phototransistors, and photoconductors (Sze, 1981). Only PINs and APDs will be considered herein, as they have demonstrated the best characteristics for telecommunications applications.

A photodiode (PIN or APD) operates in a reverse-biased mode such that incoming photons are absorbed in a depleted semiconductor region containing a strong electric field. The presence of the electric field assures that electron-hole pairs created by photon absorption are quickly separated and collected at the photodiode terminals before recombination can occur. Figure 18.2 shows an AC equivalent circuit of a basic photodiode. The current source i_s represents photocurrent resulting from the detection of an optical signal. For a PIN detector, the output current is given by

$$ i_s = \left(\frac{\eta q}{h\nu} \right) P, \tag{18.1} $$

where

P = incident optical power

η = detector quantum efficiency

q = electron charge = 1.6×10^{-19} coul

h = Planck's constant = 6.63×10^{-34} joule-sec, and

ν = optical frequency = c/λ.

A current source is an excellent small-signal model for photodiode behavior as the output current is independent of small changes in reverse-bias voltage, especially for PINs. For APDs, the avalanche gain and hence photocurrent do vary with bias voltage. However, the magnitude of the bias voltage is from tens to hundreds of volts, whereas voltage changes resulting from variations in photocurrent are on the order of millivolts and do not affect the overall bias voltage and gain appreciably in most circumstances. The capacitance C_d represents the sum of the photodiode junction capacitance and external packaging capacitance. Detector series resistance is generally small $(5-10\Omega)$ and is ignored in this model. Leakage current (dark current) is neglected for the moment but will be considered later in the section on receiver sensitivity.

18.3.2 The Quantum Limit

If one considers an ideal PIN photodiode (unity quantum efficiency and zero dark current) followed by a noiseless electronic amplifier, it is possible to arrive at the minimum received optical power required for acceptable bit-error-rate performance in a digital transmission system. This minimum received power level, known as the quantum limit, is a result of the statistically random nature of light absorption in a material medium. As described in the previous volume of this series (Personick, 1979); the probability $p(n)$ that exactly n electron-hole pairs will be created from a pulse of light of known amplitude is given by a Poisson distribution

$$p(n) = N_{av}^n e^{-N_{av}}/n!, \qquad (18.2)$$

where N_{av} is the average number of pairs produced by an optical pulse of the same amplitude, with the average being taken over a large number of trials. The unpredictability of output current from a photodiode illuminated by a steady optical signal is a phenomenon known as quantum noise or shot noise.

Personick derived a quantum limit of 21 photons per pulse for an error rate of 10^{-9}. A refinement of this calculation (Henry, 1985) notes that there is zero probability of error during the detection of an optical "zero" (no light) because no electron-hole pairs can be created. Therefore, if the density of both "ones" and "zeroes" is 50% and the average error rate is 1×10^{-9}, the allowed probability of error in the reception of "ones" is 2×10^{-9}. Setting the probability of no photons being detected when a "one" is transmitted to be 2×10^{-9}

$$p(0) = e^{-N_{av}} = 2 \times 10^{-9}, \qquad (18.3)$$

one finds that 20 photons per "one" are required, rather than 21. If 50% of the bits are zeroes, then the average required photons per bit in the quantum limit is 10.

18.3.3 The Effect of Amplifier Noise

In practice, most current optical receivers have sensitivities that are 20 dB or more from the quantum limit. The main factor preventing quantum-limited operation is noise from electronic amplifiers following the photodiode. Amplifier noise in present receivers is generally two to three orders of magnitude larger than the quantum noise discussed before. The lowest possible preamplifier noise is therefore essential for high receiver sensitivity, especially at wavelengths longer than 1 μm, where silicon becomes transparent and low-noise silicon avalanche photodiodes cannot be used.

To obtain the highest possible signal-to-noise ratio in an optical receiver, one would like the current available from the photodiode to generate the maximum preamplifier input voltage. The standard microwave approach to such maximum power transfer is to employ an amplifier that presents a conjugate impedance match to the source. Pulses as used in digital transmission, however, have a broad spectral content and, as the photodiode impedance is capacitive, exact conjugate matching is not possible over a broad bandwidth. The usual method of maximizing input voltage in optical receivers is to minimize detector and preamplifier capacitance. This capacitance can be reduced to about 0.5 pF in state-of-the-art receivers, but is still the main factor that limits preamplifier input voltage and causes amplifier noise to be the dominant limitation in optical receiver sensitivity.

18.3.4 Avalanche Photodiodes

With even the lowest noise and lowest capacitance preamplifiers available with current technology, sensitivities obtained with PIN detectors are far from the quantum limit. Sensitivity can be improved by using an avalanche photodiode with internal gain to deliver a larger output current. APDs utilize impact ionization to allow photo-generated primary electrons or holes to create additional electron-hole pairs, initiating a chain of impact-ionization events that multiplies the primary current by an average gain M. The output current is then

$$i_s = M\left(\frac{\eta q}{h\nu} \right) P. \qquad (18.4)$$

Avalanche gain increases the signal level before the addition of preamplifier noise, thereby improving sensitivity. Quantum noise is already present in the primary photocurrent and is multiplied by the same gain as the signal, hence sensitivity with APDs cannot exceed the quantum limit. In fact, APD receivers are prevented from reaching the quantum limit by the randomness of the avalanche gain process, as the gain varies randomly for each primary electron-hole pair. This random gain adds additional uncertainty to the received signal level and is termed excess noise as it increases the signal-related noise above the inherent quantum-noise level. The best silicon APD receivers have demonstrated sensitivities 13 dB from the quantum limit (Runge, 1976; Muoi, 1983).

Methods for calculating sensitivity for either PIN or APD receivers are outlined in Section 18.5.

18.4 RECEIVER FRONT ENDS

The performance of an optical receiver in terms of sensitivity and dynamic range is highly dependent on the design of the receiver front end. This section discusses receiver front-end types and noise calculations for various choices of input transistor.

18.4.1 Front End Types

Receiver front ends can be classified into three broad categories, although the boundaries between categories are indistinct and in reality there is a continuum of intermediate possibilities. The categories are 1) the low-impedance front end, where the load resistor R_L in Fig. 18.2 is small and the input RC time constant does not limit the front-end bandwidth to less than the signal bandwidth; 2) the high-impedance front end, where R_L is large for improved sensitivity but the front-end bandwidth is less than that of the signal, resulting in a requirement for subsequent equalization to compensate for inadequate front-end bandwidth; and 3) the transimpedance front end, where the load resistor is replaced by a large feedback resistor, and negative feedback around a wideband amplifier is used to obtain increased bandwidth.

18.4.1.1 Low-Impedance Front Ends. Low-impedance front ends typically consist of a photodiode (PIN or APD) operating into a low-impedance (e.g., 50Ω) amplifier, often through a length of coaxial cable or other transmission line, as shown in Fig. 18.3. A terminating resistor R_L equal to the transmission line impedance is generally included to suppress standing waves for uniform frequency response. Low-impedance front ends do not provide high sensitivity because only a small signal voltage can be developed across the amplifier input impedance and resistor R_L. Because of the ready availability of 50Ω RF and

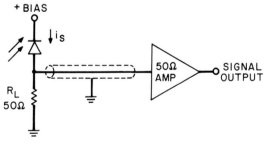

Fig. 18.3 Low-impedance receiver front end.

microwave amplifiers, such front ends are often used in cases where high sensitivity is not a prime concern. The variance of the equivalent input noise current for a low impedance front end is given by

$$\langle i^2 \rangle_a = \frac{4kTF}{R_L} \Delta f \qquad (18.5)$$

where

R_L = load resistor

F = amplifier noise figure

Δf = bandwidth

k = Boltzmann's constant = 1.38×10^{-23} joule/K, and

T = absolute temperature.

Typical values for microwave amplifier noise figures at a few GHz are from 3 dB to 6 dB (i.e., F = 2 to 4).

18.4.1.2 High-Impedance Front Ends. Because of the current-source nature of photodiodes, it is possible to increase signal voltage for a given photocurrent by operating the photodiode into an amplifier with higher input resistance. The signal voltage is therefore increased prior to the addition of noise from the following amplifier, resulting in improved sensitivity (Goell, 1974a). In a high-impedance front end, depicted schematically in Fig. 18.4, the load resistor R_L is very large and a high-impedance field-effect transistor amplifier is employed. The total input capacitance C_T is the sum of FET input capacitance, detector capacitance, and stray capacitance. As a result of the large input RC time

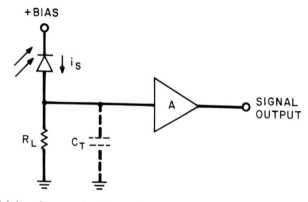

Fig. 18.4 High-impedance receiver front end.

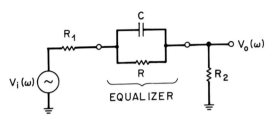

Fig. 18.5 RC equalizer for high-impedance front end.

constant, the front-end bandwidth is less than the signal bandwidth, hence for frequencies above $1/2\pi R_L C_T$ the input signal is integrated. Such integrating front ends require equalization to compensate for their lack of bandwidth.

Equalization is commonly provided by a simple RC circuit as shown in Fig. 18.5. The transfer function of this circuit is given by

$$H(\omega) = \frac{V_0(\omega)}{V_i(\omega)} = \frac{R_2(1 + j\omega RC)}{R + R_1 + R_2 + j\omega RC(R_1 + R_2)}. \qquad (18.6)$$

The numerator of this transfer function specifies a zero with a corner frequency f_L given by

$$f_L = \frac{1}{2\pi RC}, \qquad (18.7)$$

and the denominator specifies a pole with a corner frequency f_U given by

$$f_U = f_L \frac{R + R_1 + R_2}{R_1 + R_2}, \qquad (18.8)$$

which will be much greater than f_L if $R \gg R_1 + R_2$. The values of R and C are chosen such that the zero at f_L cancels the pole of the high-impedance front end, and the overall receiver bandwidth is extended to f_U.

The position of the equalizer in the linear amplifier chain must be chosen carefully to avoid a sensitivity penalty due to noise from stages following the equalizer, or reduction in receiver dynamic range due to saturation of amplifiers preceding the equalizer. Limited dynamic range is the major drawback of high-impedance receivers, as the high level of low-frequency signal components present in the integrated signal from the front end can easily saturate amplifier stages prior to equalization. For this reason, the equalizer should be placed as close as possible to the front end in the amplifier chain, where signal levels are still small, but not so close that its attenuation causes noise from the following stage to be significant compared to front-end noise.

High-impedance front ends have been used in a number of laboratory demonstrations of record receiver sensitivities (Smith *et al.*, 1982; Kasper *et al.*,

Bryon L. Kasper

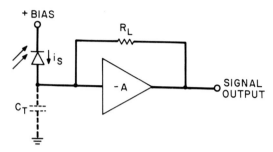

Fig. 18.6 Transimpedance receiver front end.

1985; Shikada *et al.*, 1985), and can be used in practical applications where wide dynamic range is not needed, or where the low-frequency content of signals is reduced through techniques such as coding (Brooks, 1980; Rousseau, 1976).

18.4.1.3 Transimpedance Front Ends. A configuration that provides improved dynamic range is the transimpedance front end illustrated in Fig. 18.6. The load resistor R_L is connected as a feedback resistor around an inverting amplifier with gain A. It is readily shown that the bandwidth is increased by a factor of $A + 1$ over that of a high-impedance front end with the same R_L and C_T, hence the need for further equalization is reduced or eliminated. In addition, low-frequency gain is reduced by the same factor of $A + 1$, thereby decreasing the susceptibility to saturation by low-frequency signal content and improving dynamic range.

One consideration in the design of transimpedance amplifiers is stability. The capacitive input impedance of the preamplifier causes a phase shift of 90 degrees at high frequencies, and the inverting amplifier contributes a phase shift of 180 degrees. Generally, one would like to maintain a phase margin of about 45 degrees, which allows a maximum additional open-loop phase shift of only 45 degrees up to the frequency at which the open-loop gain becomes less than unity. This small tolerable phase shift limits the gain that can be included within the feedback loop, hence it is usually not possible to obtain the desired preamplifier bandwidth if a large feedback resistor is used. (RC equalizers that could extend the bandwidth are often avoided in practice because of the need for adjustment and because of difficulty in maintaining pole-zero matching over a wide range of operating conditions.) For these reasons, practical transimpedance receivers generally give up a few dB in sensitivity by using a lower value of feedback resistor, but in exchange have greater simplicity and much wider dynamic range than high-impedance designs.

Several examples of transimpedance front-end designs are given in Section 18.7. First, however, we will consider the noise performance of various transistor types when used in receiver preamplifiers, and the sensitivities that can be achieved with either PIN or APD detectors. The noise calculations which

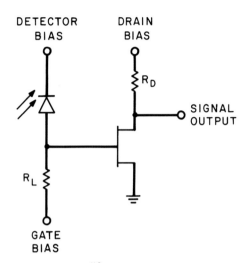

Fig. 18.7 Common-source FET preamplifier.

follow are equally valid for high-impedance or transimpedance amplifiers, since the feedback in a transimpedance amplifier, although it does change the front-end bandwidth, has the same effect on both signal and noise. The ratio of the two is not affected, hence the sensitivity of a transimpedance front end is ideally the same as that of a high-impedance front end using similar components.

18.4.2 Noise Calculations for FET Front Ends

At the present time, the noise performance of field effect transistors in optical receiver front ends is superior to that of bipolar transistors, mainly due to the development of GaAs MESFETs as low-noise microwave amplifiers.

Although preamplifiers have recently been reported using the common-drain configuration (Teare and Ulbricht, 1985; Chown and Spencer, 1986), the more usual arrangement is the common-source amplifier shown in Fig. 18.7. The principal noise sources are thermal noise from the FET channel conductance and the load resistor R_L, shot noise due to gate leakage current, and FET $1/f$ noise. The variance of the equivalent amplifier input noise current $\langle i^2 \rangle_a$ is given by (Smith and Personick, 1982; Muoi, 1984; Forrest, 1985)

$$\langle i^2 \rangle_a = \frac{4kT}{R_L} I_2 B + 2q I_{gate} I_2 B + \frac{4kT\Gamma}{g_m} (2\pi C_T)^2 f_c I_f B^2$$

$$+ \frac{4kT\Gamma}{g_m} (2\pi C_T)^2 I_3 B^3, \tag{18.9}$$

where

$$B = \text{bit rate}$$

$$R_L = \text{load or feedback resistor}$$

$$I_{gate} = \text{FET gate leakage current}$$

$$g_m = \text{FET transconductance}$$

$$C_T = \text{total input capacitance}$$

$$f_c = \text{FET } 1/\text{f-noise corner frequency, and}$$

$$\Gamma = \text{FET channel noise factor.}$$

I_2, I_3, and I_f are normalized noise-bandwidth integrals and depend only on the input-optical-pulse shape and the equalized linear-channel output-pulse shape (Personick, 1973; Smith and Personick, 1982). For rectangular NRZ input pulses, and output pulses with a full raised-cosine spectrum, the values are $I_2 = 0.562$, $I_3 = 0.0868$, and $I_f = 0.184$. The 3-dB bandwidth of the corresponding linear-channel filter in this case is 58% of the bit rate. For rectangular RZ input pulses with a 50% duty cycle, the values are $I_2 = 0.403$, $I_3 = 0.0361$, and $I_f = 0.0984$, with the corresponding filter bandwidth being 39% of the bit rate. The raised-cosine output-pulse spectra assumed above correspond to idealized output pulses that have zero intersymbol interference and are therefore convenient for theoretical analysis. Approximations to such ideal pulses can be obtained in practice with conventional low-pass filters.

Note that tighter filtering is possible for RZ pulses because their spectral content is wider than that of NRZ pulses. If the last term in Eq. (18.9) is dominant, as it would be in a well-designed receiver, then the noise advantage of RZ over NRZ is given by the ratio of the corresponding I_3 values, or $10 \log_{10}(0.0868/0.0361) = 3.8$ dB (electrical). This translates into a 1.9 dB optical sensitivity advantage for an RZ PIN/FET receiver, or somewhat less for an APD receiver. This sensitivity is based on average received power, which is less for RZ because of its 50% duty cycle. If the sensitivity were based on peak power, the RZ PIN/FET sensitivity would be reduced by 3 dB, making it 1.1 dB worse than NRZ rather than 1.9 dB better. Generally, laser transmitters are neither strictly peak-power limited nor average-power limited, but somewhere in between. Thus, from a power-budget viewpoint in transmission systems, neither RZ nor NRZ format has a distinct advantage.

The FET channel-noise factor Γ is a numerical constant that accounts for thermal noise and gate-induced noise plus the correlation between these two noise sources (Robinson, 1974; Buckingham, 1983). The thermal noise behavior of Si JFETs approaches the theoretical ideal for long-channel FETs, resulting in a low value for Γ of ≈ 0.7 (van der Ziel, 1970). GaAs MESFETs exhibit higher noise because of intervalley scattering (Baechtold, 1972) and have a value for Γ generally taken as ≈ 1.1 (Goell, 1974b; Smith and Personick, 1982); although it has been argued that the value could be as large as 1.75 (Ogawa, 1981). Short-channel Si MOSFETs are also presently noiser than the ideal due to hot-electron effects, impact ionization, and substrate effects, and have Γ values of 1.5–3 (Jindal, 1985; Ogawa, 1983).

The total capacitance C_T is given by

$$C_T = C_d + C_{gs} + C_{gd} + C_s, \tag{18.10}$$

where

C_d = detector capacitance

C_{gs} = FET gate-source capacitance

C_{gd} = FET gate-drain capacitance, and

C_s = stray capacitance (including load resistor shunt capacitance).

The 1/f-noise corner frequency f_c is defined as the frequency at which 1/f noise, which dominates the FET noise at low frequencies and has a 1/f power spectrum, becomes equal to the high-frequency channel noise, which has a white spectrum described by Γ.

The first term in Eq. (18.9) accounts for thermal noise from the load or feedback resistor R_L and can be made as small as desired by making that resistor very large (generally at the expense of receiver dynamic range). The second noise term can be minimized by choosing a FET with low gate-leakage current, and the third term can be reduced by choosing a FET with low 1/f noise. The fourth term, due to FET channel noise, is minimized by choosing a FET that gives the largest value of g_m/C_T^2. In general, this means using a device with short gate length for large f_T, and choosing the optimum gate width. As C_{gs}, C_{gd}, and g_m are all proportional to gate width, it can be easily shown by setting the derivative with respect to gate width equal to zero that the lowest noise in Eq. (18.9) is obtained when the gate width is chosen such that

$$C_{gs} + C_{gd} = C_d + C_s. \tag{18.11}$$

The above optimization of FET width is appropriate for a high-impedance front end, or a transimpedance front end with a large feedback resistor where RC equalization is used to extend the bandwidth. For the case of a transimpedance front end without additional equalization, which generally has substantial feedback resistor noise, Abidi (1988) has shown that the best sensitivity is obtained when the FET width is significantly less than that which satisfies Eq. (18.11). The lower total input capacitance allows a larger feedback resistor, giving the best compromise between low noise and adequate bandwidth.

An intuitive explanation of the bit-rate dependence of the terms in Eq. (18.9) is in order at this point. Consider a high-impedance front end. The load resistor R_L and gate leakage current I_{gate} are white-noise sources whose spectra are integrated by the receiver input capacitance. Equalization restores white spectra at the receiver output with high frequencies limited by the linear-channel low-pass filter, which has a bandwidth proportional to the bit rate. Hence, the total noise power from these two sources is proportional to B. The FET channel-noise spectrum is initially white, but is differentiated by the equalizer and is therefore modified to have an f^2 dependence. The integral of this f^2 spectrum over the low-pass filter bandwidth varies as B^3, hence the B^3 dependence of the last term in Eq. (18.9). The FET $1/f$ noise is likewise differentiated, leading to noise spectrum at the receiver output that varies as f rather than f^2, hence the integral giving the total noise power from $1/f$ noise is proportional to B^2.

A figure of merit for optimized FET receivers can be derived in terms f_T, the FET short-circuit common-source current-gain-bandwidth product, by noting that

$$f_T = \frac{g_m}{2\pi(C_{gs} + C_{gd})}. \tag{18.12}$$

At high bit rates, where the fourth term in Eq. (18.9) dominates, the minimum input noise current can be found by combining Eqs. (18.9), (18.11), and (18.12) to obtain

$$\langle i^2 \rangle_{opt} \approx 32\pi kT \frac{\Gamma(C_d + C_s)}{f_T} I_3 B^3. \tag{18.13}$$

The FET preamplifier figure of merit is then given by

$$\text{FET Figure of Merit} = \frac{f_T}{\Gamma(C_d + C_s)}. \tag{18.14}$$

FET receiver noise can be minimized by using a device with the highest possible f_T, by minimizing detector and stray capacitances and by optimizing gate width to make FET input capacitance equal to detector plus stray capacitance.

Equation (18.9) disregards noise from sources following the input FET. This assumption is generally valid if the gain of the input FET stage is high (i.e., $g_m R_D \gg 1$, where R_D is the drain load resistor). For high bit rates, a low value of R_D may be required to avoid a large interstage RC time constant and

TABLE 18.1
Typical FET Parameter Values.

	GaAs MESFET	Si MOSFET	Si JFET
g_m (mS)	15–50	20–40	5–10
C_{gs} (pF)	0.2–0.5	0.5–1.0	3–6
C_{gd} (pF)	0.01–0.05	0.05–0.1	0.5–1.0
Γ	1.1–1.75	1.5–3.0	0.7
I_{gate} (nA)	1–1000	0	0.01–0.1
f_c (MHz)	10–100	1–10	< 0.1

consequent bandwidth degradation. In this case, thermal noise from R_D and noise from following stages should be taken into account (Brain and Lee, 1985). Noise from R_D can be included in Eq. (18.9) by defining an equivalent Γ given by

$$\Gamma_{eq} = \Gamma + \frac{1}{g_m R_D}. \tag{18.15}$$

If second-stage noise is taken into account, the optimum FET input capacitance and gate width will be larger than that indicated by Eq. (18.11), since a higher value of g_m is needed to keep the signal level above noise from R_D and following amplifiers.

Typical parameter values for current FET types are given in Table 18.1.

18.4.3 Noise Calculations for Bipolar Front Ends

Silicon bipolar junction transistors are often used as preamplifiers in optical receiver front ends. Their noise is higher than that of FETs at low bit rates, but at high bit rates their performance can approach that of the best FETs.

Common-collector bipolar preamplifiers have been reported (Sibley *et al.*, 1985; Aiki, 1985; Yamashita *et al.*, 1986), but more typically, a common-emitter front end as depicted in Fig. 18.8 is used. The principal noise sources are thermal noise from the load resistor R_L, shot noise associated with the base and collector bias currents, and thermal noise from the base-spreading resistance $r_{bb'}$. The variance of the equivalent input noise current is given by (Smith and Personick, 1982; Muoi, 1984)

$$\langle i^2 \rangle_a = \frac{4kT}{R_L} I_2 B + 2q I_b I_2 B + \frac{2q I_c}{g_m^2} (2\pi C_T)^2 I_3 B^3$$

$$+ 4kT r_{bb'} [2\pi (C_d + C_s)]^2 I_3 B^3, \tag{18.16}$$

Bryon L. Kasper

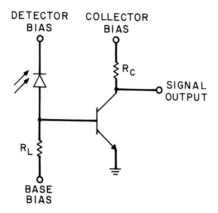

Fig. 18.8 Common-emitter bipolar preamplifier.

where

$$I_b = \text{base bias current}$$

$$I_c = \text{collector bias current}$$

$$g_m = \text{transconductance, and}$$

$$r_{bb'} = \text{base-spreading resistance.}$$

The transconductance is given by

$$g_m = I_c/V_T,$$

where

$$V_T = kT/q,$$

and the total capacitance C_T is given by

$$C_T = C_d + C_s + C_{b'e} + C_{b'c}, \tag{18.17}$$

where $C_{b'e}$ and $C_{b'c}$ are the small-signal hybrid-π-model capacitances of the transistor.

The emitter capacitance $C_{b'e}$ is the sum of a space-charge capacitance C_{je} and a diffusion capacitance, which is proportional to the collector current, such that

$$C_{b'e} = C_{je} + \frac{I_c}{2\pi V_T f_T}, \tag{18.18}$$

where f_T is the short-circuit common-emitter current-gain-bandwidth product at high I_c, where the diffusion capacitance dominates. Taking this variation of emitter capacitance into account (Muoi, 1984), it can be shown that there is an optimum collector bias current given by

$$I_c = 2\pi C_0 f_T V_T \psi(B), \qquad (18.19)$$

where

$$\psi(B) = \left[1 + \frac{I_2 f_T^2}{\beta I_3 B^2}\right]^{-1/2}$$

C_0 = total input capacitance for zero bias current

$\quad = C_d + C_s + C_{b'c} + C_{je}$, and

β = transistor current gain = I_c/I_b.

At low bit rates, $B \ll f_T[I_2/\beta I_3]^{1/2}$, the factor $\psi(B)$ has a value of

$$\psi(B) \approx \frac{B}{f_T}\left[\frac{\beta I_3}{I_2}\right]^{1/2}. \qquad (18.20)$$

The optimum bias current at low bit rates is therefore proportional to the bit rate, as suggested by an earlier analysis (Smith and Personick, 1982) that assumed that C_T remains constant. The assumption of constant C_T is valid at low bit rates, as using Eqs. (18.17), (18.18), and (18.19) the value of C_T can be expressed as

$$C_T = C_0[1 + \psi(B)], \qquad (18.21)$$

and for low bit rates $\psi(B) \ll 1$. At very high bit rates, $B > f_T[I_2/\beta I_3]^{1/2}$, $\psi(B)$ asymptotically approaches a value of 1, implying that the optimum collector current becomes constant and independent of the bit rate, and that the total capacitance approaches a maximum value of $2C_0$. The breakpoint between these high- and low-bit-rate regimes for current silicon microwave bipolar transistors with $f_T \approx 10$ GHz and $\beta \approx 100$ occurs at $B \approx 3$ Gbit/s.

Substituting the optimum bias current from Eq. (18.19) into Eq. (18.16), the minimum input-noise-current variance is found to be

$$\langle i^2 \rangle_a = \frac{4kT}{R_L}I_2 B + \frac{4\pi kTC_0 f_T}{\beta}\psi(B)I_2 B$$

$$+ \frac{4\pi kTC_0}{f_T}\frac{[1 + \psi(B)]^2}{\psi(B)}I_3 B^3 + 4kTr_{bb'}[2\pi(C_d + C_s)]^2 I_3 B^3.$$

$$(18.22)$$

At low bit rates, Eq. (18.20) can be used to obtain

$$\langle i^2 \rangle_a = \frac{4kT}{R_L} I_2 B + 8\pi kTC_0 \left[\frac{I_2 I_3}{\beta}\right]^{1/2} B^2 + 4kTr_{bb'}\left[2\pi(C_d + C_s)\right]^2 I_3 B^3.$$

$$(18.23)$$

At very high bit rates, $\psi(B) \approx 1$ and the noise is given by

$$\langle i^2 \rangle_a = \frac{4kT}{R_L} I_2 B + \frac{16\pi kTC_0}{f_T} I_3 B^3 + 4kTr_{bb'}\left[2\pi(C_d + C_s)\right]^2 I_3 B^3. \quad (18.24)$$

The bipolar amplifier at high bit rates is dominated by collector-current shot noise, and the figure of merit is given by

$$\text{Bipolar Figure of Merit} = \frac{2f_T}{C_0 + \pi f_T r_{bb'}(C_d + C_s)^2} \qquad \text{(High bit rate)}$$

$$\approx \frac{2f_T}{C_0} \quad \text{for small } r_{bb'}. \qquad (18.25)$$

The high-frequency bipolar figure of merit, neglecting base resistance, is very similar to that for FETs in Eq. (18.14), with differences being that the bipolar case has a factor of 2 in the numerator, the FET case has a factor or Γ in the denominator, and the capacitance C_0 in the bipolar case includes collector-base capacitance and emitter-base space-charge capacitance. For the same f_T, the two figures of merit are very close to one another. The optimizations for minimum noise also parallel one another. For the FET, the gate width is optimized by making the FET capacitance equal to the detector capacitance plus stray capacitance, whereas for the bipolar, the collector current is optimized when the emitter-base diffusion capacitance is equal to the sum of detector, stray, collector-base, and emitter space-charge capacitances. For equal values of f_T, the g_ms of the two devices will then be quite close. The noise performance is also similar, although the noise originates as thermal noise in the case of the FET and as shot noise in the case of the bipolar. The disadvantages of bipolars relative to FETs at high bit rates are the presence of emitter and collector space-charge capacitances, and higher base resistance than typical FET gate resistance.

At lower bit rates, bipolar performance suffers due to the presence of base current. As base-current shot noise becomes significant, the collector current must be reduced to the optimum point given by Eq. (18.19), with $\psi(B)$ as in Eq. (18.20). The noise contributed by base current is then equal to that from collector current, with each contributing half of the noise in the second term of

Eq. (18.23). Lowering bias current to reduce base-current shot noise has the effect of decreasing transconductance, which contributes to bipolar amplifiers being noisier than FETs at low bit rates. The low-frequency figure of merit for bipolar amplifiers is given by

$$\text{Bipolar Figure of Merit} = \frac{\beta^{1/2}}{C_0} \quad \text{(Low bit rate)}. \quad (18.26)$$

In addition to choosing the optimum bias current, another optimization that can be performed for bipolar receivers is the choice of optimum emitter width. Base resistance scales in inverse proportion to emitter width, whereas zero-bias capacitance $C_{je} + C_{b'c}$ is in direct proportion. It has hence been shown (Solheim and Roulston, 1987) that there is an optimum emitter width that results in minimum bipolar receiver noise.

18.4.4 Noise Comparison of GaAs MESFET, Si MOSFET, Si JFET and Si Bipolar

State-of-the-art noise performance of various transistor preamplifiers for bit rates from 1 Mbit/sec to 10 Gbit/sec is shown in Fig. 18.9. Calculations for this figure assume a detector-plus-stray capacitance of $C_d + C_s = 0.2$ pF and a load resistor R_L sufficiently large that its noise can be neglected. Noise for the Si JFET, Si MOSFET, and GaAs MESFET is calculated using Eq. (18.9) with parameter values listed in Table 18.2. Bipolar noise is calculated from Eq. (18.16) with an optimum collector bias current given by Eq. (18.19), except that the minimum collector current is limited to 0.1 mA as low bias current can cause β to be reduced (Muoi, 1984). Bipolar parameter values used in the calculations are $\beta = 100$, $r_{bb'} = 20\Omega$, $C_{b'c} = 0.2$ pF, $C_{je} = 0.8$ pF, and $f_T = 10$ GHz.

At low bit rates, all three FETs are better than the Si bipolar. Above about 50 Mbit/sec, the bipolar becomes better than the JFET, as noted previously by Smith and Personick (1982). JFETs are limited to bit rates below about

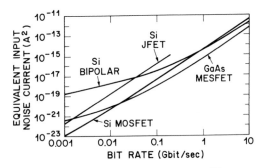

Fig. 18.9 Noise characteristics of state-of-the-art transistor preamplifiers.

TABLE 18.2
Assumed FET Parameters for Noise
and Sensitivity Calculations.

	GaAs MESFET	Si MOSFET	Si JFET
g_m (mS)	40	30	6
C_{gs} (pF)	0.38	0.8	4
C_{gd} (pF)	0.02	0.1	0.8
Γ	1.1	2	0.7
I_{gate} (nA)	2	0	0.05
f_c (MHz)	30	1	0

200 Mbit/sec because of their low values of f_T. Above about 20 Mbit/sec, the lowest noise is obtained with a GaAs MESFET, although at high bit rates the bipolar and MOSFET are not far behind. Below 10 Mbit/sec, the Si MOSFET is superior to the GaAs FET due to its lower gate-leakage current and lower $1/f$ noise. If should be noted that the curves in Fig. 18.9 are intended to give only general trends in device noise comparisons, and exact curves for particular devices will be different as a result of variations in parameter values from those assumed here.

New types of transistors with higher values of f_T than those considered have been demonstrated experimentally and should provide improvement in future amplifier noise performance. These new devices include heterojunction bipolar transistors and selectively-doped heterojunction FETs (also known as high-electron-mobility transistors).

18.5 SENSITIVITY

Given the equivalent input noise current for a preamplifier as found in the previous section, one can calculate the input optical power required for a desired bit-error rate at the receiver output. Exact error-rate calculations require a knowledge of the probability density functions (PDFs) of all noise sources present in the receiver. The photocurrent from a PIN detector is characterized by the Poisson distribution from which the quantum limit is derived. The PDF of current from an avalanche photodiode is the convolution of an initial Poisson distribution of primary electron-hole pairs created by photon absorption with the random distribution of the avalanche gain process, resulting in a distribution that is difficult to deal with analytically (Personick, 1971a; Personick, 1971b; McIntyre, 1972; Helstrom, 1984). Amplifier noise sources including thermal noise and shot noise are well characterized by Gaussian statistics. The simplest approach in calculating receiver sensitivity is to use Gaussian approximations for the signal-dependent noise produced by both PIN and APD detectors. More complex approaches providing better accuracy can be used (Personick *et al.*,

1977; Hauk *et al.*, 1978; Rocha and O'Reilly, 1982), but in general, the Gaussian approximation is sufficiently accurate for practical applications. Expressions for receiver sensitivity with either PIN or APD detectors based on Gaussian approximations are presented next.

18.5.1 PIN Sensitivity Calculations

As the sensitivity of PIN receivers is generally 20 dB or more from the quantum limit, it is possible to neglect signal-related shot noise. Following Smith and Personick (1982), noise from PIN dark current I_d can be added to amplifier noise to arrive at a total circuit noise $\langle i^2 \rangle_c$ given by

$$\langle i^2 \rangle_c = \langle i^2 \rangle_a + 2qI_dI_2B. \tag{18.27}$$

The average detected optical power $\eta\bar{P}$ required for a given bit error rate is then found from

$$\eta\bar{P} = \left(\frac{h\nu}{q}\right)Q\langle i^2 \rangle_c^{1/2}, \tag{18.28}$$

where Q is the required signal-to-noise ratio for the desired bit-error rate (Wozencraft and Jacobs, 1965). For an error probability of 10^{-9}, Q has a value of 6 implying that the distance from the signal level to the decision threshold must be 6 times the RMS noise level. Note that for a PIN receiver, which has equal noise for both the "one" and "zero" levels, the optimum decision threshold is midway between the two signal levels. Thus, if one defines the signal pulse height as the difference between the "one" and "zero" levels, then the ratio of pulse height to RMS noise must be $2Q$, or 12, for 10^{-9} BER. In decibels this signal-to-noise ratio of 12 is 21.6 dB.

18.5.2 APD Sensitivity Calculations

As described in Section 18.3.4, the internal gain of an avalanche photodiode provides sensitivity improvement by increasing the signal level prior to the addition of amplifier noise. The avalanche gain, however, also multiplies the shot noise associated with photon absorption, and adds excess noise because of randomness of the avalanche gain. The total receiver noise is the sum of amplifier noise plus APD noise. The amount of excess noise increases with gain, hence there is an optimum gain, M_{opt}, which gives best sensitivity. Below M_{opt}, the signal increases more rapidly than the total noise, whereas above M_{opt} the total noise increases faster. At optimum gain, the APD noise is approximately but not precisely equal to the amplifier noise. The method for calculating M_{opt} is described later in this section.

The dark current in an APD is comprised of two components; one flows through the avalanche region and undergoes multiplication, and another bypasses the avalanche region and is not multiplied. The dark current noise is then (Muoi, 1984)

$$\langle i^2 \rangle_d = \left[2qI_{du} + 2qI_{dm}M^2F(M) \right] I_2B, \tag{18.29}$$

where

$$I_{du} = \text{unmultiplied dark current}$$

$$I_{dm} = \text{primary multiplied dark current, and}$$

$$F(M) = \text{excess noise factor for average gain } M.$$

The excess noise factor $F(M)$ expresses the amount by which the avalanche gain mechanism increases the noise over that of an ideal, noiseless gain process. This factor is given by (McIntyre, 1966)

$$F(M) = kM + (1 - k)(2 - 1/M), \tag{18.30}$$

where k is the ratio of the electron ionization coefficient α and the hole ionization coefficient β in the detector avalanche region. The value of k is normally taken to be less than 1, hence k can represent either α/β or β/α, assuming that the avalanche process is initiated by the most highly-ionizing carrier type, as required for lowest noise (McIntyre, 1966). Typical values are $k = 0.02-0.04$ for silicon, $k = 0.7-1.0$ for germanium, and $k = 0.3-0.5$ for InGaAs APDs with InP multiplication regions.

The receiver sensitivity can be found for a given APD gain from (Muoi, 1984)

$$\eta \overline{P} = \left(\frac{h\nu}{q} \right) Q \left\{ QqBI_1F(M) + \left[\frac{\langle i^2 \rangle_c}{M^2} + 2qI_{dm}F(M)BI_2 \right]^{1/2} \right\}. \tag{18.31}$$

For simplification, the noise from unmultiplied dark current in the first part of Eq. (18.29) has been combined with amplifier noise as in Eq. (18.27) to obtain a total circuit noise $\langle i^2 \rangle_c$ that includes I_{du}. For rectangular NRZ input pulses and full raised-cosine output pulses, the Personick integral I_1 has a value of 0.548, whereas for 50% duty-cycle RZ input pulses its value is 0.500. If the multiplied component of the dark current, I_{dm}, is small enough that its noise contribution can be neglected (often not a safe assumption, especially for long-wavelength devices), then the sensitivity will be given by (Smith and Personick, 1982)

$$\eta \overline{P} = \left(\frac{h\nu}{q} \right) Q \left[\frac{\langle i^2 \rangle_c^{1/2}}{M} + qBI_1QF(M) \right]. \tag{18.32}$$

The optimum gain M_{opt} can be found analytically be setting the derivative with respect to M equal to zero, giving

$$M_{opt} = \frac{1}{k^{1/2}} \left[\frac{\langle i^2 \rangle_c^{1/2}}{q B I_1 Q} + k - 1 \right]^{1/2}. \qquad (18.33)$$

If I_{dm} is too large to be neglected, M_{opt} will be smaller than the value given by Eq. (18.33) and must be found either graphically or numerically.

18.5.3 Long-Wavelength Receiver Performance

The sensitivity of short-wavelength receivers ($\lambda < 1$ μm) has not improved substantially since the first volume in this series was written. Silicon APDs allow sensitivities in the range of 400 photons/bit to be achieved with relative ease, with the best reported result being $\bar{P} = 187$ photons/bit (Runge, 1976). Recently, much effort has been devoted to obtaining high sensitivity at wavelengths of 1.3 μm and 1.55 μm, where silica fibers have their lowest loss, but where silicon is transparent because of its wide bandgap. Appropriate detectors in this wavelength range are InGaAs/InP PINs, Ge APDs, and InGaAs/ InGaAsP/ InP APDs. The latter devices, which will be referred to as InGaAs APDs for convenience, are multilayer, heterojunction detectors in which photon absorption occurs in narrow-bandgap InGaAs, but avalanche multiplication occurs in wider-bandgap InP for reduced tunnelling dark current (Susa et $al.$, 1980; Campbell et $al.$, 1983).

Figure 18.10 shows theoretical receiver sensitivities for 10^{-9} BER at bit rates from 10 Mbit/sec to 10 Gbit/sec for three such detectors at 1.3 μm wavelength. The detector characteristics are listed in Table 18.3, and the preamplifier is assumed to be a high-impedance type with negligible load-resistor noise. The preamplifier uses a GaAs MESFET with characteristics as in Table 18.2. Stray capacitance C_S is assumed to be zero.

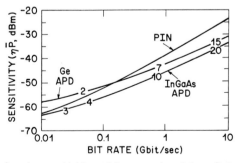

Fig. 18.10 Calculated receiver sensitivities at 1.3 μm wavelength for a GaAs FET preamplifier and either an InGaAs PIN, Ge APD, or InGaAs APD. Optimum avalanche gain is also indicated.

TABLE 18.3
Assumed Detector Parameters
for Receiver Sensitivity Calculations.

	PIN	Ge APD	InGaAs APD	Si APD
C_d (pF)	0.2	0.6	0.2	1.0
I_{du} (nA)	1	20	0	0
I_{dm} (nA)	—	20	1	0.1
k	—	0.7	0.3	0.03

At 10 Mbit/sec, the InGaAs APD and InGaAs PIN sensitivities are close to one another, whereas the Ge APD sensitivity is somewhat worse due to its higher dark current. Above 100 Mbit/sec, the Ge APD becomes better than the PIN and approaches the InGaAs APD, although the latter remains better due to its lower k-value and lower capacitance. These theoretical curves assume that there are no APD bandwidth limitations, whereas in reality APDs have a limited gain-bandwidth product, which may result in reduced sensitivity at very high bit rates (Forrest, 1984; Kasper and Campbell, 1987).

At a wavelength of 1.55 μm, InGaAs PIN and APD performance is similar to that at 1.3 μm. Ge APDs, however, suffer from a reduction in the optical absorption coefficient of germanium at 1.55 μm, resulting in lower quantum efficiency and pulse-response diffusion tails. Special Ge APD structures have been devised that alleviate these difficulties somewhat (Yamada *et al.*, 1982).

Table 18.4 is a compilation of some of the best receiver sensitivities that have been demonstrated to date with a variety of detector and preamplifier types. Receiver sensitivities are given in terms of both optical power, $\bar{P}$, and photons per bit, $\bar{N}$. Conversion from one to the other is given by

$$\bar{N} = \frac{\bar{P}}{h\nu B}$$

$$= 5.03 \frac{\bar{P}(\text{nW})\lambda(\mu\text{m})}{B(\text{Gbit/s})}.$$

(18.34)

18.6 SOURCES OF SENSITIVITY DEGRADATION

This section considers some of the factors that can cause the sensitivity of an optical receiver to be degraded from its ideal maximum. There can be many sources for such impairment. Two sources that will be considered in this section are detector dark current and transmitter extinction ratio. Other impairments that will not be discussed include noise sources other than the receiver front end, and intersymbol interference (ISI). Other noise sources could consist of laser intensity fluctuations in the transmitter, for example, or of crosstalk from other

TABLE 18.4
State-of-the-Art Receiver Sensitivities.

No.	B (Mb/s)	λ (μm)	$\bar{P}$[1] (dBm)	Photons[1] /Bit	Detector[2]	Amplifier[3]	Pulse[4] Format	Reference
1	8000	1.3	−25.8	2150	IGA APD	GA FET-HZ	NRZ	Kasper et al. (1987)
2	4000	1.51	−31.2	1440	IGA APD	GA FET-HZ	NRZ	Kasper et al. (1985)
	2000	1.51	−36.6	847	"	"	"	"
3	2000	1.54	−37.4	705	IGA APD	GA FET-HZ	RZ	Shikada et al. (1985)
	1200	1.54	−40.0	646	"	"	"	"
	565	1.54	−42.9	703	"	GA FET-TZ	"	"
4	2000	1.3	−33.4	1494	Ge APD	Si Biplr-LZ	RZ	Yamada et al. (1982)
	"	1.55	−32.0	2460	"	"	"	"
	1200	1.3	−35.9	1401	"	"	"	"
	"	1.55	−34.4	2359	"	"	"	"
	400	1.3	−40.2	1561	"	"	"	"
	"	1.55	−39.0	2454	"	"	"	"
5	1200	1.53	−36.5	1436	IGA PIN	GA FET-HZ	NRZ	Brain et al. (1984)
6	1000	1.55	−38.0	1235	IGA APD	GA FET-HZ	NRZ	Campbell et al. (1983)
	"	1.3	−37.5	1162	"	"	"	"
	420	1.55	−43.0	930	"	"	"	"
	"	1.3	−41.5	1102	"	"	"	"
7	800	1.3	(−28.0)	(12950)	IGA PIN	Si MOS-TZ	NRZ	Abidi et al. (1984)
8	565	1.3	−38.5	1635	IGA PIN	GA FET-HZ	NRZ	Smith et al. (1982)
	280	"	−43.0	1170	"	"	"	"
	140	"	−46.5	1046	"	"	"	"
9	446	1.52	−43.1	840	Ge APD	GA FET-HZ	RZ	Toba et al. (1984)
10	296	1.3	−36.3	5179	IGA PIN	Si Biplr-TZ	NRZ	Snodgrass et al. (1984)
11	274	1.3	−36.0	5995	IGA PIN	GA FET-TZ	NRZ	Lee et al. (1980)
	45	1.3	−46.7	3107	"	"	"	"
12	250	0.83	(−49.8)	(175)	Si APD	Si Biplr-TZ	BP	Muoi (1983)
13	140	1.52	−49.3	642	Ge APD	GA FET-TZ	NRZ	Walker et al. (1984)
	34	"	−55.8	591	"	"	"	"
14	50	0.85	−56.6	187	Si APD	Si JFET-HZ	RZ	Runge (1976)
15	45	1.3	(−53.3)	(696)	IGA APD	GA FET-TZ	NRZ	Forrest et al. (1981)
16	45	1.3	(−50.0)	(1453)	IGA PIN	GA FET-TZ	NRZ	Williams (1982)
17	45	1.3	(−46.8)	(3036)	IGA PIN	Si MOS-TZ	NRZ	Ogawa et al. (1983)

[1] $\bar{P}$ for 10^{-9} bit-error rate. Parentheses indicate $\eta\bar{P}$.
[2] IGA-InGaAs.
[3] Ga-GaAs. Biplr-Bipolar. MOS-MOSFET. HZ-High-impedance. TZ-Transimpedance. LZ-Low-impedance.
[4] NRZ-Non-return-to-zero. RZ-Return-to zero. BP-Biphase.

receiver circuits as a result of inadequate power supply decoupling or improper grounding. Intersymbol interference is invariably present to some degree. Sources of ISI can include inadequate transmitter, detector or amplifier bandwidth, improper receiver filtering and equalization, inadequate low-frequency response leading to baseline wander, or various distortions of optical pulse shape caused by fiber dispersion.

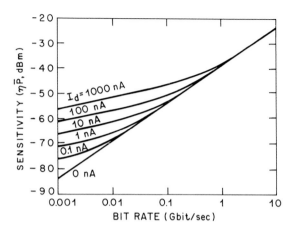

Fig. 18.11 Sensitivity versus bit rate for a PIN detector and GaAs FET preamplifier (neglecting $1/f$ noise) for various values of detector dark current.

18.6.1 Detector Dark Current

Dark current in either PIN or APD detectors can arise from a number of mechanisms and is generally sensitive to temperature (Pearsall and Pollack, 1985), thus sensitivity limitations due to high-temperature dark current must be considered in receiver design.

18.6.1.1 Sensitivity Limitations Due to PIN Dark Current. Effects of PIN dark current can be calculated from Eq. (18.27) and Eq. (18.28). Figure 18.11 gives sensitivity curves for a PIN detector with dark current from 0 nA to 1 μA at a wavelength of 1.3 μm. The detector capacitance is assumed to be 0.2 pF and the preamplifier is a hypothetical GaAs FET with most parameters as in Table 18.2. The $1/f$-noise corner frequency is taken as 0 Hz and gate-leakage current is assumed to be zero to clearly show PIN dark-current effects at bit rates as low as 1 Mbit/sec. InGaAs PINs have been demonstrated with room-temperature dark currents as low as 100 pA (Kim *et al.*, 1985), although more typical values are from 1 nA to 10 nA. It can be seen that below 100 Mbit/sec, a dark current of 10 nA or less can be the dominant receiver noise source.

18.6.1.2 Sensitivity Limitations Due to APD Dark Current. In a well-designed avalanche photodiode, the unmultiplied dark-current will be small and only the multiplied component will be significant. Dark current can be a serious limiting factor in long-wavelength APDs, especially at high temperatures. The effects of APD dark current on receiver sensitivity can be found from Eq. (18.29), with the optimum avalanche gain being obtained numerically or graphically.

Figure 18.12 indicates receiver sensitivity at 1.3 μm wavelength for an InGaAs APD with multiplied primary dark current from 0 nA to 1 μA. The

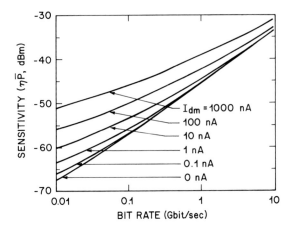

Fig. 18.12 Sensitivity at 1.3 μm wavelength versus bit rate for an InGaAs APD and GaAs FET preamplifier for various values of multiplied primary dark current.

APD capacitance is assumed to be 0.2 pF and the preamplifier is a GaAs FET as in Table 18.2. Typical room-temperature values of I_{dm} for InGaAs APDs are from 1 nA to 10 nA.

Figure 18.13 shows 1.3 μm receiver sensitivity for various primary dark current values with a Ge APD having 0.6 pF capacitance and a similar GaAs FET preamplifier. Typical room-temperature values of I_{dm} for Ge APDs are from 20 to 200 nA.

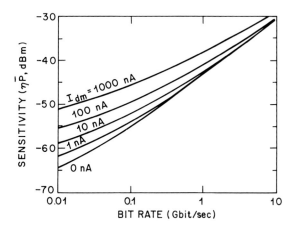

Fig. 18.13 Sensitivity at 1.3 μm wavelength versus bit rate for a Ge APD and GaAs FET preamplifier for various values of multiplied primary dark current.

18.6.2 Transmitter Extinction Ratio

The failure of an optical transmitter to turn completely off during the transmission of a "zero" causes a receiver sensitivity penalty for two reasons: 1) not all the received optical power is being modulated, and 2) the unmodulated optical power adds shot noise to the reception of both "ones" and "zeroes."

The extinction ratio r is the ratio of the average power transmitted during a "zero" to the average power transmitted during a "one." Unless r approaches 1, the shot noise will generally not be a problem for a PIN receiver and the sensitivity will be given by

$$\eta \bar{P} = \left(\frac{1 + r}{1 - r} \right) \left(\frac{h\nu}{q} \right) Q \langle i^2 \rangle_c^{1/2}, \qquad (18.35)$$

where the factor $1 + r/1 - r$ represents the penalty from unmodulated received power.

For an APD detector, the unmodulated power produces a photocurrent that acts like multiplied dark current and further reduces receiver sensitivity. The sensitivity can be calculated from (Smith and Personick, 1982)

$$\eta \bar{P} = \left(\frac{h\nu}{q} \right) \left(\frac{1 + r}{1 - r} \right) \left\{ (1 + r) \frac{Q^2 q B I_1 F(M)}{1 - r} \right.$$

$$\left. + \left[\left(\frac{Q^2 q B I_1 F(M)}{1 - r} \right)^2 4r + \frac{Q^2 \langle i^2 \rangle_c}{M^2} \right]^{1/2} \right\}. \quad (18.36)$$

As with APD dark current, the optimum value for M is most conveniently found numerically.

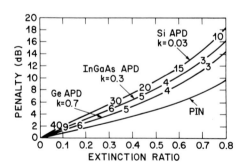

Fig. 18.14 Sensitivity penalty versus transmitter extinction ratio for a GaAs FET preamplifier at 1 Gbit/sec and either a PIN detector, Ge APD, InGaAs APD, or Si APD. Optimum avalanche gain is also indicated.

As an example of typical extinction-ratio effects, Figure 18.14 gives receiver sensitivity penalties at 10^{-9} BER versus r for a bit rate of 1 Gbit/sec and a GaAs FET preamplifier as in Table 18.2. Curves are given for four different detector types: PIN, Ge APD, InGaAs APD, and Si APD. Detector parameters are as in Table 18.3. Numbers along each curve indicate optimum values of avalanche gain. The optimum gain can be seen to decrease as the transmitter extinction ratio increases. In general, the penalty for a given extinction ratio is larger for higher sensitivity detectors (i.e., poor extinction ratio tends to equalize the performance of different detectors by degrading the best more than the worst).

18.7 DESIGN EXAMPLES

This section outlines three examples of receiver front-end design. The examples include a hybrid FET/bipolar circuit, an integrated bipolar circuit, and an integrated FET circuit.

18.7.1 Hybrid FET / Bipolar Cascode Front End

The hybrid GaAs FET/bipolar circuit originated by Ogawa and Chinnock (1979) is a transimpedance front end that has found wide usage in long-wavelength PIN/FET receivers. As shown in Fig. 18.15, the circuit contains a cascode amplifier formed by common-source GaAs MESFET Q1 and common-base pnp transistor Q2. The output stage Q3 is an npn emitter-follower from which feedback is taken back to the gate of Q1 through feedback resistor R6. The high output impedance of the common-base stage provides high voltage gain, and the use of a pnp transistor for Q2 provides level-shifting to allow DC-coupled feedback.

Receivers based on this design are available from a number of manufacturers for use at bit rates 10–400 Mbit/sec. Sensitivities reported for this type of

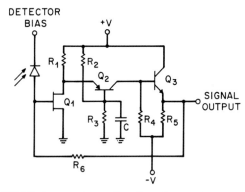

Fig. 18.15 Hybrid FET/bipolar cascode transimpedance front end.

preamplifier include -36 dBm at 274 Mbit/s and -46.7 dBm at 45 Mbit/s with a PIN detector at 1.3 μm wavelength (Lee *et al.*, 1980). At 12.6 Mbit/s, a sensitivity of -52.6 dBm and a dynamic range of 26.6 dB have been reported, along with a technique to extend the dynamic range to 44.8 dB (Owen, 1982).

Additional equalization of this transimpedance receiver is generally required at higher bit rates because of limited open-loop gain and parasitic capacitance associated with the feedback resistor.

18.7.2 Common-Emitter / Common-Collector Integrated Bipolar Front End

Bipolar integrated circuits are often used in receiver applications where reliability or low cost are critical. A popular design is the common-emitter/common-collector amplifier shown in Fig. 18.16. The high input impedance of common-collector stage Q2 allows high gain in input stage Q1. Diode-connected transistors such as Q3 provide level-shifting, and Q4 acts as an output buffer stage.

Bipolar integrated receivers of this type have been developed for submarine lightwave systems, where a sensitivity of -36.3 dBm and dynamic range of 22 dB have been obtained at 296 Mbit/s with a PIN detector at 1.3 μm wavelength (Snodgrass and Klinman, 1984). Such bipolar preamplifiers may also find application in terrestrial lightwave systems at bit rates up to 1.6 Gbit/sec or more (Ohara *et al.*, 1984).

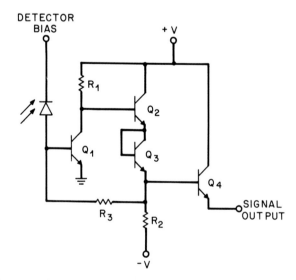

Fig. 18.16   Common-emitter/common-collector integrated bipolar transimpedance front end.

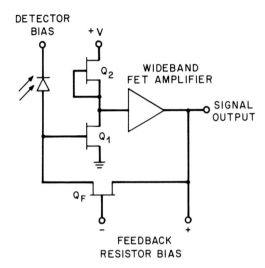

DETECTOR
BIAS   + V

WIDEBAND
FET AMPLIFIER

Q_2

Q_1

SIGNAL
OUTPUT

Q_F

–   +
FEEDBACK
RESISTOR BIAS

Fig. 18.17   Integrated FET transimpedance front end using FET feedback resistor.

18.7.3 Integrated FET Front End

One problem in the monolithic integration of FET front ends is the fabrication of a high-value, low-capacitance feedback resistor on the IC substrate. One solution to this problem, proposed by Williams (1982), is to use a small-area FET as a voltage-variable feedback resistor, Q_F, as in Fig. 18.17. High resistance and low capacitance are obtained by using a short-gate FET with a gate width of only a few microns. The use of a high open-loop-gain monolithic amplifier provides wide closed-loop bandwidth, hence further equalization is not required. Wide dynamic range can be achieved in such receivers through the incorporation of automatic gain control and the use of a second FET voltage-variable resistor that shunts photocurrent away from the input FET when high light levels are received (Williams and LeBlanc, 1986). Receivers based on this design have been implemented in CMOS technology for use in data links (Steininger and Swanson, 1986) and have potential for implementation in GaAs technology for receivers at very high bit rates.

REFERENCES

Abidi, A. A. (1988). On the choice of optimum FET size in wide-band transimpedance amplifiers. *IEEE J. Lightwave Tech.* **LT-6,** 64–66.

Abidi, A. A., Kasper, B. L., and Kushner, R. A. (1984). Fine line NMOS transresistance amplifiers. *Tech. Dig., Int. Solid-State Circuits Conf.* 1984 p. 76.

Aiki, M. (1985). Low-noise optical receiver for high-speed optical transmission. *IEEE J. Lightwave Tech.* **LT-3,** 1301–1306.

720   Bryon L. Kasper

Baechtold, W. (1972). Noise behavior of GaAs fie'd-effect transistors with short gate lengths. *IEEE Trans. Electron Devices* **ED-19**, 674–680.

Brain, M. C., Smyth, P. P., Smith, D. R., White, B. R., and Chidgey, P. J. (1984). PINFET hybrid optical receivers for 1.2 Gbit/s transmission systems at 1.3 and 1.55 μm wavelength. *Electron Lett.* **20**, 894–896.

Brain, M., and Lee, T. P. (1985). Optical receivers for lightwave communications systems. *IEEE J. Lightwave Tech.* **LT-3**, 1281–1300.

Brooks, R. (1980). 7B8B balanced code with simple error detecting capability. *Electron. Lett.* **16**, 458–459.

Buckingham, M. J. (1983). "Noise in Electronic Devices and Systems." Ellis Horwood Limited, Chichester, England.

Campbell, J. C., Dentai, A. G., Holden, W. S., and Kasper, B. L. (1983). High-performance avalanche photodiode with separate absorption "grading" and multiplication regions. *Electron. Lett.* **19**, 818–820.

Chown, D. P. M., and Spencer, M. (1986). Novel optical receiver configuration optimized for future high-speed systems. *Tech. Dig., Opt. Fiber Commun. Conf.*, 1986 p. 58.

Forrest, S. R., Williams, G. F., Kim, O. K., and Smith, R. G. (1981). Excess-noise and receiver sensitivity measurements of $In_{0.53}Ga_{0.47}As/InP$ avalanche photodiodes. *Electron. Lett.* **17**, 917–919.

Forrest, S. R. (1984). Gain-bandwidth-limited response in long-wavelength avalanche photodiodes. *IEEE J. Lightwave Tech.* **LT-2**, 34–39.

Forrest, S. R. (1985). Sensitivity of avalanche photodetector receivers for high-bit-rate long-wavelength optical communication systems. In "Semiconductors and Semimetals." (W. T. Tsang, ed.), Vol. 22, pp. 327–387. Academic Press, Orlando, Florida.

Goell, J. E. (1974a). An optical repeater with high-impedance input amplifier. *Bell Syst. Tech. J.* **53**, 629–643.

Goell, J. E. (1974b). Input amplifiers for optical PCM receivers. *Bell Syst. Tech. J.* **53**, 1771–1793.

Hauk, W., Bross, F., and Ottka, M. (1978). The calculation of error rates for optical fiber systems. *IEEE Trans. Commun.* **COMM-26**, 1119–1126.

Helstrom, C. W. (1984). Computation of output electron distributions in avalanche photodiodes. *IEEE Trans. Electron Devices* **ED-31**, 955–958.

Henry, P. S. (1985). Lightwave primer. *IEEE J. Quantum Electron.* **QE-21**, 1862–1879.

Jindal, R. P. (1985). High frequency noise in fine line NMOS field effect transistors. *Proc. Int. Electron Devices Meeting 1985* p. 68.

Kasper, B. L., and Campbell, J. C. (1987). Multigigabit-per-second avalanche photodiode lightwave receivers. *IEEE J. Lightwave Tech.* **LT-5**, 1351–1364.

Kasper, B. L., Campbell, J. C., Gnauck, A. H., Dentai, A. G., and Talman, J. R. (1985). SAGM avalanche photodiode receiver for 2 Gbit/s and 4 Gbit/s. *Electron. Lett.* **21**, 982–984.

Kasper, B. L., Campbell, J. C., Talman, J. R., Gnauck, A. H., Bowers, J. E., and Holden, W. S. (1987). An APD/FET optical receiver operating at 8 Gbit/sec. *IEEE J. Lightwave Tech.*, **LT-5**, 344–347.

Kim, O. K., Dutt, B. V., McCoy, R. J., and Zuber, J. R. (1985). A low dark-current, planar InGaAs p-i-n photodiode with a quaternary InGaAsP cap layer. *IEEE J. Quantum Electron.* **QE-21**, 138–143.

Lee, T. P., Burrus, C. A., Dentai, A. G., and Ogawa, K. (1980). Small area InGaAs/InP p-i-n photodiodes: fabrication, characteristics and performance of devices in 274 Mb/s and 45 Mb/s lightwave receivers at 1.31 μm wavelength. *Electron. Lett.* **16**, 155–156.

Maione, T. L., Sell, D. D., and Wolaver, D. H. (1978). Practical 45-Mb/s regenerator for lightwave transmission. *Bell Syst. Tech. J.* **57**, 1837–1856.

McIntyre, R. J. (1966). Multiplication noise in uniform avalanche diodes. *IEEE Trans. Electron Devices* **ED-13**, 164–168.

McIntyre, R. J. (1972). The distribution of gains in uniformly multiplying avalanche photodiodes: theory. *IEEE Trans. Electron Devices* **ED-19**, 703–713.

Muoi, T. V. (1983). Receiver design for digital fiber optic transmission systems using Manchester (biphase) coding. *IEEE Trans. Commun.* **COM-31**, 608–619.

Muoi, T. V. (1984). Receiver design for high-speed optical-fiber systems. *IEEE J. Lightwave Tech.* **LT-2**, 243–267.

Ogawa, K., and Chinnock, E. L. (1979). GaAs F.E.T. transimpedance front-end design for a wideband optical receiver. *Electron Lett.* **15**, 650–652.

Ogawa, K. (1981). Noise caused by GaAs MESFETs in optical receivers. *Bell Syst. Tech. J.* **60**, 923–928.

Ogawa, K. (1983). A long-wavelength optical receiver using a short-channel Si-MOSFET. *Bell Syst. Tech. J.* **62**, 1181–1188.

Ohara, M., Akazawa, Y., Ishihara, N., and Konaka, S. (1984). Bipolar monolithic amplifiers for a gigabit optical repeater. *IEEE J. Sold-State Circuits* **SC-19**, 491–497.

Owen, B. (1982). PIN-GaAs FET optical receiver with a wide dynamic range. *Electron. Lett.* **18**, 626–627.

Pearsall, T. P., and Pollack, M. A. (1985). Compound semiconductor photodiodes. In "Semiconductors and Semimetals." (W. T. Tsang, ed.), Vol. 22, pp. 174–245. Academic Press, Orlando, Florida.

Personick, S. D. (1971a). New results on avalanche multiplication statistics with applications to optical detection. *Bell Syst. Tech. J.* **50**, 167–189.

Personick, S. D. (1971b). Statistics of a general class of avalanche detectors with applications to optical communication. *Bell Syst. Tech. J.* **50**, 3075–3095.

Personick, S. D. (1973). Receiver design for digital fiber optic communication systems, I. *Bell Syst. Tech. J.* **52**, 843–874.

Personick, S. D., Balaban, P., Bobsin, J. H., and Kumar, P. R. (1977). A detailed comparison of four approaches to the calculation of the sensitivity of optical fiber system receivers. *IEEE Trans. Commun.* **COMM-25**, 541–548.

Personick, S. D., (1979). Receiver design. In "Optical Fiber Telecommunications." (S. E. Miller and A. G. Chynoweth. eds.), pp. 627–651. Academic Press, New York.

Robinson, F. N. H. (1974). "Noise and Fluctuations in Electronic Devices and Circuits." Clarendon Press, Oxford, England.

Rocha, J. R. F., and O'Reilly, J. J. (1982). Modified Chernoff bound for binary optical communication. *Electron. Lett.* **18**, 708–710.

Rousseau, M. (1976). Block codes for optical-fibre communication. *Electron. Lett.* **12**, 478–479.

Runge, P. K. (1976). An experimental 50 Mb/s fiber optic PCM repeater. *IEEE Trans. Commun.* **COM-24**, 413–418.

Shikada, M., Fujita, S., Takano, I., Henmi, N., Mito, I., Taguchi, K., and Minemura, K. (1985). 1.5 μm high bit rate long span transmission experiments employing a high power DFB-DC-PBH laser diode. *Proc. Eur. Conf. Opt. Commun. 11th, 1985.* Postdeadline Papers, p. 49.

Sibley, M. J. N., Unwin, R. T., Smith, D. R., Boxall, B. A., and Hawkins, R. J. (1985). A monolithic common-collector front-end optical preamplifier. *IEEE J. Lightwave Tech.* **LT-3**, 13–15.

Smith, D. R., Hooper, R. C., Smyth, P. P., and Wake, D. (1982). Experimental comparison of a germanium avalanche photodiode and InGaAs PINFET receiver for longer wavelength optical communication systems. *Electron. Lett.* **18**, 453–454.

Smith, R. G., and Personick, S. D. (1982). Receiver design for optical fiber communication systems. In "Topics in Applied Physics." (H. Kressel, ed.), Vol. 39, pp. 89–160. Springer-Verlag, Berlin.

Snodgrass, M. L., and Klinman, R. (1984). A high reliability high sensitivity lightwave receiver for the SL undersea lightwave system. *IEEE J. Lightwave Tech.* **LT-2**, 968–974.

Solheim, A. G., and Roulston, D. J. (1987). Optimum emitter aspect ratio for bipolar fiber-optic preamplifiers. *IEEE Electron Device Lett.* **EDL-8**, 437–439.

Steininger, J. M., and Swanson, E. J. (1986). A 50-Mbit/s CMOS optical datalink receiver integrated circuit. *Tech. Dig., IEEE Int. Solid-State Circuits Conf. 1986* p. 60.

Susa, N., Nakagome, H., Mikami, D., Ando, H., and Kanbe, H. (1980). New InGaAs/InP avalanche photodiode structure for the 1–1.6 μm wavelength region. *IEEE J. Quantum Electron.* **QE-16**, 864–869.

Sze, S. M. (1981). "Physics of Semiconductor Devices." Wiley, New York.

Teare, M. J., and Ulbricht, L. W. (1985). A performance comparison of two p-i-n FET receiver circuit architectures. *IEEE J. Lightwave Tech.* **LT-3**, 1307–1311.

Toba, H., Kobayashi, Y., Yanagimoto, K., Nagai, H., and Nakahara, M. (1984). Injection-locking technique applied to a 170 km transmission experiment at 445.8 Mbit/s. *Electron. Lett.* **20**, 370–371.

van der Ziel, A. (1970). "Noise, Sources, Characterization, Measurement." Prentice-Hall, Englewood Cliffs, New Jersey.

Walker, S. D., and Blank, L. C. (1984). Ge APD/GaAs FET/op-amp transimpedance optical receiver design having minimum noise and intersymbol interference characteristics. *Electron. Lett.* **20**, 808–809.

Williams, G. F. (1982). Wide-dynamic-range fiber optic receivers. *Tech. Dig., IEEE Int. Solid-State Circuits Conf. 1982* p. 160.

Williams, G. F., and LeBlanc, H. P. (1986). Active feedback lightwave receivers. *IEEE J. Lightwave Tech.* **LT-4**, 1502–1508.

Wozencraft, J. M., and Jacobs, I. M. (1965). "Principles of Communication Engineering." Wiley, New York.

Yamada, J., Kawana, A., Miya, T., Nagai, H. and Kimura, T. (1982). Gigabit/s optical receiver sensitivity and zero-dispersion single-mode fiber transmission at 1.55 μm. *IEEE J. Quantum Electron.* **QE-18**, 1537–1546.

Yamashita, K., Kinoshita, T., Maeda, M., and Nakazato, K. (1986). Simple common-collector full-monolithic preamplifier for 565 Mbit/s optical transmission. *Electron. Lett.* **22**, 146–147.

Chapter 19

Lightwave Transmitters

P. W. SHUMATE

Bell Communications Research, Inc., Morristown, New Jersey

19.1 INTRODUCTION

Sending information over fibers requires transmitters and receivers analogous to their non-optical counterparts for other media. That is, an information signal is conditioned to drive an active device that aids in matching the characteristics (e.g., impedance) of the transmission medium. At the receiving end, a weaker and/or distorted signal is detected and restored as closely as possible to a replica of the original.

19.1.1 Signaling and Modulation

When binary information is transmitted, one format may be converted to another; e.g., NRZ data may be converted to RZ or M-ary patterns, or the line rate may be raised by transmitting extra bits for framing or encoding. These modifications facilitate demultiplexing, error detection or clock recovery. For transmitting analog information, the input signal may be predistorted to compensate for nonlinearities in the light source, or digitally encoded or subcarrier-frequency modulated to circumvent nonlinearities. Coding and modulation techniques are well known (Schwartz, 1980). We will confine the treatment to circuitry specific only to the use of directly modulated lightwave devices such as light emitting diodes (LEDs) and injection laser diodes (ILDs or, simply, LDs). Furthermore, we will limit the discussion to medium- and high-speed digital applications, since the majority of current interest lies here. Analog modulation has been discussed or reviewed elsewhere (Shumate, 1980; Basch and Carnes, 1984; Gower, 1984; Sato *et al.*, 1984; Personick, 1985).

Digital transmitter circuits convert low-level electrical signals $V(t)$, usually at ECL or TTL levels, to corresponding light-intensity envelopes $L(t)$. Such *direct modulation* is intended to affect only the average optical power, and any phase or frequency information imparted to the optical carrier itself is not used at the receiver. (Coherent transmission, which utilizes such information, is discussed in Chapter 21).

19.1.2 Devices

For direct modulation, two semiconductor devices—LEDs and lasers—are suitable for use with fiber in terms of device dimensions, speed, efficiency, electrical characteristics and reliability. The devices are discussed in detail in Chapters 12 and 13. Here we compare them in terms of characteristics relevant to transmitters.

19.1.2.1 Power and Spectral Width.   LEDs are classified as either surface- or edge-emitters depending on their structure. Surface-emitting LEDs (SE-LEDs) emit light perpendicular to the contacted surface of the chip and parallel to the injected current (Fig. 19.1a). The emission pattern is approximately Lambertian (power $\propto \cos\theta$, θ measured from the beam axis, in this case perpendicular to the chip surface). Power coupled into fibers from Lambertian emitters is low: at currents on the order of 100 mA, peak power levels of -10 dBm to -20 dBm are characteristic of coupling to multimode transmission fibers, and -27 to -37 dBm into single-mode fiber. The specific value depends on the emitting area and the coupling scheme, including integral and external lenses. SE-LEDs have the widest spectral widths at full-width, half-maximum, approximately 35 to 50 nm for short-wavelength devices and 80 to 120 nm for long-wavelength devices.

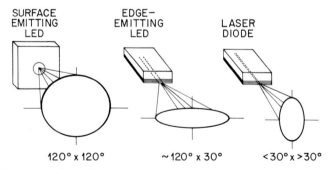

Fig. 19.1 Geometry and emission patterns of (a) surface-emitting LED, (b) edge-emitting LED and (c) injection laser.

Spectral width sets a limit on the bit-rate distance product through material (chromatic) dispersion.

Edge-emitting LEDs (EE-LEDs), structurally resembling lasers but designed to suppress gain, emit light perpendicular to the injected current from the end(s) of the chip (Fig. 19.1b). The emission pattern is more directional than a SE-LED (for many EE-LEDs, power $\propto \cos^7\theta$ perpendicular to the junction plane) due to internal waveguiding. This directionality results in a 5- to 10-dB coupling advantage over SE-LEDs. The waveguiding and subsequent self-absorption result in narrower emission spectra than SE-LEDs (30 to 90 nm), so reduced dispersion contributes to higher bit-rate distance operation than SE-LEDs.

Because of structural similarities between EE-LEDs and lasers, at high current densities or low temperatures, an EE-LED may begin to operate with net gain. This superradiant or superluminescent operation gives reduced spectral width and output beam divergence, the possibility of a laser-like threshold behavior, and a large dependence of optical output on temperature. Because of unfavorable temperature characteristics, superluminescent devices are rarely used in transmission equipment.

Injection lasers also require a waveguiding structure perpendicular to the direction of the current. For Fabry-Perot lasers, this waveguide extends between opposite edges of the chip, which, when cleaved along crystallographic planes, form mirror-like facets at opposite ends of the cavity to provide optical feedback. Alternatively, grating structures integral to the chip, such as in distributed-feedback (DFB) lasers, can be used. The output beam is highly directional (Fig. 19.1c), leading to high coupling efficiencies. Although they can approach 100%, coupling efficiencies of 15 to 50% are more typical with either multimode or single-mode fiber, depending on the coupling scheme. The resulting range of peak power levels can range from -10 dBm to over $+10$ dBm, with typical values lying near 0 dBm. The spectral width is narrow, ranging from 3 to 5 nm for typical multilongitudinal-mode lasers to $\ll 1$ nm for a single-longitudinal-mode device.

19.1.2.2 Frequency Response. LEDs are characterized by a one-pole frequency response, with 3-dB values usually 30 to 100 MHz, but with special devices extending to several hundred MHz. Device structures and doping levels can be modified to increase speed, usually at the cost of reduced output power. Laser frequency response and turn-on dynamics are complicated because they result from several coupled effects (Thompson, 1980). Equivalent 3-dB frequency response extends from about 1 GHz to over 10 GHz depending on the structure and bias point. Detailed equivalent-circuit models for both LEDs and lasers, useful for high-speed circuit modeling, have been published (Hino and Iwamoto,

1979; Descombes and Guggenbuhl, 1981; Atallah and Martinot, 1984; Tucker and Pope, 1983).

19.1.2.3 Reliability. Typical of semiconductor devices, LEDs and injection lasers also demonstrate remarkably high reliability, and detailed discussion appears in Chapters 12, 13 and 17. Short- and long-wavelength LEDs have shown extrapolated lifetimes in the 1 Mh to 1 Gh range, depending on device structure and temperature (Keramidas *et al.*, 1980; Yamakoshi *et al.*, 1981; Suzuki *et al.*, 1985; Fukuda *et al.*, 1986). Early reports on reliability testing of EE-LEDs gave extrapolated lifetimes considerably less, but recent results are very encouraging, indicating several Mh at 70°C, which is comparable to high-current-density SE-LEDs (Ettenberg *et al.*, 1984; Saul *et al.*, 1985; Fujimoto, 1986; Uji, 1986).

Lasers have a complex structure depending critically on gain. Thus, they are sensitive to more degradation mechanisms and generally, today, display one to two orders-of-magnitude shorter extrapolated lifetimes. When aged at 60°C, for example, median lifetimes (50% failure) of a few tens of kh to a few hundred kh are representative. For many applications requiring lasers (e.g., terrestrial long-distance transmission where protection switching and controlled environments are used), these lifetimes are adequate or can be made adequate by active cooling using Peltier devices. For the most demanding laser applications (e.g., submarine cable), hard screening procedures identify subsets of high-reliability devices (Gordon *et al.*, 1983; Nakano *et al.*, 1984; Nash *et al.*, 1985; Fujita *et al.*, 1985; Hirao *et al.*, 1986). Long wavelength lasers have been found more resistant to degradation mechanisms such as mirror erosion, catastrophic damage to the mirrors from high transient power levels, crystalline defects that develop with time, than short-wavelength devices.

19.1.2.4. With modest launched power and speeds, and because of broad spectral widths, LEDs are used for shorter-distance, medium-bit-rate (50 to 500 Mb/s) applications (e.g., bit-rate-length products of 100 to 2000 Mb · km/s, depending on wavelength). LEDs, having high reliability, often are required in high-temperature or uncontrolled environments such as for data links within high-speed equipment, in local area networks or in the telephone subscriber loop. These applications also employ large numbers of devices, usually without protection circuits, so failure-rate objectives are very stringent. The choice of wavelength is dictated by transmission distance, cost and availability. Compared with long-wavelength devices, short-wavelength LEDs have low temperature dependence and can be used with silicon photodetectors.

Given the 10–30 dB additional launched power levels compared with LEDs, higher speeds and narrow spectral widths, lasers are useful over a significantly wider bit-rate · length region (e.g., maximum bit-rate-length product of 100 Gb · km/s). For transmission, long-wavelength devices are nearly always used to take advantage of low fiber attentuation at 1.3 μm and 1.55 μm, as well as their higher reliability. The low cost of 780-nm lasers, driven down by the Compact Disk market, makes them possible choices for modest-temperature local-area networks. GaAs lasers can be integrated with GaAs IC transmitter circuitry, making them applicable for high-speed interconnections within equipment or even on circuit boards.

19.2 LED DIGITAL TRANSMITTERS

For high-speed two-level modulation in digital transmitters, LEDs are frequently operated with a small forward bias in the "off" state to overcome the turn-on delay associated with the space-charge capacitance of the junction. To this bias, the transmitter adds a high-speed drive current of 25 to 200 mA to reach the "on" state.

With series resistance usually a few ohms, LED drive current can be provided by simply using a line-driver IC from the logic family appropriate for the speed, as shown in Fig. 19.2. In Fig. 19.2a, a Schottky TTL driver pulls up or down the LED n-contact on transitions through an RC network conpensating the LED's complex pole, thus extending the frequency response. The off- and on-state currents are $(V_{cc} - V_f)/R_1$ and $(V_{cc} - V_f - V_{ol})/50\Omega$, respectively where V_{ol} is the gate output voltage in the low state. The circuit shown will provide current up to 60 mA (per 74S140 section) and operate to 50 Mb/s.

For higher speeds, non-saturating emitter-coupled logic, which is also compatible with GaAs ICs, is desirable. Figure 19.2b shows an ECL driver utilizing another commercial device, a 10210 circuit, which can supply currents up to 100 mA at speeds up to 200 Mb/s (Jarrett, 1974). (A related device, the higher-speed 10H209, can operate as high as 700 Mb/s NRZ (Banwell and Stephens, 1987)). In the circuit of Fig. 19.2b, the LED drive current is simply $(V_{oh} + 5.2 \text{ V})/R_1$.

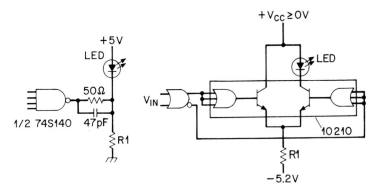

Fig. 19.2 Logic-circuit LED drivers. (a) TTL and (b) ECL input-level compatible.

The bias applied to the LED must be sufficiently positive to prevent saturation of the associated drive transistor.

By placing the LED in the collector (or drain) circuit of a current-source driver, lead inductance is overcome facilitating high-speed operation (Uhle, 1976). Even so, lead inductance must be minimized (Cooke *et al.*, 1986). Off-state bias is provided by R_1. Advantages of using logic-gate drivers include low cost, high reliability, and assurance of input logic-level compatibility and noise immunity over the full range of operating conditions.

An important characteristic of LEDs is that their optical output varies with temperature approximately as

$$P_o \propto e^{-\frac{T}{T_o}}, \tag{19.1}$$

where T_o varies from < 60 K to over 400 K depending on wavelength and structure. (Long-wavelength superluminescent devices can have T_o's as low as 8 K (Dutta, 1983).) The drivers shown in Fig. 19.2 supply approximately constant drive current leading to temperature-induced variations in output as high as $-2\%/°C$ ($T_o = 50$ K). For many applications, a power variation of 1 dB or more over the full operating temperature range is undesirable. Therefore, to stabilize the output power, transmitters are often designed to increase drive current with rising temperature (Brackett *et al.*, 1980; Fisher and Linde, 1986).

Figure 19.3 shows several alternative means of temperature compensating the current source for an emitter-coupled switch. In Fig. 19.3a, the current supplied

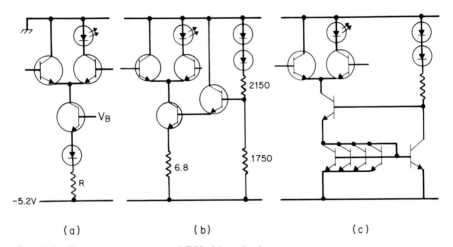

Fig. 19.3 Temperature compensated ECL driver circuits.

to the switch is

$$I_d = (V_B - V_{BE} - V_D + 5.2V)/R, \qquad (19.2)$$

where D is the lower reference p-n diode. If we take the LED output power as being linear with current $[P(I, T) \propto I \cdot \exp(-T/T_o)]$, then we find the base condition of the current-source transistor required to cancel the LED's temperature dependence to be

$$\frac{1}{P_o}\frac{dP_o}{dT} = 0 = \frac{1}{I_d}\frac{dI_d}{dT} - \frac{1}{T_o}, \qquad (19.3a)$$

which is equivalent to requiring $I \propto \exp(T/T_o)$ and leads to

$$V_B - 2V_{BE} = -2T_o\frac{dV_{BE}}{dT} - 5.2V. \qquad (19.3b)$$

The forward voltage (0.75 V) and temperature coefficient (-2 mV/$^\circ$C) of the diode have been taken the same as the base-emitter junction of the current-source transistor. With these values, and for an LED with a T_o of 90 K, we find a base bias of -3.34 V required, and $R = 7.2\Omega$ for 50 mA at room temperature. More accurate modeling of the diode and base-emitter junction, however, leads to -3.5 V and 4Ω. Either base bias is sufficiently negative to assure proper operation of the circuit, even if level shifting is used between ECL-level inputs and the bases of the emitter-coupled switching transistors. Regulation of the -3.34 V (-3.5 V) bias should be with respect to the -5.2 V rail to eliminate sensitivity to variations in the -5.2 V rail.

Figures 19.3b and c demonstrate other means for compensating an emitter-coupled pair. Figure 19.3c uses a current mirror (a widely used circuit that transfers a current level between two locations so that biasing constraints are relaxed) with a multiplier of 4 to permit regulating a high drive current with a smaller current I_{ref}. The current mirror also minimizes the voltage required between the switch and the -5.2 V rail, and is compatible with higher levels of integration that can take advantage of additional mirroring and of integrated temperature or constant-voltage (e.g., band-gap) references. One such circuit using CMOS technology was described by Fisher and Linde (1986).

The circuits just described facilitate ECL compatibility. In addition, other generic methods of inserting temperature compensation are shown in Figs. 19.4a–c, which facilitate TTL interfacing. In the figure, the box labeled T supplies a temperature-compensating current or voltage. Its design will vary depending on power-supply constraints, temperature range, device technology (e.g., availability of complementary devices) if integrated, etc.

Temperature effects can be circumvented by stabilizing the temperature of the LED using a Peltier active-cooling device. Such a thermoelectric cooler

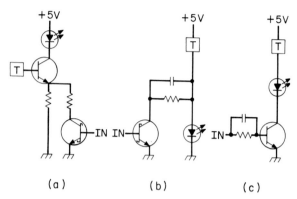

Fig. 19.4 Several means for TTL temperature compensation.

(TEC) is usually integral to the LED package. When operated at currents up to about 1A (requiring a bias the order of a volt), a miniature TEC can maintain a differential of 20 to 50 Centigrade degrees between the LED and the ambient, depending on the location and the optimization of the TEC. The TEC can also be operated in reverse to provide heating a low ambient temperatures, thus ensuring against the appearance of a threshold in edge-emitters. TEC current can be controlled under feedback to stabilize temperatures precisely, or it can be adjusted based on an open-loop measurement of the local ambient temperature.

19.2.1 High-Speed Considerations

Much attention has recently focused on operating LEDs at speeds above 100 Mb/s. Here, special compensation or "speed-up" networks are frequently used. (A "speed-up" network overcompensates the pole, spiking the drive current to help charge capacitance). High dc bias levels and/or circuits that actively remove stored charge are also used.

The circuit in Fig. 19.5 effectively combines current-source and shunt-driver circuits with a pole-zero compensation network R_1C_1. When the LED is off, a 30-mA bias is supplied by Q_3 but shunted around the LED by Q_2. Only a small residual bias, determined by the V_{DS} of Q_2 serves to charge the LED capacitance. Turn-on is a two-stage process. First, the gate of Q_2 is switched off as LED current begins to flow through the compensation network, resulting in high bias being redirected into the LED. Subsequently, the drive current appears. The result is a "pedestaled" turn-on, and the LED operates as if the large bias had been present at all times, but the light output in the off state is small. Operation at 560 Mb/s with an extinction ratio (the ratio of peak power levels in the "one" and "zero" states) of 15 dB was reported, with the circuit resulting in a 2.2X speed improvement (Chang *et al.*, 1986).

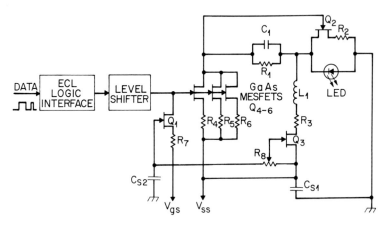

Fig. 19.5 High-speed LED driver using active turn-on and turn-off (Chang *et al.*, 1986; reprinted with permission).

Integrated driver circuits help overcome the transistor and parasitic capacitance limitations while also reducing size and lead inductance. Circuit considerations to suppress ringing in integrated LED drivers was discussed by Cooke *et al.* (1986).

Figure 19.6a shows a fully integrated bipolar current-source driver whose operation was verified at 300 Mb/s. The compensation circuit improved rise and fall times by factors of 3X, to 0.9 and 1.3 ns, respectively. No quiescent dc bias was applied in the off state (Yamashita *et al.*, 1986).

Figure 19.6b shows a GaAs IC that functions as a shunt driver, acting to remove stored charge in the zero state. The LED is driven through a pole-zero compensation network. In the off state, the output transistor is biased, so its drain voltage is adequate to maintain a small residual dc bias in the LED (extinction ratio = 13 dB) (Suzuki *et al.*, 1986a). Operation at 1 Gb/s was reported, with 0.9 and 1.0 ns rise and fall times (Suzuki *et al.*, 1986b).

An alternative to special circuitry or compensation networks is simply to use very-high-speed LEDs. A highly doped, planar 1.3 μm SE-LED was fabricated with a 12-μm-diameter p-contact, resulting in high current densities and speed. In combination with a simple GaAsFET shunt driver, open eye patterns were observed at 1.4 Gb/s RZ and 1.6 Gb/s NRZ (Suzuki *et al.*, 1984).

High-speed EE-LEDs have been fabricated. A highly doped mesa structure 8 μm wide by 150 μm long with slanted back facet to prevent feedback displayed a 3-dB bandwidth of 600 MHz. The EE-LED gave open eye patterns at 600 Mb/s without equalization or bias, and operation to 2 Gb/s with equalization and bias (Hayashi *et al.*, 1987).

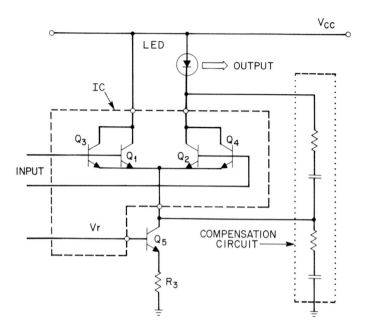

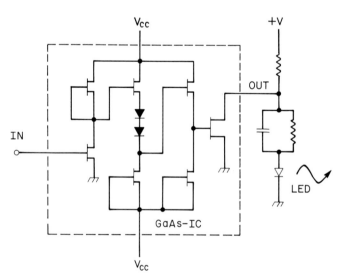

Fig. 19.6 Integrated LED driver ICs (Yamashita *et al.*, 1986; Suzuki *et al.*, 1986; reprinted with permission).

19.3 LASER DIGITAL TRANSMITTERS

For lasers, we must examine the important difference of lasing threshold and the circuitry it necessitates. This additional bias circuitry will be combined in a straightforward way, usually with current-switching driver circuits the same as those described, but often designed for lower drive current levels.

Threshold current I_{th} is the forward injection current at which optical gain in the laser cavity exceeds losses. Additional injected current is converted efficiently to light through the process of stimulated emission, resulting in light-current (L-I) transfer characteristics as in Fig. 19.7a. Threshold current near 25°C may range from less than 10 mA to about 60 mA for long-wavelength devices, although typical values lie between 20 and 40 mA. Short-wavelength thresholds typically are 20 to 30 mA higher. Below threshold, light emission is spontaneous, with characteristics (speed, spectrum and efficiency) similar to those from EE-LEDs. Above threshold, however, differential quantum efficiency (DQE) and speed are very high. DQEs range from 15% to 50% depending on the device structure and mirror characteristics.

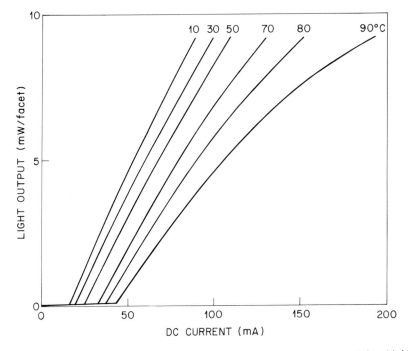

Fig. 19.7 Laser light-current transfer characteristics (a) at several temperatures and (b) with bias and drive currents superimposed.

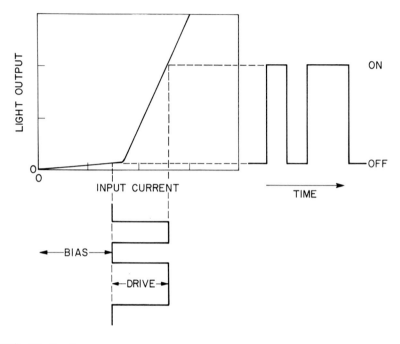

Fig. 19.7 (*Continued.*)

When a laser is turned on with a current pulse I_p, a delay in the emission of light is observed equal to

$$t_d = \tau_{th} \ln\left(\frac{I_p}{I_p - I_{th}} \right), \qquad (19.4)$$

where τ_{th} is the carrier recombination lifetime at threshold (usually about 2 ns). If the laser is dc-biased with a current I_b, however, (19.4) becomes

$$t_d = \tau_{th} \ln\left(\frac{I_p}{I_p + I_b - I_{th}} \right), \qquad (19.5)$$

and the delay is seen to vanish when $I_b = I_{th}$. A bias at or above threshold also reduces overshoot and ringing associated with laser turn-on. Excessive overshoot or ringing can degrade the output waveform, resulting in a loss of sensitivity at the receiver. As I_b is increased beyond I_{th}, laser frequency response rises. This advantage, however, is offset by reduced extinction ratio, which results in degraded sensitivity at the receiver (Personick, 1973). The design goal is to provide bias near threshold that optimizes the turn-on characteristics for the specific application. Increased frequency response is becoming less of an issue since the recent introduction of laser structures that provide current confinement without the use of reverse-biased p-n junctions (Bowers, 1986; Shen *et al.*, 1986; Zah *et al.*, 1987).

The difficulty arises not in selecting the bias, but in maintaining it. (A transmitter eye pattern, which is a bit-synchronized superposition of pseudorandom input data, can be optimized for maximum opening and minimum overshoot and jitter during assembly). Threshold current varies with temperature (Fig. 19.7a) by the same empirical relationship as (19.1), with T_o typically about 150 K for short-wavelength lasers and 60 K for long-wavelength devices. In addition to temperature, degradation of the laser with time, whether due to changes in the cavity, at the mirrors, or in current-blocking junctions, is accompanied by irreversible increases in lasing threshold. Degradation of the p-contact region (increased resistance) leads to Joule heating, which raises the threshold. Therefore, to stabilize the laser operating point with regard to both temperature and time, feedback regulation of I_b is needed.

Stabilization is accomplished by monitoring the light output from the *rear* facet or mirror of the laser with a photodiode while coupling the output of the *front* facet to the fiber. Bias adjustment using low-frequency circuitry stabilizes the photocurrent of the monitor, hence the laser operating point. The low-speed bias circuitry and the high-speed drive circuitry are connected as in Fig. 19.8. By selecting transistor Q_4 with sufficiently low C_{ob}, the inductor indicated in Fig.

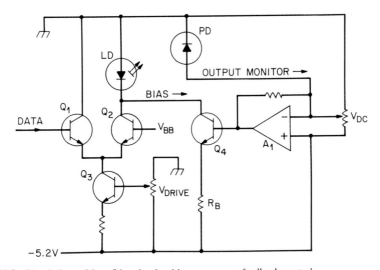

Fig. 19.8 Simple laser-driver/bias circuit with mean-power feedback control.

19.8 may sometimes be eliminated. Alternatively, a ferrite bead on the bias lead may be used. In Fig. 19.7b, we see the 50°C L-I characteristic showing combined action of the bias and drive.

19.3.1 Bias Control

We turn now to details of laser bias control, to compensate for changes in threshold, which would otherwise result in unacceptable changes in output power, extinction ratio and turn-on delay. Because from Fig. 19.7a, DQE is also seen to change with temperature, we will conclude that some cases also warrant independent control of the drive current.

It is worth noting that thermoelectric cooling can be used to *minimize* the effect of temperature, and this approach is commonly used in combination with closed-loop bias control. TE cooling essentially eliminates concern about reduced DQE by keeping the laser temperature in the 20 to 30°C range.

Simple regulation of the average power as in Fig. 19.8 is adequate when digital data are scrambled or encoded to assure an equal density of marks and spaces (50% duty cycle). Adjustment of the mean-power reference level (using V_{DC}) and peak drive current (using V_{DRIVE}) are adjusted initially at 25 °C to set the power level, while optimizing the eye for extinction ratio and turn-on delay. As mentioned earlier, the monitor photodiode usually intercepts the otherwise-unused light from the back mirror and is usually mounted inside the laser package. Alternatively, simple optics can be used to bring the back-mirror light through the laser package to a separately housed photodiode. Other monitoring schemes include placing a tap in the output fiber (Karr *et al.*, 1979) or monitoring the front-mirror light that escapes being coupled into the fiber.

Single-loop bias-control integrated circuits are used in Compact Disk players (Sharp, 1985). Designed to provide average-power regulation for continuously operated lasers, they may also be useful in transmitters that assure 50% duty cycle.

Data interrupts or long strings of one's or zero's cannot be eliminated in all applications, however, and it is sometimes undesirable to encode the data to assure 50% duty cycle. In the circuit of Fig. 19.9, the time average of the input data pattern is compared with the time average of the light output at amplifier A_1, so that power-level variations independent of the duty cycle are sensed (Shumate *et al.*, 1978). Transistors Q_4 and Q_5 are used to regenerate the logic-level data to obtain well-shaped pulses with a 0–V offset, thus minimizing drift.

The use of this form of negative feedback bias control prevents cw operation of the laser at the half-power setpoint as would be the case with Fig. 19.8 in the event of an interrupt comparable to the time constant of the bias circuit. For longer interruptions, however, the bias returns to its original set-up value, which

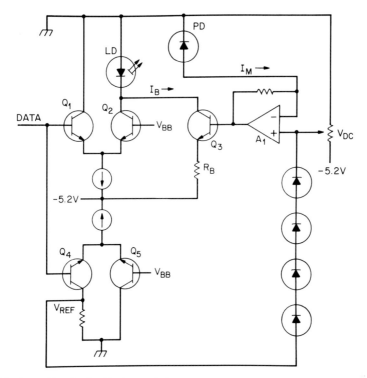

Fig. 19.9 Laser bias circuit with duty-cycle input.

may not reflect the prevailing temperature or age of the laser. When the modulation reappears, there will be a period equal to the loop time constant in which there may be serious pattern-dependent effects. The effect of interrupts can be minimized for the circuits of either Fig. 19.8 or 19.9 by using long-term memory to hold the last value of bias current when the duty cycle was 50%.

It is instructive to examine the operation of simple feedback loops such as these (Schwartz and Wooley, 1983). We can determine how effectively threshold changes are compensated and how extinction ratio changes, particularly if DQE is not constant. For simplicity, we will take the combination of A_1 and Q_3 as a current amplifier of gain A in Fig. 19.8, and treat all inputs as currents. The bias current I_B is given by:

$$I_B = A(-I_M + \beta I_{REF} + I_{DC}),\tag{19.6}$$

where I_M is the monitor photocurrent, βI_{REF} is the duty cycle β times the peak data-pattern reference current, and I_{DC} is the dc offset that establishes the set-up value of bias current. The monitor photocurrent is given in terms of the peak logical one and zero power level of the laser as

$$I_M = r\bar{P} = r[\beta P_1 + (1 - \beta)P_0], \qquad (19.7)$$

where r is the net responsivity of the detector, which is the product of the detector's actual responsivity (typically 0.5 to 1 A/W) times the fraction of the laser power intercepted by the detector.

To relate P_0 and P_1 to the bias and drive currents of the laser, the L-I transfer characteristic is modeled as two straight-line segments of slope s_0 below threshold and s_1 above threshold. For a bias current $I_B \leq I_{TH}$, P_0 and P_1 are given by:

$$P_0 = s_0 \cdot I_B \text{ and} \qquad (19.8a)$$

$$P_1 = s_0 \cdot I_{TH} + s_1(I_B + I_D - I_{TH}). \qquad (19.8b)$$

The bias current is found from the solution of (19.6)–(19.8):

$$I_B = \frac{\beta I_{REF} + I_{DC} + r\beta[I_{TH}(s_1 - s_0) - s_1 I_D]}{\dfrac{1}{A} + r\beta(s_1 - s_0) + rs_0}, \qquad (19.9)$$

which is in the form of the usual solution for a negative feedback loop. We have simplified this result by taking the forward or open-loop gain A very large (actually, the loop gain $A[r\beta(s_1 - s_0) + rs_0] \gg 1$).

The resulting expression for I_B can now be examined for its variation with the L-I parameters. For example, the partial derivative with respect to threshold current is:

$$\frac{\partial I_B}{\partial I_{TH}} = \frac{\beta(s_1 - s_0)}{\beta(s_1 - s_0) + s_0} = \frac{1}{1 + \dfrac{s_0}{\beta(s_1 - s_0)}}, \qquad (19.10)$$

which we see is unity (perfect compensation) if s_0 is zero. If s_0 is non-zero, however, (19.10) is less than unity (compensation lags changes in threshold), with the error being the order of s_0/s_1. From (19.10), we also see that, if the duty cycle β goes to zero, then (19.10) also goes to 0 (no correction), as noted in the discussion of the circuit of Fig. 19.8.

Finally we examine the extinction ratio ϵ with changes in laser parameters. With (19.9) and (19.10), this can be solved exactly leading to tedious expressions. Instead, let us use (19.8a) and (19.8b) with the result simplified by assuming $I_B \approx I_{TH}$ to get:

$$\epsilon = \frac{P_1}{P_0} = 1 + \frac{s_1}{s_0} \frac{I_D}{I_{TH}} \qquad (19.11)$$

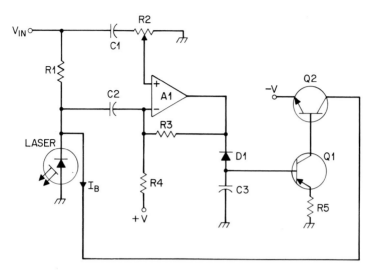

Fig. 19.10 Laser bias circuit that dynamically determines threshold from I-V characteristic (Albanese, 1978; reprinted with permission).

from which it is seen that aging- or temperature-induced increases in threshold current or sub-threshold spontaneous emission (s_0), and/or decreases in lasing slope efficiency all tend to reduce ϵ.

Circuits more complex than these can also deal with duty-cycle variations. Two interesting approaches have been reported for tracking threshold directly (rather than indirectly through mean-power measurement); one using the laser output and the other using only the laser I-V behavior. In the first method, introduced as part of a dual-loop control approach (Smith, 1978), a 1 kHz square wave ($\approx 1\% \, I_p$) was applied to the laser in the off state. Synchronous detection extracted the small 1 kHz signal from the off light and adjusted the bias to a preset value corresponding to bias just below threshold. This detection was later performed using a custom integrated circuit (Smith and Matthews, 1983).

For the second method, shown in Fig. 19.10, a low-frequency signal was also applied to the laser (Albanese, 1978). Here, however, an ac-coupled bridge measured the degree to which the laser junction voltage became clamped to the bandgap value at threshold. The bridge is formed of R_1, the left and right parts of R_2, and the laser junction plus its series resistance. If the laser junction voltage is pinned (constant), the bridge remains in balance as V_{in} is dithered. By adjusting the bias incrementally below threshold, however, V_{in} evokes a response that is amplified (A_1) and peak detected for application to the bias transistors Q_1 and Q_2. The appeal of a control scheme such as this is the elimination of the monitor photodiode, thus leading to a simpler, lower cost package.

Feedback control of bias alone usually suffices to provide proper transmitter operation. As seen from Fig. 19.7a, however, DQE is degraded at high temperatures and is more pronounced even at lower temperatures in certain laser structures (e.g., those with weak current confinement). Therefore, control of both bias *and* drive may be necessary.

Since DQE is not a strong function of temperature, one approach is simply to alter the drive current predictively (open-loop) according to the ambient temperature. The bias would be controlled by one of the conventional methods.

More-complex "dual-loop" feedback circuits have been designed that monitor the on and off light levels, or the on and average-power levels to adjust bias and drive actively. The latter is more desirable, since the off light level is usually very

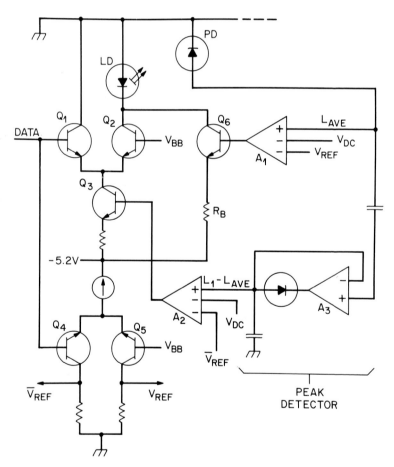

Fig. 19.11 Dual-loop laser bias and drive control circuit (Chen, 1980; reprinted with permission).

low and difficult to measure accurately (except where synchronous detection is used, as just described). Whether or not dual-loop control is employed is a question of circuit and set-up complexity and added cost, versus the need for fine control of the laser over a wide temperature range, or the need to deal with a laser structure that offers poor DQE at higher temperatures.

The circuit shown in Fig. 19.11 maintains the average and peak power constant, regardless of duty cycle (Chen, 1980).The duty-cycle reference voltage V_{REF} supplied to the control amplifiers A_1 and A_2 is derived as for Fig. 19.9. Amplifier A_2 acts with a peak detector (A_3) to compensate for changes in slope efficiency only, regulating the on level as follows: since the peak detector is ac-coupled to the photodiode, its dc (zero) reference is the average light level and its peak-to-peak input is P_1. Duty-cycle variations are compensated at A_2 using V_{REF}, so amplifier A_2 acts only on P_1, regulating it through the pulse drive current determined by the base bias A_2 supplies to Q_3. It is clear that, once P_1 is fixed, regulation of the average power by A_1 as described earlier results in bias corrections that compensate for changes in P_0, and hence threshold only.

19.3.2 High-Speed Considerations

To exploit the high speed of lasers, most driver circuits are of the emitter-coupled differential current switch configuration. Gruber (1978) has discussed high-frequency circuit considerations for 2 Gb/s. At these speeds, it becomes important to consider the interactions among all circuit elements, interconnections and packaging. An accurate high-frequency laser model was described by Tucker and Pope (1983), and packaged devices modeled in SPICE by Wedding (1987).

As with LEDs, a clear approach to combining high-speed devices with lasers, while minimizing parasitics and power dissipation, is to integrate the driver using GaAs MESFET technology. This also assures immediate logic-level compatibility with associated GaAs IC circuitry. Integrated drivers with circuit configurations similar to those in Fig. 19.6 have been described (e.g., Yamashita et al., 1986; Chen, 1986) and commercial devices are currently becoming available.

It has been widely accepted, and recently reinforced through computer modeling (Nakamura et al., 1986), that monolithic integration of the laser with the driver should result in highest-speed performance. Nakamura calculated that interconnection lead inductance severely limits laser performance at 6.3 Gb/s, and can be significantly reduced through monolithic integration.

Such optoelectronic integration with GaAs lasers combined with GaAs MESFET circuits have been reported for several years (e.g., Matsuda and Nakamura, 1984; Horimatsu et al., 1985; Horimatsu et al., 1986; Hutcheson, 1986; Nakano et al., 1986). Recent results have addressed long-wavelength

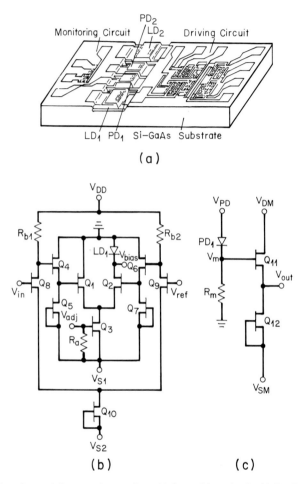

Fig. 19.12 Optoelectronic integrated transmitter. (a) Laser driver circuit. (b) Feedback bias-control circuit. (c) Layout of OEIC transmitter chip (Nakano *et al.*, 1986; reprinted with permission).

materials, using bipolar transistors, JFETs, MISFETs or heterostructure MESFETs to circumvent the difficulty in obtaining MESFETs in the InGaAsP/InP material system (e.g., Shibata *et al.*, 1984; Chen *et al.*, 1983; Koren *et al.*, 1984; Kasahara *et al.*, 1986; Mikawa *et al.*, 1987).

Figure 19.12a shows a short-wavelength (GaAs) OEIC transmitter chip, which contains the laser, a monitor photodiode, drive circuit and feedback bias control (Nakano *et al.*, 1986). (The second laser/monitor pair was included as a test structure.) The drive circuit shown in Fig. 19.12b is a conventional current switch $(Q_1 - Q_3)$ with input pulse shaping and buffering provided by the other transistors. With 1-μm gate-length GaAs MESFET and 0.8-V_{pp} input, the driver and laser were operated at 2 Gb/s. The integral monitor and amplifier/active-load (Q_{11} and Q_{12} in Fig. 19.12c) provide negative feedback control of the laser bias when V_{out} is connected to the laser bias node.

Microwave IC (MIC) hybrids, monolithic MIC (MMIC) integrated drivers and OEICs all offer alternatives for modulating lasers well into the multi-Gb/s range. At these speeds, however, laser "chirp" limits the transmission distance. Chirp is a dynamic broadening of the laser spectrum arising from transient effects during modulation. Chromatic dispersion of the broadened linewidth, which can be several tenths of a nanometer, then restricts maximum fiber length. This particularly defeats the advantage of the 1550 nm window, which is used principally for obtaining long transmission distances.

From the transmitter circuit viewpoint, drive pulses can be shaped to minimize chirp, or multiple-section lasers can be used. A dual-pulsing scheme was analyzed (Olshansky and Fye, 1984) wherein a low-amplitude pulse of width equal to the laser's intrinsic relaxation period preceded the main drive pulse. Properly synchronized with the main drive pulse, the authors calculated that transient overshoots could be eliminated, thus reducing chirp.

A resonant network inserted between the laser and driver that compensated the intrinsic relaxation of the laser was reported that reduced chirp by a factor of three (Bickers and Westbrook, 1985).

There have been reports of chirp reduction by using multiple-contact lasers and circuitry that drives one or more of the cavities out of phase with the main cavity, in such a way as to cancel chirp (Kaede, 1985; Yoshikuni and Motosugi, 1986).

Finally, external modulators can be used with cw lasers to minimize chirp. Waveguide devices such as Mach-Zehnder interferometers or delta-β-reversal switches (Schmidt, 1983) fabricated in LiNbO$_3$ eliminate chirp completely, but III-V modulators themselves induce chirp (Koyama and Iga, 1985; Suzuki et al., 1986).

19.4 SUBSYSTEM PERFORMANCE AND RELIABILITY

We have discussed the provision and control of drive and (for lasers) bias current. Laser transmitters, however, require additional circuit complexity to protect the laser from electrical transients, and often for monitoring the light output, or monitoring and control of the laser temperature if cooling is provided. These needs arise from the reliability requirements and need for fault location for laser transmitters in telephone trunking applications.

Lasers are destroyed by currents that result in optical power-density levels from the order of a few MW/cm^2 (AlGaAs) to tens of MW/cm^2 (InGaAsP) (Fukuda, 1983) but can also be degraded in more subtle ways by smaller transients (Sim *et al.*, 1984). Therefore, protection circuitry prevents the laser current from responding to ordinary power up/down operations, or, equally important, to removal or insertion of a transmitter into a powered-up equipment frame. The circuit prevents bias and/or drive from being applied until the transmitter power rails have become stable, as determined by R-C time constants, perhaps in concert with 3-terminal voltage regulators. When more than one power rail is used, it is prudent to consider carefully all turn-on/turn-off combinations (including discharge paths of capacitors) that might induce unintentional current through the laser.

The most demanding requirement of optoelectronic subsystems, and LED- or laser-based transmitters in particular, is that they operate over wide temperature ranges. For most of today's transmission equipment, this means device temperatures ranging from about $+16$ to $+50°C$ (sustained) and $+14$ to $+60°C$ (short-term). These temperatures arise from standard ambient specifications (Bellcore, 1985a) plus typical ΔTs between the ambient and the sites of devices ($12°$ in this case). Additionally, storage and transportation over the range $-40°C$ to $+65°C$ are possible.

With emphasis shifting to the telephone loop plant, the situation becomes more severe. If equipment is installed in controlled environmental vaults (CEVs), a central-office environment such as just described is assured. If, however, above-ground enclosures are used, then the sustained and short-term device ambients may be -20 to $+70°C$ and -40 to $+85°C$, respectively. Again, these arise from ambient characteristics (Olson and Schepis, 1984) plus a rise, which is typically $20°$ for more-confined loop equipment. Storage and transportation extremes remain the same.

For wide temperature ranges, all subsystems must be assured of operation at the *short-term* extremes. This means logic-level tracking, output-power regulation, etc. As mentioned previously, some edge-emitting LEDs tend to develop "thresholds" (i.e., become lasers) at low temperatures. Without proper bias, a turn-on delay will appear inducing severe eye degradation. For lasers, on the other hand, diminished DQE at high temperature will result in degraded extinction ratio without adequate control of drive current. Resistive (or thermoelectric) heating of the EE-LED package at low temperatures or thermoelectric cooling of the laser at high temperatures can offset such problems.

19.4.1 Transmitter Reliability

At the *long-term* extremes, transmitter reliability must be assured. A transmitter failure (end-of-life) can be defined as the inability to provide error-free transmission, either due to sudden death of a component or a more subtle degradation that results in inadequate optical output, unmodulated optical output, or dynamically degraded optical output (poor eye pattern) (Dean and Dixon, 1981). Since reliability of silicon devices is high, transmitter reliability is often dominated by the light source, either its intrinsic reliability or that of the coupling between the device and the fiber.

For estimating the reliability of a *system*, a model is frequently used wherein individual devices are assumed to display uncorrelated, random failures (MIL-HDBK-217D, 1980; Bellcore, 1985b). An individual (i^{th}) component with mean-time-to-failure $MTTF_i$ is therefore described by an exponential failure probability at time t of $P_{f,i} = \exp(-t/MTTF_i)$. This is more conveniently written as $\exp(-\lambda_i t)$ by defining a constant failure rate λ_i equal to the reciprocal of the $MTTF_i$.[1] The failure probability of the system at time t is the product of the $P_{f,i}$'s

$$P_f = \Pi \exp(-\lambda_i t) = \exp(-\Sigma \lambda_i t), \tag{19.12}$$

or the system failure rate can conveniently be equated to the sum of the λ_i's. It is therefore useful to work with component-level failure-rate units either reciprocally related to each $MTTF_i$ or more appropriately calculated from the failure distribution. These are usually expressed in failures/10^9h (originally called FITs for Failure unITs but often ascribed to Failures In Time), or in %/1000h. These are related as 1000 FITs = 0.1%/kh and, for random failures, both correspond to MTTF = 1Mh or about 100 years.[2]

Transmitter objectives for central-office equipment are usually in the 5000 to 10000-FIT range, and the order of 1000 FITs for loop equipment. Exact values depend on maintenance models, cost of repairs, etc. Loop equipment should have high reliability because of the large number installed, higher field-repair costs and the absence of protection circuits.

19.4.1.1 Device Reliability. As mentioned in the introduction, high-current-density LEDs, both edge- and surface-emitters, may have extrapolated 70°C MTTFs as high as 1 Mh to 10 Mh, and room-temperature values estimated as

---

[1] Semiconductor components, including LEDs and lasers, display a failure rate distributed log-normally in time; i.e., the failure rate is not constant (Peck, 1974; Jordan, 1978). It is assumed that a system can nevertheless be described as if individual component rates were constant since different failure distributions, different residual infant mortalities, different burn-in histories, etc., obscure individual details at the system level. Modifications to this model to include infant mortality have been described (Holcomb and North, 1985).

[2] Failure rates calculated using the log-normal model (Jordan, 1978) may depart substantially from this simple interrelation depending on the standard deviation (σ, logarithmic in time) of the failure distribution. For example, a small σ (e.g., a value of 0.5 indicating a tight failure peak) and a system service life much less than the median life can reduce the failure rate one or even two orders of magnitude relative to a simple reciprocal of the MTTF. Sigmas usually range from 0.5 to 1.5 for semiconductor and lightwave devices. Also see Cheng (1977).

high as 100 Mh to 1 Gh. LEDs are also characterized by low log-normal σ s. Due to the wide variety of devices available, however, many data sheets report significantly shorter MTTFs at elevated temperature. Available laser data from numerous manufacturers as well as published research results indicate 70°C MTTFs ranging from 10kh to 100kh, 30 kh to 300kh at 60°C, and 200kh to over 1Mh at 25 °C. For both LEDs and lasers, the log-normal failure rates should, in principle, be calculated for specific applications: most frequently, the reciprocal of either MTTF or median life is used which is pessimistic.

19.4.1.2 Active Cooling. Since the degradation of LEDs and lasers is thermally activated, the use of thermoelectric (or other active) cooling maintains low-temperature reliability in high-temperature ambients. The reliability of the TEC and its control circuitry must be considered, of course. Miniature TECs appear to lie in the 500 to 2000 FIT range depending on the number of elements and screening procedures (Marlow, 1975; Shumate, 1987). With the added failure rates for power-control circuitry, active cooling increases the transmitter subsystem failure rate by 1000 to 3000 FITs.

Finally, contributions for the drive circuitry and packaging are added to determine the total transmitter failure rate. Discrete semiconductors and silicon integrated circuits have individual failure rates in the 30 to 100-FIT range, depending on temperature, complexity, etc. (MIL-HDBK-217D, 1980). The failure rate of GaAs ICs has not been fully established at this time, but they are expected to be comparable to their silicon counterparts. Therefore, circuitry may contribute 100 to 1000 FITs to the transmitter.

For the network objectives given, it is clear that cooling is applicable for many lasers for achieving a net reliability gain, as seen from the following table, where it is assumed that laser failure rate at 50°C is 20k FITs:

TABLE 19.1
Laser transmitter failure rates with and without thermoelectric cooling, including reliability of TEC.
Activation energy = 0.7 eV.
Ambient temperature = 50°C.

	w/o TEC	with TEC
Laser Temperature	50°C	25 °C
Laser F. R.	20k FITs	2.5k FITs
Driver Circuit F. R.	2k FITs	2k FITS
TEC + Control F. R.	—	2.5k FITS
Subsystem F. R.	22k FITs	7k FITs

For loop equipment, however, the transmitter failure-rate objective is already on the order of the cooling components alone. Given the additional objectives of reducing both cost and power dissipation, the applicability of cooling is far less clear. Uncooled, high-reliability lasers or LEDs appear to be more appropriate solutions for the loop.

19.4.1.3 Coupling Stability and Packaging. The final aspect of transmitter reliability relates to packaging. In addition to providing high-frequency access to the emitting device, packages should provide hermetic environments in many cases and stable device-to-fiber coupling in all cases. Additionally, packages usually provide for monitor photodiodes (for lasers), TECs and temperature monitoring, and sometimes active circuitry (Maxham *et al.*, 1985) or an optical isolator to suppress reflections (Chikama *et al.*, 1986; Sugie and Saruwatari, 1986; Matsuda *et al.*, 1987). Other LED and laser packages have been described (Johnson *et al.*, 1980; Speer and Hawkins, 1980; Berg *et al.*, 1981; Berg *et al.*, 1983; Dufft and Camlibel, 1980; Kuwahara *et al.*, 1980; Kock, 1982; Khoe *et al.*, 1984; Khoe and Kock, 1985; Kawano *et al.*, 1986; Reynolds *et al.*, 1986).

Although LEDs are frequently packaged non-hermetically, hermeticity is desirable for laser packages because of concerns about die-bonding solder instabilities (Yamamoto *et al.*, 1979) and mirror erosion (Fukuda and Wakita, 1980; Morimoto and Takusagawa, 1983). If TE coolers are used with either LEDs or lasers, hermeticity is important to prevent the ingress of moisture leading to actual condensation on the cold side of the TEC, should it be run below the local dew point. Condensation can lead to TEC failure, as well as resulting in impaired coupling or facet erosion if condensation takes place on laser mirrors.

Hermetic laser packaging is often provided using either conventional or modified MIC housings (e.g., Kovar packages with glassed-through pins or butterfly feedthroughs, plus a hermetic feedthrough for the fiber), or by placing the laser chip in a cylindrical opto package and coupling to the fiber with discrete lenses. An example of MIC packaging is shown in Fig. 19.13 (Lasertron, 1986), where either a DIP-style or butterfly housing incorporates an internal cooler, a monitor photodiode, a thermistor and a hermetic fiber feedthrough.

The assurance of coupling stability over the life of the transmitter and over the expected temperature range are of primary importance. When a single-mode fiber is directly coupled to the laser, as is frequently the case, transverse displacements between the device and fiber of as little as 0.5 μm can result in 1-dB loss in coupling. (If lenses are used, a similar critical interface exists, usually between the laser and the first lens.) Multimode coupling is less sensitive by a factor of about 5.

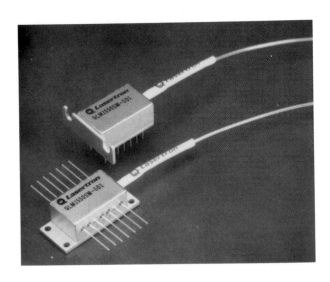

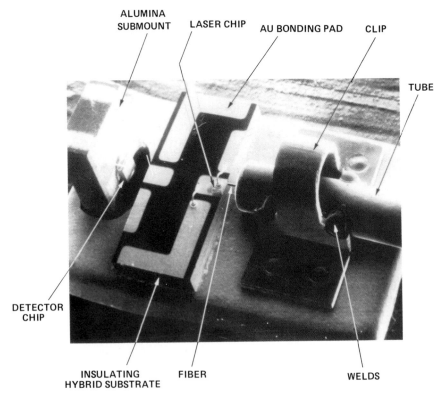

Fig. 19.13 (a) Laser packaged hermetically with internal thermoelectric cooler. (b) Details of laser-welded fiber attachment (Lasertron, 1986; reprinted with permission).

TABLE 19.2
Thermal conductivity and coefficients of linear thermal expansion
for common lightwave packaging materials.
Silica glass fiber is also shown.
Units of conductivity are W / cm · C° (near 25°C)
and expansion are ppm / C° (near 25°C).

	Conductivity	Expansion
C11000 Copper	3.9	17.0
Kovar	0.17	5.7
1010 C. R. Steel	0.5	12.6
H. D. Alumina	0.35	5.0
H. D. Beryllia	2.1	6.5
Diamond	20.0	1.2
Silicon	1.5	2.5
60Sn/40Pb Solder	0.2	23.9
Filled Epoxies	0.01–0.2	30–50
Silica Fiber	0.01	0.5

Design considerations for the laser attachment include minimizing electrical
and thermal resistance, and for the laser-fiber interface, assuring mechanical
stability and minimum differential thermal expansions (unless the interface
temperature is stabilized by active cooling). Table 19.2 gives the thermal
conductivity and expansion properties of several materials used for packaging
LEDs and lasers.

Good thermal design using workable pieceparts often results in combinations
of different materials lying between the device and the attaching point of the
fiber. For example, consider a laser die-bonded to a copper heatsink block, a
fiber attached to an alumina block, and both blocks subsequently attached to a
common heat-spreader substrate. Given the data in Table 19.2 and the figure of
0.5 μm for 1-dB coupling loss, the block heights must be ≤ 0.4 mm to assure
1-dB coupling stability over a 100-degree temperature range. Such small pieceparts are difficult to handle and align. Therefore some packages may require the
design of less-simply shaped parts or of more co-expansive interfaces.

For any given interface design, the long-term reliability (stability) may be
difficult to assess in a meaningful way. Being mechanical, deterioration in
coupling is expected to be characterized by a low activation energy. Overcoming
the low activation energy by accelerating at high temperature may be limited
when the alignment is secured with solders or epoxy. Instead, stability can be
examined by extensive thermal cycling and by storage tests at elevated temperatures. Field data then must be monitored to reinforce the results. It is clear that
an understanding of the mechanical properties of all packaging materials is
essential (low- and high-temperature solders, epoxies, etc.) and the strains
resulting from differential displacements.

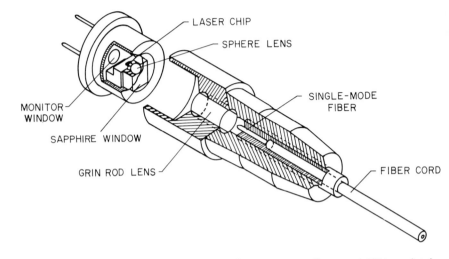

Fig. 19.14 Laser packaging using confocal lens coupling arrangement (Saruwatari, 1984; reprinted with permission).

It is considered desirable to secure the fiber alignment in MIC-style packages with high-temperature solders or spot welds, thus eliminating epoxies and questions related to mechanical strength near glass-transition temperatures, or of water and amines trapped in cured resin. For the technique shown in Fig. 19.13b, the fiber, hard-soldered into the tube, is aligned to the laser and four spot welds secure the clip to the substrate carrying the laser, monitor, etc. (Lasertron, 1987). Laser welding also minimizes heating (hence differential motion), so active alignment is possible.

For the packaging approach of Fig. 19.14 (Saruwatari, 1984), the laser and a high-index spherical lens are contained in a separate hermetic package. A cylindrical graded-index lens and the fiber are mutually aligned in a non-hermetic outer housing, which is then aligned and attached to the laser/lens combination. The displacement sensitivity in the plane between the two lenses is reduced by the confocal arrangement, so the securing of the laser in the outer housing is less critical than the other two interfaces. Also, the cylindrical symmetry reduces the susceptibility to this design to temperature-induced lateral differential displacements. For the package shown here, the laser back-mirror emission is brought out through a window to a separately packaged photodiode.

Both the MIC and cylindrical packaging schemes achieve high coupling efficiency needed for transmission, typically 25 to 40% (-6 to -4 dB). In loop

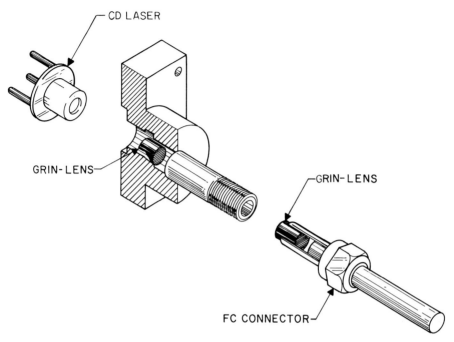

Fig. 19.15 Laser packaging using Compact Disk hermetic package and confocal lensing (Reith *et al.*, 1986).

and local-area-network applications, high power is not needed because of the shorter fiber lengths. On the other hand, lowest cost and good stability in widely varying ambient temperatures are desirable. Separately packaged lasers coupled directly to fibers using standard connectors have recently been explored, both for single-mode (Reith *et al.*, 1986) and multimode (Soderstrom *et al.*, 1986) situations.

A package design that takes advantage of a low-cost Compact Disk laser housing (but containing a 1.3 μm device instead of a 0.78 μm CD laser) is shown in Fig. 19.15. Here, the coupling efficiency is only 6% (−12 dB), but this has been traded for decreased sensitivity to both lateral and axial displacements at the lens-lens interface (Joyce and DeLoach, 1984). (*Angular* displacement must also be considered in some package designs (Reith and Mann, 1987).) Hence the output is stable over a wide temperature range, while the output power remains adequate for transmission over several km of single-mode fiber. The CD package itself is capable of operation beyond 1 Gb/s.

In summary, driver and control circuitry for both LED-based and laser-based high-speed digital transmitters have been discussed. The importance of examining the transmitter as an entire subsystem—device properties, circuit design, reliability and packaging—was emphasized. Application-specific requirements must be carefully understood in order to make the proper selection among the options available.

REFERENCES

Albanese, A. (1978), An automatic bias control (ABC) circuit for injection lasers, *Bell Sys. Tech. J.* **57**, 1533–1544.

Atallah, K., and Martinot, H. (1984), Equivalent circuit and minority carrier lifetime in heterostructure light emitting diodes, *Solid State Electron.* **27**, 375–380.

Banwell, T. (1987), private communication.

Banwell, T. C., and Stephens, W. E. (1987), Low-cost transmitter strategies using 1.3 μm CD-type laser diodes, *Proc. 13th European Conf. on Opt. Commun.*, 453–456.

Basch, E. E., and Carnes, H. A. (1984), Analog optical communications, "Fiber Optics." CRC Press, Boca Raton, 181–194.

Bellcore Technical Reference (1985a), Network equipment—building systems, TR-EOP-000063, issue 1, 4–1.

Bellcore Technical Reference (1985b), Reliability prediction procedure for electronic equipment, TR-TSY-000332.

Berg, H. M., Lewis, G. L., and Mitchell, C. W. (1981), A high performance connectorized LED package for fiber optics, *IEEE Trans. Components, Hybrids, Manufact. Technol.*, **CHMT-4**, 337–344.

Berg, H. M., Shealy, D. L., Mitchell, C. M., Stevenson, D. W., and Lofgran, L. C. (1983), Optical coupling in fiber optics packages with surface emitting LEDs, *IEEE Trans. Components, Hybrids, Manufact. Technol.*, **CHMT-6**, 334–342.

Bickers, L., and Westbrook, L. D. (1985), Reduction of laser chirp in 1.5 μm DFB lasers by modulation pulse shaping, *Electron. Lett.*, **21**, 103–104.

Bowers, J. E., Hemenway, B. R.. Gnauck, A. H., and Wilt, D. P. (1986), High-speed InGaAsP constricted-mesa lasers, *IEEE J. Quantum Electron.*, **QE-22**, 833–844.

Brackett, C. A., Hackett, W. H., and Shumate, P. W. (1980), A noise-immune 32 Mb/s optical data link, *6th European Conf. on Opt. Commun.*, York, England.

Bulley, R. M. (1986), Private communication.

Chang, G. K., Leblanc, H. P., and Shumate, P. W. (1986), Novel high-speed LED transmitter for single-mode fiber and wideband loop transmission systems, *Electron. Lett.*, **23**, 1338–1340.

Chen, F. S. (1980), Simultaneous feedback control of bias and modulation currents for injection lasers, *Electron. Lett.*, **16**, 7–8.

Chen, F. S. (1986), 4 Gbit/s GaAs MESFET laser-driver IC, *Electron. Lett.*, **22**, 932–933.

Chen, P. C., Law, H. P., Rezek, E. A., and Weller, J. (1983), Monolithic integration of laser/FET on InP, *4th Intl. Conf. on Integ. Optics and Opt. Fiber Commun.*, Tokyo, 190–191.

Cheng, S. S. (1977), Optimal replacement rate of devices with lognormal failure distributions, *IEEE Trans. Reliab.*, **R-26**, 174–178.

Chikama, T., Watanabe, S., Goto, M., Miura, S., and Touge, T. (1986), Distributed feedback, laser diode module with a novel and compact optical isolator for gigabit optical transmission systems, *Conf. on Opt. Fiber Commun.*, Atlanta, GA, paper ME4.

Cooke, M. P., Sumerling, M. P., Muoi, T. V., and Carter, A. C. (1986), Integrated circuits for a 200-Mb/s fiber-optic link, *IEEE J. Solid-State Circuits*, **SC-21**, 909–915.

Dean, M., and Dixon, B. A. (1981), Aging of the light-current characteristics of proton-bombarded AlGaAs lasers operated at 30°C in pulsed conditions, *3rd Intl. Conf. on Integ. Opt. and Opt. Fiber Commun.*, San Francisco, paper MJ7.

Descombes, A., and Guggenbuhl, W. (1981), Large signal circuit model for LED's used in optical communication, *IEEE Trans. Electron Devices*, **ED-28**, 395–404.

Dufft, W. H., and Camlibel, I. (1980), A hermetically encapsulated AlGaAs laser diode, *Proc. 30th Electron. Components Conf.*, San Francisco, 261–269.

Dutta, N. K., Nelson, R. J., Wright, P. D., Besomi, P., and Wilson, R. B. (1983), Optical properties of a 1.3-μm superluminescent diode, *IEEE Trans. Electron Dev.*, **ED-30**, 360–363.

Ettenberg, M., Olsen, G. H., and Hawrylo, F. Z. (1984), On the reliability of 1.3-μm InGaAsP/InP edge-emitting LEDs for optical-fiber communication, *J. Lightwave Technol.*, **LT-2**, 1016–1023.

Fisher, A. L., and Linde, N. (1986), A 50-Mb/s CMOS optical transmitter integrated circuit, *IEEE J. Solid-State Circuits*, **SC-21**, 901–908.

Fujimoto, N. (1986), Private communication.

Fujimoto, N., Ohtsuka, T., Taniguchi, A., Yamaguchi, K., and Nabeshima, Y. (1987), 1.2-Gb/s optical transmission experiment over 10 km of single-mode fiber using a high-speed edge-emitting LED, *Conf. on Opt. Fiber Commun.*, Reno, NV, paper MI5.

Fujita, O., Nakano, Y., and Iwane, G. (1985), "Screening by aging test for highly reliable laser diodes," *Electron. Lett.*, **21**, 1172–1173.

Fukuda, M. (1983), Facet oxidation of InGaAsP/InP and InGaAs/InP lasers, *IEEE J. Quantum Electron.*, **QE-19**, 1692–1698.

Fukuda, M., Fujita, O., Uehara, S., and Iwane, G. (1986), Reliability of InGaAsP/InP light emitting diodes, *Proc. 12th European Conf. on Opt. Commun.*, Barcelona, 113–116.

Fukuda, M., and Wakita, K. (1980), Degradation in InGaAsP/InP DH lasers in water due to facet deterioration, *Jap. J. Appl. Phys.*, **19**, 667–670.

Gordon, E. I., Nash, F. R., and Hartman, R. L. (1983), Purging; a reliability assurance technique for new technology semiconductor devices, *IEEE Electron Dev. Lett.*, **EDL-4**, 465–466.

Gower, J. (1984), "Optical Communication Systems." Prentice-Hall, London, 515–521.

Gruber, J., Marten, P., Petschacher, R., and Russer, P. (1978), Electronic circuits for high bit rate digital fiber optic communication systems, *IEEE Trans. Commun.*, **COM-26**, 1088–1098.

Hayashi, J., Fujita, S., Isoda, Y., Uji, T., Shikada, M., and Kobayashi, K. (1987), 2 Gb/s and 600 Mb/s single mode fiber transmission using a high-speed Zn-doped 1.3 μm edge-emitting LED, *Conf. on Opt. Fiber Commun.*, Reno, NV, paper PDP-17.

Hino, I., and Iwamoto, K. (1979), LED pulse response analysis considering the distributed CR constant in the peripheral junction, *IEEE Trans. Electron Devices*, **ED-26**, 1238–1242.

Hirao, M., Mizuishi, K., and Nakamura, M. (1986), High-reliability semiconductor lasers for optical communications, *IEEE J. Sel. Areas Commun.*, **SAC-4**, 1494–1501.

Holcomb, D. P., and North, J. C. (1985), An infant mortality and long-term failure rate model for electronic equipment, *AT & T Tech. J.*, **64**, 15–31.

Horimatsu, T., Iwama, T., Oikawa, Y., Touge, T., Wada, O., and Nakagami, T. (1985), 400 Mb/s transmission experiment using two monolithic optoelectronic chips, *Electron. Lett.*, **21**, 319–321.

Horimatsu, T., Iwama, T., Oikawa, Y., Touge, T., Makuichi, M., Wada, O., and Nakagami, T. (1986), Compact transmitter and receiver modules with optoelectronic-integrated circuits for optical LAN's, *J. Lightwave Technol.*, **LT-4**, 680–688.

Hutcheson, L. D. (1986), GaAs/AlGaAs monolithic optoelectronics and integrated optics, *Conf. on Lasers and Electro-Optics (CLEO)*, San Francisco, paper FO-1.

Jarrett, B. (1974), Drive fiber optic lines at 100 MHz, *Electronic Design*, **15**, July 19 issue, 96–99.

Johnson, B. H., Ackenhusen, J. G., and Lorimor, O. G., (1980), Connectorized optical link package incorporating a microlens, *IEEE Trans. Components, Hybrids, Manufact. Technol.*, **CHMT-3**, 488–492.

Jordan, A. S. (1978), A comprehensive review of the lognormal failure distribution with application to LED reliability, *Microelectron. Reliab.*, **18**, 267–279.

Joyce, W. B., and DeLoach, B. C. (1984), Alignment of gaussian beams, *Appl. Opt.*, **23**, 4187–4196.

Kaede, K., Mito, I., Yamaguchi, M., Kitamura, M., Ishikawa, R., Lang, R., and Kobayashi, K. (1985), Spectral chirping suppression by compensating current in modified DFB-DC-PBH LDs, *Conf. on Opt. Fiber Commun.*, San Diego, CA, paper WI-4.

Karr, M. A., Chen, F. S., and Shumate, P. W. (1979), Output power stability of GaAlAs laser transmitter using an optical tap for feedback control, *Appl. Opt.*, **18**, 1262–1265.

Kasahara, K., Suzuki, A., Fujita, S., Inomoto, Y., Terakado, T., and Shikata, M. (1986), InGaAsP/InP long wavelength transmitter and receiver OEICs for high speed optical transmission systems, *Proc. 12th European Conf. on Opt. Commun.*, Barcelona, Spain, 119–122.

Keramidas, V. G., Berkstresser, G. W., and Zipfel, C. L. (1980), Planar, fast, reliable single-heterojunction light-emitting diodes for optical links, *Bell Sys. Tech. J.*, **59**, 1549–1557.

Khoe, G. D., Koch, H. G., Kuppers, D., Poulissen, J. H. F. M., and deVrieze, H. M. (1984), Progress in monomode optical-fiber interconnection devices, *IEEE J. Lightwave Technol.*, **LT-2**, 217–227.

Khoe, G. D., and Kock, H. G., (1985), Laser-to-monomode-fiber coupling and encapsulation in a modified TO-5 package, *J. Lightwave Technol.*, **LT-3**, 1315–1320.

Kock, H. G. (1982), "Coupler comprising a light source and lens," U.S. Patent No. 4,355,323, Oct. 19.

Koren, U., Yu, K. L., Chen, T. R., Bar-Chaim, N., Margalit, S., and Yariv, A. (1982), Monolithic integration of a very low threshold InGaAsP laswer and metal-insulator-semiconductor field effect transistor on semi-insulating InP, *Appl. Phys. Lett.*, **40**, 643–645.

Koyama, F., and Iga, K. (1985), Frequency chirping of external modulation and its reduction, *Electron. Lett.*, **21**, 1065–1066.

Lasertron Reliability Report (1987), Reliability assessment of sources and detectors, LT86-2.

Lau, K., Bar-Chaim, N., Derry, P. L., and Yariv, A. (1987), High-speed digital modulation of ultralow threshold (< 1 mA) GaAs single quantum well lasers without bias, *Appl. Phys. Lett.*, **51**, 69–71.

Marlow, R. (1975), Reliability and failure modes of thermoelectric heat pumps, *Marlow Industries, Inc., Dallas, TX*.

Matsuda, K., Minemoto, H., Toda, K., Kamada, O., and Ishizuka, S. (1987), Low-noise LD module with an optical isolator using a highly Bi-substituted garnet film, *Electron. Lett.*, **23**, 203–205.

Matsueda, H., and Nakamura, M. (1984), Monolithic integration of a laser diode, photo monitor, and electric circuits on a semi-insulating GaAs substrate, *Appl. Opt.*, **23**, 779–781.

Maxham, K. Y., Hogge, C. R., Clendening, S. J., Chen, C. T., Dugan, J. M., Sheem, S. K., and Offutt, D. D. (1984), Rockwell 135 Mb/s lightwave system. *J. Lightwave Technol.*, **LT-2**, 394–402.

MIL-HDBK-217D (1980), "Military Handbook—Reliability Prediction of Electronic Equipment." Dept. of Defense, Washington, DC.

Mikawa, T., Miura, S., Kuwatsuka, H., Fujii, T., and Wada, O. (1987), Planar monolithic pin/FET fabricated by using an imbedded structure InP/GaInAs pin photodiode and an AlInAs/GaInAs field-effect transistor, *Conf. on Opt. Fiber Commun.*, Reno, NV, paper WG2.

Morimoto, M., and Takasugawa, M. (1983), Accelerated facet degradation of InGaAsP/InP double-heterostructure lasers in water, *J. Appl. Phys.*, **53**, 4028–4037.

Nakamura, M., Suzuki, N., and Ozeki, T. (1986), The superiority of optoelectronic integration for high-speed laser diode modulation, *IEEE J. Quantum Electron.*, **QE-22**, 822–826.

Nakamura, M., Furuyama, H., and Kurobe, A. (1987), 1 Gbit/s automatic-power-control-free zero-bias modulation of very-low threshold MQW laser diodes, *Electron. Lett.*, **23**, 1352–1353.

Nakano, Y., Iwane, G., and Ikegami, T. (1984), Screening method for laser diodes with high reliability, *Electron. Lett.*, **20**, 397–398.

Nakano, H., Yamashita, S., Tanaka, T. P., Hirao, M., and Meada, M. (1986), Monolithic integration of laser diodes, photomonitors, and laser driving and monitoring circuits on a semi-insulating GaAs, *J. Lightwave Technol.*, LT-4, 574–582.

Nash, F. R., Joyce, W. B., Hartman, R. L., Gordon, E. I., and Dixon, R. W. (1985), Selection of a laser reliability assurance strategy for a long-life application, *AT & T Tech. J.*, **64**, 671–715.

Olshansky, R., and Fye, D. (1984), Reduction of dynamic linewidth in single-frequency semiconductor lasers, *Electron. Lett.*, **20**, 928–929.

Olson, J. W., and Schepis, A. J. (1984), Description and application of the Fiber-SLC carrier system, *J. Lightwave Technol.*, LT-2, 317–322.

Peck, D. S., and Zierdt. C. H. (1974), The reliability of semiconductor devices in the Bell System, *Proc. IEEE*, **62**, 185–211.

Personick, S. D. (1973), Receiver design for digital fiber optic communication systems: II, *Bell Sys. Tech. J.*, **52**, 875–886.

Personick, S. D. (1985), Analog links for video, telemetry, i.f. and r.f. remoting, "Fiber Optics Technology and Applications." Plenum Press, New York, 194–215.

Plastow, R. (1986), LEDs for fiber optic communication systems, *Photonics Spectra*, Sept. issue, 109–116.

Reith, L. A., Shumate, P. W., and Koga, Y. (1986), Laser coupling to single-mode fibre using graded-index lenses and Compact Disc 1.3 μm laser package, *Electron. Lett.*, **22**, 836–838.

Reith, L. A., and Mann, J. W. (1987), Design parameters for a 1.3 μm connectorized low-cost laser package for subscriber loop using gradient-index lenses, *Electron. Lett.*, **23**, 520–521.

Reynolds, C. L., Jr., Nygren, S. F., Lapinsky, R. L., and Vehse, R. C. (1986), Reliable package for hermetic encapsulation of InGaAsP edge-emitting LEDs, *Proc. SPIE*, **703**, 29–32.

Saruwatari, M. (1984), Laser diode module for single-mode optical fiber, "Optical Devices and Fibers-1984." North Holland, New York, 129–140.

Sato, K., Tsuyuki, S., and Miyanaga, S. (1984), Fiber optic analog transmission techniques for broadband signals, *Rev. Elec. Comm. Lab.*, **32**, 586–597.

Saul, R. H., King, W. C., Olsson, N. A., Zipfel, C. L., Chin, B. H., Chin, A. K., Camlibel, I., and Minneci, G. (1985), 180 Mb/s, 35 km transmission over single-mode fiber using 1.3 μm edge-emitting LEDs, *Electron. Lett.*, **21**, 773–775.

Schmidt, R. V. (1983), Integrated optics switches and modulators, "Integrated Optics: Physics and Applications." Plenum Press, New York, 181–210.

Schwartz, M. (1980), Modulation techniques, "Information Transmission, Modulation and Noise." McGraw-Hill, New York, 209–316.

Schwartz, R. G., and Wooley, B. A. (1983), Stabilized biasing of semiconductor lasers, *Bell Sys. Tech. J.*, **62**, 1923–1936.

Sharp (1985), e.g., Sharp Electronics Corp., p/n IR3CO1 and IR3CO2.

Shen, T. M., Wilt, D. P., Long, J. A., Dautrement-Smith, W. C., Focht, R. L., and Hartman, R. L. (1986), Channelled substrate buried heterostructure InGaAsP/InP laser with Fe-doped semi-insulating InP base structure for 1.7 Gb/s optical communication systems, *Conf. on Lasers and Electro-Optics*, San Francisco, paper ThU1.

Shibata, J., Nakao, I., Sasaki, Y., Kimura, S., Hase, N., and Serizawa, H. (1984), Monolithic integration of an InGaAsP/InP laser diode with heterojunction bipolar transistor, *Appl. Phys. Lett.*, **45**, 191–193.

Shumate, P. W., Chen, F. S., and Dorman, P. W. (1978), GaAlAs laser transmitter for lightwave transmission systems, *Bell Sys. Tech. J.*, **57**, 1823–1836.

Shumate, P. W., and DiDomenico, M. (1980), Lightwave transmitters, "Semiconductor Devices for Optical Communication." Springer-Verlag, Berlin, 161–200.

Shumate, P. W. (1986), unpublished aging-program results: 4.5 million device hours accumulated on 17-element TECs at 70°C (hot side) without failure or significant change in performance.

Sim, S. P., Robertson, M. J., and Plumb, R. G. (1984), Catastrophic and latent damage in GaAlAs lasers caused by electrical transients, *J. Appl. Phys.*, **55**, 3950–3955.

Smith, D. W. (1978), Laser level-control circuit for high-bit-rate systems using a slope detector, *Electron. Lett.*, **14**, 775–776.

Smith, D. W., and Matthews, M. R. (1983), Laser transmitter design for optical fiber systems, *IEEE J. Selected Areas Commun.*, **SAC-1**, 515–523.

Soderstrom, R. L., Block, T. R., Karst, D. L., and Lu, T. (1986), The Compact Disc (CD) laser as a low-cost high-performance source for fiber optic communication, *Proc. FOC/LAN Conf.*, Orlando, 263–264.

Speer, R. S., and Hawkins, B. M. (1980), Planar DH GaAlAs LED packaged for fiber optics, *IEEE Trans. Components, Hybrids, Manufact. Technol.*, **CHMT-3**, 270–274.

Sugie, T., and Saruwatari, M. (1986), Distributed feedback laser diode (DFB-LD) to single-mode fiber coupling module with optical isolator for high bit rate modulation, *J. Lightwave Technol.*, **LT-4**, 236–245.

Suzuki, A., Inomoto, Y., Hayashi, J., Isoda, Y., Uji, T., and Nomura, H. (1984), Gb/s modulation of heavily Zn-doped surface-emitting InGaAsP/InP DH LED, *Electron. Lett.*, **20**, 273–274.

Suzuki, A., Uji, T., Inomoto, Y., Hayashi, J., Isoda, Y., and Nomura, H. (1985), InGaAsP/InP 1.3-μm surface-emitting LEDs for high-speed short-haul optical communication systems, *J. Lightwave Technol.*, **LT-3**, 1217–1222.

Suzuki, M., Noda, Y., Kushiro, Y., and Akiba, S. (1986), Dynamic spectral width of an InGaAsP/InP electroabsorption light modulator under high-frequency large-signal modulation, *Electron. Lett.*, **22**, 312–313.

Suzuki, T., Ebata, T., Fukuda, K., Hirakata, N., Yoshida, K., Hayashi, S., Takada, H., and Sugawa, T. (1986a), High-speed 1.3 μm LED transmitter with GaAs integrated driver IC, *J. Lightwave Technol.*, **LT-4**, 790–793.

Suzuki, T., Shikata, S., Nakajima, S., Hirakata, N., Mikanura, Y., and Sugawa, T. (1986b), GaAs IC family for high-speed optical interconnection systems, *Proc. GaAs IC Symp.*, Grenelefe, FL, 225–228.

Tachikawa, Y., and Saruwatari, M. (1984), Laser diode module for analog video transmission, *Rev. Elec. Commun. Labs.*, **32**, 598–607.

Thompson, G. B. H. (1980), Dynamic response of lasers, "Physics of Semiconductor Laser Devices." John Wiley & Sons, New York, 402–474.

Tucker, R. S. (1985), High-speed modulation of semiconductor lasers, *J. Lightwave Technol.*, **LT-3**, 1180–1192.

Tucker, R. S., and Pope, D. J. (1983), Microwave circuit models of semiconductor injection lasers, *IEEE Trans. Microwave Theory Tech.*, **MTT-31**, 289–294.

Uhle, M. (1976), The influence of source impedance on the electrooptical switching behavior of LEDs, *IEEE Trans. Electron Dev.*, **ED-23**, 438–441.

Uji, T. (1986), Private communication.

Wedding, B. (1987), SPICE simulation of laser diode modules, *Electron. Lett.*, **23**, 383–384.

Yamakoshi, S., Abe, M., Wada, O., Komiya, S., and Sakurai, T. (1981), Reliability of high radiance InGaAsP/InP LEDs operating in the 1.2–1.3 μm wavelength, *IEEE J. Quantum Electron.*, **QE-17**, 167–173.

Yamamoto, T., Sakai, K., and Akiba, S. (1979), 10,000-h continuous CW operation of InGaAsP/InP DH lasers at room temperature, *IEEE J. Quantum Electron.*, **QE-15**, 684–687.

Yamashita, K., Takasaki, Y., Maeda, M., and Maeda, N. (1986), Master-slice monolithic integration design and characteristics of LD/LED transmitters for 100–400 Mb/s optical transmission systems, *J. Lightwave Technol.*, **LT-4**, 353–359.

Yoshikuni, Y., and Motosugi, G. (1986), Independent modulation in amplitude and frequency regimes by a multielectrode distributed feedback laser, *Conf. on Opt. Fiber Commun.*, Atlanta, GA, paper TuF1.

Zah, C. E., Osinski, J. S., Menocal, S. G., Tabatabaie, N., Lee, T. P., Dentai, A. G., and Burrus, C. A. (1987), Wide-bandwidth and high-power 1.3 μm InGaAsP buried crescent lasers with semi-insulating Fe-doped InP current blocking layers, *Electron. Lett.*, **23**, 52–53.

Chapter 20

High Performance Integrated Circuits for Lightwave Systems

ROBERT G. SWARTZ

AT&T Bell Laboratories, Inc., Holmdel, New Jersey

20.1 INTRODUCTION

Integrated circuit technology lies at the heart of fiber optic communication systems, just as it does in so many other areas vital to today's technological society. But optical communications challenges the limits of IC technology as does perhaps no other application. Fiber communication requires the ultimate in speed, minimum noise, high gain, low power and as much circuit complexity under these conditions as a technology can provide. For this reason, optical systems are a ready customer for the latest and most aggressive integrated circuit technologies, and continue to be an inspiration for further rapid IC development.

In this chapter we examine the role of high performance integrated circuits in modern optical communications, and we explore the limitations of IC technology in providing the ever more demanding circuits required for new fiber systems.

We begin by reviewing the electronic components that make up a digital optical communications system. Next, we survey IC answers to the problems posed by high data rate communication requirements. Then we conclude with a closer examination of the integrated circuit technologies that have made possible the highest performance systems, and review technological and device characteristics that limit maximum circuit speed.

Before proceeding, it is appropriate for the reader to consider and to understand the benefits to be gained from the use of integrated implementations of any circuit. From the standpoint of optical communications, ICs provide more circuit features, and ultimately lower cost, than hybrid circuits by virtual elimination of hand assembly. Other advantages include compactness, lower power dissipation, and potentially greater reliability and higher performance. Disadvantages may include higher initial cost, longer lead times for component availability, and reduced performance in some applications. Particular IC technologies are suitable for some applications and eminently unsuitable for others. Integration is not the automatic answer to every electronic question, and the designer must have enough knowledge to make the correct choice where alternatives exist.

This chapter is not intended as a comprehensive general review of integrated circuits or of IC technology. Instead, it is an attempt to introduce the reader to the electronics necessary for very wideband optical communications as seen from the standpoint of an integrated circuit designer or technologist. In the course of this review we will by necessity and by design conduct an exploration of the high performance fringes of the IC world.

20.2 INTEGRATED HIGH SPEED COMPONENTS— PIECES OF THE PUZZLE

In this chapter, "high performance" is defined quite arbitrarily as 1 Gb/s and beyond. In the field of optical communications, integrated circuit components operating at data rates ranging from 500 Mb/s to 2 Gb/s are presently moving from development laboratories into production. Research facilities have demonstrated ICs functioning to 6 Gb/s, and an increase in this upper limit is assured in the future. Some functions in an optical system are more easily realized by high speed integrated circuits than others. We therefore examine the electronics in an optical system closely, reviewing their demands on an IC technology, and how existing state-of-the-art designs have met these requirements.

The electronics in a fiber system are typically assembled into either transmitter or receiver configurations, or some combination of subassemblies performing functional aspects of both (e.g., "repeaters" or "regenerators"). The digital transmitter (Fig. 20.1) assembles one or more streams of digital data into a single high data rate serial stream, and then encodes this information on the optical output of a laser or LED. The first component in this process is the digital multiplexer (mux) that, in a high speed system, receives input data at rates ranging, for example, from 45 Mb/s to 400 Mb/s. In future systems, it can be expected that data will be piped into the mux at even higher rates. The multiplexer may encode this data, or add error checking and framing bits, while performing the multiplex operation. The number of digital input channels

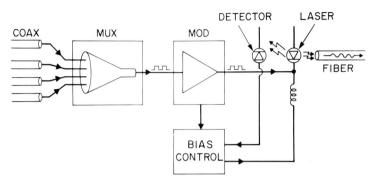

Fig. 20.1 Block diagram of optical transmitter.

required by the multiplexer typically ranges from 4 to as many as 12. As the number of of data channels increases, IC process complexity limitations and power dissipation per logic gate (with the associated problem of *cooling* that gate) become important factors in determining performance capabilities.

At the highest data rates, muxes fabricated in silicon bipolar, GaAs MESFET and fine line silicon NMOS have been reported. The fastest to date is an experimental silicon bipolar 4 to 1 multiplexer reported by Reimann and Rein (1986) with a maximum output data rate of 6 Gb/s. The architecture, illustrated in Fig. 20.2, consists of two 2:1 muxes that funnel four channels of data into a third 2:1 output mux. The input muxes are toggled by half-frequency, 90° phase-referenced clocks. The technology utilized transistors with 2 μm features and shallow junctions to achieve an f_T of approximately 9 GHz.

The architecture of Fig. 20.2 is similar to that of a more complex circuit by Liechti *et al.* (1982) who reported an 8 to 1 multiplexer fabricated in 1 μm gate-length GaAs MESFET technology that operated at a 5 Gb/s output data rate. A column of four 2:1 muxes is used to channel eight lines into four lines, and then a circuit similar to that in Fig. 20.2 is used to perform the final four to one multiplexing operation.

A high speed 4:1/1:4 mux/demux pair has been reported by Yoshikai *et al.* (1985). Fabricated in bipolar silicon "super-self aligned transistor" (SST) technology (Konaka, 1986) with 0.5 μm emitter width, and 1 μm lithography, these circuits achieved an operating speed of 5 Gb/s using transistors with an f_T of 13.7 GHz. SST is the name of a family of ultra-high performance bipolar processes reported by NTT that have been employed for a variety of optical electronic applications. This technology is examined more closely in the final section.

Among the most ambitious of the high performance circuits is a 12 channel, 3 Gb/s mux/demux chip set reported by Bayruns *et al.* (1986). Designed in a fine

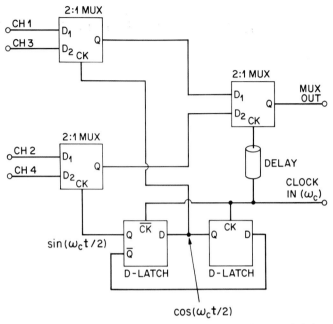

Fig. 20.2 Experimental 4 to 1 multiplexer employing staggered input and timing to achieve 6 Gb/s operation (Reimann and Rein, 1986).

line silicon NMOS technology with 1 μm features and 0.6 μm effective channel length, these chips take advantage of the high device packing density and circuit complexity afforded by MOS. The 12 channel demultiplexer also includes an on-chip barrel shifter to align output data with the proper data channel.

Other integrated circuits have combined multiplexing and demultiplexing capabilities with data encoding/decoding. These circuits, although at this time operating at somewhat lower data rates of 565 Mb/s to 1.5 Gb/s, are intended to simplify clock recovery at the receiver (Yoshikai *et al.*, 1986) and to allow provision for error monitoring and automatic output channel alignment (Mudd *et al.*, 1985).

The output of the multiplexer is delivered to the laser or LED driver circuitry. The driver's responsibility is to modulate the optical signal and to establish electronic and thermal bias conditions necessary for stable device operation. Driver circuits, particularly laser drivers, can be of considerable complexity and are currently areas of active research interest. In the highest speed laser driver circuits reported to date, the modulator and laser optical power-control circuits have been integrated separately. In this presently preferred configuration, designers optimize separately the process and circuit parameters most important for each function. For example, modulators emphasize high speed at the expense of high power dissipation, and are fabricated in state-of-the-art minimum feature size IC technologies. Power control circuits, on the other hand, rarely require high speed, but are often of greater circuit complexity than modulators. Moreover, they benefit by thermal isolation from the high dissipation modulator chip.

The fastest reported IC laser modulators utilize direct, amplitude modulation of the laser light output. [Note: Other methods include modulation of the laser wavelength, and external modulation employing a device such as a waveguide electrooptic modulator (Alferness, 1982).] Such ICs typically include a series of one or more amplifier stages followed by an output differential pair with a drain or collector connected to the laser. An example is the silicon NMOS modulator reported by Swartz et al. (1986) shown in Fig. 20.3. Advantages of a differential output stage are the separately adjustable current-setting control, minimization of transients on power supply lines, reduced sensitivity to variations in transistor threshold, and immunity to common-mode noise. This particular circuit demonstrated 120 mA of output drive at a 2 Gb/s data rate and over 200 mA of modulation current at data rates at or below 1.5 Gb/s. A chip photo is shown in

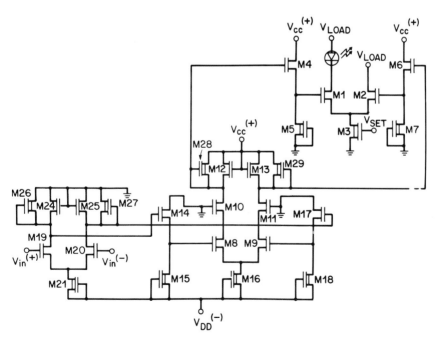

Fig. 20.3 2 Gb/s silicon NMOS laser modulator circuit schematic (Swartz et al., 1986). Circuit consists of three differential amplifier stages with intermediate source-follower buffers.

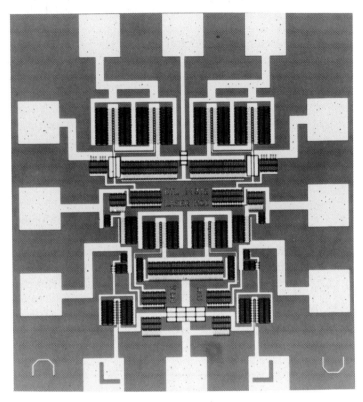

Fig. 20.4 Microphotograph of NMOS laser modulator. Note the segmented transistor structure used to obtain high power, high frequency operation.

Fig. 20.4. The segmented transistor structures visible in Fig. 20.4 consist of a large number of high performance FET modules that provide power handling capability. Even faster circuits have been reported in GaAs. Figure 20.5 shows a novel driver utilizing feedforward capacitors to speed up rise and fall times (Ueda *et al.*, 1985). This circuit was tested at data rates to 2 Gb/s with 50 mA output, and with 100 ps rise and fall times, appears capable of substantially higher top speed. The fastest reported laser modulator (Chen and Bosch, 1986) was tested to 4 Gb/s. Figure 20.6 shows an eye diagram revealing operation at that speed while supplying 80 mA of output drive.

A block diagram of a representative receiver is shown in Fig. 20.7. The first electronic components on the receiver side of the optical system are the pre-amplifier and main amplifier. The preamp can be thought of conceptually as a resistor, converting the photocurrent output of the detector to a voltage—ideally as large a signal as possible with the absolute minimum of accompanying

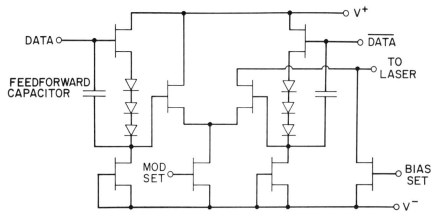

Fig. 20.5 2 Gb/s GaAs MESFET laser modulator with high frequency feedforward (Ueda *et al.*, 1985).

electrical noise. This is a particularly difficult component to implement in IC form as the requirements for broad bandwidth, low noise and high gain (A_1) are mutually conflicting. The main amplifier follows the preamplifier and amplifies its output to a level suitable for subsequent digital processing. Input signals to the preamplifier can span a wide range in amplitude because of differences in laser output, repeater spacing, and detector sensitivity, depending on the application. Consequently, the main amplifier must be configured in such a way as to amplify very small preamp voltages to full output, yet also be able to accommodate large preamplifier outputs without pulse distortion and accompanying intersymbol interference. A common solution is to use one or more variable gain

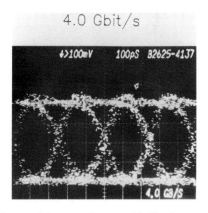

Fig. 20.6 4 Gb/s GaAs laser modulator eye diagram with 80 mA output (Chen and Bosch, 1986).

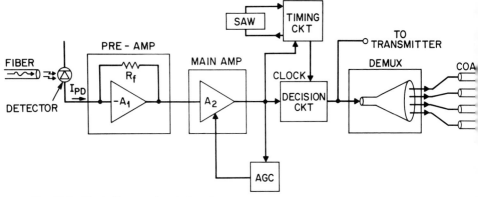

Fig. 20.7 Block diagram of optical receiver.

stages in the main amplifier with gain adjusted by an AGC (automatic gain control) feedback loop. This maintains linear circuit operation and minimizes the problems mentioned. An alternative solution is the use of high gain limiter stages without AGC. In a third configuration, the AGC function is sometimes combined with the preamplifier. In any amplifier circuit, designers must endeavor to eliminate possible oscillation and intersymbol interference arising from interstage coupling. Such coupling may be caused by inductive pick-up, parasitic coupling capacitance, or injected substrate current.

Although many examples of high quality monolithic preamplifiers and amplifiers exist in the frequency range of 600 Mb/s and below (see, for example, Aiki *et al.*, 1985; Ross *et al.*, 1986), few IC examples are reported at the highest data rates. Such preamplifiers are all "transimpedance" circuits (see Fig. 20.7) where resistive negative feedback between the amplifier input and output is used to perform the photocurrent to voltage conversion. Increasing the feedback resistor R_f improves the noise performance of the amplifier, but reduces bandwidth, although a technique proposed by Williams and LeBlanc (1986) employs an FET feedback element to improve the trade-off.

Again the familiar technological choices of silicon NMOS, GaAs MESFET and silicon bipolar are observed. Abidi (1984) has reported two preamplifiers, one with 20 dB gain and 1.26 GHz, -3 dB bandwidth, the other with 22 dB gain and 920 MHz bandwidth, both in a silicon NMOS technology with 0.45 μm effective channel length. An ACG function was implemented using resistive gate-to-drain feedback around individual amplifier stages. In this case, the "resistor" was a MOSFET with gate-controlled channel resistance. AGC amp gain was variable between 0 dB and 40 dB with corresponding bandwidths of 650 MHz and 810 MHz.

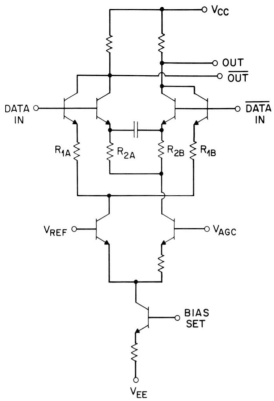

Fig. 20.8 Bipolar AGC amplifier circuit schematic (Ohara *et al.*, 1985). Variable gain is controlled by adjusting the bias current split, and therefore, the transconductance of two differential amplifiers.

Ohara *et al.* (1985) describe a two-chip silicon bipolar "equalizing" amplifier fabricated in 0.5 μm emitter SST technology. The preamplifier chip has a gain of 23 dB and a bandwidth of 3.7 GHz with a 4.5 dB noise figure. The AGC amplifier (Fig. 20.8) achieves variable gain by adjusting the bias current split between two shunt differential amplifiers with different emitter resistance ($R_{1A,B}$, $R_{2A,B}$) and therefore differing gains. The amplifier in its entirety consists of the preamp, an AGC amp, an "MGC" (manual gain control) amplifier, and a post amplifier with accompanying output buffers. The overall voltage gain is 64 dB with an AGC range of 24 dB and a −3 dB bandwidth of 1.2 GHz.

Reimann and Rein (1987) report a "limiting amplifier" alternative to the conventional AGC configuration. The amplifier they describe is a high gain circuit capable of delivering a 400 mV output in response to input signal amplitudes ranging from 1 mV to 400 mV at data rates to 4 Gb/s. These results stem from the circuit's capability of operating with little output edge jitter in either a small signal, high gain linear mode, or in a large signal, low gain "limiting" mode. Reported results include a maximum voltage gain of 54 dB and an input dynamic range of 52 dB.

GaAs technology is represented in circuits reported by Imai *et al.* (1984). Utilizing a 0.7 µm gate-length self-aligned MESFET technology, the preamplifier circuit had a 22 dB gain and a bandwidth of 2.5 GHz. The complete amplifier, including AGC, was fabricated on three monolithic chips with a total gain of 45 dB, an AGC range of 13 dB and a 1.6 GHz bandwidth.

The retiming components follow the amplifiers in an optical receiver. These consist of a timing circuit used to regenerate the system clock and a decision circuit to distinguish between digital ONE and ZERO and restore what may be a badly distorted and noise-obscured signal. In the process, the data is retimed to minimize edge jitter. Clock recovery is complicated by the non-return-to-zero (NRZ) format, which has a spectral null at a frequency corresponding to the data rate. For NRZ data, therefore, an initial nonlinear process (e.g., frequency doubling) is required to generate spectral components at the clock rate. Two techniques for accomplishing this include differentiation followed by full wave rectification, and exclusive-or'ing of the data with a half-period delayed version of itself (for example, Sugiyama *et al.*, 1984, or Stevenson and Faulkner, 1985). The resulting signal is subsequently filtered with a high Q bandpass system, typically an external surface acoustic wave (SAW) filter, before final amplification (see Rosenberg, 1984).

Monolithic integration of the clock recovery function can be accomplished using phase-locking feedback loops (PLL). For example, Hogge (1985) and Bambach and Schwaderer (1985) have reported partially integrable PLL clock recovery circuits operating to 565 Mb/s. Hogge's technique was not frequency locking, however, and required an initial frequency acquisition circuit or quartz crystal controlled oscillator. The otherwise monolithic circuit of Bambach and Schwaderer utilized a thin film voltage-controlled oscillator. Although completely monolithic PLLs have not yet been demonstrated at data rates much above 500 Mb/s, suitable individual components have been reported, and it seems likely that complete high frequency PLLs will soon appear in the literature.

Decision circuits are more easily integrated. A typical decision circuit consists of an input amplifier followed by a D-type flipflop. The regenerative (positive feedback) latching action of the flipflop causes a signal deviating very slightly from the decision threshold to be restored to a full digital ONE or ZERO. The window of logic value uncertainty around the decision threshold is referred to as the "decision threshold ambiguity width" (DTAW), and in general, the smaller it is, the better. The decision flipflop is most often implemented in an emitter-coupled logic (ECL) configuration or its GaAs equivalent. GaAs decision circuits have been reported by O'Connor *et al.* (1984) and by Ohta and Takada (1983). O'Connor *et al.* describe a circuit implemented in 1 μm MESFET technology consisting of an input amplifier, a 6-NOR D-type flipflop and output buffer with demonstrated DTAW less than 20 mV at 2 Gb/s. The GaAs "source-coupled FET logic" (SCFL) decision circuit reported by Ohta and Takada also functioned to at least 2 Gb/s with a phase margin of 150°, although with larger DTAW (250 mV) because no input amplifier was integrated with the chip. For both these circuits, test speeds were limited by measurement equipment, and operation at greater than 2 Gb/s was achievable.

Bipolar silicon alternatives are reported by Suzuki *et al.* (1984), and by Suzuki and Hagimoto (1985a). In the earlier reference, a decision circuit implemented in 0.5 μm emitter SST technology is reported. It utilizes an input amplifier, a single D-type latch regenerative circuit (as opposed to the conventional edge-triggered flipflop), and an output buffer. Operation at 2 Gb/s was obtained with a phase margin of 270° and a DTAW of less than 30 mV at that frequency. Output waveforms indicated probable operation to greater than 3.6 Gb/s. The later reference describes an alternative version using a full master-slave D flipflop implementation and fabricated in a more readily manufacturable 0.75 μm emitter version of the SST technology. This circuit demonstrated 250° phase margin and less than 10 mV DTAW at frequencies to 2.1 Gb/s.

The above circuits employ clocked D-type flipflops operating at a frequency equal to the data rate; for example, 2 Gb/s NRZ data requires 2 GHz internal clocking of the flipflop. So-called "master/slave" flipflops operate in a two-phase cycle. During the first phase, flipflops "acquire" the input data by internally amplifying the incoming data. During the second phase, the flipflop "latches" the data, i.e., it enters a regenerative mode that restores the acquired data to a full logic zero or one. During this latch phase, the flipflop ignores any new input data. Clawin and Langmann (1985) have pointed out that maximum decision circuit operating speed is often limited by the maximum toggle speed of the internal flipflop rather than by circuit output rise and fall times. In order to get around this limitation, they have proposed the architecture illustrated in Fig. 20.9. This circuit consists of two parallel master/slave flipflops, each composed of two D-type latches. The parallel decision circuits are clocked 180° out of phase with each other, and they operate on alternating bits so that while one flipflop is acquiring a new bit, the other flipflop is latching the previous bit. This effectively doubles the amount of time at any given bit rate allowed for each flipflop to regenerate the bit. The results are multiplexed together at the output using a 2:1 mux. Note that this technique is beneficial only in the NRZ format and only up to the frequency where rise/fall time limitations become dominant. The 2 Gb/s circuit reported by Clawin and Langmann was fabricated in a conventional 2.5 μm-feature silicon bipolar technology and demonstrated a 300° clock phase margin with a 500 mV input signal and 180° phase margin with a

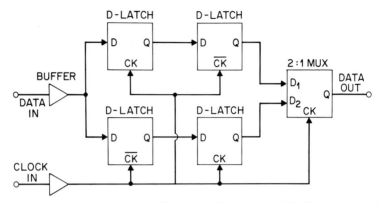

Fig. 20.9 Experimental retiming circuit (Clawin and Langmann, 1985). The upper and lower D-latch pairs each form independent decision circuits operating on alternating data bits to increase maximum circuit operating speed.

300 mV input, without benefit of an input amplifier. Their simulations indicate that 3 Gb/s operation may be obtainable in a more advanced technology now under development.

Concluding this section, we observe that with the exception of the timing recovery circuit, every other component necessary for a digital transmitter and repeater has been demonstrated in monolithic form at data rates exceeding 2 Gb/s. A major challenge for the future is clearly increased speed for even higher performance ICs. The next section takes a look at devices and technologies that are likely candidates for tomorrow's high speed optical communications systems.

20.3 HIGH PERFORMANCE IC TECHNOLOGIES: A CLOSER LOOK

The bandwidth of optical fiber is measured in terahertz, and purely from an optical standpoint, there is ample reason to suppose that 10 Gb/s, 100 Gb/s or even higher data rate systems will become common. Wavelength division multiplexing (WDM) may be one way to achieve these bandwidths. For applications other than WDM, however, it is likely that electronic component limitations will restrict the advent of such super performance systems. In this section we look more closely at today's state-of-the-art technologies, we examine some of the limitations of those technologies, and we briefly review the somewhat more exotic processes and devices that offer potential for significant advance in the future.

Today's premier high speed technologies in terms of performance, level of development and application are silicon bipolar, GaAs MESFET, and silicon NMOS. The dividing line between reality and speculation is about 6 Gb/s. As we have seen, the performance of GaAs MESFET and silicon bipolar is currently quite comparable in most repeater applications. Silicon bipolar integrated circuits have enjoyed the benefits of many years of concentrated development effort, including a virtual revolution in process technology since 1980. MESFET technology, on the other hand, offers potentially better performance from the standpoint of power dissipation and ultimate operating speed in certain applications. The third of today's competing technologies, "fine line" (very small feature) silicon NMOS, has been applied in some high performance systems. NMOS allows higher circuit complexity, although at somewhat lower maximum speeds than MESFET and bipolar integrated circuits. Furthermore, silicon CMOS, a close relative of NMOS, approaches an ideal technology for VLSI applications, where device packing density and power dissipation are paramount concerns. In the near future we may expect CMOS to find favor in medium speed (< 1 Gb/s) systems, while GaAs modulation-doped FETs and heterojunction bipolar transistors (HBT) become more prominent at the highest data rates. Other future directions include optoelectronic integration (integration of optical and electronic devices on a single monolithic chip), monochip (one-chip) repeaters, and monolithic integration of silicon and gallium arsenide to realize the advantages of both on a single chip.

Figures 20.10 through 20.12 show cross-sectional diagrams of modern silicon bipolar, GaAs MESFET, and silicon NMOS devices. The principles of oper-

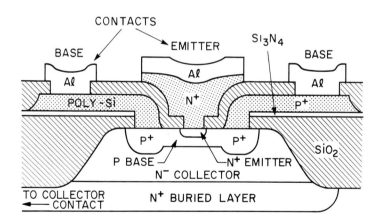

Fig. 20.10 Cross-sectional diagram of super-self-aligned bipolar transistor structure where definition of emitter, base, contacts and metallization is determined by one photolithography step (Konaka *et al.*, 1986).

ation of FET and bipolar transistors are quite dissimilar, as are their strengths and weaknesses as circuit elements. When making speed comparisons between these different device structures, simplistic assumptions based on individual device characteristics will often yield incomplete or even erroneous conclusions. *The device performance cannot be weighed outside of the circuit context in which it is found.*

The ultimate speed of a bipolar device is determined by the minority carrier transit time across the base and base-collector depletion layers, along with the time required to charge and to discharge major device capacitances such as the base diffusion and base-collector depletion capacitances and various parasitics (see Tang and Yu, 1984). A cross-sectional diagram of a state-of-the-art silicon bipolar SST transistor (Konaka *et al.*, 1986) is shown in Fig. 20.10. This device is exemplary of the revolution in bipolar transistor device design that began in the early 1980s and continues apace (see Ning and Tang, 1986 for a general review). The designation "super self-aligned" is applied to this device because the emitter and base geometry, including contact openings and metallization, are defined in one photolithographic step, allowing very small emitter and extrinsic base structures (< 0.5 μm) using comparatively relaxed photolithography. Bipolar technology has benefited from the same photolithographic advances that drive VLSI technology, so that typical minimum mask feature sizes are now approximately 1 μm. Such smaller features along with higher base doping result in a reduction of the extrinsic r_{bx} and intrinsic r_{bi} base resistances. Higher base doping is accommodated without degradation of the forward current gain by use of a polysilicon contact to the emitter region. Carefully controlled ion-implantation results in a very narrow base width (< 100 nm), which in turn minimizes the base diffusion capacitance C_D and base transit time τ_F. Oxide device isolation reduces the collector-substrate capacitance (C_{csub}) and increases device packing density, while collector resistance is minimized by a buried layer or "subcollector" in conjunction with thin epitaxy technology. Unity current gain frequencies (f_T) for this device and similar silicon bipolar transistors have been reported at 17 GHz (Konaka *et al.*, 1986; Park *et al.*, 1986) to 25 GHz (Washio *et al.*, 1987). Digital dividers operating to 10.3 GHz have also been reported (see Suzuki *et al.*, 1985b; Sakai *et al.*, 1985).

Forward current is exponentially proportional to the base-emitter voltage; therefore, the bipolar structure offers exceptionally high transconductance at any particular bias current in comparison with "square law" field effect transistors such as MOSFETs and MESFETs. This is offset by the much higher capacitance of a bipolar device arising from charge storage in the base. From a circuit standpoint, an NPN bipolar transistor process offers the designer high transconductance for driving internal and external capacitive loads, "enhancement-mode" (normally off) operation with well-controlled threshold voltage ($V_T = kT/q$) and high output impedance leading to a high gain per amplifier stage. Disadvantages of bipolar technologies include higher complexity com-

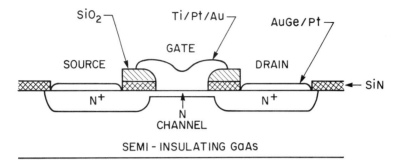

Fig. 20.11 Self-aligned gate, N-channel GaAs MESFET transistor structure with minimized extrinsic source resistance (Ishii *et al.*, 1984).

pared with an FET device in the same material system, relatively high power dissipation, and larger overall area.

A representative n-channel MESFET is illustrated in Fig. 20.11. Ultimate operating speed is determined by electron transit time across the gate-controlled channel region and by charging time (RC delay) associated primarily with the gate-source capacitance, but also with other parasitics. As one would expect, channel length is critically important in maximum operating speed and is currently about 1 μm for most technologies. Devices with channel lengths of 0.5 μm and below have reported f_t's to greater than 45 GHz (Kim *et al.*, 1985). The device in Fig. 20.11 utilizes self-aligned features to pattern the gate metallization and source/drain regions with a single photolithographic step, while ion implantation is used for the active layer and source drain contact regions (Ishii *et al.*, 1984). In analogy with silicon gate MOSFETs, this technology provides devices with well-controlled gate-to-source/drain overlap, minimized extrinsic source resistance, and good packing density for VLSI applications. Drain/source junction capacitance is reduced in comparison with silicon devices by the presence of the semi-insulating GaAs substrate. This factor, along with gallium arsenide's higher electron mobility and correspondingly greater device transconductance (g_m) results in faster operating speeds than similar (FET) structures fabricated in silicon. MESFET digital dividers operating to 17.9 GHz (Jensen *et al.*, 1986) have been reported.

When comparing MESFETs with silicon MOSFETs, GaAs enjoys other advantages as well, but comparisons with silicon bipolar are complicated considerably by the diversity of the strengths of these technologies. To the IC designer, modern GaAs MESFET technologies offer very high speed, low power, low voltage operating capabilities. The major limitations are related primarily to the comparatively immature state of the technology and include limited device threshold matching, restricted circuit complexity because of low device yield, and high cost. It can be projected that these limitations will be much less constraining in the future. Another limitation at the present time is the lack of a high performance, complementary P-channel device in GaAs. This precludes CMOS type device structures with their accompanying power advantages in VLSI applications.

Figure 20.12 shows a cross section of a high performance silicon N-channel MOSFET (e.g., Ko *et al.*, 1983). A MOSFET is very similar in topology to a MESFET except for the use of a dielectrically insulated gate (SiO_2) instead of a Schottky gate, and a conducting silicon substrate rather than a semi-insulating GaAs substrate. This particular device also uses a self-aligned gate structure, and a variety of process innovations including a sidewall gate oxide spacer for minimized gate-drain overlap (Miller) capacitance. The f_T of a silicon device of this type with an effective channel length of 1 μm is about 8 GHz, substantially less than that of a MESFET with comparable dimensions. Very short channel NMOS devices with f_T of 17 GHz have been reported along with digital dividers operating at 3.1 GHz (Fichtner *et al.*, 1985).

CMOS (complementary MOS) includes both P-channel and N-channel devices on the same circuit. It is normally slower than NMOS because of extra capacitance arising from device isolation and the added P-channel gate capacitance, and thus has had little application to date in the highest performance optical systems where VLSI-level functionality has so far been unneeded.

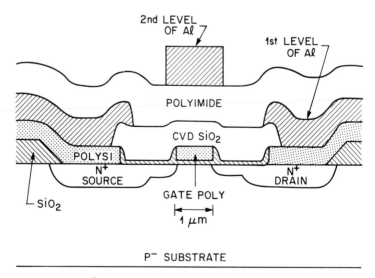

Fig. 20.12 Self-aligned poly-gate N-channel silicon MOSFET transistor structure with oxide sidewall spacer, poly source/drain contacts and minimized parasitic capacitance (Ko *et al.*, 1983).

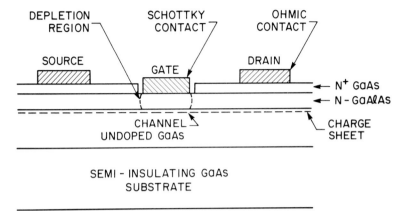

Fig. 20.13 GaAs/GaAlAs selectively doped heterostructure transistor structure.

Two other devices offering promise for the future are illustrated in Fig. 20.13 and 20.14. These are the "selectively doped heterostructure transistor" (SDHT) and the "heterojunction bipolar transistor" (HBT). The SDHT is known by a variety of other names including MODFET, HEMT and TEGFET. A comprehensive description of the operating principles of this device and its performance potential is given by Solomon and Morkoc (1984). (This review is also recommended as a general device-oriented performance comparison for various technologies.) The principles of transistor action in a SDHT are very similar to those of a conventional GaAs MESFET; that is, current flow is along a gate-con-

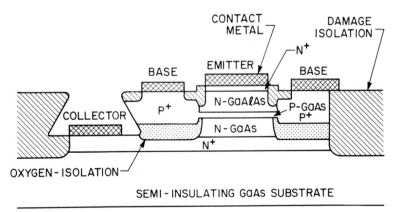

Fig. 20.14 Device cross section of a GaAs/GaAlAs heterojunction bipolar transistor (Asbeck and Miller, 1984).

trolled conducting channel between the source and drain. In a MESFET, conduction electrons are supplied by donor atoms physically located within the channel region. A consequence is that electric fields from these immobile, but ionized donor atoms interfere with electron transport through the channel by an interaction mechanism referred to as ionized impurity scattering. This results in a reduction in electron mobility and device transconductance, and ultimately in lower circuit operating speed. The SDHT circumvents this problem by physically separating the donor atoms from the conducting channel. In the device illustrated in Fig. 20.13, a doped, donor layer of AlGaAs is grown on top of an undoped GaAs layer. Electrons diffuse from the AlGaAs layer into a thin sheet (referred to as a "two dimensional electron gas") at the surface of the GaAs layer. This charge sheet, in analogy with the inversion layer of a silicon MOSFET, forms the conducting channel. Ionized impurity scattering is minimized, and the principal remaining source of electron scattering is that due to phonon (lattice) interactions, an effect much less deleterious at reduced temperatures. For that reason, liquid nitrogen-cooled SDHT circuits operating at $77°K$ have demonstrated particularly impressive results.

SDHT technology is not as well-developed as that of MESFETs, yet test circuits of considerable complexity, including a 4K bit static memory (Kuroda et al., 1984) have been reported. Compared to silicon MOSFETs, the SDHT offers the previously noted advantages of higher device transconductance and lower capacitance. In comparison with enhancement-type MESFETs, the SDHT offers higher mobility at lower forward voltages, although the depletion-type MESFET has comparable device transconductance at room temperature. At $77°K$, however, SDHT performance is unapproachable from the combined standpoint of speed and power, and for ultra-short channel devices, the SDHT may find favor even at room temperature. Devices with an f_T of 70 GHz have been reported (Mishra et al., 1985), as have digital dividers operating at 13 GHz at $77°K$ (Hendel et al., 1984). SDHT transconductance is low in comparison with bipolar devices, but its substantially smaller capacitance promises to make it a prime contender with bipolar in small-scale, ultra-high speed circuits. Furthermore, its much lower power dissipation at high operating speeds makes it potentially far superior to silicon bipolar devices in VLSI applications.

The other device currently of interest for future high speed applications is the HBT, the gallium arsenide equivalent of the silicon bipolar transistor. The cross section in Fig. 20.14 is that described by Asbeck and Miller (1984). The collector is n-type GaAs, the base is p-type GaAs and the emitter is n-type GaAlAs. The wide bandgap GaAlAs emitter creates an energy barrier to hole injection from the base into the emitter. This allows the transistor to have high current gain even when the base is much more highly doped than is permissible in a silicon bipolar device. In turn, this highly doped base gives reduced base resistance, a narrower allowable base, and minimized high electron injection (e.g., Kirk) effects. All of these factors promote higher speed operation. Furthermore, in comparison with similar silicon devices, the higher electron mobility of the GaAs base region results in reduced base transit time, and therefore, lower diffusion capacitance. This, plus a low capacitance, semi-insulating GaAs substrate translates directly to increased circuit speed.

The state of development of the HBT is less advanced than is that of the SDHT. Particular problems remain with excess carrier recombination and with device planarity (especially important in IC applications). Nonetheless, preliminary results are impressive, with device f_Ts measured to 45 GHz (Nakajima et al., 1986; Madihian et al., 1986) and digital divider circuits operating to 11 GHz (Asbeck et al., 1986).

This section so far has considered performance only with respect to pure device speed. Within a circuit context, however, the situation is considerably more complex, and a variety of factors determine ultimate circuit speed, including:

- unity gain frequency (f_T),
- minimum voltage needed for maximum f_T,
- device transconductance,
- power limitations,
- threshold control,
- magnitude and control of parameter variation with temperature, and
- circuit flexibility.

Circuit flexibility factors include:

- number of available thresholds, enhancement and depletion mode,
- availability of special circuit configurations; e.g., stacked logic, charge storage, transfer gates,
- availability of complementary devices (N-channel/P-channel, NPN/PNP), and
- output impedance: peak gain per stage.

Different technologies excel in different aspects. For example, GaAs FET technologies have the highest device f_Ts and far surpass silicon devices in that respect. Bipolar technology remains very competitive in high speed circuit applications, however, because of its higher transconductance (translating to a lesser sensitivity to fanout and capacitive loading), inherently better threshold control and suitability for stacked (current-steered) logic applications. Silicon MOS is also competitive in high complexity LSI or VLSI applications. The suitability of a particular technology for high speed applications is often evaluated by means of the ring oscillator. This circuit, while helpful in evaluating device f_T, fails to adequately characterize other important device and circuit parameters. Unfortunately, there is no single test capable of such determination, but as a complement to the usual technology comparisons, best reported digital divider operating speeds as well as f_T have been noted throughout this section. We emphasize that these benchmarks only define a starting point for circuit designers who in practical applications must design for a range of device and environmental parameters. The control of such parameters and the degree to which their variation is predictable in advance will often set limits on ultimate circuit performance.

20.4 CONCLUSIONS

High performance IC technology research is being carried out at locations all around the world. Given the demanding requirements of optical communications, it is understandable that circuits for fiber systems are frequently selected as technology demonstrations. The fastest reported integrated components function at data rates of 5 to 6 Gb/s, and it seems probable that complete systems operating at these speeds will eventually appear. Furthermore, fundamental device limits have not yet been approached, and additional speed improvements of factors of 5 or more may be ultimately achievable. One or more of the technologies described in Section 20.3 will likely be the vehicle used to reach these barely envisioned data rates. Whether that technology be bipolar or FET, silicon or GaAs, or something altogether different is difficult to predict. Moreover, that choice depends on the particular component and speed range involved. When comparing GaAs FETs with silicon FETs, and HBTs with silicon bipolar junction transistors, gallium arsenide has undeniable material advantages but has suffered principally from an inadequate state of technological development. As time goes on, it appears very likely that this technology will consolidate its material advantages in ever more complex circuits. The silicon art is not expected to stand still, however, and the lower costs and greater scale of integration in silicon may move the GaAs circuit niche to higher and higher speeds. Fortunately, as a principal customer of high performance integrated circuits, optical communications stands to benefit however the issues are resolved.

ACKNOWLEDGEMENTS

The author wishes to acknowledge and thank J. L. Hokanson, C. B. Swan, P. S. Henry, and S. Lumish for their assistance in reviewing and contributing to this manuscript. He also thanks F. S. Chen for supplying the photo used in Fig. 20.6.

REFERENCES

Abidi, A. A. (1984). *IEEE J. Solid-State Circuits* **SC-19**, 986–994.

Aiki, M., Tsuchiya, T., and Amemiya, M. (1985). *J. Lightwave Technology* **LT-3**, 392–399.

Alferness, R. C. (1982). *IEEE Trans. Microwave Theory and Techniques* **MTT-30**, 1121–1137.

Asbeck, P. M., Miller, D. L., Anderson, R. J., Deming, R. N., Chen, R. T., Liechti, C. A., and Eisen, F. H. (1984). *GaAs IC Symposium Technical Digest*, 133–136.

Asbeck, P. M., Chang, M. F., Wang, K. C., Sullivan, G. J., and Miller, D. L. (1986). *Bipolar Circuits and Technology Meeting Proceedings*, 25–26.

Bambach, W., and Schwaderer, B. (1985). *Telecommunications* **19**, 69–72.

Bayruns, R. J., Hofstatter, E. A., and Weston, H. T. (1986). *Intl. Solid-State Circ. Conf. Digest of Tech. Papers*, 192–193.

Chen, F. S. and Bosch, F. (1986). *Electronics Letts.* **22**, 932–933.

Clawin, D. and Langmann, U. (1985). *Intl. Solid-State Circ. Conf. Digest of Tech. Papers*, 222–223.

Fichtner, W., Hofstatter, E. A., Watts, R. K., Bayruns, R. J., Bechtold, P. F., Johnston, R. L., and Boulin, D. M. (1985). *Intl. Electron Device Meeting Technical Digest*, 264–267.

Hendel, R. H., Pei, S. S., Tu, C. W., Roman, B. J., Shah, N. J., and Dingle, R. (1984). *Intl. Electron Device Meeting Technical Digest*, 857–858.

Hogge, C. R. (1985). *J. Lightwave Technology* **LT-3**, 1312–1314.

Imai, Y., Kato, N., Ohwada, K., and Sugeta, T. (1984). *IEEE Electron Device Letts.* **EDL-5**, 415–416.

Ishii, Y., Ino, M., Idda, M., Hirayama, M., and Ohmori, M. (1984). *GaAs IC Symposium Technical Digest*, 121–124.

Jensen, J. F., Salmon, L. G., Deakin, D. S., and Delaney, M. J. (1986). *Intl. Electron Device Meeting Technical Digest*, 476–479.

Kim, B., Tserng, H. Q., and Shih, H. D. (1985). *IEEE Electron Device Letts.* **EDL-6**, 1–2.

Ko, P. K., Voshchenkov, A. M., Hanson, R. C., Grabbe, P., Tennant, D. M., Archer, V. D., Chin, G. M., Lau, M., Soo, D. C., and Wooley, B. A. (1983). *Intl. Electron Device Meeting Technical Digest*, 751–753.

Konaka, S., Yamamoto, Y., and Sakai, T. (1986). *IEEE Trans. Electron Devices* **ED-33**, 526–531.

Kuroda, S., Mimura, T., Suzuki, M., Kobayashi, N., Nishiuchi, K., Shibatomi, A., Abe, M. (1984). *GaAs IC Symposium Technical Digest*, 125–128.

Liechti, C. A., Baldwin, G. L., Gowen, E., Joly, R., Namjoo, M., and Podell, A. F. (1982). *IEEE Trans. Electron Devices* **ED-29**, 1094–1102.

Madihian, M., Honjo, K., Toyoshima, H., and Kumashiro, S. (1986) *Intl. Electron Device Meeting Technical Digest*, 270–273.

Mishra, U. K., Palmateer, S. C., Chao, P. C., Smith, P. M., and Hwang, J. C. M. (1985). *IEEE Electron Device Letts.* **EDL-6**, 142–145.

Mudd, M. S. J., Taylor, D. G., Wood, I. C., Childs, J. C., and Saul, P. H. (1985). *IEEE J. Solid-State Circuits* **SC-20**, 708–714.

Nakajima, O., Nagata, K., Yamaguchi, Y., Ito, H., and Ishibashi, T. (1986). *Intl. Electron Device Meeting Technical Digest*, 266–269.

Ning, T. H. and Tang, D. D. (1986). *Proceedings of the IEEE* **74**, 1669–1677.

O'Connor, P., Flahive, P. G., Clemetson, W. J., Panock, R. L., Wemple, S. H., Shunk, S. C., and Takahashi, D. P. (1984). *Conf. Optical Fiber Communication, Digest Tech. Papers*, 26–27.

Ohara, M., Akazawa, Y., Ishihara, N., and Konaka, S. (1985). *IEEE J. Solid-State Circuits* **SC-20**, 703–707.

Ohta, N. and Takada, T. (1983). *Electronics Letts.* **19**, 983–985.

Park, H. K., Boyer, K., Tang, A., Clawson, C., Yu, S., Yamaguchi, T., and Sachitano, J. (1986). *Bipolar Circuits and Technology Meeting Proceedings*, 39–40.

Reimann, R. and Rein, H. M. (1986). *IEEE J. Solid-State Circuits* **SC-21**, 785–789.

Reimann, R. and Rein, H. M. (1987). *Intl. Solid-State Circuits Conf. Digest Technical Papers*, 172–173.

Rosenberg, R. L., Chamzas, C., and Fishman, D. A. (1984). *J. Lightwave Technology* **LT-2**, 917–925.

Ross, D. G., Moyer, S. F., Eckton, W. H., Paski, R. M., and Ehrenberg, D. G. (1986). *IEEE J. Solid-State Circuits* **SC-21**, 331–336.

Sakai, T., Konaka, S., Yamamoto, Y., and Suzuki, M. (1985). *Intl. Electron Device Meeting Technical Digest*, 18–21.

Solomon, P. M. and Morkoc, H. (1984). *IEEE Trans. Electron Devices* **ED-31**, 1015–1027.

Stevenson, A. and Faulkner, D. W. (1985). *IEE Proceedings* **132**, 62–67.

Sugiyama, M., Higo, Y., and Iwakami, T. (1984). *IEEE Global Telecomm. Conf. Record*, 1139–1144.

Suzuki, M., Hagimoto, K., Ichino, H., and Konaka, S. (1984). *IEEE J. Solid-State Circuits* **SC-19**, 462–467.

Suzuki, M. and Hagimoto, K. (1985a). *Electronics Letts.* **21**, 844–846.

Suzuki, M., Hagimoto, K., Ichino, H. and Konaka, S. (1985b). *IEEE Electron Device Letts.* **EDL-6**, 181–183.

Swartz, R. G., Voshchenkov, A. M., Chin, G. M., Finegan, S. N., Lau, M. Y., Morris, M. D., Archer, V. D., and Ko, P. K. (1986). *Intl. Solid-State Circuits Conf. Digest Technical Papers*, 64–65.

Takada, T., Kato, N., and Ida, M. (1986). *IEEE Electron Device Letts.* **EDL-7**, 47–48.

Tang, D. D. and Yu, H. N. (1984). *Proceedings Intl. Electronic Devices and Materials Symposium*, Hsinchu, Taiwan, 199–205.

Ueda, D. Shimano, A., Otsuki, T., Kaneko, H., Tamura, M., Takagi, H., Kano, G., and Teramoto, I. (1985). *GaAs IC Symposium Technical Digest*, 103–106.

Washio, K., Nakamura, T., Nakazato, K., and Hayashida, T. (1987). *Intl. Solid-State Circuits Conf. Digest Technical Papers*, 58–59.

Williams, G. F. and LeBlanc, H. P. (1986). *J. Lightwave Technology*, **LT-4**, 1502–1508.

Yoshikai, N., Kawanishi, S., Suzuki, M., and Konaka, S. (1985). *Electronics Letts.* **21**, 149–151.

Yoshikai, N., Yamada, J.-I., Suzuki, M., and Konaka, S. (1986). *Electronics Letts.* **22**, 39–41.

Chapter 21

Introduction to Lightwave Systems

PAUL S. HENRY

AT&TBell Laboratories, Inc., Holmdel, New Jersey

R. A. LINKE

AT&TBell Laboratories, Inc., Holmdel, New Jersey

A. H. GNAUCK

AT&TBell Laboratories, Inc., Holmdel, New Jersey

21.1 INTRODUCTION

21.1.1 Background

The extraordinary advances in fiber and semiconductor technology described in the preceding chapters have resulted in exponential growth in the performance of lightwave communications systems. This progress is illustrated in Fig. 21.1, which shows the advance of laboratory lightwave demonstrations measured in terms of the product of bit-rate times unrepeated transmission distance for a single-wavelength (that is, non-wavelength-multiplexed) channel. The past decade shows a growth rate of roughly a factor of two per year. Today optical fibers can carry higher data rates over greater repeater distances than has ever been possible with any other transmission medium (Li, 1983; Suematsu 1983; Miller, 1979, Chs. 1 and 21; Garrett and Todd, 1982). In addition to progress in the laboratory, lightwave technology has enjoyed extremely rapid translation from basic concept to commercial application. Since the installation of a few modest circuits under the streets of Chicago in 1977, commercial lightwave systems have proliferated rapidly and now span large portions of North America, Europe and

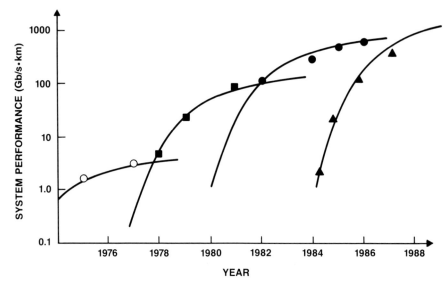

Fig. 21.1 Progress in Lightwave Transmission. The points correspond to benchmark laboratory demonstrations in the indicated year:

○ λ = 0.8 μm, multimode fiber,
■ λ = 1.3 μm, single-mode fiber,
● λ = 1.55 μm, single-mode fiber,
▲ λ = 1.55 μm, coherent detection.

The curves illustrate the evolution of each of the four generations of lightwave systems discussed in this chapter. They suggest that the sustained growth in performance derives primarily from the introduction of new technologies.

Japan (Schwartz *et al.*, 1978). The first transoceanic fiber system, TAT-8, links the continents of Europe and North America (Runge and Trischitta, 1984).

Despite its rapid growth and overwhelming commercial success, lightwave is not a mature technology—we have barely scratched the surface in terms of approaching the information-carrying capability of a fiber. Perhaps even more important, we have only just begun to investigate the application of fibers in a potentially huge market—photonic networks. Thus far, research and commercial development have focused mainly on trunk transmission technology, where fiber is used as a high-capacity conduit to carry large bundles of traffic between telephone switching offices. Only rarely does a fiber extend from a switch to an end-user. For this latter application conventional copper wire has proved to be more cost-effective than fiber. But as lightwave technology advances and costs come down, it becomes tempting to consider all-fiber communication systems, where fibers interconnect end-users directly. In these systems, lightwave technology is used not just for transmission, as in trunk systems, but also for networking.

Serving end-users directly with fiber opens up the possibility of economical wideband services (Linnell, 1986). Applications along these lines include local area networks (LANs), subscriber loops and TV distribution. This field of photonic networks, described in more detail in Chapter 26, is still evolving. Of course, the basic communications engineering principles are the same as for trunk transmission systems, but the practical constraints and possible applications are poorly defined, so our understanding is still sketchy. Nonetheless, the potential impact of photonic networks is huge. Lightwave to every home is a tremendous commercial opportunity; it is none too soon to start development of a conceptual framework for this burgeoning field.

This chapter is an introduction to the basic principles of optical fiber communications systems. Rather than study lightwave systems from a global point of view, where the goal is to elucidate a small number of very general principles, our approach will be generally historical—the development of lightwave systems is viewed in terms of evolution through four generations of technology, characterized primarily by the system operating wavelength. Most of the fundamentals are developed in the context of trunk transmission systems rather than photonic networks for two reasons: the advance of technology has been driven mainly by these applications; and explanation and analysis are much more straightforward for trunk systems. Because of unavoidable complexities and ambiguities, our discussion of photonic networks is postponed until the end of the chapter.

21.1.2 Basic Lighwave Link

The simplest lightwave communications system, shown in Fig. 21.2, consists of a transmitter, fiber transmission medium and receiver. The transmitter converts incoming binary data to on-off light pulses, which are launched into the fiber. (In Section 21.5 we will discuss more sophisticated modulation techniques.) At the receiver, the optical stream is detected and converted back into electrical

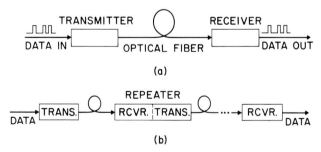

Fig. 21.2 (a) Digital lightwave transmission link. (b) Lightwave repeatered line composed of a cascade of simple links.

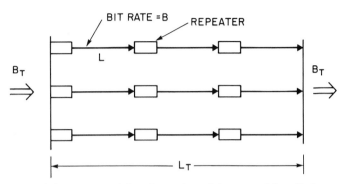

Fig. 21.3 Transmission system consisting of several parallel repeated lines. Each repeater section operates at bit-rate B over a fiber of length L.

signals. The primary measure of performance of a digital system is the bit-error rate (BER), which is the probability that an error will be made in the detection of a received bit. The maximum tolerable BER depends on the system application but for our purposes, we may use BER $\leq 10^{-9}$ as the criterion for satisfactory performance.

A useful measure of the capability of a given lightwave technology, especially in trunk applications, is the $B \cdot L$ product, where B is the bit-rate of the system and L is the length of the fiber in a single repeater section, as shown in Fig. 21.2. This simple figure of merit is consistent with the intuitive notion that both bit rate and distance are important contributors to the overall performance measure of a system. Consider Fig. 21.3, which depicts a lightwave system consisting of several repeated lines in parallel. Let the objective be to transmit a total bit-rate of $B_T \gg B$ over a distance $L_T \gg L$, where B and L are the bit-rate and fiber length of each individual repeater section. The total number of repeaters required for the system is:

$$N = \frac{B_T \cdot L_T}{B \cdot L}. \tag{21.1}$$

Thus, to minimize the number of repeaters in a system, which are frequently the most costly components, we want to use the technology that provides the greatest $B \cdot L$. Equation 21.1 shows that B and L are equally important in determining the "box count" of Fig. 21.3. We will discover, however, that with today's technology the most attractive route for maximizing $B \cdot L$ is to operate at the highest possible bit-rate, rather than strive for ultra-long repeater sections.

21.1.3 Fiber Attenuation

Except for perhaps very short links, such as might be found in intrabuilding networks, the most important consideration in lightwave system performance is fiber loss. As an optical signal propagates along a fiber, it is attenuated exponentially. The power level at a distance l kilometers from the transmitter is

$$P(l) = P_T \cdot 10^{-Al/10}, \tag{21.2}$$

where P_T is the power launched by the transmitter into the fiber and A is the attenuation constant of the fiber expressed in dB/km. It is the extremely low attenuation of optical fibers (of order 1 dB/km), coupled with their small size (~ 0.1 mm diameter), which makes them so attractive for communications purposes. The development of practical techniques for making fibers with losses only slightly greater than the theoretical minimum represents an extraordinary engineering achievement. Some early optical fibers, which were used to provide illumination in confined areas, had attenuations in the neighborhood of 5,000 dB/km (Strong, 1958). Even as recently as 1971, an attenuation of 20 dB/km was the best that could be achieved (Kapron et al., 1970). Today, fiber loss close to 0.2 dB/km is commonplace. Figure 21.4 shows the attenuation of today's high-quality (though not necessarily "champion") fused silica fibers for wavelengths between 0.8 μm and 1.8 μm, which is the region of interest for communications systems (Nagel et al., 1982; Suematsu, 1982, Ch. 3.1; Irven and Harrison, 1981). The discontinuity in the figure at 1.1 μm wavelength reflects the fact that the diagram represents losses for two different types of fiber in the wavelength regions where they are most commonly used: multimode for the shorter wavelengths and single-mode for the longer. The difference between the fibers, and the reasons for preferring one or the other, will become clear later on.

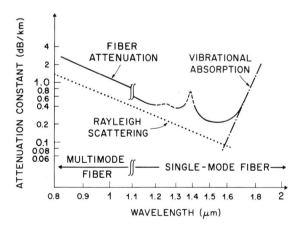

Fig. 21.4 Attenuation constant as a function of wavelength for high-quality optical fiber. Fiber loss, which is dominated by Rayleigh scattering and vibrational absorption, reaches a minimum in the vicinity of $\lambda = 1.55$ μm. OH absorption peaks are shown by dashes.

The fundamental processes that cause attenuation in optical fibers are Rayleigh scattering, which dominates at the shorter wavelengths, and vibrational absorption in silica (and other fiber constituents), which is important at longer wavelengths. The lower limit to Rayleigh scattering is set by random density fluctuations "frozen" into the fused silica during manufacture; these lead to a theoretical fiber attenuation shown by the dotted line in Fig. 21.4:

$$A \sim \frac{0.6}{\lambda^4} \text{ dB/km}, \qquad (21.3)$$

where λ is the free-space wavelength in micrometers (Okoshi, 1982). (Rayleigh scattering for multimode fibers is actually somewhat higher ($\sim 50\%$) than that given by Eq. 21.3, because of scattering from the enhanced fiber dopants.) The silica vibrational absorption peak is at $\lambda \sim 9 \ \mu\text{m}$: what is shown by the chained line in Fig. 21.4 is the short-wavelength tail of this resonance (Midwinter, 1979, Chapt. 8; Garbuny, 1965). The magnitude of the attenuation is a sensitive function of fiber composition; the behavior shown in Fig. 21.4 represents typical fiber characteristics. The important lesson to be drawn from Fig. 21.4 is that Rayleigh scattering and vibrational absorption combine to produce a low-loss "window" in silica fiber. The absolute loss minimum is near 1.55 μm, where each of the loss mechanisms contributes ~ 0.1 dB/km to fiber attenuation.

The loss peaks shown by the dashed portions of the curve in Fig. 21.4 are due to absorption by OH ions. Although this loss mechanism is not intrinsic to optical fibers, the OH ion is ubiquitous, and removing it completely from the silica during fiber manufacture is extremely difficult. High-quality fiber now in production shows excess OH attenuation at 1.39 μm in the range 0.1–2 dB/km, corresponding to 2–40 parts per billion of OH by weight (Suematsu, 1982, Ch. 2.3; Murata and Inagaki, 1981; Stone and Lemaire, 1982).

Reducing fiber attenuation is important, because a lightwave receiver requires at least a minimum amount of optical signal power P_R in order to detect a transmitted bit with an acceptable error rate. The maximum length L over which a signal can be transmitted before it is too weak to be detected is obtained from Eq. 21.2:

$$L = \frac{10}{A} \log_{10} \frac{P_T}{P_R}. \qquad (21.4)$$

The length L is called the loss-limited transmission distance. Equation 21.4, as simple as it is, makes an important point: the loss-limited transmission distance depends strongly on fiber attenuation, but only weakly on transmitter and receiver power levels. For example, moving the operating wavelength from 0.8 μm to 1.55 μm reduces A by roughly a factor of 10 (see Fig. 21.4), and thereby increases L by the same factor. A similar improvement in P_T or P_R,

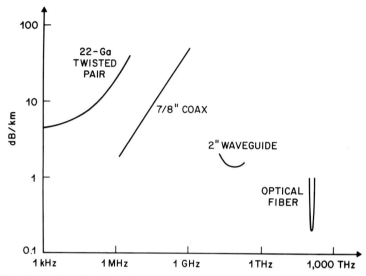

Fig. 21.5 Attenuation of various electromagnetic waveguiding media. Note that the logarithmic frequency scale obscures the enormous bandwidth advantage of optical fiber.

however, yields only a 25% improvement in L for typical system parameters. This disparity in payoff is worth remembering, especially when contrasted with the case of radio transmission, where signal attenuation obeys an inverse-square law, so a factor of 10 improvement in P_T or P_R yields more than a factor of 3 improvement in transmission distance.

The extremely low loss of optical fiber can be appreciated by comparing it with some ordinary metallic guiding media: 22-Ga. twisted pair, low-loss coaxial cable and 2-inch circular waveguide (TE_{01} mode), as shown in Fig. 21.5 (Suematsu, 1982; Irven and Harrison, 1981; Okoshi, 1982). Except possibly for wire lines at very low bit-rates (≤ 1 kb/s), optical fiber clearly shows the lowest loss. Not only is fiber superior in terms of loss, it is much smaller, making it generally easier to install, especially in existing ducts. The combination of low loss and small size is surprising at first; in metallic waveguides the general trend is toward lower loss as guide cross-section increases, since waveguide losses are dominated by resistive dissipation near the surface of the conductors, and bigger cross-sections lead to lower resistance. With fibers, on the other hand, the loss is a bulk effect in the dielectric medium, essentially independent of cross section, so fibers can be made very small without incurring added loss.

Before applying Eq. 21.4 to lightwave systems, let us gain some perspective by studying a different guided-wave medium–metallic coaxial cable. Typical parameters for a 100 Mb/s system might be $A = 20$ dB/km, $P_T = +30$ dBm (1 Watt) and $P_R = -75$ dBm, yielding $L = 5$ km (Reference Data, 1975).

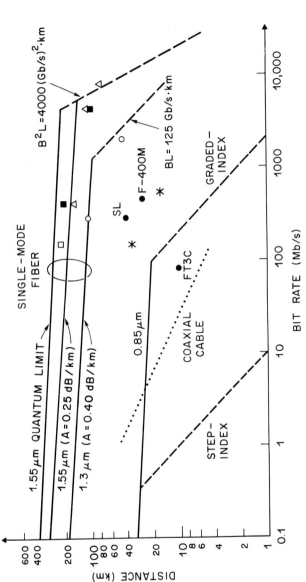

Fig. 21.6 Repeater spacing vs. bit rate for various lightwave technologies. The loss limits (solid lines) and dispersion limits (dashed lines) indicate approximately where transmission impairments become significant. The points correspond to actual achievements:

○ $\lambda = 1.3\ \mu m$,

* $\lambda = 1.3\ \mu m$, LED

□ $\lambda = 1.55\ \mu m$, controlled-dispersion fiber,

■ $\lambda = 1.55\ \mu m$, DFB laser,

△ $\lambda = 1.55\ \mu m$, C^3 laser,

● commercial systems.

The performance of coaxial cable (dotted line) is shown for comparison. The 1.55 μm Quantum Limit represents the performance that would be achieved with the ultimate lightwave receiver (ideal homodyne PSK) and a fiber with $A = 0.25$ dB/km.

Transmission length L as a function of bit rate is shown in Fig. 21.6, where it is assumed that $A \propto (\text{frequency})^{1/2}$. The transmission performance of coaxial cable depicted in Fig. 21.6 will be a useful standard against which to compare lightwave system performance to be calculated in later sections.

21.1.4 Fiber Dispersion

As the bit rate (or, more precisely, the symbol rate) is increased, dispersive effects in the fiber become more important. Fiber dispersion causes the transmitted pulses to spread and overlap as they propagate, so they become indistinct at the receiver, which leads to increased bit-error rate. The various causes of fiber dispersion will be discussed later; for now it is sufficient to consider the system shown in Fig. 21.7 and to recognize that the pulse broadening associated with dispersion manifests itself as a reduction in the input-output response of the system at high modulation frequencies. Assume an optical carrier intensity-modulated by a constant-amplitude sinewave applied to the transmitter input. As the modulation frequency is increased, the amplitude of the output signal at the receiver decreases, resulting in a low-pass system response. The deleterious effects of dispersion can be corrected after the receiver by using an equalizer to boost the attenuated high-frequency signal components. Along with the signal enhancement, however, comes an unavoidable increase in noise at the boosted frequencies, and hence a higher BER. By increasing the transmitter power, we can compensate for the enhanced noise and return the BER to the value it would have had in the absence of dispersion. The added transmitter power required to accomplish this is called the dispersion penalty, an important quantity in lightwave system design. Typically, systems operate in a regime where the dispersion penalty is small ($\lesssim 1$ dB). For this case the penalty is simply related to fiber dispersion, as we now show. Let a sufficiently narrow (in time) optical pulse be injected into a fiber; the time dependence of the pulse detected at the far end of the fiber is called $h(t)$, a readily measurable quantity.

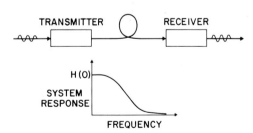

Fig. 21.7 Effect of fiber dispersion on end-to-end lightwave system response, $H(\omega)$. As dispersion increases, the bandwidth of $H(\omega)$ decreases.

The corresponding baseband frequency response is the Fourier transform of $h(t)$:

$$H(\omega) = \int_{-\infty}^{\infty} h(t) e^{j\omega t} dt. \tag{21.5}$$

For the simple case of modest dispersion and symmetric $h(t)$, a Taylor's series expansion of $h(t)$ yields

$$H(\omega) \sim H(0)\left\{1 - \frac{1}{2}\omega^2\sigma_t^2\right\}, \tag{21.6}$$

where σ_t^2 is the mean-square width of $h(t)$, sometimes called the mean-square impulse spread. As the frequency ω increases, the baseband frequency response decreases, as expected. Not surprisingly, the effect of this decrease on the error-rate performance of a digital system shows up most strongly in data sequences that have the greatest high-frequency content; that is, sequences that have lots of alternations between binary marks and spaces. Indeed, to estimate the effect of dispersion on error rate, we need only consider the "dotting" sequence: 101010.... Most of the power in this sequence is carried by the Fourier component at $\omega = \pi B$, where B is the bit-rate. To compensate for the attenuation of this component, the transmitter power must be increased by $[H(\pi B)]^{-1}$, so the dispersion penalty in dB is[1]

$$P_D \sim -10\log_{10} H(\pi B) \sim 21(\sigma_t B)^2. \tag{21.7}$$

As an approximate rule, we can say that the dispersion penalty is about 1 dB when the rms impulse spread is one-quarter of a bit period, and grows rapidly as pulse-width increases (Personick, 1973; Midwinter, 1979, Chapt. 14). For system design purposes we may use

$$\sigma_t B \leq \frac{1}{4} \tag{21.8}$$

as a condition for avoiding excessive dispersion penalty. We expect to see σ_t increase with fiber length, reflecting the fact that a pulse continues to spread as it propagates along the fiber. Thus, the condition given by Eq. 21.8 places a constraint on maximum fiber length, independent of the loss considerations discussed earlier. The maximum fiber length that satisfies Eq. 21.8 is called the dispersion-limited transmission distance. This limit, plus the loss limit defined earlier, are useful indices for comparing different lightwave systems.

The effects of dispersion can be clearly seen on an oscilloscope by means of an eye diagram, illustrated in Fig. 21.8 (Keiser, 1983). On the left of Fig. 21.8a is

---

[1]Our implicit assumption, that the receiver output signal is linear in the optical power input, is valid for most lightwave systems.

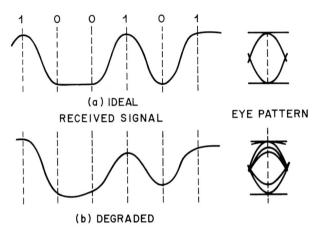

Fig. 21.8 Receiver output waveforms and corresponding eye diagrams: (a) Ideal case with no waveform distortion. (b) Distorted waveform and resulting eye closure.

an ideal received waveform, showing no dispersive degradation. The eye diagram, shown on the right, is generated by triggering the oscilloscope in synchronism with the bits of a random incoming data stream. The resulting display is a superposition of waveforms corresponding to a large number of different data sequences. In the ideal case (21.8a) the eye is fully open. When there is dispersion (21.8b), the waveform is distorted and the corresponding eye opening is reduced. The degree of eye closure is a convenient measure of system degradation.

21.2 FIRST-GENERATION LIGHTWAVE: $\lambda \sim 0.85$ μm

21.2.1 System Overview

The first generation of practical lightwave systems used AlGaAs lasers or light-emitting diodes (LEDs) as sources ($\lambda \sim 0.85$ μm), multimode fiber as the transmission medium, and silicon PINs or avalanche photodiodes (APDs) as detectors. Multimode fibers, though not capable of the lower-loss, low-dispersion performance of single-mode fibers, were (and are) extremely attractive because their large core-size (~ 50 μm vs. ~ 8 μm) makes fiber-splicing and coupling to an LED fairly easy.

We saw in Section 21.1 that fiber loss and dispersion place bounds on the maximum fiber length from transmitter to receiver. Let us now evaluate those limits for first-generation lightwave systems.

21.2.2 Loss Limit

A reasonable value for fiber attenuation at $\lambda = 0.85$ μm is $A = 2.5$ dB/km; this is slightly higher than the value shown in Fig. 21.4 to allow for the inevitable splicing losses that occur in practical systems. To evaluate the loss-limited transmission distance from Eq. 21.4, only crude estimates of P_T and P_R are needed, because the logarithm is such a weak function of its argument. For P_T a value of 1 milliwatt is reasonable. To estimate P_R, we need to know the bit rate, because higher-speed receivers need more optical power. The optical energy per bit, however, is roughly constant, independent of bit-rate. A Si APD receiver requires a photon flux of about 300 photons per bit to achieve BER $< 10^{-9}$; thus $P_R \sim 7 \times 10^{-14}$ B mW, where B is the bit-rate (Goell, 1974; Runge, 1974; Personick, 1981). The loss-limited transmission distance corresponding to these parameters is shown in Fig. 21.6 as a function of system bit-rate. Of course, a practical system design would allow some margin for component degradation, so installed distances would be somewhat shorter than those shown. For speeds above ~ 2 Mb/s optical fiber allows a longer loss-limited transmission distance than coaxial cable. The advantage of fiber grows as the bit rate is increased, because with cable the distance limit is approximately proportional to $B^{-1/2}$, whereas with fiber it decreases logarithmically with B.

The slow decrease in loss-limited transmission distance as B is increased suggests that to maximize the bit-rate distance product (Section 21.1.2), the system should be operated at the highest possible bit-rate. At high speeds, however, fiber dispersion becomes important. In the next section we estimate its effect.

21.2.3 Dispersion Limit

The simplest multimode optical fiber structure, called the step-index fiber, is shown in Fig. 21.9. The core, having index of refraction n, is surrounded by a cladding of lower index $n(1 - \Delta)$. The index step acts to confine optical energy to the fiber core. An accurate description of the propagating electromagnetic modes in this fiber requires solution of a complicated boundary-value problem.

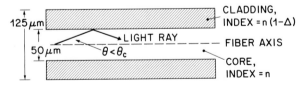

Fig. 21.9 Step-index optical fiber. The dimensions shown are for typical multimode fiber; for single-mode fiber the core diameter is much smaller, typically 8 μm.

For multimode fiber, fortunately, an approximate model based on the ray picture of geometrical optics is adequate for this discussion (Miller, 1979, Ch. 2). (Geometrical optics is applicable because we are considering a structure that supports many modes.) Light rays in the core, with directions of propagation within an angle $\theta \leq \theta_c = (2\Delta)^{1/2}$, are confined to the core by total internal reflection; rays at larger angles experience only partial reflection and are rapidly attenuated. The most important dispersive mechanism in this fiber, called modal dispersion, arises because τ, the time of flight for a ray in the fiber, is a function of θ:

$$\tau = \frac{Ln}{c \cdot \cos \theta}, \qquad 0 \leq \theta \leq \theta_c, \qquad (21.9)$$

where L is the length of the fiber and c is the free-space velocity of light. Thus, an impulse launched into the fiber will become smeared out as it propagates (unless by some mechanism the rays are confined to a negligibly small solid angle). The full width (i.e., the $2\sigma_t$ width) of the broadened impulse is approximately,

$$2\sigma_t \sim \frac{Ln}{c}\left[\frac{1}{\cos \theta_c} - 1\right] \sim \frac{Ln\Delta}{c}. \qquad (21.10)$$

Using this value of σ_t in Eq. 21.8 we find the dispersion-limited transmission distance to be

$$L \leq \frac{c}{2nB\Delta}. \qquad (21.11)$$

Equation 21.11 is plotted in Fig. 21.6 using typical parameters for step-index multimode fibers: $n = 1.46$ and $\Delta = 0.01$; the corresponding $B \cdot L$ product is ~ 10 Mb/s $\cdot$ km. For all but the lowest bit rates, modal dispersion is a much more severe limitation than fiber loss. If a way to circumvent this problem had not been found, multimode fibers would have remained an interesting, but impractical, transmission medium.

Orders-of-magnitude reduction in modal dispersion can be obtained by using a radially graded (rather than constant) index distribution in the fiber core (Gloge and Marcatili, 1973). Typically a parabolic profile is used, where the index is highest in the center of the core and gradually decreases toward the cladding. Thus, rays which stay near the core center and follow nearly straight paths, travel relatively slowly because they are in a region of high index. On the other hand, rays that make wide excursions toward the cladding traverse longer paths, but they travel faster because they spend time in lower-index material. By careful control of the index profile, the travel times for all rays can be made nearly equal resulting in an rms pulse spreading of

$$\sigma_t \sim \frac{Ln}{Bc} \Delta^2, \qquad (21.12)$$

where n is the index at the center of the core. The graded-index fiber shows a pulse spreading, which is quadratic in Δ, rather than the linear behavior shown in Eq. 21.10 for the step-index fiber. Combining Eqs. 21.8 and 21.12, we have the dispersion limit for graded-index fibers:

$$B \cdot L \leq \frac{2c}{n\Delta^2} \sim 4000 \text{ Mb/s} \cdot \text{km}. \qquad (21.13)$$

Typically, the achievable bit-rate · distance product is somewhat less than this; the dispersion limit shown in Fig. 21.6 corresponds to a $B \cdot L$ product of 2000 Mb/s · km (Cohen *et al.*, 1978; Jablonowski *et al.*, 1982). The superior performance of lightwave compared with coaxial cable is apparent in the figure. (In this discussion we have neglected mode-mixing effects, which ameliorate somewhat the modal dispersion in long fibers (Personick, 1971). These effects are important in system design, but they do not change significantly our approximate picture of dispersion in multimode fibers.)

The loss-limit and dispersion-limit curves in Fig. 21.6 represent the performance boundaries for first-generation lightwave technology; within these boundaries some highly successful systems have been developed, such as the AT&T FT3C system shown by a solid circle in the figure (Stauffer, 1983).

In the context of Fig. 21.6, progress in lightwave technology can be viewed as pushing the system performance boundaries outward, toward longer spans and higher bit-rates; in later sections we will see how much progress has been made. But before turning to this, let us briefly discuss the most important performance characteristics of first-generation sources and detectors.

21.2.4 Optoelectronic Devices

Except for our estimates of P_T and P_R for use in Eq. 21.4, we have not mentioned the operating characteristics of the semiconductor devices used as sources and detectors. In order to calculate the system performance bounds in Fig. 21.6, no further detail was necessary. But device characteristics are an important consideration in system design, and some modest familiarity with them is essential to an understanding of lightwave system capabilities.

21.2.4.1 Sources. There are two useful types of sources at $\lambda = 0.85 \ \mu$m, lasers and light-emitting diodes (LEDs); the former are the more popular for transmission distances beyond 1 km or so, but for many applications the LED is the preferred device. It is less expensive and more tolerant of environmental extremes. It is more easily coupled to a fiber, albeit with a coupling efficiency of only about 2% compared with at least 50% for the laser. In addition, because the LED does not have the temperature- and age-dependent threshold behavior of a laser, it requires only a simple driver circuit. On the negative side, the spectral width of a surface-emitting AlGaAs LED is large (25–40 nm at the 3 dB points) making chromatic dispersion a potential problem (see Sec. 21.3.2). Also, modulation speed is limited to about 1 Gb/s, with high data rates being achieved only through increased drive power (Dawson and Burrus, 1971) or at the expense of reduced output power (Harth *et al.*, 1976). Edge-emitting and superluminescent LEDs exhibit reduced spectral width, increased power, and improved coupling into fiber (Marcuse, 1977) when compared to surface-emitting devices, but have a more complicated structure, and so are costlier.

One very important use for systems incorporating LEDs and multimode fibers is in data links, where transmission distances are short (< 1 km), so the low loss and wide bandwidth of fiber may not be of primary importance (although they certainly can be). There are several attributes of fiber that make it attractive in applications where wire might otherwise be installed. Most importantly, coupling between signals in fibers and the external environment is negligible, so degradations due to crosstalk, electromagnetic interference and optical radiation are insignificant. The AT&T 5ESS® switch uses fiber data links for this reason (Martersteck and Spencer, 1985). Also, fiber is nonconductive and therefore eliminates ground loops, shorting problems, and the potential ignition of combustible atmospheres. Since it is essentially nonreactive, it can be used in many corrosive atmospheres. Its small size and weight are assets in crowded installations. Finally, fiber offers greater data security than wire, because covert taps are extremely difficult to implement.

For longer-distance applications, such as trunk lines between telephone switching offices, the laser offers higher performance than the LED. Typical lasers deliver a few milliwatts and can be directly modulated (by varying the drive current) up to several Gb/s. Achieving adequate reliability, however, was a very difficult problem, whose solution required an intensive engineering effort. With early lasers, a lifetime of an hour was sufficient for a laboratory demonstration, but for practical communications applications a mean lifetime of 10–100 years is needed (Kao, 1982). Improving the lifetime by five orders of magnitude was accomplished by painstakingly identifying failure mechanisms, especially darkline defects, and devising ways to eliminate them (Casey and Panish, 1978; Hartman *et al.*, 1977).

Although the relatively narrow spectral width of the laser (~ few nm) minimizes chromatic dispersion effects in the fiber, it is not an unmitigated blessing. The optical intensity distribution across the core of a multimode fiber is granular, rather than smooth. The fluctuations become more pronounced as the laser spectral width decreases, because they depend on coherent interference among the modes in the fiber. If there is any spatial filter in the fiber, such as an imperfectly mated connector or a directional coupler, the transmission through the fiber will depend on the details of the interference among the modes at the filter, and will exhibit random variation with time due to environmental changes. This system impairment is called modal noise; its effects are most severe in analog systems where high (40–60 dB) optical signal-to-noise ratios are required (Hill *et al.*, 1980; Petermann, 1980). In digital systems the degradation is not as great, but it is still serious enough so that lasers with extremely narrow spectral width are avoided.

21.2.4.2 Detectors. Silicon, the mainstay of the integrated-circuit industry, is also an excellent photodetector in the wavelength range 0.4–1.0 μm (Daly, 1983). As with sources, there are two choices for the detector structure: a PIN diode, or an avalanche photodiode (APD). Typically Si APD receivers enjoy a 10–15 dB sensitivity advantage over Si PIN receivers, because the APD has internal gain, which makes the noise of the amplifier stages after the detector relatively less important. Details of receiver design and performance are discussed in Chapter 18. Suffice to say that the state-of-the-art sensitivity for Si APD receivers at speeds up to several hundred Mb/s corresponds to a photon flux of roughly 300 photons per bit.

Two disadvantages of the APD are its high supply-voltage requirements ($\geq$ 50 V vs. 5–10 V for the PIN) and its temperature sensitivity. Both these drawbacks necessitate more complicated (and therefore more expensive) support and control electronics; but for high-capacity systems, the enhanced performance appears to be worth the price (Stauffer, 1983).

21.3 SECOND-GENERATION LIGHTWAVE: $\lambda \sim 1.3$ μm

21.3.1 Transition to a Longer Wavelength

Figure 21.4 shows that increasing the operating wavelength from 0.85 μm to the 1.3–1.6 μm region is a promising route to enhanced system performance. For values of P_T and P_R comparable to those used for first-generation systems, the reduced fiber loss at the longer wavelengths increases the loss-limited transmission distance by a factor of 5–10, as shown in Fig. 21.6. Only recently has it been possible to exploit this low-loss region; suitable sources and detectors based on InGaAsP and Ge had to be developed first. The practical implementation of these devices marked the beginning of second-generation lightwave technology.

Some of the earliest system designs to take advantage of reduced fiber attenuation at 1.3 μm wavelength used multimode fibers and light-emitting diodes rather than lasers (Gloge *et al.*, 1980; Martin-Royle and Bennett, 1983). The low fiber loss ($\sim$ 0.6 dB/km) roughly compensated for the modest output power of the LED, resulting in a loss-limited transmission distance comparable to that of laser systems at $\lambda = 0.85$ μm. The advantages of the LED system are simplicity of design, low cost and very high reliability. These features have made LED-based systems attractive for commercial applications. British Telecom uses LED systems in interoffice trunk lines at data rates of up to 34 Mbit/s. The original AT&T subscriber loop carrier lightwave system used LEDs at 6.3 Mbit/s over distances up to 20 km (Bohn et al., 1984). A 1.3 μm InGaAsP laser can improve system performance by injecting up to 15 dB more light into the fiber. For example, in the United States "Northeast Corridor" trunk line, wavelength multiplexing was used to add a 1.3 μm laser system to an existing 0.83 μm system (Jacobs, 1986). The 1.3 μm channel operates at 180 Mbit/s (twice the 0.83 μm system rate) and, in addition, bypasses every second repeater. As impressive as this improvement in performance is, an even bigger leap comes with the introduction of single-mode fibers, which have both lower dispersion and lower loss than multimode designs.

21.3.2 Single-Mode Fibers

A single-mode fiber, as its name implies, supports just a single, fundamental waveguide mode.[2] The simplest single-mode fiber structure is similar to the step-index multimode fiber shown in Fig. 21.9, except that the core diameter is much smaller, typically 8 μm, so that higher-order modes cannot be guided. An impulse injected into such a fiber propagates as a single waveguide mode, so modal dispersion is nonexistent. Dispersive effects are not entirely absent, however; the group velocity of the propagating mode is in general a function of the optical wavelength. Thus a pulse containing more than a single wavelength will broaden as it travels. The total dispersion of single-mode fiber is the sum of two components: material dispersion and waveguide dispersion. The former, which is the more important effect in conventional fiber designs, arises from the wavelength dependence of the index of refraction of fused silica (Gray, 1972). We postpone the discussion of waveguide dispersion until Section 21.4.2. The effect of dispersion on optical pulse propagation can be readily understood from the idealized experiment shown in Fig. 21.10. Nearly-monochromatic light impulses at different wavelengths are launched into one end of a single-mode fiber of length 1 kilometer, and their time of flight τ through the fiber measured, with results as shown (Payne and Hartog, 1977). A remarkable feature of these data is that for silica fibers τ goes through a minimum at a wavelength near 1.3 μm, denoted by λ_0. (This wavelength is the point where $d^2 n/d\lambda^2 \sim 0$.) In the vicinity of λ_0, τ varies slowly with wavelength, so dispersion effects are minimal. In general, after traveling L km through a fiber,

---

[2] In fact there are two fundamental modes, corresponding to orthogonal polarizations. This complication, for the purposes of this chapter, is not important.

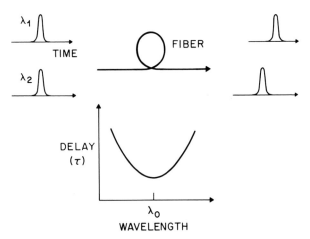

Fig. 21.10 Experiment to measure material dispersion. λ_0 is the wavelength where $d\tau/d\lambda$ (the linear material dispersion) is zero.

an impulse with center wavelength λ_c not too close to λ_0 broadens to an rms width given by

$$\sigma_t \sim D \cdot L \cdot \sigma_\lambda, \qquad (21.14)$$

where D is the linear dispersion coefficient of the fiber given by $D = d\tau/d\lambda$ evaluated at $\lambda = \lambda_c$, and σ_λ is the rms spectral width of the impulse. In principle we could imagine operating a system with $\lambda_c = \lambda_0$ so that the linear dispersion given by Eq. 21.14 would be zero, and only higher-order effects would remain (Marcuse and Lin, 1981). In most practical systems, however, the wavelength cannot be controlled that accurately, and the pulse broadening given by Eq. 21.14 is approximately correct.

From Eqs. 21.8 and 21.14 we calculate the dispersion-limited bit-rate · distance product:

$$B \cdot L \leq \frac{1}{4D\sigma_\lambda}. \qquad (21.15)$$

The behavior of the linear dispersion coefficient for a typical single-mode fiber is shown in Fig. 21.11 (Sugimura *et al.*, 1980). For conventional multifrequency InGaAsP lasers (see Section 21.3.3.1), the center wavelength can be specified to a tolerance of perhaps 20 nm, so $D \sim 2$ psec/km · nm for systems operating near λ_0. A typical laser spectral width is $\sigma_\lambda \sim 1$ nm. Equation 21.15 then yields $B \cdot L \lesssim 125$ Gb/s · km, more than an order of magnitude improvement over the corresponding result for multimode fiber given in Section 21.2, as shown in Fig. 21.6. We emphasize that this $B \cdot L$ product is not an absolute limit, but rather a general indication of the practical capability of 1.3 μm technology.

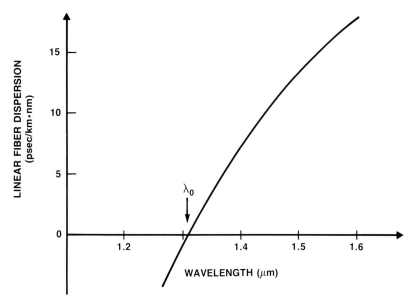

Fig. 21.11 Linear dispersion as a function of wavelength for typical single-mode optical fiber. λ_0 is approximately 1.3 μm; at $\lambda = 1.55$ μm the dispersion is approximately 15 psec/km · nm.

An additional advantage of single-mode fiber is that it has significantly lower loss than multimode fiber at the same wavelength ($\sim$ 0.4 dB/km vs. $\sim$ 0.6 dB/km at 1.3 μm). This improvement is due to the reduction of Rayleigh scattering losses because of lower doping density in the fiber core. (The index step Δ for single-mode fiber is typically 0.001–0.004, compared with $\Delta \sim 0.01$ for multimode fiber (Li, 1980; Cohen *et al.*, 1978; Nakagawa and Ito, 1979).) The implication of lower fiber loss is clear in Fig. 21.6, where the loss-limited transmission distance, Eq. 21.4, for 1.3 μm laser systems is shown. (The values used for P_T and P_R, 1 mW and 500 photons per bit, will be discussed in Sec. 21.3.3.3.) Single-mode systems at 1.3 μm enjoy a distance advantage of about 6-to-1 over first-generation systems. The two open circles shown in the figure represent benchmark demonstrations of second-generation lightwave technology (Runge *et al.*, 1982; Yamada *et al.*, 1981). Also shown are two solid circles representing commercial 1.3 μm systems: SL (AT&T Company) and F-400M (Nippon Telegraph and Telephone Public Corporation) (Abbott *et al.*, 1983; Iwahashi and Fukutomi, 1983).

21.3.3 Optoelectronic Components

21.3.3.1 Lasers. In addition to providing a reasonable amount of optical power (typically a milliwatt or so into the fiber), the lasers used in single-mode lightwave systems must have several other characteristics. Most importantly, the laser must oscillate in a single transverse mode, preferably the fundamental mode. Optical power in higher-order modes cannot be efficiently coupled into a single-mode fiber and hence is wasted. Worse still, if several transverse modes are oscillating, their relative intensities will fluctuate, creating noise in the transmitted signal. Several laser structures, e.g., buried heterostructure and channeled substrate, are capable of operation in a single transverse mode. In general these devices have a waveguiding region of small transverse dimensions with conditions favorable to laser oscillation; i.e., high optical gain and/or low cavity loss. If this region is small enough, only the fundamental transverse mode will have the net gain necessary to sustain laser oscillation.

A second important property of the laser, though perhaps not a strict requirement, is the capability of being directly modulated at high speeds. The carrier lifetime for long-wavelength lasers is 2–3 times smaller than for similar short-wavelength devices, so they are intrinsically faster. Although the earliest second-generation systems operated at speeds of a few hundred Mb/s, the highest bit-rate · distance products will be achieved only at speeds above 1 Gb/s, as shown in Fig. 21.6. As discussed in Chapter 13, not all lasers are capable of modulation at these speeds. Indeed, there is a correlation between the structure used for transverse mode control and the speed capability of the laser; some designs show poor performance at 500 Mb/s, while others operate with only moderate degradation up to at least 8 Gb/s (Linke, 1984; Koren *et al.*, 1985).

The final item in our discussion of laser characteristics is spectrum control: a laser that oscillates in a single transverse mode and has high-speed modulation capability (our first two requirements) may still not be suitable for use in single-mode systems if its spectrum is not well behaved. Under cw conditions, a typical laser spectrum might consist of a few lines spaced about 1 nm apart, as shown by the solid curves in Fig. 21.12. (These lines correspond to different longitudinal modes of the laser structure.) When the laser is modulated, the weak lines in the wings of the spectrum tend to become stronger, increasing σ_λ and reducing the maximum bit-rate · distance product given by Eq. 21.15. To ameliorate this effect the laser can be kept well above threshold for both binary symbols (marks and spaces). This technique itself causes system performance degradation (though usually minor), because the full dynamic range of the laser is not used.

Another spectrum-behavior problem is mode partition noise (Okano *et al.*, 1980; Ogawa, 1982). Even though the total output power of a laser is nearly constant, the relative intensities of the various lines in the laser's spectrum can vary considerably from one pulse to the next. The combination of these spectral fluctuations and fiber dispersion produces random distortion of the received pulses, as shown schematically in Fig. 21.13, leading to an increase in bit-error rate. Mode partition noise is a complicated phenomenon, but its effect on the

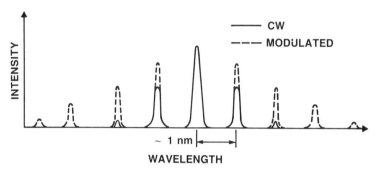

Fig. 21.12 Spectrum of a typical InGaAsP injection laser; cw (solid line) and modulated (dashed line). The spectral broadening caused by modulation leads to increased dispersion penalty.

performance of high-speed ($B \gtrsim 100$ Mb/s) lightwave systems can be summarized simply: for a fixed dispersion penalty (say 1 dB), the dispersion-limited bit-rate · distance product in the presence of mode partition noise is roughly one-half of what it would have been if the laser spectrum were not fluctuating. Thus Eq. 21.15 might be modified to read

$$B \cdot L \leq \frac{1}{8D\sigma_\lambda}, \qquad (21.16)$$

and the dispersion limit in Fig. 21.6 shifted accordingly. Various calculations and measurements of mode partition noise yield a spread of results, but most of them are within a factor of two of Eq. 21.16 (Iwashita and Nakagawa, 1982; Yamamoto *et al.*, 1982; Cheung *et al.*, 1984).

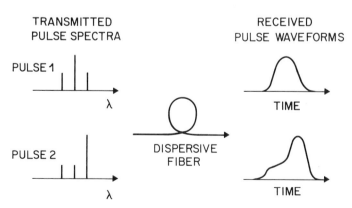

Fig. 21.13 Laser mode partition noise. Pulse-to-pulse fluctuations in the transmitted spectrum, combined with fiber dispersion, produce random distortion of the received pulses.

21.3.3.2 LEDs. Although the power that can be coupled from an LED into a single-mode fiber is quite low ($\sim$ -30 dBm), the attraction of low cost and high reliability prompts its use in second-generation systems. For example, the lifetime at room temperature for 1.3 μm devices has been estimated at 10^9 hours, and the devices continue to operate at temperatures in excess of 200°C. Edge-emitting and superluminescent LEDs are particularly suited to single-mode fiber systems, because they can couple more light than surface-emitters into the small fiber core. Laboratory results of 140 Mbit/s over 35 km and 560 Mbit/s over 15 km (Stern *et al.*, 1986) indicate the potential for this technology. These results are particularly interesting because the effect of dispersion due to the broad spectral width of the LED (90 nm full-width-half-maximum, typical for a 1.3 μm device) was minimized by optical filtering of the transmitted spectrum.

21.3.3.3 Detectors. Both PIN and APD detectors are available at $\lambda = 1.3$ μm. The most popular detector materials have been InGaAs for PINs and Ge for APDs (Brain, 1982; Yamada *et al.*, 1982). (Silicon cannot be used because it is transparent at this wavelength.) The choice of detector is a difficult one, because the Ge APD is far more temperature-sensitive and less rugged than the InGaAs PIN, but it does offer a modest sensitivity advantage ($\sim$ 3 dB at 420 Mb/s). In principle, the best features of both types of detectors can be combined in the InGaAs APD. Early attempts at building such a device were plagued by high dark current and/or slow response. A new heterojunction device appears to have overcome these problems (Campbell *et al.*, 1983). It may be the prototype for a practical InGaAs APD.

The best InGaAs APDs require about 500 photons per bit to achieve BER $\leq 10^{-9}$; this value was used for P_R in computing the loss-limited transmission distance in Fig. 21.6. The fact that APDs at 1.3 μm require roughly twice the photon flux of Si APDs at 0.85 μm (see Sect. 21.2.4.2) does not significantly reduce the distance advantage stemming from lower fiber loss at 1.3 μm.

21.4 THIRD-GENERATION LIGHTWAVE: $\lambda \sim 1.55$ μm

21.4.1 Pros and Cons of 1.55 μm Transmission

To achieve maximum loss-limited transmission distance, a lightwave system should operate at the wavelength where fiber attenuation is smallest. For fused-silica fibers (see Fig. 21.4), this occurs in the region of $\lambda = 1.55$ μm, where the attenuation constant for high-quality single-mode fiber is $\sim$ 0.25 dB/km, slightly more than half as large as the attenuation at 1.3 μm. ("Champion" fibers with $A \sim 0.16$ dB have been produced in laboratory quantities (Nelson *et al.*, 1985).) Since optoelectronic components at 1.55 μm can use the same material technology (InGaAsP and Ge) that was originally developed for 1.3 μm systems, we can reasonably expect that typical values for P_T and P_R at this new wavelength will be about the same as they were at 1.3 μm; namely, 1 milliwatt and 500 photons per bit. The loss-limited transmission distance (Eq. 21.4) will therefore be nearly doubled, as shown in Fig. 21.6.

The most serious system impairment at 1.55 μm is fiber dispersion. From Fig. 21.11, we find that at this wavelength the linear material dispersion coefficient D is approximately 15 psec/km $\cdot$ nm. Thus a lightwave system using a conventional laser with $\sigma_\lambda \sim 1$ nm would suffer a dispersion-limited bit-rate $\cdot$ distance product (Eq. 21.15) of $B \cdot L \lesssim 17$ Gb/s $\cdot$ km, which is far worse than the 88 Gb/s $\cdot$ km demonstrated at 1.3 μm. There are two promising solutions to this problem: controlled-dispersion fibers and single-frequency lasers (also called single-longitudinal-mode lasers). Third-generation lightwave systems will probably use one (or both) of these techniques.

21.4.2 Controlled-Dispersion Fiber

In the discussion of dispersion in Section 21.3.2, we remarked that for conventional single-mode fiber, waveguide dispersion was relatively small. With specially-designed fibers, however, waveguide dispersion can be enhanced and turned to advantage (Imoto et al., 1980; Croft et al., 1985). By a fortunate coincidence, the sign of waveguide dispersion is negative, which is opposite to that of material dispersion at $\lambda = 1.55$ μm. With proper choice of fiber parameters, waveguide and material dispersion can be made to cancel each other, yielding a fiber, often called a dispersion-shifted fiber, in which attenuation and dispersion are minimized at the same wavelength. The primary advantage of this approach is that it allows the multifrequency laser technology (Fig. 21.12) created for 1.3 μm systems to be used at 1.55 μm.

To develop an intuitive feeling for the behavior of waveguide dispersion, assume a step-index single-mode fiber of length L, where the core and cladding indices of refraction, n and $n(1 - \Delta)$, are independent of wavelength, so there is no material dispersion. In the limit of zero wavelength the fundamental mode of this fiber is confined to the core and travels with speed $\sim c/n$. As the wavelength is increased, the mode spreads into the cladding and the effective index decreases, becoming $n(1 - \Delta)$ in the long-wavelength limit. If this fiber were used in the dispersion-measuring experiment illustrated in Fig. 21.10, the results would be as shown in Fig. 21.14. Define λ_m to be the wavelength corresponding to the midpoint between the two limiting values of τ. The wavelength range over which the transition in τ occurs is roughly equal to λ_m. Thus the slope of the curve at λ_m, i.e., the waveguide dispersion, is approximately

$$\left(\frac{d\tau}{d\lambda} \right)_{\lambda=\lambda_m} \sim \frac{\delta\tau}{\delta\lambda} \sim - \frac{n\,\Delta L/c}{\lambda_m}, \qquad (21.17)$$

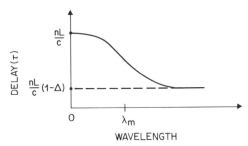

Fig. 21.14 Waveguide dispersion. Even in the absence of material dispersion, the pulse propagation delay in a single-mode fiber is a function of wavelength. With proper choice of fiber parameters this effect can be made to cancel material dispersion at the wavelength of minimum fiber loss, ~ 1.55 μm.

where $\delta\tau$ and $\delta\lambda$ are the approximate ranges of τ and λ in the transition region of Fig. 21.14. Although this formula is only a crude approximation, it shows that with reasonable values of Δ, say $\Delta \sim 0.005$, it is possible to have waveguide dispersion that is equal in magnitude (~ 15 psec/km $\cdot$ nm) and opposite in sign to material dispersion at 1.55 μm. Indeed, the experiment shown by the open square ($\square$) in Fig. 21.6 shows clearly that dispersion cancellation is more than just a possibility (Blank et al., 1985).

By using multilayer structures, rather than simple step-index designs, it is even possible to make fibers where the waveguide and material dispersion cancel at two wavelengths instead of just one (Cohen et al., 1982). These fibers typically have low dispersion across a broad wavelength range, so they are well suited to wavelength-division-multiplex applications, wherein lightwave channels at several different wavelengths are used simultaneously.

21.4.3 Single-Frequency Lasers

The second promising solution to dispersive impairments at 1.55 μm is the use of a laser with extremely narrow spectral width—the so-called single-frequency laser, which oscillates in one, rather than several, longitudinal modes (see Fig. 21.12). Such behavior has been achieved by modifying the basic Fabry-Perot laser structure in any of several different ways, as shown in Fig. 21.15 (see Chapter 13). For example, a mirror external to the semiconductor device itself has been used to provide feedback, which enhances one particular longitudinal mode and suppresses others (Fig. 21.15b) (Cameron et al., 1982; Lin and Burrus, 1983). The best performance, however, in terms of system demonstrations, has been obtained with lasers in which the actual semiconductor cavity has been modified to achieve single-frequency operation. One highly successful example of this approach is the distributed feedback (DFB) laser, which uses a periodi-

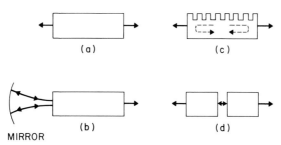

(a)

(c)

(b)

MIRROR

(d)

Fig. 21.15 Structures for single-frequency lasers: (a) Basic Fabry-Perot cavity, (b) External cavity, (c) Distributed feedback (DFB), (d) Cleaved coupled-cavity (C^3).

cally perturbed waveguide to provide frequency-selective feedback, as shown in Fig. 21.15c (Kressel and Butler, 1977; Sakai et al., 1982; Suematsu et al., 1983). Excellent single-frequency behavior, even under modulation, has been observed, with unwanted cavity modes suppressed 20–30 dB relative to the primary mode. Some system demonstrations using DFB lasers are indicated by the solid squares (■) in Fig. 21.6 (Mazurczyk et al., 1984; Gnauck et al., 1985). A practical repeater using a DFB laser transmitter has been successfully designed and tested (Nakagawa et al., 1983).

Another technique for producing single-frequency operation is embodied in the cleaved coupled-cavity (C^3) laser shown in Fig. 21.15d, which uses the interaction between two coupled laser cavities to enhance a desired mode of oscillation (Tsang et al., 1983). As with the DFB structure, the C^3 laser has shown excellent single-frequency behavior. The experiments marked with open triangles (△) in Fig. 21.6, all using a C^3 laser, established world-class standards (Kasper et al., 1983; Korotky et al., 1985; Gnauck et al., 1986). The 4 and 8 Gb/s experiments are particularly interesting, because they used an integrated-optic modulator rather than direct modulation of the laser diode itself. External modulation has several advantages over direct modulation. Since the laser does not have to be capable of being modulated at high speed, a wider choice of devices, including high-power structures, is available. Also, the effects of current modulation on the laser's spectral characteristics, to be further discussed shortly, are avoided. The penalties for external modulation are an additional system component, modulator insertion loss, and, at least with waveguide modulators, polarization sensitivity and higher drive voltage (5–10 V for gigabit rates).

The term "single-frequency laser" is a convenient but not altogether precise description of C^3 and DFB devices. For example, the spectrum of a modulated C^3 laser, shown by the solid line in Fig. 21.16, shows two significant departures from ideal single-frequency operation: a linewidth of ~ 0.1 nm FWHM, which is orders-of-magnitude greater than the "intrinsic" linewidth, and small but significant sidemodes approximately 30 dB weaker than the primary mode

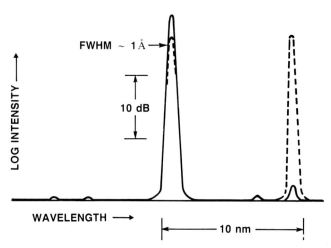

Fig. 21.16 Spectrum of a modulated C³ laser (solid line). The 1 Å linewidth leads to dispersive effects at 1.55 μm comparable to those seen with multimode lasers at 1.3 μm. In addition, mode partition fluctuations (dashed lines) can cause bit-detection errors.

(Henry, 1982). Both these features represent potentially important limitations on the performance of third-generation lightwave systems. The observed line broadening is a frequency "chirp" associated with modulation-induced changes in the carrier density, which leads to significant dispersion effects (Linke, 1985). For system applications the frequency deviation, $\Delta\nu(t)$, of a laser exhibiting chirp is closely approximated by (Koch and Bowers, 1984):

$$\Delta\nu(t) = -\frac{\alpha}{4\pi}\frac{\partial \ln P}{\partial t}, \qquad (21.18)$$

where α is the "linewidth enhancement factor" and P is the output power of one facet. The α factor is in the range of -3.5 to -5.5 for AlGaAs lasers (Harder *et al.*, 1983) and -6 to -8 for InGaAsP lasers (Schimpe *et al.*, 1986). Although materials and structures can be altered to obtain values outside these ranges, further research is needed to produce lasers that have a significantly-reduced α. At present, then, Eq. 21.18 shows that an InGaAsP laser changing power level by a factor of 10 in 200 ps, as might occur under modulation at a few Gb/s, undergoes a frequency shift of about 6.4 GHz (~ 0.05 nm). This limits system performance to second-generation levels. Chirp can be minimized, however, by reducing $\partial \ln P/\partial t$. One approach is to bias the laser so that a given modulation current does not drive it below threshold, where $\ln P$ changes rapidly. Unfortunately this leads to an extinction ratio penalty (typically several dB) at the receiver. Another approach is to damp the relaxation oscillations that can occur at turn-on and turn-off and result in large power fluctuations. This has been done by shaping the electrical drive pulses (Bickers and Westbrook, 1985). Injection locking is yet another solution (Lin and Mengel, 1984; Toba et al., 1984). At present, the most effective technique for minimizing the effects of chirp is to allow the laser to run continuously and impress data on the optical carrier with an external modulator, which might be a separate $LiNbO_3$ device, or monolithically integrated with the laser (Yamaguchi et al., 1985).

The second non-ideality of single-frequency lasers, residual sidemodes, leads to mode partition noise similar to that described in Section 21.3.3.1, but with an important difference. At 1.3 μm the difference in propagation delay between adjacent laser modes after, say 100 km of fiber is ~ 100 psec, which is a small fraction of the pulse period B^{-1}. Thus the effect of mode partition fluctuations is to cause random distortion of the received pulses. At 1.55 μm, on the other hand, the delay difference between laser modes can be several nanoseconds, so a mode partition fluctuation typically will cause some of the power in a transmitted pulse to arrive at the receiver in the wrong time-slot altogether. Even if the average power in a sidemode is extremely small, the instantaneous intensity can occasionally fluctuate up to a large value, say half the average power of the primary mode (dotted curve in Fig. 21.16), and thereby cause a bit-detection error at the receiver. The degree of sidemode suppression needed to ensure BER $\leq 10^{-9}$ is not well understood for all types of lasers, but in the case of simple Fabry-Perot devices (Fig. 21.15a), 17–20 dB suppression appears to be required (Linke et al., 1985; Henry et al., 1984). Transmission experiments with C^3 and DFB lasers suggest that substantially more than 20 dB suppression is needed.

Optical feedback from unwanted external reflections can also affect the intensity and frequency stability of lasers (Bloom et al., 1970; T. Morikawa et al., 1976; Miles et al., 1980; Temkin et al., 1986; Henry and Kazarinov, 1986). In multimode fiber systems, this degradation is reduced because reflections are distributed among many fiber modes and so couple only weakly back into the laser mode (Personick, 1985). In single-mode-fiber systems, variations in the total laser output and the output of individual longitudinal modes (which lead to mode partition noise) are enhanced by the stronger fiber-to-laser coupling. In 1.3 μm systems, however, the effect is still usually small, except perhaps at very high bit rates (see Section 21.3.3). In such cases reducing the coupling between laser and fiber may be necessary. Alternatively, an optical isolator can be installed. This is a nonreciprocal device that allows light to pass in one direction but strongly attenuates it in the other direction. With such a device, the unwanted fiber-to-laser coupling can be reduced by about 30 dB.

In systems employing 1.55 μm single-frequency lasers, reflection-induced frequency-hops and linewidth broadening present a more serious problem, because fiber dispersion is much greater than at 1.3 μm. In DFB lasers, feedback in the range of -5 dB to -30 dB results in linewidth broadening of up to 50 GHz, or 0.4 nm at 1.55 μm (Miles *et al.*, 1980; Lenstra *et al.*, 1985; Tkach and Chraplyvy, 1986; Temkin *et al.*, 1986). Similarly, C^3 lasers are known to be affected by reflections as small as about -30 dB (Olsson *et al.*, 1985). Reflections of this magnitude are common in today's systems. Especially troublesome are the fiber-to-fiber interfaces, which are inevitable in any system design. A polished single-mode-fiber end has a reflection of -14 dB due to the glass-air interface. Reflections from commercial fiber connectors can range from this level or even higher (Shah *et al.*, 1987) for noncontacting types, to -25 to -30 dB for dry contacting types, to -35 dB or better for well-index-matched connectors and splices. (Fusion splices are much better, at least -50 dB.) Other system components may also contribute reflections. Optical isolators may therefore be necessary for reliable single-frequency laser system operation.

21.4.4 Dispersion Limitations

To extract maximum usefulness from single-mode fibers, the transmitted spectral width σ_λ must be minimized. Since the smallest attainable value of the occupied bandwidth (FWHM) is roughly equal to the bit-rate B, we have

$$2\sigma_\lambda \sim \frac{\lambda^2}{c} B. \qquad (21.19)$$

To achieve this width, an external modulator must be used. If direct modulation of the laser is attempted, the resulting chirp (Eq. 21.18) will cause line-broadening significantly in excess of σ_λ above. Substituting σ_λ into Eq. 21.15 we obtain the ultimate dispersion limit:

$$B^2 L \leq \frac{c}{2D\lambda^2}. \qquad (21.20)$$

At $\lambda = 1.55$ μm where $D \sim 15$ psec/km $\cdot$ nm, $B^2 L \leq 4000$ (Gb/s)$^2 \cdot$ km, which is shown in Fig. 21.6. The 8 Gbit/s experiment of Fig. 21.6 was the first clear demonstration of this fundamental limit of third-generation technology. Modulation-bandwidth-limited spectral width at the transmitter was achieved by using external modulation of a C^3 laser. The $B^2 L$ product of ~ 4350 (Gbit/s)$^2 \cdot$ km and reported dispersion penalty of 1 dB agree well with Eq. 21.20 (Gnauck *et al.*, 1986). The intersection of the dispersion limit with the quantum limit corresponds to the highest bit-rate $\cdot$ distance product that can be achieved with the techniques described in this paper: ~ 900 Gb/s $\cdot$ km.

21.4.5 Fiber Non-Linearities

With present 1.55 μm technology, laser transmitters can launch at most a few milliwatts into a fiber. That situation may change, however; after all, AlGaAs lasers at 0.85 μm can generate close to 100 mW. Perhaps tomorrow's lasers at

1.55 μm will have similar or even greater capability. Let us examine the systems implications of such a performance improvement.

The most obvious consequence of increased laser power is, of course, greater loss-limited transmission distance (Eq. 21.4). A laser with $P_T = 1$ watt might nearly double the achievable repeater spacing in a 1 Gb/s system. If implementation of repeaters with high-power lasers were not too difficult, this added transmission distance would be a significant improvement. The flaw in this scenario is that non-linear effects in fibers make it impossible (or at least very difficult) to make effective use of high transmitter power. The most important of these effects are stimulated Brillouin scattering (SBS) and stimulated Raman scattering (SRS) (Miller, 1979, Ch. 5). Both manifest themselves as power-dependent excess fiber loss, which increases rapidly for transmitter power above a threshold level, while remaining negligible at lower power levels. For a nearly monochromatic transmitter (linewidth $\leq$ 50 MHz), SBS is the most important non-linear process; the threshold power is only about 10 mW. The impairment due to SBS can be eliminated, however, by broadening the transmitted spectrum so its spectral density is less than about 0.1 mW/MHz (Smith, 1972). This condition can be satisfied either with a noisy laser having sufficient incidental frequency modulation, or by using modulation techniques that produce the desired spectral characteristics (Cotter, 1982a, b, c). In the long run then, SBS is not likely to be a serious problem.

Unlike SBS, SRS cannot be easily eliminated. It is an incoherent process depending on total transmitted power and is relatively independent of transmitter spectral width. The power threshold at which significant added loss begins to appear is about 1 watt for a single-channel system, but can be as low as 20 mW per channel for a two-channel wavelength-division-multiplexed (WDM) system (Chraplyvy and Henry, 1983; Tomita, 1983). This may be a serious limitation for future lightwave systems, because WDM appears almost certain to be a popular technique (Olsson et al., 1985). The system power penalty due to SRS in a WDM system depends on many details, but as a guideline we can say that if the system channels are spread uniformly over a bandwidth of 40–100 nm, then significant added loss ($\geq$ 1 dB) due to SRS will be encountered when the total launched power exceeds roughly 40 mW. (More on SRS in Sections 21.5.3 and 21.6.)

21.5 FOURTH-GENERATION LIGHTWAVE: COHERENT SYSTEMS

21.5.1 The Cloudy Crystal Ball

Predicting the course of lightwave evolution beyond the 1.55 μm systems discussed in Section 21.4 is a risky business, because it depends so sensitively on technological breakthroughs yet to be made. For example, new optical-fiber materials for use at wavelengths beyond 2 μm can, in principle, have attenuation constants ranging from 0.06 dB/km all the way down to 0.0001 dB/km (Miyashita and Manabe, 1982). If these losses can be realized in actual fibers, the focus of lightwave research might shift away from 1.55 μm systems toward still longer wavelengths, necessitating entirely new research programs to develop suitable semiconductor materials. For the time being, however, the most promising direction for lightwave evolution is toward coherent systems at 1.55 μm, which we tentatively call fourth-generation technology.

The term "coherent lightwave" can be defined in a variety of ways. For our purposes we take the broadest view: the characteristic that sets coherent lightwave systems apart from earlier generations is their use of optoelectronic devices to process photons in sophisticated ways, rather than merely detect them (Yamamoto and Kimura, 1981; Okoshi, 1982; Henry, 1984; Okoshi, 1987). Not only do coherent techniques offer the possibility of improved system performance (as we shall see next), they also allow a much richer variety of system configurations and functions. One application of coherent techniques, for example, makes use of simple optical amplifiers instead of regenerative repeaters in long-distance transmission systems. Another potentially important area is multiuser photonic networks. Both of these are discussed next.

Underlying the many potential applications of coherent techniques are two fundamental processes: homodyne (or heterodyne) detection, and direct optical amplification. In this section we will develop simple models for these processes and use them to quantify the potential advantages and shortcomings of coherent lightwave systems.

21.5.2 Coherent Optical Reception

A coherent optical receiver is the lightwave analog of a superheterodyne radio set. Unlike direct detection, where the optical signal is converted directly into a demodulated electrical output, the coherent receiver first adds to the signal a locally-generated optical wave and then detects the combination. The resulting photocurrent carries all the information of the original signal, but is at a frequency low enough ($\leq$ few GHz) so that further signal processing can be performed using conventional electronic circuitry. In practice, this method offers significant improvements in receiver sensitivity and wavelength selectivity compared with direct detection. Even early implementations of coherent receivers, for example, required signal energies as low as 45 photons per bit to achieve a bit error rate of 10^{-9}, far less than the roughly 500 photons required by today's best APD receivers (Linke and Gnauck, 1988). And because of its improved selectivity, a coherent receiver might permit wavelength-division-multiplexed systems with channel spacings of only, say 100 MHz, instead of the 100 GHz required with conventional optical multiplexing technology. A further advantage of coherent reception, not often cited but potentially very important, is

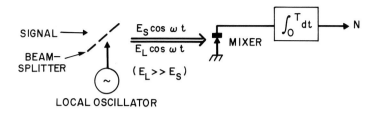

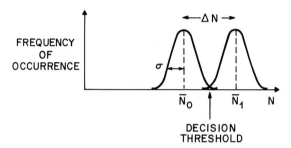

Fig. 21.17 Homodyne receiver. The incoming signal is combined with the local oscillator wave and then detected by the mixer photodiode. Shot noise from the mixer causes fluctuations in the integrator output, which can lead to bit detection errors.

that it allows the use of electronic equalization to compensate for the effects of optical pulse dispersion in the fiber. The receiver sensitivity improvement is the focus of this section.

Conceptually, the simplest type of coherent reception, though not necessarily the most practical, is achieved with a homodyne receiver, in which the local oscillator is phase-locked to the incoming optical carrier, as shown in Fig. 21.17. The optical signal is first combined with the much stronger local oscillator wave using a partially reflecting plate called a beam splitter. (A fiber directional coupler might also be used for this function.) Usually signal power is more precious than local oscillator power, so the beam splitter is made almost completely transparent, and consequently it reflects only weakly. For our idealization we will assume negligible transmission loss in the beam splitter. The combined signal and local oscillator waves illuminate a photodetector, called the mixer, whose average output current is proportional to the total optical power averaged over many optical cycles (P_{opt}). For a PIN diode mixer,

$$R = \frac{\eta P_{opt}}{h\nu} \equiv \left(E_L + E_S \right)^2, \qquad (21.21)$$

where R is the average generation rate of photoelectrons, η the quantum efficiency of the detector and $h\nu$ the photon energy. E_L and E_S, which are proportional to the envelopes of the local oscillator and signal fields incident on the photodiode, are defined in normalized units such that the simple square-law relation of Eq. 21.21 holds. (We assume these fields have the same polarization; in practice this is usually achieved by employing an adjustable polarization compensator.) The output of the photodetector is integrated for the duration of a data bit T, and the result compared with a threshold to determine if a binary "0" or "1" was transmitted.

We now calculate the sensitivity of the homodyne receiver for on-off keying (OOK), in which a binary "1" is transmitted as an optical pulse ($E_S \neq 0$), and a binary "0" is represented by the absence of optical energy ($E_S = 0$). The expected number of counts at the integrator output for the "0" and "1" signal states is

$$\overline{N}_0 = E_L^2 T$$
$$\overline{N}_1 = (E_L + E_S)^2 \cdot T \sim (E_L^2 + 2E_L E_S) \cdot T, \tag{21.22}$$

where the approximate equality is based on the assumption $E_L \gg E_S$.

When an individual bit is detected, the integrator output, of course, will not be precisely $\overline{N}_0$ or $\overline{N}_1$, because the photoelectrons are generated at random intervals. (This randomness is the basis of shot-noise, which is seen whenever current flows through a diode.) The integrator outputs will therefore be distributed around $\overline{N}_0$ and $\overline{N}_1$, in accordance with Poisson statistics, as shown by the histogram in Fig. 21.17. The error probability (BER) of the receiver is related to the fraction of each distribution on the "wrong side" of the threshold level. The narrower the distributions are relative to their separation, the lower will the error rate be. Under our assumption of a strong local oscillator ($E_L^2 T > 100$), the distributions are approximately Gaussian with peak-to-peak separation

$$\Delta \overline{N} = \overline{N}_1 - \overline{N}_0 = 2E_L E_S T, \tag{21.23}$$

and width

$$\sigma \sim \sqrt{N_1} \sim \sqrt{N_0}. \tag{21.24}$$

Thus, we can use tabulated values of the error-function to calculate error rates.[3] To achieve 10^{-9} BER requires $\Delta N/\sigma \sim 12$, so from Eqs. 21.23 and 21.24 we find

$$E_S^2 T = 36. \tag{21.25}$$

[3]For $E_L^2 T > 100$, sensitivities calculated using the Gaussian approximation are within 1% of the true values derived from Poisson statistics.

TABLE 21.1

Lightwave Receiver Sensitivity

In each box the upper entry represents the average number of photons per bit required by an ideal binary receiver ($\eta = 1$) to achieve BER $= 10^{-9}$. The lower entry gives the probability of error as a function of the average number of received photons per bit, $\overline{N}_p$. (For $x \geq 5$, erfc $\sqrt{x} \sim e^{-x}/\sqrt{\pi x}$)

Receiver		Heterodyne		
Modulation	Homodyne	Synchronous Detection	Asynchronous Detection	Direct Detection
On-Off Keying (OOK)	18	36	40	10
	$\frac{1}{2}$ erfc$(\eta \overline{N}_p)^{1/2}$	$\frac{1}{2}$ erfc$(\frac{1}{2}\eta \overline{N}_p)^{1/2}$	$\frac{1}{2}$ exp$(-\frac{1}{2}\eta \overline{N}_p)$	$\frac{1}{2}$ exp$(-2\eta \overline{N}_p)$
Phase-Shift Keying (PSK)	9	18	20	—
	$\frac{1}{2}$ erfc$(2\eta \overline{N}_p)^{1/2}$	$\frac{1}{2}$ erfc$(\eta \overline{N}_p)^{1/2}$	$\frac{1}{2}$ exp$(-\eta \overline{N}_p)$	
Frequence-Shift Keying (FSK)	—	36	40	—
		$\frac{1}{2}$ erfc$(\frac{1}{2}\eta \overline{N}_p)^{1/2}$	$\frac{1}{2}$ exp$(-\frac{1}{2}\eta \overline{N}_p)$	

But from Eq. 21.21, $E_S^2 T$ is simply the expected number of photoelectrons per bit when the mixer is used as a simple direct detector (i.e., $E_L = 0$). That is, to achieve 10^{-9} BER with OOK homodyne, the average energy of each optical pulse must be sufficient to produce 36 direct-detection photoelectrons. We can take this result a step further by noting that for antireflection-coated InGaAsP PIN diodes, the quantum efficiency η approaches unity, so 10^{-9} BER corresponds to an average received optical energy of 36 photons per pulse. For the usual case where "1s" and "0s" are equally probable, an OOK data-stream is on only half the time. Thus, the required number of photons per bit of information is half the required number per pulse, or 18, as shown in Table 21.1.

If phase shift keying (PSK) is used instead of OOK, even greater receiver sensitivity can be achieved. In this case, information is impressed on the phase of the transmitted wave and we write the two states of the received wave as $E_S \cos \omega t$ and $E_S \cos(\omega t + \pi)$. Using arguments similar to those in the OOK case, we find

$$\Delta N = (E_L + E_S)^2 T - (E_L - E_S)^2 T = 4 E_L E_S T \qquad (21.26)$$

and

$$\sigma = \sqrt{E_L^2 T}. \qquad (21.27)$$

The condition $\Delta N/\sigma = 12$ then implies

$$E_S^2 T = 9. \tag{21.28}$$

Thus for PSK homodyne detection with an ideal photodiode ($\eta = 1$), an average signal energy of 9 photons per bit is required to achieve 10^{-9} BER. (For PSK the optical signal is on all the time, so we need not distinguish between photons per pulse and photons per bit, as we did in the OOK case.) The improved sensitivity of PSK compared with OOK stems from the fact that for a given signal field E_S, the value of ΔN using PSK is twice as large as with OOK. PSK with homodyne detection provides the best sensitivity that can be achieved with the simple coherent receiver structure of Fig. 21.17. The system performance associated with this "quantum-limited" sensitivity is shown in Fig. 21.6. The transmission distance at 1 Gb/s is about 50% greater than with conventional intensity modulation and APD detection.

Homodyne receivers, though they are the most sensitive, are also the most difficult to build, because the local oscillator must be controlled by an optical phase-locked loop. The heterodyne receiver, in which the local oscillator frequency is deliberately offset from the signal frequency, is considerably easier to implement. For this case the generation rate of photoelectrons is

$$R(t) = E_L^2 + 2E_L E_S \cos(\omega_S - \omega_L)t + E_S^2, \tag{21.29}$$

where ω_S and ω_L are the optical signal and local oscillator frequencies, respectively. The desired output signal, represented by the middle term in Eq. 21.29 is centered at an intermediate frequency (IF), which is the difference between the signal and local oscillator frequencies. In most cases the IF is adjusted to be of order 1 GHz, so subsequent signal processing can be done using standard radio-frequency techniques. The price paid for elimination of the optical phase-locked loop in the heterodyne receiver is a 3 dB degradation in sensitivity compared with homodyne. A semiclassical way to view this impairment is to recognize that in a heterodyne receiver, the signal and local oscillator are constantly slipping in phase. The mixer output is most sensitive to the incoming signal when the signal and local oscillator are aligned in phase, either parallel or anti-parallel. When they are in quadrature, mixer sensitivity is negligible. The i-f signal, which carries the desired data, is an average over these "good" and "bad" conditions. With homodyne, on the other hand, the signal is always aligned with the local oscillator, so mixer response to the signal is maximized. Since the homodyne receiver takes full advantage of the incoming signal, it is not surprising that its sensitivity is greater. More precisely, for given signal and local oscillator powers, the power available from the mixer output in a heterodyne receiver is just half that available in the homodyne case. Since the shot-noise power for both is the same (because the local oscillator powers are equal), the sensitivity of the heterodyne receiver is 3 dB poorer.

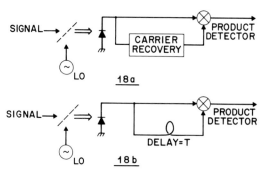

Fig. 21.18 Receivers for phase-shift keying: (a) Synchronous receiver for CPSK, (b) Asynchronous receiver for DPSK.

An important feature of heterodyne receivers, which sets them apart from homodyne designs is their compatibility with asynchronous detection. (Communications engineering textbooks frequently refer to asynchronous detection as incoherent detection; we will not use this latter term.) Asynchronous detection techniques are attractive because they are simple and robust, yet perform almost as well as their synchronous counterparts. A comparison of synchronous and asynchronous detection of phase-shift keying is shown in Fig. 21.18. The optical and IF portions of the receivers are identical. For synchronous detection (known as coherent PSK or CPSK), a carrier recovery circuit, usually a phase-locked loop, is used to generate a local phase reference. The product of this reference times the received signal yields the desired data output. In the asynchronous case (usually called differential PSK or DPSK), the carrier recovery circuit is replaced by a simple one-bit delay line. The output of the product detector indicates whether the phase of the received signal has changed from one bit to the next, and from this the desired data can be recovered.

Explicit calculation of heterodyne receiver sensitivity is considerably more complicated than for the homodyne cases discussed earlier. Nonetheless, the final results are beautifully simple. Under the usual assumptions of strong local oscillator and ideal components, sensitivities for various heterodyne configurations take on simple functional forms, as shown in Table 21.1 (Salz, 1985; Salz, 1986). The nearly equal sensitivities for the synchronous and asynchronous cases is apparent.

Another important feature of heterodyne receivers is their suitability for use with frequency-shift keying (FSK). With this modulation technique binary zeros are transmitted as light pulses at one optical frequency, and binary ones as pulses at another. FSK is attractive because it can be generated by direct modulation of the laser injection current; a separate, external modulator is not needed. The elimination of the modulator and its attendant insertion loss makes up for most, if not all, of the sensitivity penalty (shown in Table 21.1) suffered by FSK compared with PSK.

FSK receiver sensitivity can be improved by resorting to multi-frequency (non-binary) signalling, at the expense of added complexity. For example, if four optical frequencies are used instead of two, then two bits of information can be conveyed with a single pulse. Provided the frequencies are spaced widely enough, the difficulty in deciding at the receiver which one of the four frequencies was sent is only slightly greater than in the binary case. Thus, in both cases the receiver requires about the same number of photons per pulse. But each pulse in the 4-frequency system carries twice as much information, so the required optical energy per bit is cut in half! In theory at least, the sensitivity of a 32-frequency FSK receiver can surpass that of homodyne PSK.

The sensitivities shown in Table 21.1 are all roughly equal, so the choice of which modulation scheme to use in practical systems will probably depend strongly on other considerations, such as laser linewidth. Even in the absence of modulation-induced chirp, the linewidths of single-frequency diode lasers are broad, on the order of 50 MHz (Fleming and Mooradian, 1981a), as compared with values of 10s of Hz typical of gas lasers and radio oscillators. These broad lines, which are a manifestation of the relatively poor coherence of diode laser light, have made it impossible to apply radio techniques indiscriminately to coherent lightwave systems. The diode laser's small size (typically 250 μm long) is the main cause of broad linewidth. Since relatively little energy is stored in the laser cavity, there is only weak buffering against phase disruptions caused by randomly emitted spontaneous photons; broad linewidths are a direct consequence of the resulting phase instability. Linewidths as low as a few kHz have been obtained from diode lasers by incorporating them into large ($\sim$ 10 cm long) resonant cavities (Fleming and Mooradian, 1981b; Matthews et al., 1985), although the resulting external cavity lasers lack some of the most desirable features of the basic diode laser, namely, small size and ruggedness.

Some coherent communications schemes are relatively immune to degradations caused by broad laser linewidth. In particular, amplitude and frequency modulation systems can employ sources with linewidth $\Delta \nu$ comparable to the modulation speed: $\Delta \nu \leq B/10$ (Yamamoto and Kimura, 1981; Foschini et al., 1988; Garrett and Jacobsen, 1986). Systems employing phase modulation are, not surprisingly, much more sensitive to phase noise and consequently require much narrower lines: $\Delta \nu \leq B/1000$ (Kikuchi et al., 1984; Kazovsky, 1986). While the best sensitivities and longest transmission spans to date have been achieved by coherent lightwave systems employing narrow-line lasers (Iwashita et al., 1986; Linke, 1987), impressive results have also been obtained from broadline systems (Yamazaki et al., 1986; Gimlett et al., 1987), and it seems likely that such systems, owing to their simplicity, will be the first to find commercial application.

Experimental coherent transmission systems have already demonstrated nearly-quantum-limited sensitivities and ~ 300 km transmission distances at a bit rate of 400 Mb/s (Iwashita *et al.*, 1986; Linke, 1987). Beyond about 1 Gb/s, however, the observed performance advantage of coherent techniques diminishes and no coherent systems have been reported above 4 Gb/s. A possible explanation for this apparent speed-dependent degradation reveals a fundamental difficulty with coherent techniques. The output waveform of a direct-detection receiver (Fig. 21.8) is a voltage proportional to the incoming optical signal power. Any electronic nonideality (e.g., bandwidth limitation or imperfect amplifier response), which reduces the opening in the eye diagram, can be compensated by providing a corresponding increase in the received optical power. For the coherent receiver the output waveform is proportional to the field strength of the optical signal (rather than the optical power), so compensating for the same eye-closure requires a corresponding increase in the received optical field, and therefore, a larger increase in optical power than in the direct-detection case. Thus a receiver degradation that causes a 3 dB direct-detection optical power penalty will cause a 6 dB penalty in a coherent system. Since, in general, the performance of the electronic components is the limiting factor at very high bit-rates, high-speed systems are more difficult to implement using coherent detection. This problem is exacerbated by the use of heterodyne rather than homodyne techniques, since the electronic components must operate at an IF which is typically several times higher than the bit-rate. Thus it appears that the sensitivity advantage of coherent techniques is restricted in practice to speeds where electronic components are nearly ideal.

It should be emphasized that in this discussion we have assumed that the only noise in a coherent receiver is shot-noise generated by the photodiode; other noise contributions such as electronic noise in the integrator have been neglected. The justification for this assumption is that shot-noise increases with local oscillator power level; so by making the local oscillator strong enough, we can eventually overwhelm any other sources of noise. This is called shot-noise-limited operation (Abbas *et al.*, 1984; Kasper *et al.*, 1986). In practice, however, the required local oscillator power is often not available. For example, if the photodiode drives a conventional 50-ohm amplifier with a noise temperature of 300 K, the required power on the diode is roughly 4 mW (Hodgkinson *et al.*, 1983). Given normal optical coupling losses, this implies a local oscillator laser generating in excess of 10 mW, an impractically high value. With very careful low-noise design, it is possible to operate at much lower power levels. A low-capacitance PIN photodiode followed by a microwave FET amplifier, for example, can achieve shot-noise limited operation at 140 Mb/s with only about − 20 dBm local oscillator power, a level well within the range of practicality. Observe, however, that the effective noise of a FET amplifier increases as the square of the bit-rate, so we can expect increasing difficulty in achieving shot-noise-limited operation at speeds beyond a few Gb/s. Finally, we remark that even at the -20 dBm level ($\sim 10^{14}$ photons/sec), the local oscillator power is still far in excess of the expected signal power (10^{10} photons/sec), so the assumption $E_L \gg E_S$, used throughout this section, is easily satisfied.

For purposes of comparison, the right-hand column in Table 21.1 shows receiver sensitivity for on-off keying and an ideal direct-detection (photon-counting) receiver, for which BER $= \frac{1}{2}\exp(-2\overline{N}_p)$, where $\overline{N}_p$ is the average number of photons received during each bit interval. It is important to recognize that the required average flux, 10 photons/bit, is smaller than the requirement for all heterodyne and homodyne schemes except homodyne PSK. In other words, coherent techniques do not have a significant sensitivity advantage over *ideal* direct-detection schemes. In practice, however, coherent systems can perform within about 3 dB of the shot-noise limit, whereas direct-detection receivers, because of noise from the post-detection amplifiers, are typically 10–15 dB less sensitive (see Chapter 18).

The foregoing discussion of receiver sensitivity, although presented in the context of optical systems, also applies to superheterodyne radio receivers. The fundamental premise, from which everything else follows, is the existence of a radiation detector whose output current is proportional to the incident electromagnetic power. Since this is readily achieved in the radio domain with simple diode detectors, we might expect that ideal coherent radio receivers should also have sensitivities in the vicinity of 10 photons per bit. In fact, the best microwave receivers require about 10^5 photons to detect a single bit! The source of this huge degradation in sensitivity is the poor quantum efficiency of microwave diodes brought about by internal thermal fluctuations. Whereas a PIN photodetector generates roughly one photoelectron for each incident optical photon, a microwave diode requires thousands of radio photons per electron. The grossly different quantum efficiencies in these two regimes are a manifestation of a fundamental difference in the physics of optical and radio systems. In the former case, the photon energy $h\nu$ (Planck's constant times the optical frequency ~ 1 eV) is much greater than the ambient thermal fluctuation energy kT (Boltzmann's constant times room temperature ~ 25 meV). Under these circumstances, it is possible with a semiconductor diode to detect photons by direct excitation of an electron into the conduction band, while at the same time suffering negligibly small background noise (i.e., dark current). In the radio regime, on the other hand, the photon energy is much less than kT. Consequently, direct excitation of an electron across the band gap cannot be achieved. Instead detection occurs through an indirect, and much less efficient, process. To generate an electron in this case requires roughly kT of microwave energy incident on the diode or the equivalent of 10^4 photons at 1 GHz (Watson, 1969). This reduction in detection efficiency relative to the optical case translates directly into a reduction in coherent receiver sensitivity as measured in photons per bit. It must be borne in mind, however, that when sensitivity is measured in terms of energy per bit, the microwave receiver is superior, because 10^5 radio photons carry roughly the same energy as a single optical photon.

These two modes of radiation detection correspond to distinct bias states for the detector diode. For optical detection the diode is reverse-biased; current flows only when light is detected. For radio detection forward-bias is used, and the detected radiation causes an increase in diode current. The performance of coherent receivers using diode detection for either regime can be summarized by the following convenient (albeit imprecise) statement: To detect a bit of information requires about $10kT$ or $10h\nu$ of energy, whichever is greater.

Although most coherent-systems research has been directed toward improving receiver sensitivity, the selectivity advantage of coherent techniques is receiving increasing attention. Especially in the area of multiple-access photonic networks, discussed in Section 21.6, WDM is seen as a powerful tool for allocating distinct channels to many users sharing a single fiber, so the high selectivity of coherent receivers is an important feature. Consider the case where an incoming optical wave consists of a desired signal and an interferer at a nearby frequency. Discriminating between these two with optical filter techniques is a difficult task when their spacing is less than about 10 GHz. If, instead, a coherent receiver is used, the signal and interferer are both translated down to the radio domain with their absolute frequency spacing left unchanged (see Eq. 21.29). Their fractional spacing is much larger, however, so discriminating between the two is greatly simplified.

21.5.3 Optical Amplification

Coherent lightwave techniques can be used to construct long-distance digital transmission systems without regenerative repeaters. Instead of repeaters, optical amplifiers are placed at intervals along the fiber, much as conventional amplifiers are used in analog coaxial-cable systems. The difference between regeneration and amplification is not merely one of nomenclature. A regenerative repeater as currently implemented requires optoelectronic devices for source and detector, as well as substantial circuitry for pulse slicing, retiming and reshaping. The optical amplifier, on the other hand, is in principle much simpler; it is a single component that delivers at its output a linearly amplified replica of the optical input signal. In addition to simplicity, the advantage to this approach is flexibility. The same amplifier can be used for any modulation scheme at any bit-rate. Indeed, if the amplifier is sufficiently linear, a single device can simultaneously amplify several signals at different wavelengths and bit-rates.

There are two promising approaches to optical amplification: semiconductor amplifiers, which utilize stimulated emission from injected carriers, and fiber amplifiers, in which gain is provided by stimulated Raman or Brillouin scattering or fiber dopants (Mukai *et al.*, 1982; Chraplyvy *et al.*, 1983; Smith *et al.*, 1986). Both types have attractive advantages, but at present semiconductor amplifiers seem closer to practical implementation, so we will confine our discussion to them.

The structure of a semiconductor optical amplifier is similar to that of a conventional injection laser. In fact, lasers pumped just below threshold have been used to demonstrate optical amplification. The drawback to using lasers as amplifiers is that they can only achieve a relatively small gain-bandwidth product, given roughly by the reciprocal of the photon transit time within the device; typically $\sqrt{G} \cdot B \sim 50$ GHz, where G is the amplifier power gain and B is its 3 dB bandwidth (Simon, 1983). Thus to achieve a useful gain (≥ 20 dB), the bandwidth of the amplifier can be no greater than 5 GHz, or about $\frac{1}{2}$ Å at 1.55 μm, which might be uncomfortably narrow in many applications. A more promising optical-amplifier structure is the traveling-wave amplifier (Fye, 1984). Again, the device is similar to a laser, except in this case the end facets are anti-reflection-coated to eliminate cavity resonances. Photons incident on a traveling-wave amplifier do not bounce back and forth once inside the cavity; they make only a single pass through the device, stimulating enough emission during that transit to provide useful gain. The elimination of cavity resonances yields a very broad bandwidth for the traveling-wave amplifier—10 Å at 20 dB gain has been observed in an AlGaAs device.

A simple model for the noise behavior of traveling-wave amplifiers is shown in Fig. 21.19; it is valid under conditions of high signal-to-noise ratio (which is always the case with properly-functioning lightwave systems), and high gain. The model consists of an ideal noiseless amplifier of gain G preceded by an additive noise source of spectral density

$$S = K \cdot h\nu, \tag{21.30}$$

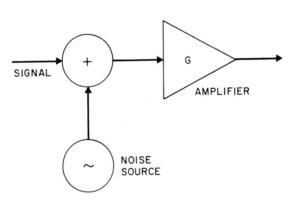

Fig. 21.19 Noise model for an optical amplifier. For typical lightwave applications, quantum fluctuations in the amplifier can be represented as simple additive noise with spectral density $\geq h\nu$.

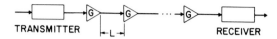

Fig. 21.20 Lightwave transmission system using optical amplifiers. Compared with a conventional regenerative repeater, the optical amplifier is simpler (it requires no high-speed digital circuitry), and more flexible (it can amplify simultaneously one or more signals at different wavelengths and bit rates).

where K, which depends on population inversion and cavity losses, is a measure of the amplifier quality and $h\nu$, as before, is the photon energy. The minimum theoretical value of K is unity, which occurs for the ideal case of complete inversion and no cavity loss. In this situation $S = h\nu$. That is, it is theoretically impossible to have a noiseless amplifier. In practice $K < 2$ has been achieved, indicating that amplifiers less than 3 dB noisier than the theoretical limit are realizable (Henry, 1986; Simon, 1986; Kaminow and Tucker, 1987; Saitoh, 1986).

The use of optical amplifiers in a long-distance transmission system is shown in Fig. 21.20 (Marshall *et al.*, 1986). After each section of fiber of length L, there is an amplifier whose gain G just compensates for fiber loss:

$$G = 10^{AL/10}. \tag{21.31}$$

As an optical pulse travels through this cascade of amplifiers, it gets noisier and noisier, because the additive noise from each amplifier is cumulative. At the end of the chain the signal-to-noise ratio is

$$\text{SNR} = \frac{P_T \cdot 10^{-AL/10}}{\left[\dfrac{L_{\text{tot}}}{L}\right] \cdot K \cdot h\nu \cdot B}, \tag{21.32}$$

where L_{tot} is the total system length, so (L_{tot}/L) is the total number of repeaters, and the noise bandwidth equals the bit rate B. For any given total system length, SNR is maximized by using an optimum link length $L \sim 4.3/A$. (This corresponds to an optimum gain $G = e$ for each amplifier.) This optimization is not useful, however, because it requires impractically short fiber links: ~ 17 km for $A = 0.25$ dB/km. But even with a sub-optimal choice of L, impressive system performance can be achieved, as we show below.

The maximum transmission distance for the system of Fig. 21.20 is reached when the accumulated noise reduces the SNR (Eq. 21.32) to a value just large enough to assure reliable bit detection at the receiver. Let us assume a 1 Gb/s system with $P_T = 1$ mW. Taking $L = 100$ km, $A = 0.25$ dB/km, $K = 4$ and SNR = 13 dB (Schwartz *et al.*, 1966), we find from Eq. 21.32 that $L_{\text{tot}} \sim 10^5$ km, which is greater than the circumference of the earth! Thus, it appears possible to join any two points on earth with a chain of non-regenerative repeaters. (A more precise calculation, which takes into account many details not mentioned here, gives maximum system lengths in the vicinity of 20,000 km —still long enough to connect any points on earth (Yamamoto and Kimura, 1981).) We can thus imagine a worldwide network of "light pipes" providing transmission of signals at essentially arbitrary bit rates between any locations in the world. But alas, standing in the way of this exciting possibility is fiber dispersion, which imposes serious limitations on system performance, as previously discussed. The "light pipe" is especially vulnerable to dispersive effects because a transmitted pulse gets progressively broader as it propagates along the amplifier chain. (With regenerative repeaters pulse-spreading does not accumulate, because the pulses are reshaped at each repeater.) Thus in the case of the "light pipe," the appropriate value of L to use in Eq. 21.20 is the entire system length, not the individual link length. For $L_{tot} = 10^4$ km, Eq. 21.20 limits B to about 600 Mb/s. This might turn out to be a significant drawback to the use of optical amplifiers in long-distance transmission systems.

21.6 PHOTONIC NETWORKS

There is little doubt that lightwave communications systems are competitive with copper wire and even satellite systems in the area of long-distance broadband transmission. As was mentioned in the Introduction to this chapter, however, photonic network technology has not yet shown similar strength. The primary attraction of the lightwave medium—extremely low propagation loss—is largely irrelevant in this application, and other fiber characteristics, such as broad bandwidth and small size, may prove to be of far greater importance. Some of the different approaches that have been suggested or tried for photonic networks are presented in Chapter 26. For purposes of this chapter, we will concentrate on the simplest architectures, namely, those in which the lightwave medium is entirely passive and static, its sole purpose being to gather and distribute optical energy among the network users in a time-invariant fashion, much as a coaxial cable does in a conventional local area network. Even in these relatively primitive networks, two problems arise in applying photonic technology that are not encountered in the point-to-point communications systems discussed earlier: (1) the optical power must be distributed to many receiving stations, and (2) switching of some form must be implemented in order to route a message to its proper destination. First we consider the power distribution problem.

The very long transmission spans attained with lightwave systems are largely attributable to the low loss of the fiber. Optical power budgets are severely limited by available transmitter power on the one hand (typically $\leq$ 1 mW) and by practical receiver sensitivities on the other. Typical lightwave system margins of 30–40 dB are dwarfed by the 80 dB or greater values achievable in radio systems. Even ideal shot-noise-limited detection would only add 10 or 20 dB to the optical margins. If we now try to share this limited optical power by, say, sequentially tapping an optical bus, we find that the signal level decreases very rapidly. Neglecting propagation loss in the fiber itself, the power available to the user at the Nth tap is

$$P_N = P_T C \left[\beta (1 - C) \right]^{N-1}, \tag{21.33}$$

where C is the fraction of power coupled out at each tap and β accounts for the excess loss in the taps and splices ($\beta \sim 0.95$, or ~ 0.2 dB loss). The exponential dependence of P_N on N is a fundamental problem of the bus architecture, which makes it very difficult with today's technology to place more than a few tens of taps on a single bus. If optical amplifiers were available, the problem could be virtually eliminated, since tap loss could be compensated by optical gain. The situation is analogous to that treated in Section 21.5.3 with tap loss replacing fiber link loss. Installing an amplifier each time tap losses reduce the bus signal by $1/e$, we could extend the bus to over a million taps.

In the absence of optical amplifiers, a star topology, in which light from each transmitter is split by a star coupler (also called a power splitter) into N equal portions and routed back to all N terminals, has a significant advantage over the bus. With an ideal N-by-N star coupler, the power reaching each receiver from each transmitter is P_T/N. In practice, of course, the transmission loss will be somewhat greater. For a star composed of elemental 2-by-2 couplers, we expect

$$P_N = \frac{P_T}{N} \beta^{\log_2 N}, \tag{21.34}$$

where β again accounts for the excess loss of each elemental coupler. For reasonable values of β (~ 0.95) Eq. 21.34 shows that the star coupler behaves almost ideally—the available power decreases essentially inversely with N, rather than exponentially as on a bus. This improved performance yields important system advantages: less transmitter power is required to assure reliable detection, and lower power levels in general mean reduced impairments due to fiber non-linearities (Section 21.4.5).

Equations 21.33 and 21.34 have profound implications for the maximum achievable performance of photonic networks, measured in terms of total throughput, $B \cdot N$. For a given available signal power at a receiver, P_N, the maximum user bit rate is

$$B = \frac{P_N}{N_{\min} \cdot h\nu}, \tag{21.35}$$

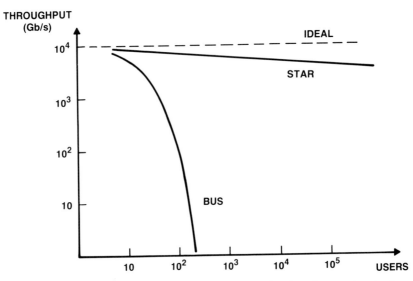

Fig. 21.21 Total traffic capacity (throughput) for representative photonic networks using bus and star topologies.

where N_{min} is the receiver sensitivity in photons per bit. Using Equations 21.33–35 with $P_T = 1$ mW, $N_{min} = 500$ and $C = 1/N$ (i.e., nominally uniform power distribution to all receivers), the throughput for bus and star topologies is shown in Fig. 21.21. Observe that $B \cdot N$ is almost independent of N for the star (a consequence of the $\sim 1/N$ behavior of Eq. 21.34), but falls rapidly for the bus as the number of users increases. This drawback of the bus architecture can be ameliorated by optimizing the coupling coefficient C, but it remains a very serious problem.

In order to establish communication links between network users, some sort of switching function must be performed. An attractive approach with photonic networks is to emulate radio: define channels on the basis of wavelength, broadcast all messages and use coherent techniques with tunable transmitter or local oscillator lasers to select a desired channel. The total tuning range needed (~ 100 nm) is comparable to the range over which InGaAsP has useful gain, and indeed external cavity lasers have been mechanically tuned over nearly this range (Wyatt and Devlin, 1983). These lasers are cumbersome, however, and not likely to be practical. A suitable electronically tunable diode laser has not been reported.

Another attractive approach to switching, less powerful than coherent techniques, but more compatible with the limitations of existing device technology, is subcarrier multiplexing (Darcie et al., 1986). In this scheme individual channels modulate microwave oscillators, which in turn are used to drive an optical source. When the optical signal is detected, the modulated microwave signal is recovered and then decoded without an optical local oscillator. Different microwave frequencies can be multiplexed onto the same laser or, using different lasers, onto the same fiber and subsequently separated at the detector using microwave frequency filters. The total bandwidth attainable in this technique is limited by the available modulation bandwidth of the sources (several GHz for semiconductor lasers) and the total number of channels is limited by the shot noise generated in the detector by the other signals. A system with 1000 channels of 1.5 Mb/s each seems possible (Darcie 1987).

REFERENCES

Abbas, G. L., Chan, V. W. S, and Yee, T. K. (1984). Cancellation of local oscillator intensity noise caused by the relaxation oscillation of GaAlAs lasers with a dual-detector heterodyne receiver. *Conf. Opt. Fib. Comm.*, New Orleans, Paper TUA2.

Abbott, S. M., Wagner, R. E., and Trischitta, P. R. (1983). SL undersea lightwave system experiments. *Conf. Rec. IEEE Intl. Conf. Comm.*, Boston MA. Paper C3.5.

Bickers, L., and Westbrook, L. P. (1985). Reduction of laser chirp in 1.5 μm DFB lasers by modulation pulse shaping. *Elect. Lett.* **21**, 103–104.

Blank, L. C., Bickers, L., and Walker, S. D. (1985). 220 km and 233 km transmission experiments over low-loss dispersion-shifted fibre at 140 Mb/s and 34 Mb/s. *Conf. Opt. Fiber Comm.*, San Diego, Paper PD-7.

Broom, R. F., Mohn, E., Risch, C., and Salathé, R. (1970). Microwave self-modulation of a diode-laser coupled to an external cavity. *IEEE J. Quantum Electron.* **QE-6**, 328–334.

Bohn, P. P., Brackett, C. A., Buckler, M. J., Rao, T. N., and Saul, R. H. (1984). The fiber SLC carrier system. *AT & T Bell Laboratories Technical Journal* **63**, 2389–2416.

Brain, M. C. (1982). Comparison of available detectors for digital optical fiber systems for the 1.2–1.55 μm wavelength range. *IEEE J. Quant. El.* **QE-18**, 219–224.

Cameron, K. H., Chidgey, P. J., and Preston, K. R. (1982). 102 km optical fiber transmission experiments at 1.52 μm using an external cavity controlled laser transmission module. *Electr. Lett.* **18**, 650–651.

Campbell, J. C., Dentai, A. G., Holden, W. S., and Kasper, B. L. (1983). High-performance avalanche photodiode with separate absorption (InGaAs), grading (InGaAsP) and multiplication (InP) regions. *Elect. Lett.* **19**, 818–820.

Casey, H. C., Jr, and Panish, M. B. (1978). "Heterostructure lasers, part B." Academic Press, New York, Chap. 8.

Cheung, N. K., Davidow, S., and Duff, D. G. (1984). A 90 Mb/s transmission experiment in single-mode fibers using 1.5 μm multi-longitudinal mode InGaAsP/InP lasers. *J. Lgtwv. Tech.* **LT-2**, 1034–1039.

Chraplyvy, A. R., and Henry, P. S. (1983). Performance degradation due to stimulated raman scattering in wavelength-division-multiplexed optical fibre systems. *Elect. Lett.* **19**, 641–643.

Chraplyvy, A. R., Stone, J., and Burrus, C. A. (1983). Optical gain exceeding 35 dB at 1.56 μm due to stimulated raman scattering by molecular D₂ in a solid silica fiber. *Opt. Lett.* **8**, 415–417.

Cohen, L. G., Kaminow, I. P., Astle, H. W., and Stulz, L. W. (1978). Profile dispersion effects on transmission bandwidths in graded index optical fibers. *IEEE J. Quant. El.* **QE-14**, 37–41.

Cohen, L. G., DiMarcello, F. V., Fleming, J. W., French, W. G., Simpson, J. R., and Weiszmann, E. (1978). Pulse dispersion properties of fibers with various material constituents. *Bell Sys. Tech. J.* **57**, 1653–1662.

Cohen, L. G, Mammel, W. L., and Jang, S. J. (1982). Low-loss quadruple-clad single-mode lightguides with dispersion below 2 ps/km · nm over the 1.28 μm–1.65 μm wavelength range. *Elect. Lett.* **18**, 1023–1024.

Cotter, D. (1982a). Observation of stimulated brillouin scattering in low-loss silica fibre at 1.3 μm. *Elect. Lett.* **18**, 495–496.

Cotter, D. (1982b). Transient stimulated brillouin scattering in long single-mode fibres. *Elect. Lett.* **18**, 504–506.

Cotter, D. (1982c). Suppression of stimulated brillouin scattering during transmission of high-power narrowband laser light in monomode fibre. *Elect. Lett.* **18**, 638–640.

Croft, T. D., Ritter, J. E., and Bhagavatula, A. (1985). Low-loss, dispersion-shifted single-mode fiber manufactured by the OVD process. *Tech. Dig., Conf. Opt. Fiber Commun.*, San Diego, Paper WD2.

Daly, J. C., Ed. (1983). "Fiber optics." CRC Press, Boca Raton, Florida, Chap. 5.

Darcie, T. E., Dixon, M. E., Kasper, B. L., and Burrus, C. A. (1986). Lightwave system using microwave subcarrier multiplexing. *El. Lett.* **22**, 774.

Darcie, T. E. (1987). Subcarrier multiplexing for lightwave multiple-access networks. *Tech. Dig. Opt. Fib. Conf.* OFC/IOOC '87, Reno, Paper MI3.

Dawson, R. W., and Burrus, C. A. (1971). Pulse behavior of high-radiance small-area electroluminescent diodes. *Appl. Opt.* **10**, 2367.

Fleming, M. W., and Mooradian, A. (1981a). Fundamental line broadening of single-mode GaAlAs diode lasers. *Appl. Phys. Lett.* **38**, 511–513.

Fleming, M. W., and Mooradian, A. (1981b). Spectral characteristics of external-cavity controlled semiconductor lasers. *IEEE Journ. Quant. Electron.* **QE-17**, 1, 44–59.

Foschini, G. J., Greenstein, L. J., and Vannucci, G. (1988). Noncoherent detection of coherent optical pulses corrupted by phase noise and additive gaussian noise. *IEEE Trans. Comm.* **36**, 306–314.

Fye, D. M. (1984). Practical limitations on optical amplifier performance. *J. Lgtwv. Tech.* **LT-2**, 403–406.

Garbuny, M. (1965). "Optical physics." Academic Press, New York, Chapter 5.

Garrett, I., and Todd, C. J. (1982). Components and systems for long-wavelength monomode fibre transmission. *Opt. and Quant. El.* **14**, 95–143.

Garrett, I., and Jacobsen, G. (1986). Theoretical analysis of heterodyne optical receivers for transmission systems using (semiconductor) lasers with nonnegligible linewidth, *J. Lightw. Tech.* **LT-4**, 323–334.

Gimlett, J. L., Vodhanel, R. S., Choy, M. M., Elrefaie, A. F., Cheung, N. K., and Wagner, R. E. (1987). 2 Gb/s, 101 km optical FSK heterodyne transmission experiment. *Tech. Dig. Optical Fiber Conf.* OFC/IOOC '87, Reno, Paper PDP-11.

Gloge, D., and Marcatili, E. A. J. (1973). "Multimode theory of graded-core fibers. *Bell Sys. Tech. J.* **52**, 1563–1578.

Gloge, D., Albanese, A., Burrus, C. A., Chinnock, E. L., Copeland, J. A., Dentai, A. G., Lee, T. P., Li, T., and Ogawa, K. (1980). High-speed digital lightwave communications using LEDs and PIN photodiodes at 1.3 μm. *Bell Sys. Tech. J.* **59**, 1365–1382.

Gnauck, A. H., Kasper, B. L., Linke, R. A., Dawson, R. W., Koch, T. L., Bridges, T. J., Burkhardt, E. G., Yen, R. T., Wilt, D. P., Campbell, J. C., Nelson, K. C., and Cohen, L. G. (1985). 4 Gb/s transmission over 103 km of optical fiber using a novel electronic multiplexer/demultiplexer. *Conf. Opt. Fib. Comm.*, San Diego, Paper PD2.

Gnauck, A. H., Korotky, S. K., Kasper, B. L., Campbell, J. C., Talman, J. R., Veselka, J. J., and McCormick, A. R. (1986). Information-bandwidth—limited transmission at 8 Gb/s over 68.3 km of single-mode optical fiber. Postdeadline Paper PDP 9, OFC '86 Atlanta.

Goell, J. E. (1974). An optical repeater with high-impedance input amplifier. *Bell Sys. Tech. J.* **53**, 629–643.

Gray, D. E. (1972). "American Institute of Physics Handbook," (3rd Ed.). McGraw-Hill, New York, Table 6b–17.

Harder, C. H., Vahala, K., and Yariv, A. (1983). Measurement of the linewidth enhancement factor α of semiconductor lasers. *Appl. Phys. Lett.* **42**, 328–330.

Harth, W., Huber, W., and Heinen, J. (1976). Frequency response of GaAlAs light-emitting diodes. *IEEE Trans. Electron. Devices* **ED-23**, 478–480.

Hartman, R. L., Schumaker, N. E., and Dixon, R. W. (1977). Continuously operated AlGaAs DH lasers with $70°$ lifetimes as long as two years. *Appl. Phys. Lett.* **31**, 756–759.

Henry, C. H. (1982). Theory of the linewidth of semiconductor laser. *IEEE J. Quant. El.* **QE-18**, 259–264.

Henry, C. H., Henry, P. S., and Lax, M. (1984). Partition fluctuations in nearly single-longitudinal-mode lasers. *J. Lgtw. Tech.* **LT-2**, 209–216.

Henry, P. S. (1984). The promise of coherent lightwave communications. *Proc. Nat. Comm. Forum*, Rosemont, Ill., 608–613.

Henry, C. H., and Kazarinov, R. F. (1986). Instability of semiconductor lasers due to optical feedback from distant reflectors. *IEEE Jour. Quant. Electron* **QE-22**, 294–301.

Henry, C. H. (1986). Theory of spontaneous emission noise in open resonators and its application to lasers and optical amplifiers. *J. Lightw. Tech.* **LT-4**, 288–297.

Hill, K. O., Tremblay, Y., and Kawasaki, B. S. (1980). Modal noise in multimode fiber links. *Opt. Lett.* **5**, 270–272.

Hodgkinson, T. G., Wyatt, R., Smith, D. W., Malyon, D. J., and Harmon, R. A. (1983). Studies of coherent transmission systems operating over installed cable links. *Proc. Globecom* **83**, San Diego, Paper 21.3.

Imoto, N., Kawana, A., Machida, S., and Tsuchiya, H. (1980). Characteristics of dispersion free single-mode fiber in the 1.5 μm wavelength region. *IEEE J. Quant. El.* **QE-16**, 1052–1058.

Irven, J., and Harrison, A. P. (1981). Single-mode and multimode fibers Co-doped with fluorine. *Tech. Dig. Intl. Conf. Integ. Opt. and Opt. Fiber Comm.*, San Francisco CA, Paper TuC4.

Iwahashi, E., and Fukutomi, H. (1983). F-400M System Overview. *Rev. Elect. Comm. Lab.* **31**, 237–243.

Iwashita, K., and Nakagawa, K. (1982). Mode partition noise characteristics in high-speed modulated laser diodes. *IEEE J. Quant. El.* **QE-18**, 2000–2004.

Iwashita, K., Matsumoto, T., Tanaka, C., and Motosugi, G. (1986). Linewidth requirement evaluation and 290 km transmission experiment for optical CPFSK differential detection. *Electron. Lett.* **22**, 791–792.

Jablonowski, D. P., Padgette, D. D., and Merten, J. R. (1982). Performance of the MCVD preform process in mass production conditions. *Tech. Dig. Top. Mtg. Opt. Fiber Comm.*, Phoenix AZ, Paper TuEE2.

Jacobs, I. (1986). Fiber-optic transmission technology and system evolution. Chapter 1 in "Digital Communications." T. C. Bartee, Ed., Howard Sams & Co., Indianapolis, IN, 1986.

Kaminow, I. P., and Tucker, R. S. (1987). Mode-controlled semiconductor lasers. In "Guided Wave Optoelectronics." T. Tamir, Ed., Springer-Verlag, New York.

Kao, C. K. (1982). "Optical Fiber Systems." McGraw-Hill, New York, Chap. 5.

Kapron, F. P., Keck, D. B, and Maurer, R. D. (1970). Radiation losses in glass optical waveguides. *App. Phy. Lett.* **17**, 423–425.

Kasper, B. L., Linke, R. A., Campbell, J. C., Dentai, A. G., Vodhanel, R. S., Henry, Kaminow, I. P., and Ko, J.-S. (1983). A 161.5 km transmission experiment at 420 Mb/s. *Ninth European Conf. Opt. Comm.*, Geneva, Paper PD#7.

Kasper, B. L., Burrus, C. A., Talman, J. R., and Hall, K. L. (1986). Balanced dual-detector receiver for optical heterodyne communication at Gbit/s rates. *Electr. Lett.* **22**, 10, 413–415.

Kazovsky, L. (1986). Performance analysis and laser linewidth requirements for optical PSK heterodyne communications systems. *J. Lgtwv. Tech.* **LT-4**, 415–425.

Keiser, G. (1983). "Optical Fiber Communications." McGraw-Hill, New York, Ch. 9.

Kikuchi, K., Okoshi, T., Nagamatsu, M., and Henmi, N. (1984). Degradation of bit-error rate in coherent optical communications due to spectral spread of the transmitter and the local oscillator. *J. Lgtwv. Tech.* **LT-2**, 1024–1033.

Koch, T. L., and Bowers, J. E. (1984). Nature of wavelength chirping in directly modulated semiconductor lasers. *Elec. Lett.* **20**, 1038–1039.

Koren, U., Eisenstein, G., Bowers, J., Gnauck, A. H., and Tien, P. K. (1985). Wide-bandwidth modulation of three-channel buried-crescent laser diodes. *El. Lett.* **21**, 500–501.

Korotky, S. K., Eisenstein, G., Gnauck, A. H., Kasper, B. L., Veselka, J. J., Alferness, R. C., Buhl, L. L., Burrus, C. A., Huo, T. C. D., Stultz, L. W., Nelson, K. C., Cohen, L. G., Dawson, R. W., and Campbell, J. C. (1985). 4 Gb/s transmission experiment over 117 km of optical fiber using a Ti:LiNbO₃ external modulator. *Conf. Opt. Fib. Comm.*, San Diego, Paper PD1.

Kressel, H., and Butler, T. K. (1977). "Semiconductor Lasers and Heterojunction LEDs." Academic Press, New York, Chap. 15.

Lenstra, D., Verbeek, B. V., and den Boef, A. J. (1985). Coherence collapse in single-mode semiconductor lasers due to optical feedback. *IEEE J. Quantum Electron* **QE-21**, 6, 674–679.

Li, T. (1983). Advances in optical fiber communications: an historical perspective. *IEEE J. Sel. Area Comm.* **SAC-1**, 356–372.

Li, T. (1980). Structures, parameters and transmission properties of optical fibers. *Proc. IEEE* **68**, 1175–1180.

Lin, C., and Burrus, C. A. (1983). CW and high-speed single-longitudinal-mode operation of a short InGaAsP injection laser with external coupled cavity. *Tech. Dig.*, Top Mtg. Opt. Fiber Comm. New Orleans LA, Paper PD5.

Lin, C., and Mengel, F. (1984). Reduction of frequency chirping and dynamic linewidth in high-speed directly modulated semiconductor lasers by injection locking. *Elec. Lett.* **20**, 1073–1075.

Linke, R. A. (1984). Direct gigabit modulation of injection lasers; structure dependent speed limitations. *J. Lgtwv. Tech.* **LT-2**, 40–43.

Linke, R. A. (1985). Modulation induced transient chirping in single-frequency lasers. *J. Quant. El.* **QE-21**, 593–597.

Linke, R. A., Kasper, B. L., Burrus, C. A., Kaminow, I. P., Ko, J.-S., and Lee, T. P. (1985). Mode power partition events in nearly single-frequency lasers. *J. Lgtwv. Tech.* **LT-3**, 706–712.

Linke, R. A. (1987). Beyond gigabit-per-second transmission rates. *Technical Digest Optical Fiber Conference OFC-87*, Reno.

Linke, R. A., and Gnauck, A. H. (1988). High-capacity coherent lightwave systems. To be publ. *J. Lgtwv. Tech.* **LT-6**.

Linnell, L. (1986). A wideband local access system using emerging technology components. *IEEE J. Sel. Area Comm.* **SAC-4**, 612.

Marcuse, D. (1977). LED fundamentals: comparison of front and edge emitting diodes. *IEEE J. Quantum Electron.* **QE-13**, 819–827.

Marcuse, D., and Lin, C. (1981). Low-dispersion single-mode fiber transmission—the question of practical vs. theoretical maximum transmission bandwidth. *IEEE J. Quant. El.* **QE-17**, 869–877.

Marshall, I. W., O'Mahony, M. J., and Constantine, P. D. (1986). "Measurements on a 206 km optical transmission system experiment at 1.5 µm using two packaged semiconductor laser amplifiers as repeaters. *Tech. Dig. 12th Eur. Conf. Opt. Comm.*, **I**, 253–256, Barcelona.

Martersteck, K. E., and Spencer, Jr., A. E. (1985). The 5ESS switching system: introduction. *AT & T Technical Journal*, **64**, No. 6, Part 2, 1305–1314.

Martin-Royle, R. D., and Bennett, H. G. (1983). Optical Fiber transmission systems in the British Telecommunications network: an overview. *British Telecommunications Engineering*, **1**, 190–199.

Matthews, M. R., Cameron, K. H., Wyatt, R., and Devlin, W. J. (1985). Packaged frequency-stable tunable 20 kHz linewidth 1.5 μm InGaAsP external cavity laser. *Elect. Lett.*, **21**, 113–115.

Mazurczyk, V. J., Bergano, N. S., Wagner, R. E., Walker, K. L., Olsson, N. A., Cohen, L. G., Logan, R. A., and Campbell, J. C. (1984). 420 Mb/s transmission through 203 km using silica-core fiber and a DFB laser. *Proc. 10th Eur. Conf. Opt. Comm.*, Stuttgart, Paper PD-7.

Midwinter, J. E. (1979). "Optical Fibers for Transmission." Wiley, New York.

Miles, R. O., Dandridge, A., Tveten, A. B., Taylor, H. F., and Giallorenzi, T. G. (1980). Feedback-induced line broadening in cw channel-substrate planar laser diodes. *App. Phys. Lett.* **37**, 990–992.

Miller, S. E., and Chynoweth, A. G. (1979). "Optical Fiber telecommunications." Academic Press, New York.

Miyashita, T., and Manabe, T. (1982). Infrared optical fibers. *J. Quant. El.* **QE-18**, 1432–1449.

Morikawa, T., Mitsubishi, Y., Shimoda, J., and Kojima, Y. (1976). Return-beam induced oscillations in self-coupled semiconductor lasers. *Electron. Lett.* **12**, 435–436.

Mukai, T., Yamamoto, Y., and Kimura, T. (1982). S/N and error rate performance in AlGaAs semiconductor laser preamplifier and linear repeater systems. *IEEE J. Quant. El.* **QE-18**, 1560–1568.

Mukai, T., Saitoh, T., Mikami, O., and Kimura, T. (1983). Fabry-Perot cavity type InGaAsP BH-laser amplifier with small optical-mode confinement. *Elect. Lett.* **19**, 582–583.

Murata, H., and Inagaki, N. (1981). Low-loss single-mode fiber development and splicing research in Japan. *IEEE J. Quant. El.* **QE-17**, 835–849.

Nagel, S. R., Walker, K. L., and MacChesney, J. B. (1982). Current status of MCVD: process and performance. *Tech. Dig.*, Top Mtg. on Opt. Fib. Comm., Phoenix, Paper TuCC2.

Nakagawa, K., and Ito, T. (1979). Detailed evaluation of an attainable repeater spacing for a fibre transmission at 1.3 μm and 1.55 μm wavelengths. *Elect. Lett.* **15**, 776–777.

Nakagawa, K., Ohta, N., and Hagimoto, K. (1983). Design and performance of an experimental 1.6 Gb/s optical repeater. *Intl. Conf. Int. Opt. and Fiber Comm.*, Tokyo, Paper 29C2-1.

Nelson, K. C., Brownlow, D. L., Cohen, L. G., DiMarcello, F. V., Huff, R. G., Krause, J. T., Reed, W. A., Shenk, D. S., Sigety, E. A., Simpson, J. R., and Walker, K. L. (1985). Fabrication and performance of long lengths of silica core fiber. *Conf. Opt. Fib. Comm.*, San Diego, Paper WH1.

Ogawa, K. (1982). Analysis of mode partition noise in laser transmission systems. *IEEE J. Quant. Elect.* **QE-18**, 849–855.

Okano, Y., Nakagawa, K., and Ito, T. (1980). Laser mode partition noise evaluation for optical fiber transmission. *IEEE Trans. Comm.* **COM-28**, 238–243.

Okoshi, T. (1982). "Optical Fibers." Academic Press, New York.

Okoshi, T. (1982). Heterodyne and coherent optical fiber communications: recent progress. *IEEE Trans. Micr. Theory and Tech.* **MTT-30**, 1138–1148.

Okoshi, T. (1987). Recent advances in coherent optical fiber communication systems. *J. Lgtwv. Tech.* **LT-5**, 44–52.

Olsson, N. A., Hegarty, J., Logan, R. A., Johnson, L. F., Walker, K. L., Cohen, L. G., Kasper, B. L., and Campbell, J. C. (1985). Transmission with 1.37 Tbit · km/sec capacity using ten wavelength division multiplexed lasers at 1.5 μm. *Conf. Opt. Fib. Comm.*, San Diego, Paper WB6.

Olsson, N. A., Tsang, W. T., Temkin, H., Dutta, N. K., and Logan, R. A. (1985). Bit-error-rate saturation due to mode-partition noise induced by optical feedback in 1.5 μm single longitudinal-mode C^3, and DFB semiconductor lasers. *Jour. of Lightwave Tech.* **LT-3**, 215–218.

Payne, D. N., and Hartog, A. H. (1977). Determination of the wavelength of zero material dispersion in optical fibers by pulse-delay measurements. *Elect. Lett.* **13**, 627–629.

Personick, S. D. (1973). Receiver design for digital fiber optical communication system—Part I. *Bell Sys. Tech. J.* **52**, 843–874.

Personick, S. D. (1981). "Optical Fiber Transmission Systems." Plenum Press, New York, Chapter 3.

Personick, S. D. (1971). Time dispersion in dielectric waveguide. *Bell Sys. Tech. J.* **50**, 843–859.

Personick, S. D. (1985). "Fiber Optic Technology and Applications." Plenum Press, New York, p. 125.

Petermann, K. (1980). Nonlinear distortions and noise in optical communication systems due to fiber connectors. *IEEE J. Quant. Electr.* **QE-16**, 761–770.

"Reference Data for Radio Engineers" (1975). Sams, H. W., Indianapolis, IN, Chap. 24.

Runge, P. K. (1974). A 50-Mb/s repeater for a fiber-optic PCM experiment. *Conf. Rec., Int. Comm. Conf.*, Minneapolis, MN, Paper 17B.

Runge, P. K., Brackett, C. A., Gleason, R. F., Kalish, D., Lazay, P. D., Meeker, T. R., Ross, D. G., Swan, C. B., Wahl, A. R., Wagner, R. E., and Williams, J. C. (1982). 101-km lightwave undersea system experiment at 274 Mb/s. *Tech. Dig.*, Top. Mtg. on Opt. Fiber Comm., Phoenix AZ, Paper PD7.

Runge, P. K., and Trischitta, P. R., Eds. (1984). Joint Special Issue on Undersea Lightwave Communications. *J. Lgtwv. Tech.* **LT-2**.

Saitoh, T., and Mukai, T. (1986). Low-noise 1.5-μm GaInAsP traveling-wave optical amplifier with high-saturation output power. *Tech. Dig.*, IEEE Intl. Semicon. Laser Conf., Kanazawa, Paper PD-5.

Sakai, K., Utaka, K., Akiba, S., and Matsushima, Y. (1982). 1.5 μm range InGaAsP/InP distributed feedback laser. *IEEE J. Quant. El.* **QE-18**, 1272–1278.

Salz, J. (1985). Coherent Lightwave Communications. *AT & T Tech. J.* **64**, 2153–2209.

Salz, J. (1986). Modulation and detection for coherent lightwave communications. *IEEE Communications Magazine* **24**, 38–49.

Schimpe, R., Bowers, J. E., and Koch, T. L. (1986). Characteristics of frequency response of 1.5 μm InGaAsP DFB laser diode and InGaAs PIN photodiode by heterodyne measurement technique. *Elec. Lett.* **22**, 453–454.

Schwartz, M., Bennett, W. R., and Stein, S. (1966). "Communication Systems and Techniques." McGraw-Hill, New York, Esp. Chap. 7.

Schwartz, M. I., Reenstra, W. A., Mullins, J. H., and Cook, J. S. (1978). The Chicago lightwave communications project. *B.S.T. J.* **57**, 1881–1888.

Shah, V., Young, W. C., and Curtis, L. (1987). Large fluctuations in transmitted power at fiber joints with polished end-faces. *Tech. Dig. Opt. Fib. Comm. Conf.* OFC/IOOC '87, Reno, Paper TuF4.

Simon, J. C. (1983). Semiconductor laser amplifier for single mode optical fiber communications. *J. Opt. Comm.* **4**, 51–62.

Simon, J. C. (1986). Present status and future of optical amplification for communication. *Tech. Dig. 12th European Conference on Optical Communication*, III, Barcelona, pp. 39–45.

Smith, R. G. (1972). Optical power handling capacity of low loss optical fibers as determined by stimulated Raman and Brillouin scattering *Appl. Opt.* **11**, 2489–2494.

Smith, D. W., Atkins, C. G., Cotter, D., and Wyatt, R. (1986). Application of Brillouin amplification in coherent optical transmission. *Tech. Dig. Opt. Fib. Comm. Conf.*, Atlanta, GA, Paper WE3.

Stauffer, J. R. (1983). FT3C—a lightwave system for metropolitan and intercity applications. *IEEE J. Sel. Areas Comm.* **SAC-1**, 413–419.

Stern, M., Gimlett, J. L., and Cheung, N. K. (1986). Dispersion-limited transmission and spectral filtering in light-emitting diode single-mode fiber communication systems. *Optics Letters* **11**, 584–586.

Stone, J., and Lemaire, P. J. (1982). Reduction of loss due to OH in optical fibers by a two-step OH- > OD exchange process. *Electron. Lett.* **18**, 78–80.

Strong, J. (1958). "Concepts of Classical Optics." Freeman, San Francisco, Appendix N.

Suematsu, Y. (1983). Long-wavelength optical fiber communication. *Proc. IEEE* **71**, 692–721.

Suematsu, Y., Ed. (1982). "Optical Devices and Fibers." North-Holland, Amsterdam.

Suematsu, Y., Arai, S., and Kishino, K. (1983). Dynamic single-mode semiconductor lasers with a distributed reflector. *IEEE J. Lgtwv. Tech.* **LT-1**, 161–175.

Sugimura, A., Daikoku, K., Imoto, N., and Miya, T. (1980). Wavelength dispersion characteristics of single-mode fibers in low-loss region. *IEEE J. Quant. El.* **QE-16**, 215–225.

Temkin, H., Olsson, N. A., Abeles, J. H., Logan, R. A., and Panish, M. B. (1986). Reflection noise in index-guided InGaAsP lasers. *IEEE Jour. Quant. Electron*, **QE-22**, 2, 286–293.

Tkach, R. W., and Chraplyvy, A. R. (1986). Regimes of feedback effects in 1.5 µm DFB lasers. *Jour. Lightwave Tech.* **LT-4**.

Toba, H., Kobayashi, Y., Yanagimoto, K., Nagai, H., and Nakahara, M. (1984). Injection-locking technique applied to a 170 km transmission experiment at 445.8 Mbit/s. *Elec. Lett.* **20**, 370–371.

Tomita, A. (1983). Cross talk caused by stimulated raman scattering in single-mode wavelength division multiplexing systems. *Top. Mtg. Opt. Fiber Comm.*, New Orleans, LA, Paper TUH5.

Tsang, W. AT., Olsson, N. A., and Logan, R. A. (1983). High-speed direct single-frequency modulation with large tuning rate and frequency excursion in cleaved-coupled-cavity semiconductor lasers. *App. Phys. Lett.* **42**, 650–652.

Watson, H. A., Ed. (1969). "Microwave Semiconductor Devices and Their Circuit Applications." McGraw-Hill, New York, Chap. 12.

Wyatt, R., and Devlin, W. J. (1983). 10 kHz 1.5 µm InGaAsP external cavity laser with 55 nm tuning range. *Electron. Lett.* **19**, 110–112.

Yamada, J. I., Machida, S., and Kimura, T. (1981). 2 Gbit/s optical transmission experiments at 1.3 µm with 44 km single-mode fibre. *Electr. Lett.* **17**, 479–480.

Yamada, J., Kawana, A., Miya, T., Nagai, H., and Kimura, T. (1982). Gigabit/s optical receiver sensitivity and zero-dispersion single-mode fiber transmission at 1.55 µm. *IEEE J. Quant. El.* **QE-18**, 1537–1547.

Yamaguchi, M., Emura, K., Kitamura, M., Mito, I., and Kobayashi, K. (1985). Frequency chirping suppression by a distributed-feedback laser diode with a monolithically integrated loss modulator. Paper WI3 OFC '85, San Diego, *Tech Digest* pp. 103–104.

Yamamoto, S., Sakaguchi, H., and Seki, Norio (1982). Repeater spacing of 280 Mbit/s single-mode fiber-optic transmission system using 1.55 µm laser diode source. *IEEE J. Quant. El.* **QE-18**, 264–273.

Yamamoto, Y., and Kimura, T. (1981). Coherent optical fiber transmission systems. *IEEE J. Quant. El.* **QE-17**, 919–935.

Yamazaki, S., Emura, K., Shikada, M., Yamaguchi, M., Mito, I., and Minemura, K. (1986). Long-span optical FSK heterodyne single-filter detection transmission experiment using phase-tunable DFB laser diode. *Electron. Lett.* **22**, 5–6.

Chapter 22

Interoffice Transmission Systems

STEVEN S. CHENG

Bell Communications Research, Inc., Morristown, New Jersey

ERIC H. ANGELL

AT&T Bell Laboratories, Inc., Ward Hill, Massachusetts

22.1 INTRODUCTION

Between 1980 and the end of 1985, several million kilometers of optical fiber had been installed in the United States, effectively doubling the transmission capability of the telecommunication networks. Point-to-point optical links are being used in ever increasing applications ranging from equipment interconnection between computers, digital switches to high capacity interoffice and intercity backbone transmission systems. In many metropolitan areas, fibers are being used to provide interconnection services between various local area networks scattered in geographically separated areas. Use of fibers in local loops and its far reaching impact in the information age is generally considered as the most challenging problem in telecommunication for the balance of this century.

Following the feasibility demonstration of a 44.7 Mb/s optical trunk transmission system in Atlanta in 1976 (Jacobs, 1976), and a successful field experiment using the same equipment in downtown Chicago in 1977 (Schwartz *et al.*, 1978), the stage was set for wide deployment of multimode fiber transmission systems in the metropolitan and suburban areas between central offices. The first generation fiber optics system was thus designed, engineered, installed and maintained as a cost-effective replacement for the then prevailing 1.544 Mb/s T1 interoffice trunking system. Especially important to the metropolitan environment is its high bandwidth, low loss and small size. The high bandwidth matched the large circuit demand between large central offices. Low fiber loss resulted in longer repeater spacing, which often removed the need of placing intermediate repeaters in manholes. Small cable size conserved the crowded duct space underground. High capacity and longer repeater spacing also drastically reduced the maintenance activities often required for the T1 systems with the same capacity and distance. The largest multimode fiber trunking system was installed in 1983–84 between Richmond, Virginia and Cambridge, Massachusetts. Known as the North East Corridor system, it has 80,000 km of fiber along a 1250 km route. The system initially ran at 90 Mb/s but was designed to be upgradable to support two short wavelength channels (0.8–0.9 μm) and one long wavelength channel (1.3 μm) simultaneously on each fiber.

By 1979–1980, the rapid progress in single-mode fibers and long wavelength lasers (1.3 μm region) indicated that the cost and performance of a single-mode fiber transmission system could soon surpass that of a multimode fiber system. This was demonstrated in several system experiments in Japan (Ito *et al.*, 1983) and the United States (Cheng *et al.*, 1983). By operating lasers at the essentially zero dispersion wavelength of fiber at 1.3 μm, the single-mode transmission system has negligible dispersion at the highest bandwidth possible as a consequence of power and noise considerations. The transmission speed was constrained by the availability of the fast electronics in transmitter, regenerator circuitry and digital multiplexer. The lower cable loss at 1.3 μm, typically 0.5 dB/km installed, was responsible for increasing the engineered repeater spacing to 40 km at a bit rate of 400–600 Mb/s. The nearly two orders of magnitude increase in bit rate-distance product of a single-mode system over the initial 44.7 Mb/s multimode fiber system and the potential for substantial upgrades are mainly responsible for its mercurial acceptance in interoffice network. By 1986, the majority of telephone companies were installing single-mode fibers for interoffice as well as feeder routes.

In Section 22.2, the adaptation of the revolutionary fiber technology into the interoffice transmission network is described, including a discussion of the equipment, service configuration and the compatibility with the existing digital transmission hierarchy.

Until very recently, because of the nature of point-to-point high capacity transmission, there is little distinction in the design and the operation of a long haul intercity fiber system and that of a short haul interoffice fiber system except for the number of regenerators required. This is changing, however, as a result of the network partitioning after divestiture, the rapid progress in VLSI technology and the desire to provide customers with flexible and timely network services through software control. In the near future fiber optics will become the driving force and a key ingredient in the design of a wideband, integrated and intelligent transport for the metropolitan environment. Section 22.3 explores the desirable characteristics of such a transport and its impact on network topologies.

The last section discusses several promising emerging lightwave technologies that are being explored for interoffice transport system for possible use in the next 6–10 years. They may greatly enhance the network functionalities and increase the speed of transmission while evolving the fiber based network to become an all photonic network.

22.2 INTEROFFICE TRUNKING SYSTEMS

22.2.1 Interoffice Trunking Environment

This section discusses the present and anticipated applications of lightwave transmission systems to interoffice needs. Interoffice as used in this chapter refers to the trunking needs between central offices. This application is driven by different considerations than other applications, discussed in other chapters of this book, such as undersea, intercity, and distribution. For example, the primary cost in long haul is in the repeaters and cable, which connect the two distant end points together, and not the multiplexers or office circuits, which occur only at the ends of the long haul system.

Contrast this with the interoffice applications where the distance is short and optical repeaters are almost unnecessary because of the close proximity of the offices. For this case, the primary cost is usually in the multiplexers and office circuitry. Another factor important to interoffice applications is the fact that each office interconnects with more than one other office. Again, this is in contrast with long haul, which by its very nature tends to connect two distant offices together (the traffic, of course, connects to more than these two offices, but these offices are the gateways between long haul and interoffice systems). These differences lead to the implementation of different fiber systems, which each match the specific characteristics of the different environments to yield cost effective solutions. In order to have end-to-end network integrity and efficient operation, however, it is imperative for the intercity and interoffice systems to function harmoniously together.

22.2.2 Digital Transmission Systems

Interoffice circuits were originally carried on copper pairs, with analog amplification for the longer spans. One voice circuit required two pairs of wires, one for each direction of transmission. Since the 1960s, with the advent of T-carrier, a conversion of interoffice network circuits to digital transmission started. The driving force at that time was the cost of providing connectivity between offices. Since T1 carrier provided 24 voice circuits on two pairs of wires, the savings was due to "pair gain," or the savings of not needing as many copper pairs. This allowed more circuits to traverse a given copper cable, avoiding the need for new or larger copper cables. The added cost of putting the voice signals into the digital format and multiplexing them together was subtracted from this pair gain to yield the net savings. The distance at which these factors were equal (zero net savings), was the break even point or "the prove-in distance" (Fig. 22.1) Thus, the offices with greater distances between them and with larger circuit demands were the offices that initially converted to digital circuits. This also resulted, however, in the addition of outside plant equipment in the form of regenerators and manholes to house them. These were required at intervals of about 1 or 2 kilometers. More time and effort were necessary to maintain this outside plant equipment, which in the interoffice environment was often below the busy city streets in congested ducts. Other factors influencing the decision to add T-carrier include projected circuit growth, office wiring congestion and requirements for digital connectivity.

One of the advantages of deploying T1 carrier systems was that they were robust enough to work over almost any of the existing interoffice trunk cables. This avoided the expense of replacing the cables below the city streets. Later, other T-carrier systems, in the form of T1C and T1D were introduced. These systems reduced cost per channel by carrying 48 channels each, but maintained the same repeater spacing as the T1 system. At higher bit rates, the crosstalk in the cables became more severe, and T1C could only be put on a subset of total T1 applications and required replacement of apparatus cases. New, more complicated T1D repeaters were introduced to maintain the same regenerator spacing, apparatus cases, and cable usage as T1. However, the savings realized still allowed a reasonable prove-in distance. The cost of the added circuits to process the more complicated coding methods went up rapidly as the bit rate increased. Hence, the bit rates for the higher level signals, such as T1C and T1D were kept to a minimum by not including a lot of overhead bits (non-traffic bits) at each multiplex stage.

At the DS3 rate (44.7 Mb/s) or above, however, the balanced copper pairs could no longer be used due to the high loss and crosstalk, so coaxial cable was necessary. A coaxial cable based digital transmission system at DS4 rate of 274 Mb/s was developed in mid-1970, but because of its size and cost, it was deployed in a very limited application. (European deployment of digital trunk transmission came along at a later time and developed a digital hierarchy similar to but not identical to the North American hierarchy just described. Both are depicted in Fig. 22.2).

The strategy of deploying a transmission medium with high bandwidth relative to current needs and then upgrading it with higher bit rates in the future is a sound one if the cost of the medium is not too high. With the advent of optical fiber and its promise of high capacity plus low loss, the interoffice trunking area seems to have found the next generation of transmission medium

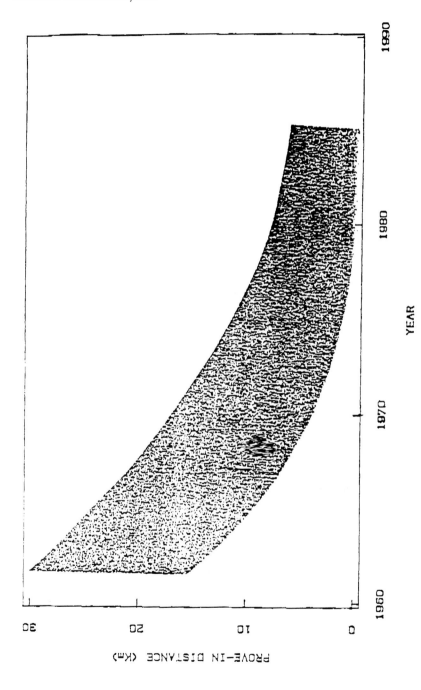

Fig. 22.1 T-Carrier prove-in distance.

Digital Signal Level	1.544 Mb/s Based Hierarchy	2.048 Mb/s Based Hierarchy
0	64 Kb/s	64 Kb/s
1	1.544 Mb/s	2.048 Mb/s
1C	3.152 Mb/s	
2	6.312 Mb/s	8.448 Mb/s
3	44.736 Mb/s	34.368 Mb/s
4E	139.264 Mb/s	
4	274.176 Mb/s	139.264 Mb/s

Note: Both Hierarchies continue to evolve with time and are not the only hierarchies used worldwide.

Fig. 22.2 Digital Hierarchies.

that meets these same needs. The use of optical fibers often saves the much higher expense of installing additional duct, which would have been necessary to meet the now much higher circuit requirements with balanced copper cable or coaxial cable.

Optical fiber has been used in the interoffice trunking area since the early 1980s. However, since the inherent capacity was high and the initial cost of the fiber was also relatively high, multiplexing more circuits together helped bring the cost per circuit down. Since multiplexers were already in existence to reach the DS3 level (developed for the DS4 coaxial cable transmission system), these were typically used and the input to the fiber system itself was the DS3 level. (Some early systems integrated an M13 into the lightwave terminal and were functionally identical.) One or more DS3s were further multiplexed together in a

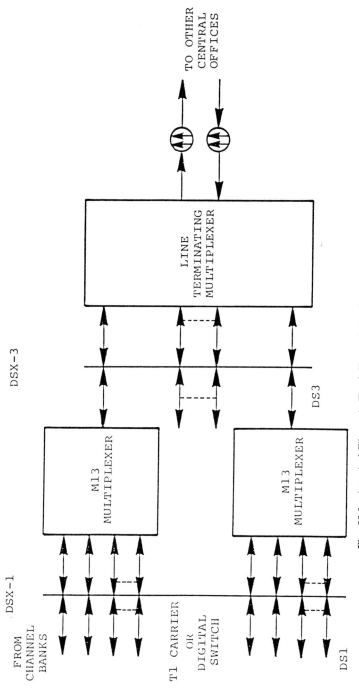

Fig. 22.3 A typical Fiberoptic Trunk Transmission System.

839

lightwave terminal before being sent out optically (Fig. 22.3). The use was motivated by the smaller duct space required, the longer spacing of repeaters (hence less outside plant electronics required), and the desire for a medium that could be exploited later for higher capacity without requiring replacement. The predominance of optical fibers in place by the mid 1980s in the interoffice trunking area were DS3 based.

More recently, several other factors have entered the forefront of thinking in meeting the needs of interoffice facility planning. In particular, one new factor that has entered into the picture is the recognition that a whole network is involved and not just point-to-point transmission links. Secondly, from this network view, it can be seen that the needs of each office are to send digital traffic to a number of other offices. This leads to the notion of some traffic on each fiber dropping in the office, some being added and some going through to other offices. Using the current DS3 format, this is straightforward but generally requires back-to-back M23 and M12 multiplexers (sometimes combined in an M13). Since the traffic is usually in DS1 bundles, it is not possible to extract DS1s in the present DS3 format without demultiplexing the full DS3 to DS2s. Then each DS2 with a desired DS1 must be fully demultiplexed. One of the new approaches to overcome this is found in the proposal of a new DS3 format (identified as SYNTRAN) (REF T1X1.4/85-050) using the same 45 Mb/s bit rate, but using a new format. This does give the advantage of allowing direct add/drop of DS1s from the DS3, but significant processing is required mainly because the frame length of DS1 and DS3 do not coincide.

Another approach being taken is to define new formats that are not constrained by the older hierarchy, which was chosen when bandwidth was a very expensive commodity. Because incremental bandwidth is not the expensive constraint on fiber that it was on balanced copper cable and coax cable, it is now possible to use the bandwidth for new rates and formats, which provide direct DS1 observability for advantages in add/drop realizations and also provide overhead for maintenance functions. In other words, the inherent bandwidth capabilities of fiber allow a rethinking of network architectures and maintenance features to achieve the most economical interoffice network.

Another set of factors that has aided the deployment of optical fiber systems in the interoffice area is the rising cost of craft time and the increasing complexity available with lower cost electronic circuits primarily through the advent of VLSI. These factors favor the use of fiber, which has offered longer repeater spacing than copper cable T-carrier systems, and hence less outside plant maintenance time from the craft. The VLSI circuit cost reductions have decreased the multiplexing costs and have led to a desire for electronic crossconnecting of circuits at intermediate offices to decrease the time required and cost necessary for rearranging circuits manually.

22.2.3 Multimode Fiber System Design

In order for the multimode fiber to achieve the best performance in bit rate-distance product, a GaAlAs laser operating at 0.8–0.9 μm and a silicon APD were used in the design of the initial interoffice transmission system, achieving a repeater spacing of 7–10 km for 44.7 Mb/s (DS3) transmission. For shorter distances, lower bit rate systems and higher reliability considerations, LED and pin detectors had been used as alternative sources and detectors. Figure 22.4 shows a plot of repeater spacing vs. bit rate achievable using short wavelength laser and APD detectors on multimode fiber. Good engineering practices dictate that the optical link be designed to operate essentially in the loss limited region, i.e., impairment due to dispersion limited to 1 dB or less. The laser will typically launch 0 dBm optical power into a 50 μm core, 0.2 NA fiber. A silicon APD is typically 10–15 dB more sensitive than a pin detector. At 44.7 Mb/s, the APD detector sensitivity is about -50 dBM. The 50 dB difference in transmitter and receiver power is allocated to accommodate optical path loss, regenerator impairments and system margin, which includes environmental effects, aging and repair margin. A typical service configuration including standard electrical interfaces and optical interface is illustrated in Fig. 22.3.

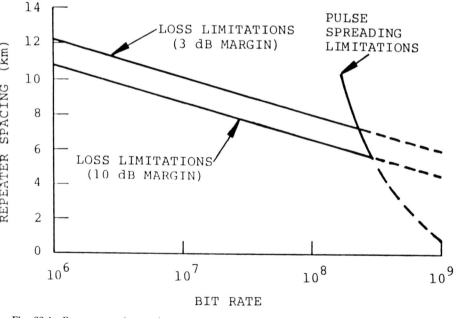

Fig. 22.4 Repeater spacing vs. data rate—short wavelength laser system with APD detector and multimode fiber.

This would enable the fiber trunking system to operate in a T1 environment. A DSX-1 cross-connect frame terminates signals from T1 sources such as channel banks, digital loop carriers and digital switch interfaces. These signals are forwarded to a M13 digital multiplexer, which combines up to 28 DS1 signals to a 44.7 Mb/s DS3 signal as described in 22.2.2. The 44.7 Mb/s signals from M13 multiplexers and other DS3 sources will appear in a DSX-3 cross-connect frame. For fiber trunk that operates higher than 44.7 Mb/s, a line terminating multiplexer is employed to combine multiple DS3 signals to a high bit rate signal that directly modulates a laser transmitter. Line terminating multiplexers with two, three, four, nine and twelve DS3 inputs are now available commercially. Automatic protection switching is usually built into the line terminating multiplexer to protect multiplexer and repeatered line failures. Microprocessor based maintenance subsystem enables failures to be rapidly detected, sectionalized and localized for ease of repair. Optical cable distribution frame provides a flexible point to patch outside plant fibers to in-office fibers. Because of the 7–10 km repeater spacing, most of the repeaters can be housed in central offices instead of manholes.

With the availability of long wavelength lasers, lower fiber loss and smaller dispersion in the 1.3 μm region, multimode fiber systems can now handle 180 Mb/s (4 DS3) with a repeater spacing of 25 km. Taking advantage of these advances, the 90 Mb/s North East Corridor system is being upgraded to triple its capacity by adding 180 Mb/s in the 1.3 μm region, while skipping one intermediate repeater location. Increasing capacity while eliminating repeaters is a truly remarkable achievement unique only to fiber transmission technology. However, 180 Mb/s is about the highest bit rate that can be transmitted without incurring significant modal dispersion penalties. Further improvement will have to come from single-mode fiber technology, which became commercially available in 1983.

22.2.4 Single-mode Fiber System Design

In the U.S., as soon as they became available, 1.3-μm long wavelength lasers and single-mode fibers were first incorporated with the existing 90 Mb/s and 180 Mb/s transmission systems as a means to extend the repeater spacing to over 40 km at these rates. World-wide efforts to harness the potential of single-mode fiber technology actually started even before the first multimode fiber system became commercially available. The coincidence of zero dispersion and low fiber loss at 1.3 μm region (Chapter 21) provided a broad wavelength window to engineer a very high bit rate trunking system with long repeater spacing suitable for the long haul intercity environment. The system transmission rate of 400–600 Mb/s was limited by the availability of fast electronics rather than by the performance of fiber, laser or photodetector. With a transmitted power of typically -2 dBm and a receiver sensitivity of -34 dBm at 500 Mb/s, the available system gain is 32 dB of which 21 dB is allocated for optical path loss and the remainder for regenerator impairments and system margin. With an installed cabled fiber loss, including occasional single-mode fiber splices of 0.4–0.5 dB/km, the repeater spacing of 40 km is achievable. When this system is installed for interoffice transmission in a metropolitan area, the need for intermediate repeaters between offices is virtually eliminated. This advantage and the potential for upgrade to operate at higher bit rates or by the use of wavelength division multiplexer channels quickly made it the preferred choice for trunking expansions.

The remarkable rapid changeover from multimode fiber systems to single-mode fiber systems for interoffice transmission in mid-1980s does not change the service configuration described in section 22.2.2. For the initial single-mode system, the line terminating multiplexer typically handles nine to twelve DS3s with a line rate of 417 Mb/s to 680 Mb/s. Many of these systems are designed and engineered for upgrade to double or quadruple the initial capacity.

While single-mode fiber is being established as the preferred choice of media in virtually all areas of telecommunications, the cost, reliability and need to use long wavelength lasers is coming under scrutiny for feeder and loop applications and for short and dedicated trunks for corporate users in the metropolitan area. With the recent encouraging experimental results of LEDs for single-mode fiber transmission (Shumate *et al.*, 1985), increasing numbers of such applications will turn to LEDs as sources wherever the cost and performance can justify it.

22.2.5 Field Experience to Date

The first major Bell System deployment of fiber optic transmission systems in the interoffice arena took place in Smyrna, Georgia in 1980. Several more successful system applications followed, both in the U.S. and other countries. The initial applications were on multimode fiber and fairly low bandwidth usage, with thoughts of WDM or increased bit rate as a way of allowing future upgrades. These systems were successful for the intended application, but the expected improvements in technology with time developed sooner than initially anticipated and have voided some of the potential upgrade scenarios as single-mode fiber has taken over the market. As might be expected with deployment of new technology, there were some problem areas such as short laser lifetimes, aging of optical cable splices, and lack of cable robustness. Despite these problems, no loss of service was experienced and the systems met the expectations for overall reliability, through the use of automatic protection switching.

As the manufacturing techniques matured, quality systems became the common end result of optical fiber applications in the interoffice arena. End users, especially the larger more sophisticated ones with their own telecommunications departments, started recognizing that fiber systems gave superior performance to copper systems. Reports in the literature indicate that fiber system bit error rates were extremely good, in the range of about 99.9999% error free seconds (Tse, 1986, LeGall, 1986). The net result has been a demand from end users to specify that their circuits be carried via fiber whenever possible. Operating company T Carrier lines have to be measured and carefully selected for providing DATAPHONE® Digital Service (DDS).[1] With the performance achieved via optical fiber systems, this measuring and selecting procedure for DDS is avoided. These factors have further accelerated the deployment of fiber systems.

22.3 DESIGN TOWARD AN INTEGRATED TRANSPORT IN EXCHANGE NETWORKS

22.3.1 Fiber Impact on Interoffice Network Topology

Fiber deployment in the interoffice arena has not only resulted in excellent span relief strategies, which tended to be point-to-point in an overall mesh network, but also allowed a reexamination of topologies. Optical fiber bandwidth being very large, and the proportionally increasing cost of craft and maintenance becoming a larger factor, have led this thrust. The potential of installing a medium that would allow future upgrades for added bandwidth, both for meeting increasing traditional service needs and the introduction of new services to take advantage of the low cost bandwidth, is important. These new services include seasonal and time of day offerings, which require more remote software control to avoid manual craft rearrangements. These factors lead to careful reexamination of the interoffice topology.

With the bandwidth available on fiber, a pure hub topology (Fig. 22.5a) seemed appropriate because the cost of backhauling and misrouting of circuits was offset by the ability to use an automated crossconnect facility in a larger hub office. Several drawbacks to a pure hub topology, however, became apparent. First, the seemingly unlimited bandwidth of fiber is in reality limited at any point in time, and while bit rates are moving upward, the use of the very latest technology does pose cost and reliability concerns. The bandwidth needed to connect all circuits through a single hub in the interoffice arena is the sum of all the individual office demands, and while each office has fairly limited demands, the number of offices is usually large. Because the fiber capacity is still large relative to a single office demand, several offices could share a single fiber headed for the hub office. The end result is a topology that demands very high bandwidth even relative to today's fiber systems, and in addition, poses potential problems with respect to reliability of service. A single fiber cable cut or system

---

[1]DATAPHONE is a trademark of AT & T.

outage would result in cutting off service to a large portion of the network, certainly a very risky situation.

Another approach is to use a combination network with a hub topology and additional direct paths along with automated add/drop at intermediate offices (Fig. 22.5b). This allows for a flexible higher reliability topology and a much more cost effective network. Such a system is currently offered by at least one company and is also being discussed in the industry standards arena under the name of SONET (T1X1.4/85-017). It is expected that eventually this effort will allow interconnection of different vendor's optical systems at an optical midspan meet (see also 22.4.2). The new formats used, along with the internal synchronization to a single interoffice standard clock, also allow for very cost effective crossconnecting. These features implemented with flexible software provisioning and control architectures appear to be the next step in the evolution of the

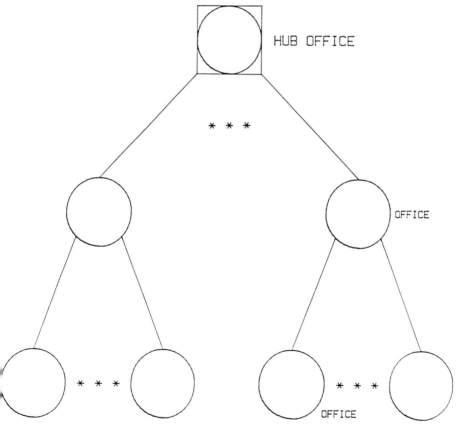

Fig. 22.5 (a) HUB topology.

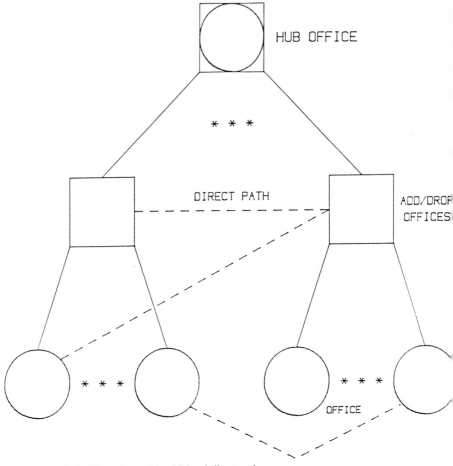

Fig. 22.5 (b) HUB topology with additional direct paths.

network. Some of the implications of these new approaches are described in the paragraphs that follow.

22.3.2 Synchronous Transmission and Multiplexing

Currently, most of the interoffice fiber systems in North America interface with DS3 signals in the end offices. These DS3 signals are first multiplexed together with added overhead bits and then converted to an optical signal to be launched onto the fiber. The approach used is to assume that all of the signals entering the fiber system as DS3s are nominally running at the same bit rate, but are in fact asynchronous and running at slightly different rates, each being defined by a different clock. In order to multiplex them together, they must all be synchronized to a common clock. In current systems this is usually the transmit clock of the optical system, which for cost reasons is usually not a high stability clock.

Each of the input signals is bit stuffed up to this transmit clock rate before multiplexing; a drift in either the incoming DS3 rates or the transmit clock rate are thus accommodated. (See Transmission Systems for Communications, 1982, for bit stuffing explanation.) At the receive end of the system, a clock is recovered from the incoming bit stream and all signals are demultiplexed, destuffed and smoothed before going on to a DSX3 crossconnect for connection to the next system. This method of bit stuffing is cost effective for point-to-point systems with asynchronous input signals. For the interoffice network, newer systems with internally synchronous formats may be more cost effective. This results from the need to interconnect many optical routes within one office and requires crossconnecting and add/dropping arrangements. Since these systems are not synchronized to each route's transmit clock, but rather to a single interoffice reference (such as the Basic Synchronization Reference Frequency [BSRF]), they may be rearranged without the added expense of demultiplexing and desynchronizing. In the future, as more of the traffic is generated with bit rates that are synchronized directly to the master clock in the network, the need to bit stuff will disappear. In the meantime, bit stuffing for asychronous inputs is required. An effective transition vehicle appears to be bit stuffing for these asynchronous inputs, followed by bit interleaving at the higher levels. Bit stuffing requires more processing circuits than bit interleaving. At the lower bit rates, where VLSI is available, this allows flexibility and does not appear to be a cost penalty. As bit rates increase, however, both bit and byte interleaving methods may be considered. Other considerations become important in synchronous networks, such as wander and synchronous boundaries. These topics are being addressed in the Exchange Carrier Standards Association (ECSA) committee T1 work and are beyond the scope of this book.

22.3.3 Facility Switching

As discussed in Section 22.3.1, the very large bandwidth on fiber results in a concern about the reliability of service if there is a loss of a fiber or cable. This can be addressed in a number of ways. Automatic protection switching of the electronics and the associated fiber spans is normally employed to increase reliability. As the volume of traffic increases, however, this method alone may not be enough. If the protection switching routes traffic on the same fiber cable, and the cable is cut, all service would be lost. One way to avoid this cable cut problem is to route the protection fiber in a different cable, preferably along a different physical path. In some cases this is possible; however, the added cost of diverse routing of all protection fibers is often prohibitive.

Another approach to increasing the reliability of service is to provide facility switching. This means that when service is lost along the primary path, it is restored over another path. When there are several paths in the network to connect the desired end points together, and the system involved has the intelligent software control to utilize it, service may be restored in this fashion. There may be several levels of multiplex and other equipment involved between the end points and these may all be included within the switched path. This method provides a highly reliable network and in addition, leads to other features, as discussed in the next section.

The ability to allocate bandwidth dynamically with facility switching on trunks, could, in 6 to 10 years, lead to a restructuring of tandem switch functions. While it is too early to give exactly what form this restructuring will take, it is clear that this is a step in the evolving interoffice network.

22.3.4 Dynamic Network Control

The previous sections have indicated that the interoffice network is evolving toward an optical fiber intelligent network to meet the current needs. In addition, this same network will provide features that will enable new services to be added in a cost effective manner to enhance the revenues generated for the operating companies. A key to this future is the concept of dynamic network control. Dynamic network control implies that the interoffice network has the intelligence under software control to allow it to be monitored and controlled remotely. This flexibility, without attendant craft activity spread over many offices, will allow the network to be tailored at any one time to meet the needs of individual users. In some cases, this will lead to time of day or seasonal services scheduled in advance. Changes in traffic patterns, with perhaps Interexchange Carriers requesting points of presence at different offices over time, can be accommodated in this way. In addition, end users may find service that they wish to take advantage of will be most beneficial without advance planning. These services will then be handled by providing network bandwidth on demand. The combination of a high bandwidth medium in fiber optics, coupled with synchronous internal format and software control of the network, will allow these advanced feature packages with network resources on demand to become the cornerstone in the continuing evolution of interoffice network.

22.3.5 Metropolitan Area Networks (MAN)

The proliferation of local area networks (LAN) that support office automation, factory automation and computer-aided design provides an excellent communication, computing and processing environment in a small geographical area typically spanning several hundred to several thousand feet in an office building or a campus environment. Packet communication with various contention handling protocols are implemented to efficiently share the common resources within LAN. However, the interconnection of LANs in geographically separated areas poses a new challenge to the data communication and telecommunication industries. This is often encountered by banks and many large corporations with branches scattered in a metropolitan area. The use of statistical data multiplexers and dedicated private data lines provides only partial solution to the low speed data services. It does not address, for instance, the high speed file and image transfer problems that often require variable high speed transmission. The bottleneck is real and its magnitude is growing.

At the time of this writing, IEEE 802.6 committee is considering a MAN network standard using a slotted ring for distances up to 50 km at 43 Mb/s, while the Accredited Standards Committee (ASC) X3J9 is considering an enhancement version of 100 Mb/s Fiber Distributed Data Interface (FDDI) to include circuit switching capability for distances up to 100 km. The successful resolution in the standards arena as well as the telecommunications industry's ability to adapt these standards into the existing or the upcoming intelligent transport facility described previously are keys to provide low cost, efficient and integrated inter-connection services between LANs and PBXs in a metropolitan area. It will enable the business community to take full advantage of the integrated broadband services promised by the fiber and pave the way for ultimate deployment of fiber to residences.

22.4 EMERGING TECHNOLOGY AND APPLICATIONS

As fibers are being installed in many parts of the telecommunications network, the optical technologies are also advancing rapidly. In this section, we will examine several areas of advances that might have potential impact on the exchange network in the next decade. These include dense wavelength division networks, photonic switching and a brief discussion of the promises of integrated optoelectronics.

22.4.1 Dense Wavelength Division Multiplexing Network

Conventional WDM techniques of packing two to four channels in each of the three wavelength windows (0.8–0.9, 1.2–1.3 and 1.5–1.6 μm) have long been advertised as a means to upgrade an existing transmission system. In practice, however, they are seldom used because of the need to install additional regenerators at each repeater site. To date, upgrades have been usually accomplished by increasing the speed of transmission wherever possible. In the last several years, however, with the installed fiber loss of as low as 0.3 dB/km in the 1.5–1.6 μm region and the use of stabilized, narrow linewidth single frequency sources such as DFB lasers, experimental networks based on dense WDM principle became an interesting area of research (Brackett, 1986; Payne et al., 1985; Goodman et al., 1986). Two such networks are depicted in Fig. 22.6. Each local office is assigned a unique wavelength for transmission purpose. The wavelength spacing

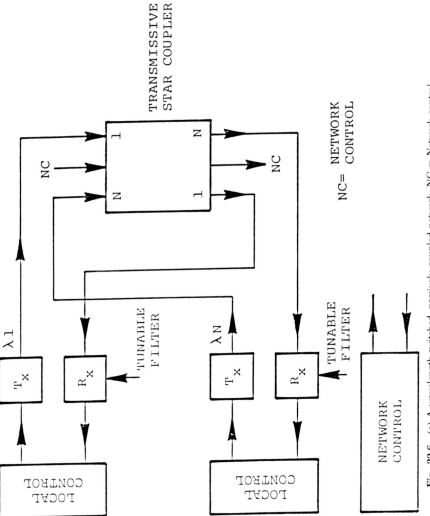

Fig. 22.6 (a) A wavelength switched, passively coupled network. NC = Network control.

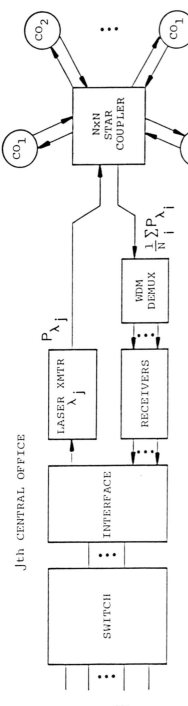

Fig. 22.6 (b) A multiwavelength passive star exchange network.

of 1–2 nm between channels would accommodate 32–128 channels in the 1.5–1.6 μm region. At the hub site of the network, a passive star coupler combines all input signals and redistributes them to all receiving offices, thus performing the function of a WDM as well as a broadcasting star. Since signals from all parts of the network are present optically at each receiving office, the office can use a wavelength sensitive tunable filter to filter out the portion of traffic designated for it as depicted in Fig. 22.6a, or optically demultiplex all signals and then rely on an array of optical receivers and fast electronics to further process the signals (Fig. 22.6b). The former approach, a form of wavelength switching at edges of the network, requires the technology of wavelength sensitive fast tunable filters and a coordinated network control and synchronization scheme. The latter approach provides a transparent network with broadcast capability. However, it suffers from requiring an optical wavelength division demultiplexer with N outputs, N optical receivers and N electronic processors for traffic sorting at each office, where N is the number of central offices in the cluster. The large number of wavelengths and the distance (20–30 km) that can be accommodated by this scheme matches closely to the characteristics of the hubbed interoffice environment. Further progress in optical amplifiers, integrated optoelectronic receiver arrays, and optical processing of network control signals might bring this type of network closer to realization.

22.4.2 Photonic Switching

The benefits of fiber optics have so far been confined mainly to transmission. With the proliferation of fibers in many parts of telecommunications networks, it becomes increasingly apparent that photonic switching and optical signal processing, including optical multiplexing, will play an important role in the network evolution into an all photonic network.

As we survey the technological trends suitable for small and simple facility switches to large and complex circuit and packet switches, only the mechanical and small size LiNbO$_3$ switch technology is available now. They tend to be useful for facility switching purposes, i.e., the entire bit stream on a fiber is switched. A typical size might be 4 × 4. For a larger switching matrix, the crosstalk and insertion loss become the dominating factors. Hence, reduced crosstalk level and optical regeneration will have to be designed into large switching matrix. For more sophisticated photonic circuit or packet switches, the switch control signals usually run at a much lower speed than the signals on fiber. These signals can be extracted optically and converted into electrical form, so a hybrid photonic switch may offer advantages over an all photonic switch. Recently, simple optical time slot interchanger using optical bistability and laser amplification had been demonstrated (Sakaguchi et al., 1985). It appears that innovations in non-linear optics, material research, and new network architecture are needed for photonic switching to thrive in the evolving optical network.

For near term applications of photonic switching in the exchange network, the adoption of a synchronous optical network (SONET) standard will accelerate the need for automatic optical crossconnect system, very high speed synchronous optical multiplexing and optical add/drop. An optical crossconnect system controlled electronically could provide a flexible point for interconnection of different vendors' optical equipment at midspan, a large number of loop fibers to trunk fibers in a local office and among various trunk fibers in a hub office. It is transparent to bit rate and format. Synchronous optical add/drop will enable more efficient and flexible use of single-mode fibers. Together they will provide the basic ingredient of fiber-on-demand and bandwidth-on-demand capabilities.

22.4.3 Integrated Optoelectronics

For the interoffice network, increased integration and greater functionality are the key to increased applications of fiber optics. Advances are being made in making array lasers and array detectors on the same semiconductor substrate. Furthermore, it is possible to fabricate combined optical and electronic circuits on a monolithic chip. The integrated optoelectronic technology may hold the key for low cost and high functionality subsystems such as laser transmitters. Many functions that are traditionally implemented by discrete, passive and active components can be integrated on the same chip with laser sources. These include WDM, modulator, power splitter, tunable filter, optical digital multiplexer and optical amplifier. Such integration has the promises to reduce the physical size, power consumption, simplify optical interconnection between various functional sections and more importantly, may lend itself for mass production. Progress in these areas will surely move the ultimate vision of bandwidth on demand at affordable cost closer to reality.

REFERENCES

Brackett, C. A. (1986). A view of the emerging photonics network. *ICC*, 1730–1733.

Cheng, S., Gardner, W. B., and McGrath, C. J. (1983). Single mode lightwave transmission experiments at 432 and 144 Mb/s. *Proceedings Integrated Opt. and Opt. Fiber Comm. Conf.*, 28C2-3.

Goodman, M. S., Kobrinski, H., and Loh, K. W. (1986). Application of wavelength division multiplexing to communication network architectures. *International Conf. on Comm.*, 931.

Ito, T., Shinohara, S., Ishida, Y., and Uchida, N. (1983). Results and experience of the field trial for the first fully engineered single-mode fiber cable transmission system at 400 Mb/s. *Proceedings Integrated Opt. and Opt. Fiber Comm. Conf.*, 28C2-2.

Jacobs, I. (1976). Lightwave communications passes its first test. *Bell Labs Record*, 291–297.

LeGall, Marc, and Auffret, René (1986). Transmission systems on optical fibers used in digital broadband networks. *ICC*, 1706–1709.

O'Neill, E. F. (1985). A history of engineering and science in the Bell System: transmission technology (1925–1975).

Payne, D. B., and Stern, J. R. (1985). Wavelength switched, passively coupled single mode fiber optical networks. *Proceedings Integ. Opt. and Opt. Fiber Comm. Conference*, 585.

Roohy-Laleh, E., and Ross, N. (1986). Network design for survivability: procedure and case study in a dynamic network architecture. *ICC*, 923–930.

Sakaguchi, M., and Goto, H. (1985). High speed optical time-division and space-division switching. *Proceedings Integrated Opt. and Opt. Fiber Comm. Conf.* 2, 81.

Schwartz, M. I., Reenstra, W. A., Mullins, J. H., and Cook, J. S. (1978). Chicago lightwave communication project. *Bell System Tech. J.* 57, 1981,

Shumate, P. W., Gimlett, J. L., Stern, M., Romeiser, M. B. and Cheung, N. K. (1985). Transmission of 140 Mb/s signals over single mode fiber using surface and edge-emitting 1.3 μm LEDS. *Elec. Letters* 21, 12, 522.

Stauffer, J. R. (1983). FT3C—A lightwave system for metropolitan and intercity applications. *IEEE Journal on Selected Areas in Communications* SAC-1, 3, 413.

Tam, R. M. (1986). Network transport services. *ICC*, 910–915.

Tse, K. A., and Smith, D. E. (1986). Transmission performance of the AT & T FT series G lightwave system. *ICC*, 1534–1537.

The International Telegraph and Telephone Consultative Committee Red Book (1985). Vol. III-Fascicle 111.3, Geneva.

Transmission Systems for Communications (1982). Bell Laboratories, Inc.

T1X1.4/85-050 (1985). Exchange Carrier Standards Association, Proposed Draft American National Standard for Synchronous DS3 Format (SYNTRAN).

T1X1.4/85-017 (1985). Exchange Carrier Standards Association, Proposed Work Project of Fiber Optic Network Interfaces (SONET).

Chapter 23

Terrestrial Intercity Transmission Systems

DETLEF C. GLOGE

AT&T Bell Laboratories, Holmdel, New Jersey

IRA JACOBS

Virginia Polytechnic Institute & State University, Blacksburg, Virginia 24061

23.1 INTRODUCTION

23.1.1 Outline

This chapter considers performance requirements, system design, technology, and applications factors for intercity fiber optic transmission systems. It is an expansion and update of a paper (Jacobs, 1986) on "Design Considerations for Long-Haul Lightwave Systems."

A simplified block-diagram of an intercity fiber optic transmission system is shown in Fig. 23.1. The system consists of terminal locations in which there is access to and egress from the system, and intermediate repeater stations to regenerate the signal. Since maintenance functions are performed at the terminal sites,[1] the portion of the system between adjacent terminals is termed a maintenance span. End-to-end performance requirements may be translated into performance requirements on each maintenance span (Section 23.2). The overall architecture of the system is contained in the functional design of the terminal (Section 23.3). Key technology issues relate to the design of the regenerators (Section 23.4) that are contained in the intermediate repeater

---

[1]Maintenance may also be remoted from the terminal sites to a centralized maintenance center.

855

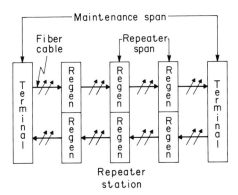

Fig. 23.1   Fiber optic transmission system block diagram.

stations as well as in the terminals. Applications considerations (Section 23.5) include the relative costs of the various parts of the system (terminals, cable, and regenerators), as well as upgrade capabilities to higher capacities. Future trends (Section 23.6) relate not only to technology advances, but also to possible architectural advances.

23.1.2 Evolution

Fiber optic transmission systems were initially applied largely for intracity (metropolitan area) trunking, where there was already extensive application of digital transmission on wire-pair carrier systems. It was difficult initially to economically prove-in fiber optic systems for long distance intercity transmission, where existing analog microwave radio and coaxial cable installations could be upgraded in capacity. However, the advent and wide deployment of digital switching for intercity telecommunications afforded considerable terminal cost savings when the switch interfaced with digital rather than analog transmission facilities. Fiber optic systems became the economic choice on high growth routes, particularly when the distance between terminal cities was relatively short. A prime example was the AT & T Northeast Corridor System between New York and Washington, D.C., for which construction approval was sought in January 1980, and which began service in January 1983 (Stauffer, 1983).

Initial intercity applications in the 1979–1983 time period in the U.S. (Cook and Szentisi, 1983), in Europe (Moncalvo and Tosco, 1983) and in Japan (Ishio, 1983) were at short wavelength on multimode fiber, at bit rates of 45 and 90 Mb/s in the U.S., at 34 and 140 Mb/s in Europe, and at 32 and 100 Mb/s in Japan, with repeater spacings of about 10 km. It was generally recognized (Midwinter, 1981), however, that widespread application of fibers for intercity transmission required increases in both bit rate and repeater spacing. Rapid technology improvements have been and continue to be made, and intercity systems installed world-wide since 1984 all utilize single-mode fiber, with transmission generally at a wavelength of 1310 nanometers, with bit rates typically between 400 and 600 Mb/s, and with repeater spacings of about 40 km. Field experiments (Fishmann *et al.*, 1986) have been performed at 1.7 Gb/s, and laboratory experiments (Gnauck *et al.*, 1986) have been performed at 8 Gb/s. Upgrade of existing systems into the gigabit-per-second range, is anticipated in the 1987–1988 time period.

23.2 SYSTEM PERFORMANCE REQUIREMENTS

There are two types of performance requirements: system requirements (e.g., cross-connect specifications, outage, and error performance) and component subsystem requirements (e.g., the performance of a regenerator). The former, typically contained in standards documents (Bell Communications Research, 1985), are intended to assure proper interworking with other facilities and appropriate performance for subscriber services; e.g., data services have more demanding error performance requirements than voice services. The latter (subsystem requirements) are imposed by the designer (and, not infrequently, by the purchaser) to assure that the end-to-end requirements are met.

This section considers only two of the many performance criteria for intercity fiber optic systems—viz., outage and error performance. Other performance criteria include protection switching performance (switching thresholds and switching times), transient response (recovery times), and jitter generation and accommodation. These criteria are more related to the electronics than to the fiber optics portion of the system and are associated with a level of design detail outside the scope of this chapter.

23.2.1 Outage

An outage is said to occur "when the bit error rate (BER) in each second is worse than 10^{-3} for a period of ten consecutive seconds" (CCITT, 1985). In the U.S., short haul systems (up to 400 km in length) are required to have an outage (measured at the 1.544 Mb/s DSX-1) of no more than 0.02%. Seventy-five percent of the outage is allocated to the DSX-3 to DSX-3 transmission link and is prorated by length; viz., 0.0000375% per km.

For long haul systems, the same fractional outage, 0.02%, is used for a 6400 km system. This full outage is allocated to the DSX-3 to DSX-3 transmission link and is prorated by length giving 0.000003% per km, a more stringent requirement than for short haul systems.

More stringent requirements imposed on long haul than on short haul facilities are likely a carry-over from the time when long haul transmission facilities were of higher quality than short haul facilities (e.g., coaxial cable compared to paired cable). Such distinctions may not be meaningful for fiber optic facilities.

High capacity fiber optic systems invariably employ automatic protection switching to meet outage requirements. This is the case even if the outage requirements could be theoretically met without automatic protection, since the consequences of an outage at high capacity are so severe, and the cost of protection is relatively low.

If there are N working lines and M protection lines in a protection group, then the fractional outage per working line is given by

$$\text{Outage} = \frac{(N+M)!\,P^{M+1}}{N!\,(M+1)!}, \tag{23.1}$$

where P is the outage in the absence of protection

$$P = \frac{\text{Mean Time To Repair (MTTR)}}{\text{Mean Time Before Failure (MTBF)}}. \tag{23.2}$$

Typically, $M = 1$, in which case Equations 23.1 and 23.2 reduce to

$$\text{Outage} = \left[\frac{N+1}{2}\right]\left[\frac{\text{MTTR}}{\text{MTBF}}\right]^2. \tag{23.3}$$

Equation (23.3) provides the requirements on the reliability (MTBF) of the equipment in a protection span to achieve a given outage for a given protection ratio (N) and a given MTTR (typically assumed to be four hours). In Fig. 23.2, the MTBF to achieve an outage of 10^{-5} is shown. This outage corresponds to the requirements for a 320 km protection span in a long haul system. Assuming 40 km spacing between regenerators, there are 16 regenerators in the two-way span. If half of the outage is allocated to the regenerators, then for a protection

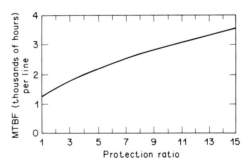

Fig. 23.2 Mean time before failure (MTBF) of a two-way line between switch points to achieve 10^{-5} outage probability. (Assumes two-way protection switching and 4 hour MTTR).

ratio of 7, each regenerator should have a MTBF of at least 80,000 hours (no more than 12,500 FITS).[2] Although this is adequate to meet the outage requirement, higher reliability is desired to minimize repair activity.

If protection switching were not used ($M = 0$), then each regenerator would have to have a MTBF of at least $1.3(10)^7$ hours (no more than 80 FITS) to meet the outage requirement in the example. This would be extremely difficult to achieve.

23.2.2 Error Performance

There are three common measures of error performance:

i) Long term BER
ii) Percent error-free-seconds (EFS) and
iii) Severe-errored-seconds (SES).

In the U.S., both long haul (6400 km) and short haul (400 km) BER requirements are typically $< 10^{-8}$. These are allocated by mileage; thus a 40 km short haul system would have a BER requirement $< 10^{-9}$. Assuming randomly distributed errors, it makes no difference whether the BER requirement is applied to a tributary input rate or to the line rate.

In the case of EFS, the requirement is, of necessity, dependent on the bit rate at which it is measured. For example, a 1 Gb/s system operating at a 10^{-8} BER would have essentially zero EFS measured at the line rate, since each second would contain, on average, 10 errors. Requirements on EFS typically apply to end-to-end digital services, measured at 1.5 Mb/s or below. Establishment of similar requirements at 45 Mb/s is currently under study. An errored second at 45 Mb/s (DS3) can affect from 1 to 28 DS1s. If only 1 DS1 is affected (isolated random errors), then the DS3 EFS requirement may be relaxed by a factor of 28 relative to the DS1 requirement, i.e., the percent errored seconds could be a factor of 28 higher. On the other hand, if all 28 DS1s are affected (burst error), then the DS3 EFS requirement is the same as the DS1 requirement.

A severe errored second is generally defined as a second in which there are more than 100 bit errors. A less precise, but somewhat more meaningful, definition is that a SES contains a sufficiently large burst of errors such that *all* tributaries are affected and are considered severely errored. Thus, by definition, severe errored seconds are independent of whether they are measured at the line or tributary rates.

SES requirements are typically in the range of, on the average, no more than 4 to 8 SES per day, where the lower figure corresponds to short haul (400 km) systems, and the higher to long haul systems (6400 km).

---

[2]FIT is a unit of failure rate. 1 FIT corresponds to 1 failure in 10^9 hours.

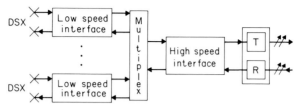

Fig. 23.3   Fiber optic transmission system terminal block diagram.

How do these requirements impact the design of an intercity fiber optic system? Clearly, individual repeater sections have to be designed for BER considerably below 10^{-8} (typically $< 10^{-11}$). Also, the design should assure that events causing severe burst errors (e.g., external interference or internal intermittents such as amplifier "dropouts") occur less than of the order of one per day. Fiber optic systems should operate essentially error free.

23.3 SYSTEM ARCHITECTURE (TERMINAL DESIGN)

A simplified functional block diagram of the terminal of a fiber optic transmission system is shown in Fig. 23.3. Each element of this block diagram, with the exception of the regenerator (designated T/R in the figure), which is discussed separately in Section 23.4, is discussed next.

23.3.1 Cross-Connect

Intercity fiber optic systems interface at a standard digital cross-connect (DSX). In the U.S. this has been almost universally at the DSX-3 (44.736 Mb/s third level of the North American digital hierarchy). In Europe, the interface is at the CEPT-4 (139.264 Mb/s) level,[3] and in Japan at the 397.20 Mb/s 5th level of the Japanese hierarchy. Table 23.1 compares the North American, European, and Japanese hierarchies.

Early digital transmission systems operated at capacities corresponding to that of the defined levels in the digital hierarchies, and separate multiplexers were used to provide the transition between levels. Largely for economic reasons (but also for technology and applications reasons as noted in the following sections), the trend in fiber optic systems (particularly in the United States) has been to incorporate multiplexing into the fiber optic system terminal. Thus, the capacity of the fiber optic system may be an integral multiple of the cross-connect capacity, and is not constrained to be equal to a cross-connect capacity. For example, in the United States, fiber optic systems exist or have been announced

---

[3]Some European systems have been applied in the U.S. utilizing a separate multiplexer (factor of three) between the U.S. DSX-3 and the CEPT-4 levels.

TABLE 23.1
Digital Hierarchies

		North American	European (CEPT)	Japanese
Level 1	Bit rate (Mb/s)	1.544	2.048	1.544
	Voice Circuits	24	30	24
Level 2	Bit Rate (Mb/s)	6.312	8.448	6.312
	Voice Circuits	96	120	96
Level 3	Bit Rate (Mb/s)	44.736	34.368	32.064
	Voice Circuits	672	480	480
Level 4	Bit rate Mb/s	274.176	139.264	97.728
	Voice Circuits	4032	1920	1440
Level 5	Bit Rate (Mb/s)	Not defined	565.148	397.20
	Voice Circuits		7680	5760

with capacities the following multiples of the DSX-3 cross-connect capacity: 1, 2, 3, 4, 6, 9, 12, 18, 24, and 36.

23.3.2 Low-Speed Interface

The principal functions of the low speed interface are to provide the coding and decoding (codec) for the cross-connect interface,[4] synchronization and desynchronization (syndes) for the subsequent multiplexing and demultiplexing, and the addition (and removal) of bits for performance monitoring and order wire and telemetry channels. The advantage of adding these bits at the low speed interface rather than at the line rate[5] is that it reduces the amount of high speed signal processing required, facilitates add-drop multiplexing (see following section), and facilitates the upgrade to higher capacity. The disadvantage is that it is somewhat more wasteful of bandwidth (not of primary concern in fiber optics transmission), and it requires demultiplexing of the line signal to access these bits (not difficult with synchronous multiplexing, see following section). The advantages tend to outweigh the disadvantages.

23.3.3 Multiplexing

If the signal processing is all performed at the low speed input level, then the multiplexing and demultiplexing can be simply synchronous bit interleaving and separation. In the demultiplexing, proper identification of the tributaries may be achieved by complementing one of the tributaries. Apart from this identification

---

[4] The cross-connect interface specifies not only pulse rate, but also a specific pulse code format for intra-office transmission. For example, the DSX-3 in the North American hierarchy specifies a B3ZS (bipolar with three zero substitution) code.
[5] In some systems the bits are added at a rate intermediate between the interface and line rates.

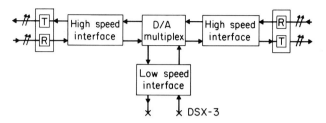

Fig. 23.4 Drop-add terminal block diagram. The D/A Multiplex provides a high-speed through path as well as provision for drop-add of a small number of DS3s.

of which tributary is "channel 1," each of the tributaries is identical and completely self-contained (i.e., all of the maintenance and performance monitoring information is contained within the tributary).

At terminal locations where access is required of only a small number of tributaries, low speed interfaces may be provided only for those tributaries accessed, with the remainder of the signal passed directly through (see Fig. 23.4). An alternate and somewhat simpler (but less flexible) arrangement is to have a given location serve as a terminal for some of the fiber pairs in the cable, and as an intermediate repeater location for other pairs. In this case, terminals are provided for the protection pair at such locations. If there is a failure in a through system, then the individual protection spans making up this system are concatenated.

23.3.4 High Speed Interface

The primary function of the high speed interface is to introduce appropriate statistics into the line signal. This function may, in principle, be implemented by appropriate parallel processing of the input tributaries, prior to multiplexing, rather than serial processing after multiplexing. However, this may necessitate aligning the frames in each of the tributaries, which makes the dropping and inserting of tributaries more difficult.

Most systems in the United States use unrestricted binary transmission, with scrambling to guard against low frequency patterns occurring with probability greater than predicted by random statistics. This format has been used by all of AT & T's high speed (45 Mb/s and above) lightwave systems, as well as a previous high speed coaxial cable system (Rubin, 1975).

In Europe, balanced block codes are generally used (Windus *et al.*, 1985). This allows a higher low frequency cut-off in the regenerator circuits, and facilitates timing recovery and performance monitoring. The use of block codes with binary transmission, however, results in a higher line rate. For example, the 5B6B code (translating a 5 bit binary input sequence to a 6 bit binary line signal), widely used in European systems, requires a 678 Mb/s line rate for a

565 Mb/s line capacity. This puts an added burden on high speed regenerator circuitry.

Ternary (three-level) transmission, commonly used on metallic systems, has not generally been used on fiber optic systems, owing to the higher required signal-to-noise ratio,[6] and the greater sensitivity to eye degradations (see Section 23.4.7). One such European system has been announced (Drupsteen, 1985), however, utilizing a 4B3T code. This allows achieving a 565 Mb/s capacity using the same 423 Megabaud transmission employed in European digital coaxial cable systems. However, it does reduce eye margins in half.

In Japan, long haul systems use a 10B1C code in which a complementary bit (to the previous bit) is added after every 10 information bits (Iwahashi and Fukotomi, 1983), which assures no more than 11 bit in a row without transitions.

There does not appear to be convincing arguments in favor of any particular coding arrangement.

23.3.5 Implementation

In microwave radio, coaxial cable, and early fiber optic intercity transmission systems, it was common to separate the maintenance and transmission functions. Thus, a separate monitor and switch bay (or a separate shelf within the transmission bay) would contain the protection switches and a common monitor processor serving all of the channels within the system. This had the advantage of spreading common system costs over a large number of channels, which was advantageous from the standpoint of fully-loaded costs, but disadvantageous from the standpoint of getting started costs.

Both technologic advances (VLSI making per-channel processing more economic) and the growing importance of first-costs have led to distributed architectures and implementations that minimize common equipment. This is illustrated by the photograph (Fig. 23.5) of the AT&T FT-Series G 417 Mb/s terminal. This 13 inch high by 26 inch wide by 12 inch deep shelf contains all of the circuitry necessary for interfacing nine DS3 signals to and from a 417 Mb/s optical line. In addition to all of the functions depicted in Fig. 23.3, it contains the power supplies, monitor and control, and protection switch cards that serve this shelf. The use of a dedicated power supply, rather than a common power supply serving an entire bay, avoids the need for redundant power, since the power units may be contained within the protection switching. Thus, a failure of either the transmission or power circuit packs will result in a switch to the protection shelf.

---

[6]On metallic systems, reduced bandwidth offers the advantage of lower cable loss. This does not occur in fiber optic systems. Also, on baseband metallic systems, balanced transmission with $+$, 0, and $-$ voltage levels are used. With incoherent fiber systems, ternary transmission entails 0, half, and full intensity transmission levels.

Fig. 23.5   The Terminating Muldem (Multiplexer-Demultiplexer) assembly of the AT&T FT-Series G System.

The nine tributary line cards, and the corresponding switch card located above each line card, need not all be equipped initially, but may be added as needed. Up to eight such assemblies may be combined with one serving as protection for the other seven. There are nine protection switching buses (one for each of the nine tributaries) such that a tributary failure detected by the monitor will result in that tributary's signal being switched onto the protection bus. Thus, for example, simultaneous failure of different tributaries (whether or not they are on the same shelf) may be accommodated.

The distributed architecture also facilitates the upgrade to higher speed. For example, in FT-Series G, a four-shelf configuration, providing four 417 Mb/s lines, is converted to 1.7 Gb/s by removing the 417 Mb/s regenerators and adding a fifth shelf containing the additional multiplexing and 1.7 Gb/s regenerators.

23.4 REGENERATOR DESIGN AND PERFORMANCE

Figure 23.6 shows the key components needed to regenerate an optical signal. The signal enters the avalanche photo diode (APD) through a connectorized single fiber cable. The resulting photocurrent is amplified to an input-independent signal level. This signal is used to extract a clock or timing signal at the baud rate. The timing wave clocks a decision circuit, which completes the regeneration of the data stream. The regenerated signal is reconverted to an

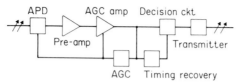

Fig. 23.6 Regenerator block diagram.

optical signal in the transmitter and leaves the regenerator through another optical connector.

All of these functions are generally contained on a single circuit pack; although sometimes the transmitting and the receiving and regenerating portions are separated into two circuit packs, particularly in terminal locations. Note that a regenerator interfaces with two fibers (via the transmitter and receiver connectors). Thus, at intermediate repeater locations, a regenerator serves as a one-way line repeater, whereas at terminal locations, the same circuit pack serves as a two-way terminal repeater. As illustrated in Fig. 23.7, these different functions may be accommodated with a common circuit pack, using appropriate backplane wiring in the terminal and line repeater bays.

Not shown in Fig. 23.6 is the part of the regenerator associated with fault location and performance monitoring. This usually involves a microcomputer, which collects analog data from certain test points (photocurrent, AGC voltage, loss of clock, laser bias current, laser temperature, etc.) and converts these into a

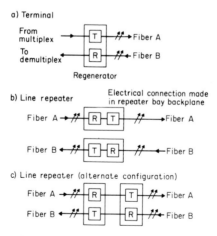

Fig. 23.7 Repeater configurations. (a) Terminal Regenerator. Line repeaters are generally config-ured as in (b) such that each regenerator circuit pack (T/R pair) serves a single fiber path. The configuration shown in (c) may be used if, for example, different technologies are employed on the spans to the left and to the right.

data signal, which is then transmitted to the maintenance interface. Since the maintenance aspects tend to be proprietary to each vendor's system, the following discussion will focus on the technological aspects of the functions depicted in Fig. 23.6.

23.4.1 Receiver

The detector and the first low noise photocurrent amplifier are usually combined in the receiver package to minimize parasitic and circuit capacitances. (The reader is referred to Chapter 18 for details on receiver design and optimization.) Simple PIN photodetectors are often used in regenerator implementations in the interest of low cost. Better sensitivity can be achieved using avalanche photodiodes (APDs), which provide photocurrent multiplication in the device. Such APDs have been made from Ge and InGaAsP for the wavelength regions of interest at 1.3 and 1.5 micron (see Chapter 14). The first amplifier stage must be chosen for low noise and large dynamic range. This is achieved by selecting an active feedback circuit, which gives a broadband high-sensitivity response (Williams, 1986). In such receivers, the parasitic feedback capacitance can be minimized by using a specially designed micro-FET as a feedback resistor. Some implementations of these active feedback IC receivers include an input shunt AGC circuit, which extends the dynamic range by diverting excess photocurrent away from the input of the basic receiver.

For a given error rate there is an optimum APD gain to minimize the required optical power. At smaller gains, the preamplifier noise dominates; at larger gains, the multiplication noise of the APD is the dominant noise. For APDs made from Ge or InGaAsP, the optimum gain is in the range between 10 and 20 depending on the bit rate of the received signal.

23.4.2 AGC Amplifier

The AGC amplifier consists of several variable gain stages providing a maximum gain of 40 to 50 dB, a fixed gain power amplifier and a low-pass filter to limit noise. The low-pass filter shapes the frequency response of the linear channel to limit noise without introducing much intersymbol interference. An optimum design minimizes the detected power for a given error rate. The high frequency roll-off of the AGC amplifier should be at a frequency higher than the cut-off of the low-pass filter in order that bandwidth changes at the onset of AGC do not affect the overall bandwidth of the linear channel. The low-frequency roll-off should be low as possible to avoid pattern-induced dc-wander in the decision circuit. The roll-off is determined by the coupling capacitors. In reality, there is a limit to the size of these capacitors so that practical low-frequency roll-offs are in the range of a few kHz causing a dc-wander of a few percent at bit rates of several hundred Mb/s.

The level of the output signal is monitored by a peak detector and compared against a reference. The response time of the peak detector is a fraction of a second. The voltage at the output of the comparator controls the gain of the AGC amplifier and to a lesser extent that of the APD. The total AGC range is in the vicinity of 40 dB. Note that this corresponds to a dynamic range of 20 dB of the detected optical power since the APD is a square-law detector. For low detected power, the AGC amplifier gain is varied and the APD gain is at its optimum value. For very large detected power, the amplifier AGC saturates and the APD bias voltage is reduced to decrease its gain.

23.4.3 Decision Circuit

The decision circuit compares the conditioned signal from the AGC amplifier with a threshold and decides, at sampling times determined by the recovered clock, whether the signal is a "one" or a "zero." In the absence of errors, the decision circuit output is a logic-level version of the lightwave signal originally transmitted. To obtain the extremely low probability of error required in an individual regenerator, it is essential to have adequate margins against decision errors. The phenomena involved can be approximately represented on an eye diagram, which is a superposition of all the possible pulse waveforms that can appear in any time slot. Figure 23.8a shows an ideal decision process for a non-return-to-zero (NRZ) signaling format. The decision is made at the moment when the "zero" and "one" levels are maximally separated. Figure 23.8b shows

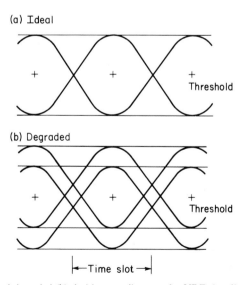

(a) Ideal

+   +   +
                         Threshold

(b) Degraded

+   +   +
                         Threshold

|←—Time slot—→|

Fig. 23.8 Ideal (a) and degraded (b) decision eye diagrams for NRZ signaling.

in a simplified way how the margins against errors are reduced by an assortment of perturbations. In reality, the degraded signal is not quite symmetric since the "light on" state has more noise associated with it than the light off state. For this reason, the optimum decision level is not exactly at the center of the eye, but closer to the off state. In addition to noise, typical perturbations are jitter, timing offset and dc-wander. There is also a certain range of ambiguity (typically a few tens of mV) around the decision level, in which the decision circuit makes decisions on a statistical basis. A typical decision circuit consists of an input amplifier, a clocked D flip flop as the decision and reshaping element, and an output buffer amplifier for off-chip interfacing. Either master-slave flip flop arrangements (Maione *et al.*, 1978) or edge-triggered flip flops (O'Connor *et al.*, 1984) are being used.

23.4.4 Timing Recovery

The timing recovery circuit has the purpose of extracting a timing wave at the symbol rate from the conditioned signal pulse stream provided by the AGC amplifier. This is complicated by the fact that the spectrum of the pulse train lacks a component at the symbol rate when NRZ signals are received. In this case, the message spectrum is first passed through a high-pass filter, which symmetrizes the spectrum about a frequency equal to half the symbol rate. It suppresses much of the low frequency jitter components that would otherwise perturb the timing wave. Full-wave rectification of "squaring" of the filtered signal then creates a discrete line at the symbol rate. In the time domain, the output of the squarer is a series of current pulses corresponding to the transitions of the original input signal. The narrow spectral line at the symbol rate is superimposed on a broad continuum related to the message pulse shape and statistics. To remove most of the continuum, the signal is now either processed by a phase-locked loop (Maione *et al.*, 1978) or filtered by a narrow band resonant filter. These are surface acoustic wave (SAW) filters (Rosenberg *et al.*, 1983) in the case of most modern regenerators operating at hundreds of Mb/s. The bandwidth of the filter is a compromise between the need to limit jitter (and therefore to remove as much of the continuous spectrum as possible) and the need to minimize the phase slope of the filter in order to limit the effect of filter detuning. Detuning will generate a static phase error and therefore a timing error in the decision process as a result of the unavoidable slope in the phase-frequency characteristic. The causes of detuning include aging and temperature variations of the filter and the adjacent electronics and tuning errors remaining after manufacture and phase adjustment. Most of the detuning mechanisms are randomly distributed among the regenerator population. Those that are deterministic, like the temperature characteristics of the filter and the electronics, can be offset against each other for overall compensation. More specifically, one can choose the crystal orientation of the filter substrate in such a way that phase slope with temperature of the SAW filter compensates the phase slope of the timing recovery electronics.

23.4.5 Transmitter

The transmitter accepts the logic-level signal from the decision circuit and provides either an RZ or NRZ optical signal of a fraction of a mW. It consists of an optical source, a backface monitor to regulate the laser output level over time and temperature and electronic circuits to drive the laser current and to activate the regulating feedback loop. A more complete description of the transmitter is given in Chapter 19.

In most modern regenerators, the laser source is an InGaAsP double hetero-structure design emitting either in the 1.3 micron or in the 1.5 micron wavelength range. If used in the 1.5 micron range at high bit rates, the laser must emit in one single longitudinal mode to avoid dispersion effects on typical terrestrial cables, which have substantial dispersion in this range. In the 1.3 micron range, conventional multimode lasers can be used if operation is restricted to a narrow range around the minimum dispersion wavelength. To extend the lifetime of the device, the laser can be mounted on a thermo-electric (Peltier) cooler, which generally controls the temperature to within a few degrees.

The laser is biased near threshold. A positive going input pulse brings the laser from below to above threshold. Through the feedback circuit, the prebias current is automatically established at a value required to maintain the light output of the laser constant. As part of this control circuit, a PIN photodiode senses the light output from the back face of the laser. By deriving a reference from the input signal, the control is insensitive to the "ones" and "zeros" density in the data. The output power ratio between "ones" and "zeros" (extinction ratio) must be at least 15 to avoid excess noise and loss of sensitivity at the receiver.

23.4.6 Electronic Technology

More than anything else, electronic circuit capability has an influence on the bit rate achievable. The technology of choice from 45 Mb/s to maybe 600 Mb/s is small-to-medium scale integrated silicon emitter-coupled logic. MOS technology is making inroads at 45 Mb/s because of lower power consumption.

GaAs FETs and GaAs integrated circuits have become very popular as first-stage photocurrent amplifiers because of low input capacitance and high current gain. If equipped with a specially designed micro-FET as a feedback resistor and shunt AGC circuit (Williams, 1986), these amplifiers have a broadband high-sensitivity response and very good dynamic range.

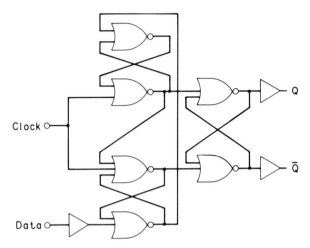

Fig. 23.9 Logic diagram of GaAs decision circuit.

Above 1 Gb/s, hybrid integrated circuits using discrete GaAs or Si FETs are adequate for most of the analog functions (Fishman *et al.*, 1986). The more complex logic functions are better developed in monolithic integrated form to achieve high speed performance, high reliability, low power consumption and good temperature compensation. Using a self-aligned technology (Nakagawa *et al.*, 1984), one can achieve bipolar integrated transistors with a unity gain frequency of 12 GHz. At these frequencies, GaAs competes with Si bipolar technology for all electronic functions both in the form of discrete and small-scale integrated circuits. Figure 23.9 shows the circuit diagram of a 2 Gb/s decision circuit (O'Connor *et al.*, 1984). Depletion and enhancement mode FETs are used in a direct coupled technology, which minimizes power consumption. The circuit is part of the 1.7 Gb/s regenerator which was field tested (Fishman *et al.*, 1986) in September 1985 as part of the AT & T Gigabit lightwave system called FT Series G.

23.4.7 Regenerator Performance

The effects of regenerator degradation are best described as a reduction in "eye opening" (see Figure 23.8) resulting from phenomena such as intersymbol interference, signal dependent noise, decision threshold offsets, timing offsets and degradations in laser on-off ratios.

Consider a signal as in Figure 23.8 in which the ideal eye height is unity, in which there is an initial eye degradation D_o, and in which we wish to allow for a subsequent degradation D. The additional power margin required to accom-

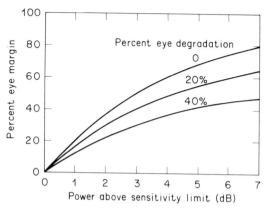

Fig. 23.10 Percentage eye margin resulting from increases in power above the sensitivity limit, with initial eye degradation as a parameter. (The sensitivity limit is also dependent on the initial degradation).

modate an eye margin D is given by

$$L_x = 10 \log_{10} \left[\frac{1 - D_o}{1 - D - D_o} \right]. \tag{23.4}$$

The relation between eye margin and power margin is shown in Fig. 23.10. The power margin is dependent on the initial eye degradation. Thus, if it is desired to have a 50% eye margin for subsequent degradations, the power margin needs to be 3 dB if there are no initial degradations, but needs to be 5 dB if there are initial degradations totaling 25%. Eye margin, rather than power margin, is a more appropriate measure of regenerator performance. The degradation allowance, L_x, should thus be based on an analysis of eye degradations and on a *measured curve* of power margin. Regenerator dynamic range should be specified as the range of input power over which the eye margin exceeds a pre-assigned threshold.

One source of eye degradation that impacts on subsystem specifications, is chromatic dispersion, which results in both intersymbol interference and in mode partition noise (Ogawa, 1982). To limit the equivalent eye degradation from dispersion to no more than 10% requires that the dispersion be less than 0.1 of a pulse period. For systems operating at the wavelength of minimum chromatic dispersion (λ_o), this sets requirements on both the laser spectral width, and on the off-set of the laser center wavelength from λ_o. For operation far from λ_o (e.g., at 1.55 μm on non-dispersion-shifted fiber), there are much more stringent requirements on laser spectral width (see Table 23.2).

TABLE 23.2
Spectral requirements resulting from chromatic dispersion and mode partition noise for single-mode fiber (minimum dispersion at 1.3 μm) and 40 km repeater spacing. ($\Delta\lambda$ is the rms spectral width of the laser, and $\Delta\lambda_o$ is the difference between the laser center wavelength and the minimum dispersion wavelength of the fiber. The mode partition noise k-parameter is assumed to be 0.5).

BIT RATE	$\lambda = 1.3\ \mu m$	$\lambda = 1.55\ \mu m$
417 Mb/s	$(\Delta\lambda)(\Delta\lambda_0) = 40\ nm^2$	$\Delta\lambda = 0.2\ nm$
1668 Mb/s	$(\Delta\lambda)(\Delta\lambda_0) = 10\ nm^2$	$\Delta\lambda = 0.05\ nm$

23.4.8 Regenerator Spacing

The loss budget for a regenerator span is calculated by taking the difference between the transmitter power P_T and receiver sensitivity P_R, subtracting for connector losses (L_C), and allowing for various degradations and margin (L_X)

$$L_1 = P_T - P_R - L_C - L_X. \qquad (23.5)$$

The loss of the connectors within the regenerator are usually included in the definitions of P_T and P_R; L_C is intended to accommodate additional connectors (and jumpers) within terminal or repeater sites. For example, an optical patch panel may be desired to permit flexible interconnection and rearrangement, and also serves as an appropriate point for the installation of wavelength division multiplexing (WDM) filters.

This maximum loss allowance, L_1, is then allocated to the installed fiber cable loss. Since there are statistical variations in cabled fiber loss and in splice loss, characterization of these variations allows a statistical treatment that is less pessimistic than a worst case analysis. The 2σ outside plant loss is given by (Meskell, 1984)

$$L_2 = \mu_f S + \mu_s M + 2 \left[\frac{\sigma_f^2 S^2}{M - 1} + \sigma_s^2 M \right]^{1/2}, \qquad (23.6)$$

where μ_f and μ_s are mean fiber and splice losses (including any temperature effects), σ_f and σ_s are the corresponding standard deviations, S is the regenerator spacing, and M is the number of splices including one at each end of the span. Thus, $M - 1$ is the number of statistically independent fiber sections.

The outside plant loss L_2 must be less than or equal to the loss allowance L_1. A representative calculation for the AT&T FT-Series G system, with a loss allowance of 22 dB leading to a repeater spacing of 44 km, is provided in table 23.3.

TABLE 23.3
Regenerator Spacing Calculation

Representative Example for FT Series G

$P_T = -3$ dBm
$P_R = -33$ dBm (PIN receiver, BER $= 10^{-9}$)
$L_c + L_x = 8$ dB (combined in a statistical analysis)
$L_1 = 22$ dB
$S = 44$ km
$M = 23$ (2 km between splices)
$\mu_f = 0.4$ dB/km, $\qquad \sigma_f = 0.1$ dB/km
$\mu_S = 0.1$ dB, $\qquad \sigma_S = 0.1$ dB

23.5 APPLICATIONS CONSIDERATIONS

23.5.1 Comparison with Other Media

For intercity transmission, terminals are much further apart than for metropolitan and local area systems, and the line haul costs (installed cable and repeaters) are the dominant cost element. Also, unlike metropolitan systems where wire pairs were the dominant transmission medium, intercity transmission was dominated by microwave radio and, to a lesser degree, coaxial cable.

Fiber has completely supplanted coaxial cable as the new cable medium for intercity transmission. Fiber costs are typically less than that of coaxial units, transmission capacity is higher, and most importantly, distance between repeaters is much greater.

Comparisons with microwave radio are more difficult to make. Microwave radio system capacities are limited by spectrum allocations, and fiber systems are advantageous to meet growth on high cross section intercity routes. Also, whereas maximum utilization of the limited radio bandwidth typically results in analog (single sideband) modulation, fiber systems are well matched to digital transmission.

It is expected that fiber and microwave radio systems[7] will coexist to meet intercity transmission needs. The 1970s were a period of rapid growth of microwave radio, leading by 1980 to microwave radio carrying about 75% of the equipped circuit capacity of the AT&T-Communications (then known as AT&T-Long Lines) network. The period of the 1980s is seeing a rapid growth of intercity fiber facilities, and the percentage of intercity transmission on fiber is expected to grow rapidly.

---

[7]For domestic applications, satellite systems tend to be more advantageous to meet broadcast rather than point-to-point transmission needs.

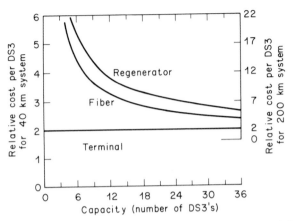

Fig. 23.11 Illustrative relative material costs for high-speed fiber optic transmission systems. The scale on the left corresponds to an unrepeatered 40 km system. The scale on the right corresponds to a 200 km span between terminals (four intermediate repeaters) in a long-haul system. Cost assumptions are given in Table 23.4.

23.5.2 Relative Costs

System design choices for fiber optic systems are strongly influenced by economic factors, but detailed economic considerations are outside the scope of this chapter. However, simple relative cost arguments may be both illustrative and informative.

There are three principal materials cost components of a fiber optic system—the fiber cable, the terminals, and the regenerators. A highly simplified example, depicted in Fig. 23.11, illustrates the costs of these components as a function of system capacity and length. Normalized costs for this example are provided in table 23.4.

TABLE 23.4
Relative cost assumptions and calculation of system cost.
(Unit relative cost corresponds to about \$2000).

Terminal cost per DS3 (one-end, excludes regenerators)	1
Regenerator cost	$\sqrt{n}$
Fiber cost (regenerator section)	7.5
Cost per DS3 for system capacity of n DS3 and containing M regenerator sections	$C(n) = 2 + 2M/\sqrt{n} + 15M/n$

Note that the cost of the terminal per DS3 is taken to be independent of the number of DS3s. The reason is that much of the terminal cost is associated with the low-speed interface, and hence the overall cost of the terminal is proportional to the number (n) of DS3s. Although the high speed circuit costs are shared over more DS3s as n increases, the cost of these circuits also increases.

Owing to the added complexity and power requirements of higher speed circuits, we have assumed a $\sqrt{n}$ dependence of regenerator costs. For simplicity, we have also assumed that the regenerator spacing is independent of n. This requires either increased transmitter power or improved receiver sensitivity (smaller number of photons per bit) as the bit rate increases. This is further reason for assuming regenerator cost increases as n increases, although the $\sqrt{n}$ dependence is admittedly arbitrary.

It follows from Fig. 23.11 that the capacities below 9 DS3, fiber rapidly becomes the dominant system cost element. At capacities in the gigabit per second range, fiber and regenerator costs are comparable. Also, as capacities increase, terminal costs become a larger fraction of system costs. It should be noted, however, that much of the terminal costs may be deferred until the circuits are implemented, whereas all of the fiber costs, and a larger fraction of the regenerator costs are paid up front.

23.5.3 Upgrade

There are two principal techniques for extending the capacity of a lightwave system upon exhaust—either increasing the capacity by higher speed serial transmission (TDM) or by wavelength division multiplexing (WDM).

The economics favor higher speed serial transmission since the cost of a single high speed regenerator is generally less than that of multiple lower speed regenerators and WDM filters. For example, with the assumptions of the previous section, a factor of m increase in capacity with TDM results in regenerator costs a factor $\sqrt{m}$ less than that with WDM.[8] Note that the terminal and fiber costs are essentially independent of whether the capacity increase is obtained by TDM or WDM, although the fiber loss and dispersion characteristics (as well as the regenerator transmitter spectral width) need to be chosen appropriately to support WDM.

Although the economics favor TDM, there may be a point at which speed limitations on both regenerator and terminal circuitry, as well as bandwidth limitations on sources and detectors, makes WDM a more viable option for upgrading capacity than continuing to increase serial transmission speed.

---

[8] If added rather than total costs are considered, then the advantage is a factor $(m - 1)/\sqrt{m}$.

23.6 TRENDS

23.6.1 Technology

The two principal technology trends in fiber optic transmission systems, higher speed transmission and longer repeater spacing, have particularly benefited intercity systems, where there is both a high concentration of traffic and relatively long distances to traverse. It may be noted, however, that transmission speeds have increased much more rapidly than the capacity of installed systems. For example, a 144 fiber cable with 45 Mb/s transmission has the same capacity as a 16 fiber cable with 405 Mb/s transmission. However, the higher speed transmission at long wavelengths has resulted in a 9-fold reduction in the number of fibers, and about a 50-fold reduction in the number of regenerators relative to that required in early short wavelength systems. This in turn has greatly increased the number of intercity routes on which fiber optic systems have economically proved-in.

Further increases in transmission capacity per fiber will be needed to continue to upgrade the capacity of installed systems. Advances in electronic integrated circuits, both silicon and GaAs, will continue to extend the speed capability of serial transmission. Integrated optics offers the option for more economic wavelength division multiplexing (WDM) configurations, so that WDM may become an economically viable option relative to higher speed serial transmission. Even with higher speed serial transmission, optical processing may well begin to supplant electronic processing for the final multiplexing stages. Also, optical regenerators, not requiring back-to-back O-E and E-O converters, may well be achieved.

In addition to capacity increases, further advances in regenerator spacing are also anticipated. This may be more significant for new routes than for existing routes where repeater spacings are quantized by existing terminal and repeater sites. Of itself, 1.5 μm technology does not allow a doubling of repeater spacing relative to systems at 1.3 μm but may well do so combined with advances in transmitter power and receiver sensitivity.

23.6.2 Architecture

For fiber optic systems, the distinction between intercity, intracity, and local area systems are blurring. For example, an intercity carrier may desire to provide direct access to a business customer located along an intercity fiber route. Thus, the ability to economically add-and-drop small amounts of traffic from a high capacity line are important in intercity as well as in metropolitan and local area networks.

Short distance systems are more terminal cost intensive than long distance systems. On the other hand, as repeater distances increase, and once the fiber is already in the ground, the primary cost element in the long distance system is also the terminal.

A further trend, affecting both long and short distance systems, is the merging of transmission and switching technologies. The desire to switch signals optically may well impact on the formats used for transmission. For example, the bit interleaving high speed multiplexing used in present-day intercity systems, may be supplanted by "burst multiplexing" permitting direct optical interfaces between transmission lines and switches.

Intercity transmission needs have driven the fiber optics technology to higher speed and longer distance capability. Many of these advances are now being applied to shorter distance systems as well. Whereas the long distance systems have driven the technology, it is likely that short distance (local and metropolitan area) networks will drive the architecture. Such architectural advances may be expected to find application to long distance as well as short distance networks. Indeed, the existence of long-haul nationwide fiber networks, interconnected to both metropolitan and local fiber networks, provides not only the promise of meeting present telecommunication needs, but offers the opportunity of shaping future needs.

REFERENCES

Bell Communications Research, Inc. (1985). Single-mode inter-office digital fiber optic systems. Technical Advisory TA-TSY-00038 Issue 2.

CCITT Red Book (1985). Recommendation G. 821. Error performance of an international digital connection forming part of an integrated services digital network.

Cook, J. S., and Szentisi, O. I. (1983). Northern American field trails and early applications in telephony. *IEEE J. Selected Areas in Communication* **SAC-1**, 393–397.

Drupsteen, J. (1985). A high-capacity 565 Mb/s optical system bridges long distances. *Philips Telecommunication Review*, **43**, 31–42.

Fishman, D. A., Lumish, S., Denkin, N. M., Schulz, R. R., Chai, S. Y., and Ogawa, K. (1986). 1.7 Gb/s lightwave transmission field experiments. *Conference on Optical Fiber Communication* (OFC '86), PDP-11.

Gnauck, A. H., Korotky, S. K., Kasper, B. L., Campbell, J. C., Talman, J. R., Veselka, J. J., and McCormick, A. R. (1986). Information-bandwidth-limited transmission at 8 Gb/s Over 68.3 km of single-mode optical fiber. *Conference on Optical Fiber Communication* (OFC '86), PDP-9.

Ishio, H. (1983). Japanese field trials and applications in telephony. *IEEE J. Selected Areas in Communication* **SAC-1**, 404–412.

Iwahashi, E., and Fukutomi, H. (1983). F-400M system overview. *Review of the ECL, NTT*, **31**, 237–243.

Jacobs, I. (1986). Design considerations for long-haul lightwave systems. *IEEE J. Sel. Areas Comm.* **SAC-4**, 1389–1395.

Maione, T. L., Sell, D. D., and Wolaver, D. H. (1978). Practical 45 Mb/s regenerator for lightwave transmission. *Bell Syst. Tech. J.* **57**, 1837–1856.

Meskell, Jr., D. J. (1984). Fiber optic system engineering. *Proceedings of the National Communications Forum*, **38**, 142–148.

Midwinter, J. E. (1981). Studies of monomode long wavelength fiber systems at the British Telecom Research Laboratories. *IEEE J. Quantum Electronics* **EQ-17**, 911–918.

Moncalvo, A., and Tosco, F. (1983). European Field Trials and Early Applications in Telephony. *IEEE J. Selected Areas in Communication* **SAC-1**, 393–397.

Nakagawa, K., Iwashita, K. Ohara, M., and Horiguchi, S. (1984). 1.6 Gb/s optical transmission experiment with monolithic integrated circuits. *International Communications Conference*, Amsterdam, Holland.

O'Connor, P., Flahive, P. G., Clementson, W., Panock, R. L., Wemple, S. H., Shenk, S. C., and Takahashi, D. P. (1984). A monolithic multigigabit/second DCFL GaAs decision circuit. *IEEE Electron. Device Letters* **EDL-5**, 226.

Ogawa, K. (1982). Considerations for single-mode fiber systems. *Bell Syst. Tech. J.* **61**, 1919–1931.

Rosenberg, R. L., Ross, D. G., Trischitta, P. R., Fishman, D. A., and Armitage, C. B. (1983). Optical fiber repeatered transmission systems utilizing SAW filters. *IEEE Trans. on Sonics and Ultrasonics* **30**, 119.

Rubin, P. E. (1975). The T4 digital transmission system—overview. *IEEE International Conference on Communications*.

Stauffer, J. R. (1983). FT3C-A lightwave system for metropolitan and intercity applications. *IEEE J. Sel. Areas Comm.* **SAC-1**, 413–419.

Williams, F., and Leblanc, H. P. (1986). Active feedback lightwave receivers. *Lightwave Technology* **LT-4**, 1502–1508.

Windus, G. G., Jessop, A., and Hemsworth, A. D. (1985). Considerations in the design of a 565 Mbit/s optical fibre system. *Proceedings of 3d International Conference on Telecommunication Transmission*, *IEE*, London, 146–150.

Chapter 24

Undersea Cable Transmission Systems

P. K. RUNGE

AT & T Bell Laboratories, Inc., Holmdel, New Jersey

NEAL S. BERGANO

AT & T Bell Laboratories, Inc., Holmdel, New Jersey

24.1 INTRODUCTION

If the information age is our present era then our ability to communicate is the fiber that joins us. As the information age expands about us, so does our need for telecommunication capacity and functionality. These increasing demands are being met with communication systems that traverse the land, sea, and outer space.

Today there is little doubt that fiber optics will play a key role in the proliferation of telecommunication services in the next decade. The application of fiber optics in the terrestrial market will produce a greatly expanding array of digital services, which in turn will stimulate the market for these services over international circuits (Thomas, 1986; Kochman, 1986). Digital lightwave submarine cable systems along with digital satellite systems will provide the communication links for these new services.

Present international development of undersea lightwave systems calls for the installation of three major systems to be installed in the 1980s. These systems will be shared by several North American, European, and Asian countries. The major systems and suppliers are shown in Table 24.1. The first major system to be installed will be the eighth transatlantic telephone cable (TAT-8), which will be supplied by AT&T, STC, and SUBMARCOM (Fig. 24.1). Following the

TABLE 24.1
First Generation of Long-haul Undersea Lightwave Systems

System	Suppliers	Transmission* Capacity	Service Date
TAT-8 United States to England and France	AT&T STC SUBMARCOM	560 Mb/s (2 × 280 Mb/s)	1988
TPC-3 Hawaii to Japan and Guam	AT&T KDD	560 Mb/s (2 × 280 Mb/s)	1988
HAW-4 California to Hawaii	AT&T	560 Mb/s (2 × 280 Mb/s)	1988

*One-Way Capacity

installation of TAT-8, the TPC-3 and HAW-4 systems will be installed in the Pacific basin (Fig. 24.2) (Ishikawa and Jeffcost, 1986).

These cable systems represent a radical departure from previous intercontinental cables. TAT-8 will be the first digital submarine system, it will employ optical transmission, integrated circuits in the repeaters, an undersea branching capability, and a supervisory system capable of both monitoring and controlling the undersea repeaters.

In this chapter we will explore fiber optic telecommunications in the undersea arena, by highlighting the major differences between terrestrial and submarine

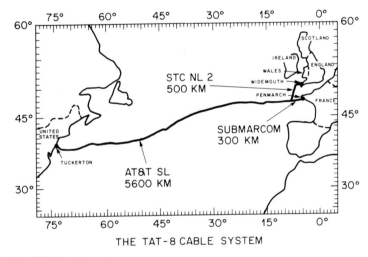

Fig. 1   The TAT-8 cable system.

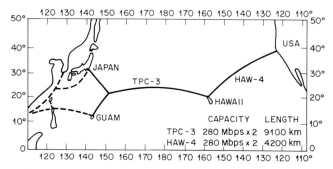

Fig. 2   The TPC-3 and HAW-4 cable systems.

systems. As we shall see, the driving force behind the special designs of lightwave hardware is the need for high reliability. This need affects all aspects of the system. A review of the growth of undersea communication traffic in the past three decades is given, along with a comparison of cables and satellite systems. The special aspects of fiber cables and undersea components are covered in two separate sections. Repeater diagnostics, transmitter sparing and fault location are included in the supervisory section. We conclude the chapter with a look at system span engineering of present and future lightwave systems.

24.2 RELIABILITY

Probably the single most important issue that influences the design of an undersea lightwave system is its need for high reliability. In contrast to its terrestrial counterpart, each component and subsystem of the undersea re-peatered line is extremely inaccessible. This fact precludes the flexibility of being able to design for inexpensive repairs for failed components. Inaccessibility necessitates that the system be capable of working with a minimum of actual repairs over its life time (Easton *et al.*, 1985). Planned undersea cable systems such as TAT-8, HAW-4, and TPC-3 are designed for 25-year service life. During a system failure of the undersea communications link, there are few options for re-routing communication traffic. Other driving forces for high system reliability are the large cost of repair and lost revenue during system down time. Each at-sea repair is an elaborate, time-consuming operation that has to rely on unpredictable elements such as the weather conditions and location of the fault with respect to an available repair ship.

   The stability of the eventual resting place at the bottom of the ocean does have some reliability advantages. The constant low temperature on the ocean bottom is advantageous for long life operation of the semiconductor lasers and electronics. In the deep water sections most of the cable and repeaters are insulated from the human and environmental perturbations of the terrestrial world.

During the design of modern analogue undersea systems there existed a rich knowledge base of reliability data to draw from for critical components such as high frequency Si transistors, and coaxial cable. This made the qualification of the analogue systems easier than the lightwave designs, since actual long-term data was available. Since fiber optics is a relatively new technology, the associated reliability issues are still in a transient state. Practical considerations preclude the use of real time testing of system components to establish reliability. Therefore, the major challenge to the system designers is to design a highly reliable system with a 25-year lifetime using relatively new technology that has little or no reliability knowledge base.

A reliability strategy that addresses the special needs of such a system is shown schematically in Fig. 24.3 (Amster and Hooper, 1986). Starting with the basic system architecture and lifetime specifications, a reliability budget is constructed. This budget allocates an allowable steady state failure rate to each repeater and cable section. From this it is possible to estimate the allowable span failure rate. For example, consider an undersea system with 75 repeater sections. The probability that a section fails in its 25-year lifetime must be less than 0.024 in order to limit the number of ship repairs to three or less, with 90% confidence. From this probability of failure it is possible to estimate the allowable span failure rate. The failure rate allocated to the repeatered section is then further subdivided to assign a reliability goal for the optoelectronics, fibers, integrated circuits, and mechanical fixtures.

In reliability budgeting, the parameter often used to describe failure rate is a FIT, or failure in 10^9 hours. Assuming that a population of devices that have been operating from time $t = 0$, as time progresses some of these devices will fail. If $N(t)$ is the number of devices still operating within specifications at time t, then we can define an instantaneous failure rate $h(t)$

$$h(t) = -\frac{1}{N(t)} \frac{dN(t)}{dt}. \tag{24.1}$$

For a wide class of devices this failure rate will follow the "bath tub" characteristic, which is shown in Fig. 24.4. From this model we can identify three regions: the beginning of life, where the failure rate is decreasing; the mid-life, where failure rate is relatively constant; and the end of life, where the failure rate is increasing. Assuming that we are able to implement screening procedures to "weed out" the early failures, and that the time to wear out is much longer than the system life time, we can bound the mid-life failure rate to an acceptable level. We express this bound, or constant failure rate in terms of FITs. A device with a failure rate of F fits has a probability of failure during H hours of

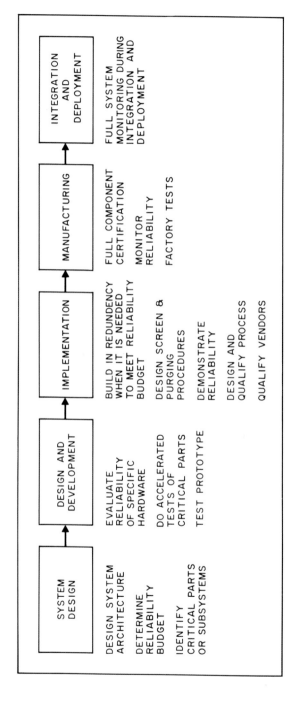

Fig. 3 An undersea reliability strategy (adapted from Amster and Hooper, 1986).

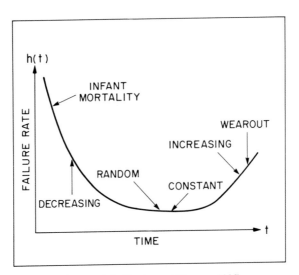

Fig. 4 Bathtub failure rate characteristic (Amster and Hooper, 1986).

operation given by

$$Pr \left(\text{failure during } H \text{ hours}\right) = FH \times 10^{-9} \qquad \text{for } FH < 10^9. \qquad (24.2)$$

In our 75 repeater example, the steady-state failure rate for each repeater/cable section is 110 FITS.

During the system design and development, the reliability study has to identify those components that are critical to the system's performance, but which do not have a sufficient reliability knowledge base to qualify their use. The most critical components in the undersea lightwave system are the semiconductor lasers and photodiodes. The approaches used to establish high reliability for these components are, where possible, eliminating all of the known failure mechanisms from the design and making maximum use of accelerated testing and purging, then designing redundancy into critical areas. As previously stated, real time testing is not feasible; therefore, an indirect approach that employs component stressing is used to accelerate the aging time. Typical stress conditions for semiconductors are high temperature, high voltage, and high humidity operation (Miller, 1984). The risk involved in this procedure is that the stressed conditions might not accelerate *all* of the failure mechanisms equally so that the actual time to failure might not be actually predicted.

The accelerated life tests for the critical components progress systematically through a series of tests, and analysis designed to determine the component population is uniformly accelerated by stress condition. These procedures lead to an understanding of the accelerated factors, or activation energies of the stress procedure, which in turn allow the failure rate "bath tub" parameters to be bounded (Nash *et al.*, 1985). The TAT-8 system will use a multilongitudinal mode buried heterostructure InGaAsP laser diode, operating at a nominal wavelength of 1.3 μm. The AT&T TAT-8 laser's accelerated life tests have estimated the device's probability of failure in 25 years at 10°C at about 0.3%. This level of reliability translates into a failure rate of about 15 FITS (see Chapter 17 for more details). From the experience gained during the testing procedure, it is possible to design tests to screen out the devices that will suffer from infant mortality. These procedures, known as purging and burn-in, could include operating the unscreened devices in the laboratory under field conditions until that failure rate becomes constant, or falls below some acceptable value, or becomes unmeasurably small.

Using the estimated steady-state failure rate of the critical components along with the system reliability budget and system architecture, it is possible to determine the required level of redundancy. Studies show that incorporating redundancy into a system can greatly increase the allowable number of FITs given to an individual component. As an example consider the 75 repeater system with two working lines. The reliability budget allocates 32 FITS to the transmitter (Easton, 1986). If one spare transmitter was employed per regenerator, then the FIT count for the individual transmitters could be increased to 385 FITS, and if two spares were used, then the count could be increased to 900 FITS.

All of the system's critical components must be qualified for high reliability operations. The qualification testing program is a series of tests performed on a product to assure developer, manufacturer, and customer that the product will meet its detailed specifications and that it will do so over a predetermined set of environmental, mechanical, and electrical stresses (Alles, 1986). Qualification tests are performed as a matter of course on new products during the initial manufacturing period. The devices used in the qualification tests are selected at random from early or pre-production lots. An attempt is made to select devices over a period long enough to insure that the devices include a representative sample of several semiconductor and piece-part lots. During the production of systems quality components, devices are chosen at random and are subjected to extensive autopsies. This "surveillance" program adds assurance that the process procedures are within the system's specifications.

Each component and subsystem to be included in the undersea portion of the system is 100% certified. Functionally, this means that every fiber, electronic component, and metal fixture is certified to be undersea quality. Each critical component is subjected to screening, purging and burn-in procedures that were developed in the early design stages. For each critical component, many parameters are measured and compared to the population to determine if the device fits the norm. This data is then reviewed to determine if the component is suitable for undersea use. During the system integration, when the repeaters are joined with the cable sections, the system is constantly tested for proper operation. This monitoring procedure is carried out through system deployment to insure proper system performance.

In addition to *all* of the subsystems requirements for high reliability, the implementation of the system must be qualified reliable. This means that the procedures used to manufacture and deploy the system must also have reliability checks in place. For example, if epoxy is to be used in the repeater housing, it must not evolve any gases that would contaminate the repeater environment for its 25-year service life time. Another example, during the deployment of the system, great care needs to be taken so that the cable does not exceed its specification for minimum bend radius.

24.3 TRANSMISSION NEEDS

The introduction of fiber optics to the undersea arena was a timely solution to an otherwise difficult telecommunications problem. The growing needs for international communications were pushing the limits of conventional analogue cable systems. AT&T's first trans-Atlantic telephone cable in the North Atlantic was TAT-1, commissioned in 1956 and provided 48 voice channels.[1] Since this cables installation, the number of voice circuits in the North Atlantic has been increasing exponentially, at an annual growth rate between 20 and 27 percent (Anderson *et al.*, 1980; Wagner, 1984). Figure 24.5 shows a plot of the cumulative number of transatlantic telephone circuits that have been installed by AT&T. In the late 1970s, system planning indicated that a cable with ~ 8000 voice channels would be needed in the 1988 time frame to follow this growth characteristic. An analogue system capable of meeting these needs would have required a cable greater than 3 cm in diameter, and repeater spacing of about 5 km. In the Pacific the rate of growth of international cable circuits has been about 15–20% (Yamamoto, 1986) (Fig. 24.6). Based on the transmission capabilities of TPC-1 and 2, a new system with a throughput similar to the Atlantic system will be required in the same 1988 time frame. Unlike the analogue systems, technologically the digital fiber systems could easily support the channel capacity needed for the end of this decade, and the vast amount of fiber bandwidth allows for future growth.

The needs of international communications cannot be met by cable systems alone. The other major technology that entered the public telecommunications market in 1965 is oversea satellite. Fiber optic cable systems and satellite technology have overlapping markets where point-to-point links that have heavy

---

[1] All references to the number of voice circuits are quoted before time assignment interpolation (TASI) or digital compression techniques.

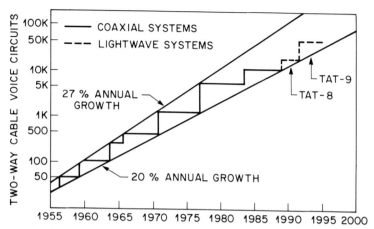

Fig. 5 Cumulative trans-atlantic telephone circuits installed by AT&T.

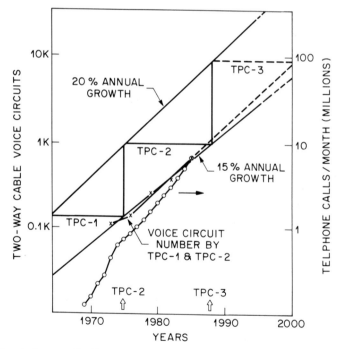

Fig. 6 Cumulative trans-Pacific telephone circuits (Yamamoto *et al.*, 1986).

traffic utilization are needed. The advantages that the undersea system has over its satellite counterparts are security, system availability, minimal time delay, and system lifetime. The nature of the light guiding properties of single-mode fiber coupled with the remote cable location makes the undersea system very secure against covert eavesdropping and immune to RFI and EMI. The intrinsic 0.6 second round trip time delay associated with satellite communications is a nuisance in two-way conversations and data transfers. System availability is an advantage of submarine cables compared to satellite systems, which can be affected by sunspots or factors of nature. This advantage is exemplified by the TAT-8 owner's requirements for system availability. For this system the mean bit error ratio (BER) measured over any 24-hour period must be less than 4.4×10^{-8}. There must also be less than 8.2 minutes per day with BER $> 10^{-6}$, and less than 2.56 seconds per day with BER $> 10^{-3}$. This system has been designed to be available greater than 99.996% of its lifetime. Planned international systems such as TAT-8 and TPC-3 will have expected service life times of 25 years, opposed to 10 years for INTELSAT VI for example. On the other hand, satellites provide route and facility diversity, they afford economic advantages in supplying communications to remote or landlocked countries, and they give countries a measure of control over their own means of communications (Thomas, 1986).

24.4 FIBER AND CABLES

The fiber optic cable has to simultaneously satisfy a multitude of system requirements including reliability, low optical loss, protection to the single-mode fibers, and a power path for the repeaters. In this section we shall review the special requirements for single-mode fibers used in the submarine environment, and then explain how these fibers are incorporated into the undersea cable.

Starting with single-mode fiber that exhibits suitable transmission characteristics for a lightwave system, additional mechanical requirements such as tensile strength and fatigue life time must be satisfied in order to consider its use in a submarine system. The production of undersea quality fiber in a transatlantic system might require a 40,000 km fiber inventory. The successful production of such long fiber lengths necessitates an optimized design, along with a low-loss, high strength splice procedure. The special manufacturing procedures for undersea quality fiber include the selection of high quality starting materials, delicate handling and surface preparation of preforms, clean adherent coating materials, manufacturing in a clean room environment, and prooftesting (see Chapter 4).

At ambient temperature and moisture conditions, the intrinsic tensile strength of short sections of silica-based fiber is approximately 800 ksi (Nagel, 1984). This translates to a breaking force of 15 lbs for a 125-μm diameter fiber. For comparison, a high strength plow steel wire of the same diameter would have a

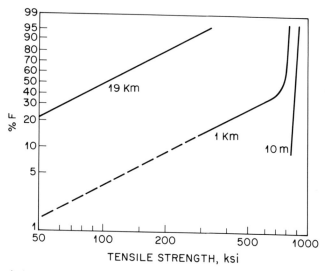

Fig. 7 Cumulative plot of fiber breaking strength for several gauge lengths (Nagel, 1984).

breaking strength of about 4 lbs (Chin and Mechtly, 1985). The fiber strength specifications for the first generation of undersea system is 200 ksi (2% strain).

Figure 24.7 shows a cumulative probability plot of fiber breaking strength for several gauge lengths (Nagel, 1984). The tensile strength of silica fiber is reduced from its intrinsic level because of defects or flaws that are randomly distributed along the fiber (Sakaguchi, 1984). Such defect centers can lead to catastrophic failure when the fiber is stressed. This results in a bi-modal strength distribution for fiber lengths longer than tens of meters, since the probability of encountering a weak spot increases with fiber length. Considering this bi-modal behavior and that repeater span is in the 50–70 km range, we need a prooftesting procedure to ensure fiber integrity.

Careful prooftesting of the entire fiber inventory limits the remaining crack or defect size in the system fiber. In order to achieve the TAT-8 strength requirement, fiber flaws must be limited in size to hundredths of microns (DiMarcello and Krause, 1986). During prooftesting, the fiber passes through a set of belts and pulleys that subject the fiber to a predetermined set of stress conditions. Great care must be given to the design of the testing apparatus for high strength prooftesting, such that the testing procedure itself does not weaken the fiber. This is accomplished by designing a prooftest that rapidly unloads the applied stress to the fiber, which limits crack growth in the post testing region.

To obtain long lengths of high strength fiber for use in an undersea system, it is necessary to have a splicing procedure that gives both high strength and low loss. This has been accomplished using a combination of chemical stripping, flame fusion splicing, and over coating techniques (Krause and Kurkjian, 1983). These techniques can produce low-loss splices with splice strength similar to the intrinsic strength of the fiber. Splice losses in the range of 0.01–0.1 dB are routinely made with fibers that possess a high degree of core to cladding concentricity (Nagel, 1984).

Stability of the fiber attenuation has a large impact on the system planning for long life. During the system planning it is important that any added loss mechanisms be understood so that they can be either designed out of the system, or accounted for in the systems optical loss budget. Two mechanisms that increase the fiber attenuation over time have been identified in undersea fiber cables. These are hydrogen and radiation induced losses. Hydrogen gas can be generated by electro-chemical reactions between metal in the cable and trapped moisture (Mochizuki et al., 1984). Hydrogen can increase the loss of GeO_2 doped fibers through two different mechanisms (Stone, 1987). The first effect, which is reversible, is caused by molecular hydrogen diffused into the glass being polarized by high electronegativity of oxygen in SiO_2 molecule. An absorption peak at 2.42 μm occurs because energy is transformed to the vibrational state of the polarized H_2 molecules. This effect has an overtone at 1.24 μm, thus increasing the loss in the 1.3 μm band. The second effect, which is irreversible, is caused by a hydrogen reaction that forms OH radicals. This hydroxyl formation causes an increase in the loss peak at 1.41 μm, which induces a broad band loss increase.

The attenuation of optical fibers can also be increased by exposing them to ionizing radiation (Friebele, 1984; Mies and Soto, 1985). This effect has been correlated to phosphorus concentrations in the light guiding regions of the fiber. In the North Atlantic, the background sea bottom radiation dose has been estimated at 100 mrad/year (Schulte, 1985). With this dose rate, a 100 km 1.55 μm system using a high phosphorus fiber could experience a loss increase of ~ 0.25 dB per 100 km span in a 25-year lifetime.

The cable used in submarine lightwave systems has to protect the fiber from damaging stresses, provide a satisfactory environment to ensure fiber reliability, and provide power to the undersea regenerators. During the deployment of the submarine system the cable must be able to support its own weight in water without over straining the fibers. In the maximum depth of 5-1/2 km in the TAT-8 route there might be as much as 9 km of cable suspended between the ocean bottom and the cable ship. For deep water cable, the maximum strains occur near the water's surface during a recovery operation (Adl et al., 1984). In this situation the cable tension is a combination of the static loading of the cable's weight in water and the dynamic loading caused by the ship's motion and cable recovery speed. For the AT&T TAT-8 cable it is estimated that the combined maximum static and dynamic tensions encountered during cable recovery will be 78 kN, which could result in a 0.78% cable strain.

The mechanical design of undersea cables is influenced by the environmental conditions of the ocean bottom as well as the practical considerations of cable handling (Gleason *et al.*, 1978; Adl *et al.*, 1984). The design window of cable strength vs. weight in water for AT&T undersea cable is shown in Fig. 24.8. This characteristic assumes a cable diameter of 21 mm. The cable's design region is bounded by five different constraints. The maximum strength and weight are dictated by the cable ship. AT&T's cable ship C. S. LONG LINES has a tension capability of about 72 kN. In the maximum system depth of 5.5 km there might be 9 km of cable suspended between the ship and the ocean bottom; therefore, the upper limit on weight in water is about 72 kN/9 km or 8 kN/km. To protect the ship's cable handling equipment, and for safety reasons, the maximum strength of the cable is set to 180 kN, which is the rating of the cable engine's braking apparatus. Referring to the figure, the vertical line at 2 kN/km gives the minimum weight so that a 21 mm cable would have a horizontal sinking speed greater than 0.97 km/hr. This minimum sinking speed is required to minimize cable suspensions between ocean bottom irregularity. Within this rectangular design region, there are two additional strength restrictions. The lower boundary is given to ensure that the cable can support its own weight and any dynamic loading stress during its recovery. Cable experience shows that a minimum strength requirement of the equivalent weight of 19 Km of cable in water is sufficient. Finally, as a practical matter the strength of the cable will be less than the strength attainable from an all-steel cable of the same

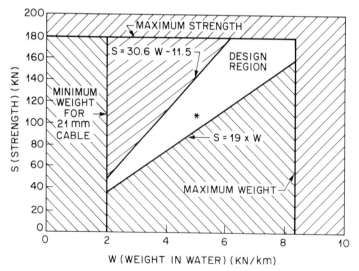

Fig. 8 Design window for AT&T undersea cable (Adl *et al.*, 1984 and Gleason *et al.*, 1978).

diameter. The actual properties for the deep water SL cable are 4.9 kN/km weight in air, and minimum breaking strength of 107 kN, (shown as an asterisk on Fig. 24.8).

An important function served by the cable is to supply power to the submarine repeaters. In contrast to the analogue coaxial undersea cables the fiber cable has only one conduction path. Therefore, the cable's dielectric layer has to provide proper electrical insulation from the sea water (which is at earth ground) and mechanical insulation so that the cable is protected from abrasions. Power is supplied to the repeaters by a d.c. current on the cable's conductor. Assuming that the cable's resistance is between 1/2 and 1 ohm/km with repeater spacings of 40–60 km and a 1.5 amp supply current, the voltage drop across each section of cable could be from 30 to 90 volts. This voltage could easily exceed the voltage drop across the repeater itself. As an example, consider a 6000 km system with 120 repeaters each needing 1.5 amps at 20 volts. If the cable resistance is 0.75 ohm/Km, the terminal power supply would need to bias the cable with a voltage drop of about 10 kV. Typically this is achieved using a bipolar supply. One end of the cable is biased with +5 kV and the other end with −5 kV.

The design philosophy adopted by the major manufacturers of undersea cables is to build a basic structure capable of deep water operation, with the flexibility of providing additional armor protection for shallow water or high risk areas. Five deep water cable designs are shown in Fig. 24.9 (Iwamoto, 1986). In most of the designs, the fibers are placed in a helix shape and are buffered from the center wire and the cables strength members. This helix structure serves to average the stresses applied to the cable in the vicinity of a bend since stress on any fiber will alternate from compression to elongation. Combinations of steel wires and copper sheaths are used to supply high strength and low resistance. Typically polyethylene is used as a dielectric. Water blocking compounds are used to fill the air spaces between strength members, which inhibits water propagation down the length of the cable in the event of a cable failure.

Cables deployed close to the shore in shallow water run the greatest risk of being damaged by either human intervention or natural causes. To protect the cable the basic structure is surrounded by armored steel wires. This entire structure may be buried on the ocean bottom near the shore end and extend several kilometers out into the ocean.

24.5 UNDERSEA COMPONENTS

The design of the optics and electronics used in the undersea systems poses a special set of reliability and performance problems. The system's inaccessibility dictates that reliability needs to be the primary requirement. This reliability requirement necessitates the use of integrated circuits for the regenerator elec-

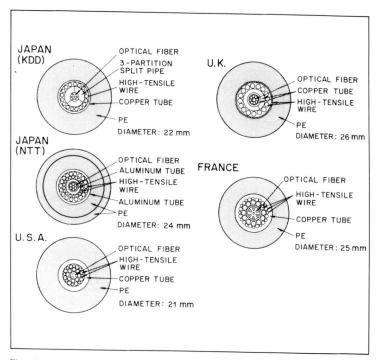

Fig. 9   Five deep water cable cross sections (Iwamoto, 1986).

tronics and the use of redundancy in the form of stand-by transmitters, and/or repeatered lines. High performance requirements are necessary from both an economic and reliability standpoint. The cost of the undersea repeaters is a significant percentage of the overall system expense; therefore, maximizing the repeater spacing could reduce the system cost. Additionally, reducing the number of repeaters could in turn ease the reliability allocation for any individual span. An integral part of the undersea electronics is a remote monitoring and control system. This supervisory system monitors the performance of the remote electronics and controls laser sparing operations.

The major functional blocks within the regenerator are the optical receiver, automatic gain controlled (AGC) amplifier, timing recovery circuit, decision circuit, transmitter and supervisory interface. A typical block diagram showing the relationship of these components is shown in Fig. 24.10 (Ross *et al.*, 1984). The functionality of the receiver and transmitter are described in detail in Chapters 19 and 20. The AGC amplifier adjusts the amplitude of the signal from the receiver to some optimal level for the decision circuit. The retiming circuit extracts a clock signal from the data stream that is phase coherent with the data. The decision circuit accepts the recovered clock, and data stream and

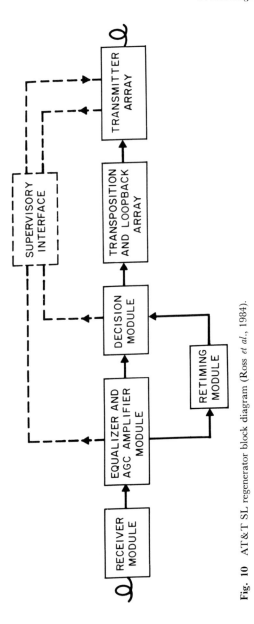

Fig. 10 AT&T SL regenerator block diagram (Ross *et al.*, 1984).

then reconstructs the data with precise amplitude and transition spacing (Reinold and Ross, 1986). The supervisory circuitry provides the necessary monitoring and control functions in the repeater (see next section).

One measure of repeater performance is its effective gain, which is the difference between the transmitter output power and the receiver sensitivity. Maximizing this parameter is more important in an undersea application than in a terrestrial system. For land-based systems, it is convenient to install the new lightwave systems along established intercity trunking routes. These routes were originally designed to accommodate coaxial wire transmission equipment where span lengths were shorter than the fiber systems. Since the possibility of reusing the established equipment vaults exist, and the fact that new equipment housings are not a major expense, maximum repeater spacing is not imperative for terrestrial systems. In an undersea system such as TAT-8, however, an increase in repeater gain could have a large effect. For example, an increase of 2 dB in repeater gain could increase the repeater span in a 1.3 μm system by 5 km, which might reduce the repeater count in a transatlantic system by as many as 6 repeaters.

The effective repeater gain is dominated by the regenerator's receiver and transmitter. Submarine receiver designs employ very low noise front end amplifiers with either a p-i-n or avalanche photodiode detector, which is similar to terrestrial systems. Uncommon to land-based systems are the high reliability designs of the devices and the packaging schemes. For example, the AT&T TAT-8 receiver uses a special common package for the InGaAs p-i-n detector and Si IC (Snodgrass and Klinman, 1984). This allows for high temperature bakeout and purge for the common package before it is installed in the receiver housing, which is important for establishing high reliability. This special design also reduces the total input capacitance at the receiver's input, which is important for performance optimization. The NTT F-400M receiver uses a Ge APD that has the ohmic contact made of Au/Pt/Ti rather than the standard Al metal. (Chino and Fukuda, 1983). This alleviated a failure mechanism that was due to Al metal penetration into the Ge depletion layer.

Laser transmitters designed for use in the submarine environment possess most of the functionality of terrestrial units, with special emphasis placed on optical coupling stability, low thermal resistance to the repeater housing and a high degree of integration for the bias control, and high speed electronics. Transmitter reliability is strongly affected by the lifetime of the laser diode, which in turn is affected by its operating temperature. Terrestrial laser package designs may incorporate thermoelectric coolers for active temperature control (Asous et al., 1984). For reliability and powering reasons, however, thermoelectric coolers are not used in the submarine systems. Fortunately, the cold temperature at the bottom of the ocean (which is nominally 2°C in deep water) offers a good heat dissipation environment. Submarine transmitters are designed to provide a low thermal resistance path from the laser stud to the repeater body. It is estimated that the stud temperature of a working TAT-8 transmitter will be from 7°C to 12°C above the transmitter base plate (Bosch *et al.*, 1984).

As previously stated, the required system reliability of the first generation undersea systems necessitate the use of transmitter redundancy. Each regenerator has its primary transmitter along with one or more inactive transmitters that can be switched on in the event the primary unit becomes marginal in performance. To accommodate these needs the transmitter output must be connected to the transmission fiber through an optical sparing device, such as an optical relay or a passive optical coupler. Optical relays are made of short lengths of single-mode fiber that are supported in fixtures that allow the core of the launching fiber to be moved from one output to the next (Kaufman *et al.*, 1984). These optical relays have been realized in 1 by 2 or 2 by 4 arrangements. The 1 by 2 version has been qualified for undersea use and will be used in TAT-8 and TPC-3. The sparing function could easily be performed by a passive coupler, however, couplers suffer an intrinsic 3 dB loss for a 1 by 2 device. If the system reliability budget allows for only one spare transmitter, then a passive coupler using polarization redundancy could be used to couple two sources to one fiber with no intrinsic loss (Tsutsumi *et al.*, 1984). In this case the fiber pigtail is made from polarization maintaining fiber. By using a polarization beam splitting device the two pigtails launch light to the transmission fiber in orthogonal states of polarization.

24.6 REPEATER SUPERVISION

The repeater supervisory system is essential to the integration, deployment and operation of the submarine lightwave system (Anderson and Keller, 1984). Monitoring and remote control of the repeater hardware is also an essential part of the system reliability program. The monitoring functions of the supervisory system will provide valuable data for determining the repeater's long-term performance, such as transmitter aging. As previously stated, the long-term reliability of the system is achieved by incorporating redundancy in the submarine repeaters. Redundancy management is administered using the supervisory system to communicate with, and control the repeaters.

For the majority of the system's life, the supervisory system will work in the in-service mode. Since the system is fully operational at these times the functionality of the supervisory system is limited to a repeater monitor. One of the key in-service indications of the repeater's performance is a measure of the channel's bit error rate. Since an actual in-service measurement of the error rate is impractical, different estimation schemes are used. One possible method that will be employed by AT&T (Anderson and Keller, 1984) and KDD (Wakabayashi *et al.*, 1986) in the TPC-3 system is to monitor parity block errors.

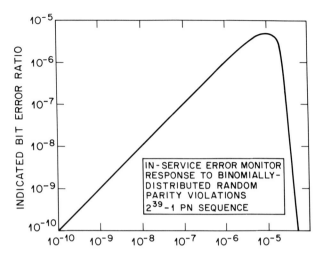

Fig. 11 Bit error rate as measured by an in service error monitor vs. actual bit error rate (Anderson and Keller, 1984).

Experience has shown that this method yields good agreement with the actual error rate for randomly distributed errors for bit error rates from 10^{-10} to 10^{-6} (Fig. 24.11). Depending on the particular implementation, a multitude of other monitoring functions can be designed to operate in-service and/or out-of-service. These monitored parameters include logic states such as the position of optical relays, electronic relays, and transmitter states. Analogue parameters also require monitoring such as the transmitter laser bias current. Long-term bias monitoring will provide useful aging data. At the front end of the regenerator the received optical power can be estimated by measuring either the detector bias current or the AGC amplifier control voltage. This measurement gives valuable information on the span's sensitivity margin.

The different levels of redundancy used in undersea systems such as TAT-8 and TPC-3 are multiple transmitters per regenerator and spare regenerator/fiber pairs per repeatered span. The normal mode of operation of these systems will be two working fiber pairs, and one stand-by line (Fig. 24.12a). In the event of a transmitter failure a spare could be switched using the electrical and/or optical switching circuits. The spare regenerators and fiber pairs would be utilized in the event of a span failure, other than the transmitter. In this case the failed component could be circumnavigated by a transposition operation, which is shown in Fig. 24.12b. Here, the high speed data is switched to a spare regenerator and fiber, then after passing the failed section, the data would be switched back to the primary fiber.

In the event of a failure, the system operators must be able to identify the nature of the failure and then take action to correct the situation, this is made

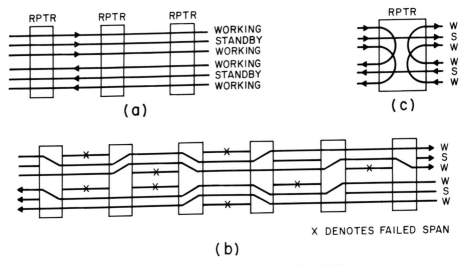

Fig. 12   TAT-8 loop-back and transposition (Anderson and Keller, 1984).

possible by the loopback function (Fig. 24.12c). The loopback operation involves sending a test signal, say in the east to west direction and then instructing the repeater to redirect the data stream back to the originating end on another fiber. By successively polling each repeater to perform the loopback function, while monitoring the repeater's vital signs, it is possible to locate and to identify a fault, such as a cable break or repeater outage.

The supervisory system remotely initiated monitor and control functions require a two-way communication channel from the terminal equipment to the undersea repeaters. In an all optical supervisory system this two-way link has to be transmitted as either high speed dedicated bits, or as a modulation impressed on the high speed data. The in-service channel selected for the TPC-3 and HAW-4 systems is a mixture of the two. In these systems the terminal to repeater link is accomplished by periodic parity violations. These parity violations are detected by the regenerator's error rate monitor and are decoded by the supervisory logic. The response channel is accomplished by phase modulating the data stream. This low speed phase modulation does not interfere with the high speed signal and can be recovered by demodulating the clock signal at the terminal site. These techniques can be used for both in- or out-of-service operation. Other communication channels are possible for out-of-service operation. In the KDD section of the TPC-3 systems the mark to space ratio is modulated to communicate with the repeaters. In the French TAT-8 section, special alarm words are transmitted in the out-of-service mode (Lacroix, 1984).

24.7 SYSTEM SPAN ENGINEERING

So far we have discussed the requirements of a variety of lightwave components and subsystems used in the submarine environment. Now we address the problem of putting all of the pieces together in order to have a high speed, long haul lightwave link. Here we shall explore the different system options that utilize technologies that are either presently available, or should be available for submarine applications in the next five to ten years.

We start our discussion with a list of assumptions that guides us through the multitude of available options. First, we limit our discussion to include only direct detection systems. In the future, coherent optical systems will become practical for commercial lightwave systems; however, in the next ten years these systems will probably be limited in the undersea market to special long unrepeatered applications. Second, we assume that the system requirements call for maximum reliability at minimum cost. Often these constraints translate into a requirement for long repeater span lengths.

The two transmission wavelengths to be considered are 1.3 μm and 1.55 μm. These wavelengths have the obvious advantage of occupying the low loss regions in silica-based fibers. The technology for 0.8 μm GaAs lasers exists today; however, the higher fiber losses at this wavelength would not allow for the maximum span length criteria. The wavelength band beyond 1.55 μm is not considered since little knowledge of long wavelength fibers and optoelectronics is available.

The optical sources that we considered are Fabry-Perot (F-P) semiconductor lasers and distributed feedback (DFB) lasers. The 1.3 μm multi-longitudinal mode F-P type semiconductor laser is the workhorse of today's terrestrial and undersea long haul market. Lightwave transmitters are now being manufactured with output power of approximately 1 mW and root mean square (RMS) spectral width of about 2 nm. The reliability of these devices has been established and it is believed that it is possible to extend the technology to 1.55 μm. Future systems requiring a single frequency source will probably rely on the DFB laser structure. Both 1.3 μm and 1.55 μm DFB lasers are now commercially available, and it is thought that the high reliability required for submarine applications will be demonstrated in the years to follow.

The fiber types to be considered are the step-index profile that have a zero dispersion wavelength (λ_0) in the 1.3 μm region, and the triangular index profile dispersion shifted fibers (DSF), which have a λ_0 around 1.55 μm. The step-index designs include the conventional dual window fiber types that can be useful at both 1.3 μm and 1.55 μm (Nagel, 1984), and the single window fibers optimized for low loss at 1.55 μm. The optimized fibers have demonstrated the lowest losses for single-mode fiber, which is about 0.16 dB/km at 1.55 μm (Csencsits et al., 1984).

The bit rate at which the lightwave system is specified has an obvious impact on the optics and optoelectronics. For comparison, we will consider five different ranges of bit rates. The distinction between these ranges is somewhat arbitrary, however, they will be useful in describing the needs of systems.

The different options for the optics and optoelectronics are:

Wavelengths:	1.3 μm	1.55 μm
Laser Types:	F-P	DFB
Fiber Types:	Step-Index	DSF
Bit Rates:	0.1 < B < ~ 2Gb/s.	

From this list it is possible to make many combinations of technically feasible systems. Table 24.2 shows these options, with the most attractive highlighted. Some would make a usable combination; however, the complications that are involved are not required for the particular bit rate (i.e., the system would be overkill). These systems are labeled "NR." Others are not technically feasible, these are labeled "NA" (not applicable).

Referring to the low bit rates the most attractive system would use 1.55 μm multifrequency lasers, with step index fibers. This system architecture gives the maximum span length of any system that we will consider. The large amount of chromatic dispersion of the step-index fibers at 1.55 μm would be tolerable because of the low bit rates. System experiments in excess of 200 km have been demonstrated as being loss limited at 3 Mb/s (Mazurczyk, 1984). As part of a U.S. Government application, AT&T has installed a 150 km system in the Pacific that uses this architecture (McNulty, 1984). Although this type of system is technically feasible, its application would be limited to short haul and/or unrepeated systems. The high cost of installing a long undersea system often necessitates higher information rates so as to be cost competitive with other carriers.

The first generation of true long-haul undersea lightwave cable systems fall into the medium speed range. These systems will use 1.3 μm buried heterostructure F-P semiconductor lasers with step-index fiber designs. Large lightwave systems that are in this category include TAT-8 and TPC-3 which will operate at 295.6 Mb/s. The higher bit rates of these systems require an architecture that can control the effects of chromatic dispersion (see Chapter 21), this is accomplished by placing the mean wavelength of the multimode lasers near the λ_0 of the fiber.

The second generation of undersea lightwave systems will probably incorporate an increased bit rate. Indeed preliminary negotiations between North American and European countries have specified the TAT-9 system to operate in the 600 Mb/s range. The choices of the system architecture could be one of four possibilities. These include the present 1.3 μm F-P systems, 1.55 μm F-P

Undersea Lightwave System Options

λ	1.3 μm	1.55 μ (2nd Generation Systems)			
	Multimode F-P	Multimode F-P		Single-mode DFB	
Laser Structure	Step-index	Step-index	DSF	Step-index	DSF
Fiber					
B < 10 Mb/s	• Medium span length (> 110 km) • Short-haul, specialty system applications	• Longest span length (> 150 km) • Short-haul specialty system applications	NR	NR	NR
10 Mb/s < B < 300 Mb/s	• 70–110 km spans • First generation long-haul systems. e.g, TAT-8, TPC-3, and HAW-4	• 70–110 km spans • Applications in specialty systems for bit rates < 45 Mb/s • NA for B > 45 Mb/s	• 100–150 km spans • Moderate tolerance on λ_0, $\bar{\lambda}$ and σ	• 110–170 km spans • Wide Tolerance on λ_0 and $\bar{\lambda}$	NR
300 Mb/s < B < 1 Gb/s	• 55–70 km spans • Small tolerance on λ_0, $\bar{\lambda}$ and σ	NA	• 70–100 km spans • Small tolerance on λ_0, $\bar{\lambda}$ and σ • Dispersion limited	• 90–110 km spans • Wide Tolerance on λ_0 and $\bar{\lambda}$ • Requires Low Chirp Laser	• 80–95 km spans • Wide tolerance on λ_0 and $\bar{\lambda}$ • Most expensive system
1 Gb/s < B < 2 Gb/s	• Dispersion limited spans • Very Strict Tolerance on λ_0, $\bar{\lambda}$ and σ • Short-Haul, specialty system applications	NA	NA	• 65–90 km spans • Wide Tolerance on λ_0 and $\bar{\lambda}$ • Requires very low chirp laser	• 55–80 km spans • LModerate Tolerance on λ_0, $\bar{\lambda}$ and laser chirp
B > 2 Gb/s	NA	NA	NA	• Wide Tolerance on λ_0 and $\bar{\lambda}$ • External Modulator Required	• Maximum Bit Rate Distance product • Direct Modulation May Be Possible

Notes. NR Not Required. NA Not Applicable

type system using dispersion shifted fiber, a 1.55 μm distributed feedback laser system using the step-index fiber or a 1.55 μm distributed feedback laser system using dispersion shifted fiber.

1) 1.3 μm F-P/STEP
2) 1.55 μm F-P/DSF
3) 1.55 μm DFB/STEP
4) 1.55 μm DFB/DSF

The advantages of remaining with the 1.3 μm system is that the high reliability of lasers and fiber technologies could be directly transferred to the next generation. The disadvantages would be that the higher bit rates would result in shorter span lengths and would require tighter tolerances on the laser's spectral content. The main advantage of moving to the 1.55 μm region would be increased repeater span length that would result from the lower fiber losses. This added repeater span length would come at the cost of a development program (including reliability assurance) for the 1.55 μm F-P/DSF and/or 1.55 μm DFB/STEP systems. The development effort needed for the 1.55 μm F-P laser would require less effort than the 1.55 μm DFB laser because of similarities to the 1.3 μm F-P laser used in TAT-8, TPC-3, and HAW-4. The 1.55 μm F-P/DSF system, however, would have similar wavelength tolerances to the 1.3 μm arrangement. These wavelength tolerances will become narrower as the bit rate and span length are increased. The tolerance problems could be relaxed in a 1.55 μm DFB system. The narrow spectral line width of the DFB laser could control the chromatic dispersion effects that the multimode F-P suffer from. The major disadvantage of a submarine system based on a DFB laser is the effort to qualify their reliability and spectral stability. However, once the reliability has been established the DFB laser will probably be the favored light source for future submarine transmitters.

Systems operating at bit rates that exceed 1 Gb/s will probably rely on single frequency lasers. As bit rate is increased, the tolerances in a F-P laser system would become stricter. In the multi-gigabit region these diminished tolerances will make a long-haul multi-longitudinal mode submarine system impractical. Therefore, these high bit rate systems will require single-longitudinal mode lasers in conjunction with either conventional fiber or dispersion shifted fiber. At these high bit rates, the modulation induced spectral broadening or laser chirp, will need to be considered. This effect in conjunction with the high bit rates, and large amount of chromatic dispersion of the step-index fibers at 1.55 μm will lead to a dispersion limitation. To circumvent this problem, DFB lasers could be used along with dispersion shifted fiber. It is believed that this type of system could yield the maximum bit rate distance product of the systems measured.

REPEATER SPACING

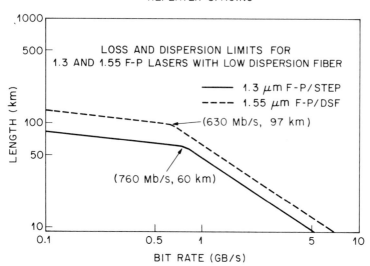

Fig. 13 Repeater spacing vs. bit rate for 1.3 μm and 1.55 μm multi-frequency lasers, with low dispersion fiber.

Figures 24.13 and 24.14 show the repeater spacing for these four systems, as a function of bit rate. The assumptions used to generate this figure are given in Appendix A. In these curves we observe a discontinuity in the slope, which corresponds to the boundary between the repeated span being loss limited, and dispersion limited. For bit rates less than 600 Mb/s, all four systems are loss limited.

For the F-P laser systems, the dispersion limit was fixed at 10% pulse spreading due to chromatic dispersion that occurs from the mean laser wavelength ($\bar{\lambda}$) mismatched from the zero dispersion wavelength (λ_0) of the fiber. The two multimode break points occur at 630 Mb/s and 760 Mb/s for the 1.55 μm and 1.3 μm devices respectively. For both systems we have assumed an 11–13 nm difference between the λ_0 of the fiber and mean operating laser wavelength $\bar{\lambda}$. Since the slope of the dispersion characteristic is steeper for the step-index fiber, it yields a larger chromatic dispersion for the same wavelength offset. Even though the dispersion limitation for the 1.55 μm system is displaced toward the higher bit rates, the break point occurs at a lower bit rate. This results from the longer span lengths afforded by the lower fiber loss at 1.55 μm.

For the 1.55 μm single frequency systems the dispersion limit is due to dynamic spectral broadening or laser chirp. The effects of chirp pose a potential span limitation for long-haul digital fiber optic systems (Linke, 1985). A chirping DFB laser in the presence of chromatic dispersion can cause a sensitiv-

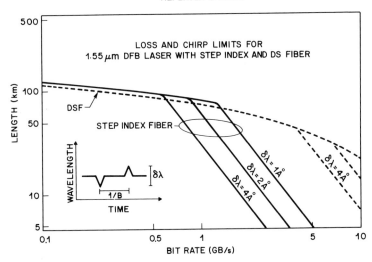

Fig. 14 Repeater spacing vs. bit rate for 1.3 μm and 1.55 μm DFB lasers with step-index and dispersion shifted fiber.

ity penalty in the regenerator arising from inter-symbol interference due to pulse distortion. Several recent works have treated the system impact of chirp (O'Reilly and Silva, 1987; Yamamoto, et al., 1987; Corvini and Koch, 1987). Typically, the metric used to quantify the chirp phenomenon is the change in the mean wavelength of the modulated laser. The top curve in figure 24.15 shows how the mean wavelength of a modulated DFB laser changes in time during NRZ modulation. The bottom trace shows the laser's intensity waveform. The chirp or the change in the mean wavelength is plotted for the time segments when the intensity is greater than 20% of its maximum value. This laser exhibits the characteristic blue shift at the leading edge of the pulse and red shift at the trailing edge, which has been reported in many references. The chirp was measured using a wavelength discriminator method (Bergano, 1988).

We relate chirp and regenerator sensitivity penalty using an empirical model given in appendix A (24.A5). The repeater spacing vs. bit rate for a 1.55 μm DFB laser using conventional step-index fiber (Figure 24.14) was generated by limiting the chirp penalty to 1 dB and using the parameters listed in table 24.3. The breakpoint between the loss and dispersion limit varies between 600 Mb/s and 1.3 Gb/s for a chirp value of 4 Å to 1 Å respectively. Therefore, the usefulness of this type of system in the 1–2 Gb/s region will be independent on the development of a low chirp DFB laser.

If a low chirp lasers can not be realized, then high speed next generation systems will rely on dispersion shifted fiber. A 1.55 μm DFB system using DS

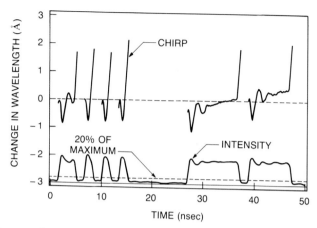

Fig. 15 Mean wavelength and intensity vs. time for a DFB laser modulated with a 600 mb/s NRZ signal.

fiber is shown to be loss limited up to 4 Gb/s assuming 4 Å chirp. For lasers with lower chirp, the breakpoints occur at even higher bit rates.

24.8 SUMMARY

We have completed our examination of fiber optic telecommunications in the submarine environment. As we have seen, the systems' need for high reliability caused several design departures from the terrestrial systems. The first submarine application of fiber optics for long-haul communications will be the TAT-8 system, which is scheduled to be commissioned in 1988. As more and more systems are added to the international digital network, we will move closer to achieving the Global "Ring Main" built from fiber optic submarine cables (Kochman, 1986).

The reader who is interested in a more detailed understanding of undersea fiber optic systems is directed to two collections of works. The first entitled "Undersea Lightwave Communications" (Runge and Trishitta, 1986) contains a collection of papers originally written for a special edition of the *Journal of Lightwave Technology* by the same name. The second source is the meeting proceedings from the International Conference on Optical Fiber Submarine Telecommunications Systems (Suboptic).

APPENDIX A

Theoretical calculations for the loss and dispersion limits of four different lightwave systems are presented. The first configuration is a TAT-8-like system that uses a multilongitudinal mode laser at 1.31 μm with step-index fiber. The second system uses a 1.55 μm multimode laser and dispersion shifted fiber (DSF). The third system uses a single frequency laser at 1.55 μm with step index fiber. The maximum bit rate distance product would be possible with the fourth system, which uses a single frequency laser at 1.55 μm, with DSF.

System	Abbreviation
1)	1.3 μm F-P/Step
2)	1.55 μm F-P/DSF
3)	1.55 μm DFB/Step
4)	1.55 μm DFB/DSF

Our findings are presented in Fig. 24.13 and 24.14, which show the repeater spans, as a function of bit rate.

Loss Limit. The loss limit calculation is relatively straightforward. The number of decibels to be allocated to fiber loss is calculated by subtracting the receiver sensitivity and margin from the launched power. The margin is assumed to be 10 dB in all cases, independent of bit rate. To obtain the bit rate dependence, we assume that the receiver sensitivity changes at 3 dB/octave below 1 Gb/s and at 4 dB/octave above 1 Gb/s. We index the receiver sensitivity to be -31.5 dBm at 1 Gb/s, so that receiver sensitivity becomes;

$$\text{Receiver Sensitivity} = \begin{cases} -31.5 + 10\log(B) & 0.1 \leq B \leq 1 \text{ Gb/s} \\ -31.5 + \dfrac{40}{3}\log(B) & 1 \leq B \leq 10 \text{ Gb/s} \end{cases} \tag{24.A1}$$

For the single frequency systems, we assume that the average launch power of the transmitter will be 3 dB less than their multilongitudinal mode counterparts. The span length is calculated by dividing the available loss by the average fiber loss.

Multilongitudinal mode dispersion limit. For the dispersion limitation we set the pulse spreading to be one tenth of a time slot, i.e.,

$$mBz\sigma = \frac{1}{10} \tag{24.A2}$$

or

$$z(B) = \frac{1}{10mB\sigma}. \tag{24.A3}$$

The chromatic dispersion for step-index fiber operated at 1.55 μm is approximated at 17 ps/km/nm. For operation near the zero dispersion wavelength of DSF and step-index fiber, we approximate the dispersion to be

$$m = S|\bar{\lambda} - \lambda_0|_{\max} = S\Delta\lambda. \tag{24.A4}$$

TABLE 24.3
Parameters Used in Span Length Calculations

System	P_t (dBm)	S (ps/km-nm^2)	$\Delta\lambda$ (nm)	m (ps/km-nm)	σ (nm)	$\delta\lambda$ (Å)	α (dB/km)
1.31 μm F-P/STEP	0	0.089	11	0.98	2.0	—	0.38
1.55 μm F-P/DSF	0	0.05	13	0.65	2.5	—	0.24
1.55 μm DFB/STEP	−3	—	—	17	—	1, 2, 4	0.22
1.55 μm DFB/DSF	−3	0.05	13	0.65	—	1, 2, 4	0.24

Notes-

P_t Launch power
S Dispersion slope at λ_0
$\Delta\lambda$ Wavelength separation between λ_0 and $\bar\lambda$
m Dispersion
σ RMS spectral width of F-P lasers
$\delta\lambda$ DFB laser chirp
α Fiber attenuation

Our assumptions for the dispersion slope (S), maximum wavelength offset ($\Delta\lambda$), dispersion (m), linewidth (σ), loss available to the cable (L), and fiber loss (α) are given in Table 24.3.

Single-longitudinal mode dispersion limit. A chirping DFB laser in the presence of chromatic dispersion can cause a sensitivity penalty in the regenerator arising from inter-symbol interference due to pulse distortion. A simple empirical relationship between chirp and regenerator sensitivity penalty is:

$$penalty \approx kmz\delta\lambda B^2 \qquad (24.A5)$$

where $k \approx 5\, dB \cdot nsec$ and $\delta\lambda$ is laser chirp. For the dispersion limits in figure 24.14 we assume a regenerator sensitivity penalty of 1 dB. Our assumptions for m, and $\delta\lambda$ are given in Table 24.3.

REFERENCES

Adl, A., Chien, T-M, and Chu, T-C. (1984). Design and testing of the SL cable. *JLT* LT-2, No. 6. Alles, D. S. (1986).

Amster, S. J., and Hooper, H. H. (1986). Statistical methods for reliability improvement. *AT&T Technical Journal*, **65**, Issue 2.

Anderson, C. D., Gleason, R. F., Hutchison, P. T., and Runge, P. K. (1980). An undersea communication system using fiber guide cables. *Proceedings of the IEEE*, **68** No. 10.

Anderson, C. D., and Keller, D. L. (1984). The SL supervisory system. *JLT* LT-2, No. 6.

Asous, W. A., Palmer, G. M. Swan, C. B., Scotti, R. E., and Shumate, P. W. (1984). The FT4E 432 Mb/s lightwave transmitter. "Links for the future, science, systems & services for communication." Dewilde, P. and May, C. A. (editors). IEEE/Elsevier Science Publishers. B. V. (North-Holland).

Bergano, Neal S. (1988). A Wavelength Discriminator Method of Measuring Dynamic Chirp in DFB Lasers. *To be submitted for publication in the Journal of Lightwave Technology.*

Bosch, F., Palmer, G. M., Sallada, C. D., and Swan, C. B. (1984). Compact 1-3 μm laser transmitter for the SL undersea lightwave system. *JLT* **LT-2**, No. 6.

Chin, G. Y., and Mechtly, E. A. (1985). "Reference data for engineers: radio, electronics, computer, and communications." Howard W. Sams & Co., Inc., pp. 4–43.

Chino, K., and Fukuda, K. (1983). Degradation mechanism in breakdown of Ge APD. *Tech. Dig.*, Fall Meeting IECE, Japan, p. 96.

Corvini, P. J., and Koch, T. L. (1987). Computer Simulation of High-Bit-Rate Optical Fiber Transmission Using Single-Frequency Lasers *JLT* **LT-5**, No. 11.

Csencsits, (1984). In the optical fiber conform proceedings, 1984.

DiMarcello, F. V., and Krause, J. T. (1986). Advances in high-strength fiber fabrication. *Technical Digest of the Optical Fiber Communication conference*, paper TUE1, Atlanta, Georgia.

Easton, R. L. (1986). Redundancy to tailor system design to establish component reliability. Comm. 6.1 Suboptic, 86 Février, France.

Easton, R. L., Hartman, R. L., and Nash, F. R. (1985). Introduction to assuring high reliability of lasers and photodetectors for submarine lightwave cable systems. *AT&T Tech. J.* **64**, No. 3.

Friebele, E. J. (1984). *Nucl. Inst. Meth. Phys. Res.* **B1**, 355.

Fukinuki, H., Takeshi, I., Aiki, M., and Yoshihivo, H. (1984). The FS-400M submarine system. *JLT*, **LT-2**, No. 6.

Gleason, R. F., Mondello, R. C., Fellows, B. W., and Hatfield, D. A. (1978). Design and manufacture of an experimental lightguide cable for undersea transmission systems. *Proc. Int. Wire and Cable Symp.* (Cherry Hill, NJ).

Hakki, B. W., Fraley, P. E., and Eltrongham, F. (1985). 1.3 (μ) laser reliability determination for submarine cable systems. *AT&T Tech. Journal* **64** No. 3.

Ishikawa, Y., and Jef/cost, C. M., (1986). HAW-4/TPC-3: System planning and progress. Comm 3.2, Suboptic 86, Février, France.

Iwamoto, Y. (1986). Optical fiber submarine communications and its evolution. *Anritsu News* **6**, No. 27.

Kaji, H. (1985). Trans-Pacific ocean cable project—history and future development. *Proc. of KTCIF '85*, Seoul, Korea.

Kaufman, S., Reynolds, R. L., and Loefler, G. C. (1984). An optical switch for the SL undersea lightwave system. *JLT* **LT-2**, No. 6.

Koch, T. L., and Linke, R. A. (1986). Effect of nonlinear gain reduction on semiconductor laser wavelength chirping. *Appl. Phys. Lett.* **48**(10).

Kochman, Z. (1986). Global "Ring Main" for fiber optic submarine cable networks. Comm. 12.1 Suboptic '86, Février, France.

Krause, J. T. and Kurkjign, C. R. (1983). Improved high strength flame fusion single mode splices. *Tech. Dig.* 4th IOOC (Tokyo, Japan) **29A**, pp. 4–6, 96–97.

Lacroix, J-C. (1986) Dialogue channels used for remote supervision and remote control of the underwater plant in submarine digital telephone links using optical fibers. "Undersea Lightwave Communications," edited by Peter K. Runge and Patrick R. Trischitta, IEEE Press.

Linke, R. A. (1985). Modulation Induced Transient Chirping in Single Frequency Lasers. *IEEE JQE* **QE-21**, No. 6.

Mazurczyk, V. J. (1984). 202 Km transmission spans at 155 nm with multilongitudinal mode lasers. *Tech. Dig.* OFC '84, Post-deadline paper WJ8 (New Orleans, LA).

McNulty, J. J. (1984). A 150 Km repeaterless undersea lightwave system operating at 1.55 μm. *JLT* **LT-2**, No. 6.

Mies, E. W. and Soto, L. (1985). Characterization of the radiation sensitivity of single-mode optical fibers. *Tech. Digest of IOOC-ECOC '85* Venezia, Italy.

Miller, L. E. (1984). Ultra-high reliability ultra-high speed silicon integrated circuits for undersea optical communications systems. *JLT* **LT-2**, No. 6.

Mochizuki, K., Namihira, Y., and Kuwazuru, M. (1985). Optical fiber transmission loss due to hydrogen. KDD Konnichiwa, No. 57, pp. 20–22.

Mochizuki, K., Namihira, Y., Kuwazuru, M., and Nunokawa M. (1984). Influence of hydrogen on optical fiber loss in submarine cables. *JLT* **LT-Z**, No. 6.

Nagel, S. R. (1984). Review of the depressed cladding single-mode fiber design and performance for the SL undersea system application. *JLT* **LT-2**, No. 6.

Nash, F. R., Joyce, W. B., Hartman, R. L., Gordon, E. I., and Dixon, R. W. (1985). Selection of a laser reliability assurance strategy for a long-life application. *AT&T Tech. J.* **64**, No. 3.

O'Reilly, J. J., and Silva, H. J. A. (1987). Chirp-Induced Penalty in Optical Fibre Systems. *Elec. Lett.* **23**, No. 19.

Reinold, G. A., and Ross, D. G. (1986). Highly integrated repeaters for the SL undersea lightwave transmission system. Comm. 7.1 Suboptic '86 Février, France.

Runge, P. K. and Trischitta, P. R. (1984). The SL undersea lightwave system. *JTL.* **LT-2**, No. 6.

Runge, P. K., and Trischitta, P. R. (1986). "Undersea lightwave communications." IEEE Press, New York.

Sakaguchi, S. (1984). Drawing of high-strength long-length optical fibers for submarine cables. *JLT* **LT-2**, No. 6.

Sakaguchi, S., and Nakshara, M. (1983). Strength of proof tested optical fibers. *J. Amer. Ceram. Soc.* **66**, No. 3. pp. C-46–C-47.

Schulte, H. J. (1985). *Technical Digest*, OFC '85 Conference (San Diego, Ca) Paper TUQ2.

Sharifi, M. H., and Arozullahi, M. (1986). Comparison of satellite and fiber optics technologies for intercity and intercontinental communications. I.C.C. 26.5. 1-5.

Snodgrass, M. L., and Klinman, R. (1984). A high reliability, high sensitivity lightwave receiver or the SL undersea lightwave system. *JLT* **LT-2**, No. 6.

Stone, J. (1987). Interactions of hydrogen and deuterium with silica optical fiber: a review. *JLT*, **LT-5**, No. 5, 712.

Thomas, D. G. (1986). Interconnectivity the oceans: progress for terrestrial lightwave systems and universal information services. Comm. 12.2 Suboptic '86, Février, France.

Tsutsumi, S. Ichihashi, Y. Sumida, M. and Kano, H., (1984). LD redundancy system using polarization components for a submarine optical transmission system. *JLT* **LT-2**, No. 6.

Wagner, R. E. (1984). Future 1.55 μm undersea lightwave systems. *JLT* **LT-2**, No. 6.

Wakabayashi, H., Ashida, T., Oda, H., and Yano, H. (1986). The OS-280M repeater supervisory system. Communication 8.1, International Conference on Optical Fiber Submarine Telecommunication Systems, February 18-21, 1986.

Worthington, P. (1984). Cable design for optical submarine systems. *JLT* **LT-2**, No. 6.

Yamamoto, S., Sakaguchi, H., Nunokawa, M., Namihira Y., and Isamuta, Y. (1986). Next generation optical fiber submarine cable systems and TDM submarine branching systems. Comm. 13.6, Suboptic '86, Février, France.

Yamamoto, S., Kuwazuru, M., Wakabayashi, H., and Iwamoto, Y. (1987). Analysis of Chirp Power Penalty in 1.55 μm DFB-LD High-Speed Optical Fiber Transmission Systems *JLT* **LT-5**, No. 10.

Chapter 25

Optical Fibers in Loop Distribution Systems

PATRICK E. WHITE

Bell Communications Research, Inc., Morristown, New Jersey

L. S. SMOOT

Bell Communications Research, Inc., Morristown, New Jersey

25.0 INTRODUCTION

This chapter provides an overview of the application of optical fiber technologies to the portion of the telephone network known as the "local access network" or "loop distribution network" or simply "the loop."

Section 25.1 will identify the unique characteristics of this portion of the network. Section 25.2 will briefly outline the historical application of digital transmission technologies in the loop. Section 25.3 describes the introduction of fiber optic carrier systems in the loop, and the effect of the loop environment on their design requirements. Section 25.4 provides motivation for the current research on the Broadband Integrated Services Digital Network (BISDN), the emergence of which will substantially alter the subscribers' view of telecommunications. In the fifth section, some of the current thinking concerning architectures for the BISDN is discussed. Finally, Section 25.6 highlights some of the BISDN standards activities taking place around the world. The definition of mutually agreed-upon BISDN standards in a timely fashion will speed this technology to the marketplace.

25.1 THE LOCAL DISTRIBUTION NETWORK

The structure of the telephone network is illustrated in Fig. 25.1. The principal elements of this network are the Central Offices (COs), interexchange trunks, and the loop distribution plant.

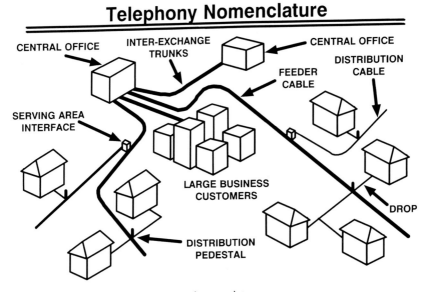

Fig. 25.1   Telephone network structure and nomenclature.

The COs provide the voiceband switching function in each of the local geographic areas they serve. They, in turn, are connected by an inter-exchange "trunk" network to other offices in a hierarchical manner such that calls may be completed within a local area (by the local office) or via intermediate offices in the hierarchy, from one locale to another. It is the function of the loop plant (the term loop derives from the fact that the two wires comprising a telephone circuit form an elongated loop) to connect customers to the local Central Office. One can make note of the fact that the facilities in the traditional loop plant are dedicated on a per subscriber basis. This is in contrast to the routes in the inter-exchange network, where the norm is for facilities to be shared (trunks, switching facilities) over many subscribers.

The loop distribution plant is generally deployed as a concatenation of cables of decreasing size as the distance from the CO increases. Large (hundreds of individual wire pairs) "feeder" cables emanate from the CO and are connected to smaller "distribution" and "drop" cables so as to extend service to individual customers in an efficient manner. The typical CO is capable of supporting copper loops with direct-current resistances up to 1500Ω. For longer loop lengths, cable gauges are increased as needed to remain within this limit, and load coils (series inductors) are added to equalize the frequency response.

25.2 DIGITAL LOOP CARRIER SYSTEMS

In the early 1970s, a trend toward the increased use of electronics in the outside plant was begun with the introduction of digital loop carrier (DLC) systems. Digital multiplexing systems were economically justified for use in the highly shared trunk plant, before their introduction in the dedicated loop plant. Decreasing electronics costs, increasing copper cable costs, and the labor associated with cable installation, however, led to the economic prove-in of digital transmission systems for the loop. These systems are based on the T1 digital transmission system introduced in 1961 for the interexchange trunk plant. The T1 system has a transmission bit rate of 1.544 Mb/s and is capable of carrying 24 voice circuits. Each voice circuit, $\approx$ 3 kHz bandwidth, is sampled at an 8 kHz rate; one byte is used to encode the amplitude of each sample, and one bit of overhead is added each 125 μs (24 bytes) to define a "frame."

With the new digital carrier systems, the feeder plant could be used more efficiently since 24 customer lines could be multiplexed onto one pair for each direction of transmission. Thus, these systems are said to introduce "pair gain"; that is, the number of feeder pairs is fewer than the number of distribution/drop pairs they serve. Although early systems were nonstandard, the generic DLC has evolved to the view shown in Fig. 25.2 It consists of two primary subsystems, a Central Office Terminal placed at the Central Office and a Remote Terminal interconnected to it via one or more T1 lines. Individual subscribers are connected to the Remote Terminal within a loop resistance limit specified by the equipment manufacturer. The line side of the Remote Terminal provides the functions necessary to interface with analog telephone sets; talking battery, ringing voltages, etc. The T1 carrier side of the Remote Terminal provides the

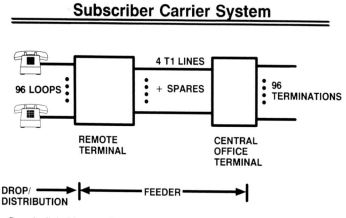

Fig. 25.2 Generic digital loop carrier.

analog-to-digital conversion, and the remaining T1 line interface functions. The Central Office Terminal terminates the T1 line and demultiplexes the 24 channel signal into separate subscriber switch interfaces. When the cable loss at the T1 rate exceeds an acceptable margin, digital line repeaters are added at regular intervals along the T1 cable in order to regenerate the digital signal. Depending on the application, typical repeater spacings are on the order of 1.5 km.

25.3 FIBER CARRIER SYSTEMS

In a fiber carrier system, an optical fiber cable is used to replace the T1 copper cable in the link between the Remote Terminal and Central Office Terminal. There are several motivations for this replacement. First, the longer unrepeatered spans possible on optical facilities eliminate the need for repeaters, one of the larger causes of DLC system failure. In particular, a 1983 loop survey showed that 97% of all loops in the United States were less than 10 km. Since currently available fiber transmission systems permit repeater spacings on the order of 25 km, nearly all fiber subscriber carrier systems are unrepeatered. Finally, the larger bandwidths and the smaller cross-sectional areas of optical fibers alleviate duct and manhole congestion and help to minimize "churn-induced" outages; that is, outages inadvertently caused by telephone company craft when additional capacity is being provided or when nominally unrelated repairs are being made on adjacent facilities.

25.3.1 Fiber Loop Carrier Design Requirements

The first commercial Fiber Loop Carrier (FLC) [Olson and Shepis, 1984] systems were introduced in the early 1980s and were essentially an outgrowth of the Digital Loop Carrier systems, which preceded them. It is instructive to examine the requirements for such systems to gain insight into the effect that designing for the loop environment will have on future broadband local access systems.

The remote terminals of the FLCs might be expected to operate in outside plant enclosures with temperature extremes of from $-40°C$ to $+65°C$. In addition, humidity levels can be high, and there is significant risk of contamination. Since craftspersons must be dispatched when outside plant failures occur, there is significant economic penalty for poor reliability.

Due to these requirements, some of the critical optical devices in such systems were hermetically sealed and required to have extremely high reliabilities; corresponding to extrapolated lifetimes on the order of over a million hours at $70°C$. This level of reliability was viewed to be consistent with the large deployment levels anticipated and corresponded to an approximate 1% field dropout rate/year.

Since injection laser diodes (ILDs) and avalanche photo detectors (APDs) were not capable of this type of performance, the systems were based on the more reliable P-type-intrinsic-N type (p-i-n) photodetectors and light emitting diodes. Since LEDs suffer an approximate 15 dB penalty compared to lasers ($\approx$ O dBm coupled power) when launched into "standard" (50 μm core 0.2 numerical aperture fiber), *and* p-i-n diodes do not supply the approximate 5 to 10 dB of increased receiver sensitivity of APDs; the system loss budget for these early FLCs was critical.

In order to offset this limited loss budget, special fibers were engineered (62.5 μm, 0.29 NA) to couple an additional 5 dB of optical power from an LED. In addition, the optical receivers used for such systems were required to be highly sensitive, thus implying high impedance designs but with large dynamic ranges to accommodate the uncertainty of FLC Central Office Terminal to Remote Terminal span lengths. Special AGC techniques were used in these early receivers to achieve high optical dynamic range (on the order of 40 dB) but with low sensitivity penalties (on the order of 0.5 dB for a 45 Mb/s receiver). Ultimately, these systems evolved to the use of single-mode fibers employing lasers with thermo-electric coolers.

An important requirement for these fiber loop carrier systems was that a single failure should not affect more than a few subscribers. Consequently, the optical line interface units were duplicated and two sets of fibers used in the link between the Central Office Terminal and Remote Terminal. During normal operation, both optical line units send information on both sets of fibers. Specialized circuitry in the far-end optical line interface unit monitors one of the fibers, the primary link, for errors. When these errors exceed a preprogrammed threshold, a link switch is initiated, causing reception to be derived from the spare fiber and optical line interface unit. In this manner, telephone calls are largely unaffected by single failures in principal subsystems.

25.4 BROADBAND ISDN

The next generation of fiber optic systems in the loop plant will probably not seek to improve on the use of optical fiber transmission equipment in its role as a replacement for T1 lines. It will be as part of a more sweeping concept, which will totally change the nature of the existing network.

The concepts for this new network (actually a combination of fiber optic and emerging electronic technologies) are lumped under the heading of the Broadband Integrated Services Digital Network (BISDN). The ISDN portion of the terminology indicates end-to-end digital connectivity with a standard set of interfaces for integrated services delivery. It can be viewed as an extension of the (narrowband) ISDN concepts being field trialed at the time of this writing (1987). "Basic" ISDN service (Fig. 25.3) would provide subscribers access to two

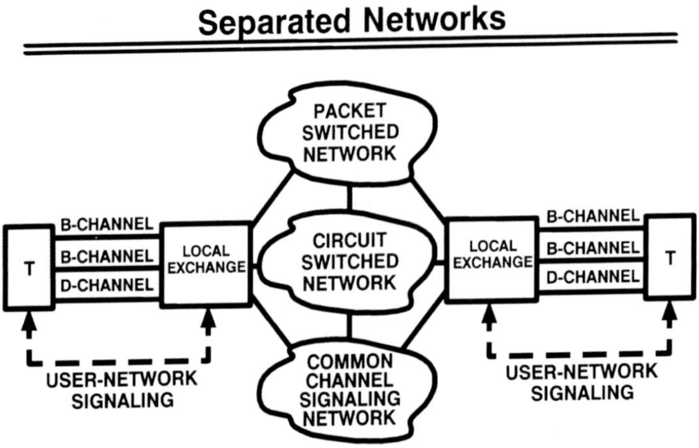

Separated Networks

Fig. 25.3   ISDN network topology.

64 kb/s digital circuit switched "B" channels and a 16 kb/s packet switched "D" channel on regular copper loops. Although many modern Central Offices employ digital switching internally (with peripheral analog to digital and digital to analog conversion), this is the first time that the subscriber will be able to access the underlying digital switch fabric directly. The "D" channel is used for subscriber packet data transport and to control the two "B" channels, which can be used for digitized speech transmission (using a subscriber's on-premises CODECs) or for relatively high-speed data transport. A "primary rate" access is also supported essentially giving customers direct T1 rate access. The key here is that the delivery of these services on copper loops imposes an inherent bandwidth bottleneck.

Broadband ISDN Services, on the other hand, would be delivered via optical fiber terminating directly on subscriber premises and in contrast to narrowband ISDN, would offer bit rates measured in the hundreds of Mb/s. The BISDN research being conducted at the time of this writing is being spurred by the perceived need in the residential and business communities for broadband communication services, i.e., "market pull" and the emergence of technologies required to realize such systems or "technology push."

For the residential subscriber, the primary market may be for the delivery of entertainment video services, which includes the distribution of commercial broadcast television (or cable TV-like services), "video on demand" (i.e., the ability to select a movie or other material for immediate viewing) based on the development of third-party vendors of "on-line" stored video sources, as well as "narrow casting." This latter category includes the broadcast of video material to small groups of selected subscribers such as college football games to alumni, musical concerts to paying fans, etc. It is also believed that residential subscribers may be interested in full-motion high-resolution video text for applications such as catalog shopping, where the catalog is a full-motion high-resolution video display, remotely selectable and controllable by the subscriber (Armbruster, 1986).

Business services are usually grouped under a "communications umbrella," which includes the interconnection of Local Area Networks; multimedia voice/data/video teleconferencing, the broadcast of high resolution documents to multiple remote sites, and the distribution of company training materials (a form of video on demand).

25.4.1 BISDN Bit Rates and Channel Structure

Since one of the primary "market pulls" for BISDN is the distribution of cable TV-like video information; this service must be of high quality and delivered in sufficient "quantity" so as to serve the needs of the network subscribers. The initial thinking is to provide an "extended quality" (EQTV) type digital video transmission channel, this means, for instance, that in contrast to commercial (off air) video, which is NTSC encoded (color information band limited and frequency-division-multiplexed into the black and white information via a color subcarrier), the video delivered by the BISDN would be component encoded. In this system, the color information leaving a video camera would in essence be separately digitized, time-division multiplexed, and sent through the network to a subscriber where it would be converted back to an analog representation of the (still separate) original color information. This means of video transmission offers a perceived picture improvement but requires more bandwidth > 100 Mb/s) than digitally encoded NTSC (< 100 Mb/s).

While "EQTV" addresses the quality issue, the quantity is determined by the fact that in the United States the majority of residences have at least two television sets and a rising percentage of households own a video cassette recorder. These devices are "sinks" of broadband video services; thus, a reasonable network should provide for the delivery of 3–4 digital video sources. In order to hold coding, and especially digital video *decoding* expenses down, the BISDN channel bit rates should be as high as possible, consistent with optical technology limits and with economical technologies, i.e., complimentary metal oxide-semiconductor (CMOS) VLSI for processing the digital video. After considering these inputs, it would seem that a bit rate of from 100 to 150 Mb/s per video channel would be appropriate.

In addition to these circuit-switched video channels, the forward looking ISDN network must provide a means to transport the large variety of emerging data rates associated with high-speed LANs and Metropolitan Area Networks (MANs). This channel should be maximally flexible and synergistic with the transport of the other broadband (video channels). It is for this reason that a fourth packet-mode channel whose line rate is also around 150 Mb/s is postulated. In addition to its subscriber packet data conveyance function, this channel serves to handle subscriber-network signaling information to control logical channels within the high-speed packet channel, as well as the other circuit switched digital video channels. The digital video channels and the packet channel would be time-division-multiplexed for transport leading to an aggregate delivered bit rate of from 500 to 600 Mb/s.

25.5 BISDN ARCHITECTURAL CONSIDERATIONS

Although debate continues as to the selection of the appropriate optical network topology for the BISDN, there is some agreement on the factors, that should be used to evaluate alternative choices. These factors include the cost of the network, particularly the cost of subscriber access to the network; the reliability of the network, including the susceptibility of the network to fraudulent use; the privacy and security of customer data on the network; and the ease of upgrading the network to satisfy unforeseen customer demand.

25.5.1 BISDN Revenue Potential and Willingness to Pay

The issue of cost and revenues has been studied extensively. Based on an analysis of current residential subscriber "willingness to pay" for services similar to that planned for the BISDN, it is estimated that the per-subscriber line costs of the BISDN should be in the range $1,500 to $2,000 to become a viable alternative distribution technology. Table 25.1 shows an updated version of material, which first appeared in (Bellcore Network Equipment, 1986).

TABLE 25.1
BISDN Willingness to Pay—A Scenario

Residential Broadband Services	$/Month
Plain old telephone services	15.00
Video broadcast	20.00
Video recorder/tapes	11.00
Quality/Convenience Premium	5.00
Total video	$36.00
Energy management	2.00
Meter reading (Utility Co. Paid)	2.00
High resolution videotext	5.00
Video phone	0.50
TOTAL	$60.50

For an investment of $1500, if we ignore tax effects, assume an interest rate of 7 percent, a straight line book depreciation schedule of 10 years, and a return on investment of 12 percent, then the monthly charge would be about $70. Tax depreciation effects, including the investment tax credit (if one exists in 1990) should serve to reduce this revenue requirement. This simple analysis also ignores operations savings, which could also reduce the revenues needed to justify residential BISDN services.

25.5.2 Reliability

Telephone network reliability is a critical consideration. For example, depending on usage, and component reliabilities, alternate facilities must be provided when the number of customers affected by a single failure exceeds a given threshold, typically 64 in today's network. In addition, as a common-carrier network provider, a telephone company must provide a network that protects the privacy and security of customer data. Mechanisms must also be provided to limit the capability of individual network subscribers to affect the level of service provided on the network. Routing and transmission cannot be based on a LAN or shared-media type of architecture, since the network provider cannot rely on customers accurately identifying themselves to the network. Protecting the privacy of subscriber data on a shared media also poses serious technical problems. Similarly, the routing and transmission integrity of the network must be resistant to failures or customer actions that might cause the continuous transmission of erroneous data from a subscriber terminal into the network. These requirements, together with the above cost constraints, tend to limit the topologies which may be considered viable for the BISDN to switched-star type architectures, or their off-shoots, where the network provider maintains control of the network and subscriber data is not interaccessible.

25.5.3 The Double Star Topology

The network topology currently receiving the most research attention is the double star architecture illustrated in Fig. 25.4. It has a structure similar to the typical telephone distribution introduced previously. The first star in this network, the Remote Electronics, is analogous to the remote terminal in the subscriber carrier system. It provides "fiber gain" via multiplexing and concentrating (i.e., by demand assignment) of customer traffic from several customers to feeders with less capacity than the maximum aggregate of subscribers access bandwidth office (Kneisel, 1986). Therefore, as in the subscriber carrier systems, the costs of the feeder, trenching, cable, and feeder transmitter/receiver electronics would be shared. Furthermore, in actual systems, only the drop fiber and trenching (if buried) would be dedicated to the particular subscriber. The trenching and fiber cable used in the distribution would also be shared among

Preliminary BISDN LATA Architecture

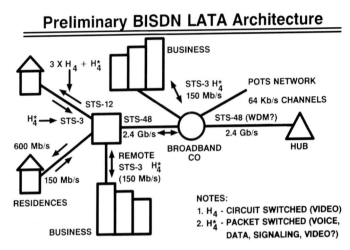

Fig. 25.4 Preliminary BISDN local access architecture. Notes: H_4-circuit switched (video); H_4^*-packet switched (voice, data, signaling, video?).

several subscribers. Thus, theactual implementation of the double-star architecture might be, for example, as shown in Figs. 25.5 and 25.6. The feeders shown in Fig. 25.5 would serve 384 customers, which, as shown in Fig. 25.6, could be further organized into four zones of 96 customers served by distribution and drop fibers. This particular model assumes an "upscale" residential neighborhood with homes on ≈ 1 acre lots (D^2 in Fig. 25.6). Thus, four homes are served from a Distribution Panel. Fibers in the link between the Distribution

Loop Plant Layout

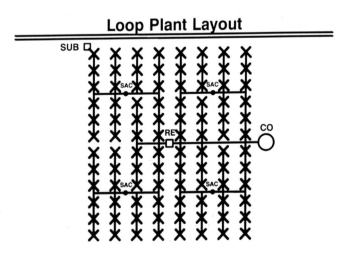

Fig. 25.5 Loop plant topology.

Serving Area Distribution

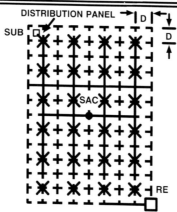

Fig. 25.6 Serving area distribituion topology.

Panel and Serving Area Crossconnect Panel (SAC) share a common trench and are organized in groups of 12. At the Serving Area panel, the size of the group expands to 72, sharing a common trench in the link to the Remote Electronics, which is placed in an outside-plant enclosure such as a Controlled Environment Vault or "Hut."[1]

Since customers are each delivered up to 600 Mb/s of digital bandwidth, in order to realize basic "fiber gain," the feeders must operate at a higher line rate than the individual subscriber to RE fibers. The present technology limits set a rate of around 2 Gb/s for a viable field-mounted system. This provides 16 150 Mb/s channels/feeder such that subscribers' 150 Mbs channels have access to these shared feeder channels via high-speed circuit switching within the RE. This circuit allocation is on a demand basis and the 150 Mb/s bit rate is within the capabilities of economical fine-line CMOS VLSI crosspoint switches (Linnell, 1986). Thus, for the 384 subscribers, a maximum of 24 fibers would be required.[2] Depending on the traffic, however, this could be reduced further.

Bell Communications Research, based on a survey of the Bell Operating Telephone Companies, has proposed that single-mode fiber (rather than multimode fiber or coax) be used for the drop and distribution, in addition to the feeder (Kaiser, 1985). There are several reasons for this. The ubiquitous use of single-mode fiber allows for common installation and maintenance practices

---

[1]This design was proposed by L. J. Scerbo, Bell Communications Research.

[2]This follows if we assume each subscriber has a 150 Mb/s upstream link to the Central Office and that each feeder operates at 2.4 Gb/s. For broadcast video (downstream) traffic, even greater levels of concentration can be achieved.

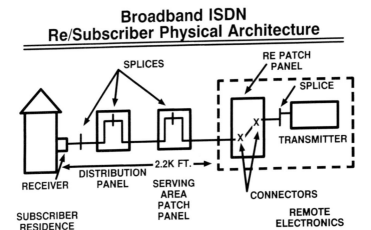

Fig. 25.7   Broadband ISDN remote-electronics/subscriber physical architecture.

across the trunk and loop plants. This fiber offers higher transmission band-widths in direct detection systems due to the elimination of modal dispersion and is also synergistic with various integrated optic technologies and coherent transmission. Since fibers in the drop and distribution are dedicated to individual subscribers, this region represents a significant portion of the costs of a broadband system; therefore, it is prudent to install the most robust technology. Figure 25.7 shows a physical design for the Remote Electronics/subscriber link of the double-star architecture. We concentrate on this segment of the architecture since it has the greatest impact on cost for the new broadband network (components are largely dedicated to individual subscribers) as shown in Section 25.3. The deployment of fiber in the feeder segment can be justified in some areas by operations savings for today's voice and low-speed data services.

24.5.4 BISDN Loop Fiber Costs

The design example shown in Fig. 25.7 shows splices in the outside plant occurring at the network termination at the subscriber premises, the distribution pedestal (panel), the Serving Area Patch Panel, and the Remote Electronics line card. Two connectors are located at the Remote Electronics line card. Two connectors are located at the Remote Electronics patch panel to provide flexibility in connecting subscribers. This design is used for purposes of illustration and more optimized designs (fewer splices) might be used for final system deployments.

Table 25.2 summarizes the projected 1995 fiber and component costs associated with this design.

TABLE 25.2
BISDN Loop Fiber Costs

Item	Quantity	Unit Cost ($)	Total ($)
Fiber Cable	2,200 ft	0.03/ft	66
Splices	4	25	100
Connectors (labor/matl)	2	10	20
Drop Placement	200	1/ft	200
Distribution Placement	2000	0.02/ft	40
Grand Total			426

25.5.5 BISDN Remote Electronics Optical Link Technologies

The double-star structure of the BISDN utilizes relatively short fiber links between the Remote Electronics (RE) and the subscriber premises and longer feeder links between the Central Office and the Remote Electronics. This structure allows the use of low-power, medium-speed, and potentially more economical transmitters (and receivers) at the subscriber premises and at the Remote Electronics end of the subscriber link, where costs are dedicated. The Remote Electronics, however, must be linked to the Central Office with higher speed, higher output power and more expensive transceiver systems. However, the greater cost of these systems is offset by the fact that they are shared across all the subscribers connected to the Remote Electronics.

We first take note of the fact (mentioned earlier) that some 97% of all telephone loops in the former Bell System are less than 10 km in length. Since BISDN will ultimately serve the same subscribers that are now receiving conventional telephone service, we can use the 10 km limit as a reasonable upper bound on the length of the feeders connecting the Remote Electronics to the Central Office. If we further assume that the subscriber "loop" (i.e., the distribution and drop) can be held to under 2 km, we have bounded the problem sufficiently to begin to consider the technologies for the links.

25.5.6 The Subscriber to Remote Electronics Link

For the residential application, approximately 600 Mb/s of aggregate bandwidth in the downstream, and 150 Mb/s in the upstream direction is assumed. For business applications, 150 Mb/s bidirectional transmission might be required. Thus, the loss budgets are determined by the 600 Mb/s rates. Two fibers will be used to support bidirectional transmission in both applications; although, this can be relaxed if wavelength-division multiplexors prove more cost effective. We will note that the use of single-mode fiber for transmission at these bit rates is not absolutely required (a multimode fiber system would also be possible), but that the single-mode system is selected for the reasons stated earlier.

There are several choices for transmitters. A conventional solution might be to use a tightly coupled injection laser diode operating at the 1.3 μm wavelength. This wavelength is chosen since it corresponds to the zero material dispersion point of the silica fibers now widely deployed in the network. The laser would couple a large amount of power, but there would be severe cost penalties associated with 1987 "transmission quality" lasers. These devices typically cost around $1000 and would not be expected to reach a sufficiently low cost by the mid-1990s when BISDN deployment is contemplated.

Laser reliability would also be of concern. Devices suitable for the loop application have reliabilities similar to devices used in present fiber loop carrier systems, that is, approximately 1 M hours MTBF at 70°C. Lasers with this level of reliability may appear in the required time frame but this is by no means certain.

A second choice would be to use a laser mounted in a relaxed-tolerance (also known as "low cost") package. These devices sacrifice coupled optical power and trade the lost power for ease of assembly. Low-cost laser packages making use of grin rod lens coupling systems, as well as other simple, low tolerance alignment schemes have been reported (Shumate et al., 1986). They may use diodes with less-stringent requirements (i.e., lower allowable kink powers) than the "transmission quality" lasers. These devices typically launch 10 to 20 dB less optical power than their highly coupled counterparts. This may not present a serious problem for the short loop lengths being considered here, but there is still concern over long-term reliability.

A final emitter choice would be the LED recently proposed by Bell Communications Research and other laboratories (Gimlett et al., 1985). This research was stimulated by the search for a low-cost highly-reliable emitter for loop application. LEDs, when coupled to single-mode fiber, have a 15 dB additional penalty compared with connection to a multimode fiber. However, given the short distance between the Remote Electronics and subscriber, it was believed that there was sufficient coupled power to make this concept feasible. Accordingly, experiments have been performed at 150 Mb/s using surface and edge-emitting LEDs over distances of 4.5 km respectively (Shumate et al., 1985). Edge-emitters have been employed at 600 Mb/s over 20 km of fiber (Fujimoto et al., 1986) and at up to 2 Gb/s (Fujita, 1987).

As shown in Fig. 25.7, a field deployable system may have four splices in the RE to subscriber link. These splices would be of the fusion or mechanical types (non-reconnectable) with losses of about 0.2 dB each. In addition, there would be two connectorized mounts at the Remote Electronics patch panel with losses of about 0.4 dB each. These losses are summarized in Table 25.3.[3] They show a total loss of about 2.8 dB.

---

[3]The data in this table were obtained from discussions with S. D. Personick, P. Kaiser, and N. K. Cheung, Bell Communications Research.

TABLE 25.3
RE / CO Power Budget

Item	Loss (dB)
2 km Fiber @ 0.5 dB/km	1.0
4 Fusion/Mechanical Splices @ 0.25 dB	1.00
2 Connectors @ 0.4 dB	0.8
TOTAL	2.8

TABLE 25.4
Transmitter / Receiver Power Budget

	Power/Sensitivity Budget (dBm)		
	RE/Subscriber		RE/Central Office
Sources/Sinks	150 Mb/s	600 Mb/s	2.4 Gb/s
Surface Emitter	−30	−30	NA
Edge Emitter	−22	−22	NA
Laser Diode	−10	−10	−6 (DBF?)
p-i-n	−38	−32	−26
APD	−45	−39	−33

Table 25.4 shows the power available as a function of bit rate for the various transmitters that may be considered for the Remote Electronics/subscriber link of the BISDN, and also the sensitivities of the various receivers.[4]

The table shows that the surface and edge-emitting LEDs are capable of coupling −30 and −22 dBm respectively into the single-mode fiber. On the other hand, the low-cost laser diode, which may be used in the loop, might couple about −10 dBm. The table also shows typical receiver sensitivities ranging from −32 dBm at 600 Mb/s for the p-i-n to about −39 dBm for the APD when operated at the same bit rates.

The results presented in Tables 25.3 and 25.4 cast serious doubt on the use of the surface-emitting LED for this application. If a p-i-n detector is used at the 600 Mb/s level, the available margin is 2 dB, which is insufficient to handle the 2.8 dB fiber insertion loss. The APD improves the surface-emitter margin to 15 dB and 9 dB at 150 Mb/s and 600 Mb/s. Hence, this combination could be used in the 600 Mb/s application.

The edge-emitting LED has a MTBF comparable to the surface-emitter (see Chapter 19), and it is also similar in cost. However, it is capable of coupling −22 dBm. Hence, the available power would be in the range of 16 dB at 150 Mb/s and 10 dB at 600 Mb/s with the p-i-n receiver. Thus, at the 600 Mb/s bit rate, 7.2 dB would be available after fiber insertion. Furthermore, in field deployable systems, extra loss should be allocated to handle miscellaneous

---

[4] N. K. Cheung supplied the data shown in Table 25.4.

considerations, such as an additional 1 dB for cable breakage (fusion splicing), 1.5 dB for aging misalignment of the optical receiver, and as much as 3 dB for variations in LED power output due to aging. Therefore, even when these additional 5.5 dB of loss are considered, the edge-emitter/p-i-n combination provides sufficient margin when it is operated at the 600 Mb/s rate.

The laser diode provides the most margin, 35 dB, and 26 dB with the APD receiver at 150 Mb/s and 600 Mb/s respectively. As stated earlier, however, application of the laser diode in the loop is dependent on an improvement in reliability and a reduction in cost.

25.5.7 Remote Electronics / Central Office Link

Now let us consider the fiber feeder technology choices. We have already established that this link need not be longer than 10 km. The high bit rates required to provide economical fiber sharing (2.4 Gb/s) dictate that this link must employ laser diodes. A simple power-budget analysis of this link shows that there is more than sufficient margin. For example, with 10 km of fiber, the 10 required splices ($\approx$ 2.5 dB) and four connectors (1.6 dB), (2 connectors are assumed to be at a Remote Electronics patch panel, and 2 are assumed to be at the Central Office main distribution frame); the total loss is about 6.6 dB, while the available gain should be about 20 to 27 dB depending on the choice of detector.

25.5.8 Network Connecting and Terminating Electronics

The NCTE is the equipment required at the subscriber's end of the RE-subscriber network to provide for the optical interconnection of customer equipment to the BISDN. It might consist of a 600 Mb/s optical transceiver, 600 Mb/s to 150 Mb/s bit interleavers and deinterleavers, and 150 Mb/s interfaces to subscriber circuit switched equipment (digital video feeds to video equipment and 150 Mb/s packet interface equipment). A possible near-term implementation for a deployable NCTE (Fig. 25.8) might be based on the use of hybrid integrated circuit technologies for the complex analog functions contained in the receiver. The 600 Mb/s interleavers/deinterleavers and transmitter could be accomplished with LSI emitter-coupled logic technologies and an external emitter. All 150 Mb/s functions: packet assembly/disassembly (PAD), framing, digital phase alignment, tributary scrambling, video coding/decoding may be accomplished by the use of fine-line CMOS VLSI. It is expected that the ongoing optical-electrical IC research will allow large cost reductions in the very-high-speed processing electronics and allow the optical transceivers and MUX/DEMUX to be integrated on one or two OEICs. Additionally, advances in integrated optics may lead to low-cost WDM devices, allowing a single fiber to support up- and downstream traffic. These advances indicate that an ultimate NCTE cost might be in the neighborhood of $200.

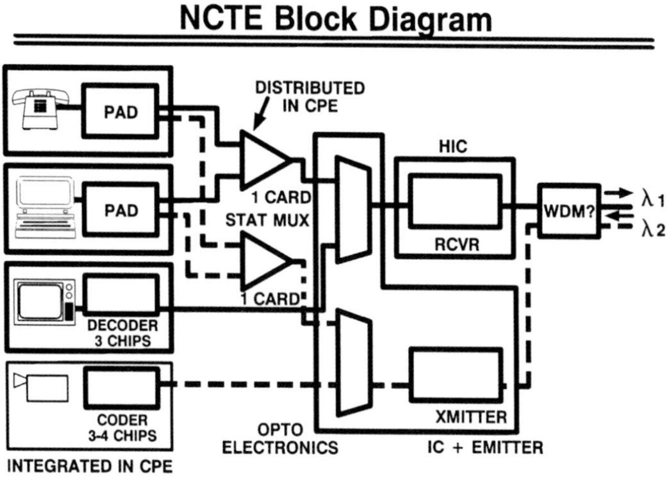

NCTE Block Diagram

Fig. 25.8 Network connecting and terminating electronics block diagram.

25.6 BISDN STANDARDS ACTIVITIES

Although BISDN is still in the research phase, substantial efforts at standardization, principally in CCITT Study Group XVIII, have begun (ISDN Task Group, 1986). The current plan appears to be to issue preliminary recommendations in 1992 at the end of the current CCITT plenary, with final standards to follow in 1996. Some administrations, however, most notably the Federal Republic of Germany and the U.S., appear anxious to accelerate this schedule to be more consistent with their plans to deploy trial systems by 1990.

The details of the BISDN subscriber access are some of the more hotly debated topics currently facing the CCITT. There is general agreement that the BISDN should evolve from the narrowband ISDN. Some administrations argue that the network-subscriber signaling protocols supported by BISDN should be compatible at least at layers 2 and 3 of the International Standards Organizations Open Systems Interconnect model.

Indeed, they argue that the differences between the protocols should be limited to specific compatible improvements required for the BISDN. Other administrations argue that since the bit rates and physical interfaces to the BISDN and narrowband ISDN are different, it is not necessary that the higher layer protocols be strictly compatible, providing that the adaptation function is simple (i.e., permitting a single-chip realization, for example). Accordingly, the customer interface proposals before the CCITT fall into two main classes. The first consists of purely circuit-switched subchannels with rates at 16 kb/s, 64 kb/s, 2.048 Mb/s (1.544 Mb/s, 30 Mb/s (32 Mb/s, 45 Mb/s)), and a "video rate channel" with a rate less than or equal to the maximum payload of a 139.264 Mb/s signal, the European CEPT 4 transmission rate (CCITT, 1985).

TABLE 25.5
International Transmission Hierarchies

Level	N. America		Europe	Japan
	Asynch	SONET		
1	1.544		2.048	1.544
2	6.312		8.448	6.312
3	44.736	49.920	34.368	32.064
4	274.176	149.760	139.264	97.728
5		599.040	565.148	400.352

The second main proposal is to use packet switching (described earlier in this chapter) for all services with bit rate requirements less than 45 Mb/s, and circuit switching for the remaining services. The first proposal can be implemented with today's technology, while the second requires new developments in high-speed, low-delay packet switching, such as reported in Beckner *et al.*, 1987. However, the packet approach has significant advantages: a common user/network interface for a variety of services, and a more systematic network structure, eliminating the need for bit-rate specific network components (Wu and Huang, 1986).

25.6.1 H4 Channel

The administrations argue that the so-called "H4" channel rate, which will be used primarily to transport entertainment video signals for residences, should be "high enough" to minimise coding complexity and cost (Bellisio, 1986). The bit rate should also be compatible with the transmission hierarchy currently in place or planned; see Table 25.5 for an outline of current international hierarchies.[5] The bit rates in the column labeled SONET (Synchronous Optical Network) are based on a relatively new channel structure proposed by Bell Communications Research. SONET provides a hierarchy of bit rates, which are all exact integer multiples of an atomic STS-1 (49.920 Mb/s) rate. SONET offers significant equipment and operations savings for the telephone network, since its frame format and multiplex structure simplify access to individual bytes in a frame.

Preferably, the H4 rate should be within the capabilities of moderately aggressive, i.e., 1 to 2 μm CMOS technologies, such that the costs of customer premises and network equipment can be reduced through the use of high levels of integration. These requirements suggest that a rate for the United States of 50 Mb/s, 100 Mb/s, or 150 Mb/s may be acceptable depending on the level of video quality that is finally offered. Indeed, this is a fundamental impediment to standards agreement. There is currently insufficient research on the tradeoffs

---

[5] Note that with the exception of SONET, the hierarchies are asynchronous.

TABLE 25.6
Proposed Broadband Channel Rates

Channel	Nation	Rate (Mb/s)
H4	USA	144.768
	Germany	138.240
	France	136.704
	Italy	135.168
	Japan	92.16
H2	USA	44.160
	Germany	33.792
	France	34.176
	Italy	33.792
	Japan	30.720

between perceived video quality, transmission bit rate, coding complexity, and signal processing in the camera, and receiver. The failure to reach international agreement (1986) on the adoption of the Japanese High Definition TV proposal (i.e., 1125 line, 5.33:3 or 16:9 aspect ratio image) was based in part on the lack of clear consensus on the best mechanism for improving the quality of today's television picture. In the interim, it is proposed by the United States that a bit rate of 144 Mb/s, the information payload of the SONET STS-3 signal, represents an optimal choice (Bellcore Bit-Rate Proposal, 1986).

Table 25.6 summarizes the standards positions of some of the major countries participating in the June 1986 meeting of CCITT Study Group XVIII. Note that two broadband channels were identified: an H4 channel operating at a rate beyond 90 Mb/s and an H2 channel with a rate less than 45 Mb/s. The major discussion was on the H4 channel. Observe that in all cases the general intent was to standardize the domestic transmission hierarchy.

Further, because of the need to provide several video channels and the high-speed packet channel, the United States proposes an aggregate downstream video rate of around 600 Mb/s (4 bit-interleaved SONET STS-3 channels). As described earlier, a 150 Mb/s upstream channel using a high speed slotted packet protocol is also proposed.

25.6.2 High-Speed Packet Proposals

Two different approaches for packet switching have been proposed; one termed Deterministic Time Division Multiplexing (DTDM), and the other, Asynchronous Time Division Multiplexing (ATDM) (LeBris and Servel, 1986). In both proposals, minimal error correction is provided by the protocol, the bulk being provided end-to-end by the application itself (this error treatment is justified since fiber links are capable of operation at less than 10^{-9} bit error rates). The main difference between the two proposals is that DTDM has fixed

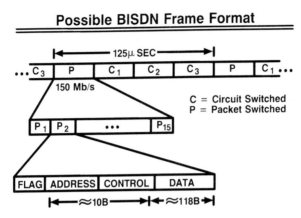

Fig. 25.9 Possible BISDN frame format.

length slots, synchronous to the frame, while ATDM slots are asynchronous. Thus, DTDM offers some efficiencies in locating and processing packets once framing is accomplished. With DTDM, the subscriber access could have the structure shown in Fig. 25.9.[6] Here it is assumed that the bit rate of the broadband channel is $\approx$ 150 Mb/s, and that the size of the DTDM slot is 133 bytes (the payload of the SONET 3rd level frame is $\approx$ 0.8 kb). Thus, $\approx$ 15 DTDM slots would be located in each 125 μs frame. Furthermore, each slot would have a header, which would identify the type of packet, contain source and destination address, logical channel number, CRC, and other protocol functions.

The statistical demultiplexer at customer premises, after receiving framed data from the multiplexer at 150 Mb/s, would locate the header of the first slot (in the first few information bytes). The header of the second slot would be 133 bytes further in the frame and so on. The demultiplexer would use the destination address in the header to deliver it to the appropriate terminal. Alternatively, if privacy and security are not important for the application (i.e., all destination terminals belong to the same subscriber), the packet demultiplexing function could be distributed. That is, incoming slots would be broadcast to all terminals with final selection accomplished by the application (e.g., as a function of supplementary address, packet type—video to monitor, data to a PC, etc.).

25.7 SUMMARY AND CONCLUSIONS

We have reviewed the impact of fiber optic transmission systems on the local access network and noted that their use so far has generally been as a part of the general trend toward the increased application digital loop carrier systems. We

---

[6]This is under discussion and undoubtedly will change before BISDN standards are published.

described the benefits that the optical transmission media brings to these systems: higher digital bandwidths resulting in large numbers of customers being multiplexed onto relatively few fibers, and straight-forward system capacity upgrades, small fiber cross section allowing better duct-space utilization and the elimination of outside plant repeaters due to low media losses. We then pointed out that the next generation of fiber transmission systems in the loop plant will be as part of a larger concept known as the BISDN. This new broadband communications concept will deliver large digital bandwidths, via optical fiber, directly to individual subscribers and will offer digital video and high-speed packet services previously impossible or prohibitively expensive using ordinary copper telephone loops.

REFERENCES

Armbruster, H. (1986). Applications of future broadband services in the office and home, *IEEE JSAC* **SAC-4**, No. 4, 429–437.

Beckner, M. W. (1987). A protocol and prototype for broadband subscriber access to ISDN, submitted to International Switching Symposium 1987.

Bell Communications Research (1986). A broadcast channel (H4) bit rate and format proposal, T1D1.1/86–046.

Bell Communications Research (1986). Technological and market obsolescence of telephone network equipment.

Bellisio, J. A. (1986). Television coding for broadband ISDN, *Proceedings, Globecom '86*, **2**.

Federal Republic of Germany (1985). Interface structures at the S/T reference point for the broadband user-network interface. *CCITT D*, 306 XVIII, Kyoto.

Fujimoto, N., Ohtsuka, T., Kondou, R., Yamaguchi, K., and Nabeshima, Y. (1986). 600 MBIT/S optical transmission experiment over 20 KM of single-mode fiber using edge-emitting LED. First Optoelectronics Conference (OEC'86) Post-Deadline Papers Technical Digest, Tokyo.

Fujita, J. (1987). 2 Gbit/s and 600 Mbit/s single-mode fibre transmission experiments using a high-speed Zn-doped 1.3 μm edge-emitting LED. *NEC Electronics Letters* **23**, No. 12.

Gimlett, J. L. (1985). Transmission experiments at 560 and 140 Mb/s using single mode fiber and 1300 nm LEDs, *Electronics Letters* **21**, 1998–1200.

Kaiser, P. (1985). Single mode fiber technology for the subscriber loop. *ECOC II Proceedings*, Venice, Italy.

Kneisel, K. E. (1986). Broad-band communications systems in Germany. *IEEE JSAC* **SAC-4**, No. 4.

LeBris, H., and Servel, M. (1986). Integrated wideband networks using asynchronous time division techniques, *ICC '86 Conference Record* **3**, 1720–1724.

Linnell, L. R. (1986). A wide-band local access system using emerging technology components. *IEEE JSAC* **SAC-4**, No. 4.

Olson, J. W., and Shepis, A. J. (1984). Description and application of the fiber SLC/carrier system. *JLT* **LT-2**, No. 3.

Shumate, P. W. (1985). Transmission of 140 Mbit/s signals over single-mode fibre using surface- and edge-emitting 1.3 μm LED. *Electronics Letters* **21**, No. 12, 522–524.

Shumate, P. W., (1986). Laser coupling to single-mode fiber using graded-index and compact disc 1.3 μm laser package. *Electronics Letters* **22**, No. 16, 836–838.

Working Party XVIII/1 (1986). Draft status report of the task group on ISDN—Broadband aspects. *TD 16* (P)-E, Geneva, Switzerland.

Wu, L. T., and Huang, N-C (1986). Synchronous wideband network and interoffice facility hubbing network, *Proceedings 1986 International Zurich Seminar on Digital Communications*, 33–39.

Chapter 26

Photonic Local Networks

IVAN P. KAMINOW

AT & T Bell Laboratories, Inc., Holmdel, New Jersey

26.1 INTRODUCTION

Lightwave technology for long-haul telecommunications is well established, as the preceding chapters amply demonstrate. It is now widely believed that the next frontier is the application of this technology to the distribution of high-speed digital signals within a local region, such as a metropolitan area, neighborhood, campus or building. Since the subscriber distribution network in the U.S. represents about 90% of the telephone plant and some 10^8 subscriber "loops," or telephone connections to the central office, it is clear that the local distribution network is a rich market that, unlike the long-haul market, offers the prospect of large numbers of components with the attendant economies of scale. Bringing optical fibers to homes and businesses will permit subscribers to enjoy "integrated digital services" including digital voice, data and video. The data and video services might be either interactive or one-way broadcast.

At the same time, computers, terminals and peripherals have escaped from the computer center and are now widely dispersed in offices and sub-centers. These components are currently being linked by wire-based local area networks (LAN). A LAN can be characterized as a data network within a local environment (say within a 10 km diameter) that provides interconnection on either a random access basis or a switched circuit basis, as in a private branch exchange (PBX). Users' appetites for high-speed links among these components have grown beyond the slow rates available from telephone modems. Optical fiber networks offer the prospect of distributed computer networks in which many users can share programs, files, databases and computing facilities at very high speeds (~ 1 Gb/s) with sophisticated graphics and real-time video.

The conjunction of these telecommunications and computer applications of lightwave technology for the distribution of bits at high speed in a local setting can be described by the trendy, high-tech apellation, photonic local networks. We are concerned here with the physical aspects of photonic networks. A few random access protocols are discussed briefly as they relate to the design of a physical network, but the details of higher-level protocols and communication theory are not our concern. The International Standards Organization (ISO) has developed an open system interconnection (OSI) reference model that separates the design of a LAN into seven layers that ideally fit together at their interfaces with standard protocols, independent of the inner workings of each layer (Tanenbaum, 1981). The top or seventh layer, the application layer, provides the software interface with the user. The lowest layer, the physical layer, encompasses the hardware, which is our main concern. The next higher layers, the data-link and network layers, provide the protocols for interconnecting users and sharing the network. Thus, a given physical facility can operate with a wide variety of network protocols.

Lightwave technology has already appeared in the form of point-to-point links both in the subscriber network and in computer local area networks, particularly within a computer center for interconnecting high-speed processors and input/output (I/O) devices (Crow, 1985). For the most part these "data links" have employed GaAlAs LEDs, silicon photodetectors and multimode fiber. Their bandwidth, convenient size, electromagnetic interference (EMI) immunity and competitive cost make these lightwave connections superior to copper wire pairs or coaxial cable. In fact, certain applications would not be feasible with copper links. Still, these applications do not stretch the capability of lightwave point-to-point links as demonstrated in long-haul experiments (Henry et al., 1987), for example, at 8 Gb/s over a span of 70 km, nor do they represent a conceptual advance over a copper point-to-point link. The more sophisticated uses of lightwave in photonic local networks are only just being explored and one of our tasks here is to review some of this research. We also describe several conventional LANs that employ lightwave data links, with the objective of seeing how they might be extended to much higher speeds by using all aspects of lightwave technology more fully. The use of lightwave links in the subscriber distribution network has been covered in an earlier chapter (White, 1988).

Since our emphasis is on future high-speed photonic networks, we focus on operation with lasers and single-mode fiber at a wavelength of 1.3 μm, near the dispersion minimum, or possibly at 1.55 μm, near the loss minimum. While LEDs and lasers made with GaAlAs and operating at 0.8 μm and multimode fiber have technical advantages in special applications and cost advantages in the short-run, many believe that cost, performance, "upgradability" and compatibility with the long-haul network will favor single-mode fiber at the longer wavelengths in the future. Indeed, any permanent installations of fiber in buildings or in the network should be as flexible as possible in allowing upgrades of terminal technology without replacing the transmission medium. In order to establish working parameters for a future photonic network, we assume a region within a 10 km diameter containing at least 100 terminals operating at speeds greater than 100 Mb/s with a network capacity in excess of 1 Gb/s. Digital transmission is assumed throughout to allow integrated services, to obtain good sensitivity and to avoid non-linear effects in optical components. The type of traffic will determine the network configuration: Voice and real-time video suggest circuit-switched networks while "bursty," interactive data communication suggests packet-switched networks. The type of traffic also determines what a user is willing to pay: Entertainment distribution, such as digital TV, must be less expensive than interactive video, which has more immediacy and intrinsic value.

Cost and performance are the key criteria for the introduction of any new technology. Terminal cost is an especially sensitive factor in local networks because it is not shared among many users as in the case of a long-haul repeater. The litany of lightwave performance advantages over conventional wire technology is familiar:

- Optical transmission bandwidth is readily available and cheap—but electrical terminal bandwidth is still a bottleneck.
- Fiber dispersion is small (at 1.3 μm).
- Fiber loss is negligible (1.3 to 1.55 μm).
- Fiber is cheaper than coax at data rates greater than 10 Mb/s.
- Fiber is easier to install than coax because of its small size and weight.
- Electromagnetic interference and ground loops are negligible.
- Free-space lightwave propagation can be directed between buildings or within a building without safety hazard (at 1.5 μm), without frequency allocation constraints and without complex wiring.

The first point is paramount in order to exploit the unique performance capability of lightwave networks. It is difficult for optical components such as switches and multiplexers to compete directly with electronic integrated circuits (IC), where the ICs are available. At speeds where ICs are not available, however, optical processing is an essential ingredient for sophisticated photonic networks. In subsequent sections, we will explore the opportunities and limitations set by the availability and physical realizability of novel optical components. For example, it has been noted that the low-loss window in optical fibers between 1.4 and 1.55 μm offers an enormous potential baseband width of 20 THz; but the challenge is to develop inexpensive, single-frequency, tunable lasers that can be closely-spaced in frequency to provide a comb of well-controlled optical carriers. The final point listed reflects the high cost of building wiring that might be reduced in special situations by indoor infrared network concentrators (Yen, 1985; Chu and Gans, 1987; Chu and Gans, 1988).

26.2 TOPOLOGY AND ACCESS PROTOCOLS

26.2 Introduction

The possible layouts of a photonic network can be classified as a ring, bus, mesh or star network or combinations and variations of these such as a tree, hub, clover leaf, or Manhattan street network. These topologies are encountered in wire networks. The physical realization of the optical fiber topology, however, differs from that of the wire topology in a fundamental way because, in the bus and star, the dimensions of the required taps are greater than a wavelength, $\lambda = c/nf$, in the optical case and less than λ in the wire case. Here c is the velocity of light, n the refractive index in the transmission line and f the frequency. Optical taps and stars will be discussed in later sections. Another distinction is that a repeater in the optical ring, star or bus networks involves optical-to-electrical (O/E) and electrical-to-optical (E/O) converters, which may be expensive.

In the following discussions, we will use the four basic topologies as a basis for describing network access protocols that have been associated with them in commercial radio or wire-based LANs or in LANs that utilize simple optical data links. We are concerned with the lower, physical levels of these protocols. At the high data rates contemplated for photonic networks, we will see that these media access protocols all have their own shortcomings.

26.2.2 Ring Network

A ring (or loop) network is illustrated in Fig. 26.1. In its usual application using a token-ring protocol for media access, as in the IBM 4 Mb/s token ring network (Strole, 1987; Bux et al., 1983; Bux et al., 1982), a repeater is required at each station. A token, i.e., a "1" or a "0" bit, or group of bits, is propagated in one direction from station to station. If a station has a packet to send to another station, it adds the address of the receiving station in a header and holds the combined packet in a buffer. It reads the tokens as they go by until an empty one, a 0, is received. It converts the 0 to a 1, a busy token, and appends the packet. The intermediate stations repeat the bits in the packet and also "listen" for their own addresses. If a station recognizes its address in the packet header, it copies the packet. When the packet returns to the sender, it serves as

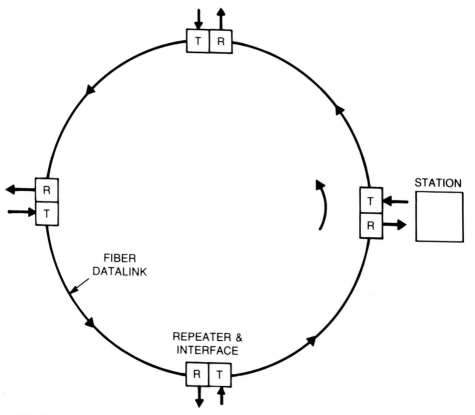

Fig. 26.1 Unidirectional ring network.

an acknowledgement and the sender removes it from the ring after converting the token back to 0. If the physical length of the packet is greater than the perimeter of the ring, the sender can convert the token and remove the packet bits as they arrive. If the packet is correct, he can eliminate it from the buffer, otherwise he can retransmit.

Commercial token rings use wire interconnections or optical data links to join the stations at rates in the 10 Mb/s range. The actual utilization of the network is less than 10 Mb/s because of the time it takes an empty token to pass around the ring. This transit-time delay increases linearly with the number of stations and includes the propagation delay between stations and the delay at each station, which must examine the token of every packet before repeating the bits to the next station.

A token ring architecture is not especially attractive for a high-speed (~ 1 Gb/s) optical network because of the cost of high-speed repeater optoelectronics at each station and the delay due to packet processing. In addition, at high bit-rates, the packet time may be much shorter than the propagation time around the ring, unless a packet contains an unusually large number of bits. Efficient use of the ring with relatively short packets may call for multiple tokens, which can lead to complex protocols. Reliability, if one station is disabled or if the fiber breaks, is a problem in both fiber and wire rings. A double-ring optical network, like the Columbia University MAGNET (Patir et al., 1985) or the NTT OPALnet-II (Tokura et al., 1985), both of which operate at 100 Mb/s, or the Fujitsu 200 Mb/s TDM loop (Minami et al., 1985), address these reliability problems by providing for bypass of defective stations and loop-back around a fiber break. Each repeater has two inputs and two outputs connected to two rings operating in opposite directions, which, of course, increases cost. The TDM frames in these implementations can be divided dynamically to provide both circuit-switched service for voice and video, and packet service for bursty data. A 1.2 Gb/s TDM ring, operating with multimode fiber at 1.3 μm, has been reported by NEC (Goto et al., 1985) and assigns 0.4 Gb/s to multiaccess packet switching and 0.7 Gb/s to circuit switching.

The Fiber Distributed Data Interface (FDDI) (Ross, 1986; Burr, 1986) is a proposed American National Standard for a 100 Mb/s double-ring TDM LAN that employs 1.3 μm multimode-fiber and LED data links between stations. It is designed to provide backbone services, interconnecting lower speed LANs, and backend services, interconnecting mainframes, mass storage systems and other high-speed peripherals. Although it has not been implemented, the standard system is designed to operate with components that were commercially available in 1986 at a low cost. Detailed protocols are specified for the OSI physical and data link layers. Even the optical connector receptacle on the wall is considered. The standard can provide both packet-switched and circuit-switched services. As many as 500 stations can be connected, with a maximum of 2 km between stations and maximum perimeter of 100 km.

26.2.3 Bus Networks

26.2.3.1 Ethernet—CSMA / CD. A bus network is based on a single main transmission line that does not close on itself. It may follow a straight, serpentine or spiral path. Stations tap onto the bus to transmit or to receive signals that are broadcast to stations on the bus.

One of the best known protocols for gaining access to the bus is the Ethernet or carrier sense multiple access with collision detection (CSMA/CD) protocol. In the original Ethernet, which operated at 3 Mb/s and was later upgraded to 10 Mb/s, a coaxial cable served as the bus and a tap consisted of a small antenna probe inserted through a hole in the outer conductor of the bus coax. This antenna is part of the transceiver, which physically attaches to the coax on one side and to the station on the other side, by means of a transceiver cable.

The antenna, which is much smaller than $\lambda \approx 30$ m, radiates bidirectionally on the bus. The probe penetration is small so that reflections on the bus are small and the shunt impedance is large. Present-day taps are a bit more sophisticated but behave similarly. Even with this small coupling, station amplifiers can restore the signal strength to the sensitivity of electronic receivers (about -60 dBm). Problems with reflections, attenuation and dispersion would begin to appear at rates of ~ 100 Mb/s.

A station with information to send collects and forms a packet containing the data and header with the receiver address. It senses the bus, by testing for a dc carrier level, to see if any other station is transmitting, if not, it sends its packet. All stations listen as the packets go by. If a station recognizes its address, it copies the packet, which finally gets absorbed at either end of the bus.

If a station, call it A, starts sending a packet at $t = 0$, another station, call it B, a distance L away will not sense A until $t = t_1 = L/(c/n)$ due to the electromagnetic propagation delay. And, if B starts to send its packet just before $t = t_1$, then A will not sense B until $t = 2t_1$. If the packet length T is greater than $2t_1$, then A will detect the collision (CD) before completing the packet transmission. It then stops transmitting. B will have detected a collision at $t = t_1$ and stopped transmitting. After a random delay, each will again try to send its packet. After successful transmission, the packet is removed from the storage buffer. During periods of heavy demand, many collisions and retransmissions may occur, resulting in poor utilization of the bus. A critical constraint on the CSMA/CD network is that the packet duration, T, must be greater than the maximum round-trip propagation delay, i.e.,

$$T \geq 2L/(c/n), \tag{26.1}$$

where L is the bus length, in order that all collisions be detected before packet transmission is completed.

A number of difficulties arise in trying to extend CSMA/CD to fiber optic networks. These include the limitations of optical taps, which restrict the number of stations as discussed next, and the difficulty of sensing a collision with a remote station whose signal will be weak, in the presence of the strong optical signal from the local station. A fiber CSMA/CD bus has been proposed by (Tseng and Chen, 1983) and the 1.2 Gb/s ring (Goto $et\ al.$, 1985) uses CSMA/CD at 0.4 Gb/s. An experimental 10 Mb/s fiber Ethernet that sidesteps the tap and collision detection problems by employing an active star is described in Section 26.2.4. At higher speeds (~ 1 Gb/s), the requirement of long packets containing many bits, according to Eq. (26.1), is a strong constraint. As an example, consider $L = 1$ km, $n = 1.5$ and a bit rate $B = 1$ Gb/s, then we require $T \geq 10^{-5}$s and the number of bits per packet is given by

$$N = BT$$

$$\geq 10^4 \text{ bits/packet}. \tag{26.2}$$

Such a large number of bits/packet requires a large high-speed buffer and places restrictions on applications. For example, if a large number of bits were continuously available, as in a large file transfer or real-time video, then (26.2) would not be restrictive; but, if only occasional bursts of data containing fewer than N bits were available in time T, as in some interactive applications, then a packet defined by (26.2) might carry many empty bits.

26.2.3.2 Optical Taps and Directional Couplers. An optical tap may take the form of a beam splitter (partial mirror) or a directional coupler, which are functionally equivalent 4-port devices, as illustrated in Fig. 26.2. In either case, light is coupled by a traveling-wave interaction from an incoming path to an

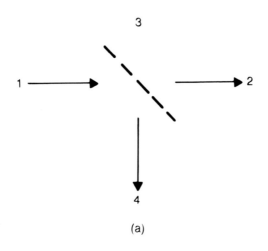

(a)

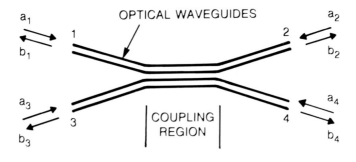

(b)

Fig. 26.2 (a) Beam splitter. (b) Directional coupler.

adjacent path over a coupling distance of many optical wavelengths. The coupling is unidirectional. On the other hand, a tap on a wire transmission line at 10 MHz, where the wavelength is 30 m, behaves like a bidirectional point source. Since the optical directional coupler is an essential component of most photonic networks, it is worthwhile to understand its behavior in depth.

Multiport devices can be represented by a scattering matrix S_{ij} (Ramo et al., 1984) connecting waves a_j entering the junction on port j with outgoing waves b_i on port i. The normalized waves a_j, b_i are defined in terms of the effective optical fields E_j^+ and E_i^- for incoming and outgoing waves, respectively, averaged over the waveguide cross section A_{ij} as

$$a_j = \frac{E_j^+}{\sqrt{Z_{oj}/A_j}}, \qquad b_i = \frac{E_i^-}{\sqrt{Z_{oi}/A_i}}, \qquad (26.3)$$

where $Z_{oi,\,j}$ is the characteristic waveguide impedance. Hence the average power flowing into port k is

$$(W_k)_{av} = \frac{1}{2}(a_k a_k^* - b_k b_k^*), \qquad (26.4)$$

which is the difference between the incident and reflected powers. The scattering matrix S_{ij} relates outgoing waves to incoming waves

$$b_i = \sum_j S_{ij} a_j. \qquad (26.5)$$

Reciprocity demands S_{ij} be symmetrical,

$$S_{ij} = S_{ji}. \qquad (26.6)$$

Reciprocity reflects the symmetry of the wave equation, which is quadratic in time t, for time reversal $t \to -t$. Reciprocity does not hold in the presence of a dc magnetization M unless the direction of M is also reversed, i.e., $M \to -M$ as $t \to -t$. For a loss-free junction, conservation of energy demands that S_{ij} be unitary,

$$\sum_j S_{ij} S_{jk}^* = \delta_{ik}, \qquad (26.7)$$

independent of Eq. (26.6), where the asterisk denotes the complex conjugate and δ_{ik} is the delta function.

The traveling-wave nature of the interaction permits the design of an ideal lossless 4-port junction (Ramo et al., 1984) as illustrated in Fig. 26.2b with the following additional constraints on S_{ij}: If ports 2 and 4 are terminated in matched impedances Z_{02} and Z_{04} so that $a_2 = a_4 = 0$, then ports 1 and 3 are not coupled, i.e., $S_{13} = S_{31} = 0$. Also, if power is incident on port 1, all the power should be divided between 2 and 4 with no reflection on 1, i.e., $S_{11} = 0$. Similar constraints hold for permutations of the ports,

$$S_{ii} = 0, \qquad S_{ij} = S_{ji}, \qquad S_{13} = S_{24} = 0, \qquad (26.8)$$

assuming a reciprocal junction. For a lossless junction, the unitary condition Eq. (26.7) leads to

$$S_{23} = -S_{14} = \sqrt{f} \qquad (26.9a)$$

$$S_{12} = S_{34} = \sqrt{g} \qquad (26.9b)$$

$$f + g = 1, \qquad (26.9c)$$

where $\sqrt{f}$ and $\sqrt{g}$ can be made real by the proper choice of reference plane. Thus, Eqs. (26.8) and (26.9) define the performance of an ideal lossless directional coupler in which unit power incident on port 1 without reflection produces an output power $(1 - f)$ on the straight-through path 2, f on the coupled path 4 and zero on the reverse path 3. A coupling ratio of $f = 1/2$ corresponds to a -3 dB coupler and $f = 1/10$ to a -10 dB coupler.

Practical couplers can be fabricated by bringing two optical waveguides, either fibers or integrated optics guides, close together so that their wavefunctions overlap over an extended coupling length. The coupling ratio f can be adjusted in the range $0 \leq f \leq 1$ by varying the coupling length. In practical couplers (Tomlinson, 1988), the energy conservation condition Eq. (26.9c) may not hold exactly, because some power may be reflected in the input or reverse ports and/or some power may radiate out of the coupling region, appearing as a loss (or an extra port). In good quality, polarization-maintaining, -3 dB couplers made by fusing two single-mode fibers together (Yokohama et al., 1986), the fractional excess loss, h, or the equivalent dB loss, $10 \log_{10}(1 - h)$, may be quite small, where

$$f + g + h = 1, \qquad (26.10)$$

with $h \approx 0.02$ $(-0.1$ dB). The excess loss may vary with f and the coupler design. Note that for coupling ratios $f < 0.1$, f and h are of the same order in an actual coupler. For illustrative purposes in the next section we will assume that for unity input power at port 1, the fraction f will exit at 4, $(1 - f - h)$ at 2 and zero at 3, corresponding to

$$f' = f, \qquad g' = 1 - f - h. \qquad (26.11)$$

Coupling ratios $f' \leq h$ are usually not practical because the useful power coupled off the bus is less than the power dissipated in the excess loss.

26.2.3.3 Optical Bus. The directionality of optical taps and the linear topology of the bus require a separate coupler for transmitter and receiver, as illustrated in Fig. 26.3 for a folded bus with N stations (see e.g., Limb and Albanese, 1985; Tseng and Chen, 1983). Several options are possible: a repeater may or may not be provided at the head end, as in Fig. 26.3, and the couplers may be identical or their coupling ratios may be tailored so that they increase with the distance from the head end in such a way that the power at each receiver is the same.

The case of a repeatered bus with tailored coupling ratios (Altman and Taylor, 1977; Auracher and Witte, 1977; Limb, 1984) gives the maximum number of stations for a given power margin R. The power margin is defined as,

$$R = 10 \log_{10} P_T / P_R, \qquad (26.12)$$

where P_T is the available optical transmitter power and P_R is the minimum receiver power that provides a tolerable bit-error rate (BER). A tailored-coupling bus, however, does not appear to be a practical solution for several reasons: For a typical allowed number of stations, $N = 32$ (Limb, 1984), the couplers at the head end would have to have coupling ratios less than $1/32 = 0.03$, which is comparable with the excess coupling loss. Also, the precise coupling ratios may be difficult to manufacture and expensive to stock in the wide variety required. Finally, it is not possible to add new stations without rearranging all the coupling ratios on the bus; since present-day fused-fiber couplers are not field adjustable, such a network would be difficult to modify. An experimental tailored-coupling bus with 15 single-mode, fused couplers, with minimum coupling ratio at the center of the bus, was operated at 500 Mb/s (Villarruel *et al.*, 1985).

A more practical arrangement is to employ identical couplers, or a small number of different couplers (Limb and Albanese, 1985), but in this case about half as many stations can be supported and the receivers must also have a wider dynamic range. As a simple example of the restriction on the number of stations imposed by the optical coupler, we consider a receiver bus with identical couplers, corresponding to the lower half of the bus of Fig. 26.3 with a head-end repeater, and P_T, the repeater output, applied directly on the bus while $P_R = P_{RN}$, is the output of the coupler at the Nth (last) station. Then N is limited by the power margin,

$$\frac{P_R}{P_T} = f(1 - f - h)^{N-1}(1 - \alpha)^{N-1}.$$

$$\approx f \exp[-(N - 1)(f + h + \alpha)], \qquad (26.13)$$

for couplers of the type described by Eq. (26.11), where we assume

$$(f + h + \alpha) \ll 1, \qquad (26.14)$$

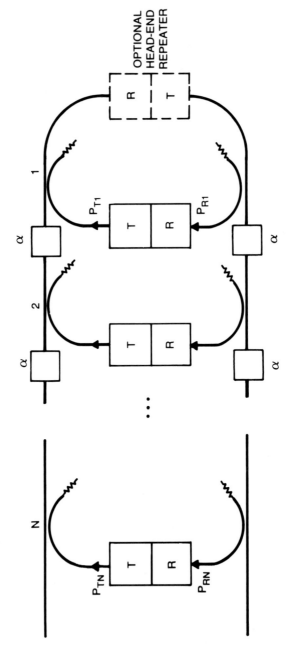

Fig. 26.3 Folded unidirectional bus. The dangling ends of couplers are terminated in matched impedances. Loss due to fiber and connectors is represented by α. A repeater at the head-end may be provided in an active bus.

with α the loss between couplers due to fiber attenuation and connector loss, using the approximation $\exp x \approx 1 + x$. Taking the log of Eq. (26.13) and setting the derivative $dN/df = 0$, we find that the maximum number of stations occurs for

$$f = (N - 1)^{-1}, \qquad (26.15)$$

where $N > 1$ to assure $f \leq 1$.

In the case of an unrepeatered bus with identical couplers, where now the transmitter power is applied via a directional coupler and $P_T = P_{TN}$, the power transmitted from the last station, and $P_R = P_{RN}$, the power received by the last station,

$$\frac{P_R}{P_T} = f^2 \exp[-2(N - 1)(f + h + \alpha) - \alpha], \qquad (26.16)$$

and the maximum number of stations again occurs for

$$f = (N - 1)^{-1}, \qquad (26.17)$$

but we now have $2N$ couplers for N stations (Villarruel et al., 1981). The optimum number of stations in either case can be found by substituting (26.15) or (26.17) into (26.13) or (26.16), respectively, for a given power margin. Taking $R = 40$ dB, which is reasonable for a good receiver at 100 Mb/s, $P_T \approx 0$ dBm in the fiber, $h = 0.023$ (-0.1 dB excess coupler loss), $\alpha = 0.045$ (-0.2 dB connector and fiber loss), we find for the repeatered case, Eq. (26.13), $N = 64$ stations and $f = 0.016$, (-18 dB) and in the unrepeatered case, Eq. (26.16), that $N = 16$ stations and $f = 0.067$ (-12 dB). In either case, the power coupled to the receiver turns out to be of the same order as the power lost in the coupler, connector and fiber.

These estimates indicate that a simple unrepeatered bus is limited to tens of stations, which may be satisfactory for some networks or for segments of larger networks with mixed topologies. Of course, if repeaters are allowed, the number of stations can grow considerably. An even more attractive approach might be to include optical amplifiers at each station or after the number of stations that reduces the signal by an optimum fraction (Henry et al., 1988). Semiconductor optical amplifiers (Kaminow and Tucker, 1988) offer the prospect of providing gains of 10 to 20 dB with reasonable noise figure at lower cost and complexity than a repeater.

A "token bus" protocol is employed in several folded unidirectional bus networks, such as Fasnet (Limb and Flores, 1982), Express-Net (Frattos et al., 1981) and D-Net (Tseng and Chen, 1983). The folded bus in Fig. 26.3 must be modified so that each station has an additional coupler on the transmit side oriented to sense signals transmitted from the upstream direction. In simplified terms, the station N transmits a "locomotive" packet and appends a data packet if it has one. The next station senses the locomotive and the end of the data "train," and then adds its packet if it has one. The locmotive collects packets along the transmit side of the bus and then returns along the receive side, where stations can read packets destined for them. When the end of the train reaches station N, station N generates a new locomotive and the process repeats. As in the token ring, there is no contention and retransmission, but the delays and inefficiency can be substantial. If station N is the only one with a packet of duration T_p to send and the propagation delay to the head-end is T_d, then the utilization is only $T_p/(2T_d + T_p)$. Even if all the stations have packets of length T_d/N, the utilization is only $1/3$. We have neglected the delays between sensing of a signal and the start of transmission, which can be substantial in terms of bits at high data rates.

26.2.4 Active Star Networks

26.2.4.1 Introduction. The star topology lends itself naturally to many physical situations such as the wiring of a building or neighborhood, in which it is often necessary to follow many diverse paths to all offices or homes. Further, it is often convenient to have a central node to place a large main-frame computer, file server and network controller. Thus, although in ideal cases, it might seem that a region within radius r containing N stations could be covered with shorter total lengths of fiber for a ring or bus than for a star, in practice the ring or bus might have to be sufficiently deformed to require fiber lengths similar to a star.

The active star, in which all incoming optical signals are converted to electrical and then converted back to optical for outgoing signals, permits central control as opposed to distributed control in the ring and bus. Packets may then be broadcast to all stations from the central node with appropriate address headers, after any contention protocol has been applied.

A passive star has similar topological properties but no active processing or regeneration takes place at the node; an incoming signal from a given station is merely divided among all N stations. Passive stars will be treated later.

26.2.4.2 ALOHA. ALOHA is one of the simplest random access protocols (Abramson, 1973). It was devised at the University of Hawaii to allow a number of stations to access a central computer by radio. A similar protocol can be used in any situation in which many stations are connected to a central node and/or interconnected via the node to other stations, as illustrated in Fig. 26.4.

A station with a packet to transmit sends it and waits for an acknowledgment from the node. If none is received after a time corresponding to the expected round-trip delay, it retransmits the packet at a random later time. No acknowledgment is sent if two packets overlap at the node. Acknowledgments are sent over a separate (radio frequency) channel. While simplicity is its main attrac-

STATION

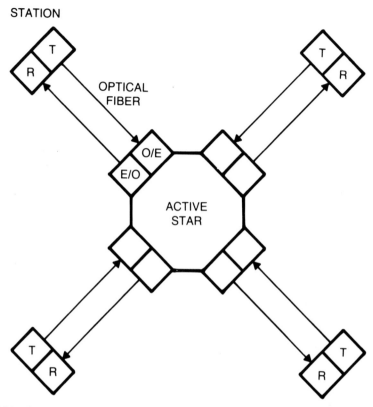

Fig. 26.4 Active star network. Optical to electrical (O/E) and E/O converters must be provided at the star.

tion, the shortcomings of ALOHA are the need for a central controller and the poor utilization of the transmission medium due to acknowledgment delays and packet retransmissions. In the standard ALOHA, the maximum fractional usage of the transmission medium is 0.18, and, in slotted ALOHA, in which packets may be sent only at fixed intervals so that all packets arrive at the node simultaneously, the utilization is 0.36. Although it has not been used in optical networks, ALOHA may indeed be suitable for bursty, low-traffic applications (Maxemchuk, 1986).

26.2.4.3 Fiber Ethernet.   The problems associated with a fiber-optic bus topology for implementing the CSMA/CD protocol, as outlined in Section 26.2.3, can be avoided with an active star as in the Xerox Fibernet II (Schmidt *et al.*, 1983; Rawson, 1985). The main problem with the bus topology is the limited number of users allowed with directional coupler taps. Further, the difficulty in detecting collisions between weak and strong packets occurs in the passive star as well as the bus.

In the Fibernet II CSMA/CD, Ethernet-compatible transceivers are linked, by means of dual data links that operate with GaAs LEDs at 10 Mb/s, to the active star repeater. Collision detection and broadcast retransmission are provided electrically by the star. Stations are notified of collisions by means of a 1 MHz square wave optical signal that cannot be confused with 10 Mb/s packets. In case of collisions, the stations retransmit after a random delay.

26.2.4.4 Datakit® Virtual Circuit Switch. The AT&T Datakit VCS protocol (Fraser, 1983) and the similar ISN (information system network) protocol (Acampora *et al.*, 1983; Acampora and Hluchyj, 1984) provide virtual circuit switches (VCS) in that a reliable data path is setup for each session and packet retransmission due to collisions is not required. The concept is illustrated in Fig. 26.5. Remote stations that may consist of mainframe computers, concentrators that bring together many terminals, or gateways to other networks are connected by 8 Mb/s fiber-optic datalinks to individual electronic modules at the node. These modules plug into two electronic buses that are short (~ 1 m) compared to a packet length (16 bytes). Packets are formed and buffered in the module with a header containing the source address. When the packet is complete, either by achieving the full number of bytes or by waiting a fixed time for added bytes, the module transmits its binary address on the contention bus while listening for bits transmitted by others. If it transmits a 1 and hears a 1, it transmits the next address bit; if it transmits a 0 and hears a 0, it transmits the next bit; and, if it transmits a 0 and hears a 1, it stops transmission, having lost the contention. This process is equivalent to an OR operation and assigns the contention to the highest address. The winner transmits the packet on the contention bus in the next time frame and the switch replaces the source address with the destination address and transmits the packet on the broadcast bus, where it is recorded by the appropriate module and sent to the remote station over the fiber link. The switch establishes a correspondence between source and destination at the beginning of a session, as in a circuit switch, so that source modules need not know the bus position of destination modules; the switch has a directory of positions and terminal names. Various protocols have been proposed (Acampora *et al.*, 1983) to set user priority and also to allow both voice (synchronous) and data (asynchronous) traffic on the bus.

If one were to go to very high bit-rates, the buses would no longer be short compared with a packet and collisions due to delays might upset the "perfect scheduling" of packets. Although methods for overcoming this limitation have been proposed (Acampora and Hluchyj, 1984), the cost of electronic circuits and electrical reflections on the bus would limit effectiveness of a centralized bus at very high frequencies.

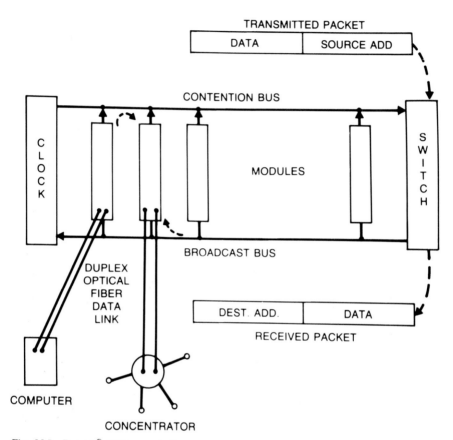

Fig. 26.5 Datakit[®] VCS network. Remote stations are connected to the electrical node by 8 Mb/s data links (after (Acampora *et al.*, 1983)).

26.2.5 Passive Stars and Splitters

An $N \times N$ passive optical fiber star is illustrated in Fig. 26.6. In the ideal case, all N input and N output ports are equivalent. If power $P_T(k)$ is transmitted by station k, then $P_T(k)/N$ will appear at each of the output ports (and none at the input ports) and will be broadcast to all stations, including k, equally. In practice, of course, the power division may not be equal and the output distribution may vary with the input port. In addition, there will be some excess power loss, ΔP, such that

$$\frac{\Delta P(k)}{P_T(k)} = 1 - \frac{\sum_j^N P_R(j,k)}{P_T(k)} \tag{26.18a}$$

$$\approx 1 - \gamma, \tag{26.18b}$$

where $P_R(j, k)$ is the power received at j due to $P_T(k)$. For simplicity we may assume independence of input and output ports and define a constant γ such that,

$$\frac{P_R(j, k)}{P_T(k)} \approx \gamma/N, \qquad (26.19)$$

for all j and k with $(1 - \gamma)$ the excess loss factor, and $10 \log_{10}\gamma$ the excess loss in dB.

A simple multimode star can be made by butting N input and N output fibers to a block of glass, in analogy with the diagram in Fig. 26.6. Since each multimode fiber contains ~ 100 modes when properly excited and the glass block supports about 100N modes, the power distribution across the output facet of the block can be nearly uniform, providing a near uniform output to all the fibers. However, the limited filling factor introduced by the core/cladding ratio, ρ, of the fibers gives an excess loss factor $\gamma \approx \rho^2 \approx 1/4$ (-6 dB) for $\rho = 1/2$. An alternative multimode star structure consists of a bundle of fibers twisted and fused together to replace the block of glass.

Single-mode stars are more complicated. They can be fabricated by arrays of -3 dB directional couplers interconnected by fibers that must cross (making an integrated optic version difficult). As illustrated in Fig. 26.7a, the -3 dB coupler is itself a 2×2 star. Four of these couplers can be combined as in Fig. 26.7b to form a 4×4 star. In a similar way, an $N \times N$ coupler can be made by cross-connecting m $n \times n$ couplers with n $m \times m$ couplers (Hermes $et\ al.$, 1985), as shown in Fig. 26.7c, with

$$N = nm = 2^M, \qquad (26.20)$$

and M an integer. The interconnection shown in Fig. 26.7c is not unique. In going from any input port to any output port one must pass through

$$M = \log_2 N \qquad (26.21)$$

-3 dB couplers. The total number of -3 dB couplers is

$$Q = \frac{1}{2}MN = \frac{1}{2}N \log_2 N. \qquad (26.22)$$

Single-mode passive stars can also be built up using arrays of 3×3 couplers, which are made by fusing 3 fibers together to produce 3×3 stars (Villarruel $et\ al.$, 1985).

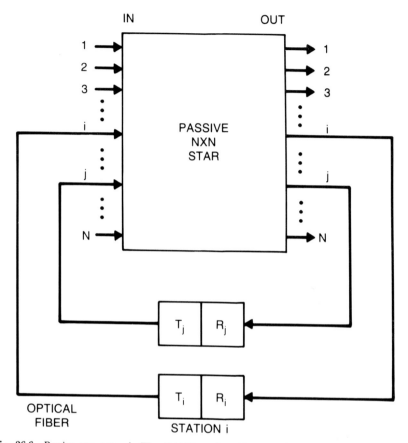

Fig. 26.6 Passive star network. The $N \times N$ star has N optical fiber input ports and N optical output ports. As in the active star, two fibers connect a remote station with the star.

The excess loss in dB for a passive star, $10M \log_{10}(1 - h)$, increases logarithmically with N, where h is the excess loss factor for each -3 dB coupler as in Section 26.2.3.2. On the other hand, the maximum loss in a linear transmit/receive bus with N stations, $20N \log_{10}(1 - h)$, is directly proportional to N. Hence, the passive star can support a larger N for a given power margin and also provides a uniform distribution of power among all stations. However, the star requires $\frac{1}{4}\log_2 N$ times as many -3 dB couplers.

In Section 26.3, we discuss network applications of the single-mode passive star. We also need to consider the single-mode $1 \times N$ splitter, which equally divides a single input among N outputs. Of course, the $N \times N$ star with only one input excited produces the same effect. Fewer -3 dB couplers, however, are required for the branching $1 \times N$ splitter illustrated in Fig. 26.8. In this case, the number of ranks M through which each output must pass is the same as

(a) 2 × 2

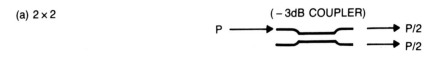

(b) 4 × 4

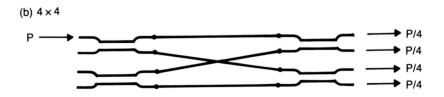

(c) N × N   (N = mn)

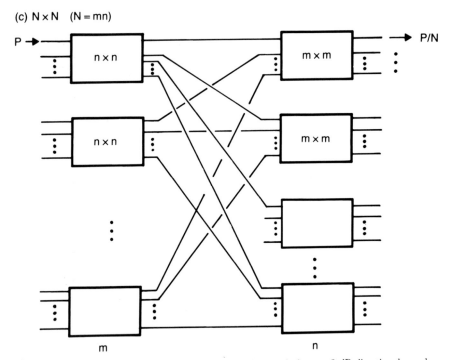

Fig. 26.7 Single-mode fiber $N \times N$ passive star couplers made from -3 dB directional couplers. (a) 2×2, (b) 4×4, (c) $N \times N$, with $N = mn$.

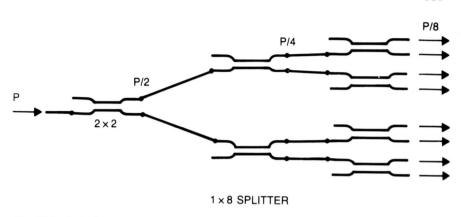

Fig. 26.8 A single-mode fiber 1 × 8 splitter. The dangling ends of the directional couplers are normally terminated by matched impedances.

given in Eq. (26.21). However, the number of -3 dB couplers is reduced to

$$Q' = N - 1 \qquad (26.23)$$

compared to the larger value in Eq. (26.22).

If power P is incident on the single port on the left in Fig. 26.8, P/N will appear at each of the N output ports. Since the $1 \times N$ splitter is a reciprocal device, if power P is incident on any one of the N ports on the right only P/N will appear at the single fiber on the left. The remainder will appear at the unconnected ports of the directional couplers. These dangling ports would normally be terminated by matched impedances to prevent reflections. Thus, if an $N \times N$ coupler were formed by connecting an $N \times 1$ combiner to a $1 \times N$ splitter by means of one single-mode fiber, the power splitting ratio would be $1/N^2$ instead of $1/N$ as for the star of Fig. 26.7.

Thus far, we have been concerned with power division and have not considered interference between coherent waves. If two or more coherent sources are incident on different input ports of an $N \times N$ star or an $N \times 1$ combiner, the outputs will be given by the appropriate complex scattering matrices having dimensions $2N$ or $N + 1$, respectively, as discussed in Section 26.2.3.2 with complex input amplitudes a_j. We will be describing network applications of these devices in time division (TDM) and wavelength division (WDM) multiplex systems in Section 26.3. Interference effects are not normally present in these applications since only one coherent input signal is present at a given instant.

26.2.6 Mesh Networks

A mesh network consists of a distribution of active nodes connected by point-to-point links. Unlike ring or bus networks, more than two links connect each node in general. The network of telephone central offices is an example. This network has a tremendous bandwidth even though each node and link have rather small bandwidths. Furthermore, there are many paths between nodes to assure reliable communication links despite failures of several nodes and links. In the telephone network, of course, many of these links are now lightwave.

A regular mesh network, known as the Manhattan street network (Maxemchuk, 1985), has been proposed as a topology for a packet LAN. Each active node has two inputs and two outputs, as in the bidirectional ring. In the conceptual topology, loops of fiber running north and south interconnect columns of nodes, and loops running east and west interconnect rows of nodes. Nodes are placed at the intersections of avenues and streets as in Manhattan. The topology has considerable flexibility in joining communities of interest in closed loops. Since packets may arrive at a node simultaneously, provision must be made for storage and control at each node. Alternatively, the need for storage can be avoided by the following protocol: if two packets enter the node simultaneously, each is assigned one of the outputs, even if it is not the one leading to the shortest route; nevertheless, the packet is eventually sent to its destination after effectively being stored *en route*. A dual ring network has the reliability rerouting feature of the mesh, but, in the ring, each repeater must operate at the network bandwidth whereas, in the mesh, each repeater operates at the terminal bandwidth.

While a mesh network has large throughput and good reliability, it may prove to be very expensive to implement where high-speed terminals and large, fast buffers are required at each node.

26.3 CIRCUIT-SWITCHED NETWORKS

26.3.1 Introduction

The media access protocols described in Section 26.2 provide packet communication between stations by means of "distributed switching." That is, packets with destination addresses are broadcast from the source to all stations and recorded by the destination station. And various methods are employed to share the transmission medium on a statistical basis.

It is more common in telecommunications, as opposed to local data networks, to dedicate a permanent path between two stations for the duration of a session. The multiple access (MA) paths may be defined by dividing the medium according to any of its dimensions: space division (SDMA), time division (TDMA), wavelength division (WDMA) or equivalently frequency division (FDMA), subcarrier frequency division (SFDMA), or code division (CDMA, or spread spectrum). An important advantage of the dedicated link for high-speed communication is that buffers, retransmissions, and acknowledgements are not required within the network.

Two types of bandwidth may be defined for a network: (a) the bandwidth of a terminal transmitter and receiver, and (b) the network bandwidth, which is the sum of bandwidths of all terminals that may transmit simultaneously. In the public telephone network, which is an extensive circuit-switched mesh network, each telephone has a narrow terminal bandwidth (~ 64 kb/s) but many million calls can be handled simultaneously to give an enormous network bandwidth. In an Ethernet random access network, on the other hand, only one terminal can use the medium at one time (and, because of the access protocol, less than one on average) so that the terminal bandwidth (10 Mb/s) for packet transmission is greater than the network bandwidth. Since wideband (> 100 Mb/s) terminals are expensive and many are required in a network, but the bandwidths of the transmission medium and network controller might be relatively cheap per terminal, the terminal costs may be dominant, and then, a circuit-switched network would be superior to a packet-switched network for certain types of traffic. Real-time video, graphics and high-speed transfers of very large (~ Gbyte) files favor the circuit-switched network; occasional bursts of short (~ Mbyte) packets, as in interactive video, might favor the packet-switched approach.

Research efforts in circuit-switched photonic networks are summarized in the following sections. It will be seen that a high level of sophistication has not yet been realized and that critical components limit the performance of these networks. Further photonic network progress depends upon the development of inexpensive, single-mode, high-frequency devices such as connectors, optical amplifiers, tunable lasers, optical AND gates, optical memories, and large optical or electronic space division switches. As we have seen elsewhere in this book, wideband (~ 10 Gb/s) transmitters, receivers and transmission lines for lightwave telecommunications are already available.

26.3.2 Space Division

The crosspoint switch shown in Fig. 26.9 is the simplest means for connecting any set of N inputs to any set of N outputs. In addition, it is possible to rearrange any subset of interconnections without disconnecting any other connections, i.e., it is strictly "non-blocking." Furthermore, it is possible to broadcast or narrowcast the same message from one input to all or several outputs. To realize this flexibility, however, the crossbar switch needs N^2 crosspoint switches. A number of other configurations that require fewer switches is possible at the expense of more complexity and less flexibility (Hill, 1986). For example, a 2×2 switch, that has two states (bar and cross), as indicated in Fig. 26.10a, can be used as a crosspoint in various switch matrices, including the crossbar. Only $N(N - 1)/2$ of them are required to produce a "rearrangeably non-blocking" $N \times N$ switch (Taylor, 1974), which means that adding a new interconnection

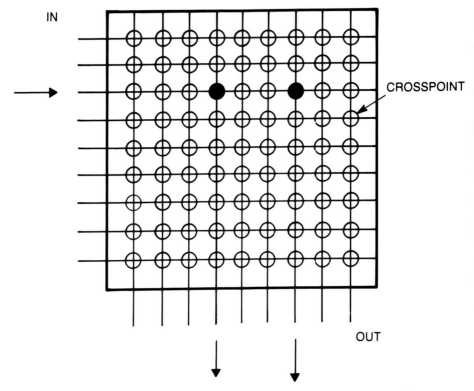

Fig. 26.9 Crossbar matrix switch. The full circles indicate closed switches in this example.

may require changing some paths through the switch for existing connections, temporarily interrupting service. Broadcasting cannot be supported in this switch. A 4×4 version is illustrated in Fig. 26.10b. The special constraint satisfied by this $N \times N$ switch is that connections between 2×2 elements do not cross. Thus, it is suitable for use in planar integrated optics applications. If cross-overs are allowed, the number of elements required can be reduced to order $N \log_2 N$. An electrooptically controlled directional coupler, for which the coupling ratio switches between 1 (cross) and zero (bar), can serve as the 2×2 switch element (Kaminow and Li, 1979).

Various $N \times N$ switches have been realized electronically using silicon integrated circuits and optically using titanium-diffused lithium niobate ($Ti:LiNbO_3$) directional coupler switches (Schmidt and Buhl, 1976; Korotky and Alferness, 1988). Two speeds must be defined for these matrix switches: (a) the rate at which the network can be rearranged, and (b) the bit-rate of signals that can be transported through the network without distortion, i.e., the

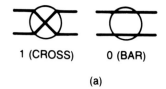

1 (CROSS)   0 (BAR)

(a)

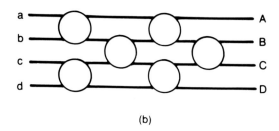

(b)

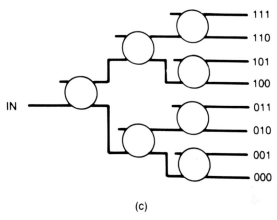

(c)

Fig. 26.10 (a) A two-state 2×2 directional-coupler switch. (b) A 4×4 switch. (c) 1×8 branching switch. Output addresses correspond to states of 2×2 switches in each rank (after (Kaminow and Li, 1979)).

throughput rate. The throughput rate is practically unlimited in the optical switches, since the refractive index dispersion for path lengths of ~ 10 cm is negligible at 10 GHz. For ICs, on the other hand, the capacitative loading introduced by the metallic paths between switches produces significant dispersion. The rearranging rate in both types of device is limited by transit times and RC time constants in the switch elements themselves. (The speed of the electrooptic effect is $\sim 10^{12}$ Hz). Individual Ti:LiNbO$_3$ switches have operated at rates of ~ 10 GHz; (Korotky and Alferness, 1988), and individual silicon transistors operate at ~ 1 GHz, while discrete GaAs transistors operate at ~ 10 GHz (Swartz, 1988). However, low-capacitance circuits to drive all the switch-points in the IC matrix may be more difficult to achieve than for the optical matrix.

The IC arrays have a fabrication advantage for large N since large-scale integrated circuit technology is well established and each transistor occupies a small chip area (~ 10 μm $\times$ 10 μm), although heat dissipation can be a problem for large packing densities. Since electrical pulses can be reformed and retimed on the IC chip, large N $\times$ N arrays can be built up from smaller sub-arrays. The pulse regeneration may take place at each crosspoint, or, for simple gates, at the edge of the sub-array. On the other hand, each optical directional coupler switch requires a large, elongated area (~ 20 μm $\times$ 1 cm) but available substrates are of limited extent so that large arrays are difficult to realize. Also, optical pulse reforming is not possible. The required large substrate diameter also contributes to control circuit capacitance and loss.

Thus, with current technology, the high rearrangement-rate applications of electronic and optical switches are complementary. For small N (< 16) and high throughput rate (> 500 Mb/s), optical switches may have the advantage. And, for large N at low throughput rates, electronic switches may be preferable.

An experimental 16 $\times$ 16 silicon IC crossbar switch, employing 2 μm CMOS transistors at the crosspoints, operated at a throughput rate of 240 Mb/s and a reconfiguration rate of ~ 3 Mb/s (Hayward et al., 1987). An 8 $\times$ 8 GaAs IC matrix switched a 1.2 Gb/s signal in 5 ns (Hayano et al., 1987). A 4 $\times$ 4 GaAs matrix switched a 2 Gb/s signal in 1 ns (Nakayama et al., 1986), and was incorporated in a three-chip optoelectronic circuit with 4 input and 4 output optical fibers that switched 560 Mb/s data with low crosstalk (Iwama et al., 1987). The latter experiment highlights the difficulty to be encountered in bringing many fibers on and off a chip.

Several state-of-the-art Ti:LiNbO$_3$ crossbar switch matrices have been reported. The emphasis has been on dimensionality, crosstalk, optical loss and switching voltage, but not speed. We may assume that large throughput rates (> 10 Gb/s) will not be a problem. Since each crosspoint switch contains 3 electrodes, however, it may prove difficult to realize the ultimate rearrangement rate, limited by the crosspoint itself, due to circuit parasitics. The largest switch reported is 8 $\times$ 8 and contains 64 directional coupler switches operating at 1.3 μm with 26 V switching voltage and -30 dB extinction ratio, when properly biased, and end-to-end throughput loss of 7 dB (Granestrand et al., 1986). The extinction ratio, which is a measure of crosstalk, is the fraction of power at a given input port that appears at any output port that is not meant to be connected to the given input. A matrix switch of this type is required for the

LOCNET LAN proposed by Heinrich Hertz Institute (Hermes, 1985). The active area of the substrate is about 0.6 cm × 6 cm. A 4 × 4 matrix with 16 switches operating at 1.3 μm with 13 V, -35 dB extinction and 5 dB insertion loss was also reported (Bogert *et al.*, 1986). It has permanently attached fiber pigtails. Another 4 × 4 matrix, but with two directional coupler switches per crosspoint (i.e., 32) to improve crosstalk, had crosstalk of -34 dB (not better than the single crosspoint switch above) and loss of 7 dB (Kondo *et al.*, 1985). A 1 × 16 branching switch array, similar to the 1 × 8 illustrated in Fig. 26.10c (Kaminow and Li, 1979), has also been fabricated with 15 Ti:LiNbO$_3$ switches with a throughput loss of 3 dB (Watson, 1986). It has permanent pigtails and is insensitive to optical polarization.

If the throughput rate of a space-division switch is greater than the ultimate terminal speeds required, then the dimensionality of the switch can be increased with a "time-slot interchanger" (TSI). The signal entering each port of the N × N space switch is time-division multiplexed with p channels destined for p users on one or more output ports, each assigned one of the p time slots. In general, the sequence of input channels will not correspond to the destination channels; thus, the source at the input assigned slot i on input port m may wish to communicate with the terminal assigned slot j on output port n. The TSI on the input port m reorders the pulses, by means of delays introduced by switched-in lengths of fiber (Thompson and Giordano, 1987), to shift the i slot to the j slot, and the space switch connects m with n. The next bit, in slot $(j + 1)$, may be sent to a different output port by the space switch. For flexibility, it is convenient to also have a TSI on the output side of the space switch. As discussed in the following section, a TSI can also be made using bistable lasers.

26.3.3 Time Division

In a TDMA network, a frame of duration T is divided into p time slots of duration T/p containing optical pulses of duration $\tau \le T/p$, as illustrated in Fig. 26.11. The length of fiber corresponding to the duration can be found by multiplying the time by the propagation velocity in fiber: 2×10^{10} cm/s $= 20$ cm/ns $= 200$ μm/ps. Each interconnection from one station to another is assigned a particular time slot by a central controller. The transmitter sends a pulse in that time slot at the frame rate $F = 1/T$ and the receiver selects that time slot. Any of the topologies mentioned earlier may be used to broadcast the frames. Since the receiver must discern its pulse from adjacent pulses, the receiver bandwidth must be pF, which is the network throughput rate. Similarly, the transmitter must be capable of generating pulses with this large bandwidth. Hence, a network with $p = 100$ stations each operating simultaneously at $F = 100$ Mb/s would require terminals with 10 Gb/s bandwidth, which is

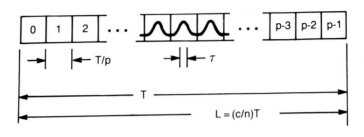

TDMA FRAME

Fig. 26.11 TDMA frame with p slots and frame rate $F = 1/T$.

difficult to realize inexpensively by electronic means. The challenge is to do the multiplexing and demultiplexing optically.

Synchronization of the receiver with its assigned time slot is also a challenge when $T/p < 1$ ns. Both the clock frequency and clock phase must be precisely synchronized between transmitter and receiver for synchronous detection of long data streams. For short packets, the clocks need not be precisely synchronous (asynchronous detection) and the phase may be recovered at the receiver during the packet header. Established methods for timing recovery between remote stations used in electronic data networks are adaptable to the higher speeds of photonic networks (Tucker *et al.*, 1987). Alternatively, optical clock pulses could be transmitted from a central node (Prucnal *et al.*, 1986a) to the communicating stations.

A straightforward optical TDMA network based on a dual-ring and repeater technology is described in Minami *et al.* (1985). It operates at 200 Mb/s with a frame, generated by a supervisory node, that has 70 time slots, which can be assigned to one or more stations depending on the need. The ring is exactly one frame in perimeter and a timing bit synchronizes the stations.

An optically sophisticated TDMA experiment used a combination of Ti:LiNbO$_3$ 1 × 4 branch switches and bistable laser diodes, which are fabricated by applying a split electrode to a 1.3 μm DCPBH laser, to produce a 4-channel time-slot interchanger (Suzuki *et al.*, 1986). A 256 Mb/s TDM signal enters the 1 × 4 switch, which scans the four outputs in sequence at a 64 Mb/s rate. Bistable lasers are located at each output port. If a 1 is present, the bistable laser is switched on, otherwise it remains off. The bistable lasers are reset every frame. The output 4 × 1 switch can be scanned in any sequence to read the bistable switches and thereby generate a reordered TDM bit stream. This element operates as a TSI and might serve as the basis for interconnecting a set of transmitters assigned given time slots to a set of receivers assigned their own slots. It does not alleviate the bandwidth requirements on the terminals, how-

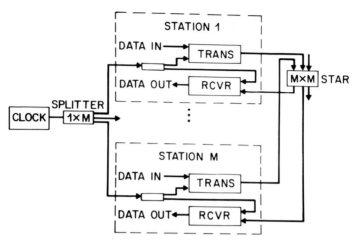

Fig. 26.12 An optical TDMA network based on a central clock and passive star (Prucnal *et al.*, 1986b).

ever. Interestingly, this array of bistable lasers could also serve as an optical memory with storage time determined by the reset signal.

An attempt to use optical techniques to reduce the electronic bandwidth requirements is contained in a proposal by (Prucnal *et al.*, 1986a) and by a demonstration experiment (Prucnal *et al.*, 1986b). As illustrated in Fig. 26.12, a centralized clock is the focus of the network. In the experiment, the clock is a pulsed ($\tau = 2$ ns) 1.3 μm laser operating at $F = 10$ MHz. A $1 \times M$ passive splitter sends the clock to each station where it is divided between the transmitter and receiver. The transmitter is a Ti:LiNbO$_3$ gate that passes the coded optical bits to a passive M $\times$ M star, where it is distributed to all the receivers. Suitable optical delays are provided so that each of the M stations produces a bit at the M $\times$ M star in the proper time slot. A suitable optical delay is also provided in the receiver so that the clock and assigned time slot coincide at the receiver. The correct signal pulse coincides with the clock pulse and produces a higher level in the photodetector than other signal pulses, corresponding roughly to a logical AND operation. In the experiment, 50 10 Mb/s time slots were available and 15 were occupied. A mode-locked laser (Tucker *et al.*, 1983) in place of the pulsed laser could produce very short clock pulses for a higher speed network. A sensitive, high-speed optical AND gate would alleviate the need for a high-speed electronic receiver. A pair of photoconductive switches in series serves as an optical AND gate (Desurvire *et al.*, 1988). Alternatively, a titanium-diffused LiNbO$_3$ electrooptic TDM demultiplexer, based on the waveguide tree structure of Fig. 26.10c, can be employed (Tucker *et al.*, 1987). A difficult problem in any optical TDM configuration is the maintenance of time-slot alignments within a frame because of delay variations caused by differences in source laser wavelengths and thermal changes in optical path lengths.

26.3.4 Frequency and Wavelength Division

In a WDMA network, the available spectral bandwidth Λ is divided into p segments. An optical carrier at wavelength λ_k is located at the center of each segment and adjacent carriers are separated by Λ/p. Since optical sources have been traditionally described in terms of wavelength, it is customary to speak of WDM. From the engineering standpoint, however, it is simpler to deal with optical frequencies. Thus, with c the velocity of light in vacuum, $f_k = c/\lambda_k$, and increments in frequency are given by

$$df/d\lambda = -c/\lambda^2 = \{178 \text{ GHz/nm@}1.3 \text{ } \mu\text{m}, 125 \text{ GHz/nm@}1.55 \text{ } \mu\text{m}\}.$$
$$(26.24)$$

We will use WDM for widely-spaced carriers (say, greater than 1 nm) and reserve FDM for closely-spaced carriers (say, within a few times the modulation rate). A recent long-haul WDM experiment employed 10 free-running distributed feedback (DFB) single-frequency laser carriers each modulated at 2 Gb/s (Olsson et al., 1984). The closest channel spacing was 1.35 nm or 180 GHz at 1.5 μm. For network applications, we will be concerned mainly with the FDMA domain in order to make efficient use of the available spectrum.

Optical fibers have losses (≤ 0.5 dB/km) low enough for local communications over an enormous bandwidth, ~ 1300–1550 nm, which is more than 20 THz. The network bandwidth, however, will be limited by the terminal equipment, particularly, a tunable laser. At present, an external grating laser (Favre et al., 1986) can be mechanically tuned continuously (without mode hopping) over about 15 nm or 3 THz at 1.3 μm by simultaneously translating and rotating the grating. If mode hopping between continuous ranges is allowed, larger regions, ~ 55 nm or 7 THz at 1.5 μm, can be covered (Wyatt and Devlin, 1983). If continuous tuning at high speed is required, however, as it would be for a heterodyne system, the tuning must be electronic. A tunable grating, distributed Bragg reflector (DBR) laser (Tohmori et al., 1985; Yamaguchi et al., 1985) was reported with a tuning range of 775 GHz at 1.55 μm (Kotaki et al., 1988, Murata et al., 1987), and a split-electrode distributed feedback (DFB) laser was reported (Yoshikuni, 1986) with a tuning range of 280 GHz at 1.5 μm.

A FDMA network might consist of a number of stations connected to a passive star or bus. A comb of optical carriers spaced apart by Λ/p, or equivalently F/p, with $F = c\Lambda/\lambda^2$ must be maintained, as illustrated in Fig. 26.13. The spacing F/p might be six times the terminal bitrate, B. With $F = 600$ GHz, one could expect as many as one hundred 1 Gb/s channels. A

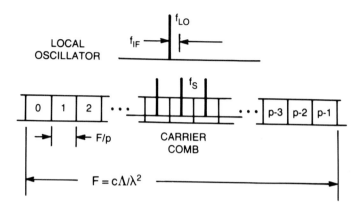

FDMA BAND

Fig. 26.13 FDMA band with p carriers at optical frequencies $f_k = c/\lambda_k$. For heterodyne detection with $f_{IF} = F/2p$, it is necessary to employ an image rejection filter; otherwise, f_{IF} must be greater than F.

central controller would assign a frequency channel to interconnect a transmitter with a receiver. The receiver could be a tunable filter with width $\leq F/p$ and tuning range F, followed by a direct detector (Mallinson, 1985). At present, piezoelectrically tunable fiber Fabry-Perot (FFP) filters with a finesse $\mathscr{F}$ of 200 have been demonstrated (Stone and Stulz, 1987). A 200 μm long FFP can tune over 500 GHz (the free spectral range) to select individual channels and provide a hundred 1 Gb/s channels. Electrooptically tuned filters (Korotky and Alferness, 1987) may be useful in providing fast tunability, although at present their resolution is poor (> 20 GHz) for this application. A higher sensitivity, at the expense of greater complexity, can be realized using optical heterodyne or homodyne detection methods, which require a tunable local oscillator laser and polarization control at the receiver. In the case of either a filter or heterodyne receiver, one of the most difficult challenges for a FDMA network is to establish a stable, equi-spaced comb of carriers produced by a large number of independent transmitter lasers that, in general, will not be co-located. One proposal is to have each transceiver servo periodically under microprocessor-based control to sense the nearest-neighbor carriers and maintain equi-spacing (Williams, 1986). Another scheme is to lock each laser carrier to a mode of a fixed FFP (Glance et al., 1987); one of the lasers can also be locked to an atomic resonance (Chung and Tkach, 1988) as an absolute reference. Various experimental schemes are described next.

A simple WDMA network was demonstrated (Payne and Stern, 1985) using an 8×8 single-mode passive star and tunable-filter receivers. Multi-longitudinal mode lasers in the 1.2 to 1.6 μm range and tunable holographic-grating filters with ~ 15 nm bandwidths were employed. In the LAMBDANET^{TM} experiment (Kobrinski et al., 1987), 18 DFB lasers operating near 1.5 μm and separated by 2 nm are modulated at 1.5 Gb/s and the channels are demultiplexed with a diffraction grating. Because of the wide carrier separations, no elaborate carrier-comb stabilization was required.

A two-channel FDM demonstration (Toba et al., 1986) employed several 1.5 μm DFB lasers separated by 11 GHz and modulated at 450 Mb/s by an external electrooptic intensity modulator. The lasers were independently stabilized to Fabry-Perot resonators and temperature controlled to $\pm .01$C. A tunable Mach-Zehnder waveguide interferometer was used as a demultiplexer (Inoue et al., 1988).

The tunable Fabry-Perot in the single-mode fiber configuration (FFP) in conjunction with a star coupler lends itself nicely to a simple and potentially inexpensive FDMA network (Kaminow et al., 1987, 1988a). Frequency-shift keying (FSK) is a modulation format that takes advantage of intrinsic semiconductor behavior. An analysis of such a network predicts and an experiment (with $B = 45$ Mb/s channels) supports a minimum channel spacing of $f_c = B/6$, similar to that for heterodyne systems (Kazovsky, 1987). Then, the maximum number of channels is

$$N = \mathscr{F}/6, \qquad (26.25)$$

independent of B. It is expected that $\mathscr{F}$ can be improved by better mirror coatings to ~ 1000. Another approach is to employ two FFPs in a tandem Vernier configuration (Saleh and Stone, 1988), which allows the effective finesse to be many times that of a single FFP while reducing the channel spacing to $f_c = B/3$ giving,

$$N = \mathscr{F}/3 \qquad (26.26)$$

due to the sharper filter characteristic (Kaminow et al., 1988a). A stabilized tandem FFP experiment gave an effective finesse of 680 with individual $\mathscr{F} \approx 50$ (Kaminow et al., 1988b). Ideally, this FDMA-FSK-star-filter network configuration can support up to 1000 high bit-rate channels with f_c comparable to that of a heterodyne network but without the need for a local oscillator or polarization control.

A rudimentary heterodyne network with three channels separated by 5 GHz was demonstrated (Bachus et al., 1985) using a novel centralized carrier stabilizing scheme. One of the optical carriers at 825 nm is locked to a fiber ring resonator. An additional control laser has its frequency swept linearly in time. A heterodyne technique produces IF pulses that are temporally spaced in proportion to the carrier frequency separation. A feedback circuit maintains equal spacing. The three carriers, spaced by 1 GHz with 70 Mb/s modulation, were fed into a star coupler; an upstream channel was also provided. A 10-channel FDMA heterodyne system with 70 Mb/s data (Bachus et al., 1986) had unstabilized carriers spaced by 6 GHz and used ordinary (Fabry-Perot) GaAlAs lasers whose frequencies are controlled by temperature and injection current.

A heterodyne FDM-FSK network experiment with $B = 45$ Mb/s showed that f_c could be reduced to 300 MHz with negligible crosstalk (Glance et al., 1988). The capability of separating two channels spaced-apart by 1 GHz using optical heterodyne techniques was demonstrated (Smith et al., 1985) at 1.5 μm using an external-grating laser as the tunable local oscillator and a 566 MHz intermediate frequency. The problem of rejecting the image band if more than two carriers were present, however, was not addressed experimentally.

In order to implement heterodyne detection of one of the p channels in a multicarrier comb with minimum carrier spacing using an IF frequency on the order $F/2p$ rather than F, an optical image rejection filter is necessary (Glance, 1986). Since F may be on the order of 100 GHz, $f_{IF} \approx F$ would not be practical. With the image rejection filter, a well-known component in microwave radio, only one of the components satisfying $f_S = f_{LO} \pm f_{IF}$ is selected, where f_S, f_{LO} and f_{IF} are the optical signal, local oscillator and intermediate frequencies, as illustrated in Fig. 26.13. This device was demonstrated (Darcie and Glance, 1986) at 1.3 μm with $f_{IF} = 1.7$ GHz. Alternatively, if every other channel is left empty, then it is possible to use $f_{IF} = F/2p$ without image rejection.

26.3.5 Subcarrier Frequency Division

A proposed subcarrier frequency division multiplexed (SFDM) network (Darcie, 1987) avoids the need for high-speed receivers inherent in TDM and the needs for inter-carrier stabilization and optical filtering or optical heterodyning inherent in FDM. The price is a lower network sensitivity margin and restricted bandwidth, which may be tolerable in a local environment. In operation, digital or analog information from individual channels is modulated onto separate microwave sinusoidal subcarriers, which may be added in a standard microwave combiner. The sum of these subcarriers modulates the current in a single semiconductor laser. Alternatively, each subcarrier can be modulated onto a separate laser, whose wavelength need not be controlled, and these channels can be combined optically. In either case, after distribution on a fiber bus or star, the multiplexed optical signals are photodetected and the channels are demultiplexed at radio frequencies by standard, inexpensive microwave filters or microwave heterodyne methods. The receiver electronics for a given channel need only have a bandwidth, centered at the subcarrier frequency, corresponding to the channel width and not the network bandwidth as in TDM. Laser

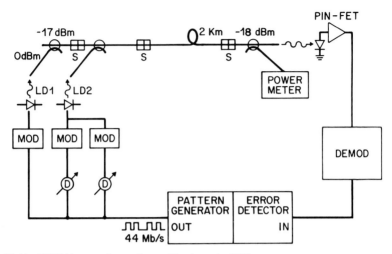

Fig. 26.14 SFDMA network experiment (Darcie *et al.*, 1986).

linearity (Darcie *et al.*, 1985) is an important constraint: in certain applications the subcarrier band must be limited to an octave to avoid second harmonic distortion, and in broadcast networks the subcarrier optical modulation depth must be limited to avoid intermodulation distortion. In a network experiment (Darcie *et al.*, 1986) at 1.3 μm, two lasers modulated in the 3.5 to 4.0 GHz band were coupled to a bus as shown in Fig. 26.14. One laser supported two 44 Mb/s channels and the other one 44 Mb/s channel. The two subcarriers on a single laser were spaced by 150 MHz and the channels were demultiplexed and detected with a bit-error rate of 10^{-10}. The SFDM approach was used to transmit 60 FM analog TV channels over 18 km of single-mode fiber (Olshansky and Lanzisera, 1987).

26.3.6 Code Division

In code division multiplexing (CDM), or so-called "spread spectrum," each bit is sub-divided into a number of "chips" consisting of 1s and 0s. The chip sequence constitutes a code that permits a bit stream broadcast on a network to be selected by means of a correlation process at the receiver destination. A large number of chip codes can be assigned to different users; however, the number of orthogonal codes that can be used simultaneously on the network is substantially less than the number of chips so that, perhaps, less than a quarter as many channels are available as for a TDM system with equivalent network bandwidth. Still CDMA offers an alternative means for network access. At microwave frequencies, the method has the advantage in satellite communications that the chip modulation spreads the band over which microwave power is distributed and reduces interference with other users of the band.

An asynchronous CDMA network with a passive star for distribution and a combination of fiber delay lines for the correlation has been demonstrated (Prucnal *et al.*, 1986c). It operates at a terminal rate of 3 Mb/s with a 32 chip code to give an equivalent TDM network bandwidth of 100 Mb/s. A single channel was detected from among five coded bit streams.

26.4 CONCLUSIONS

The configurations of future photonic networks will depend on the applications and types of traffic to be distributed. The applications will depend upon the unique capabilities of photonic networks. And the capabilities will depend upon novel optical, electronic and optoelectronic components that permit enhancements in performance and cost beyond that available by present-day methods.

Passive distribution networks require the development of higher power laser transmitters and optical amplifiers in order to support more stations. The implementation of optical random access and packet switched networks requires high-speed and high-capacity buffer memories and shift registers. FDMA networks require precisely tunable lasers and stabilization techniques. TDMA networks require fast, sensitive optical AND gates to demultiplex signals cheaply. Since terminal costs cannot be shared, the cost of terminal components, such as single-mode connectors and optoelectronic interfaces, must be reduced substantially by clever design, by integration of optical and electronic circuits and by large-scale production.

Electronic integrated circuit technology is much further advanced than photonic device technology; and it is advancing faster because of the greater effort being devoted to ICs. Thus, electronic solutions to a switched star network or to high-speed multiplexing and demultiplexing will compete strongly with the ambitious photonic solutions being proposed. The ultimate result may be a high-speed electronic node and electronic terminals with single-mode fiber optical data links joining them. Advances in photonic and electronic circuits during the next few years will determine the outcome.

The brief history of long-haul lightwave systems teaches us that the demand for high-bit-rate services grows faster than we can imagine today. And the history of sophisticated consumer optoelectronics appliances, such as Compact Disc players, teaches us that costs can be driven down dramatically without sacrificing performance if the market is sufficiently large.

REFERENCES

Abramson, N. (1973). The ALOHA system. In "Computer-Communication Networks." N. Abramson and F. Kuo, eds, Prentice-Hall, Englewood Cliffs, N.J., Chap. 14.

Acampora, A. S., Hluchyj, M. G., and Tsao, C. D. (1983). A centralized-bus architecture for local area networks. *IEEE International Conference on Communications*, ICC '83, Boston, MA, pp. 932–938.

Acampora, A. S., and Hluchyj, M. G. (1984). A new local area network architecture using a centralized bus. *IEEE Communications Magazine* 22, 12–21.

Altman, D. E., and Taylor, H. F. (1977). An eight-terminal fiber optics data bus using tee couplers. *Fiber and Integrated Optics* **1**, 135–152.

Auracher, F., and Witte, H.-H. (1977). Optimized layout for a data bus system based on a new planar access coupler. *Applied Optics* **16**, 3140–3142.

Bachus, E.-J., Braun, R.-P., Eutin, W., Grobmann, E., Foisel, H., Heimes, K., and Strebel, B. (1985). Coherent optical-fiber subscriber line. *Electronics Letters* **21**, 1203–1205.

Bachus, E.-J., Braun, R.-P., Caspar, C., Grossman, E., Foisel, H., Heimes, K., Lamping, H., Strebel, B., and Westphal, F.-J. (1986). Ten-channel coherent optical fibre transmission. *Electronics Letters* **22**, 1002–1003.

Bogert, G. A., Murphy, E. J., and Ku, R. T. (1986). Low crosstalk 4×4 Ti:LiNbO$_3$ optical switch with permanently attached polarization maintaining fiber array. *J. Lightwave Technology* **LT-4**, 1542–1545.

Burr, W. E. (1986). The FDDI optical data link. *IEEE Communications Magazine* **24**, 18–23.

Bux, W., Closs, F., Janson, P. A., Kummerle, K., Muller, H. R., and Rothauser, E. H. (1982). A local-area communication network based on a reliable token-ring system. In "Local Computer Networks." P. C. Ravasio, G. Hopkins and N. Naffah, Eds, North-Holland Publishing Company.

Bux, W., Closs, F. H., Kummerle, K., Keller, H. J., and Mueller, H. R. (1983). Architecture and design of a reliable token-ring network. *IEEE J. Selected Areas in Communications* **SAC-1**, 756–765.

Chu, T. S., and Gans, M. J. (1987). High speed infrared local wireless communication. *IEEE Communications Magazine* **25**, 4–10.

Chu, T. S., and Gans, M. J. (1988). Indoor infrared beam data link. *Conference on Lasers and Electro-Optics*, April 25, 1988, Anaheim, CA, Paper THM42.

Chung, Y. C., and Tkach, R. W. (1988). Frequency stabilization of a 1.3 μm DFB laser to an argon line using the optogalvanic effect. *Electronics Letters* **24**, 804–805.

Crow, J. D. (1985). Computer applications for fiber optics. *IEEE Communications Magazine* **23**, 16–20.

Darcie, T. E., and Glance, B. (1986). Optical heterodyne image-rejection mixer. *Electronics Letters* **22**, 825–826.

Darcie, T. E., Dixon, M. E., Kasper, B. L., and Burrus, C. A. (1986). Lightwave system using microwave subcarrier multiplexing. *Electronics Letters* **22**, 774–775.

Darcie, T. E., Tucker, R. S., and Sullivan, G. J. (1985). Intermodulation and harmonic distortion in InGaAsP lasers. *Electronics Letters* **21**, 665–666.

Darcie, T. E. (1987). Subcarrier multiplexing for multiple-access lightwave networks. *J. Lightwave Technology* **LT-5**, 1103–1110.

Desurvire, E., Tell, B., Kaminow, I. P., Qua, G. J., Brown-Goebeler, K. F., Miller, B. I., and Koren, U. (1988). High contrast GaInAs:Fe photoconductive AND gate for time-division demultiplexing. *Electronics Letters* **24**, 396–397.

Favre, F., Le Guen, D., Simon, J. C., and Landousies, B. (1986). External-cavity semiconductor laser with 15 nm continuous tuning range. *Electronics Letters* **22**, 795–796.

Fraser, A. G. (1983). Towards a universal data transport system. *IEEE J. Selected Areas in Communications* **SAC-1**, 803–816.

Fratta, L., Borgonovo, F., and Tobagi, F. A. (1981). The express-net: a local area communication network integrated voice and data. *Int. Conf. Data Comm. Syst: Performance Applications*, Paris.

Glance, B. (1986). An optical heterodyne mixer providing image-frequency rejection. *IEEE J. Lightwave Tech.*, **LT-4**, 1722–1725.

Glance, B., Stone, J., Fitzgerald, P. J., Pollack, K. J., Burrus, C. A., and Stulz, L. W. (1987). Frequency stabilization of FDM optical signals originating from different locations. *Electronics Letters* **23**, 1243–1245.

Glance, B., Pollack, K. J., Burrus, C. A., Kasper, B. L., Eisenstein, G., and Stulz, L. W. (1988). WDM coherent optical star network. *J. Lightwave Technology* **6**, 67–72.

Goto, H., Akashi, F., Hirosaki, B., and Shimizu, H. (1985). A 1.2 GOPS optical loop LAN for wideband office communications. *IEEE Globecom '85*, 462–467.

Granestrand, P., Stolz, B., Thylen, L., Bergvall, K., Doldissen, W., Heinrich, H., and Hoffmann, D. (1986). Strictly nonblocking 8 × 8 integrated optical switch matrix. *Electronic Letters* **22**, 816–818.

Hayano, S., Nagashima, K., Asai, S., Maeda, T., and Furutsuka, T. (1987). A GaAs 8 × 8 matrix switch LSI for high-speed digital communications. *IEEE GaAs IC Symposium*, Portland, Oregon, October 13, 245–248.

Hayward, G. A., Gottlieb, A. M., Jain, S. K., and Mahoney, D. D. (1987). CMOS VLSI applications in broadband circuit switching. *IEEE J. Selected Areas in Communications* **SAC-5**, 1231–1241.

Henry, P. S., Linke, R. A., and Gnauck, A. H. (1988). Transmission systems principles. "Optical Fiber Telecommunications II." S. E. Miller and I. P. Kaminow, eds, Academic Press, New York, Ch. 21.

Hermes, T., Hoen, B., Saniter, J., and Schmidt, F. (1985). LOCNET—a local area network using optical switching. *J. Lightwave Technology* **LT-3**, 467–471.

Hill, A. M. (1986). One-sided rearrangeable optical switching networks. *J. Lightwave Technology* **LT-4**, 785–789.

Inoue, K., Takato, N., Toba, H., and Kawachi, M. (1988). A four-channel optical waveguide multi/demultiplexer for 5-GHz spaced optical FDM transmission. *J. Lightwave Technology* **6**, 339–345.

Iwama, T., Oikawa, Y., Yamaguchi, K., Horimatsu, T., Makiuchi, M., and Hamaguchi, H. (1987). A 4 × 4 GaAs OEIC switch module. *Optical Fiber Communication Conference Digest*, Reno, January 19, 161.

Kaminow, I. P. and Li, T. (1979). Modulation techniques. "Optical Fiber Telecommunications." S. E. Miller and A. G. Chynoweth, eds. Academic Press, New York, Ch. 17.

Kaminow, I. P., and Tucker, R. S. (1988). Mode-controlled semiconductor lasers. In "Guided Wave Optoelectronics." T. Tamir, ed., Springer-Verlag, New York.

Kaminow, I. P., Iannone, P. P., Stone, J. and Stulz, L. W. (1987). FDMA-FSK star network with a tunable optical filter demultiplexer. *Electronics Letters* **23**, 1102–1103.

Kaminow, I. P., Iannone, P. P., Stone, J. and Stulz, L. W. (1988a). FDMA-FSK star network with a tunable optical filter demultiplexer. *J. Lightwave Technology* **6**, September.

Kaminow, I. P., Iannone, P. P., Stone, J. and Stulz, L. W. (1988b). Frequency division multiple access network demultiplexing with a tunable Vernier fiber Fabry-Perot filter. *Conference on Lasers and Electro-optics Digest*, Anaheim, CA, April 25, 164–167. (submitted to Electronics Letters).

Kamiya, T., Tanaka, I., and Kamiyama, H. (1987). GaAs integrated optoelectronic grating circuit suitable for 10-Gb/s demultiplexing. *Conference on Lasers and Electro-optics Digest*, Baltimore, April 26, paper MB2.

Kazovsky, L. G. (1987). Multichannel coherent optical communication systems. *J. Lightwave Technology* **LT-5**, 1095–1102.

Kobrinski, H., Bulley, R. M., Goodman, M. S., Vecchi, M. P., and Brackett, C. A. (1987). Demonstration of high capacity in the LAMBDANET architecture: a multi-wavelength optical network. *Electronic Letters* **23**, 824–826.

Kondo, M., Takado, N., Komatsu, K. and Ohta, Y. (1985). 32 switch-elements integrated low-crosstalk LiNbO$_3$ 4 × 4 optical matrix switch. *IOOC-ECOC '85*, Venice, 361–364.

Korotky, S. K., and Alferness, R. C. (1988). Active optical components. "Optical Fiber Telecommunications II." S. E. Miller and I. P. Kaminow, eds. Academic Press, New York, Ch. 11.

Kotaki, Y., Matsuda, M., Ishikawa, H., and Imai, H. (1988). Tunable DBR laser with wide tuning range. *Electronic Letters* **24**, 503–504.

Limb, J. O., and Flores, C. (1982). Description of fasnet—a unidirectional local-area communications network. *Bell Syst. Tech. J.* **61**, 1413–1440.

Limb, J. O. (1984). On fiber optic taps for local area networks. *Proc. ICC '84*, Amsterdam, pp. 1130–1136.

Limb, J. O., and Albanese, A. (1985). Passive undirectional bus networks using optical communications. *Proc. IEEE Globecom '85*, New Orleans, pp. 1190–1194.

Mallinson, S. R. (1985). Fibre-coupled Fabry-Perot wavelength demultiplexer. *Electronics Letters* **21**, 121–122.

Maxemchuk, N. F. (1985). The Manhattan street network. *IEEE Globecom '85*, New Orleans, pp. 252–261.

Maxemchuk, N. F. (1986). Linear, unidirectional networks and random access strategies in fiber-optic local distribution systems. *Globecom '86*, Houston, pp. 29–33.

Minami, T., Yamaguchi, K., Nakagami, T., Takanashi, H., Fujino, N., Hamano, H., Suyama, M., Iguchi, K., and Yamada, I. (1985). A 200 Mbit/s synchronous TDM loop optical LAN suitable for multiservice integration. *IEEE J. Selected Areas in Communications* **SAC-3**, 849–858.

Murata, S., Mito, I., and Kobayashi, K. (1987). Over 720 GHz (5.8 nm) frequency tuning by a 1.5 μm DBR laser with phase and Bragg wavelength control section. *Electronics Letters* **23**, 403–405.

Nakayama, Y., Ohtsuka, T., Shimizu, H., Yokogawa, S., Kameo, K., and Nishi, H. (1986). A GaAs data switching IC for a gigabits per second communication system. *IEEE J. Solid-State Circuits* **SC-21**, 157–161.

Olshansky, R. and Lanzisera, V. A. (1987). 60-channel FM video subcarrier multiplexed optical communication system. *Electronics Letters* **23**, 1196–1197.

Olsson, N. A., Hegarty, J., Logan, R. A., Johnson, L. F., Walker, K. L., Cohen, L. G., Kasper, B. L., and Campbell, J. C. (1984). 68.3 km transmission with 1.37 Tbit km/s capacity using wavelength division multiplexing of ten single-frequency lasers at 1.5 μm. *Electronics Letters* **21**, 105–106.

Patir, A., Takahashi, T., Tamura, Y., Zarki, M. E., and Lazar, A. A. (1985). An optical fiber-based integrated LAN for MAGNET's testbed environment. *IEEE J. Selected Areas in Communications* **SAC-3**, 872–881.

Payne, D. B., and Stern, J. R. (1985). Single mode optical local networks. *Proc. IEEE Globecom '85*, New Orleans, pp. 1201–1205; also SPIE, Fiber Optic Broadband Networks, **585**, pp. 162–169.

Prucnal, P. R., Santoro, M. A., and Sehgal, S. K. (1986a). Ultra-fast all-optical synchronous multiple access fiber networks. *IEEE J. Selected Areas in Communications* **SAC-4**, 1484–1493.

Prucnal, P. R., Santoro, M. A., Sehgal, S. K., and Kaminow, I. P. (1986b). TDMA fiber optic network with optical processing. *Electronics Letters* **22**, 1218–1219.

Prucnal, P. R., Santoro, M. A., and Fan, T. R. (1986c). Spread spectrum fiber-optic local area network using optical processing. *J. Lightwave Technology* **LT-4**, 547–554.

Ramo, S., Whinnery, J. R., and Van Duzer, T. (1984). "Fields and Waves in Communication Electronics." 2nd ed., John Wiley, New York, p. 535ff.

Rawson, E. G. (1985). The fibernet II ethernet-compatible fiber-optic LAN. *J. Lightwave Technology* **LT-3**, 496–501.

Ross, F. E. (1986). FDDI—a tutorial. *IEEE Communications Magazine* **24**, 10–17.

Saleh, A. A. M., and Stone, J. (1988). Two-stage Fabry-Perot filters as demultiplexers in optical FDMA LANs. *J. Lightwave Technology* **6**, accepted.

Schmidt, R. V., and Buhl, L. L. (1976). Experimental 4 × 4 optical switching network. *Electronics Letters* **12**, 575–577.

Schmidt, R. V., Rawson, E. G., Norton, R. E., Jackson, S. B., and Bailey, M. D. (1983). Fibernet II: a fiber optic ethernet. *IEEE J. Sel. Areas in Commun.* **SAC-1**, 702–711.

Smith, D. W., Hodgkinson, T. G., Malyon, D. J., and Healey, P. (1985). Demonstration of a tunable heterodyne receiver in a two channel FDM experiment. *IEE Colloquium Digest* No. 1985/30, London.

Stone, J., and Stulz, L. W. (1987). Pigtailed high-finesse tunable fibre Fabry-Perot interferometers with large, medium and small free spectral ranges. *Electronics Letters* **23**, 781–783.

Strole, N. C. (1987). The IBM token-ring network: a functional overview. *IEEE Network Magazine* **1**, 23–30.

Suzuki, S., Terakado, T., Komatsu, K., Nagashima, K., Suzuki, A., and Kondo, M. (1986). An experiment on high-speed optical time-division switching. *J. Lightwave Technology* **LT-7**, 894–899.

Swartz, R. G. (1988). Integrated circuits for lightwave systems. "Optical Fiber Telecommunications II." S. E. Miller and I. P. Kaminow, eds, Academic Press, New York, Ch. 20.

Tanenbaum, A. S. (1981). "Computer Networks." Prentice-Hall, Englewood Cliffs, New Jersey.

Taylor, H. F. (1974). Optical-waveguide connecting networks. *Electronics Letters* **10**, 41–43.

Thompson, R. A., and Giordano, P. P. (1987). An experimental photonic time-slot interchanger using optical fibers as reentrant delay line memories. *IEEE J. Lightwave Technology* **LT-5**, 154–162.

Toba, H., Inoue, K., and Nosu, K. (1986). A conceptional design on optical frequency-division-multiplexing distribution systems with optical tunable filters. *IEEE J. Selected Areas in Communications* **SAC-4**, 1458–1467.

Tohmori, Y., Komori, K., Arai, S., Suematsu, Y., and Oohashi, H. (1985). Wavelength tunable 1.5 μm GaInAsP/InP bundle-integrated guide distributed bragg reflector (BIG-DBR) lasers. *The Transactions of the IECE of Japan* **E-68**, 788–790.

Tokura, N., Oikawa, Y., and Kimura, Y. (1985). High-reliability 100-Mbit/s optical accessing loop network system: OPALnet-II. *J. Lightwave Technology* **LT-3**, 479–489.

Tomlinson, W. J. (1988). Passive and low-speed active optical components for fiber systems. Optical Fiber Telecommunications II, S. E. Miller and I. P. Kaminow, eds, Academic Press, New York, Chapter 10.

Tseng, C.-W., and Chen, B.-U. (1983). D-Net, a new scheme for high data rate optical local area networks. *IEEE J. Selected Areas in Communications* **SAC-1**, 493–499.

Tucker, R. S., Eisenstein, G., and Kaminow, I. P. (1983). 10 GHz active mode-locking of a 1.3 μm ridge-waveguide laser in an optical-fibre cavity. *Electronics Letters* **19**, 552–553.

Tucker, R. S., Eisenstein, G., Korotky, S. K., Buhl, L. L., Veselka, J. J., Raybon, G., Kasper, B. L., and Alferness, R. C. (1987). 16 Gb/s fiber transmission experiment using optical time-division multiplexing. *Electronics Letters* **23**, 1270–1271.

Villarruel, C. A., Moeller, R. P., and Burns, W. K. (1981). Tapped tee single-mode data distribution system. *IEEE J. Quantum Electronics* **QE-17**, 941–945.

Villarruel, C. A., Wang, C.-C., Moeller, R. P., and Burns, W. K. (1985). Single-mode data buses for local area network applications. *J. Lightwave Technology* **LT-3**, 472–478.

Watson, J. E., Milbrodt, M. A., and Rice, T. C. (1986). A polarization-independent 1×16 guide-wave optical switch integrated on lithium niobate. *J. Lightwave Technology* **LT-4**, 1717–1721.

White, P. E. (1988). Distribution transmission systems. "Optical Fiber Telecommunications II." S. E. Miller and I. P. Kaminow, eds, Academic Press, New York, Ch. 25.

Williams, G. F. (1986). Wavelength devision multiplexing optical communication systems. U. S. Patent 4,592,043.

Wyatt, R., and Devlin, W. J. (1983). 10 kHz linewidth 1.5 μm InGaAsP external cavity laser with 55 nm tuning range. *Electronics Letters* **19**, 110–112.

Yamaguchi, M., Kitamura, M., Murata, S., Mito, I., and Kobayashi, K. (1985). Wide range wavelength tuning in 1.3 μm DBR-DC-PBH-LDs by current injection into the DBR Region. *Electronics Letters* **21**, 63–65.

Yen, C.-S. and Crawford, R. D. (1985). The use of directed optical beams in wireless computer communications. *IEEE Globecom '85*, New Orleans, pp. 1181–1184.

Yokohama, I., Kawachi, M., Okamoto, K., and Noda, J. (1986). Polarisation-maintaining fibre couplers with low excess loss. *Electronics Letters* **22**, 929–930.

Yoshikuni, Y., Oe, K., Motosugi, G., and Matsuoka, T. (1986). Broad wavelength tuning under single-mode oscillation with a multi-electrode distributed feedback laser. *Electronics Letters* **22**, 1153–1154.

Index